Handbook of
Automotive Body Construction
and Design Analysis

To Paul, Peter and Denise

Handbook of Automotive Body Construction and Design Analysis

by

John Fenton

Professional Engineering Publishing Limited
London and Bury St Edmunds, UK

First published 1998

ISBN 1 86058 073 4

A CIP catalogue record for this book is available from the British Library.

Printed in Great Britain by Redwood Books, Trowbridge, Wiltshire

CONTENTS

Chapter 6:
Structure design: applying classic theory

Chapter 7:
Open/closed structures

Chapter 8:
Safety under impact

Chapter 9:
Occupant protection/restraint

Chapter 10:
NVH

Chapter 11:
Structural applications

PREFACE

A revolution has taken place, world-wide, in vehicle body starting with the influence of J Edward Deming on Japanese manufacture which was set world-class benchmark standards to emulate. The philosophy of manufacture has changed as much as the techniques, both of which are addressed in the first half of the book.

The widespread availability of computing power to designers and engineers has created a renewed interest in predictive techniques in automotive structural analysis which are outlined in the second half of the book. Analytical methods that were once considered too laborious and time-consuming can now be handled with considerable speed by programming in design calculations for solution over a wide range of input variables.

Weight reduction continues to drive many areas of automotive design and development. However efficient the means of converting energy of the fuel into tractive force at the wheel-to-ground contact area, the fuel itself has a limited energy capacity. So the 'potential' energy of the fuel, if equated to the kinetic energy (again proportional to weight) of a vehicle, moving at a given speed, reveals a finite energy value which no fuel economizing device can ever exceed.

The case for reducing fuel consumption by reducing mechanical component and equipment weight, in turn reducing translation and rotational inertial drag on acceleration, as well as rolling resistance and gradient resistance, has been well argued. The configuration and assembly of components and structural elements is also of importance from a weight reduction standpoint.

As excess weight at one part of a vehicle has to be supported by extra weight added elsewhere, conversely a weight reduction in one part may result in a meaningful saving in others. Also the fitting of a lighter system, such as a smaller engine, in one place could result in a mechanical change in another, smaller brakes, for example, so there is a chain reaction. A lighter body structure may result in smaller wheels, and so on. The overall system design is thus of major importance.

A quantitative measure of the effect of weight saving on fuel economy is indicative in studies carried out to investigate automatic gearboxes by a UK manufacturer. These showed that the proportions of fuel used to accelerate the mass of a typical European car, when driven according to the USA Environmental Protection Agency 'Highway', 'Urban', and 'EEC 15' cycles, were 18, 38, and 25 per cent, respectively. Thus it was deduced that a one-third reduction in vehicle weight may be expected to yield some 6-12 per cent fuel economy. This comparatively modest yield meant that new component development for lighter weight was normally accompanied by design for lower cost in material and labour.

Vehicle weight also has a close relationship with body size so compact vehicles reap rich rewards in fuel savings, not only in weight saving but with proportionally reduced air drag as cross-section area is reduced. Now engineers are looking at the body structures themselves to see if the weight in them is efficiently distributed to react applied loads with the minimum of redundancy, to reduce 'skin' drag by smoothing surfaces and to provide passenger protection in accidents.

In this review of technical literature the analytical techniques are summarized which are necessary to design the structure, as a preliminary stage, prior to finite-element analysis. The analytical processes within the text will be of particular value to students in course levels between Technical T-5 and second year degree in vehicle engineering. Parts of the book will further be of value in the final year degree and post graduate studies in vehicle engineering.

Until recent years students and apprentices in automotive engineering still entered the industry with a background of mechanical engineering training in such subjects as thermodynamics, theory of machines, hydraulics, mathematics, mechanical design, and strength of materials. The teaching of theory of structures had been more common in courses of aeronautical engineering, civil (structural) engineering, and naval architecture. Within the vehicle manufacturing industry other trends are throwing a new emphasis on body structural design: these intensify the need for wider training in structural theory for body designers. At the same time there are strong commercial pressures to shorten the lead times of new model introductions and to meet the projected standards of structural safety, emission control, ride/handling refinement and reduced fuel consumption, all without resort to extensive road proving programmes.

The increasing application of analytical techniques to the solution of these problems demands a greater knowledge of automotive-engineering theory from those about to enter vehicle design/construction, and from those already employed in vehicle body design and manufacture. The third category of reader, the specialist vehicle builder, will also find much to attract him in the techniques applied both in his own and other related specialities.

The book is an introduction to vehicle design analysis in the sense of providing readers with an insight into analytical methods given in a wide variety of published sources such as technical journals, conference papers, and proceedings of engineering institutions, for which a comprehensive list of references is provided. The analyses are therefore not necessarily fully developed or rigorously evaluated; recourse to the original references is necessary particularly in order to understand the limiting assumptions on which the analyses are based.

Some of the earlier works involved analysis techniques developed from empirical experimentation and predate the widespread introduction of SI units. In such cases conversion to SI units, and reconstruction of the related graphical illustrations was felt to be inappropriate. Much of the analytical work is centred around impending legislation and, where this is quoted in the text, it is for illustration only and it is, of course, important to examine the latest statutes in the locality concerned. The list of references given at the end of each specialist chapter is a key element of the publication, providing where possible a link to the original publication source. Where the original publication is no longer in print many of the sources can be found in such libraries as those of the Institution of Mechanical Engineers, London or the Motor Industry Research Association, in Nuneaton, as well as the British Library

JOHN FENTON

Chapter 1: Body fabrication, jointing and assembly

Recent fabrication trends; achieving world-class; fasteners, snap-fit, spring-clip and special springs; specialised fastening techniques; bolting and riveting systems; arc and gas welding techniques; spot, seam and projection welding; laser-welded tailored blanks; adhesive bonding, structural adhesives and welding of plastics; 'welding' ceramics; design for automated and robotic assembly

While most vehicle makers have divested themselves of machining operations, with the passing down of mechanical system construction to the tier-one systems suppliers and their subcontractors, forming and fabrication of metals and composites used in body construction is often still an in-house activity. Ever developing construction techniques mean that both product and production engineers must be always vigilant of changing methods in order to maintain high quality at economical price.

Recent fabrication trends

The 'oil-shock' of 1974 was the impetus for the thinking which has led to the new era of motor manufacture. The need for vehicles of much lighter weight, and manufacturing methods of much higher productivity, began a period of major reinvestment and reappraisal of manufacturing systems which is even now still under way.

Hartley[1] plotted the changing scene in car manufacture by 1977, which was already well advanced by that time, pointing out that the skill of the car worker then was well above the traditional tasks required of body fabrication and vehicle assembly; workers thus had to put up with dull, repetitive tasks under the techniques of volume production pioneered by Ford. With the new assembly methods being introduced the workers appreciated most the simple fact that they were not mechanically paced, but had some control over the rate at which they could work.

In body-in-white shops the reappraisal principle is that mechanisation has been adopted where work has proved particularly arduous. Fiat showed the way by introducing robots in the darkest parts of its Mirafiora factory, where the work was tiring and many disputes seemed to arise. There were initial difficulties, naturally, and Fiat found it could only use one multi-welder prior to the Unimate spot-welding robot operations, Fig 1, and weld-flange widths had to be increased by 3 mm for the 131 car. The company also started to assemble mechanical components to the body structure automatically, and automated the handling of sheet metal and pressings between the forming and fabrication machines.

In the automatic assembly operation the mechanical units were first assembled manually on fixtures, and then the fixture and the necessary fasteners were inserted in a machine which offered the fixture to the body, tightened the fasteners and then lowered the fixture. Each fixture was carried on a Digitron trolley which could follow tracks laid under the floor, with control by radio-telemetry. The mechanical units – engine, transmission, suspension and steering assemblies – were laid on the Digitrons, Fig 2, as they move slowly along; then each Digitron was guided into one of 12 bays where the horizontal or inclined fasteners were installed and tightened, the engine mounting bolts, bolts for the suspension pivots and fixings for the steering rack being typical examples. Fiat adopted group, or team, assembly at this stage in the sequence, there being five men to each team.

When this part of the work had been completed, the Digitron carried the components to the assembly machine, the fixture being transferred to a platen in which the fixing bolts for the components have already been positioned. Assembly of the me-

Fig 1: Spot-welding with Unimate robots

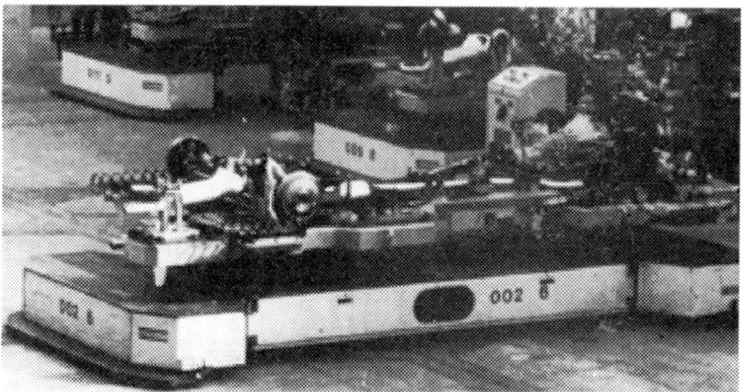

Fig 2: Use of Digitron automated guided vehicles

chanical units took place automatically, the four-man team needing only to fix the top of the suspension struts and the rear dampers to the body. Since the cycle time was about six minutes, the operations involved were carried out at a reasonable pace, while working conditions were infinitely better than when operatives had to crouch beneath the bodies, trying to line up bolts as the track moved along.

This was all relatively simple, of course, compared with the dramatic Volvo factory at Kalmar, which consisted of hexagonal areas to give the impression of a number of small workshops – Volvo recognising that people like to belong to a small enough unit with which to identify themselves. The cars were assembled on trolleys working on the same principle as the Digitron. A team of about 20 men could either work as separate smaller teams, with the trolleys stopping at a number of stations, or each team could complete assembly of a car in one bay or siding, workers deciding for themselves which jobs they did and how they rotated jobs.

Hartley explained that the door-hangers dispute at Ford UK in Autumn 1976 showed why mechanical pacing should be avoided. The men objected to the introduction of new fixtures because they then had to work at the same spot all the time and could not work their way up the line in order to get ahead – in other words, they had learned to pace themselves to gain a break when they needed it, and now they could not do so. The actual work rate was not affected. To assess the value of various replacement systems of working Renault set up a department to 'restructure working methods', headed by Pierre Tarriere, and this department has analysed all the obvious combinations – from the standpoints of economics and workers' preferences. The systems at Renault installed at Le Mans and at Douai passed the test from these points of view. Tarriere believed that the new methods improved productivity as well as working conditions. He pointed out that, with a conventional line, if 20% of the workers were absent the production level dropped by 50%; with any form of group assembly a similar rate of absenteeism cut production by only 20% — and absenteeism at Renault then averaged about 12%.

Ingersoll Manufacturing Consultants has assessed workers' preferences for various systems in terms of five factors: 1. Variety, indicated by the number of functions involved in the job and the length of the task; 2. Autonomy, or the freedom the worker has to organise his task; 3. Responsibility for reaching his production target without mechanical pacing; 4. Interaction, or the degree of social contact that can be built up among the workforce; 5. The completeness of the task, being an indication of whether the operations produce a sub-assembly in which the worker can take some individual pride. Engineers on the IMC project team found that workers demonstrated their need for these factors as they worked. For example, when they pre-assemble washers to bolts, or bolts to housings, they show the need for variety and autonomy, while the storage of a few components under the bench indicates responsibility.

Significantly, IMC found that both individual and team assembly were theoretically much preferred by the workforce to conventional lines, but that individual assembly would allow little interaction, which seems a strong desire among manual workers. The value of the work is shown by its explanation of why the Fiat workers at Termoli did not like individual assembly. IMC concluded that as far as engine assembly is concerned, for high throughputs about 30% of the operations should be automated, leaving the men to assemble the remainder in groups — perhaps with three or four groups on each 'line' carrying out part of the assembly.

Fabrication techniques

Although such a system can be adopted in most existing factories for engines and transmissions, more space is needed for body-in-white and complete car assembly; but, of course, space is at a premium in most car factories therefore some form of assembly in which the lines are divided into sections by buffer stores is preferable. The other technique is to sub-assemble as much as possible off the line, since this can be done by groups or individuals without mechanical pacing. This concept was pioneered by Ford where operatives assembled the complete Fiesta fascia before installation into the car.

Clearly there is plenty of scope for adopting new assembly methods, although it is essential to gain an indication first of the workers' preferences in some detail — and these, Hartley explains, will be the innate, unconscious preferences, not the opinions they express. Secondly, experience indicates that workers must be given at least six months to adapt to a new method before any conclusions can be reached. But the evidence does indicate that these desires can be met without reducing productivity; in some cases, they help to increase both productivity and quality, two vital factors for success in manufacturing.

Fig 3: Body framing for the Austin Metro

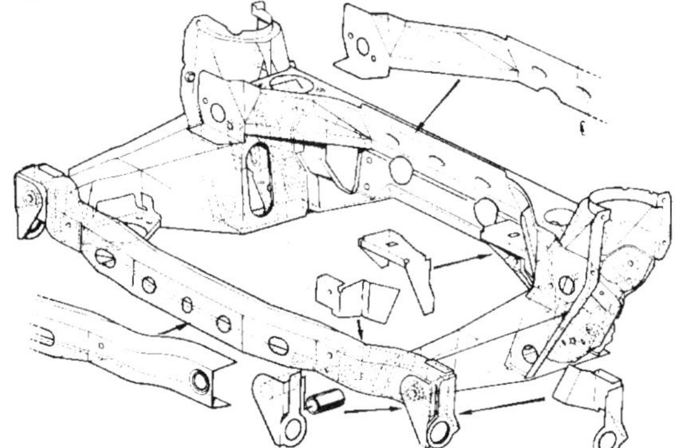

Fig 4: Metro front-suspension subframe

Automated body framing In the UK an early application of automated body fabrication was to the Austin Metro, now the Rover 100, in 1980. Underframe, front-end, monoside and tail-door apertures passed by conveyor to tagging lines (tags provide a loose skeleton for the initial panel dimensional location prior to welding) before moving on to two Sciaky ABF lines which added the roof and constructed the body shell. Shells were framed and welded on two 110 metre long lines each incorporating three multi-welding stations and 14 computer-controlled robot welders which were electronically synchronised.

At station one the skeleton was clamped on to an accurate body jig and 72 key spot welds made. With the roof being added at an intermediate stage, two further multi-welding stations applied nearly 140 further spot welds. The body was then a secure structure which passed to a fourth station for audit, 24 critical dimensions being electronically probed, the results being shown on a display panel with automatic print-out for quality inspection. Failure of this test results in stopping of assembly and halting progress of the body to the robot lines. The bank of 14 Unimate welding robots on each line, two of which are 'flexible' to allow for breakdown, perform about 250 further welds on the shell in an average cycle time of 23 seconds (Fig 3). Some robots travel up to nine metres to weld inside and outside the shell.

Subframe production was carried out in separate plants on fully automated press lines. The front suspension sub-frame, Fig 4, was produced in a row of five presses at 600 units/hour. Blanks pre-cut from 2.5 mm thick coil were fed into the first press and after the first blanking operation parts were drawn, pressed and pierced in three stages and finally, trimmed. Small parts were made at rates up to 400/minute from 2.5 mm thick strip. After initial sub-assembly using hanging spot-weld guns, the parts were resistance tack-welded together on a jig. Assemblies were then placed on carousel conveyors where CO_2 MIG welding was carried out, before passing to six-stage drilling and reaming machines.

1980 also saw the first large-scale automated body fabrication of Ford's Escort model, where 39 robots were used to make some 543 welds per body, Fig 5a. Body-sides were made in separate complexes by multi-welding, Fig 5b. Both multi-welding and robotic-spot welding was used for joining rear floor, centre-floor and engine-compartment sections of the

Fig 5a: Escort body framing

Fig 5b: Body-side fabrication

underbody. Two years later Ford invested some £500 million in its UK plant for Sierra body production, resulting in 88 % of spot-welds being automated, Fig 6. Benefits to end-users included a 78% length reduction in panel joints and a reduction in total spot-welds from 5155 to 4213. There were 75 fewer panels used in the shell and a single body side outer panel was employed. High-strength steels featured in 36.5 kg of panel weight out of the 264 kg total and a 6.2 kg saving in BIW shell weight was achieved.

Body-side assembly The outer body panel was formed on a line of six 1000 tonne presses. The bodyside assembly complex consisted of two lines, each approximately 120 metres long, and utilised a total of 54 robots, 52 for spotwelding and 2 for sealer operations. These lines which accommodated both saloon and estate car derivatives, had at the head of each line, a rotary fixture. The outer skin and the reinforcing components were manually loaded onto the fixture which then turned through 180 degrees for tack welding by two Kuka robots (Fig 7). The resulting sub-assembly was automatically transferred by an overhead translift conveyor to the main welding line. The sub-assembly was then finish-welded by a single floor-mounted Kuka robot and four overhead Nimak robots. Sealer was then applied to the wheelarch by an ASEA robot prior to the introduction of the wheelhouse components. The assembly then passed down the line for further welding operations undertaken by three floor mounted Kuka and 14 overhead Nimak robots.

Engine compartment and underbody assembly This assembly complex featured a total of 31 robots (26 Kuka and five ASEA) and five major 4-post multiwelders. The principal sub-assemblies that made up the complete underbody were the engine compartment (comprising apron and side-member, dash panel and upper and lower cross-members), the front floor, and rear underbody assembly (comprising side-member, cross-member and rear floor). The largest concentration of robots in this complex was devoted to the construction of the engine compartment. In the apron and front side-member area four ASEA robots undertake the preliminary MIG welding while 10 Kuka robots apply spotwelds to complete the welding operations. This sub-assembly was then transferred by overhead conveyor to the engine compartment complex. Together with the upper and lower front cross-members and the dash panel (which had been stud-welded by an ASEA robot), the apron and front side-member were loaded into a fixture and transferred automatically on to a four stage turntable for tack welding. A shuttle beam conveyor system transferred the engine compartments in pairs through a total of 10 Kuka robots for finish welding. The completed assembly was then conveyed to the underbody marriage station. Meanwhile the front floor assembly was being welded in a 4-post multiwelder before being automatically transferred to the same station. At the same time the rear floor was being assembled and welded in a similar facility before being automatically transferred to the rear underbody assembly area. The rear side-members were welded by a static multiweld fixture (using two Kuka robots in a handling mode) and then matched with a rear cross-member and loaded manually onto a turntable. Four Kuka robots performed the spotwelding task and the resulting rear floor frame ('H' frame)

Fig 6: Sierra body framing

was married with the rear floor. It then passed through a 4-post multiweld press to produce the rear underbody assembly, which was then automatically transferred to the main marriage station to meet up with the completed front floor and engine compartment. A shuttle beam took the assemblies through two 4-post multiwelder presses and the completed underbody was then transferred to an overhead conveyor.

Body framing/finish-welding At the preload stage the bodysides and underbody were assembled using tabs, together with the cowl top (stud-welded by an ASEA robot), front and rear headers and lower back panel. The tabbed shell was then loaded into the first stage framing buck where multiwelding was carried out. It then entered the 'inter stage' where further spotwelds were applied prior to the addition of front wings and roof. In this area, the welding was fully automated, using 6 Nimak robots (two robots on each of three lines). Following the interstage, the shell entered the second stage framing buck, was multi-welded and emerged as a complete bodyshell ready for finish welding. This facility consisted of two lines, approximately 50 metres in length, each with four robot welding stations using 12 Cincinnati hydraulic robots. The bodyshells were automatically placed onto a beam transfer system which conveyed them along the line of robots. At the end of the line the complete shells were unloaded automatically onto a conveyor system prior to metal finish and the fitting of bonnet, doors and tailgate. At a separate location hinges were welded to front and rear doors. These were two stations each with a rotary table, with four fixtures, and two ASEA robots which carried out MIG welding of the hinges to the doors. For direct glazing (Fig 8), a PU adhesive sealed and bonded the glass to the body, correct location being achieved by a finisher pre-assembled to the glass.

Fig 7: Sierra body-side fabrication

Achieving world class

Burman[2] asserts that there are a number of 'benchmarks' which can be used to identify where a company stands in the race to achieve and maintain world class status: production costs; quality performance; non-productive time; quality standards; manufacturing cycle time; time-to-market and customer satisfaction. The technique of benchmarking is now well established and has been defined by McKinsey's Steve Walleck as follows: it first identifies the manufacturing and managerial processes that a company needs to improve, then selects other companies that are known to perform analogous processes with outstanding results, and measures in detail how they perform them. Unlike competitive analysis, which is usually a staff exercise, benchmarking generally requires participation of line personnel, so it becomes more than an analytical process; it can be a tool for encouraging change.

Benchmarking is a way to go backstage and watch another company's performance from the wings where all the stage tricks and hurried realignments are visible. When a benchmarking exercise has been carried out, the gaps identified and measured, and the weaknesses causing these gaps pin-pointed, plans can be prepared to move the company towards world class status. Such plans should generally be based on

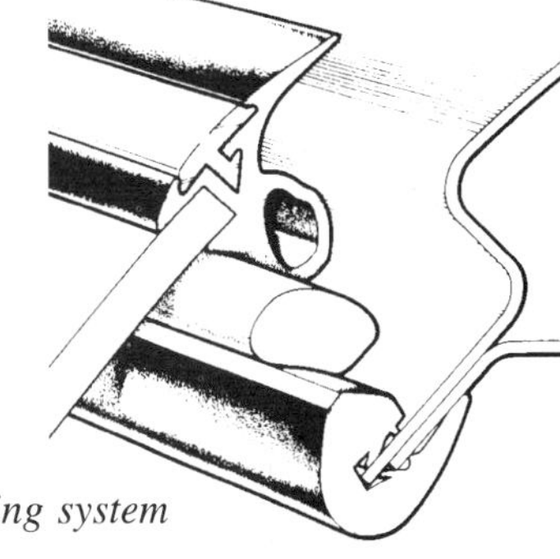

Fig 8: Direct glazing system

the concept of continuous improvement (the Japanese Kaizen) and the first step must be to ensure that all personnel, from top to bottom, accept the need for, and are properly prepared for, personal change.

According to a team from the Institute of Production Engineering in Aachen, Germany, continuous improvement means relentless efforts on the part of all employees to improve each individual process. In this context, continuous improvement applies not only to products and services which focus on product orientation, but also to the way in which machines are operated, how people work or how systems and guide-lines are implemented

Psarouthakis[3] used an example of three men of equal ability, working on identical machines, to explain the concept of continuous improvement. Each man set himself a target: the first set out to be perfect and his slogan was 'I will accept nothing short of perfection'. He failed because perfection is an ideal to be aimed for but never achieved. The second set out to be 'number one', better than all the rest. His slogan was 'Why not the best?'. He failed because others also had the same slogan and there can only be one best. The third had a much more practical slogan 'To perform better than I did yesterday'. He succeeded.

Manufacturing systems consultants Lucas Engineering & Systems say their core objective is 'to enable businesses to attain global competitiveness by implementing the world's best practices'. To achieve this they believe that gradual change is not good enough, and therefore apply their own innovative 'step change' method, which reorganises basic business processes to achieve a sharp rise in performance (step) before installing IT systems to consolidate gains (Fig 9).

When a new vehicle body is specified, that specification sets out a series of performance criteria, which may be conflicting, for example: passenger and storage capacity; overall dimensions and weight; performance and fuel economy; also price. The design must not only meet the specification but also achieve objectives such as: attractive appearance; resistance to corrosion; good all-round visibility; high level of safety; low wind noise and NVH; simplicity of manufacture; short time-to-market and, of course, be better than previous models. VW/Audi Group use a graph showing the continuous improvement concept in action. The options for body materials are also studied by the group and Fig 10 shows the changes in composition from 1985 to what was predicted for 1995, as an example.

Design for manufacture DFM principles can be summarised as the need for the designer and other members of the concurrent engineering team to recognise the size of the company's investment in production facilities (machines, tooling, robots and people skills). This recognition does not mean that change cannot be considered but merely that true

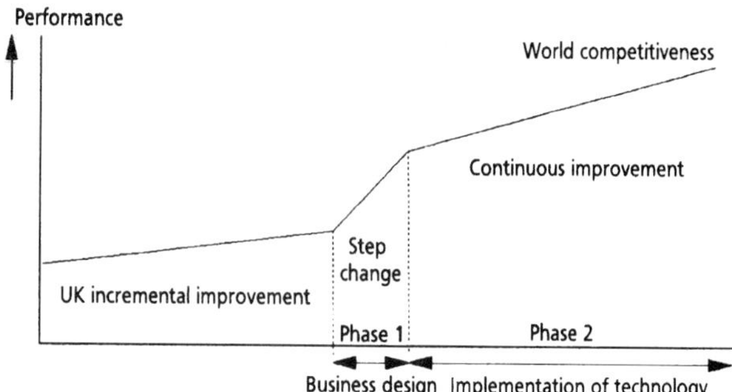

Fig 9: Step-change in improvement

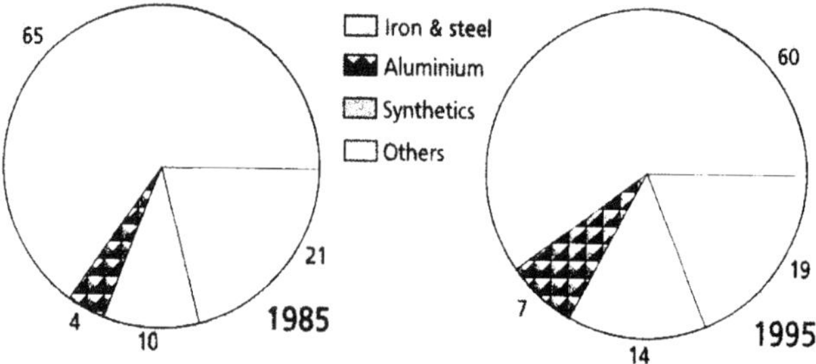

Fig 10: Changes in body materials over a decade

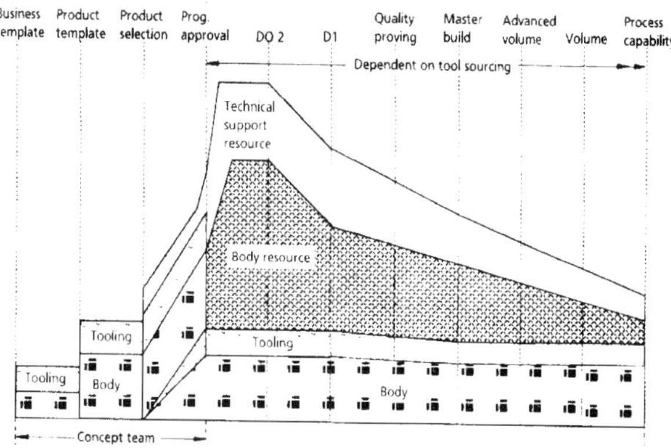

Fig 11: 'Front-end-loading' of a project

costs of such changes must be properly evaluated and justified. To achieve world class, a manufacturer must be organised in such a way that such variations as production volume, the number of derivatives, and the rate at which these factors are changing, do not adversely affect his ability to develop a design within three years of conception

Rover Group researchers[4] recently published a report which reviews the Body Engineering function within the Body and Pressings unit of the Group. This report also discusses the performance levels achieved over the previous five years and investigates potential future strategies. The need for the reorganisation described in this report was based on a new company strategy to move up-market, to increase product diversity, to reduce time-to-market and to ensure that the incidence of late and costly design changes was eliminated by front-end loading of the project resources as shown in Fig 11.

Formerly a slow and involved process surrounded the production of a new boy-in-white, working through a series of development stages and covering a period which could be as long as seven years but was never much less than five. Now the trend is to three years or possibly less. It has been suggested that the current methods of car body construction are based on the traditional craft of the coach builder, refined and developed out of all recognition, but still reliant on the assembly of a large number of manageable components which must then be put together by spot welding. It is therefore, Burman[5] argues, reasonable to question whether — given the facilities now available — a totally new approach should be considered.

This should put the questions: why so many components?, why spot welding?, why steel? and most importantly, what are the alternatives and constraints? For example, is it not possible to replace some groups of parts with a single composite part which has, moulded into it, the necessary mountings, brackets, ducts, pipes and cables? Is it not possible to develop structural adhesives which will allow such components to be assembled quickly and securely? Is there a way of making large sheet metal panels of complex shape without using presses? Can composite panels be made on a volume basis? Can aluminium extrusions be used to a much greater extent as structural components?

The answer to all these questions is probably 'yes' or perhaps it is more likely to be 'yes but...' The automotive industry and its suppliers have committed vast amount of investment to current body build technology and this could be wiped out if a new approach were to be adopted. But surviving in a competitive market requires that continuous improvement takes place, typified by new approaches to handling body panels between pressworking operations. The Fanuc pendulum arm robot, Fig 12, is designed to speed up parts transfers and tool changes in tandem press lines. The company has evaluated the economics and production rates of various types of transfer system: manual, pick-and-place shuttles,

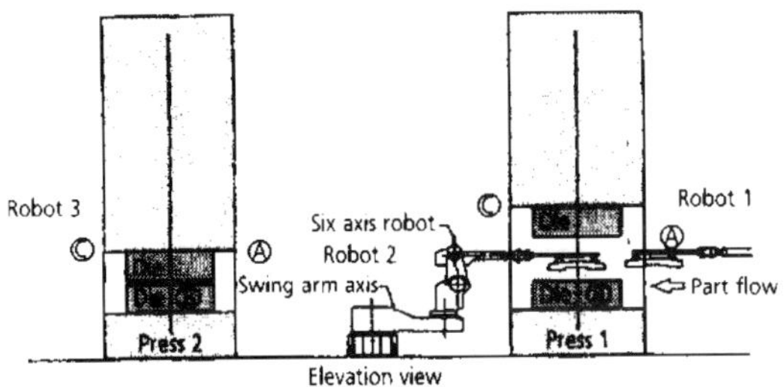

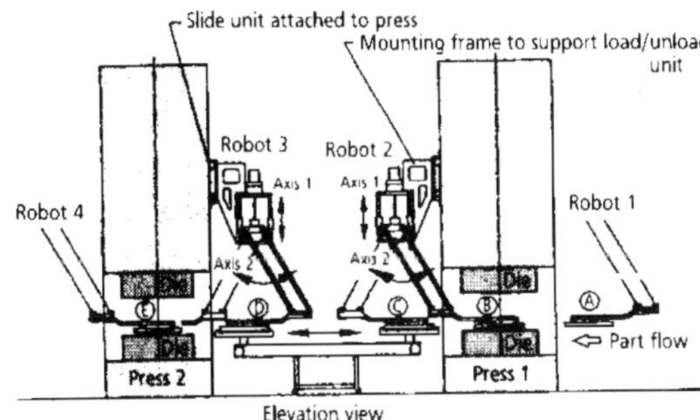

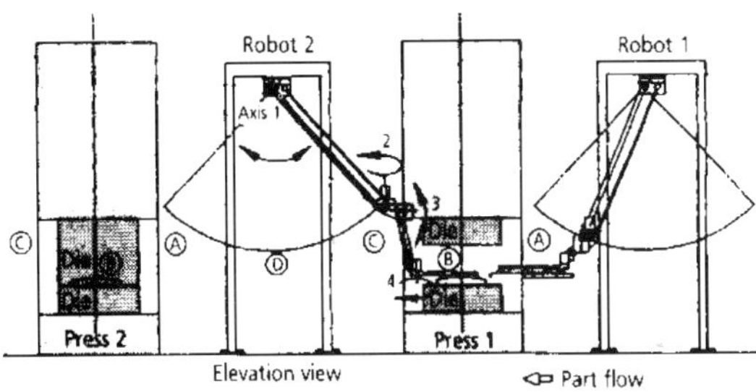

Fig 12: Fanuc pendulum-arm robot

robotic arms and the pendulum arm. From the results of these comparisons it is concluded that the pendulum robot is superior to the others because, against manual parts transfer, output is increased by up to 100 parts per hour; accuracy of placement reduces scrap, downtime and tool damage; cost per part is reduced.

Against pick-and-place and swing-arm, installation can take place without closing the line; die change-over times are reduced because the pendulum does not have to be removed from between the presses; parts are transferred by a single gripping operation instead of two; the pendulum arm has automatic EOAT (End of Arm Tool) change facility to reduce change-over times.

Joining technology for vehicle body panels

Joining techniques for body panels and components continually evolve and new forms of old techniques can soon challenge some of the newer ones.

Those who consider riveting, for example, as pre-punching holes, hand-insertion of rivets and fixing by special machine automatically assume spot-welding to be a cheaper and faster process. With the introduction of self-piercing rivets by such companies as Henrob, this assumption must be questioned. Insertion tools can be robot-mounted and each hydraulically-actuated riveting tool has its own bandolier of rivets and a fully-automated process thus result. Benefits can include a reduction in the number of joint-points, higher confidence in joint quality and lower costs. Even strength improvements over spot-welding are possible, Fig 13, and self-monitoring sensors constantly check quality parameters.

Laser-welded tailored blanks are another example of ever-changing joining technology. Fig 14 shows a door pillar, by Volvo, produced from different thicknesses of sheet. Butt-welding by laser causes appreciably less joint thickening then seam welding and there is a smooth transition from one thickness to another and very high welding speeds can be obtained. In the Volvo application, current from a 40 kHz generator is delivered to a coil placed over the joint. Eddy currents induced in the sheet metal cause heating of the weld area prior to the panels being force-butted together. The material is raised almost to melting point, but solid-phase welding is in fact maintained and there is a controlled amount of upsetting. Typically the operation is completed in six seconds.

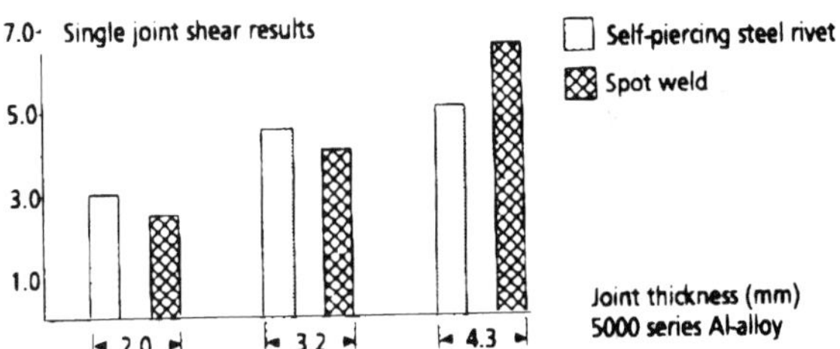

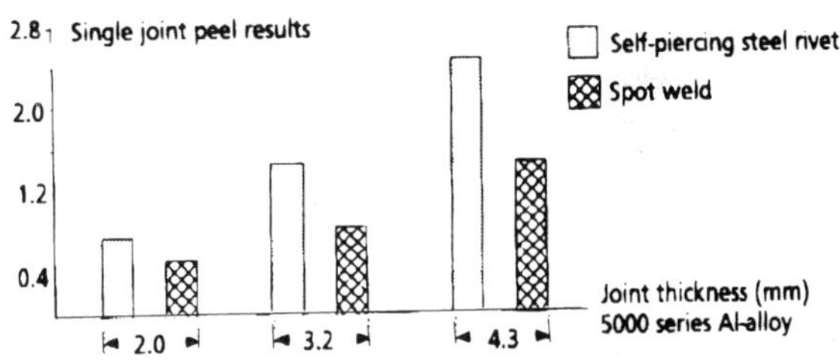

Fig 13: Strength comparison between spot-welding and self-pierce riveting

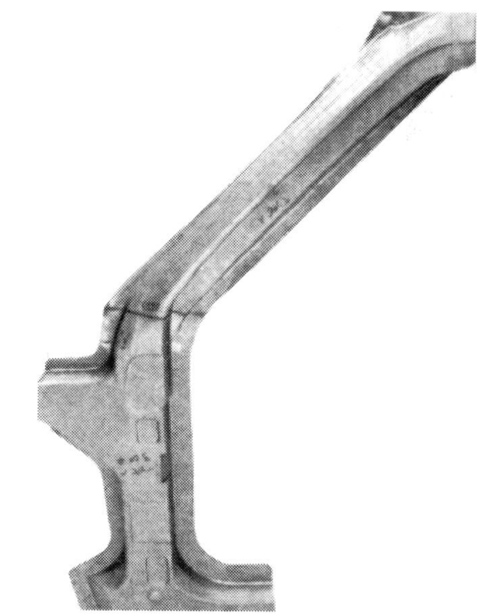

Fig 14: Door pillar produced by laser welded tailored blanks

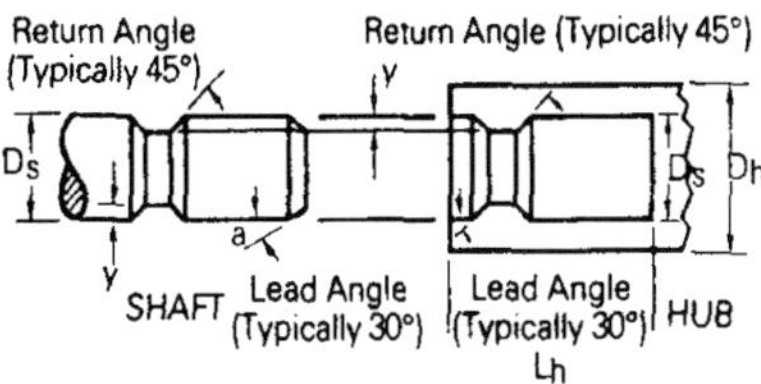

Fig 15: Cylindrical snap-fit

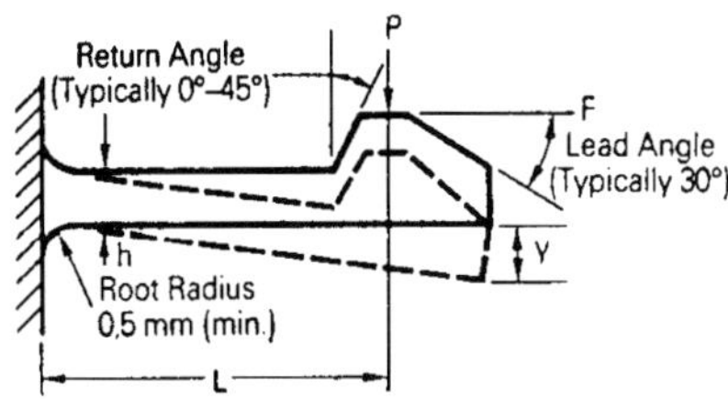

Fig 16: Lug-type snap-fit

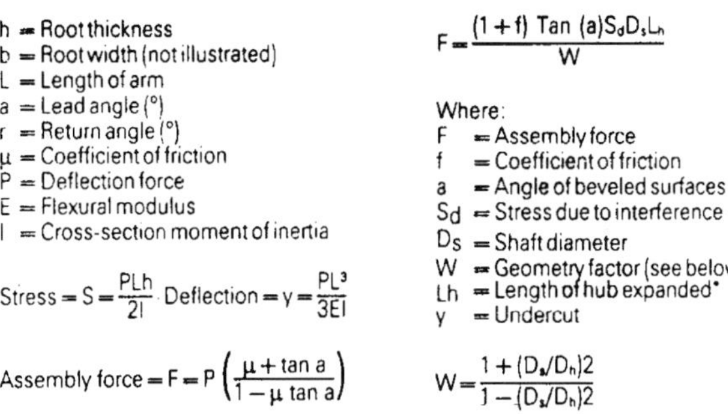

Fig 17: Assembly force calculation

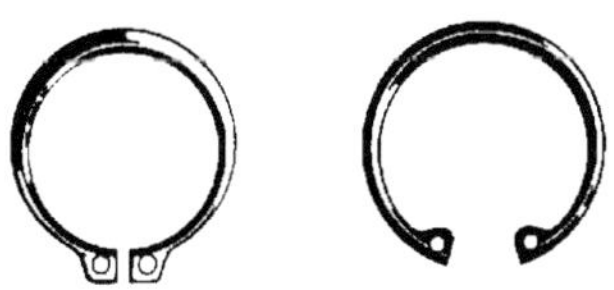

Fig 18: Internal/external retaining rings

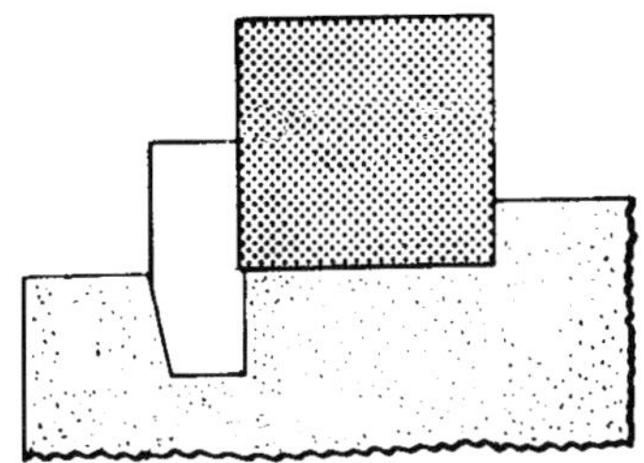

Fig 19: Bevel-edge retaining ring

Fastening techniques

Fabrication of vehicle and body systems can be considerably eased by the correct choice of fastening technique, of which threaded, riveted, tabbed and spring-clip mechanical fasteners are among the alternatives. There are also, of course, a wide range of welding/brazing options. Sometimes the design of the part to be joined can be arranged to give it an inherent fastening 'ability for mating with the adjacent part.

Non-threaded fasteners

Snap fits It is possible by careful design of plastic mouldings to avoid the use of separate mechanical fastenings or adhesives — by using the snap-fit potential of the materials. According to Du Pont, all snap-fits are variations on two basic designs, the cylindrical (Fig 15) or cantilevered lug (Fig 16). They work by using an undercut in one piece and on the other, a lip or hook. Unlike press fits they do not involve permanent deformation since the parts return to an unstressed state after fixing. Their main possible disadvantage is the additional tooling cost for making the constituent parts. If the cylinder of the joint is slotted then effectively the joint can be considered as a series of lugs. The lugs can be made easy to separate, or not, by varying the return angle. Undercuts of 90 degrees can if necessary be moulded by using the side cores of corresponding slots in the mating part. Lugs should be designed with thickness or width tapering from the root to give uniform distribution of stress — and an adequate bending length must always be allowed. The basic strength and deflection equations recommended by the company are given in Fig 17.

Spring clips These fasteners can simplify and speed the assembly of body and vehicle systems in a way that is difficult to match by threaded fasteners. By careful attention to design rules, reliable performance can be obtained for the assembled system.

Retaining rings are possibly the best known form of spring clip fastener and these may be used to replace machined shoulders, set collars, rivets, cotter-pins, nuts and other bulky fastening devices — as well as the associated machining operations. Rings divide between internal and external locking types, Fig 18, and bevelled edge types can be used to perform precision locking jobs such as the retention of bearings, as shown in Fig 19.

Other clips are of the push-in or nut retainer type and a large number of hybrids have been developed incorporating barbs, tags, riveting or other facilities. A selection are shown in Fig 20 from the designs developed by Salter. At (a) is a fastener which is attached to a body by a single rivet and used to retain a circular moulded emblem which is snapped in place, without tools, and conceals the fixing. At (b), another variant snaps into a rectangular hole and studs engage the fastener, on the part to be fixed. Triangular, arched-shape clips at (c) are intended for metal assemblies requiring a high retaining force. At (d) large clips used in pairs, to this design, are employed in retaining a loudspeaker on a trim panel, typically. The clip at (e) is designed to mate with a circular stud and can withstand an axial load on the stud of some 175 lb; it can be quickly assembled with a push-on tool or arbor press. That shown at (f) is for securing flexible materials, the upturned ends preventing digging into the soft surfaces. Like (e) in function, (g) is used to retain against forces on a larger diameter shank but with minimal radial operating clearances and might typically be used to secure an indicator lamp in a dash panel. Type (h) is used for retention within a cavity ranging in size from 3/8 to two inch diameter.

Metal-to-metal assemblies have been known to use a variety of spring clips, some of which are typified in Fig 21. They exert a firm grip on the edge of panels and can be positively anchored by piercing a small hole in one member to receive a tongue. One of this last type is also shown (left) at (b) developed for attaching weatherstrip. The complementary part, (right), snaps in place over the abutting corners of the strip. Captive types are illustrated at (c) which snap into place and provide some latitude for engagement with slightly off-position fixing screws. A special captive clip is seen at (d) which was developed as a squab adjuster for the rear seat of a car; the appropriate mating bolt is merely 'zipped' into place to achieve the required positional adjustment.

Spring-pin fasteners are another resourceful assembly aid — some applications devised by Firth Cleveland Fastenings being shown in Fig 22. That company's Rollpins are essentially slotted, chamfered, cylindrical spring-pins which are heat-treated and tempered to achieve optimum toughness, resilience and shear strength. These also can be used to make inexpensive handles and latches as shown in

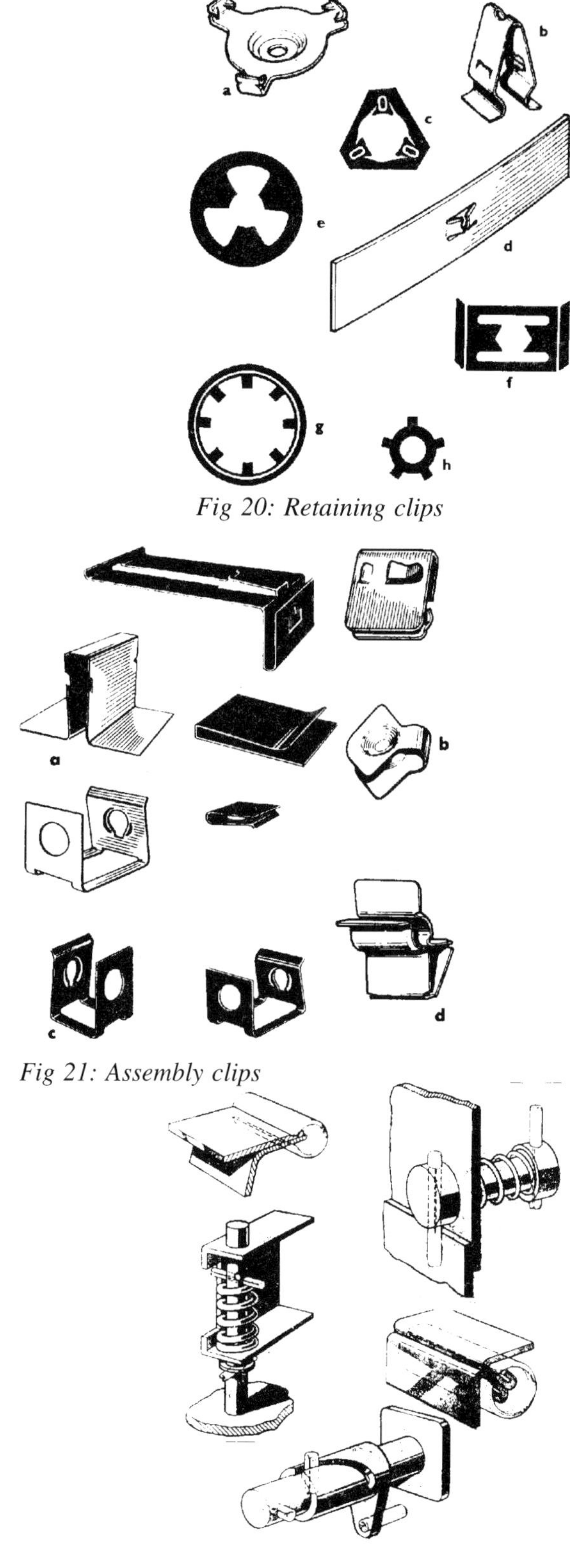

Fig 20: Retaining clips

Fig 21: Assembly clips

Fig 22: Spring-pin applications

the figure as well as securing knobs, joining shafts or guiding control cables. Other examples shown are for spring retention and cam driving.

High-torque blind cage nuts are another application for the spring clip, that shown in Fig 23a being produced by Lintite. A variant shown alongside has a torque arm projecting from the cage to increase torque capacity. Projections at the end of the arm spring into location holes. Lintite also make the Rapid series of spring clip fasteners shown at 23b.

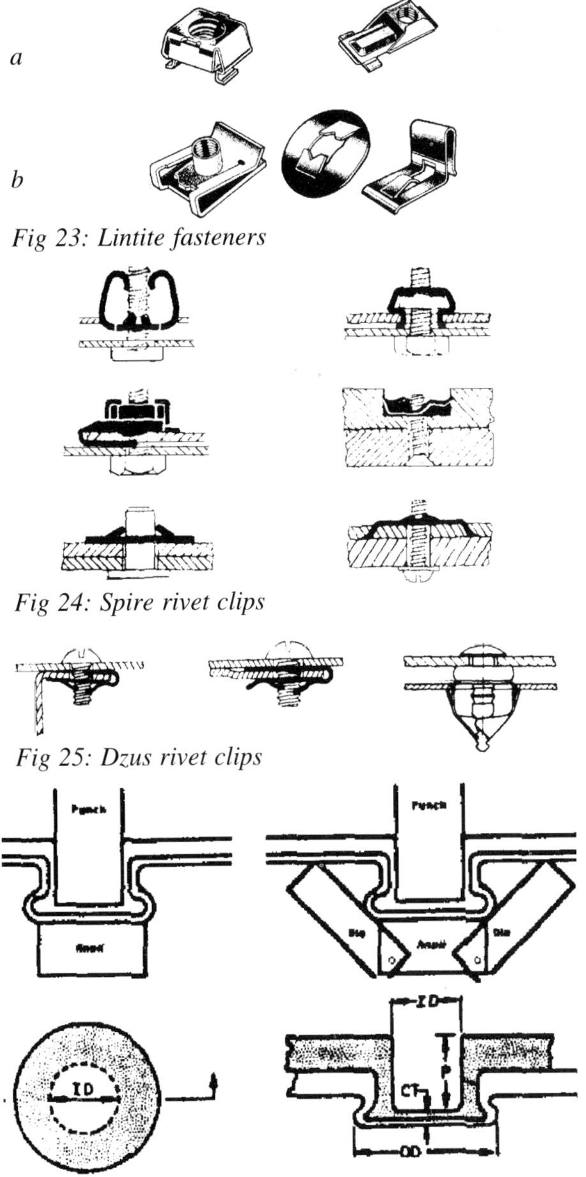

Fig 23: Lintite fasteners

Fig 24: Spire rivet clips

Fig 25: Dzus rivet clips

Fig 27: Punch, die and anvil

Clips used in conjunction with rivets, studs and bolts are shown from the Spire range in Fig 24 while an interesting variant in Fig 25 is produced by Dzus. It is called Pilot and comprises a sprung receptacle for a mushroom-headed stud. The clip is inserted in a panel while the stud is collared to hold a grommet which serves to retain the stud in the closure panel, when open, as well as providing a cushion between the mating parts. The end of the stud is laterally grooved to allow the receptacle to grip it in the closed position.

Combination fastening systems Some fasteners share some of the attributes of both threaded and non-threaded systems. Lintite's Quarter-turn captive nut, Fig 26a, comprises two rectangular plates set above one another, at right angles, on a U-shaped clip. The lower plate which can be slid up and down the clip has a central threaded hole to form the nut. It is inserted in a rectangular hole and rotated so the bottom and top plates secure the nut, then a screw is inserted to effect the fastening. The company also now offer a high-torque version, Fig 26b, of its classic cage-nut. The unit has an extended torque-reaction arm, projections on the end of which are sprung into a location hole. A fastener which is half screw and half rivet is the Scrivet from Jet Press, Fig 26c, designed for

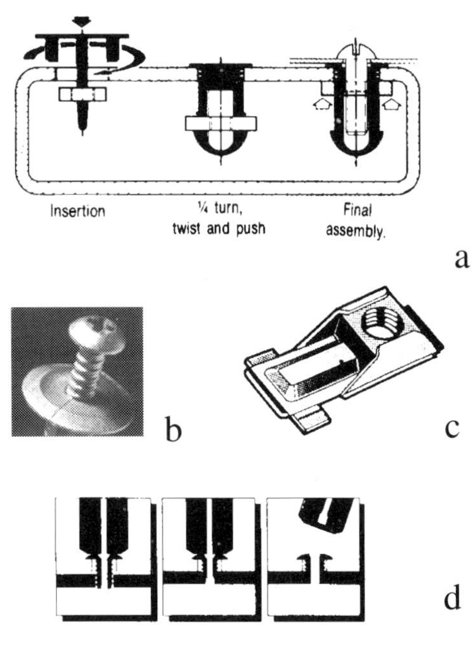

Fig 26: Combination fasteners

applications where frequent removals are likely but still allowing rapid assembly. The assembly can be fitted from one side of a panel and requires only a hammer blow to insert into a pre-punched hole. Meanwhile, Avdel's Rivscrew, Fig 26d, is a rivet which is both removable and reusable. It is a blind hollow rivet which is expanded into its hole by drawing a mandrel through it. However, as the rivet expands, a spiral groove around the shank deforms the material into which the rivet is inserted, creating a screw-thread. Furthermore the mandrel is hexagonal in section so that an internal form is created, the correct size for an Allen wrench, for removing the rivet.

Press-joining An interesting mechanical joining technique which is a possible alternative to spot-welding, riveting or bolted connection is the Eckold press-joining process. The joint is made with relatively simple equipment and heat-treatment properties of the substrate sheets are unaffected.

The method has been called H-forming and the tooling consists of a punch and shearing/forming die, Fig 27. The centre of the die has a convex anvil which has cutting inserts set on either side. The punch forces the top sheet, and in turn the bottom sheet, on the cutting edges to form a step. As the material is pushed down it is forced out sideways by the anvil. The inner edges of the cutting blades are thus forced to spring out slightly whilst constraining the material so that it flows sideways under the bottom sheets and the joint is wider than the cut-step.

The form of the top part of the joint is determined by the contours of the punch but in any case is leak-proof. Fig 28 shows the characteristic dimensions of the joint: degree of penetration and thickness of the cold-formed web beneath the joint. Penetration depth *DT* is determined by distance between the top of the shearing blades and the bottom of the die while web thickness is a function of the sheet characteristics. Fig 29 shows the final form of the joint from which it can be seen that the shape varies in the X and Y planes as the die is open-ended in the X-direction.

Thickness in the Y-direction depends on punch width and separation of the cutting leaves. Forming of the inner joint element is crucial to the strength. Fig 30 relates the stages of forming the joint with the force/penetration plot The H-joint has the advantage over spot-welding that non-conductive surfaces can be applied to the sheets and power requirements are considerably less to form each joint. Sometimes the joint can be made within the forming operation of the sheet parts. The higher cost of the punch-and-die set than spot-welding electrodes is offset by their life which is usually 200,000 joints before re-machining. This is 20 times greater than electrode life and 100 times greater in the case of welding galvanised steels.

The process had its origins in the Tog-l-loc joint developed by the US BTM Corporation, and was referred to as a button-type joint. During its development it was found that to insure the punch

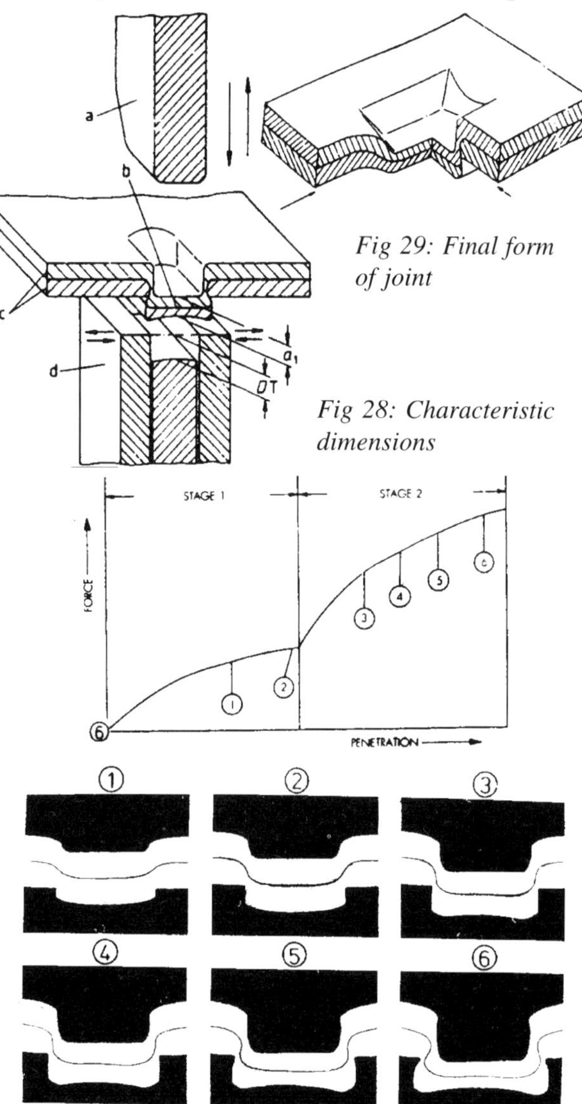

Fig 29: Final form of joint

Fig 28: Characteristic dimensions

Fig 30: Stages of forming

reaches a positive stop, and will not stall during the process, a press of at least 2 ton capacity was required for joining mild-steels up to 0.03 inch thickness. The extent the button protrudes from the sheets should be such that anvil depth - thickness = 0.005 inch for sheets up to 0.05 inch thickness. Less formable steels may be joined with a smaller anvil depth, to eliminate tearing at the internal surface of the joint. Strength of the joint was found mainly to be a function of the cap thickness of the button.

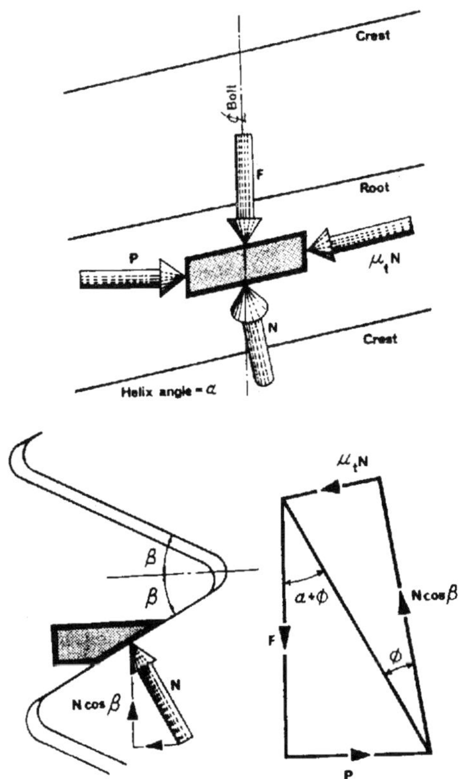

Fig 31: Forces at thread interface

Threaded fasteners

Critical bolted joints on vehicles require a knowledge of the theory of threaded fasteners so the exact capacity of the joint can be calculated. This is crucial to braking, steering and suspension components, for example, where the safety of road users can easily depend on the clamping effectiveness of the joint.

The forces acting on a threaded fastener are shown in Fig 31 for the thread interface. The torque applied to a threaded fastener not only has to overcome friction at the thread interface but also has to overcome bolt head to clamping surface friction and to do work in inducing the clamping tension into the bolt. It is usually assumed, for calculation purposes, that annular friction torque acts at the mean radius of the bearing faces and that the thrust N between two threads is on a helix. The total tightening torque can then be approximated to

$$FD[\mu_a(R_i + R_o)/2D + \mu_t(R_t sec\beta/D) + (P/2\pi D)]$$

where D is bolt diameter, R thread helix radius (inner/outer) and F bolt tension. The three bracketed terms are represented by K_a, K_{tf} and K_u in the table of Fig 32 which gives numerical values for head, thread and induced friction constants. In practice some 90% of the total applied torque does no more than overcome friction (which can vary by up to 20% between fasteners in practical systems). Use of molybdenum disulphide lubricants can reduce the mean K value to about 0.15. It is also advisable to make the bearing face of the nut and mating surface flat to within +/- 0.001 inch.

Choice of threaded fastener should resolve itself into determination of expected applied tension load and the selection of bolt which will withstand at

NOMINAL DIA. (in)	K_a		K_{tf}				K_u			
	BSW and BSF	UNC and UNF	BSW	BSF	UNC	UNF	BSW	BSF	UNC	UNF
1/4	0·696	0·688	0·488	0·509	0·504	0·522	0·032	0·025	0·032	0·023
5/16	0·669	0·649	0·497	0·512	0·510	0·528	0·028	0·023	0·029	0·021
3/8	0·651	0·624	0·505	0·517	0·515	0·536	0·027	0·021	0·026	0·018
7/16	0·657	0·643	0·505	0·516	0·515	0·534	0·026	0·020	0·026	0·018
1/2	0·660	0·624	0·503	0·518	0·519	0·540	0·027	0·020	0·025	0·016
5/8	0·654	0·626	0·510	0·523	0·523	0·545	0·023	0·018	0·023	0·014
3/4	0·651	0·626	0·516	0·523	0·527	0·547	0·021	0·018	0·021	0·013
7/8	0·622	0·625	0·518	0·527	0·529	0·547	0·020	0·017	0·020	0·013
1	0·620	0·625	0·519	0·528	0·523	0·546	0·020	0·016	0·020	0·013

Fig 32: Values of K_a, K_{tf} and K_u

least the same load in the pre-stressed condition. A safety margin of 80% of the yield strength is further added as a protection against over-stretching in the plastic range. By measuring the length change in the bolt during the tightening process, the stress, and hence the induced load, can be calculated from the elastic modulus.

Critical bolts should be tightened with torque spanners to prescribed values, remembering that only 10% will be inducing clamp load. The table in Fig 33 gives tightening torques and approximate tensions for a range of bolt sizes.

Anti-fatigue bolting

Critical mounting points to the body structure, for powertrains and suspensions, can be subject to fatigue loading. With the Boesner 'Y-con-bolt', developed mainly for conn-rod assembly but applicable to many vibrating joints, a fastener is available which absorbs larger dynamic stresses, according to research findings regarding the influence of elasticity ratios between the bolt and the part tightened by it.

At a given dynamic stress of a bolted connection, the proportion of the dynamic load felt by the thread (stress amplitude *A*) becomes smaller with the reduced elasticity of the bolt and the increased elasticity of the tightened parts of the connection, Fig 34a. In bolt connections with eccentrically acting momentum F_A, a distinct dependence on prestressing force of the bolt stress F_{SA} results so that it seems important to provide a bolt which, despite great elasticity, possesses a high yield point and which consequently can be highly prestressed, Fig 34b.

In vibration-prone applications, Fig 35, a bolt has to be specified whose curve in the stress-elonga-

Size	A QUALITY (28 tonf/in²) Torque (lbf.ft)	A QUALITY (28 tonf/in²) Tension (lbf)	R QUALITY (45-55 tonf/in²) Torque (lbf.ft)	R QUALITY (45-55 tonf/in²) Tension (lbf)
THREADS				
10	0·104	946	0·168	151
9	0·146	118	0·233	187
8	0·227	158	0·368	256
7	0·339	209	0·543	335
6	0·074	261	0·768	422
5	0·730	346	1·165	560
4	1·045	441	1·680	703
3	1·545	579	2·490	935
2	2·360	771	3·800	1240
1	3·420	985	5·470	1580
0	4·760	1273	8·000	2100
1/4 in	6·05	1460	9·55	2300
5/16 in	11·70	2260	19·00	3680
3/8 in	2·10	3375	33·80	5440
7/16 in	33·75	4650	54·70	7520
1/2 in	50·60	6110	81·20	9860
5/8 in	101·00	9720	163·00	15 780
3/4 in	181·00	14 550	283·00	22 800
7/8 in	284·00	19 600	458·00	31 600
1 in	430·00	25 900	688·00	41 600

Fig 33: (left) Tightening torques and approximate tensions

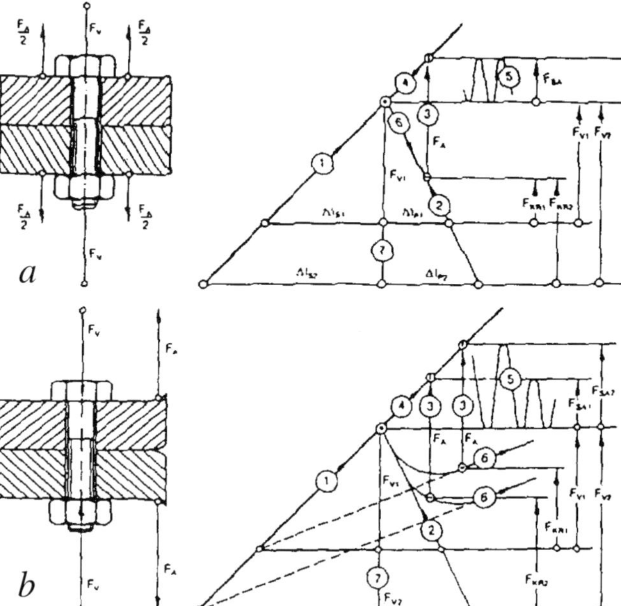

Fig 34: Tightening diagrams for co-axial (top) and eccentric load cases

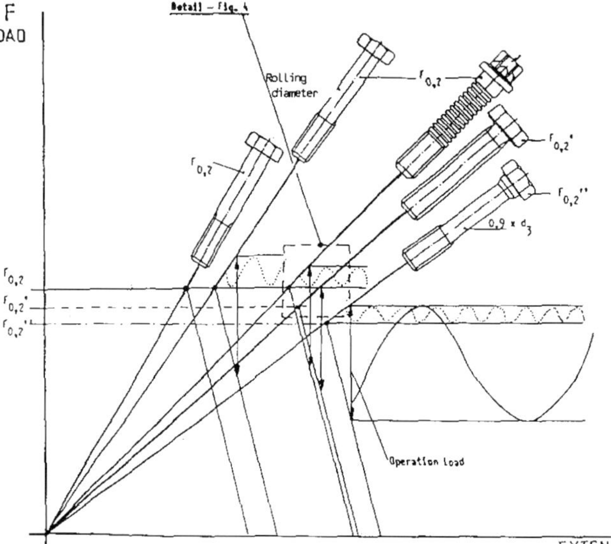

Fig 35: Shaft geometry effect on bolt resilience

tion diagram, besides having a good elastic resilience, must possess a high yield point R_{el} as well as high uniform elongation A_g. In the prestress-controlled tightened Y-con-bolt, with integrated fitted shank comprising concentrically arranged anti-fatigue rings, a high degree of prestress, up to the yield point of the computed thread dimension, and strength within the hyperelastic range, is available. The bolt body acts in a resilient manner so that the proportion of the dynamic load felt in the bolted connection is reduced and a long service life thus ensured, say Boesner.

The stress cross-sections of both the fastening thread A_S, as well as of the integrated anti-fatigue rings, A_k are uniform over the entire length of the bolt body, while the outer diameters of the fastening thread and of the fitted anti-fatigue shank differ, Fig 34b. The outer diameter of the integrated fitted shank is larger than the dimension of the fastening thread so that it is possible for the components to be connected to be brought together and retained in a snug fit. The

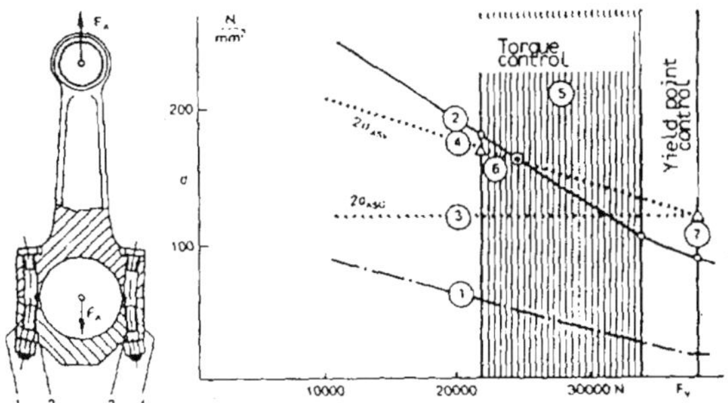

Fig 36: Reduction in alternating stress with rise of prestress-force

same stressing cross-section over the gripping length ensures, when speedily tightened and while subject to prestress within the elastic and hyperelastic range, no elongation crack occurs and that, in the stress-elongation diagram within the hyperelastic range, the uniform elongation A_g is increased, Fig 36.

In this case, connecting rod and cap are tightened to the greatest possible extent by means of the connecting rod bolts; a yield-control mode or angle-control mode prestressing of up to 0.2% yield point is effected. This requires that the bolts possess high uniform elongation in stress-elongation, Fig 37, satisfied by the Y-con-bolt, with both high elastic elongation and high tensile strength R_m. It is thus possible when tightening or prestressing up to 0.2% yield point, or within the hyperelastic range while subject to an additional operating load, for the elastic-plastic linear change to be absorbed by the bolt prior to the tensile strength R_m and the incipient constriction of the bolt reached. Also, unforeseen changes in the ratio of tightening torque to prestressing force (wide spreading of the coefficient of friction or deformation of the separating line) are absorbed by the high uniform elongation of the bolt — so that the fracture will not take place prematurely.

Those bolts with snug fit and waist, or rolling diameter, within the shank area are, during prestressing within the hyperelastic range, subject to shortcomings, say Boesner, Fig 38. For example, the static stability under load of the waisted bolt ($A_T = 0.9d_3$) is smaller due to the tapering anti-fatigue shank, compared with the fastening thread. The cross-section of the waisted shank A_T amounts to only 90% of the

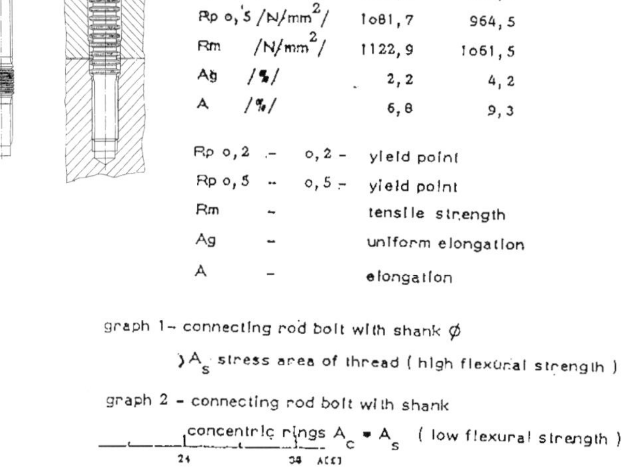

n	1	2
Rp o,2 /N/mm²/	1o59,3	939,1
Rp o,5 /N/mm²/	1o81,7	964,5
Rm /N/mm²/	1122,9	1o61,5
Ag /%/	2,2	4,2
A /%/	6,8	9,3

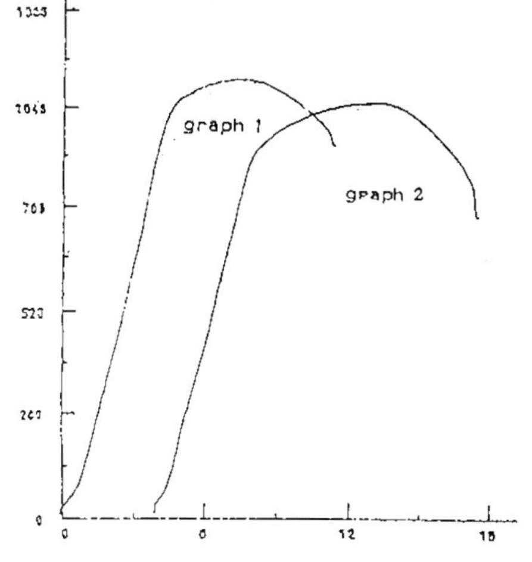

Fig 37: Stress (N/mm²) vs percentage elongation

cross-section of the thread core diameter A_{d3} of the fastening thread — so that the yield point level of the waisted bolt and thus the possible maximal prestressing are reduced by approximately 20 times in comparison with the Y-con-bolt, Fig 39. It has also been shown that, despite the waisted anti-fatigue shank, only an exceedingly small uniform elongation exists in the stress-elongation diagram. On account of the structurally contingent length of two short waisted shanks interrupted by the cylindrical snug fit, the position of the constriction is already predetermined when prestressing within the hyperelastic range so that, immediately after reaching 0.2% yield point, a drop in the tensile strength in the stress-elongation diagram results in a fracture. This bolt is not suitable for a double or multiple prestressing to yield point. Furthermore, in the event of a dynamic stress, a mating surface corrosion occurs on the cylindrical snug fit which produces a molecular surface destruction. In the Y-con-bolt, with concentric rings, this is prevented due to its special geometry, Fig 40. The slender-shank connecting-rod bolt (shank diameter = thread flank diameter d_2) does exhibit an equally high static stability under load but it is not resilient and possesses a smaller uniform elongation and, from a design point of view, cannot be usefully provided with a snug fit so that rod-to-cap guiding means have to be additionally provided, with a set-pin or fitted sleeve.

In eccentrically tightened and eccentrically stressed connections, where the majority of the bolted

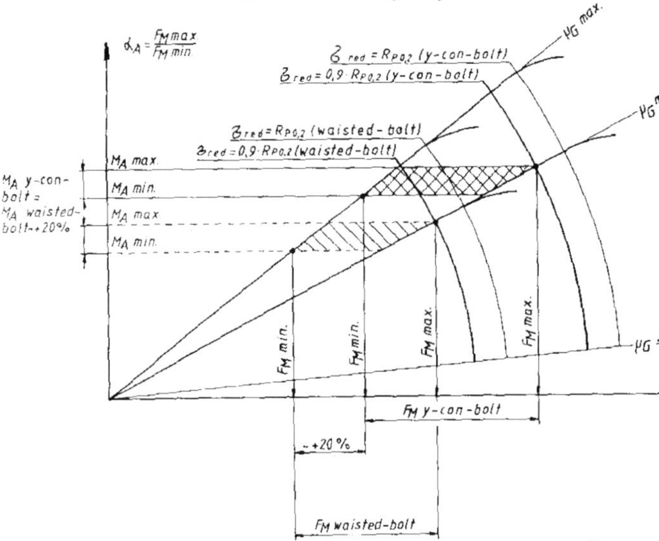

Fig 39: Tightening vs prestressing torques with friction and tightening torque spreads

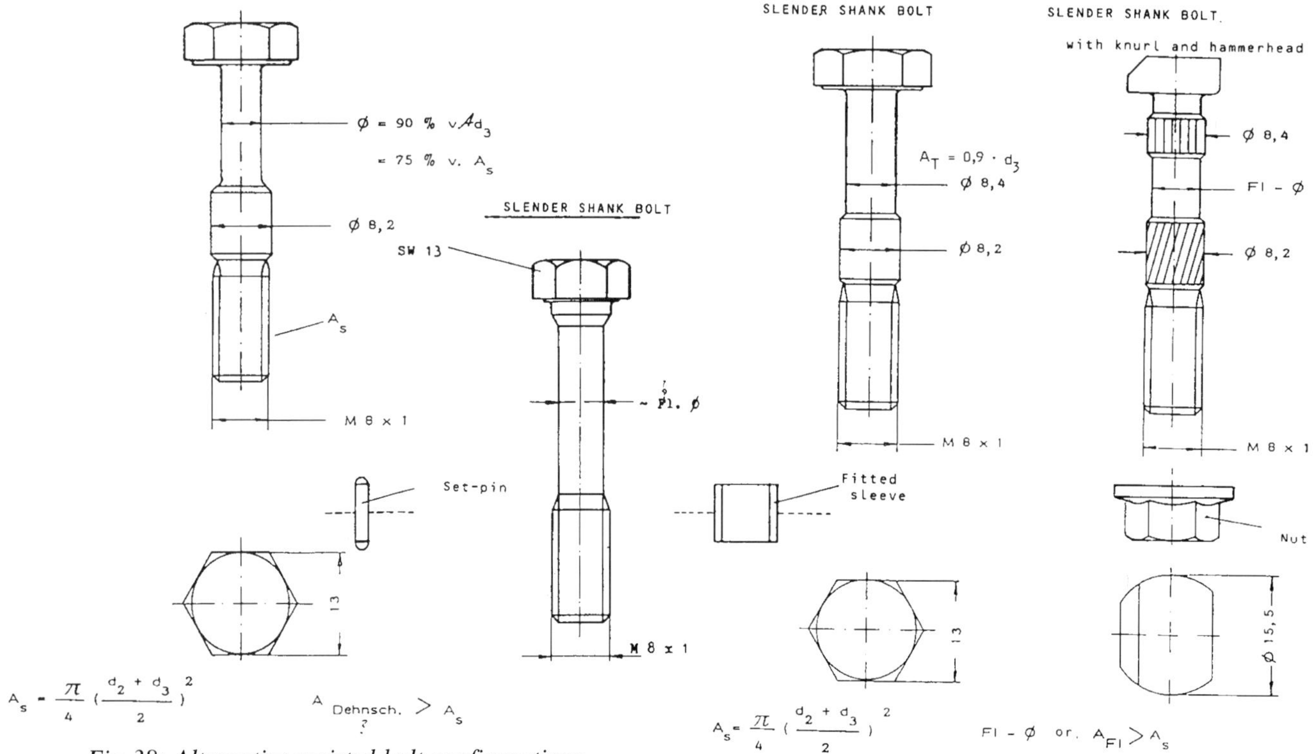

Fig 38: Alternative waisted bolt configurations

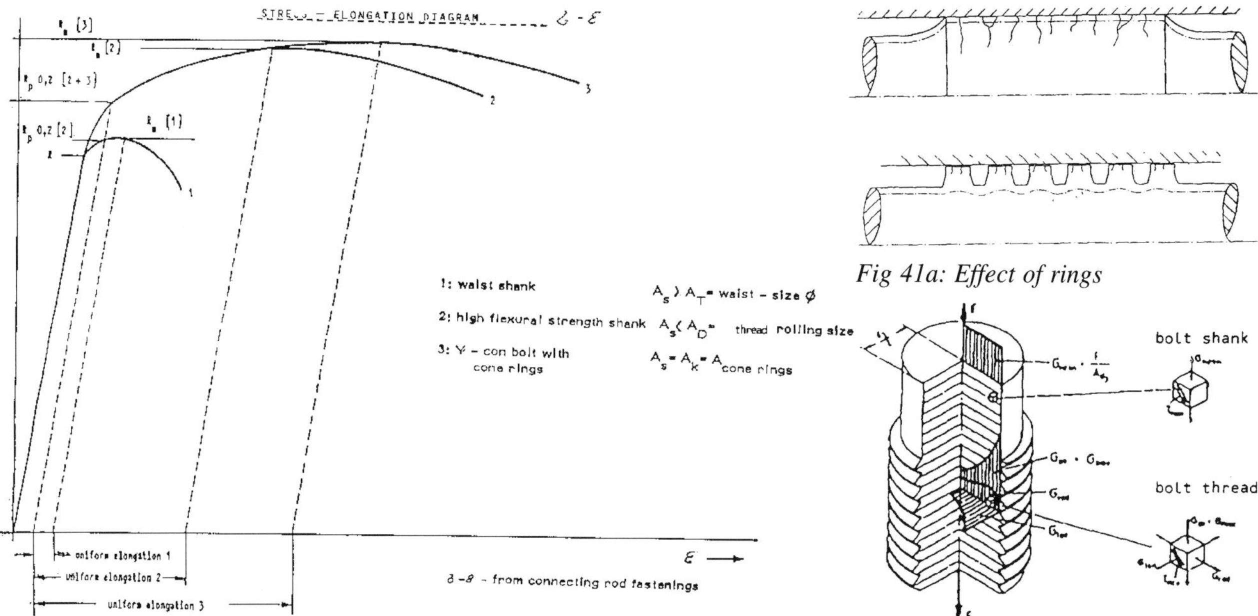

Fig 40: Stress vs elongation diagram for Y-con-bolt

Fig 41a: Effect of rings

Fig 41b: Stress conditions in fast-stressed threaded bolt

connection is stressed and/or tightened outside the centroidal axis, the bolting mechanics exhibit a quite clear dependence on prestressing force for the bolt stress. Owing to the greater prestressing force of the Y-con-bolt, the service life in such applications is distinctly improved. A further constructional feature is that, when subject to tensile stress, the transverse contraction of the bolt is prevented. Axial, tangential and radial stresses over the grooved shank are thus produced, Fig 41a. In materials that possess adequate platicizing capability, this stress condition results in an increase in the yield point and tensile strength, subsequent to a stress relief, when the connecting member had already been prestressed to the elastic limit of 0.2%, Fig 41b. Consequently, the bolt is particularly well-suited for connections which are twice or three times hyperelastically prestressed. The elastic limit of 0.2%, in the second tightening while a uniform elongation of the bolt is maintained, is higher than after the first tightening within the elastic limit range, Fig 42. Due to the induced stress in the grooved shank following the first hyperelastic prestressing, an improvement in fatigue strength is achieved, analogous to the final rolling of a thread.

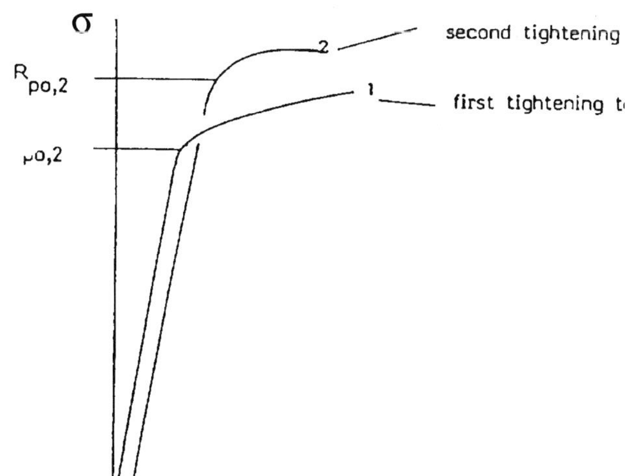

Fig 42a: Stress elongation diagram for Y-con-bolt with cone rings showing first/second tightenings-to-yield

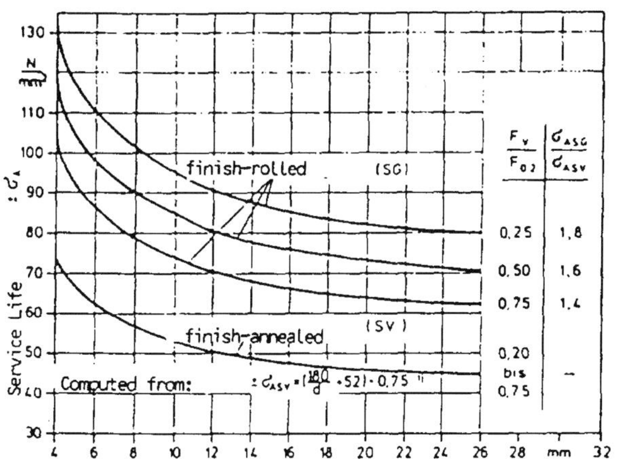

Fig 42b: Service life of threads

The internal radius in the groove root of the fitted anti-fatigue shank is greater than in the core of the fastening thread, alternatively the groove root of the anti-fatigue shank is constructed in the form of a double radius which is two to three times larger than the thread core radius. The stress concentration factor, or fatigue notch factor, is thus smaller than in the fastening thread. Therefore, when subject to prestress, higher stress peaks occur in the freely loaded thread portion than in the anti-fatigue shank of the concentric rings.

In order to ensure a close coaxiality tolerance between the fit-dimension of the integrated anti-fatigue shank and the fastening thread of the bolt, both the fastening thread as well as the ring abutments are manufactured in combination. This is done with the aid of a special tool, involving an 'equalizing' rolling, in one operation. Through a light re-rolling, with a smooth pair of rollers, the dimension of the fit, up to a tolerance of 0.01 or 0.009 mm, is produced. With this manufacturing method, both the precise fit relative to the fitted bore, as well as the close coaxiality tolerance between the fastening thread and the diameter of the fit on the crest of the concentric fitted anti-fatigue rings, is ensured, Fig 43.

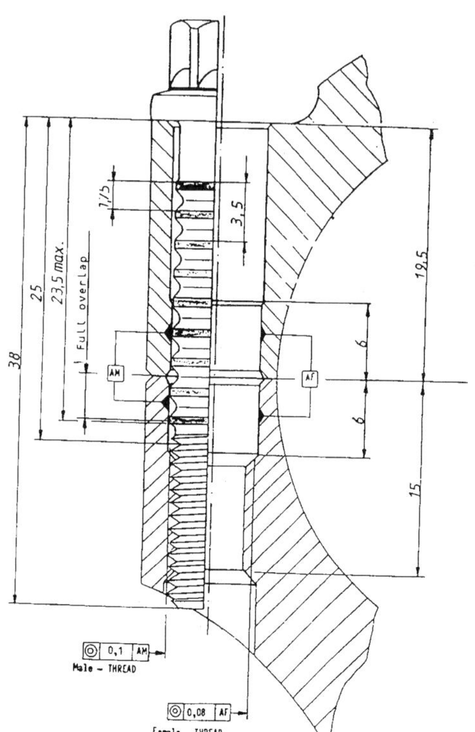

Fig 43: Ensuring precision abutment of mating parts

Riveting systems

As possibly the most common body panel jointing technique in commercial vehicle construction, riveting is governed by standard codes of practice which can lead to improved efficiency. A number of variations on the rivet are available for special purposes and for configurations where substantial loads are being transferred, stressing procedures can be adapted to desktop computer programs which allow optimum layouts to be predicted with the minimum of fuss.

British Standard BS 4620 covers solid rivets for general purpose and shows the standard heads as in Fig 44. Part (a) refers to cold-forged and (b) to hot-

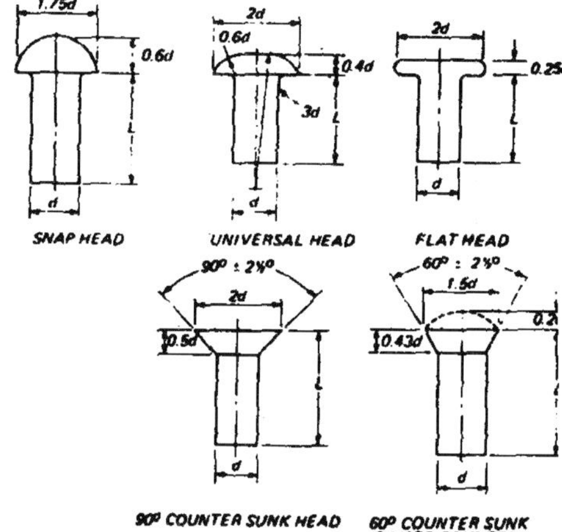

Fig 44: Types of common rivet

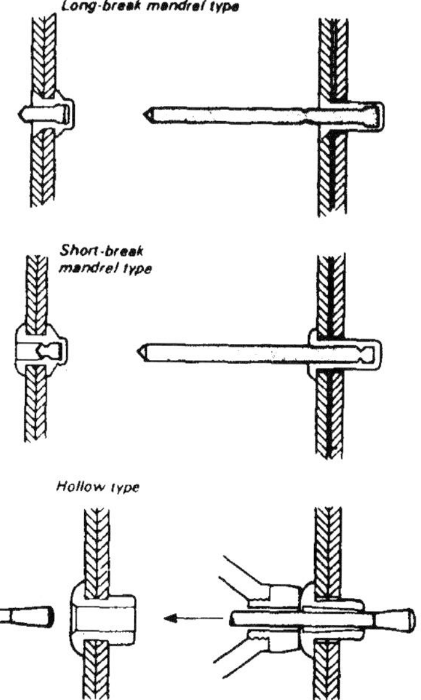

Fig 45: Types of pop-rivet

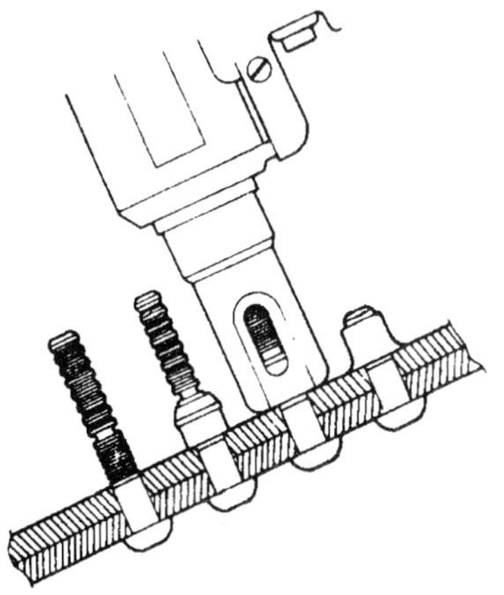

Fig 46: The Huck-bolt

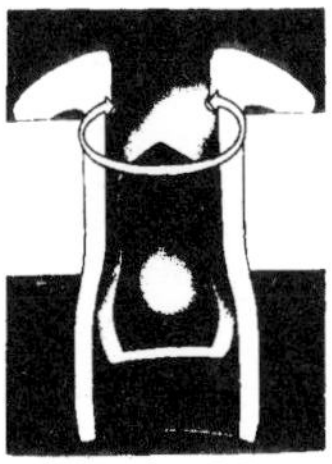

Fig 47: The Magna-Lok

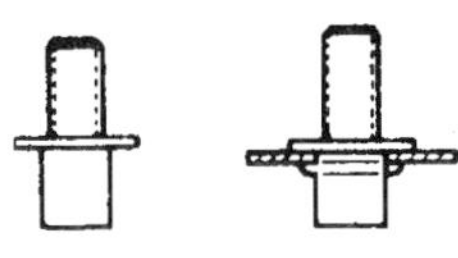

Fig 48: Riveting captive nut

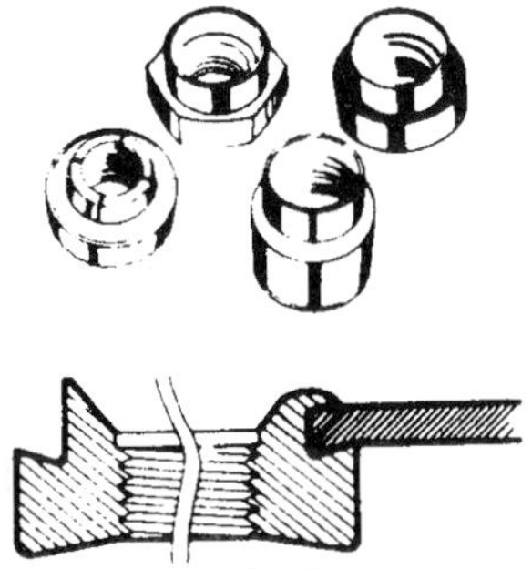

Fig 49: Alternative nut fixing by Delta

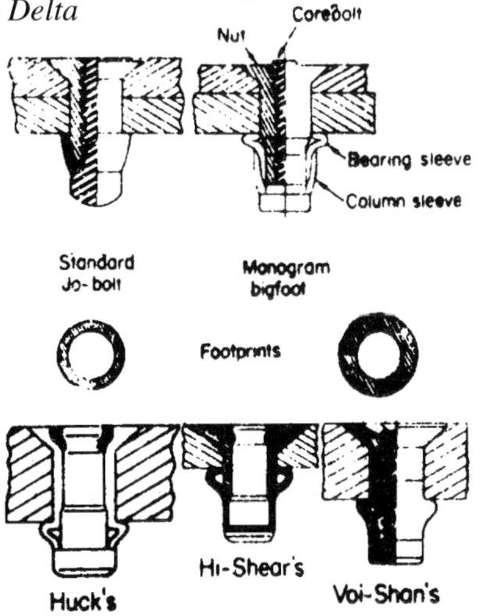

Fig 50: Systems to avoid pull-through

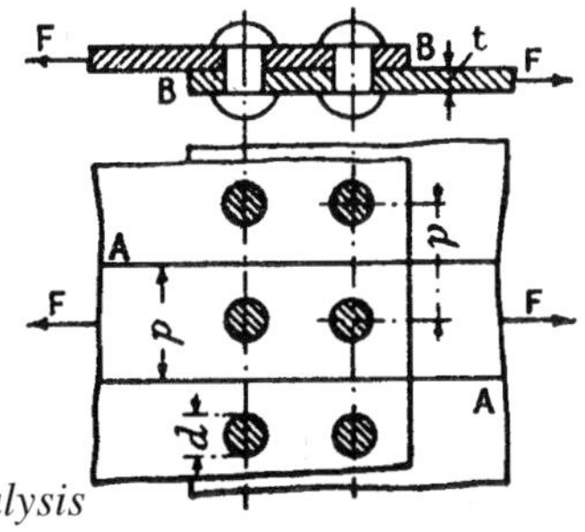

Fig 51: Lap-joint analysis

forged rivets, usually applying to larger sizes. The sizes involved are from 1mm to 39mm diameter; small sizes only are covered in a further standard, BS 641:1951, heads for these being snap, pan, mushroom, flat and countersunk to 60, 90 and 120 degrees. Semi-tubular and bifurcated designs are covered in BS 4895:1981 and 4894:1973 respectively.

Pop-rivets widely used in body construction fall into the tubular category and are classified as shown in Fig 45. It is recommended that rivets for joining aluminium alloy be made from NR6 or HR30 material — alloys suitable for cold driving. Accepted practice is to use rivets with a diameter of one to three times the thickness of the thinnest member through which they pass — and pitched at three to six times their diameter. Pitch is usually increased by an amount depending on the thickness of the materials being joined and the need to avoid separation of the sheets between rivets.

Falling between the categories of 'rivet' and 'bolt', the Huck-bolt (Fig 46) has a positive-lock collar which results in a vibration-resistant fastening. The clamp force applied during the fixing results in a predictable uniform preload and a combination of the high shear resistance of the rivet is combined with the tensile resistance of the bolt. Installation rate is as high as 25/minute; the joint is waterproof and can be undone with a special tool. A high-tensile version, the HMP suits special applications like the mounting of concrete pumps on heavy-duty chassis. Greater fatigue resistance is claimed against comparable bolted systems due to the relative smoothness of the contoured grooves.

Another variation on the rivet is the Furma self-piercing fastener which can be used to join even pre-painted panels. A variant on the pop-rivet is the Unitac Multigrip blind riveting system. Here the shank of the rivet has peripheral grooves which collapse in a manner such as to expand the shank in the hole as the mandrel is pulled through them. The mandrel ultimately breaks to form a solid head on the blind side of the workpiece, thus sealing it. Huck's Magna-Lok fastener (Fig 47) is claimed to achieve a similar result. The riveting principle can also be used to fix a captive nut or threaded stud as in the Fastener Techniques and Delta Metal Products systems shown in Figs 48 and 49. For softer panel materials, provision for high pull-through resistance, and avoidance of 'volcano' effect, can be obtained with the systems shown in Fig 50.

Riveted joint load distribution When calculating the stresses and strains on riveted joints, it is usually assumed the load is transmitted entirely by the rivets — friction between the connecting plates being ignored. When the centre of gravity of the rivets is on the line of action of the load, all rivets are assumed to carry equal parts of the load. If the line of action of the load is eccentric to the centre of gravity of the rivet group then the torque due to eccentricity must be added to the direct load in computing individual rivet shear forces.

Lap joints such as that in Fig 51 can be analysed by considering just a pitch width strip of the joint and noting that the minimum cross sectional area of the sheet occurs between the rivet holes. Three main failure criteria for the joint are tearing of the sheet, shear of the rivets or crushing of the sheet/rivet at the interface. For the forces causing these three failure modes to be equal

$$(p - d)t\sigma = \pi n \tau d^2/4 = ndtb$$

where p is rivet pitch, d diameter, t sheet thickness, n number of rivets, s limiting tensile stress in sheet, t limiting shear stress in rivet and b limiting compressive stress in sheet/rivet. Usually the third term is ignored as for most materials too large diameter of rivet would theoretically be required if the clamping action of the rivet is not taken into account. The diameter of the rivet is therefore normally chosen empirically and the pitch determined from the first two terms. A corner or flange joint as in Fig 52 often involves torsional forces as the load would not pass through the centroid of the joint. In the case shown each rivet would be subject to load Qd^2/sds in the direction of loading together with a force $QLr_i d^2/sr_i d^2$ perpendicular to the line between the rivet and the joint centroid, due to the torque of the applied load about the centroid.

Welding technology and systems

For bodywork welding, the SMMT Bodybuilders Code of Practice requires adherence to British Standard BS 5750:Part 2 in which it is pointed out that welders should be provided with laid-down procedures and that finish welds should be inspected in accordance with BS 5289 in the case of fusion welds. There is thus a new focus on weld quality and the need to understand the basic processes involved.

Resistance welding, by spot or seam techniques, is appropriate to a number of relatively thin sheet materials — while for plate, tube and sections, fusion welding is often more appropriate, with melting temperatures obtained either by oxy-acetylene or electric-arc. Different conditions are required for

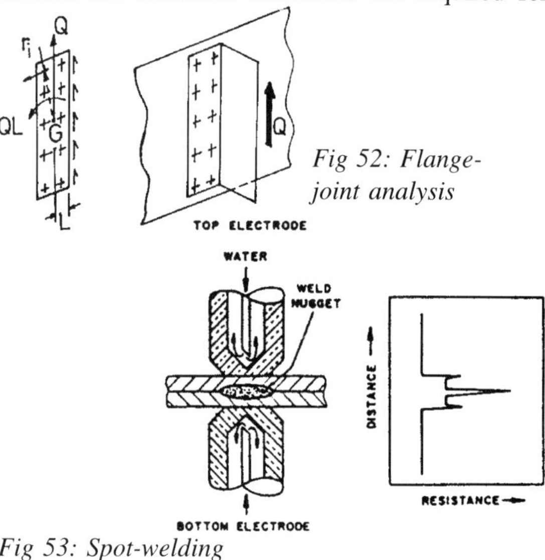

Fig 52: Flange-joint analysis

Fig 53: Spot-welding

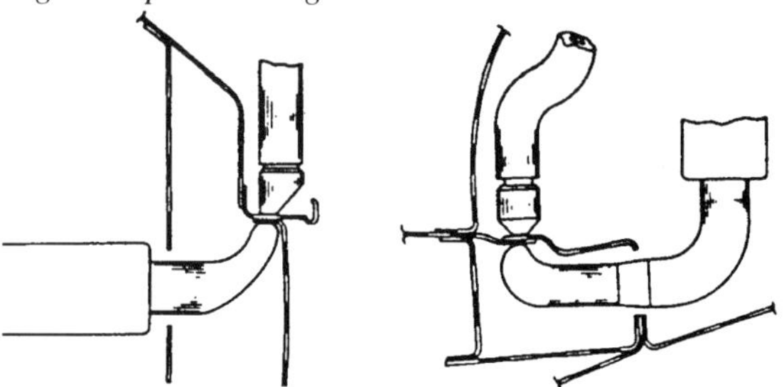

Fig 54: Reaching inaccessible flanges

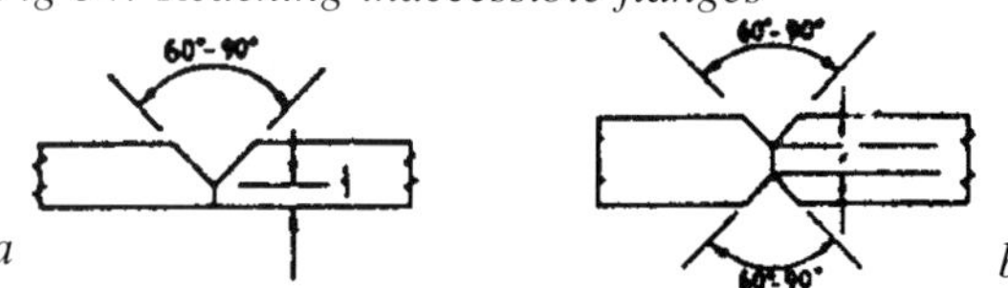

Fig 55: Single and double vee fillet welds

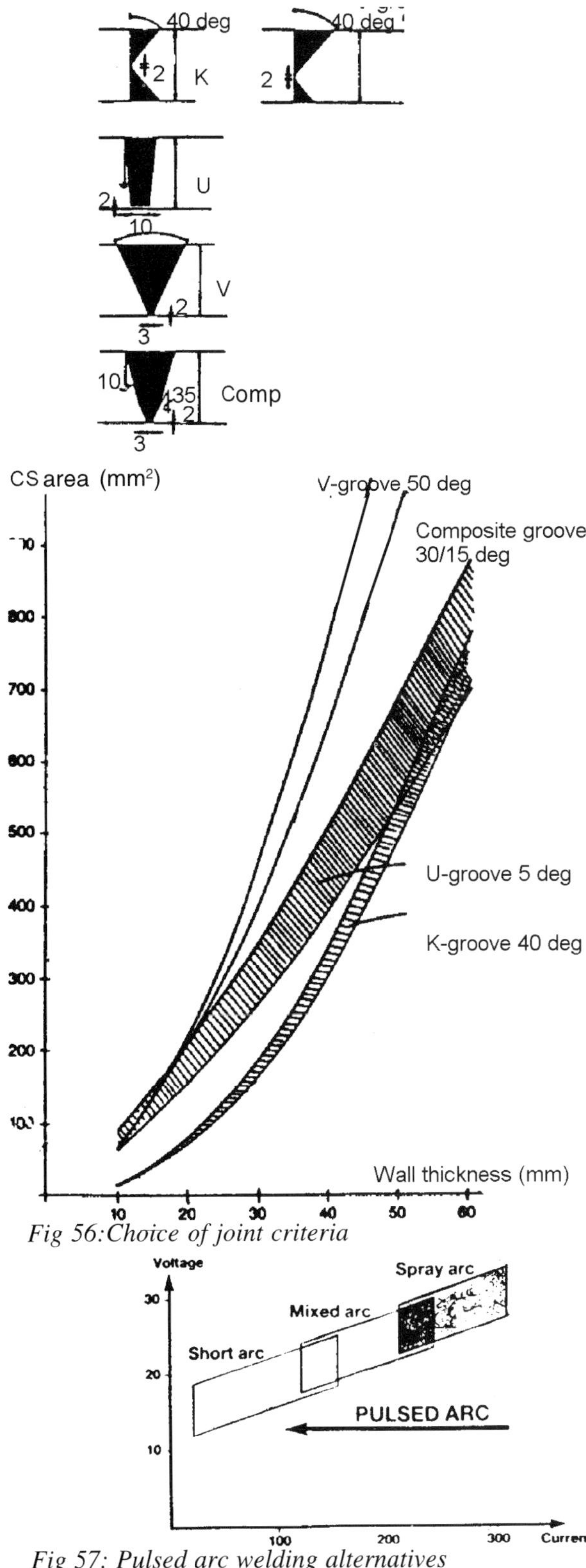

Fig 56: Choice of joint criteria

Fig 57: Pulsed arc welding alternatives

welding steel, rather than aluminium alloys, and for welding heat-treated, extra high strength, materials.

Spot-welding (Fig 53) has advantages of high application speed and relatively low cost (after capital costs are defrayed) but very careful control of welding parameters is important to the achievement of satisfactory quality. There is also the problem of surface deformation on aesthetically critical areas. Part fit-up is important — particularly when using very high strength alloys. In the case of coated materials, such as zinc/steel, longer weld times are involved in 'breaking through' the coatings. Accessibility of relatively hidden flanges is possible with careful electrode design, Fig 54.

Fusion welding of steel can involve a variety of processes. Arc welding may be metal inert gas (MIG) or tungsten (TIG) shielded. Gas metal arc welds (GMAW) can be made in all positions given control over arc current and voltage; electrode wire and shielding gas feed, also torch, speed. Advantages are high productivity and reliability coupled with low cost. Disadvantages include the bulkiness and unwieldiness of the equipment. A similar process is flux cored arc welding (FCAW), with shielding provided by gas formed in burning the flux. This uses simpler plant but can involve welding slag. Another fusion process is electron beam welding in which the weld is created by focusing a high energy stream of electrons on to the workpiece — with the advantage of providing a deep narrow weld zone, offset by the cost and difficulty of providing vacuum conditions.

Aluminium alloys can be welded using oxy-acetylene gas welding up to thicknesses of 0.125 inches when nearly pure aluminium is involved. Generally TIG or MIG arc welding is used in bodybuilding — where structural alloys are the rule — but it is important to note that any strength of the parent material obtained by heat treatment, in the joint region, will be lost. Minimum thickness for welding by MIG is usually 1/8 to 3/16 inch but there is practically no upper limit. A fine-wire process extends the lower limit down to 14 SWG and even lower if support is provided by jigging incorporating backing chills.

TIG is not quite so economic as MIG for aluminium alloys. As the arc is struck between an argon-shielded non-consumable electrode, and the material to be welded, filler wire is added separately as in gas welding. It can be used for thicknesses from

20SWG to 3/16 th inch. From 3/16 to 1/4 inch either TIG or MIG can be employed but above 1/4 inch, MIG is usually more economical and effective. TIG is a slower process than MIG, thus degree of control over weld shape is high. This is useful in welding extruded body members.

Butt welds in aluminium alloy require a temporary backing strip with MIG but not with TIG. Material up to 3/16 inch can be welded without a bevel, in one pass, by both MIG and TIG. From 3/16 to 3/8 inch, a single 70 degree included angle vee bevel is required with root face equal to 1/3 of the plate thickness, Fig 55a. Above 3/8 inch, a double-vee, Fig 55b, is required. Fillet welds do not require bevelling of abutting edges unless severe dynamic loading is required of the finish part. However, a leg length must be deposited equivalent to the member thickness. Where dynamic loading is envisaged, excessive convexity of the fillet should be avoided. It is possible to compensate for some of the strength loss, mentioned above with heat treatable alloys, by using a permanent fillet welded backing to a butt joint. With non heat treatable alloys a weld strength over 90 percent of the annealed strength can be expected.

Mechanised welding, in which the welding is done by machine, but not as in automatic or robot welding where the operator makes little input, involves constant supervision and manual correction of the welding parameters and joint tracking. According to ESAB, the opportunity could arise of replacing a highly trained welder with a machine operator. Key to success is planning of the production project, it is claimed. Choice of joint, Fig 56, should depend upon process, welding position and thickness. Another proposal by the company is for improving the welding of stainless steel components, using the pulsed-arc process. Here a low background current is used just enough to keep the arc alive without migration of metal filler. Pulses of higher current then cause successive deposition of the metal — but without the short-circuiting that takes place in conventional short-arc welding. Advantages are minimum spatter, more uniform welding and a lower heat input than with the alternative spray arc process, Fig 57.

Friction stir welding Because friction welding normally entails relative motion between the two parts being joined, it is unsuitable for butt welding of large pieces of sheet material. In friction stir welding, however, the parts are held motionless throughout the process, which consequently is applicable to sheet and plate. This is the process that TWI has developed recently for welding aluminium alloys and strong prospects have recently been held out for commercial welding of sheet. Among applications being investigated is tailored blanks.

In friction stir welding, the probe of a rotating or reciprocating tool is inserted between the abutting edges of the parts being joined. A plasticized region forms round the probe. When the tool is drawn along the junction, the plasticized region behind the probe immediately coalesces to form a solid-phase bond. There is no theoretical limit to the length of the weld. Among advantages are low distortion and good surface appearance. Friction stir welding is also applicable to other non ferrous alloys as well as to plastics.

A rotating probe, Fig 58, which is harder than the workpiece is the crux of the process from TWI known as friction stir welding. It could revolutionise the welding of sheet-metal parts having complex geometries which are normally outside the scope of spotwelding and so introduce greater design freedom from the aesthetically-dreaded flange. No axial displacement is necessary, a tool is merely held in close contact and a portion of the probe is then entered into the workpiece to create a plasticised region around its immersed part, which solidifies on withdrawal.

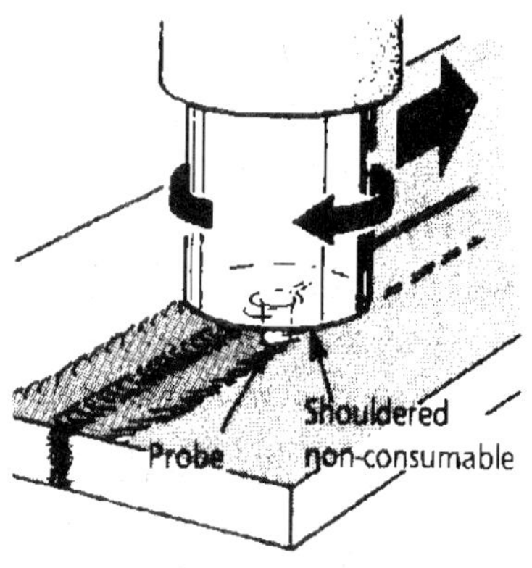

Fig 58: Friction stir welding

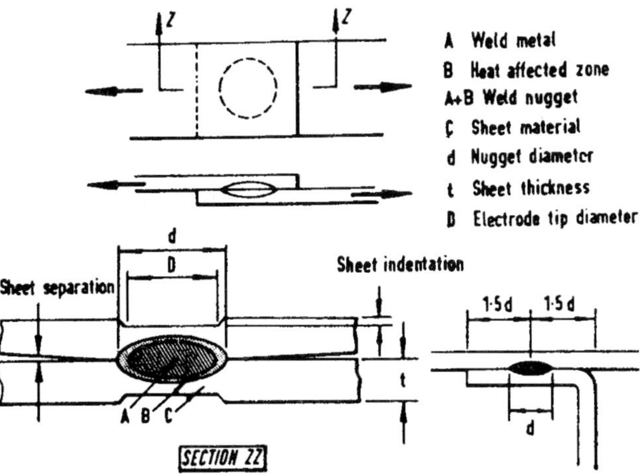

Fig 59: Spot-welding parameters

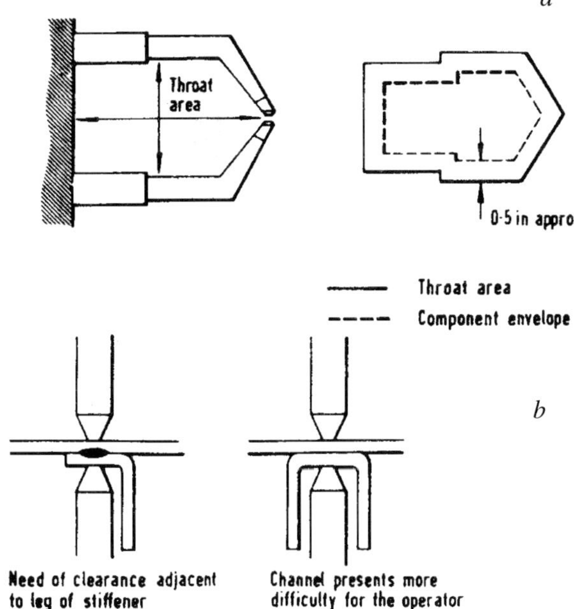

Fig 60: Spot-weld access and clearance

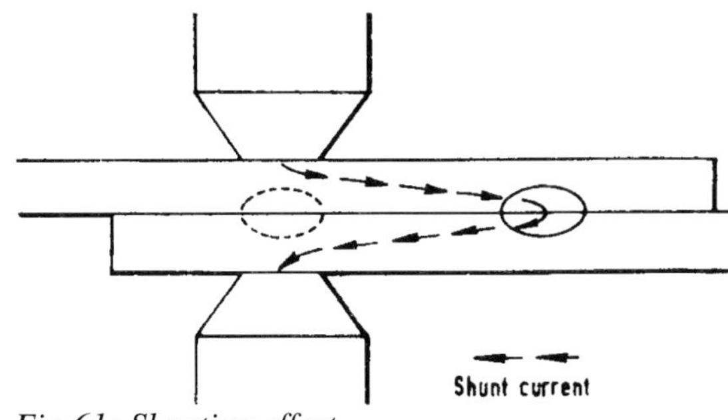

Fig 61: Shunting effect

Spot-welding technique

The basic advantages of spot welding have been listed above. Here the techniques and limitations of spot welding are considered in detail and guidelines suggested for design of flanges and other joint configurations. According to Cranfield University researchers[6] the heat and pressure required for the spot weld both need careful control to obtain optimum joint properties. Fig 59 is useful for identifying the important parameters affecting joint performance. In the case of welding mild steel sheet, electrode tip diameter is usually chosen to equal the square root of the sheet thickness in contact. Thus, where different thicknesses are being joined, different electrode sizes are required either side of the joint. The resulting nugget diameter is closely related to the electrode tip diameter and determines the strength of the joint. Sheet indentation and separation should be minimised for best performance of the joint.

Distribution of the spots on the workpiece may be limited by the throat area of the welding gun and only in the case of very high volume production is it economical to use different arm lengths to obtain different 'reaches' on, say, one particular vehicle body fabrication. Local difficulties of electrode access should also be considered, Fig 60a, as should sufficient edge distance to contain the weld nugget, Fig 60b. The two principal failure modes for a spot-welded joint are shearing through the nugget and pull-out of the nugget. While load values corresponding to these two modes can be found from tests on single joints, care is needed in extrapolating loadings for multiple-spot joints due to the so-called 'shunt' effect, Fig 61, in which it is seen that a portion of the current can pass through an adjacent, previously

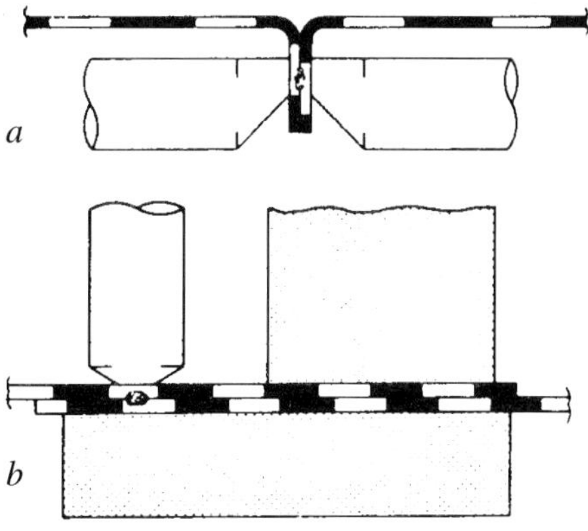

Fig 62: Concealed flange and lap joints

made, weld. Thus load per spot can also reduce with decreasing spot pitch. To optimize the appearance of spot welded structures, it is normal to conceal or partially shield the joints. A relatively inexpensive way to join panels is to use the coach joint of Fig 62a with flanges turned towards the inside of the body. The visible line of the joint is preferably made to correspond with a styling feature; a PVC plastisol sealer can be used between the flanges to obtain weather proofing. If, by contrast, a lap joint is necessary the spot-weld can be made less visible by using a flat area of copper to replace the electrode on the show side of the joint, Fig 62b.

Fatigue testing of spot-welded joints is gaining in acceptance as a design proving technique, Fig 63, due to AISI, showing a comparison in performance between carbon steels and high-strength-low-alloy steels. Such tests have shown that there is a critical weld diameter above which failures behave in a normal fatigue crack mode. The crack grows around the nugget and pull-out failure occurs. Below the critical diameter, however, rather than pull-out, an inter-facial failure occurs. Fig 64 gives a table showing weld capacity in terms of ultimate shear strength per spot, V_{su}. Special set-ups can be designed for spot-welding pre-coated steels which are painted on one side only. Galvanised steels can also be welded, but with higher current and longer weld time required to overcome the interface resistance and shunting effects due to current flow through the higher conductivity surface coating. Special procedures are also available for welding aluminium and lead/tin coated steels. In situations where only one side of a jointed sheet can be accessed, the technique of argon-arc fusion spot welding can sometimes be employed to advantage. In so-called puddle welding the heat source is an arc struck across a pre-determined gap between a non-consumable tungsten electrode (negative) and the parts to be welded (positive). A shielded argon atmosphere is provided and the upper component is melted through to the lower one. Stud-welding is another related process which uses short duration of arc (38 milliseconds) to produce stud attachments to panels without causing any back-marking. A small projection on the weld face, Fig 65, first contacts the mating part and a very high current passes which disintegrates the projection. Current continues to flow across the resulting gap, in the form of an arc, while the stud is driven home by spring or air force.

Thickness of thinnest outside sheet (inches)	V_{su} (kips)	Thickness of thinnest outside sheet (mm)	V_{su} (kN)
0.010	0.130*	0.25	0.58
0.020	0.437	0.51	1.94
0.030	1.000	0.76	4.45
0.040	1.423	1.02	6.33
0.050	1.654	1.27	7.36
0.060	2.282	1.52	10.15
0.070	2.827	1.78	12.57
0.080	3.333	2.03	14.82
0.094	4.313	2.39	19.18
0.109	5.987	2.77	26.63
0.125	7.200	3.18	32.03
0.188	10.000*	4.77	44.48
0.250	15.000*	6.35	66.72

Fig 64: Weld capacity

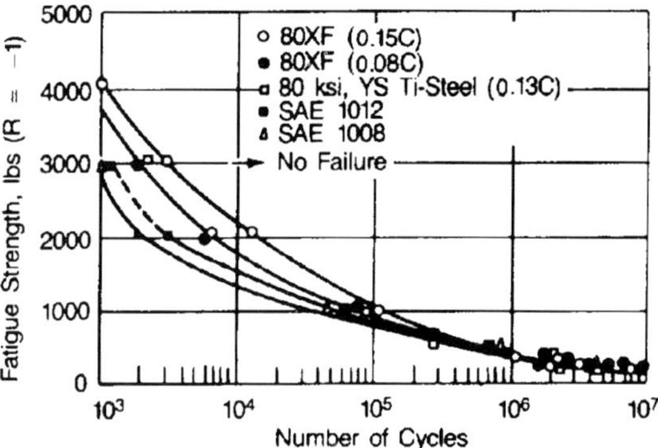

Fig 63: Fatigue strength of spot welds

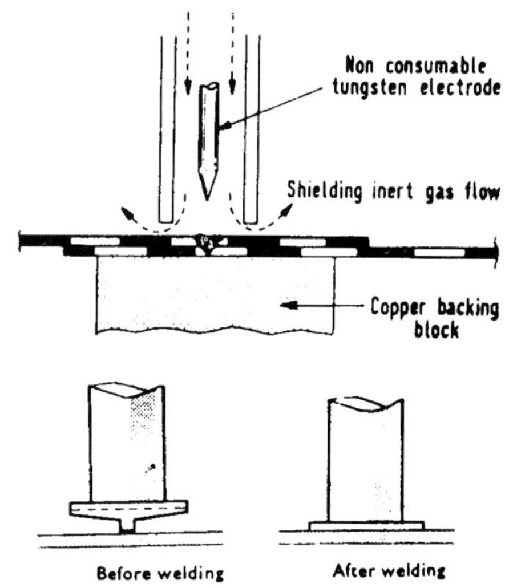

Laser-welded tailored blanks

Researchers at Isoform and Sollac[7], point out that weight reduction of the body-in-white; reduced usage of steel and reduced tooling cost, as well as increased stiffness of the fabricated body-shell: all are advantages of laser-welded tailored blanks. There are also benefits in fatigue resistance, corrosion resistance and crash performance.

In their paper, they demonstrate the advantages with the example of the car body side panel shown as Fig 66, involving A and B pillars plus quarter panel stamped in one piece. Typical current gauges of sheet steel are 0.8 mm for the main panels and 1.5 mm for local pillar reinforcements. By adopting the tailored blank route, Fig 67, the percentage of load-bearing sheet in the structure rises from 30 to 50%. In place of the reinforcements, local panel-thickness increase is applied: by 1.3 mm for A pillar, 1.4 mm for B pillar and 0.65 mm for quarter panel.

With such an approach the body-side unit can be reduced in weight by 15% and, to provide improved crash performance, the body steel used for A and B pillars is substituted by 260 MPa yield strength bake-hardening steel. For this purpose a finite-element computer program called ISOPUNCH has been developed by Sollac which will evaluate blank formability and another called RADIOSS for computing crash performance.

In applying welded blanks throughout the body-in-white, the table of Fig 68 has been prepared. Key parameter is weld length related to space between parts. Blanks can be coated or uncoated and thickness ratio from 1:1 to 1:3. The table of Fig 69 shows main part categories which can benefit from the process. Due to the small welding zone, a small gap between sheets and good alignment with the welding beam is needed. Tests have shown the gap between two sheets must be lower than 10% of the thinnest sheet of the fabricated assembly.

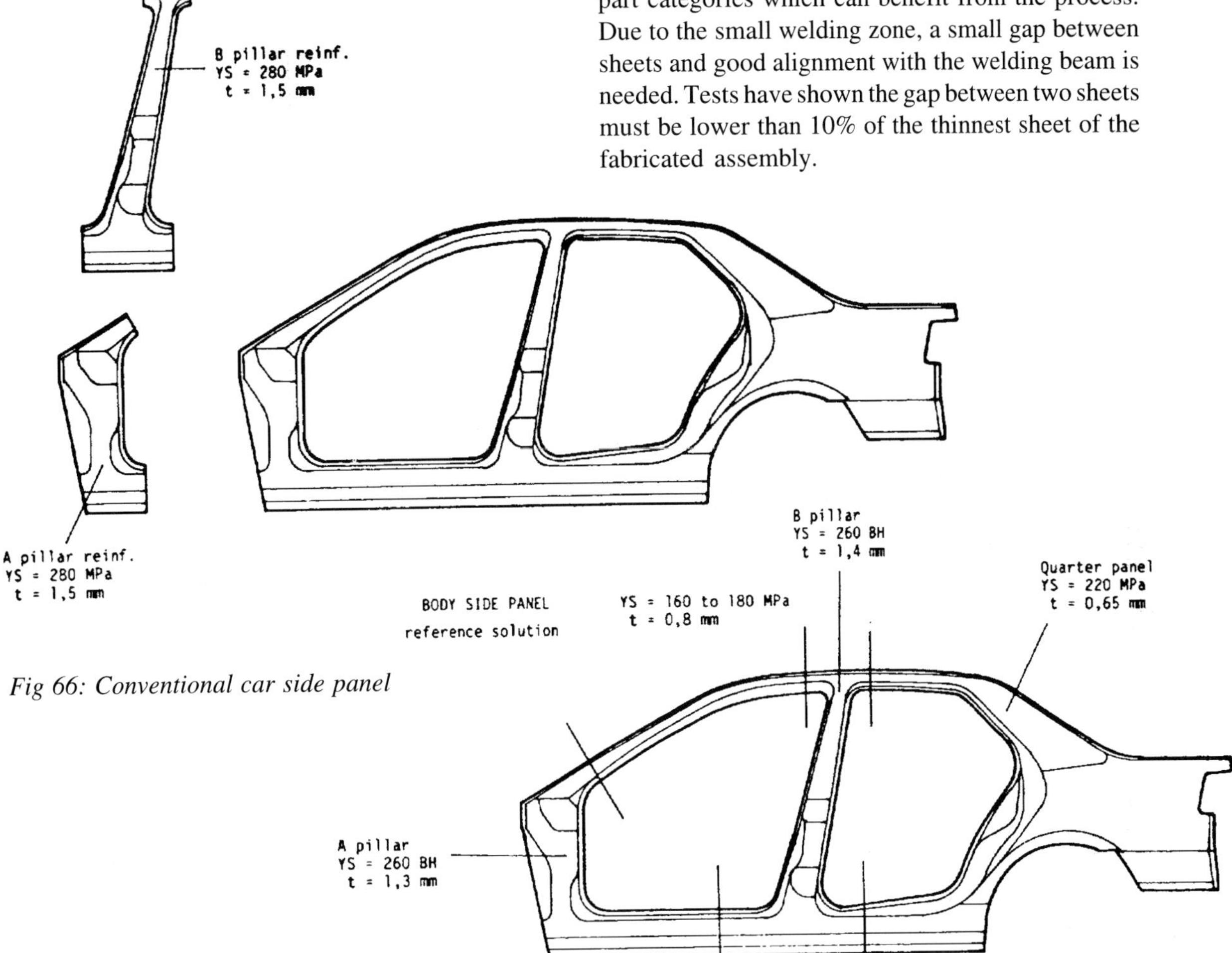

Fig 66: Conventional car side panel

Fig 67: Side-panel with tailored blanks

Fig 70 shows how gap is due to either lack of rigidity in the shearing equipment or the cut blanks having curled edges there is also the problem of transit damage, to cut edges, in moving from one machine to another. During welding, the gap can also vary due to heat expansion effects; generally, over a 500 mm seam, with 2 mm sheet thickness, thermal opening exceeds the maximum permissible gap of 10% sheet-thickness. To allow for this, welding speed must be reduced, to increase seam width, or a cooling system introduced. Clamping integrity is also important.

As the welding zone is less than 1 mm, this must be accurately positioned on the contact line of the sheets, typical allowable tolerance being +/- 0.1 mm, remembering the higher the welding speed the thinner the welded zone, Fig 71. The offset of the surfaces must also not be out of plane with the main sheet by more than 10% of the thickness of the thinnest sheet.

Sollac have patented a device for reducing the need for such high precision to the welding line. This introduces lateral pressure for providing elastic deformation of the sheet to reduce the gap between

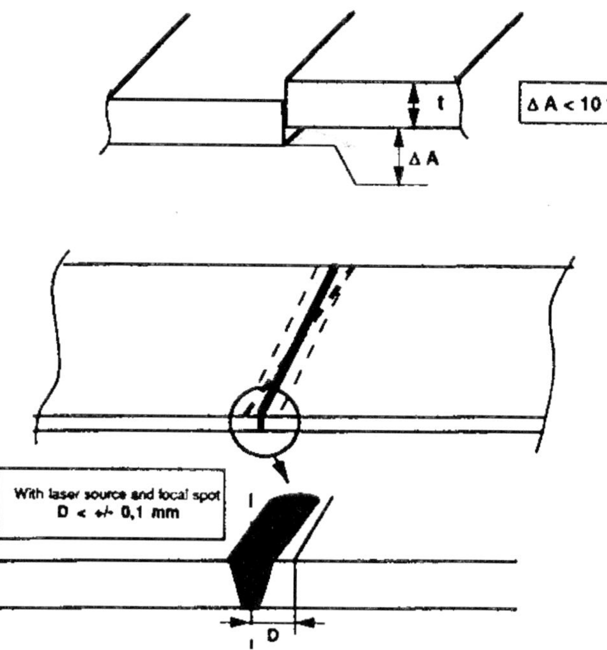

Fig 70: Gap aspects

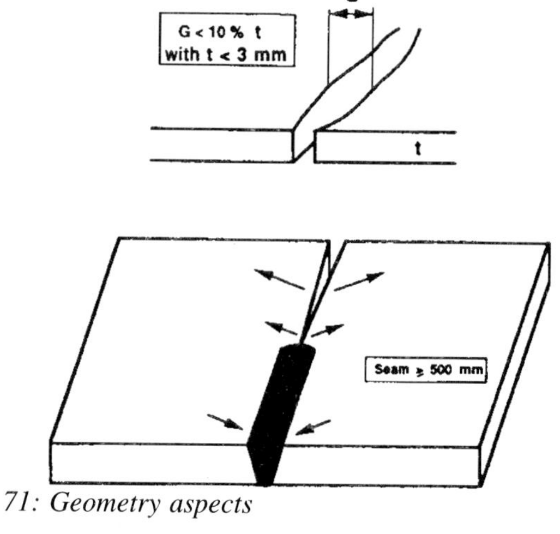

Fig 71: Geometry aspects

Kind of blanks	Parts	Kind of welds
1. Blanks with a small weld length	Rails Cross members Pillars Rocker panels	Linear welds with a length < 1 m
2. Blanks with a big weld length	Doors inner Wheelhouses Floor panels Dash panels Roofs	One or two linear welds with a total length > 1 m
3. Big sized blanks	Body side panels Doors inner	Non parallel linear welds. Discontinuous welds with seam < 50 % part width.

Fig 68: Application of steel tailored-blanks

	Safety	Weight reduction	Cost	Part sizes
Rails, cross members, pillars	•	•		
Deck lids, door frames			•	
Rocker panels			•	
Deck lids inner, doors inner, wheelhouse.		•		
Floor panels	(*)	•		(*)
Body side panels		•	•	
Truck cabs (roof outer)				•

Fig 69: Categories to benefit from steel tailored-blanks

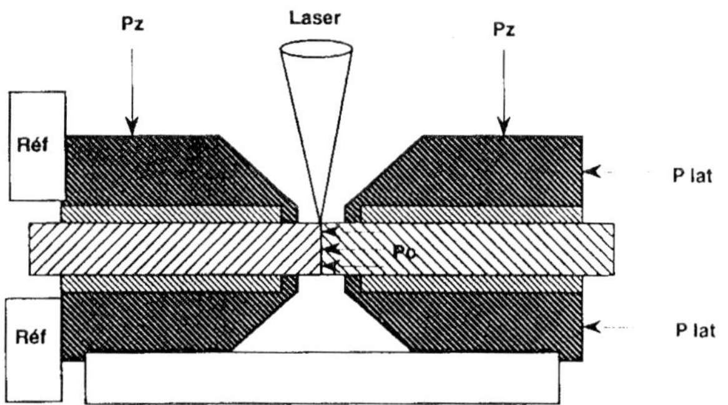

Fig 72: Applied pressures

sheets. By applying 30 MPa pressure, gap between cut sheets in traditional shearing methods can be made lower than 0.02 mm. There is also vertical pressure applied to prevent sliding of one sheet over the other, Fig 72, and a high friction base is required which must be sufficiently flexible to avoid any panel deformation, as well as being able to withstand pressure of 10 N/mm^2.

During production trials, maximum gaps that could be tolerated were 0.8 mm. Some 5000 panels were produced from three blanks, with two welds, reject parts being less than 2%. By suitably automating the process the company are confident that short enough cycle times can be achieved.

relief in the blank holder in the area of the welding seams

upper blank holder

tailored blank

bottom blank holder

consideration of the changing thickness in the blank holder and punch

Fig 73: Varying sheet thicknesses in the tool

Aluminium tailored blanks This topic was tackled by VAW Aluminium researchers[8], who consider a 40% weight saving is possible by applying the process to their material. Feasibility studies have been carried out on AA 5182 alloy (Al Mg 4.5 Mn 0.4) for body inner panels of thickness range 0.8 to 1.5 mm. Trials were made with both TIG and CO_2 laser welding, in both cases using SG-Al Mg 4.5 Mn Zr fillers. Both weld quality and formability of the welded blanks were measured.

Tests were carried out in two segmented tools, involving relieved blank-holders and punch, to achieve appropriate surface pressure over the blank holders, Fig 73. Test specimens were cups having two seams, the second having an angle of 20 degrees with the material flow. Test programme for the two welding processes is illustrated in the table of Fig 74.

To evaluate fatigue behaviour, welded flat specimens with different sheet thicknesses were

material :	→ AA 5182 = AlMg 4.5 Mn 0.4 w27	
combination of sheet thicknesses :	→ 0.8 + 1.0 mm → 0.8 + 1.5 mm → 1.0 + 1.5 mm	
welding process :		
- Laser :	→ 5 kW CO_2 - Laser (RS 5000 ; Rofin Sinar)	
	→ welding speed :	v = 7.2 m/min
	→ welding filler :	SG - AlMg 4.5 MnZr
- TIG :	→ Tungsten - Inert - Gas,	DC (Electrode : negative)
	→ inert gas :	Argon - Helium (50:50)
	→ welding speed :	v = 2.0 m/min
	→ welding filler :	SG - AlMg 4.5 MnZr

Fig 74: Test programme

Fig 75: Welded blank specimens after stretch forming

loaded, with a stress ratio of 0.5, and while the TIG specimens reached nearly the same fatigue endurance limit as the base material, but lower total cycles were achieved. A steeper hardness gradient on the laser welded specimens led to strengthening of the weld seam and a reduction in fatigue strength.

Formability tests showed the deep-drawing behaviour of the welded panel to be as good as that for the blanks but stretch-forming behaviour was 30% lower. The Bulge test specimens are shown in Fig 75 to reveal no difference between the welding processes and the position of the seam relative to the grain had no influence on stretch forming; see Fig 76.

In the smaller of the two segmented tools, blanks with one seam perpendicular to the punch edge were formed into cups, the segments of the pressure pads being stepped, according to the sheet thickness increments, so as to obtain correct surface pressure. Maximum drawing depth was 70 mm, Fig 77. In the other tool, three-segment blanks were used, joined by two seam welds, one of them at 20 degrees to the punch edge and maximum draw depth was 120 mm. While first tryouts led to cracks at the bottom of the tubs, indicating rectangular geometry to be inappropriate, a thinner sheet and optimised blank, Fig 78, reduced the draw ratio and frictional forces on the flange; crack/fold-free cups could then be formed.

Tests on welded tailored blanks, Fig 79, showed that both TIG and laser welded blanks could be formed to a depth of 120 mm but there was some folding on the flanges. Generally there was an increased blank preparation expense required with laser welding but this is offset by the higher speed attainable than TIG welding.

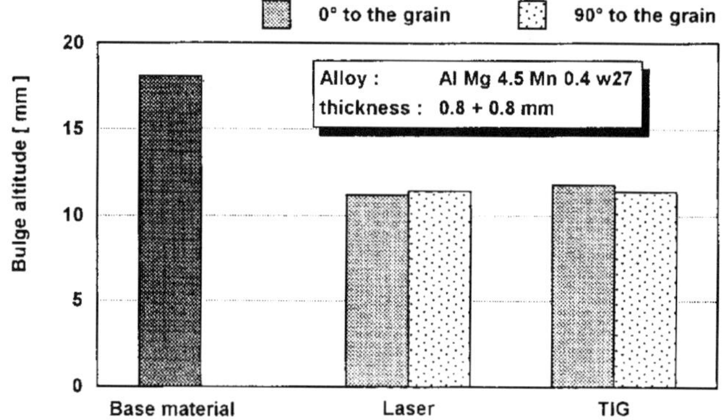

Fig 76: Characteristic forming values

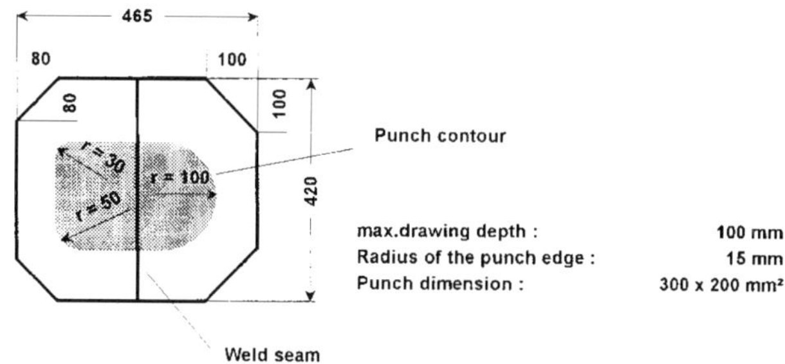

Fig 77: Optimised geometry of 2-piece blank

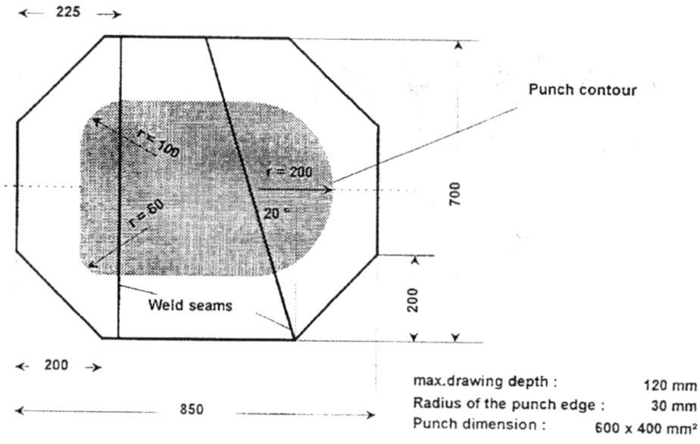

Fig 78: Optimised geometry of 3-piece blank

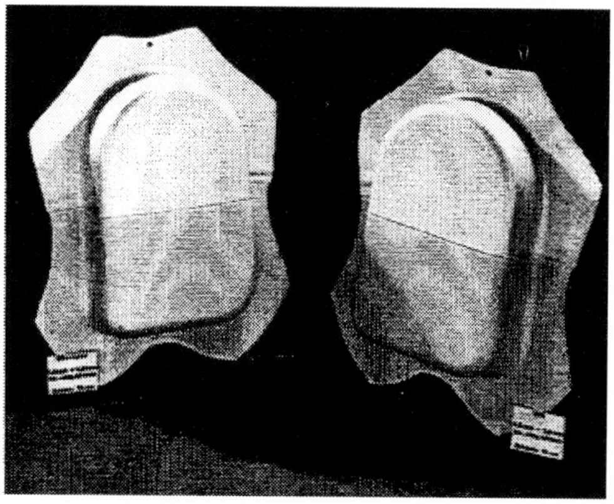

Fig 79: Deep-drawn 2-piece blank

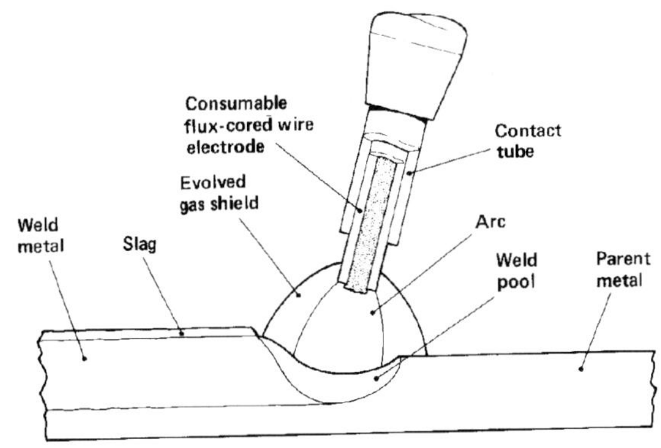

Fig 80: Flux-cored arc-welding

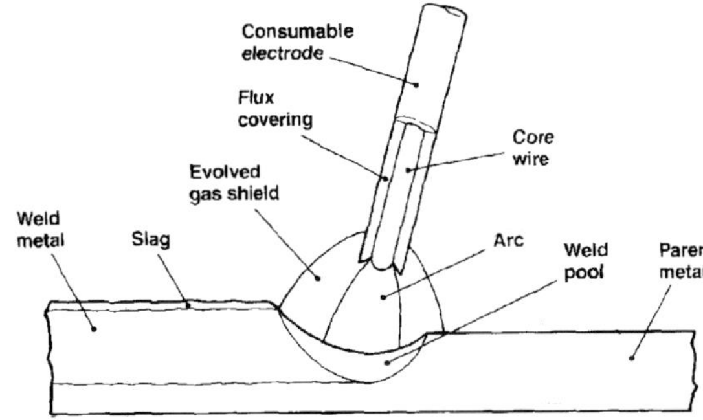

Fig 81: Manual metal arc -welding

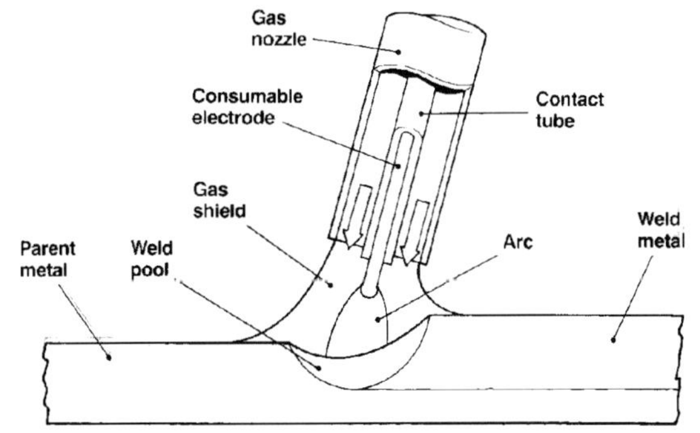

Fig 82: MIG welding

Electrode size (mm)	(SWG)	Typical current in amps	Length of electrode (mm)
1.6	16	25	250
2.0	14	45	350
2.5	12	65	350
3.5	10	115	450
4.0	8	145	450
5.0	6	215	450
6.0	4	265	450
6.3	(1/4 in.)	285	450
7.0	2	320	450
8.0	(5/16 in.)	360	450

Fig 83: Typical welding current data

Arc-welding techniques

While arc-welding is a manual skill which must be learnt by practice, a number of rules should be observed to optimise the performance of the resulting joint. There are also various factors which affect the particular technique of arc-welding which are discussed here.

In flux-cored arc-welding, a hollow electrode filled with flux provides shielding gases, deoxidisers, alloy additions and slag formers. Provided there are enough materials in the core there may not be need for external shielding gas provision and such wires allow welding which is resistant to draughts and is said to be self-shielding; however, a higher capacity power source will be needed for the heavier currents needed. High deposition rates and deep penetration is attainable but greater fuming can be a penalty. Heavier work such as might be met in construction vehicles is the usual application. The principle of the technique is seen in Fig 80, due to Houldcroft[9] of TWI, compared with the manual metal arc process in Fig 81 and metal inert gas (MIG) welding in Fig 82.

In the MIG process a continuous solid electrode is used, rather than a fixed-length rod with MMA, thus intermittent 'pools' of weld, corresponding to the electrode length, are avoided. A consistent arc length is maintained by self-adjustment such that the constant voltage power source causes current changes which oppose arc-length changes. Shielding gas and other parameters can be set to give desired metal transfer rate from the electrode.

Temperatures in MMA welding reach up to 6000C, the heat at the upper end of the arc melting the electrode while that at the lower melts the parent metal being welded. The arc is struck by tapping down the electrode to make contact then quickly raising it to the required arc gap — usually equal to the electrode diameter — then steadily feeding down the electrode as it burns away. As heating is more concentrated, higher welding speeds are possible compared with gas welding. Fig 83, due to Gibson and Smith[10], shows typical welding current data for a mild steel electrode. When a wide bead of weld is needed this can be obtained by weaving the electrode from side to side as it is advanced forward, Fig 84 showing some typical weave patterns.

The basic terminology for arc-welding is seen in Fig 85, due to Hicks[11]. This also differentiates between butt and fillet welding. Only if the metal is thin, less than 2 mm, is it possible to make a full penetration weld between abutting edges. Otherwise a gap between edges has to be left which can be minimised by bevelling the edges almost to full depth. With thicker metal the metal can be bevelled on both sides. To minimise distortion it is good practice to distribute welds equally about the neutral axis of a section and intermittent welding can also be helpful in this respect.

Strength of welds can be calculated on the throat thickness, Fig 86a, or in the case of a fillet weld, Fig 86b. In the latter case the stress on the weld is obtained by dividing the load by the throat area. Normally it is preferred to have a chart such as that in the figure giving strength of fillet welds in load per unit length.

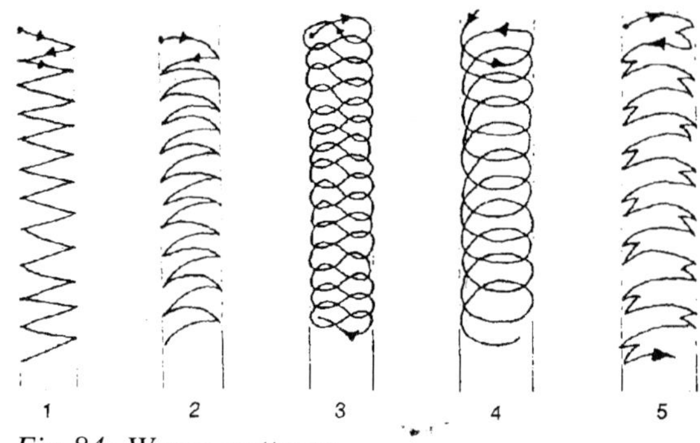

Fig 84: Weave patterns

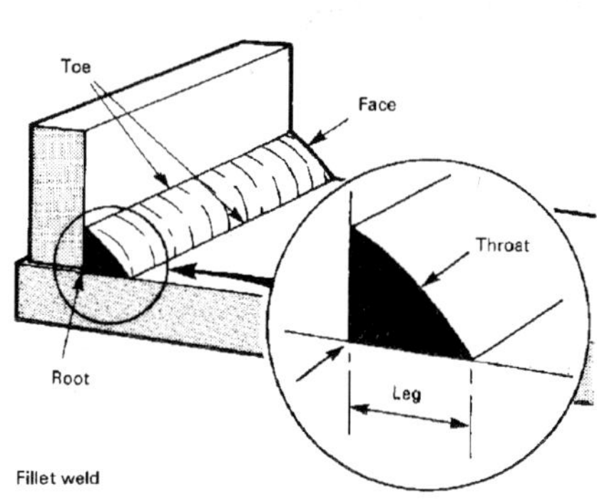

Fig 85: Welding terminology

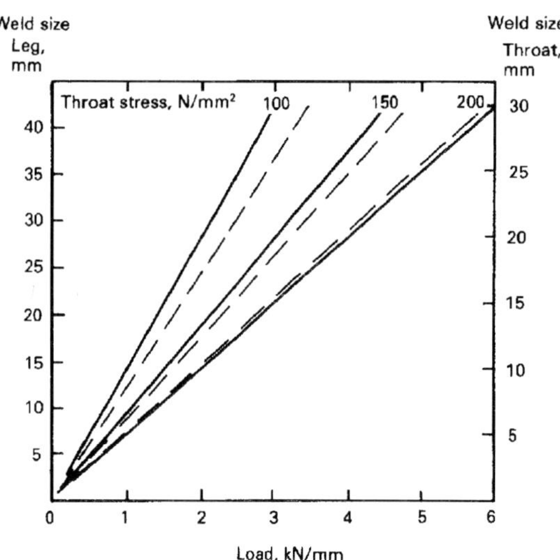

Fig 86b: Fillet weld design chart

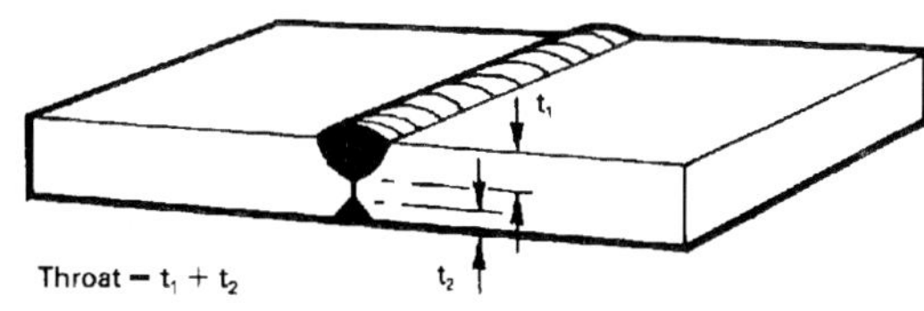

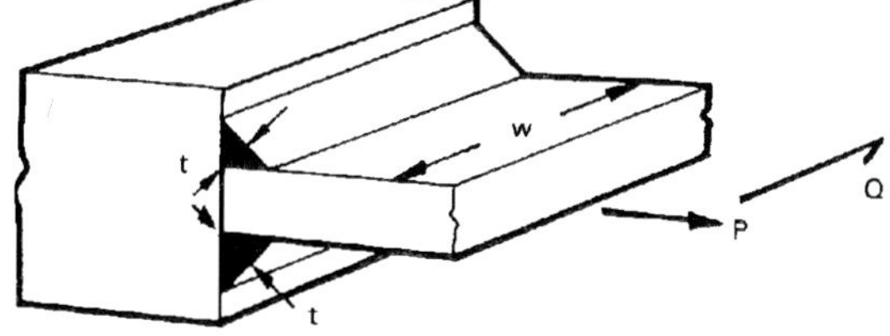

Fig 86a: Stress calculation on butt and fillet weld;tw = weld throat area; P/2bw = weld stress for load P and Q/2tw weld stress for load Q

Welding 'difficult' alloys

Aluminium alloys While long-established techniques for welding mainstream aluminium alloys have ensured the material's widespread use in both passenger and commercial vehicle structures, in striving for higher specific strengths, available with the more exotic alloys care must be taken to meet the welding requirement of the particular alloy[12]. Fig 87 shows designations for wrought aluminium alloys and their principal alloying elements.

In the non-heat-treatable alloys, material strength depends on the effect of work-hardening and solid-solution hardening of alloying elements mainly found in the 1xxx, 3xxx and 5xxx series. When welded, these alloys may lose the effects of work-hardening and the heat-affected zone (HAZ), adjacent to the weld, will become softened. With the heat-treatable alloys the solution treatment, quenching and artificial ageing processes produces a fine dispersion, which enhances hardness and strength, of the alloying elements mainly found in the 2xxx, 6xxx, 7xxx and 8xxx series. Fusion welding redistributes the hardening constituents in the HAZ, which reduces strength in this region. High strength alloys (7010 and 7050) and most of the 2xxx series are not recommended for fusion welding because they are prone to liquidation and solidification cracking.

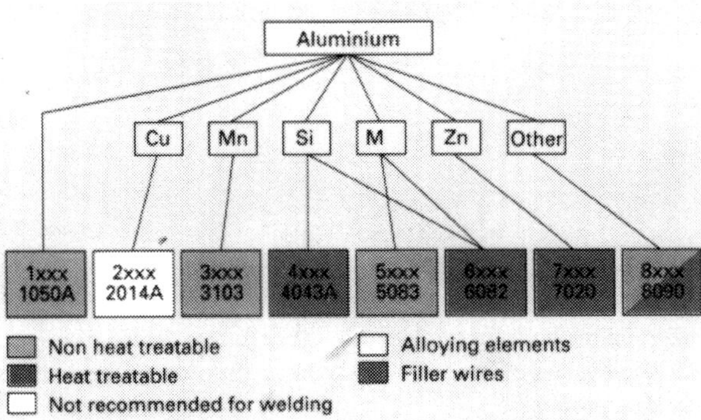

Fig 87: Aluminium alloy classification

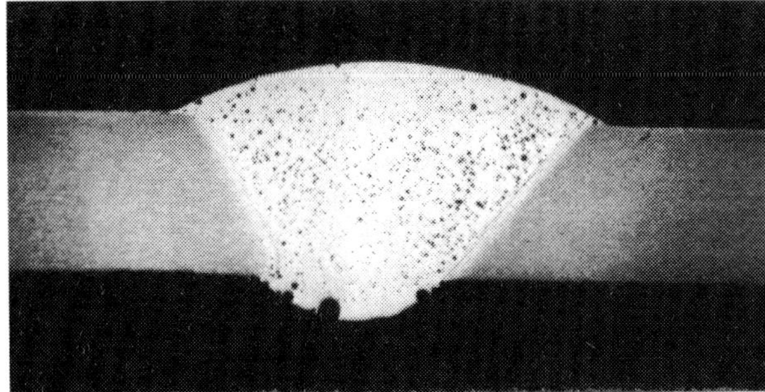

Fig 88: Porosity in a TIG weld

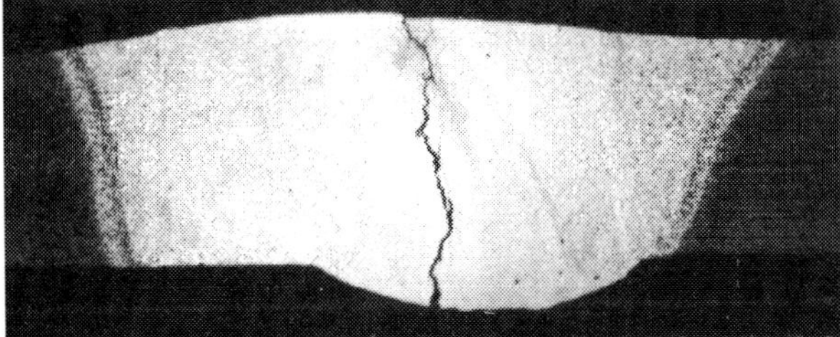

Fig 89: Solidification crack in TIG weld

Filler metal composition is determined by weldability of the substrate, minimum mechanical properties of the weld metal, corrosion resistance and anodic coating requirements. For the non-heat-treat alloys, nominally matching fillers are used while for the heat-treat alloys, non-matching fillers prevent solidification cracking. British Standards BS 3019 Pt1 and 3571 Pt1 give fillers for TIG and MIG welding, respectively.

Appropriate precautions must be taken against imperfections in fusion welds: porosity, cracking and poor bead profile. Porosity, Fig 88, is usually caused by absorbed hydrogen from hydrocarbons and moiscation. They may form in the centre of the weld, Fig 89, or in the weld crater at the end of the operation. Usual causes are incorrect filler, wrong weld geometry or restraint conditions on welding. Using a non-matching crack-resistant filler can solve the first problem but it may have lower strength and not respond to heat-treatment. Weld bead must be thick enough to withstand the contraction stresses and appropriate edge-preparation/joint-set-up can minimise the degree of restraint on the weld.

Liquation cracking occurs in the HAZ, Fig 90, when low-melting-point films are formed at the grain boundaries which cannot withstand the contraction stresses. Alloys most susceptible are 6xxx, 7xxx and 8xxx series. A lower melting-point filler can be used to alleviate the problem but do not use 4xxx filler with magnesium-element alloys as a silicide can form which worsens the situation.

Cast irons Iron-based alloys containing more than 2% carbon, 1-3% silicon and up to 1% manganese form this classification. They are grouped according to their structure, influencing mechanical properties and weldability, as in Fig 91. Grey cast irons contain 2-4.5% carbon and 1-3% silicon and have a structure of graphite flakes joined in either a pearlite, ferrite or combined matrix, Fig 92a; the flakes form weak-planes which result in a weaker material than steel. The graphite shape can be modified to remove the planes to produce so-called spheroidal irons., Fig 92b. By reducing carbon and silicon content, white cast irons can be produced, with carbon retained as iron carbide, for extra hard but brittle castings. Alternatively, heat treatment of white irons to give extra ductility in so-called malleable irons. These fall into three groups: whiteheart which are inhomogeneous with a decarburised surface skin and a higher carbon core; blackheart, which are annealed to produce a homogeneous mixture of carbon in a ferrite matrix for improved mechanical properties; finally, perlitic-malleable which have higher strength but lower ductility than ferritic blackheart irons.

Weldability of grey iron is poor because its brittleness cannot stand the contraction stresses on solidification. Spheroidal Graphite irons, by contrast, have greater ductility so are more weldable. White cast iron is considered unweldable. In general bronze welding is frequently employed, to avoid cracking, but extra-clean surfaces are required. For fusion welding, low heat input conditions, extensive preheating and slow cooling are required to avoid HAZ cracking.

Welding of plastics

Linear vibration welding, hot plate welding and induction implant welding are processes being examined by TWI for use by vehicle manufacturers. Linear vibration welding can cope with panels up to 650 x 650 mm, from 1-4 mm thick, and a weld time of 5-10 seconds is involved at 100 Hz. With induction-implant, a resistive implant involves a conductive layer of carbon fibre within a PEEK resin. Generally, TWI consider that welding processes available can be divided into two groups: processes involving mechanical movement: ultrasonic, friction (spin, angular and orbital) and vibration (linear friction) welding; processes involving external heating — hot plate, hot gas and resistive and inductive implant welding.

Ultrasonic welding is a bonding process which uses high frequency (ultrasonic) mechanical vibrations for joining. The parts to be assembled are held together under pressure and then subjected to ultrasonic vibrations of usually 20-40 kHz at right angles

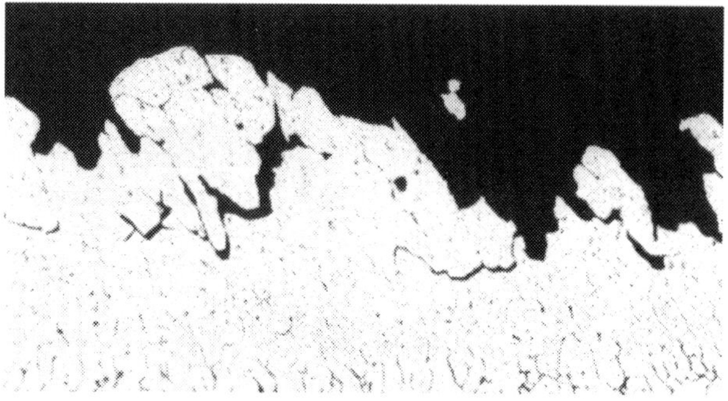

Fig 90: Liquation cracking

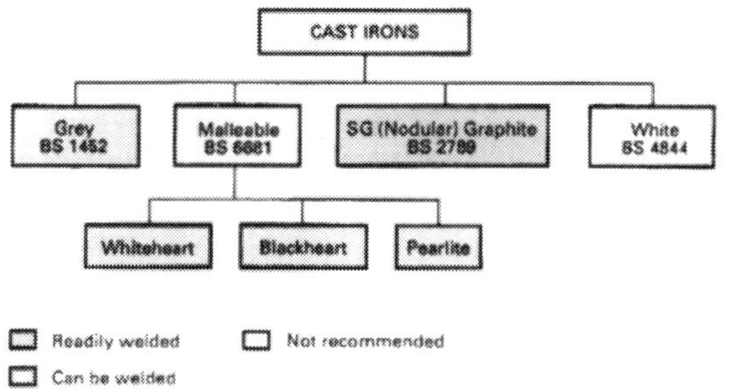

Fig 91: Cast-iron classification

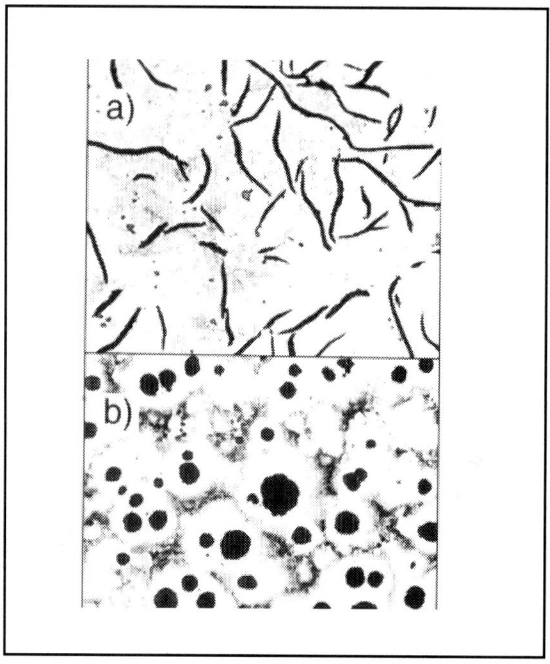

Fig 92: Pearlitic, a, and Nodular, b, structure of cast irons

to the contact area, so that longitudinal vibrations are transmitted through the component. The consequent alternating high frequency stresses produce heat in the material and, if the components are properly designed, this heat can be selectivity generated at the joint interface. Heat generation is by a combination of surface and intermolecular friction.

The welding equipment consists of five basic components: power supply; converter; booster horn; horn and assembly stand. The ease with which material can be welded depends on its ability to transmit high frequency vibration. As a simple rule, plastics having a high modulus of elasticity also possess low internal losses for ultrasonic vibration and hence allow maximum transmission of energy to the joint. This means that rigid materials weld more readily than flexible ones.

The most important factors in joint design are a loose fit and the provision of an 'energy director'. A slip fit is essential since the welding process depends on movement between the two parts as well as friction and pressure. In its simplest form the energy director is a small triangular ridge, typically 0.4-0.8 mm high, moulded on one of the mating parts. This concentrates energy to provide rapid melting of the material contained in the director. This molten material is dispersed throughout the joint area to form, in effect. an adhesive which binds the parts together.

Ultrasonic welding is well suited to the needs of mass production because it is capable of a high level of automation and can achieve very high production rates. Typical weld times are in the order of 0.5-1.5 second. Tooling is expensive, so that large volume production runs are preferred. The maximum output power of machines is limited at present and so small components only can be welded with maximum weld lengths of several centimetres. For large components multi-head machines are used but they are complex and expensive.

Ultrasonic energy can also be used to produce joints which do not involve welding. Plastics can be staked or riveted to other plastics or metals or threaded metal inserts can be pushed into plastics using ultrasonics to soften the plastics material.

Friction welding Friction or spin welding of thermoplastics involves basic principles as for metals, although there is a fundamental difference as metals are welded without melting but for plastics melting occurs at the joint interface. However, this molten material is substantially displaced from the joint into the weld flash. Relative motion can be spin, angular or orbitial. Orbital has the advantage that it can weld non-circular components. The major welding parameters are rotational speed, friction pressure, forge pressure, weld time and burn-off length. Thetype of plastics influences these parameters. In general, rotational speeds are in the range 1-20 m/sec, friction pressures in the range 0.8-1.5 bar, and forge pressures 1-3 bar with average weld times in the range 2-10 seconds.

The advantages of friction welding are said to be high weld quality, simplicity and reproducibility of the process. For many applications little end preparation is required. The main disadvantage is that in its simplest form it is only suitable for applications in which at least one of the components is circular and requires no angular alignment. Although such welds can be made by orbital welding, the process is more complex.

Vibration welding Vibration (linear friction) welding has become increasingly used for welding various mass produced thermoplastic components. As with conventional rotary friction welding, frictional heat is generated by relative movement between the two parts to be welded, which are held under pressure. The difference between the processes is that, with vibration welding, movement consists of linear oscillations. Once molten material has been generated at the joint interface, vibration is stopped then parts are aligned and the weld core solidifies on cooling. The process is rapid (weld times of 1-5 seconds are generally used), the vibration is typically 100-240 Hz at 1-5 mm amplitude, while pressures are similar to those used in rotary friction and ultrasonic welding (14 N/mm^2).

The main advantage of vibration welding is its ability to weld large complex linear joints at high production rates. Other advantages include: capability of welding a number of components simultaneously; simplicity of tooling; and ability to weld almost all thermoplastic materials, whether injection moulded, extruded, blow moulded, thermoformed, foamed or stamped. The technique is also said to be particularly suitable for crystalline plastics such as acetal, nylon, polyethylene and polypropylene, which are not easily joined by ultrasonic welding or by solvent bonding.

Heated tool welding There are several techniques which involve use of a heated tool. The simplest uses a flat hot plate but more complex tools are also used. A heated plate is clamped between the surfaces to be jointed until they soften. The plate is then withdrawn and the surfaces are brought together again under controlled pressure for a specific period. The fused surfaces are allowed to cool, forming a joint which normally has at least 90% of the strength of the parent material.

The welding tool or heating element normally has built in electric heaters, although other heat sources have been used, and has surfaces coated with PTFE to prevent sticking. Typical temperatures are between 180-230C depending on type and thickness of material to be welded. The main advantages of the process are the simplicity of the equipment and the fact that virtually all thermoplastics can be welded. Examples of assemblies which lend themselves to this process are those where strong liquid tight joints are essential such as brake fluid reservoirs, batteries, and lamp assemblies.

The main disadvantage of the process is that it is relatively slow. Weld times range from 10-20 seconds for small items up to 30 minutes for very large pipes.

Hot gas welding This method of fabrication is somewhat similar to oxyacetylene welding of metals, the open flame being replaced by a stream of hot gas. The general technique is the same for most plastics (Fig 93). The thermoplastic components to be joined are first cleaned and a rod of composition identical to the parent material is used. Welding rod and weld bead are then heated simultaneously by a hot gas stream from the welding gun. Typical temperatures are 200-300C and flow rates 15-60 litre/min. As the surfaces soften, the filler rod is continuously forced into the weld.

In common with metals, the welding rods are normally circular in section with a diameter of about 3 mm. For thicker sheets several weld runs along the joint are employed. It has been suggested that one disadvantage of conventional round rod is that air pockets tend to be trapped in the weld run during multiple weld runs, resulting in weakness. This can be overcome by using the more recent triangular section welding rods.

Gases used include compressed air, nitrogen, hydrogen, oxygen and carbon dioxide. Nitrogen is used for oxygen sensitive plastics such as polyethylene. Whilst oxygen results in higher weld strengths, compressed air is popular since it gives satisfactory results for many purposes and is cheap. Typical plastics which can be welded include PVC, polyethylene, acrylics, polycarbonates and nylons.

The principal advantage of hot gas welding is that large, complex fabrications can be constructed. However, it is slow and weld quality is entirely dependent on the manual skill of the welder.

Implant welding This type of welding is based on the principle of trapping a metal insert between the two parts to be joined and then heating the insert by induction or resistance heating so that the plastics material around the implant fuses to form the joint. With induction heating a high frequency electromagnetic field (2-30 MHz) is used to induce an electric current in the metal implant, which resistively heats it. A development of this technique uses ferromagnetic particles (usually iron oxide) dispersed in a thermoplastic matrix which is preplaced along the joint in the form of a paste or tape. In resistance implant welding, conductive wire is trapped between the components and directly resistance heated by an electric current. Currents as high as 150A may be used.

Implant welding is a simple process which has been applied to complicated joints in large components such as vehicle bumpers and electrically driven vehicle bodies. Welding times are short, up to 20 seconds for the largest components. The process can be used to join almost all thermoplastics, but the presence of the insert limits the joint strength.

Welding composites The basic plastic welding processes are now being supplemented by others which may one day make the welding together of

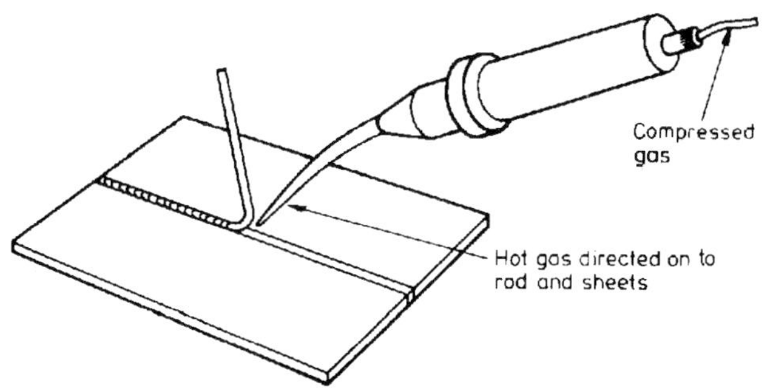

Fig 93: Hot-gas welding

structural composite vehicle body panels of the future as easy as joining steel ones. One such process is focused infrared welding, Fig 94, which TWI has demonstrated to have potential for body panels. It is an extension of hot-plate welding in which the hotplate is replaced by an infrared lamp and water-cooled elliptical reflector. The angled copper mirror splits the focused single heat source and directs it to each welding surface of the two parts to be joined. A typical lamp would be powered at 2400 W, sufficient to provide 16W per mm of weld length. The surfaces are pushed together, on reaching the correct temperature, and the lamp/mirror simultaneously withdrawn. The surfaces heat in about one second and welding is complete after a further three seconds. Tests have already been carried out on filled and unfilled polypropylene — also two polycarbonate alloys.

Fig 94: Focused infrared welding

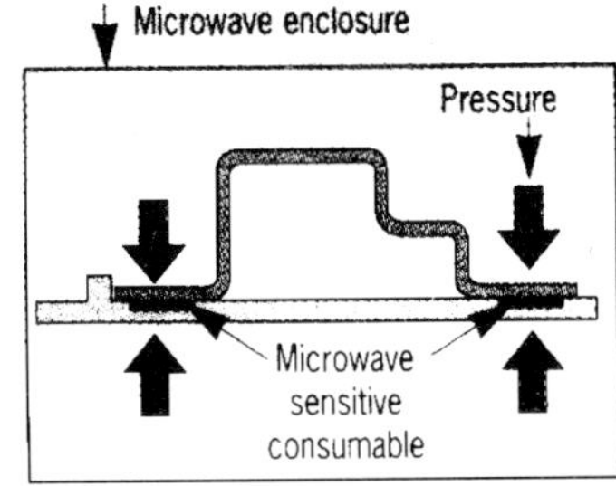

Fig 95: Microwave welding

Some time ago, TWI carried out trials to show that welding techniques applied to join unreinforced thermoplastics could also be applied to composites, the results being contained in Bulletin Volume 28 for hot-plate, ultrasonic, vibration and induction-heating techniques. Comparable work on hot-gas welding is not mentioned although it is widely used in the automotive repair trade for rectifying damaged bumpers and the like. In recent years the TWI have stated that the process suffers from a lack of standardisation and a widely accepted procedure for fabricators. They believe most thermoplastics can be hot-gas welded provided care is taken to optimise welding conditions. Long term creep behaviour should be careful watched, when structural parts are considered.

A recently introduced composites welding process is applicable to thermoplastics, including those filled with glass and Kevlar. The process is performed in an oven operating at the same frequency as a domestic microwave unit, 2.45 GHz. Components to be joined are heated indirectly by an implant inserted at the joint, Fig 95. This implant is made of a microwave-sensitive consumable material. The selection of the appropriate material for a specific application is crucial to the success of the process. Otherwise, the technology is relatively simple, and said to be reliable and easy to use in production.

In the oven, microwave energy melts the consumable implant to form the welded joint. The process is fast, and the weld is made in only 10 sec. During this time, multiple joints can be welded simultaneously. It can be a continuous, automated process and there is no need to provide access to joints for weld tooling. As a result, there is freedom in component design, and complex assemblies can be joined in one operation. The process is applicable to large assemblies, too. Body panels, dashboards, load floors and bumpers are some possible applications.

TWI is also investigating the use of microwave welding for under-bonnet assemblies, and reinforced nylon manifolds are an example. A less expensive method of manufacture is to produce the manifold as two mouldings which are then welded together. However, if vibration welding is used, the joint must lie in a flat plane, and this imposes limitations on the design. By adopting microwave welding, this restriction would be avoided, and the part could be made to the optimum shape.

Projection, stud and seam welding

While spot-welding is probably the most important category of resistance welding, seam, stud and projection welding are very useful alternatives for particular applications. One of the most famous automotive applications of stud welding was the US Ford wire wheel which had spokes welded at either end to the rim and hub of the wheel.

In projection welding, current and pressure are localised at the weld section by means of an embossed upset on one or both sides of the work. The operation is carried out on a press-welder with the workpiece placed between copper platens which are within the single winding of a large transformer. It is particularly suited for fastening attachments such as brackets, spigots and weld-nuts to sheet metal which can only be accessed from one side.

Electrodes are flat and so can contact the parts over a wide area allowing more accurate aligning of parts. Use of several projections within the electrode area also allows several welds to be performed simultaneously and individual welds can be closer together than with spot-welding as the shunting problem does not apply. Even with a single weld the conditions of the weld guarantee a higher quality than that obtained from spot-welding and it is sometimes possible to weld in an area inaccessible to spot-welding electrodes. This is particularly true of parts too small to spot-weld with a narrow electrode without burning.

In joining sheet material a round form of projection is ideal, Fig 96, having radius 1-2 times sheet thickness. For curved surfaces elongated projections are recommended, Fig 97. Low projections give minimum marking but high ones (not exceeding 90% metal thickness) give greater strength. Where different metals are being welded the one with the higher conductivity should carry the projections. Aluminum alloys need special attention. Controlling slope to suppress the first few cycles of welding current is the secret — so as to reduce current density while the projection is collapsing and so avoiding metal expulsion.

A special case of projection welding is stud-welding, Fig 98, which is used extensively for attaching fastenings for such applications as soft trim attachments. Sheets down to 26 gauge in steel and 22 in aluminium alloy can have studs attached without back-marking. When the firing current is closed, a current of up to 3500 A is discharged through the stud thus disintegrating the stud. Current continues to flow across the gap between stud and parent metal while the stud is driven home by spring or pneumatic pressure.

In seam welding, Fig 99, the electrodes are replaced by wheels whose peripheral form controls the nugget shape. Either a continuous seam can be produced, for such applications as the flanges of fuel tanks, by the roll-spot technique, or the method can be used to produce intermittent welds. The continuous seam is generated by overlapping spots. Welding current is usually supplied in pulses, with wheel rotation being continuous. In some cases the work is driven through by the wheels while in others the wheels are idle and the work is fed through by an external force. Material thicknesses for seam welding are usually in the range 0.5 to 3mm and most common metals and alloys can be handled. It is a fast method of welding and readily controllable for quality. A higher current than that used for spot-welding is required as the previously made nugget provides a shunt path for the current flow.

Fig 96: Projection for flat sheet

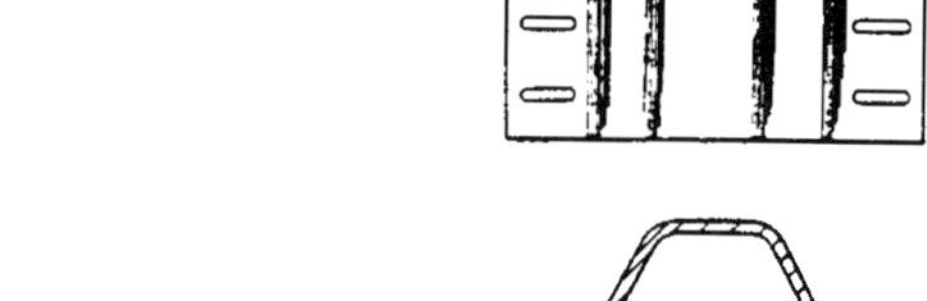

Fig 97: Projection for curved sheet

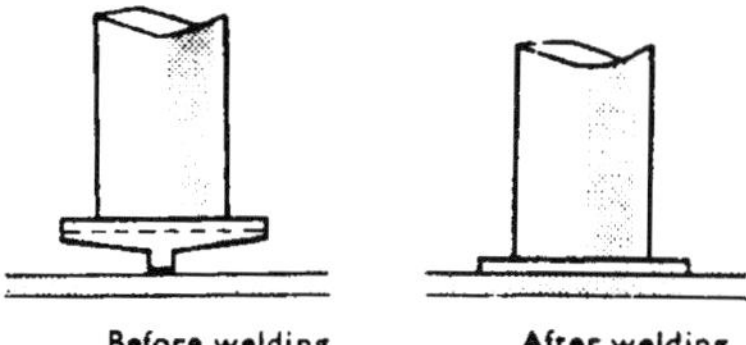

Fig 98: Stud welding

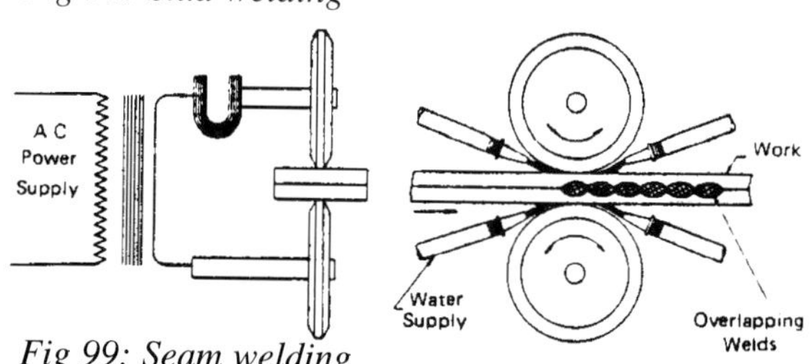

Fig 99: Seam welding

Friction jointing

Assembling one part to another in a light interference fit has been refined into a process called Trib-jointing, which can be used to joint dissimilar metals, for example. These joints use the familiar process of seizure to bind and lock rubbing surfaces in the presence of a friction-enhancing fluid called an 'anti-lubricant'. Thus it produces rapid galling, a solid phase cold-welding that leads to seizure.

The anti-lubricant can be one of several chemicals that act to increase friction and cause cold welding between rubbing metal surfaces. The fluids are thought to act as oxygen scavengers at asperity contact, when trapped and degraded due to high local mechanical forces and flash temperatures, and when subject to the catalytic influence of exposed unreacted surfaces, all resulting from initial plastic deformation. The oxygen scavenging action delays the re-oxidation of the damaged asperity surface and reduces oxides immediately surrounding the contact point, to enlarge the contact by solid phase welding. Surface disruption spreads rapidly as welds are broken and reformed, due to continued rubbing, until sufficient mechanical coupling is achieved between the rubbing surfaces. There is no heat-affected area or curing period and the joints are as strong as the joined parts but cost less than most other methods. Applications being considered include prop-shafts; starter and alternator rotors; clutch/brake parts, gears fitted to shafts, water and oil pumps; composite crank and camshafts.

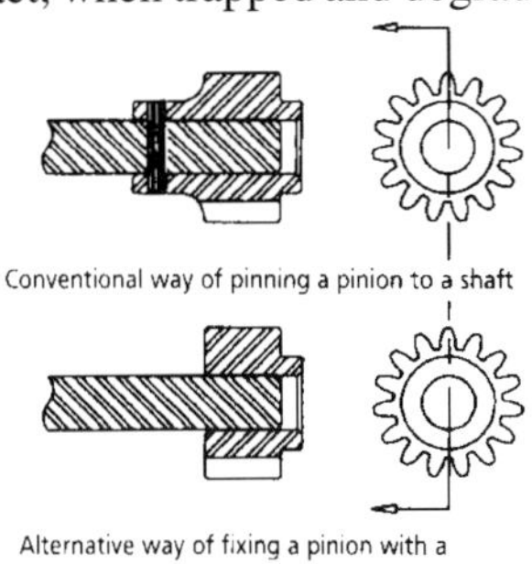

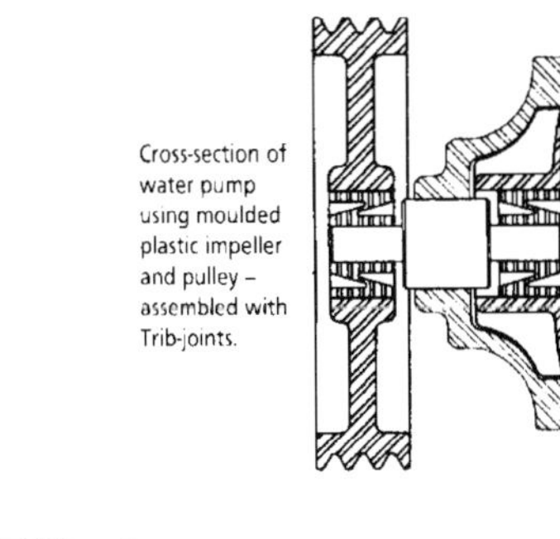

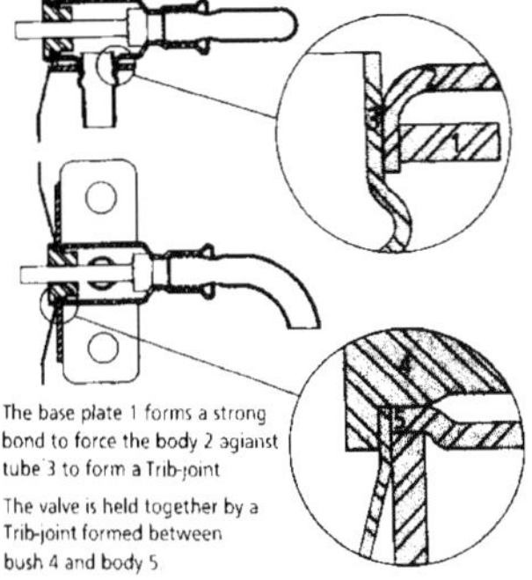

Fig 100: Gear securing, water pumps and EGR poppet valves: three potential uses for Trib-joints

The University of Hertfordshire has conducted research into the mechanics of the system, under contract to Trib-Tech Group. This has resulted in a mathematical model of the basic joints so that parameters such as push-together, pull-apart and torsional strengths can be predicted. A computer program incorporating this model enables simple iterative tests to be made to optimise joint dimensions.

Three potential uses are shown in Fig 100. Where a gear is secured to a shaft, a traditional interference fit often cannot be relied on, so a pin is introduced. Using a Trib-joint the parts are simply coated with anti-lubricant and pushed together to provide a joint as strong as the pinned version. No surface preparation is needed and normal machine finishes are acceptable.

A typical water pump assembly is also shown in section, for which it is common to use interference fits to retain the impeller and pulley. Here the Trib-joint allows use of a low cost injection-moulded plastic impeller and pulley in place of the usual metal parts. Also shown is a light-weight stainless steel poppet valve used in exhaust gas recirculation to control cross flow of exhaust gas back into the induction manifold. Traditionally these valves use a cast iron body bolted to the engine block, and need a stainless-steel valve seat to be set into the cast iron body. Here the valve body with seat can be formed by cold working a lightweight stainless-steel tube, the assembled valve being attached to the engine via a mounting bracket trapped by the poppet bush as the valve is assembled by simply pushing it together.

Adhesive bonding

The ability to bond a whole surface rather than a few points is one of the main attractions of adhesives. Advances in materials and applicator technology are also making bonding more price and time competitive than bolting riveting and spot welding. And as load bearing structures become more refined the need to maintain structural continuity at the joint line will further the preference for adhesives.

Provided the ground rules of designing for bonded structures are obeyed, high load capacities are available from modern adhesives. It must be remembered that bonded joints perform best under pure tension, compression or shear, less well under cleavage and poorly under peel loading, Fig 101. Of the many available joint configurations, a number of which are shown in Fig 102, the tapered and double-lap joints are better than the single lap but, of course, more costly to produce. Fig 103 shows a number of ways to reduce peeling tendency which might otherwise destroy a joint. These are some of the ideas put out to support the sale of Araldite epoxy adhesives by Ciba Geigy

Very broadly, adhesives divide between the 'natural' animal (casein), vegetable (gum), mineral (bitumen) and the 'synthetic', elastomers, thermoplastics and thermosets. These divisions are by chemical source and a more useful classification follows the main uses which have arisen from the commercially developed forms. In that respect 10 major families have been defined by Permabond Adhesives as follows:

Anaerobic	Plastisol
Cyanoacrylate	Polyurethane
Epoxy	Solvent rubber
Hot melt	Tape
Phenolic	Toughened types

The last category sub-divides again into anaeorobic, acrylic and epoxy adhesives. Application varies according to viscosity, obviously, with liquids applied by brush or spray, at a price, but a little ingenuity can often save money in this area. Solvent based adhesives normally involve a time penalty in waiting for the solvent to dry either between application and jointing, or with porous surfaces, after jointing. Film adhesives are often used particularly when bonding the skins of sandwich panels to honeycomb cores. This usually involves a mixed adhesive and requires an elevated temperature to cure. As well as the question of joint configuration and load capacity, key factors also to be considered are type of material and surface condition; relative expansion coefficients and joint tolerances. For working loads above 3.5 MN/m^2 (500 lb/in^2), some experts recommend that only structural adhesives of the thermoset type be used.

For an epoxy, such as Araldite, recommendations are to clean, degrease and abrade surfaces;

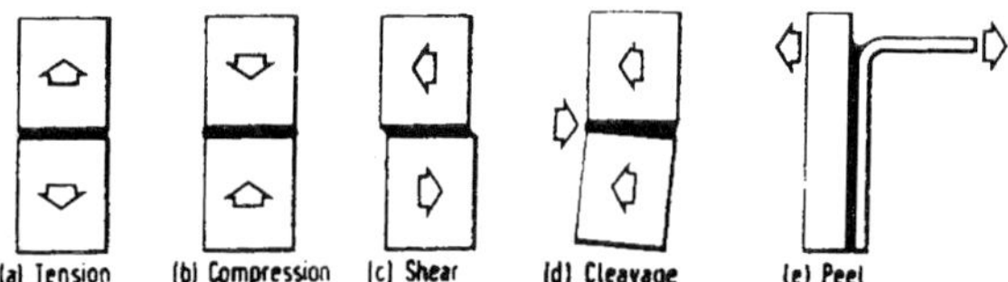

Fig 101: Bonded joint failure modes

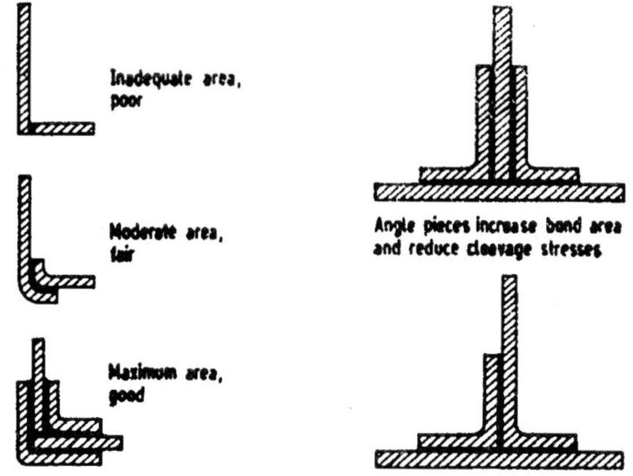

Fig 103: Ways to prevent peeling

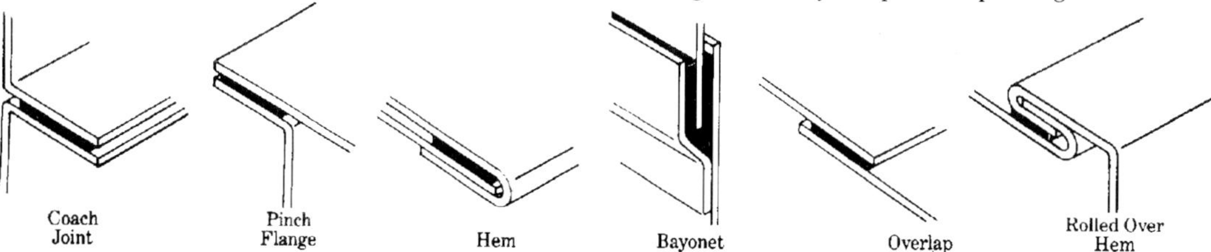

Fig 102: Joint configurations

observe curing time specified and apply clamping pressure; avoid operation above 100 C. Select viscosity according to joint condition thin for close-fitting, thick for gap-filling; seek special advice when jointing materials which have grossly differing expansion coefficients and choose a low modulus grade where high vibration or impact loads are expected; observe 5 per cent weight tolerance in the mix components of the adhesive and for large quantities, mix in shallow trays to avoid heat build up which can shorten life at this stage. Heat applied after jointing, during the cure period, can enhance joint strength. The table in Fig 104 gives joint strength ratings, key for which is (1) moderate, (2) less strong than jointing materials, (3) bond strength of 2000 lb/in^2 on standard test specimen.

Fig 105: Part-bonded centre section

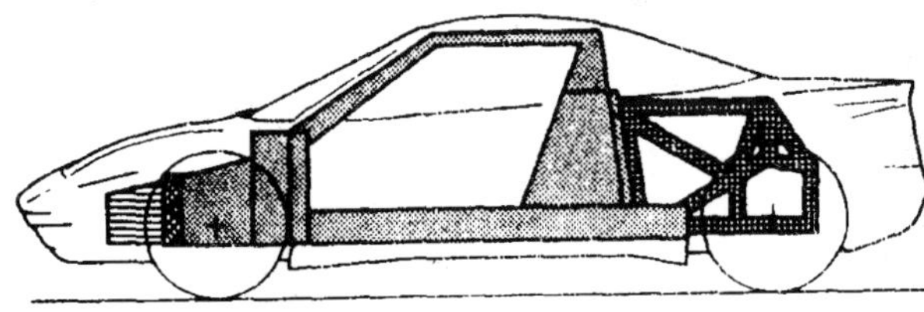

Fig 106: Assembly of sections

Alcan International recently completed a bonded structure project on the Ferrari 408 research vehicle using a toughened epoxy adhesive heat cured to 180 C. The part-riveted/part-bonded centre-section (Fig 105) connects the front and rear end-frames (Fig 106) and has torsional stiffness of 10.4 kNm/deg. This part of the structure is made up of parts which are folded on simple press brakes. Assembly technique is to put the parts together with self-tapping screws in the rivet holes to check for fit and then bond them with the adhesive which is specially formulated to have good balance between slump-resistance and extrudability. Ordinary paint stoving ovens can be used for the cure.

Structural adhesives

The work at Alcan associated with the above example has led to some manufacturing recommendations formulated by the company. Adequate peel resistance of the adhesive is, of course, the prime require-

JOINTING MATERIAL	LIN EXPAN COEFF × $10^{-6}/^{\circ}C$	OPTIMUM JOINT STRENGTH (a)	ACID ETCH TREATMENT	JOINT STRENGTH RATINGS (see chart key)						
Aluminium	24	4750	sulphuric and chromic	3	3	3	3	2	2	3
Brass	19	1000–1500	—	2	2	1	1	1	1	2
Ceramics	2·5-4	ceramic breaks	—	3	3	3	3	3	3	3
Chromium plate		1500–2000	hot dilute hydrochloric	3	3	3	3	1	1	3
Copper	17	1000–1500	ferric chloride	2	2	1	1	1	1	2
Glass	8·5-10	glass breaks	—	3	3	3	3	2	2	3
Glasscloth laminates epoxy	12-25	laminate breaks	—	3	3	3	3	3	3	3
polyester	12-35	laminate breaks	—	3	3	3	3	3	3	3
Graphite and carbon	6-7	carbon breaks	—	3	3	3	3	3	3	3
Lead, tin and solder	22-27	500–1000	—	2	2	2	2	2	2	2
Magnesium	27	2000	chromate	3	3	3	3	3	3	3
Nickel	13	2000	—	2	2	2	2	1	1	2
Nylon	90	—	—	2	2	1	1	1	2	2
Paper laminates—melamine and phenolic	15–35	laminate breaks	—	3	3	3	3	3	3	3
'Perspex'	90	800	—	1	1	1	1	2	2	1
Polystyrene	60-80	—	—	1	1	1	1	2	2	1
Polyethylene and polypropylene	150-300	1000	chromic acid or flame	1	1	1	1	1	1	1
P.t.f.e.	60-70	1000	(b)	2	2	2	2	2	2	2
P.v.c.	50-60	1000	trichloroethylene	2	2	2	2	2	2	2
Rubber	130	rubber breaks	conc. sulphuric or nitric for 5 min	3	3	3	3	3	3	3
Mild steel and iron	11	2000–3000	dil. phosphoric	3	3	3	3	2	2	3
Stainless steel	18	2000–3000	oxalic	2	2	2	2	1	1	2
Wood (c)	3–60	wood breaks	—	3	2	2	3	3	3	3
Zinc	26	500–700	—	2	2	3	2	1	1	2

Fig 104: Joint strength ratings

ment and this, say the company, must stand up over a temperature range from -40 to +120 C. As part of the Aluminium Structured Vehicle Technology (ASVT) programme, the next most important requirement is precision of placement to avoid contamination of jigs and fixtures. Application speeds from 10 mm/sec, for complex joints, to 500 mm/sec must be sustainable, with bead diameters from 1 to 5 mm on inclined surfaces from horizontal through to vertical. Robotic application of adhesive is seen as a way of obtaining predictable control of quality — with a careful integration of adhesive, robot and dispensing system being required.

As well as the structural performance requirement of the adhesive, viscosity, thixotropy, filler abrasivity and slump resistance have each to be controlled. A 10 kg 6-axis electric-drive robot is suggested but there is a special requirement for the dispenser to adjust flow exactly in proportion to the speed of the nozzle over the workpiece.

Weld bonding has become a familiar way of using structural adhesives in which just a few spot-welds help to locate the joint elements and hold the structure together prior to cure of the adhesive. The cost configurations for the ASV and conventional vehicle, comparing weld-bonded and plain spot-

Fig 108: Ferrari 408 bonded structure

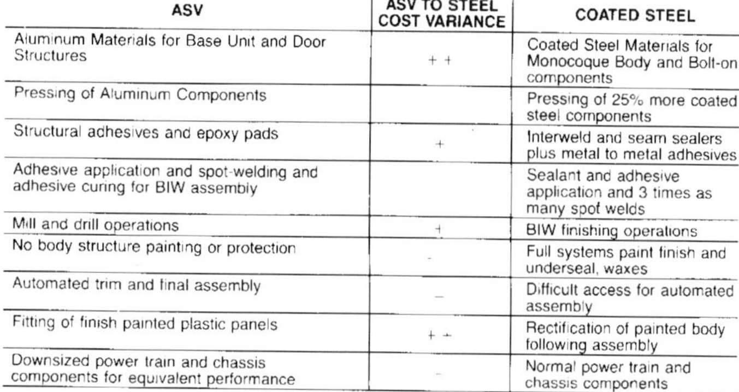

ASV	ASV TO STEEL COST VARIANCE	COATED STEEL
Aluminum Materials for Base Unit and Door Structures	+ +	Coated Steel Materials for Monocoque Body and Bolt-on components
Pressing of Aluminum Components		Pressing of 25% more coated steel components
Structural adhesives and epoxy pads	+	Interweld and seam sealers plus metal to metal adhesives
Adhesive application and spot-welding and adhesive curing for BIW assembly		Sealant and adhesive application and 3 times as many spot welds
Mill and drill operations	+	BIW finishing operations
No body structure painting or protection	-	Full systems paint finish and underseal, waxes
Automated trim and final assembly	–	Difficult access for automated assembly
Fitting of finish painted plastic panels	+ +	Rectification of painted body following assembly
Downsized power train and chassis components for equivalent performance	-	Normal power train and chassis components

Fig 107: Weld bonding and spot welding compared

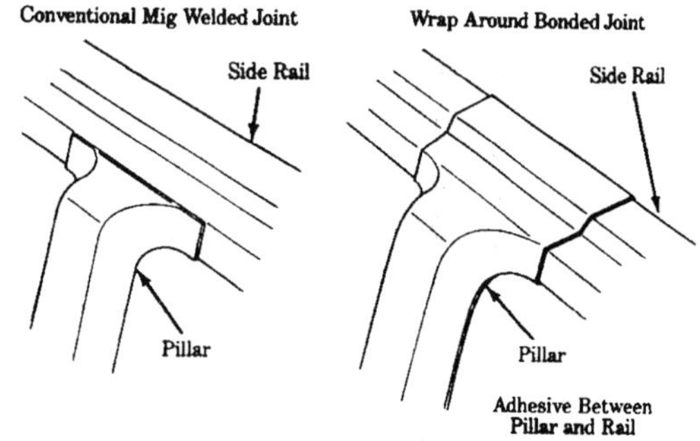

Fig 112: Joint modification for bonding

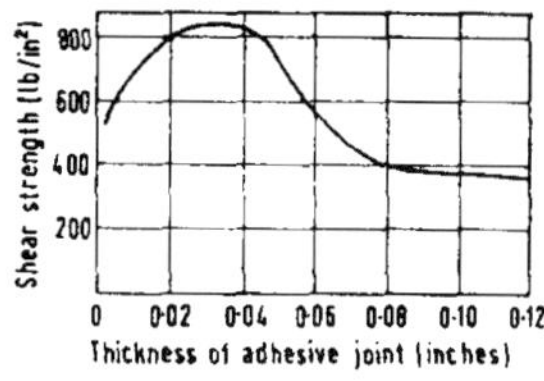

Fig 109: Strength variation with thickness

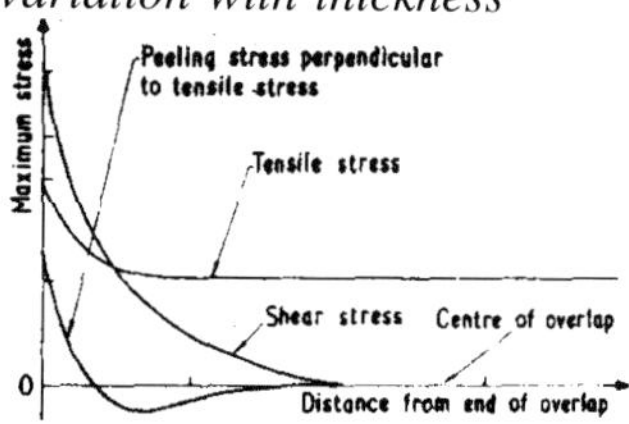

Fig 110: Stresses in single lap joints

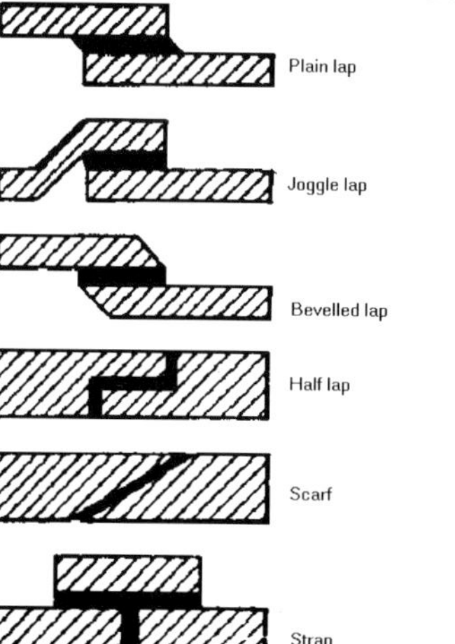

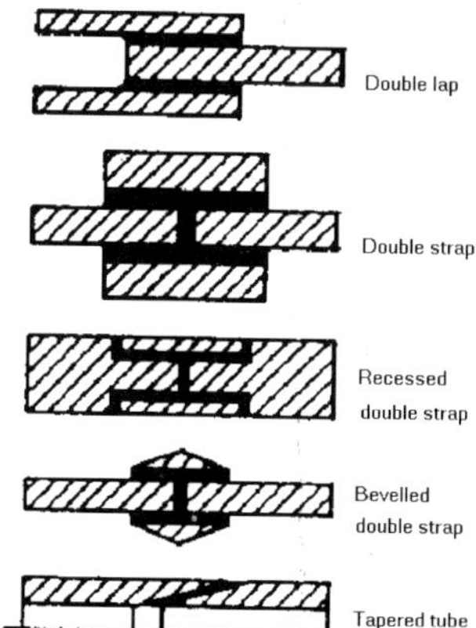

Fig 111: Joint configurations

welded lap joints is shown in Fig 107. There is also the advantage of improved corrosion resistance. The weld bonded structure is also said to be typically 25% stiffer than its spot-welded counterpart.

The Ferrari 408 structure exemplified in the previous section has become a mobile laboratory for the assessment of bonded structures in aluminium alloy, Fig 108. It is now recognised that the advantage of bonding can only be realised if the early concept design of the vehicle allows for this construction technique. Design should be based on the need to exploit the ability of the adhesives to exploit their highest strengths in tension and shear; this implies the minimisation of peel and cleavage forces. Gap thickness is also critical and Fig 109 shows the variation in shear strength of a PVC plastisol adhesive with thickness of the jointing material. Conventional weld-type flanges should be avoided and it should be noted that in the common over-'lap' joint, while bond strength increases linearly with joint width, this is not the case with overlap length. A joint factor has been proposed, to allow for this effect in calculation, of $(\text{thickness})^{1/2}$ / lap-length. Stresses in a single lap joint are shown as Fig 110. Of the joints shown in Fig 111, tapered and double-lap joints are better than the single lap.

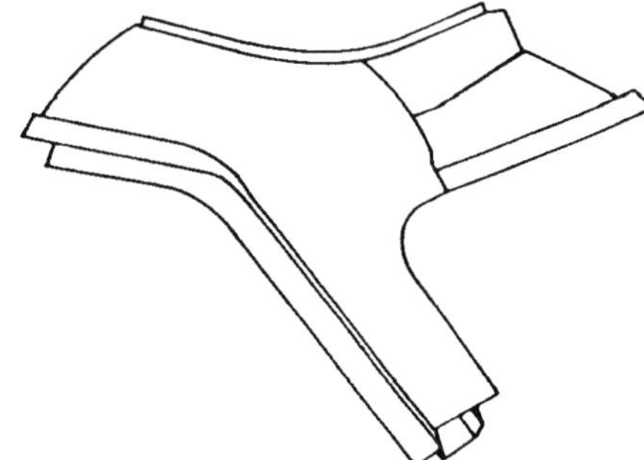

Fig 114: Cant rail to header joint

FAILURE LOADS FOR LAP AND T-JOINTS LOADED IN TENSION FOR BONDED ALUMINUM AND SPOT-WELDED STEEL

	LAP JOINT		T-JOINT	
	FAILURE LOAD (N)	NOMINAL METAL STRESS (MPa)	FAILURE LOAD (N)	NOMINAL METAL STRESS (MPa)
ALUMINUM 1.6mm 5251-0 BONDED	5960	149	2150	54
SINGLE SPOT-WELD 1.0mm (SOFT) SPOT-WELD	8030	320	984	39

Fig 115: Lap and tee geometry characteristics compared

A suggested body joint modification for bonding steel panels, due to AISI, is shown in Fig 112. To prevent undue squeeze out of adhesive, either formed dimples or recessed flanges can be used, Fig 113, for prevention of distortion by intermittent spot-welds. In the case of aluminium alloy, finite-element analysis work by ASVT on the typical A-post/cant-rail/header joint of Fig 114 showed that a bonded aluminium alloy joint could equal the spot-welded steel one in terms of structural stiffness.

A single-part heat-cured toughened epoxy material was used in the ASVT work plus occasional spot-welds, blind-rivets or self-drilling drive screws for tacking. Fig 115 shows typical strength values for the lap and tee geometries of Fig 116 to compare 1.6 mm thick bonded aluminium with 1 mm thick spot-welded steel. This does not take into account the discontinuous naturc of thc spot-wclded joint and with a 40 mm pitch spot joint the aluminium/steel figures for lap and tee configurations were 238/200 and 86/25 N/mm.

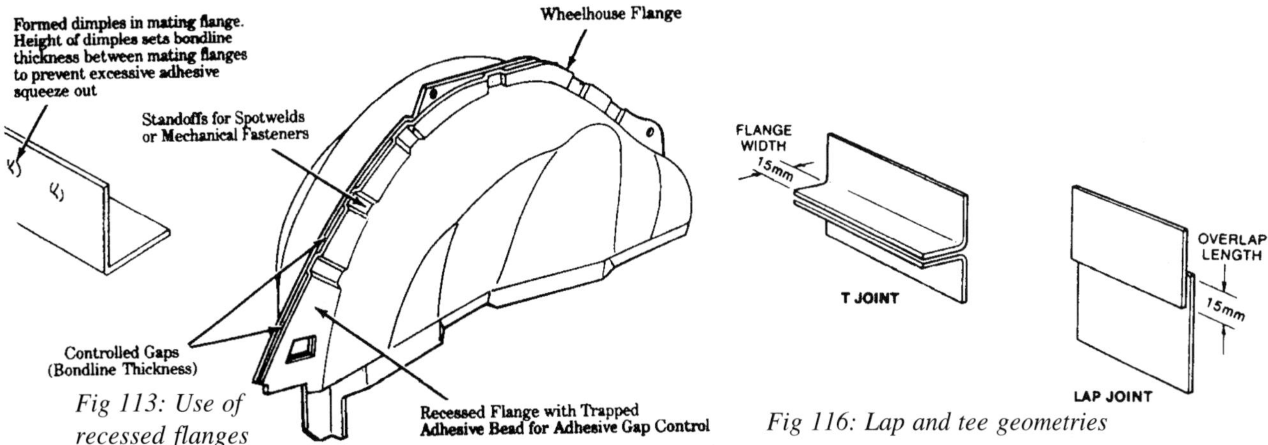

Fig 113: Use of recessed flanges

Fig 116: Lap and tee geometries

According to Alcan researchers[13], single part epoxies now show the most promise as adhesives. They are suitable for curing in the paint-bake oven when the whole structure is assembled, unlike phenolic adhesives, which, despite their known durability on aluminium, are ruled out due to their need of pressure during cure. The limitations of single-part epoxies have been poor peel-strength at sub-zero temperatures, and unsatisfactory creep performance at high temperatures. Work has now been done in conjunction with Ciba Geigy to address these areas of concern.

To help engineers attain structural optimization in an aluminium-bodied vehicle, the company have developed a joint design approach for adhesively bonded and spot welded structures[14]. The approach is based on a combination of full FE modelling of the vehicle structure, and experimentally determined joint performance envelopes. Its development showed that the metal gauge had a much greater influence on the strength of the spot-welded joint than the bonded one. Other factors affecting joint strength include the fillet size in T-peel joints, defined in Fig 117. The effect of changing this can be seen in Fig 118.

The design approach developed consists of three stages: modelling the joint line areas within the FE model; extraction of the important joint loads from the FE model and comparison of the joint loads with joint failure envelopes. Joint-line elements are created automatically in the FE mesh of the vehicle structure by a ANSYS and NASTRAN-compatible piece of software that has been developed, Fig 119.

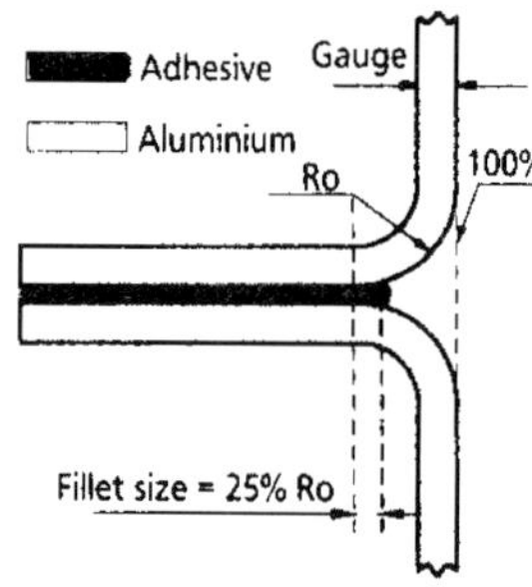

Fig 117: Definition of fillet size of adhesive bond

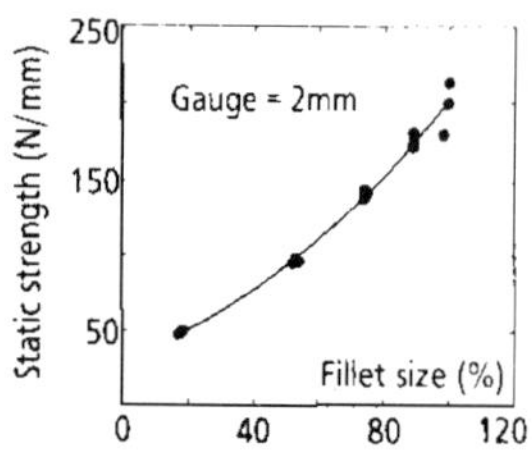

Fig 118: Effect of changing joint fillet size on static strength for adhesively-bonded T-peel joint

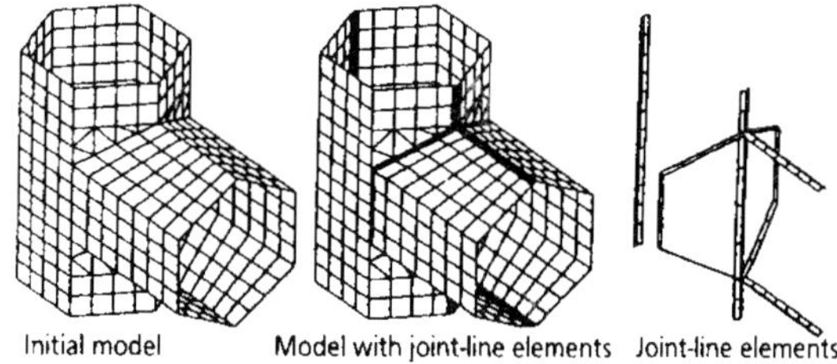

Fig 119: Automatic joint-line element creation is possible with an Alcan-developed program

Mechanically-assisted bonded joints

According to University of Warwick research[15] Adhesively bonded joints, the T-Peel joint offers a solution to the problem of minimising peel-forces, while withstanding cure temperatures, and is convenient to manufacture using existing spot welding equipment. The research has examined various combinations of joining techniques, different adherend thicknesses (16swg and 14swg) and widths (the width being determined by the joining technique) and variations of fastener pitch. Techniques appraised included adhesive bonding, riveting; riveting and bonding, spot welding; spot weld and bond; press joining (Eckold and Tox) also press joining and bonding.

For fastener/bonded joints: initially the tensile force rises steeply until the bond fails: the deforma-

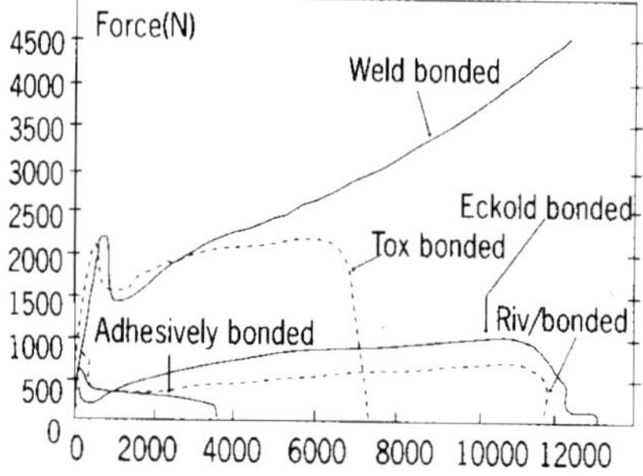

Fig 120: Joint stiffness curves

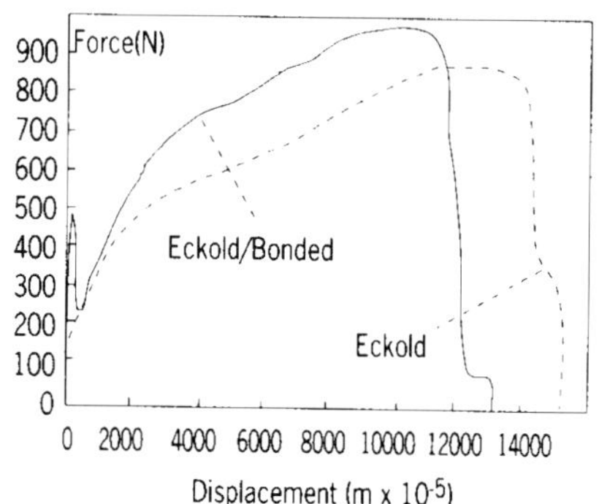

Fig 121: Eckold and eckold/bonded joint

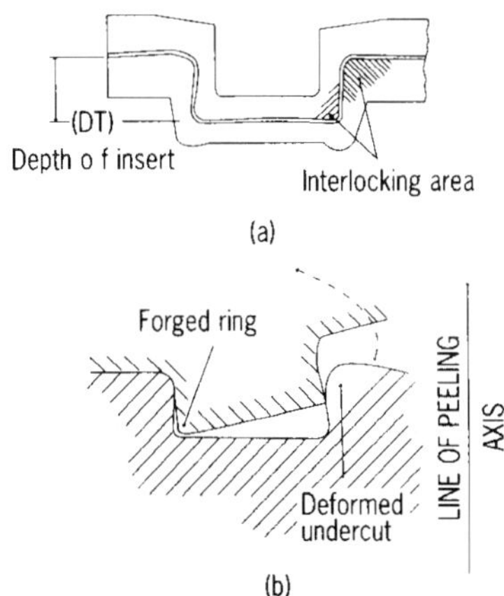

Fig 122: Tox joint — a. interlocking features; b. insert fastener relative movement under peel load

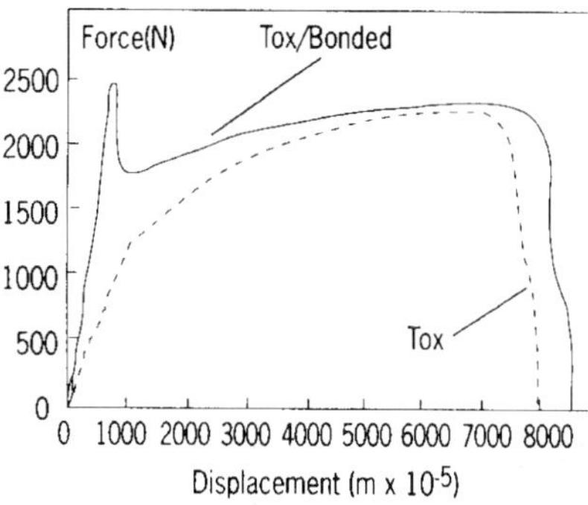

Fig 123: Tox press and Tox-bonded joints

tion is finally determined by the behaviour of the fasteners (Fig 120). The failure load is proportional to the bond width and these joints can withstand a failure load of approximately 364N/mm of joint width. The stiffness of a fastener/bonded joint is characterised by the super-position of the adhesive's stiffness and the fastener's stiffness. For adhesively bonded joints the stiffness is determined at 9.2kN/mm, and the steady joint displacement after failure is caused by the progressively breakdown of the adhesive as the adherends peeled apart. Stiffer adherends ensure a more uniform stress distribution to the whole of the adhesive layer, producing a slight reduction in edge stress, an increase in the tensile stress and reduction in peeling stress.

Riv/bonded joints exhibit a linear, high stiffness and a force at failure equal to the ultimate strength of the riveted joint. Addition of an adhesive produces a slight increase in joint strength, and greatly enhances the peel stiffness. In spot-welded joints, the addition of an adhesive increases the initial joint stiffness but has little effect on the strength. The final stiffness is produced by the welds tearing from the adherend. As the weld pitch increases the peak loading separates into two distinct peaks of lower amplitude. This is caused by joint imperfections which cause the peel stresses to be distributed unevenly to the welds. As the welds were created manually and individually, the eccentricity varies for each weld. With a significantly large difference in eccentricity, the welds fail separately and therefore generate two peak loads. Also, the second weld is inferior to the first because of electrical shunting.

In a weld-bonded T-peel joint the majority of the load is transferred by the spot weld. Although the joint requires a higher tearing force after failure of the welds, the benefit is not as great as for the equivalent lap joint. The benefit gained from the use of adhesive (load spreading and reduction in stress concentration) is negated by the T-peel geometry. The peeling stress is initially concentrated along the bondline edge and eventually around the welds less joint distortion is created since the stress concentration around the weld is reduced by the presence of the adhesive.

The Eckold/bonded joint gives a higher ultimate strength, but less elongation than the Eckold joint. The higher strength in an Eckold/bonded joint is induced by extra adesion forces provided by the adhesive and enhanced frictional force created between the pressed surfaces of the joining technique since the adhesive fills gaps created by the joining technique, Fig 121. Rupture force required for the Eckold fastening technique is increased by the presence of the adhesive but the allowable strain for the adhesive is much lower than that for the Eckold fastener. Stiffness of Tox press joints subjected to peel loads is characterised by the deformation of the interlocks (Fig 122a). The resistance to separation is partially provided by a frictional force, induced when the forged nng (Fig 122b) skids over the undercut. The energy consumed for joint separation can be increased by increasing the depth of insertion (DT) or increasing the surface friction of the joined parts. The peeling stiffness for the TOX/bonded joint is greater than that for the TOX joint (Fig 123). As with the Eckold joints, improvement in joint strength can be achieved by filling the gaps inherent in the joining technique and so increasing the force required for skidding.

Weld-bonding of sheet steel

According to research at British Steel's Welsh Technology Centre, considerable interest is now being shown in adhesive bonding as a complementary assembly technique to spot-welding. An evaluation programme has been carried out to determine optimum welding conditions, establish weld growth mechanisms and evaluate performance of weldbonding, related to adhesive-bonding and spot-welding as independent processes.

For a given weld current, with uncoated mild steel, it has been found that weld size, obtained with weldbonding involving a range of adhesives, was greater than that observed with spot-welding alone. This is due to the higher contact resistance applicable with weldbonding giving rise to earlier heat build-up. The same was also found to be true for zinc-coated steel and adhesive filler composition seemed to have little effect on the results. In fact it is the viscosity of the adhesive which is the principal factor affecting weldability, there being a limited value above which adhesive cannot be displaced from the interface to allow passage of weld current. By increasing the weld current progressively from a no-weld to a splash condition, the effect of adhesive on weldability was thus determined as in Fig 124. Gaps in the joints to represent poor panel fit-up were also simulated but these had little effect, with high electrode forces.

By increasing the weld time in intervals of one cycle of current flow, up to 15, it was possible to obtain comparative growth rates for both spot-welded and weldbonded joints, Fig 125 showing the more rapid growth for the latter process. Further tests allowed an examination of the growth mechanism. Peel testing of joints made with up to three cycles of current flow showed little or no bonding to have taken place. When sections were examined it was found that there was progressive increase in heat development with grain growth. After four to five cycles of current flow, evidence of crystalline fracture was observed but there was some solid phase bonding. The original interface, in such areas, had been eliminated by grain growth and the structure transformed, on cooling, to a ferrite/cementite structure. After six cycles, there was pronounced directionality in the grain structure, corresponding to solidification, melting having occurred at the interface contact zone. After seven cycles, rate of weld growth was limited. Contact area studies showed that prior to current initiation, the adhesive had been entirely displaced by electrode impact. However, after six cycles of weld current, a fully fused weld nugget was formed. While similar growth patterns were observed between weldbonding and spot-welding, heat development, and resultant weld growth occurred earlier in weldbonding.

Structural tests carried out on 0.5 metre long box-hat sections, Fig 126, showed similar torsional strengths for both weldbonded and spot-welded specimens. However, torsional stiffness developed by a bonded structure was 25% greater than for spot-welded specimens having 25 mm spot weld pitch, Fig

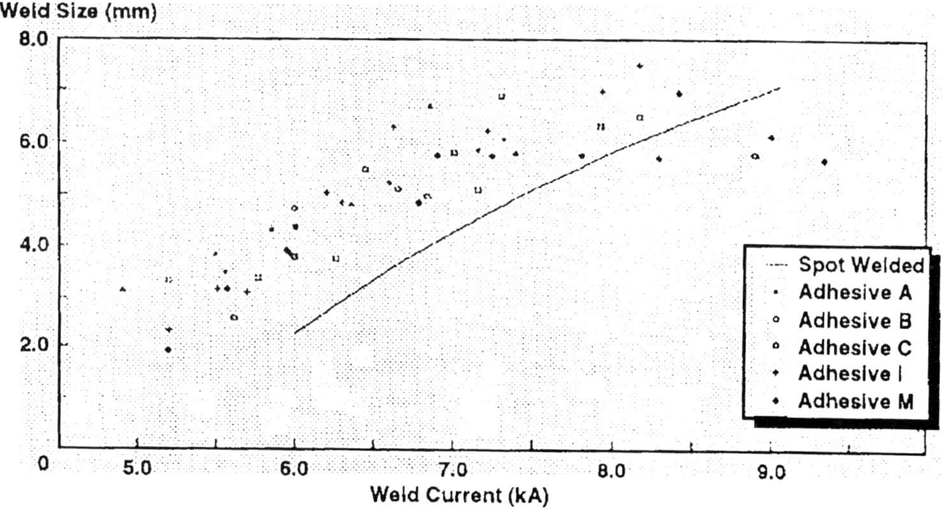

Fig 124: Weld growth comparison for weldbonded joints with one-part adhesives and spot-welded joints

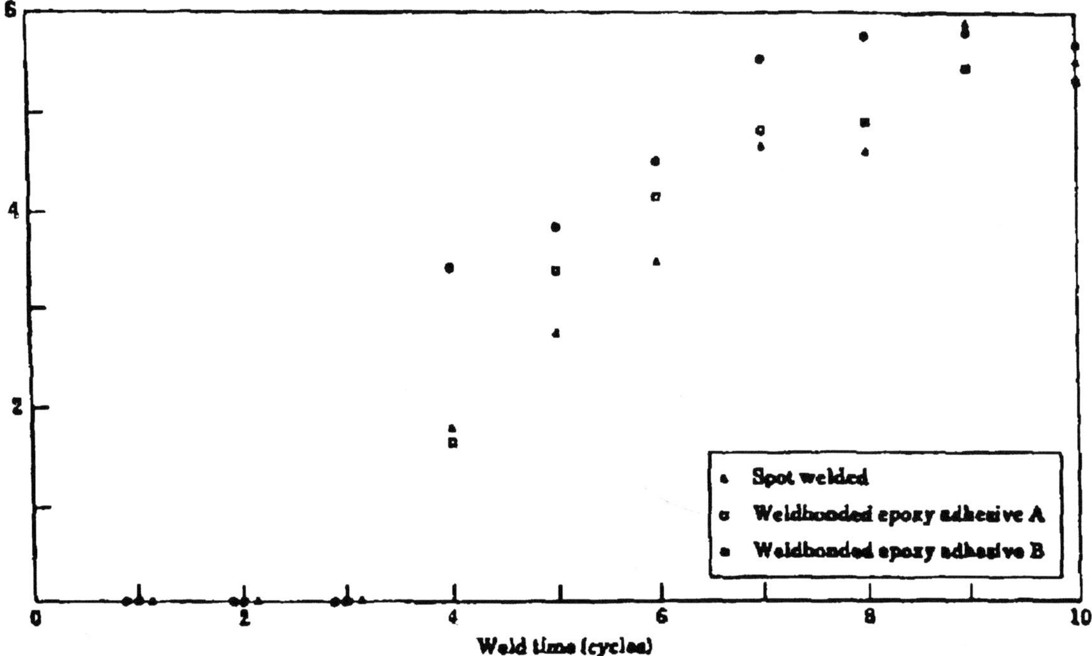

Fig 125: Influence of weld time on weld size for both spot welded/weldbonded joints in 1.2mm mild steel

127. Adhesive bonded structures were prone to instability under end impact loads owing to a tendency to flange separation. Spot-welded structures collapsed without weld failures under these conditions. By combining both techniques the spot-welds in a weldbonded structure tended to act as crack arresters which limited flange separation. Thus greater stability was obtained throughout the collapse process and a reduced degree of collapse was obtained than with either of the independent fastening techniques. Fatigue strengths, too, were enhanced significantly by weldbonding when designs which minimise peel stress were used, Fig 128.

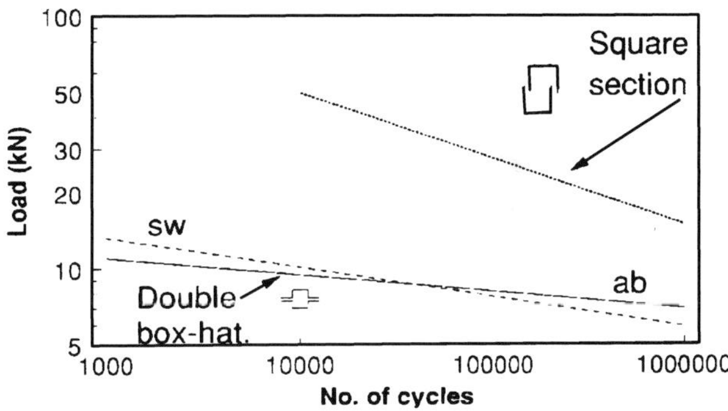

Fig 127: Comparison between torsional stiffness of adhesive bonded and spot-welded top-hat structures

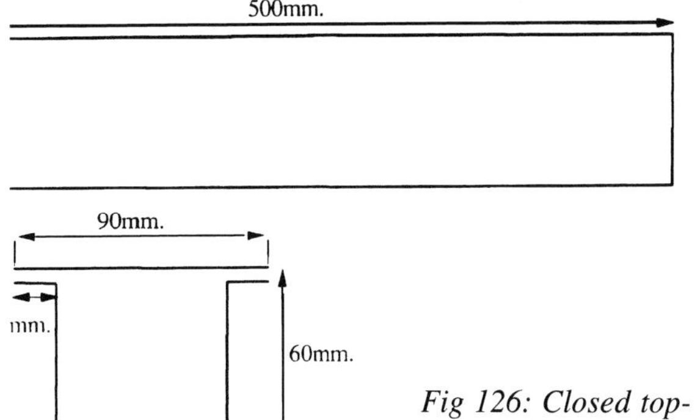

Fig 126: Closed top-hat section dimensions

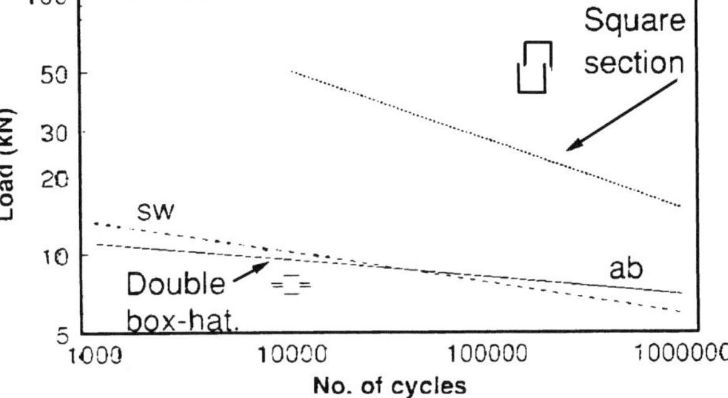

Fig 128: Influence of section shape on the fatigue properties of adhesive bonded structures in 1.2 mm mild steel

Mechanical bonding

Hook and loop fasteners are providing a valuable alternative 'bonding' technique which is of particular value to interior trim applications. The pioneering Velcro brand can be woven, knitted or moulded into plastics and composites.

Peel strengths range from 0.35 to 3.5 lb/in^2 (246-2460 kg/m^2), tensile strengths from 2 to 24 (14-168) and shear strengths from 6.5 to as high as 120 lb/in^2 (45.5-840 N/m^2). Higher performances can be obtained when the system is used in other than planar overlap configuration. Examples are curved surfaces and interference shear traps, both of which combine the various modes of separation in order to enhance joint strength. There is a tendency in shear separation for the mating surfaces to move laterally and so an interference shear set-up as in Fig 129 substantially increases the pull-out force. Another technique is the so-called manipulative shear trap, in which the configuration works as a hinge which closes to secure the joint for high shear force though relatively low peel force is required to break the joint, Fig 130.

In general the hook is a monofilament and the loop a multifilament fibre. To form the hook the fibre goes through a cutting process in which the hook eye is positioned between a set of comb teeth with cutting blades at every other tooth. Separation of hooks and loops occurs over a single line of engagement and tension forces are slightly higher than peeling ones because a greater area of hook engagement is involved. In shear the hooks and loops are being raked over each other, thus increasing the separating force yet further.

A special development used to close two panels on a flush surface without any gap is called anti-

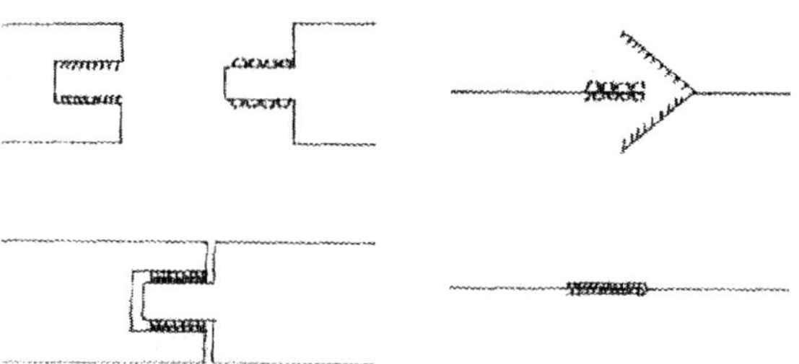

Fig 129: Interference shear *Fig 130: Manipulative shear*

TAPE PRODUCTS	TYPE/BRAND	DESIGNED FOR
	Nylon	General Purpose
	Polyester	Water and Chemical Resistant
	HI-AIR® (NOMEX*)	Flame Retardant
	HI-GARDE® (Stainless Steel)	Very High Heat Resistance
	MID-TEMP®	Moderate Heat Resistance
	HI-MEG®	Electrically Conductive or for R.F.I. Shielding
	ASTRO-VELCRO®	"Non-Burning" "Non-Igniting"
	Molded Nylon Hook MVH Series	Extra Strength
	Molded Nylon Arrowhead MVA Series	Specialty Fastening

Fig 131: Velcro tapes

SPECIALTY PRODUCTS	TYPE/BRAND	INSTALLATION
	VELSTRAP® (VST)	Reuseable, for High Strength Cinching (with Ring)
	BACK TO BACK STRAP	Reuseable Self-Wrapping Tape
	FACESTRAP (VFS®)	Adjustable Reuseable Cinching Strap (without Ring)
	VELSTICK®	Semi-Rigid Screw, Nail or Rivet
	INSTANT VALANCE	Screw, Nail or Rivet
	EDGECLIP®	For Slipping on Open Edges
	VELTRACK® BRAND	Slotted Insertion
	VELWASHER	Surface Mounted Screw, Nail or Rivet

Fig 132: Speciality products

backlash closure. This involves a flexible component, to create flush attachment, which collapses under closing pressure. Another new departure is moulding of the fastener as an integral part of the closure panels. This eliminates secondary attachment operations, of course, and avoids piercing the substrate with atttachment holes.

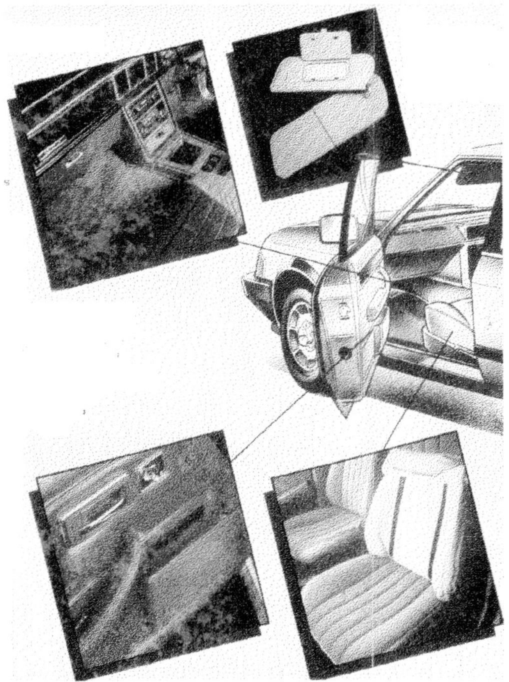

Fig 134: Velcro application areas

A benefit of hook-and-loop fastening which is not so obvious is the ability to absorb impact over a jointing surface. This has already been put to use by one motor-manufacturer, in the fixing of a lightweight bumper requiring less than normal energy-absorbing properties to meet the statutory 5 mph impact test. Compared with discrete metallic fasteners, moulded-in Velcro also overcomes the problem of pull-out caused by differential thermal expansion rates between a composite substrate and its fasteners.

Specialist Velcro fasteners, in the range supplied by Selectus Ltd are available as tapes shown in Fig 131 and speciality products in the table of Fig 132. Row 1 includes types for securing convertible hoods, truck cab headlings and floor carpeting. Row 2 types would be used for jobs like speaker grille attachment. Row 3 is for 'spot' fastening applications where a double faced foam pressure-sensitive backing is required. Row 4 applications include boot carpeting, car headliners and door panel assemblies. Row 5 has uses in secondary dashboard attachments and those for truck cab interiors. Row 6 covers removeable valances and Row 7 for securing acoustic materials. Fig 133 shows technical data for automotive industry products and Fig 134 shows key application areas.

	STANDARD VELCRO	TYPE 19P	MVA 8 HOOK	MVA 10 HOOK
MATERIAL	Nylon 6.6.	Polyester	Injection moulded nylon with standard loop	Injection moulded nylon with standard loop
SHEAR STRENGTH (LENGTHWISE) KG/CM²	0.84	1.70	1.30	1.30
PEEL STRENGTH (LENGTHWISE) KG/CM²	0.21	0.16	0.45	0.45
TENSION/ LATCHING	0.42	0.84	0.67	0.91
NO. OF OPENINGS/ CLOSINGS	In excess of 5,000	3,500 max	350	1,000
TEMPERATURE RANGE HIGH –	200°F – 70°F 93°C – 56°C	200°F – 70°F 93°C – 56°C	200°F – 70°F 93°C – 56°C	200°F – 70°F 93°C – 56°C
AVAILABLE WIDTHS MM	10, 15, 20, 25, 30, 50, 100	15, 20, 30	8, 12.5, 25	8, 12.5, 25
AVAILABLE COLOURS	20	natural, black	black, white	black, white
FINISHES AVAILABLE	Pressure sensitive adhesive, Welding (U.F. & R.F.) Solvent reactivated	Pressure sensitive adhesive, Welding (U.F. & R.F.) Solvent reactivated	Pressure sensitive adhesive, Welding (U.F. & R.F.)	Pressure sensitive adhesive, Welding (U.F. & R.F.)

Fig 133:Technical data for Velcro tapes

COMPONENTS AND COLOURS

SOFT HARDWARE® COMPONENT	PLASTIC SUBSTRATE COLOUR	VELCRO FASTENING TAPE	TAPE COLOUR(S)
VELSTICK FASTENERS	Black White Black	Woven nylon hook & loop Woven polyester hook and loop Moulded nylon hook	Black White Black
VELSTICK VELWASHER FASTENERS	Black White Black	Woven nylon hook & loop Woven polyester hook & loop Moulded nylon hook	Black White Black
VELSTICK VELCOIN FASTENERS	Black White Black	Woven nylon hook & loop Woven polyester hook & loop Moulded nylon hook	Black White Black
VELTRACK FASTENERS	Light grey Light grey Light grey	Woven nylon hook & loop Woven polyester hook & loop Moulded nylon hook	Light Grey White Grey
EDGECLIP FASTENERS	Clear Clear Clear Clear Clear Clear	Moulded nylon hook Moulded nylon hook Moulded nylon hook Woven nylon hook & loop Woven nylon hook & loop Woven nylon hook & loop	Black White Beige Black White Beige
INSTANT VALANCE FASTENERS	White White	Woven nylon hook & loop Woven polyester hook & loop	White White
VELSTUD FASTENERS	Black Black Black	Moulded nylon hook 8	Black

Joining by metal injection

Injected Metal Assembly (IMA) is a process of joining parts together which applies die casting technology to small component assembly. It cost-effectively replaces many assembly methods such as crimping, press fitting, shrink fitting, staking, swaging, riveting, adhesive bonding, soldering, brazing and welding. It also overcomes many of the problems associated with these assembly methods. From two to 10 components can be assembled in one operation.

The components to be assembled are positioned in their correct relationship to each other and held firmly by the assembly tool portion of the system. Molten alloy is injected into a cavity located at the intersection where the components are to be joined. The molten alloy flows in and around physical features on the components such as grooves, knurls, keys or other locking shapes .

The injected metal solidifies in milliseconds, shrinking slightly toward the theoretical centre, creating a strong, permanent lock. Additional shapes can also be cast as part of the injected metal hub, in many cases eliminating one, or more components. The system ejects the finished assembly, which is ready to use with no secondary operations required.

The components to be assembled are either hand fed or automatically loaded from a combination of vibratory feeders, hoppers, magazines or reels of wire, Fig 135. Assembly rates can exceed 1,000 per hour. In an automated system, the cycle is controlled by a programmable logic controller. To change operating sequences from one assembly to another, an appropriate EEPROM memory module is simply plugged into place. Systems can be dedicated to a single assembly operation, or interchangeable inserts can be used to produce variations.

Fig 136 shows a small gear and shaft assembly fabricated by the IMA process. Here advantage can be taken of the metal shrinkage on solidification to obtain rigidity and allow applied torques to be reacted. For zinc alloys used in the process, shrinkage is 0.7%. Small grooves can be incorporated into the components to be joined to enhance these effects. Fig 137 shows how combined functions can be obtained by the ability of the process to incorporate useful shapes into the final assembly without adding other components or operations. In evaluating the amount of injected metal required to fill the proposed cavity, and form additional shapes required, this is mainly a

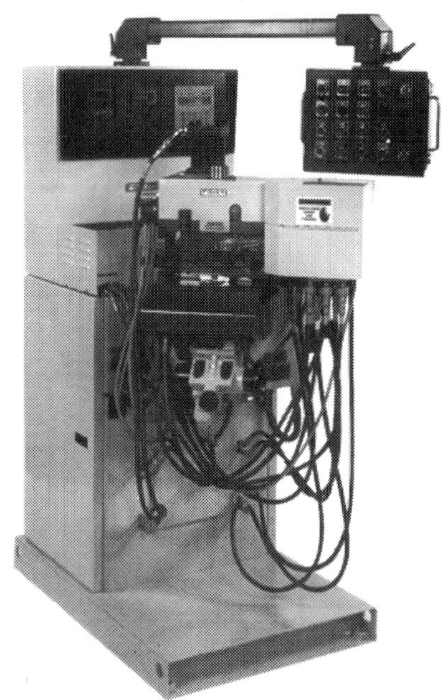

Fig 135: The assembly machine

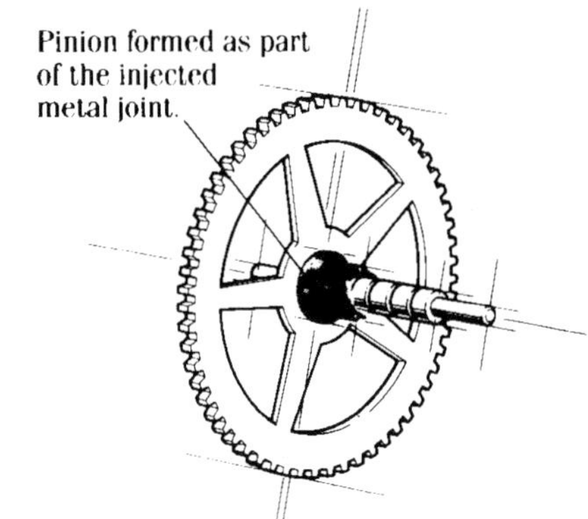

Fig 137: Obtaining combined functions

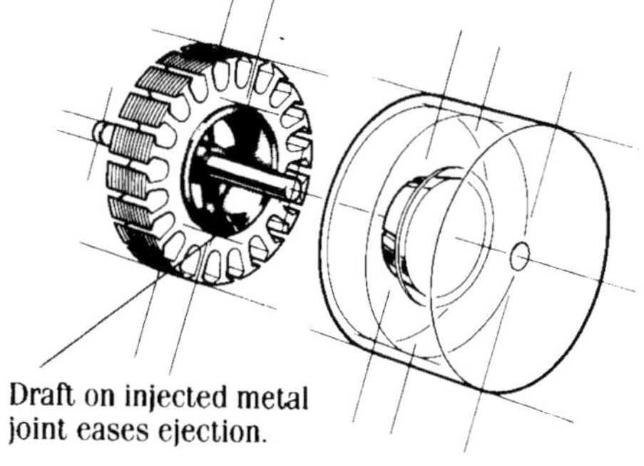

Fig 138: Easing ejection after joining

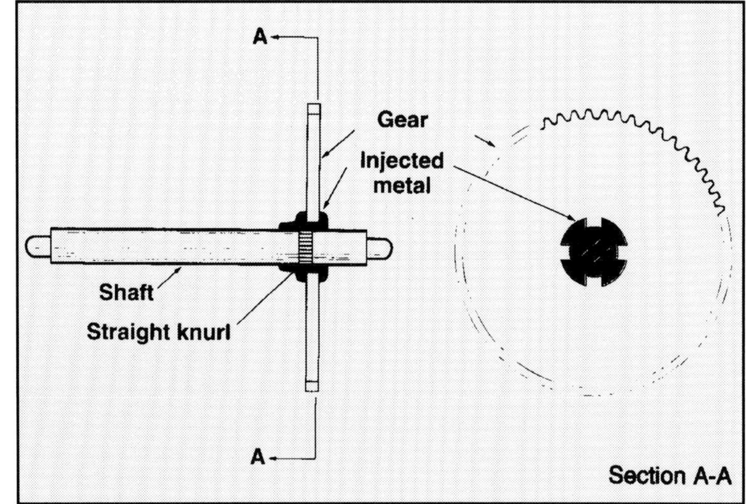

Fig 136: Gear and shaft assembly joined by IMA

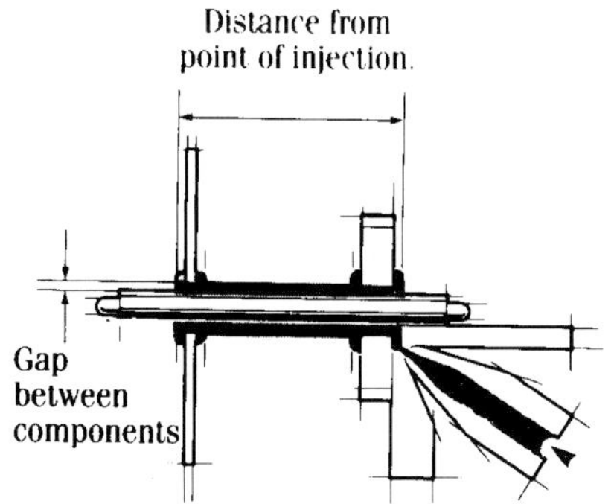

Fig 139: Easing metal flow

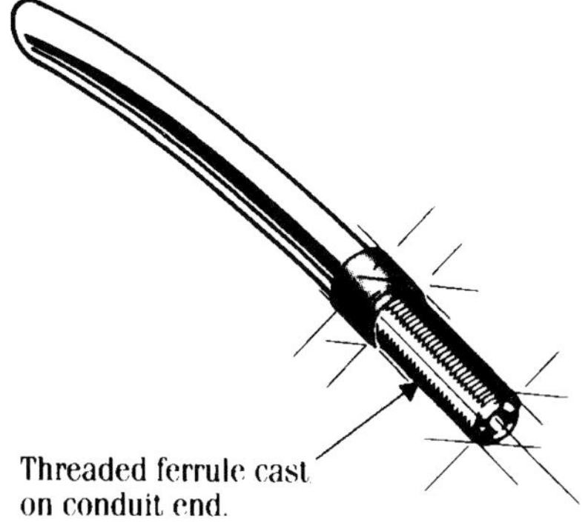

Fig 140: Forming external threads

function of the size of the hub required to withstand the application loads on the assembly. Using zinc alloy, average tensile strength is 41,000 lb/in^2 and compression strength 60,000 lb/in^2. Brinell hardness is 82 and Charpy impact strength (1/4 x 1/4 inch bar) is 43 lb-ft.

Fig 138 shows how a slightly tapered cavity or core is needed to allow the assembly to be easily ejected after joining. It is best to ensure the molten metal travels over as short a distance as possible to avoid any porosity occurring and sufficient clearance must be provided between the components to ease metal flow, Fig 139. Normally 0.75 mm is required on each side, with 0.13 mm as an absolute minimum. Typically the IMA process is used to assemble components which are 150 mm or less in their largest dimension, although this can be increased by the use of specialised equipment. At Manufacturing Week 96, the assembly of a 30.5 mm long by 2.3 mm diameter steel shaft was demonstrated, to which a 38.1 mm diameter by 2.03 thick die cast gear was attached.

It is possible to form external threads as part of the joint, Fig 140, but as a parting line must run the full length of the thread, diametrically opposed flats to root diameter depth should be allowed whenever possible. Fishertech application engineers can make recommendations on ways to enhance the assembly process using IMA. The process is a proprietary technique of Fishertech - Division of Fisher Gauge Limited, at PO Box 179, Peterborough, Ontario, Canada, K9J 6Y9.

Welding ceramics

Future automotive parts, and some already used in racing cars, are likely to be combinations of ceramics, metals and polymer composites. Techniques for joining them may not strictly be welding but they have a good deal in common with early pressure-welding techniques.

The Welding Institute (TWI) have showed that the successful exploitation of material properties can often be achieved by using newly-developing joining techniques to combine certain properties of one material, or composite, with the complementary properties of those of another. Thus, it is argued, whereas materials like ceramics, with attractive thermal and wear performance, have often been compromised by a brittleness of the material, this might be overcome by joining it to a tougher substrate. TWI have shown already that even a ceramic faced engine tappet is feasible with such techniques.

Hot isostatic pressing is one such method; it can also be used to make ceramic parts from silicon carbide and nitride. It differs from conventional hot-pressing in that pressure is applied on the three perpendicular axes, rather than in one direction. The pressing involves a closing up of pores and consequent slight loss in volume, Fig 141. The process can be carried out with or without a secondary encapsulating material of, say, glass or other ceramic. Diffusion-bonding of ceramics to metals would involve the sort of pressure distribution shown in Fig 142. The TWI's facility offers pressure up to 2000 bar and temperatures to 2200 C and component size limits are 100 mm diameter by 150 mm high.

Ceramic materials can also be sprayed on to substrates using plasma or high velocity oxygen fuel spraying (HVOF), Fig 143. The plasma process uses a DC arc struck between non-consumable electrodes, an inert gas being fed into the arc to form a high-temperature plasma (15,000 C) which is compressed and accelerated by the torch nozzle. The coating material is fed into the plasma in powder form, the whole being ejected at velocities up to 600 metres/sec. Oxyfuel spraying, already used for ceramic coating of piston crowns, involves combustion of a fuel gas with oxygen at high pressure and/or high flow rates. A high velocity gas flame thus created, propels powdered coating materials on to a substrate, Fig 144.

Interestingly, the welding together of dissimi-

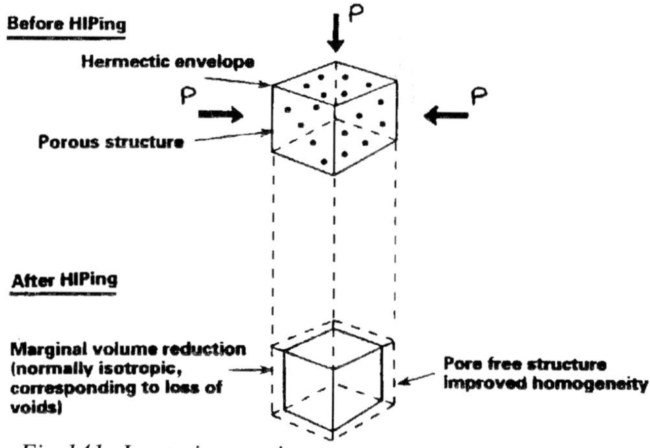

Fig 141: Isostatic pressing

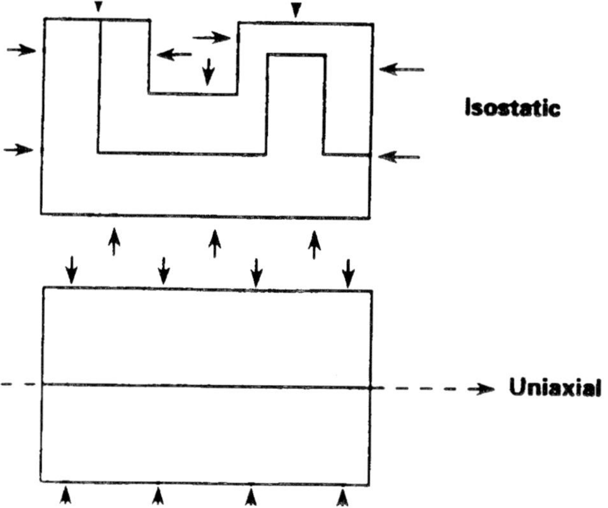

Fig 142: Pressure distribution

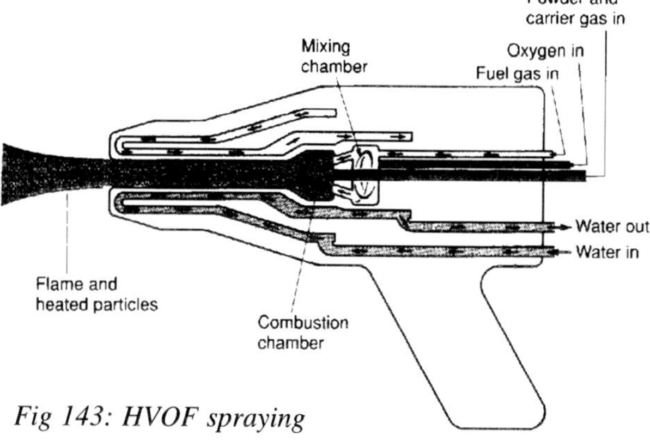

Fig 143: HVOF spraying

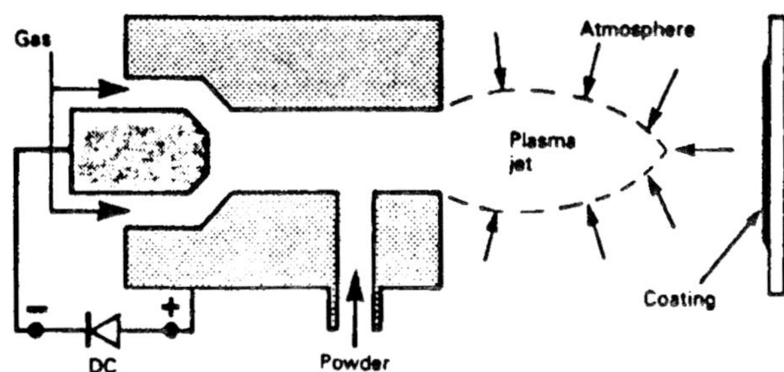

Fig 144: Plasma spraying

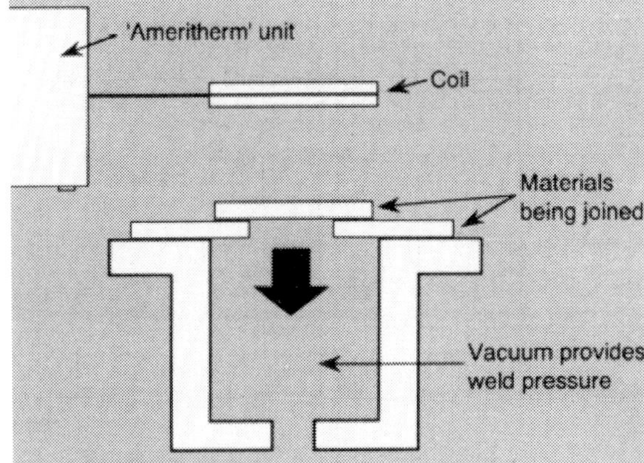

Fig 145: PCM welding

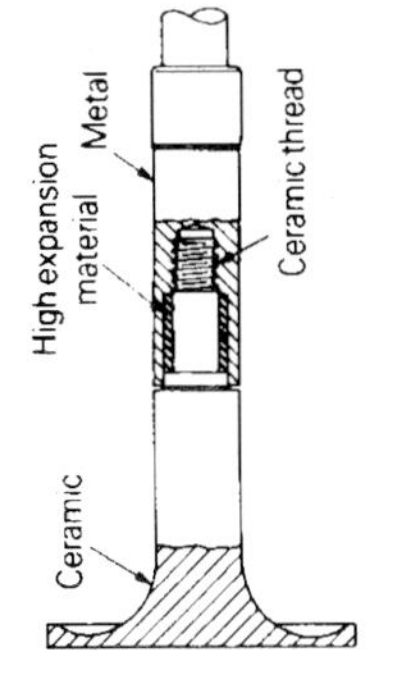

Fig 146: Brazing combined with screw-attachment

lar materials can be aided by coating techniques, too. In so-called Polymer Coated Material (PCM) welding, developed by TWI, carbon-fibre thermoplastic composites can be joined to aluminium alloy, for example. One component is pre-treated with a thermoplastic coating, after which a proven plastic welding technique such as induction welding, Fig 145, can be used to join the parts.

American researchers[16] have described Ford USA work on structural ceramics, in which investigations into adiabatic diesels has led the company to identify a critical area of exploiting structural performance as the connection between ceramic and metal support. For the turbocharger rotor-assembly both brazing of the ceramic rotor to the metal shaft and a screw attachment, Fig 146, have been tried. In the latter a trapped high bulk expansion material at the interface accounts for differential thermal expansion. The three phase approach to the adiabatic engine is also illustrated, Fig 147, to show first, ceramic-coatings (left) second, inserts (centre) and then the all-ceramic-engine (right); in all these, key ceramics property is good thermal shock resistance balanced with high temperature dimensional stability; chemical resistance to combustion products; low thermal conductivity and expansion coefficients, high strength and Weibull modulus. Engine testing over 100 hours on the ceramic engine has been achieved without incident for the SiN_4 cylinder, piston and gudgeon pin.

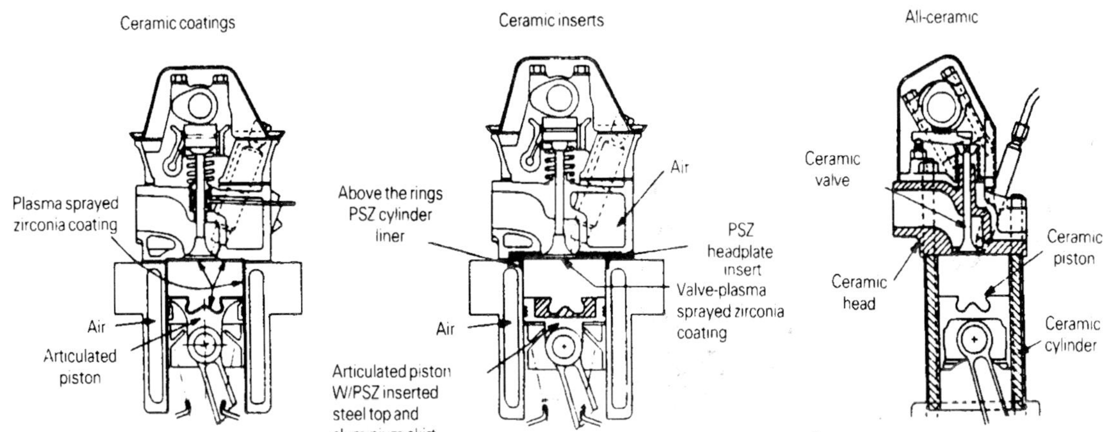

Fig 147: Ceramic jointing in the steps towards the adiabatic engine

Automated assembly

This important new discipline has added benefits to plastics and composites moulding for it can avoid many of the common fastening techniques by consolidation of individual parts into a composite whole. Bayer's Technical Applications Division has shown that by consolidating as many parts as possible into the individual mouldings, substantial economies should be possible.

The case study considered here uses an instrument panel, Fig 148, the commercial success of which, as a design, depends on the consideration, at the early concept stage, of such factors as: flowability, shrinkage, tolerances and wall thickness to rib ratios; also of surface quality and visibility of flow lines; finally, aspects relating to the mould such as position of the gate, number of hot-runner nozzles, demouldability and general cycle feasibility must be considered.

Computer programs are now available for simulating the mould filling process without actually constructing the mould. With knowledge of material properties, flow lengths can even be predicted for a particular wall thickness. A general relationship is shown in Fig 149 for different injection times — and for wall thickness of 2.6 mm. When making corrections to the mould, shrinkage of the material must be determined both during and after the moulding process. This relates to pressure drop in the cavity according to material characteristics — as shown for typical polycarbonate/ABS blends in Fig 150.

If bosses to take screws are required on the backs of visible surfaces, the standard boss (left) should be replaced by the design (right), in Fig 151, to provide a wall-thickness ratio of 1.2 to 1.3, required for best visual results. The same requirements apply to ribs as well as bosses, in these situations. Another surface effect is particularly noticeable with

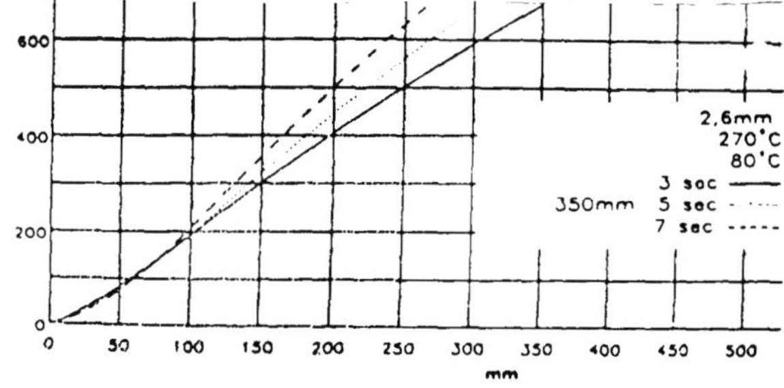

Fig 149: Flow-length/pressure relationship

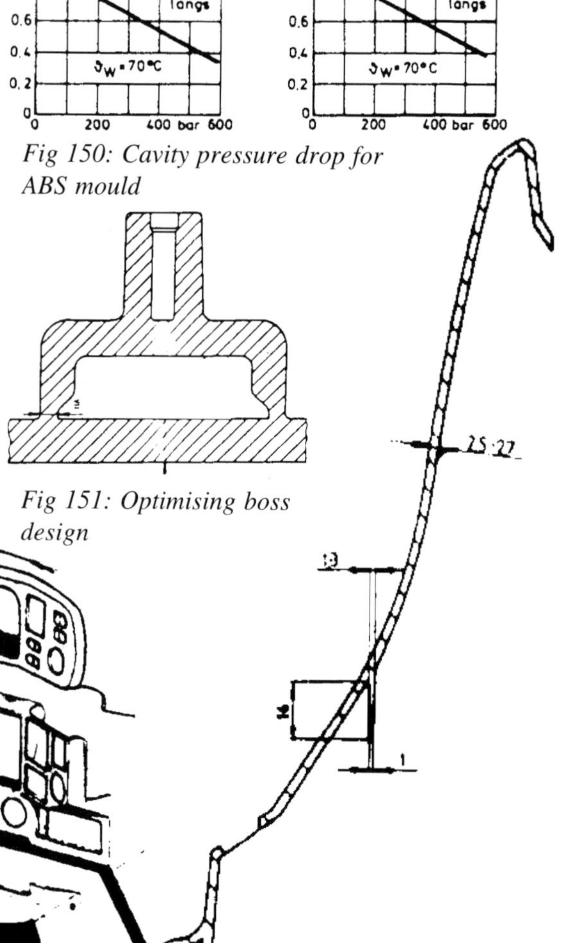

Fig 150: Cavity pressure drop for ABS mould

Fig 151: Optimising boss design

Fig 152: Ribs on a sloping side

Fig 148: Dash panel case study

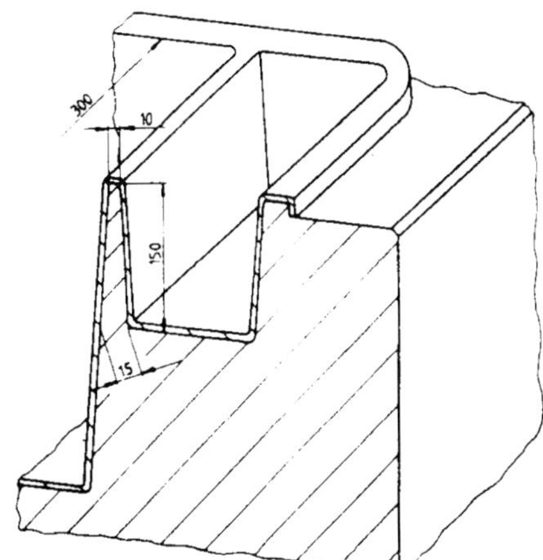

Fig 153: Double-wall area of mould

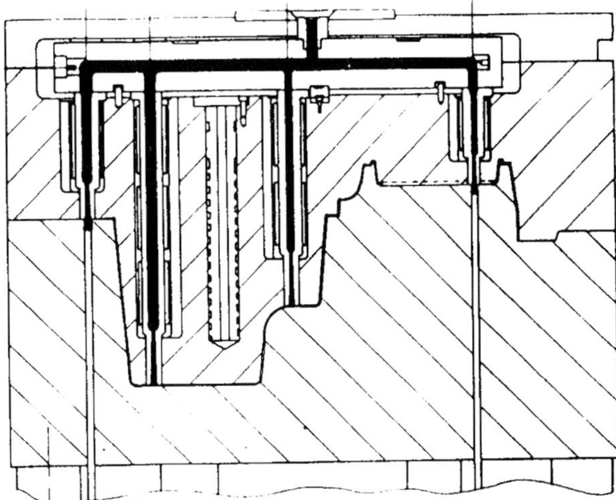

Fig 154: Hot-runner system

grained surfaces (obtained by photo-etching the inside of the mould's cavity); this is the weld line, the position of which can be varied, to find a less visible area, by repositioning the gate. To obtain an evenness of gloss, moulded parts which are located in one and the same visible area should all be made of the same material.

The grained surface can also be disfigured if inadequate draft is provided. Generally there should be one degree per 20 microns mean roughness in addition to the normal one degree of draft. Another problem area is ribs positioned on surfaces of the moulded parts that slope inside the mould (Fig 152). This is because the tip of the core that forms in the mould can be deformed by the injection pressure. This, in turn, can exert pressure on the grained surface after cooling. Double-walled areas of mouldings are also critical (Fig 153). Cores of this type are readily deflected through pressures acting on one side during the mould filling process. This leads to problems in demoulding; also it is no longer possible to cool the cores — which themselves determine the length of the cycle. In tall moulds, the importance of core centring, with respect to the direction of mould opening, must also be stressed.

Plastics mould parameters As a more detailed part of the case study, the glove locker section of the overall panel is considered as a separate moulding, shown in Fig 154 with its hot runner system designed by computer simulation. The principal visible surface makes it necessary for the core to be located on the fixed half of the mould — because of the injector pin marks. Despite central gating, with which the melt is injected from the inside, outwards, elastic deflection could still be expected with this layout which would influence the filling process. Thus the core must be centred.

The computer drawn filling pattern predictions in Fig 155 help to determine where flow lines occur and if any trapped air is likely. Provision was made for two types of gating in this example — both on visual grounds and to suit the mould. In the compartment area there are short sprue gates and in the parting surface region for the openings there are rectangular plates. The simulation also helps to see whether all parts near the sprue are filled uniformly and simultaneously. Inspection shows the paths taken by the fill are very different. To even these out further calculations are necessary to balance pressure according to temperature profile. In this example runner diameters vary from 11 to 24mm depending on their length and the melt throughput.

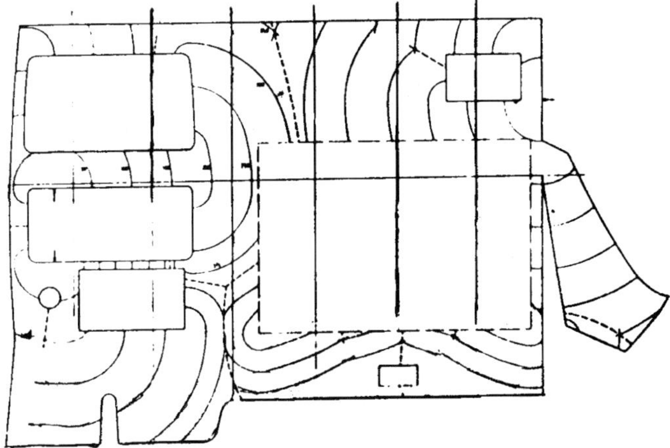

Fig 155: Computerised filling-prediction

Design for robotic assembly

While basic considerations of automated assembly were considered in the last section with the particular case study of a large plastic moulded instrument panel showing the benefits of integrating separate parts into a single moulding, here techniques are examined for easing the assembly of those parts which cannot be integrated.

When considering the wider implications of robotic assembly benefits are available even to the small-series builder if a sufficiently flexible robot is programmed for a wide variety of assembly operations. Considerable design changes may be needed to the vehicle in order to exploit the technique fully, however. The volume sector teaches us, for example, that the basic body shell may need to have a detachable front-end in order to place more easily the major mechanical assemblies into the vehicle.

At a more elemental level careful planning of the stations in an assembly system can reap rewards in the automated feeding of small parts for a larger assembly, Fig 156. In the illustration, a spindle assembly is considered and Fig 157 is the associated process model used in planning the procedure.

Minimising part count is an equally valuable discipline in simplifying the assembly process. With sheet metal parts it is obviously an advantage to create features from bending the parts rather than welding on stand-offs, inserts or snap features, Fig 158. The deformable properties of plastic can often be used to take up tolerances within an assembly, Fig 159, and integral 'springs' can be created to eliminate parts, Fig 160, or tabs in the case of sheet metal.

By designing a product so that subsequent parts are located on previously assembled ones, reorientation can often be avoided by use of features on the earlier part to locate the subsequent one. Self-locating features can be used in this context, Fig 161, such as the appropriate creation of a chamfer. In restricted areas features can be used to positively locate a part before it is released, Fig 162, or even adding a dimple can facilitate the alignment of blind features, Fig 163.

Parts should also be made symmetrical where possible to reduce orientation and if this is not possible it may be feasible to design obvious asymmetrical features into the part to facilitate orientation, Fig 164. Obviously it is wise to avoid parts that might nest or tangle and adjustments can often be reduced by using a long positioning hole on one side of a two-screw fixing, for example, Fig 165.

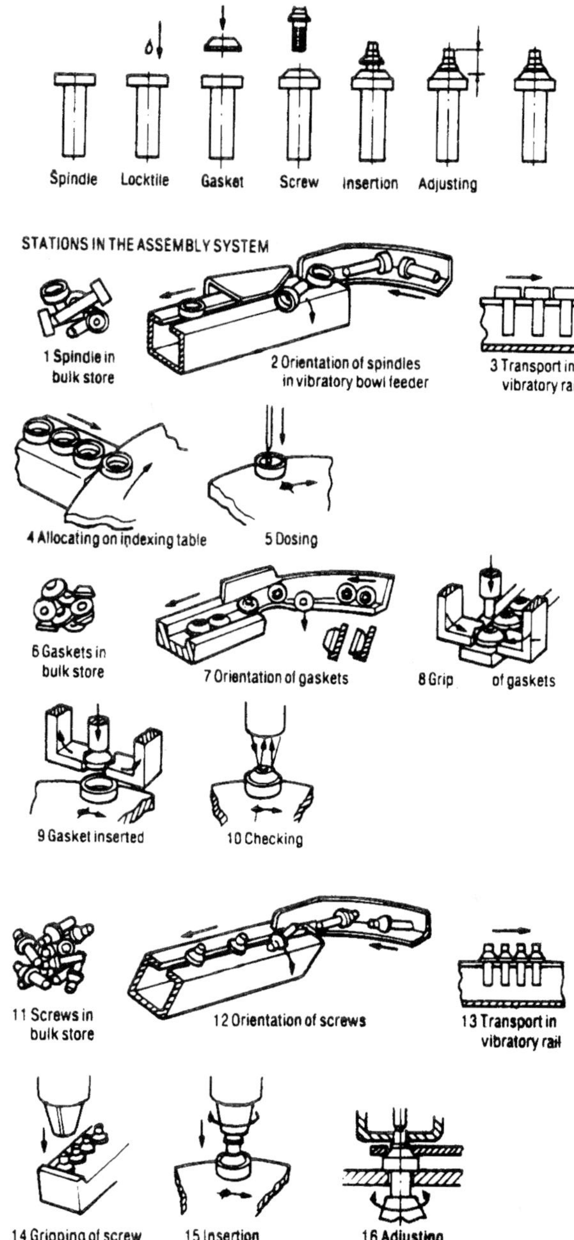

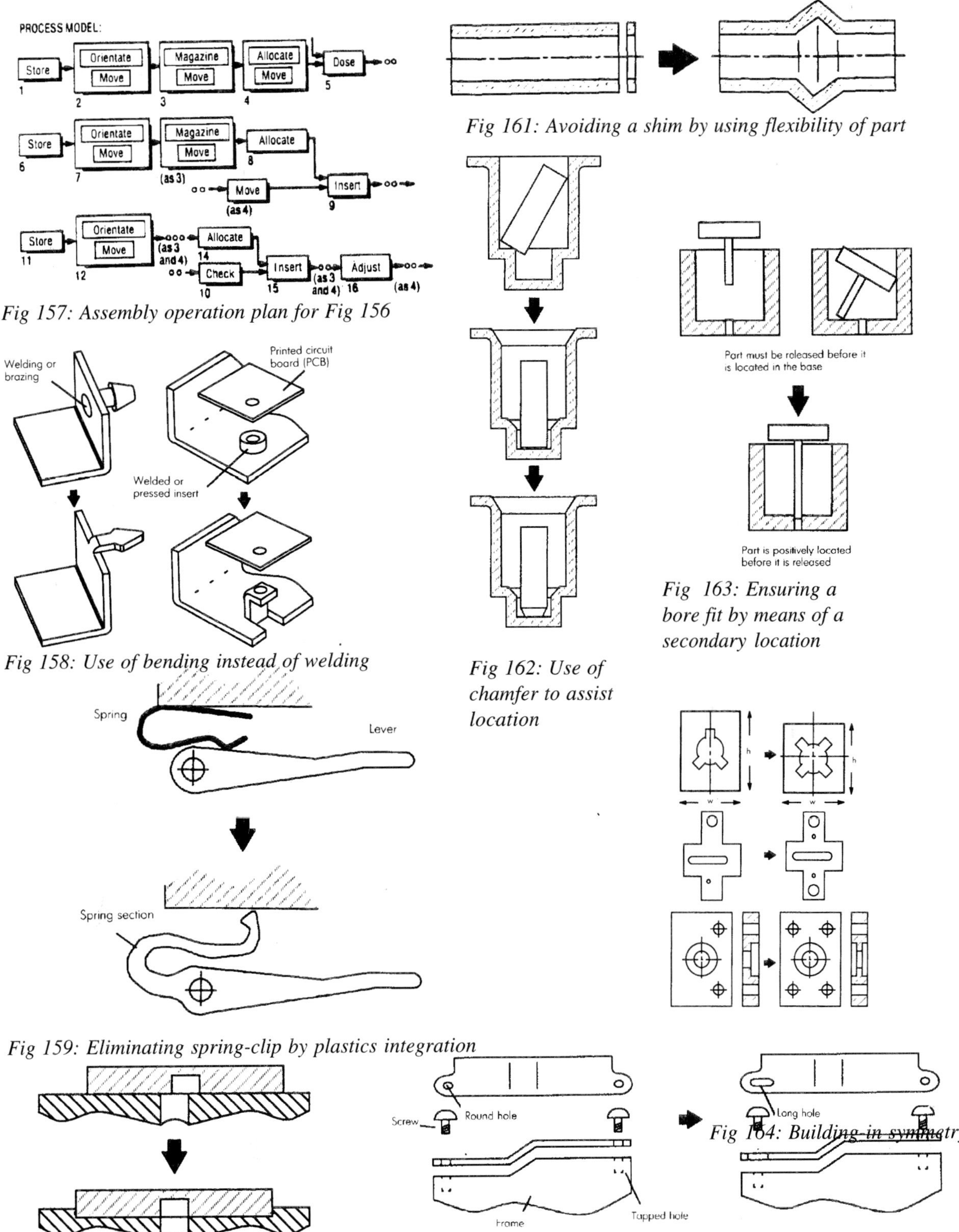

Fig 157: Assembly operation plan for Fig 156

Fig 158: Use of bending instead of welding

Fig 159: Eliminating spring-clip by plastics integration

Fig 161: Avoiding a shim by using flexibility of part

Fig 162: Use of chamfer to assist location

Fig 163: Ensuring a bore fit by means of a secondary location

Fig 164: Building-in symmetry

Chapter 2: Panel cutting and forming

Press-working state-of-the-art; metal-forming basics — shearing, bending and profiling; flame and jet cutting; press-working basics and advanced developments; deep-drawn press-work and superplastic forming; roll-forming, flex-forming and spinning/flow-forming

Pressworking state-of-the-art

Several events[1] mark the arrival of the present-day sophistication in pressworking technology. Breakthrough for presswork simulation in Japan came in 1988 when Honda engineers analysed the single-action stamping process — using CAD surface data for the tool and elastic-plastic membrane FE elements for the blank sheet. New trends in car design also resulted in a reduction in number of panels, exemplified by the floor panel in Fig 1 which was once made from five separate parts and then reduced to only two. The same trend was found in car side panels. The wider use of HS steels and aluminium alloys was also presenting new problems to die-designers and there was the effect of the introduction of laser-welded blanks and the need to predict springback and surface deflection effects in the drive to greater accuracy of assembly. Tryout times had also to be shortened for meeting the shorter overall lead-times and improved working conditions were needed to attract young recruits into the trade.

One of Japan's first steps in presswork simulation was the development of ITAS 3D code to simulate sheet bending, using solid elements. Shortly after, shell elements were introduced to simulate the wrinkling phenomenon in metal sheets. This was followed by a development of the code for the description of metal-to-tool contact behaviour, involving point, mesh and patch descriptions. By 1993 the code was able to predict springback, as exemplified by the truck frame member of Fig 2. Next stage was the modelling of draw beads and an improved description of friction. However, by then the limitations of using membrane theory were evident and a new code called ITAS-Dynamic was developed. The resulting code is now state-of-the-art and has already been used by commercial companies.

In a review of the German industry[2], a consideration of pressworking techniques, and the advent of four point cushion systems, is shown as a further

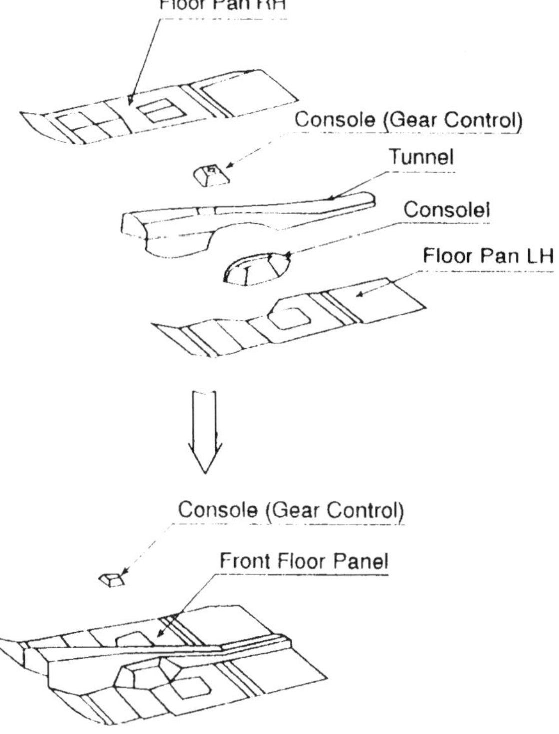

Fig 1: Panel integration

milestone. Forming of large area auto parts generally involved a combination of stretch forming and deep drawing, Fig 3. With non-axis-symmetrical parts the problem is to steer the flow of material between upper and lower binders so that wrinkles and tears are avoided. Its author points out that both friction and bending forces involved with the use of draw-beads hinders flow.

Friction can also be influenced by elastic deformation of the blankholder. To control material flow by elastic bending of the blankholder careful consideration is needed of manner of load application, die elastic behaviour and binder separation distance. On double-acting presses it is possible to adjust blankholder forces at the four pressure points of the blankholder ram. The technique is developed already for hydraulic presses but still has some way to go for mechanical double-acting presses. Fig 4 shows a hydraulic 4-corner cushion system developed by the Technical University Berlin. This is a passive system in which the pressure in the four cylinders increases by compression as the lower binder is forced downwards. It is thus possible to steer the pressure in the cylinders by servo valves adjusted to each cylinder.

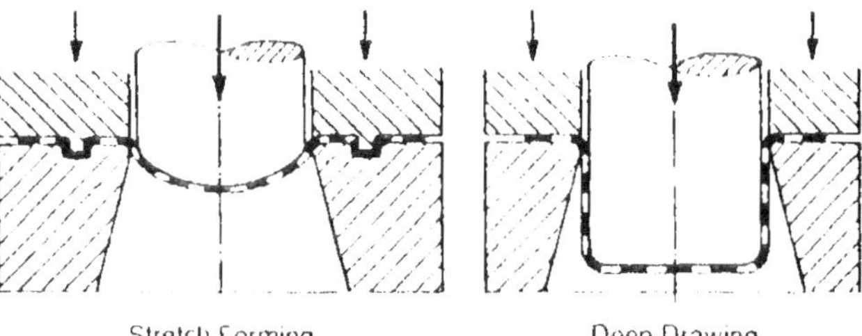

Fig 3: Drawing of irregular large-area parts

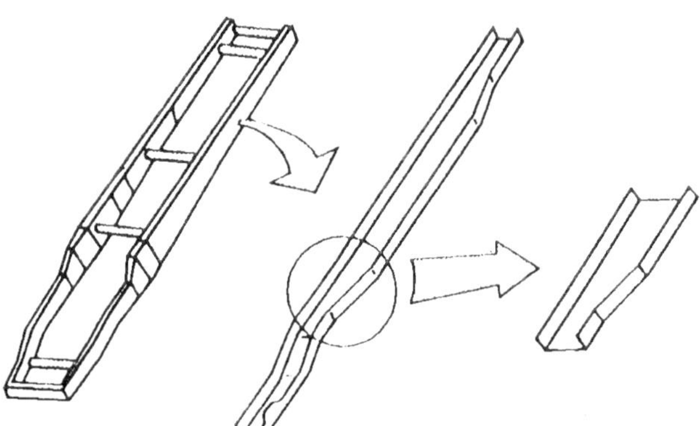

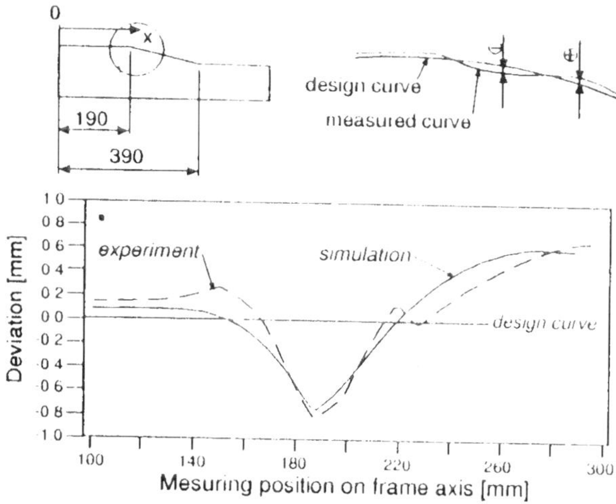

Fig 2: Comparison of flange bending between simulation and experiment

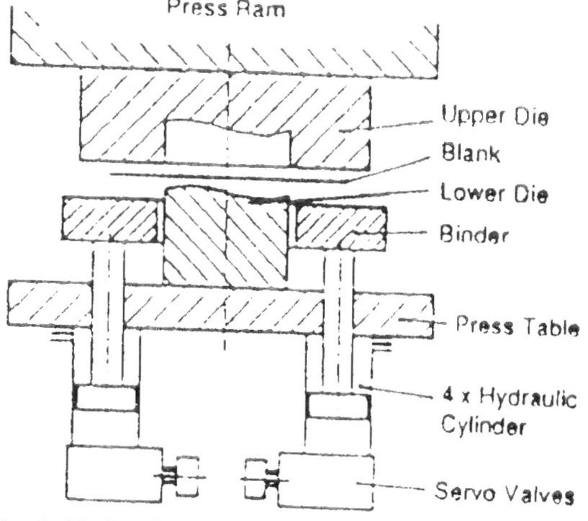

Fig 4: Hydraulic 4-corner cushion system

Shearing, forming and profiling

The profiling, and cutting, of sheet metal has been markedly changed by the introduction of programmable computer-numerical-control techniques. It has affected most of the various ways of cutting sheet-metal including blanking, shearing, gas-cutting and laser/water-jet profiling — providing greater accuracy, repeatability and the ability to set up quickly from drawing.

The latest CNC punch presses are permitting the ready inclusion of simple forming operations into the metal-blanking cycle. Stiffening ribs and louvres are among the possibilities, without removing the workpiece from the press. Turret-mounted tooling allows quick change between operations. Such devices as quick set-up tool-length adjustment are also valuable — in changing between one job and another. Tool height is obviously critical in operations like swage-forming and such devices mean that by the twist of a knurled collar, the height can be stepped by small discrete increments.

Another useful device is the positive stop, shown as Fig 5, which is the product of USA tooling specialists, CE Tooling. This is useful for such operations as coining and making push-out blanked-holes to make subsequent wiring apertures in parts. The stop causes the tool to lightly crimp the part near the end of the stroke. It can also be used to speed the set-up of such operations as swage-forming, by progressively bringing the tool-height up to tolerance while carrying out test-strikes on the part. This is important because the higher cycle speeds possible with CNC blanking mean that changes in design for swage-forming might be necessary.

The cost of blanking tools can often be reduced by thoughtful dimensioning of the part-design[3]. In Fig 6, for example, none of the radii shown may be necessary for the performance of the part. Although the tool for producing blank B would cost more than that for blank A, the tool for producing blank C would cost less than that for blank D — the reason being that far fewer punch replacements would be necessary during the production run.

Blanking and folding

The use of CNC front-gauging can make considerable improvements in part accuracy. Mazak Nissho Iwai make a typical state-of-the-art machine, the 50/

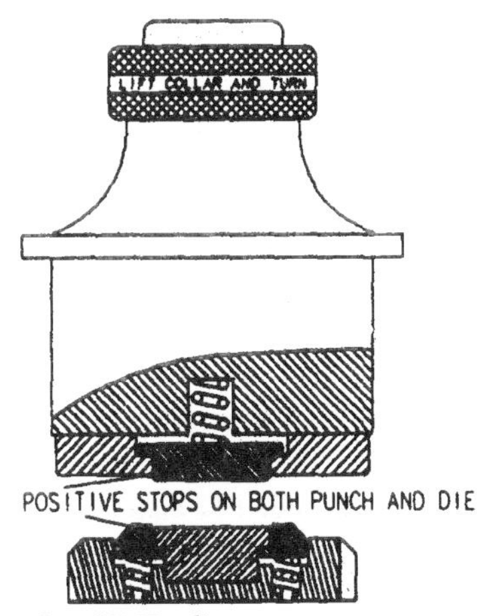

Fig 5: Positive stop by CE Tooling

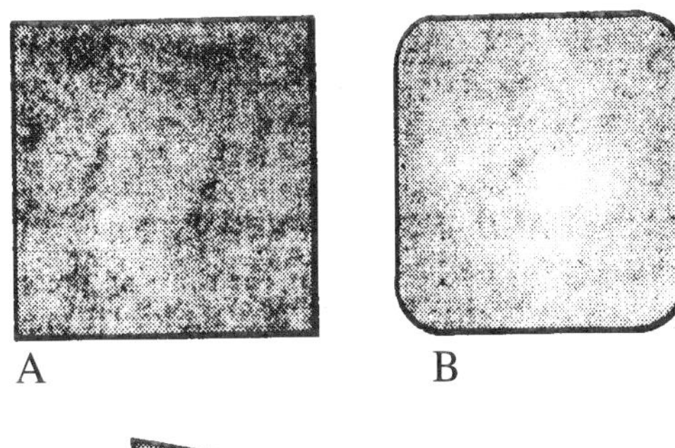

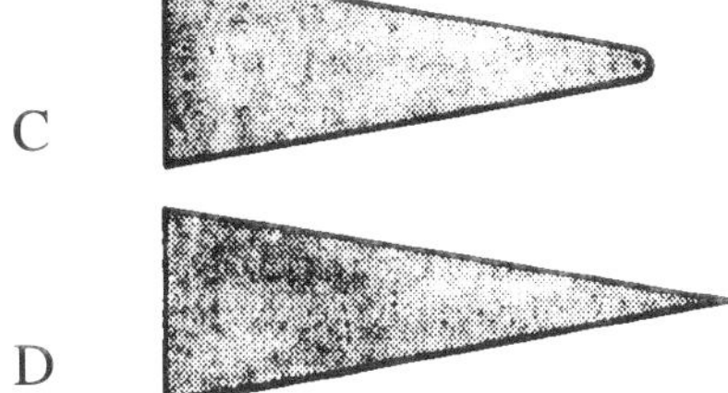

Fig 6: Attention to detail design

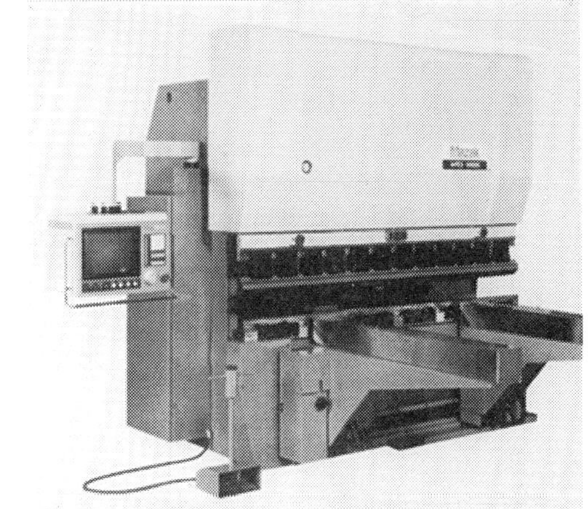

Fig 7: Hydraulic press-brake

100/150 tonne Apex CNC-controlled hydraulic press-brake, shown in Fig 7, which has ram positioning effected by a linear scale feed-back system giving accuracy to 20 microns. CNC back-gauging, and automatic die-clamping are also available. Fold lengths of 1.25-2.0, 2.5-3.1 and 3.1-4.0 are attainable, according to the machine specified. A 20 mm/sec fold rate is possible in an opening height of 420 mm.

Gas, plasma, laser and water-jet CNC profile cutting is also speeding the cutting out of more complex shapes. Among recent machine introductions is the Technicut Por-Tech 20 from Electro-Craft. This is a 20 foot capacity unit using plasma-flame to cut sheet from 16 to 20 swg. It uses DC servo motor/tachometers to cut at 10 m/min with 0.1mm accuracy on receipt of instructions from a CAD workstation.

The introduction of lasers with power in excess of 1kW has made possible the high speed cutting of thick plate and both articulated arm and gantry machines have been built to exploit the process. A recent introduction by Laser Lab is said to have made an important advance in the reduction of cost of such systems. The Contour, Fig 8, is an integrated design such that the operator controls the machine from one working position, the control taking care of both the motion system and the laser. A dual-pallet table means that plate loading down-time can be minimised. Power ramping and two or three axis motion are also optionally available.

Flame and jet cutting Gas cutting is applicable to ferrous metals for providing rough cuts in plate adequate for subsequent welding in fabrication work. The oxy-acetylene torch uses a central jet of oxygen which is under the control of the operator, the cutting action being a combination of chemical action and erosion by the narrow, high velocity, jet. Material is locally heated to ignition temperature by the oxy-acetylene flame, and the metal subsequently oxidised by the flame. The width of cut is about 3/16th of an inch and, with mechanically guided torches, a dimensional tolerance of about 1/32nd of an inch can be held for material thicknesses up to two inches.

In plasma cutting, a gas-stabilised electric arc passes through an orifice to generate intensely high temperature, high velocity, ionised flow of gas — the plasma stream which melts and blows the metal away. A higher quality of cut, and faster rates,

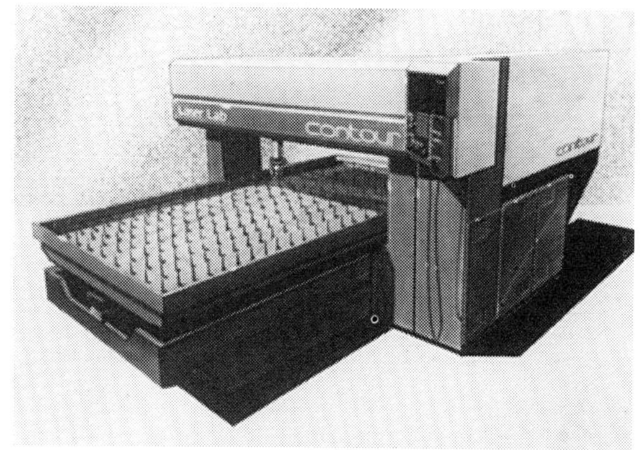

Fig 8: Laser cutter

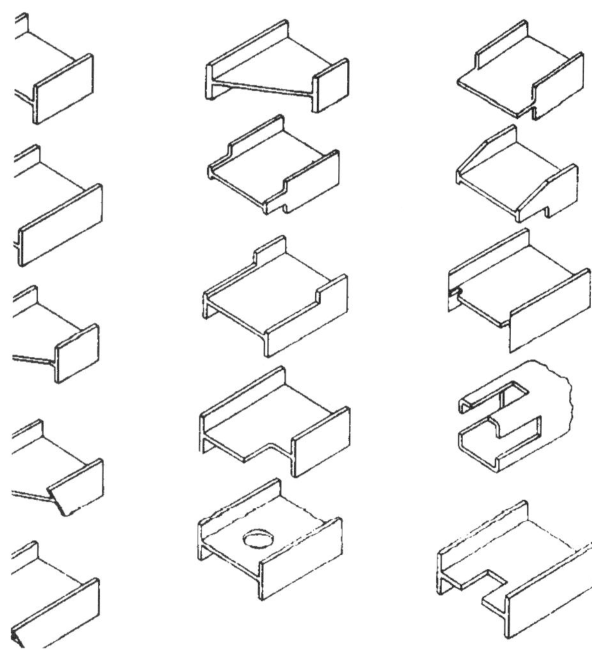
Fig 9: Flame-cut sections

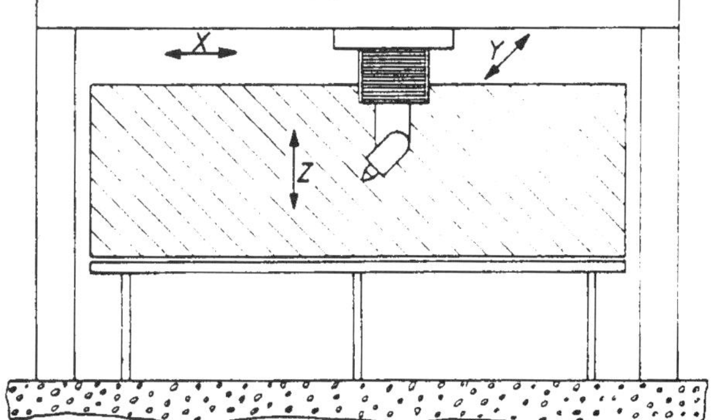

Fig 10: Gantry laser system

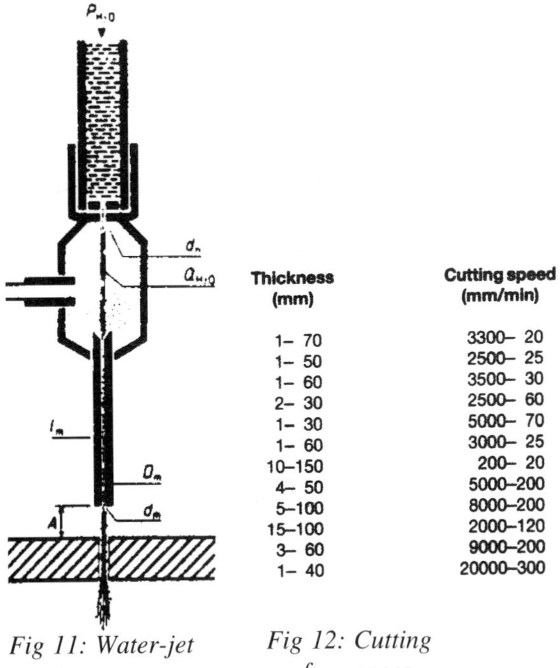

Thickness (mm)	Cutting speed (mm/min)
1– 70	3300– 20
1– 50	2500– 25
1– 60	3500– 30
2– 30	2500– 60
1– 30	5000– 70
1– 60	3000– 25
10–150	200– 20
4– 50	5000–200
5–100	8000–200
15–100	2000–120
3– 60	9000–200
1– 40	20000–300

Fig 11: Water-jet cutting

Fig 12: Cutting performances

Fig 13: Portal-type cutter

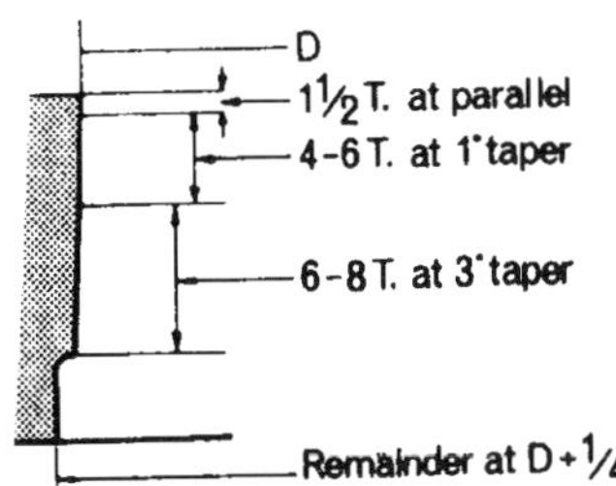

Fig 14: Die clearance considerations

than oxy-fuel cutting can thus be achieved. A wider range of materials can be cut, too, when mixed-gas techniques are used.

According to ESAB Automation, robotised cutting is gradually coming on stream for cutting steel structural sections. Fig 9 shows some sections involved in a recent collaborative contract. The robot uses a 3-D sensor to measure the actual dimensions of the profile and these are matched with stored design data in order to download the guidance program to the controller on set-up. Ten-fold productivity increase in comparison with manual cutting is claimed. Fig 10 shows a typical overhead gantry laser system used for cutting. The company have also reported progress in cutting-nozzle development for oxy- and plasma cutting which have resulted in cutting speeds being raised by 20%. For oxy-cutting a Laval socket has been incorporated into the oxygen cutting channel; this results in a mach 2.5 flow velocity and the cutting of thin sheets has been particularly improved.

In abrasive water-jet cutting, involving the addition of abrasive to the water jet stream, Fig 11, the kinds of performance occur as listed in the table in Fig 12, according to ESAB. Advantages claimed for the process include no heat distortion, a narrow cut (kerf) and no mechanical deformation. Pure water-jet cutting, without abrasives, is useful for cutting carpets, plastics and grp composites. For gas-cutting, the ESAB-Hancock Ultralex UXD-P portal-type cutting machine is capable of speeds of 12 metres/ minute. Shown in Fig 13, it is made in two sizes for cutting widths of 1500 and 2000 mm. It is numerically controlled and can be programmed quite easily for fabrication work.

Presswork techniques

A number of conflicting requirements between the optimum design, on the one hand, and production on the other, apply to pressed parts. It is important that form design meets the problems posed by both the toolmaker and pressworker before new shapes are committed for sheet metal working.

A major panel can be affected by something as basic as strip-mill rolling widths: these might govern the depth of draw for a panel-van roof panel, for example. Despite this, the need to use as large a single-piece panel as possible applies to almost every pressing situation. To produce a typical panel it may be necessary to use a blank and pierce tool; a crop or notch tool; a preform and then a flange tool, finally a setting tool. The need to ensure that a thickness reduction beyond 10 per cent does not occur is likely to require more than one 'hit' on all but the simplest deep drawn parts. In view of the number of operations involved, it is of course desirable that the minimum number of panels be used in all cases.

Cutting clearance between punch and die[4] is important in avoiding blank dishing and high loading requirement on the one hand, or, on the other, a break in the blank edge that might induce crack propagation. Die clearance, Fig 14, is also important — with 'backing off' chosen to avoid either distorted blanks or a burst die. In these connections it is necessary for the product engineer to be aware of metal gauge variation due to 'mill' tolerances; also to any variation in temper and grain run in relation to bend direction. Elastic recovery to the flat state will vary with temper and the appropriate overset to compensate for this effect must be determined. Fig 15 shows the reason for using a pressure-plate to stop wrinkle formation during deep drawing.

The illustration shows a sector of a flat circular blank before and after deep drawing. During deformation R1 must reduce to R2 and L1 to L2. Displacement of the material shown shaded sets up hoop stresses in the annular region outside the die aperture, buckling and direct stresses from which influence the amount the blank diameter can be reduced in one draw.

Press-forming modes

A complex part may involve the use of several forming operations. In the stretching operation the blank is clamped at the so-called die ring either by hold-down pressure or lock beads and a contoured punch pushed through the die opening. All deformation takes place in the metal over the area of the die opening only, the bi-axial tension resulting in a thinning over the total 'dome'. In the deep draw situation, the blank is drawn into a die, circular as shown in Fig 16a, by a flat bottom punch, b, which pulls the flange radially towards the die opening and the deformation as shown in each case occurs. More common, though, is plane bending as shown for 'V' and 'U' dies alongside a 'wiping' bend, Fig 17, in which one edge is clamped while the punch wipes the free end down. In all cases one or more edges swing through space and the swing must be calculated to

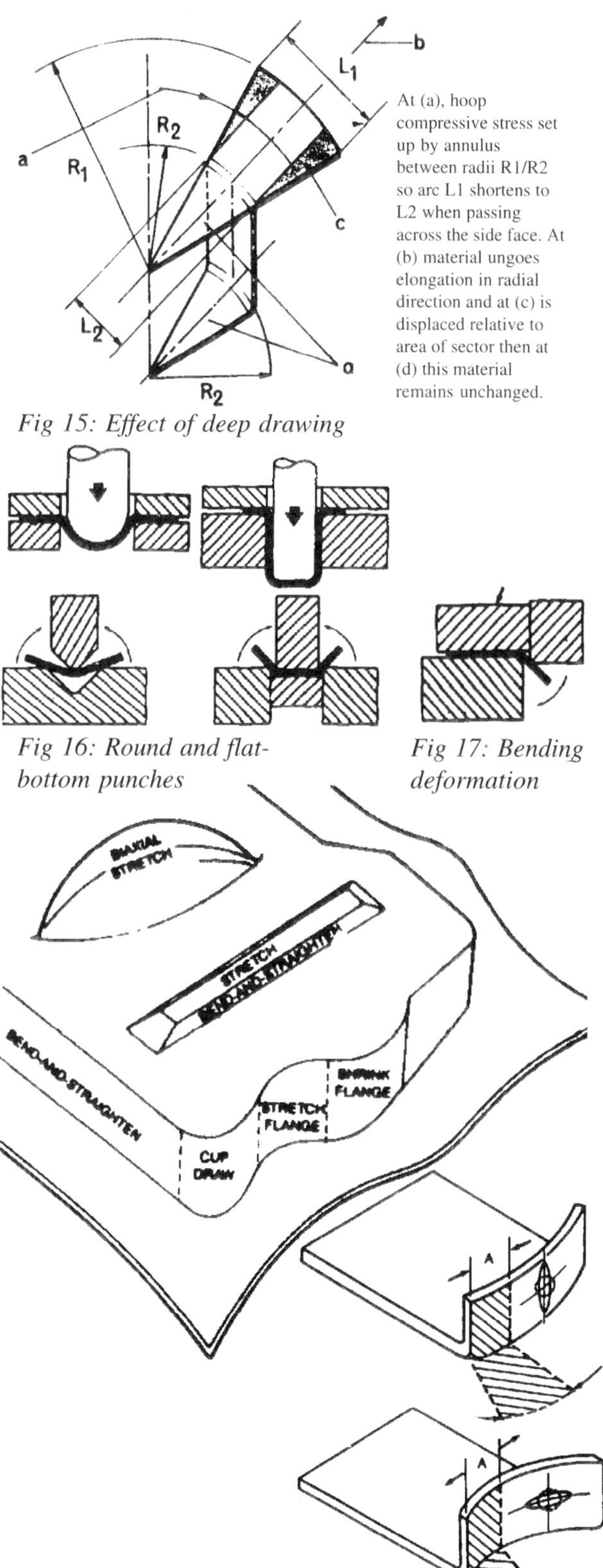

At (a), hoop compressive stress set up by annulus between radii R1/R2 so arc L1 shortens to L2 when passing across the side face. At (b) material ungoes elongation in radial direction and at (c) is displaced relative to area of sector then at (d) this material remains unchanged.

Fig 15: Effect of deep drawing

Fig 16: Round and flat-bottom punches

Fig 17: Bending deformation

Fig 18: Effect of curved hinge line

allow sufficient room in the tooling; the metal outside the 'plastic hinge' is, of course, unstrained. When the hinge line is changed from straight to curved, the metal outside the hinge no longer remains unstrained and either shrink or stretch flanging takes place as shown in Fig 18.

To achieve the same shape as that generated by the above bending operation, a 'bend-and-straighten' can be used which will give the finish product different characteristics. Here the swing of the metal is prevented by the blank-holder and the bottom radius formed round the punch by a bend operation. A simultaneous radius is formed around the die and thereafter each individual element in the final wall begins in the flange and is pulled towards the radius zone. As the die radius line is straight, no compression is created along it and upon entering the die radius zone, the element is bent to form the radius contour. On leaving it, the element is likewise unbent or straightened to conform again to the wall, hence the name of the operation. The main difference between it and 'bending' is that in the latter case, the final wall remains in the unworked state whereas in the former, the bending and unbending sequence hardens the metal and reduces subsequent formability.

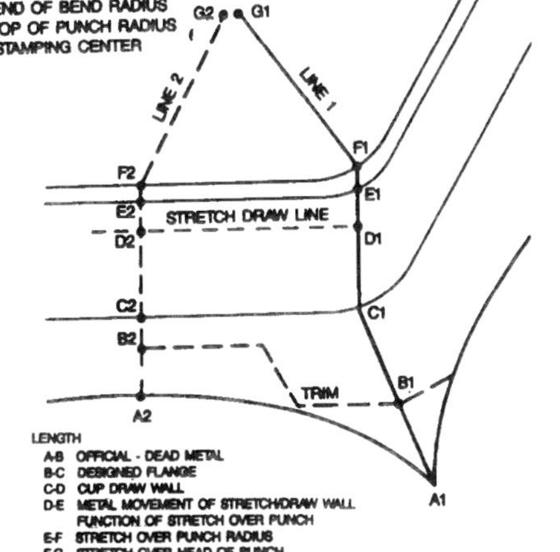

Fig 19: Line length analysis technique involving blank edge, trim line, die radius, stretch-draw line, end-of-bend radius, top-of-punch radius and stamping centre

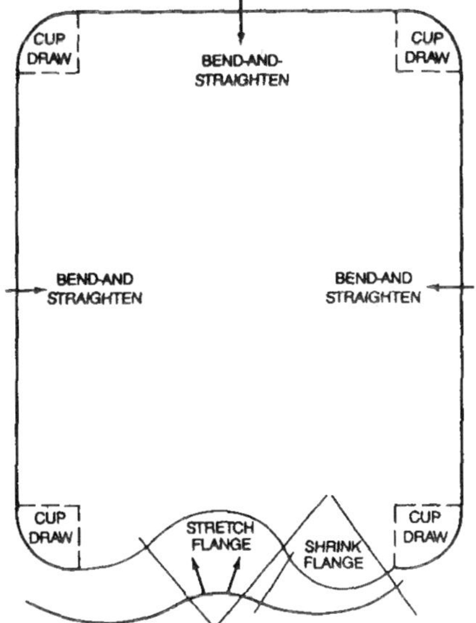

Fig 20: Plan view of pressing

Fig 19 shows the combination of forming modes which could go into one part — indicating the normal function breakdown required before analysis of a pressing. One analysis method is the length-of-line technique, illustrated, in which the salient stages are listed. Line 1 is generated by a radial or cup drawing mode obtained by examining a plan view of the pressing, Fig 20, where it can be seen that the four corners can be combined to form a cup created by deep drawing. Line 2 is generated by bending and straightening, in contrast. In the secondary forming operations, trimming and reverse flanging may present problems if thickness remains constant over Line 2 and increases, to 40 per cent say, over Line 1 in the corner cup draw. One solution may be to increase blank width to provide an additional flange, after trimming, to meet subsequent welding requirements. This will restrict metal flow from the hold-down area and increase the depth of the pressing required by stretching over the punch. If such action created a failure over the punch, such a trend could be reversed by increasing the die radius to allow easier flow of the extended blank into the die cavity. If, however, the die radius is now too large then a restrike operation may be required to sharpen the radius.

Metalworking terminology Shearing is a straight cut across the coil width of the delivered metal to form a square or rectangular blank. More often, 'blanking' will be required where the contour is both straight and curved. Metal can be saved in blanking by the best outline for 'nesting' within the strip or coil, Fig 21, in which sharp corners are seen sometimes to be preferred to more aesthetically curved shapes. Contouring should also be arranged to match as near as possible the die opening. This will encourage more uniform metal flow into the die cavity preventing buckling of the flange and reducing drag of extra flange metal behind critical zones. Finally, contouring should be arranged to minimise any post-forming trim operations.

Preforming is an additional stage for gathering metal into a zone for later use before the edge metal is restrained. This allows the necessary length of line to be generated without excessive tension that might cause breakage due to metal thinning over tight radii. Thereafter, blank insertion, positioning and closing the hold-down ring are the next operations. The ring may be formed in one or two planes, in its function of preforming the blank closer to the contours of the punch. This avoids metal being trapped under the punch and balances the forces on it in the interest of avoiding skidding of the character line. Punch action and retraction takes place in the main forming operation. Ejection and transfer of the part then usually complete the operation.

Simple bending is affected considerably by relation of thickness, t, to bend radius. The latter dimension can range from $0.5t$ for soft materials up to about $7t$ for full temper steels. If bending is with the metal grain the radii can increase by up to a factor of two. In calculating the developed length, Fig 22,

$$l = b + c + 1.5708(t/3 + r)$$

this value could alter with ductility and grain direction, however. It would only be correct for the open-vee tool but not the side-drawing tool illustrated in Fig 23. Draw dimensional rules are also much variable with conditions. About two degrees of draft is helpful on most work to avoid wrinkling and in general it is helpful if depth of draw does not exceed 1.5 x punch diameter for each strike. Radii at the bottom of the part should be as large as possible to avoid the situation of Fig 24. A 'square' component can be drawn to a depth at least equal to its width providing the ratio of depth to corner radius is no more than five. Die aperture radii should generally be within four to 10 times metal thickness and the punch nose radius should not fall below twice metal thickness.

Pressworking dos and donts[5] In Fig 25, raising of a flange where the blank contour passes acutely across the bend line should be avoided. Otherwise the metal diminishes in width towards the bend line and where it fails to meet the die wall an unsightly bulge results,(a). When beads, stiffening ribs or depressions are required in a panel their root radius should be as generous as possible consistent with the strengthening effect required; allow ample clearance around swages to allow possible easing of the radii (b).

For a 'clean' finish panel, work must be carried out at all points or the remaining, unworked, metal will remain soft and easily distort; for a straight drawn shell an adequately strong bottom is only possible by stiffening in the form of beads or a radiused die line at the closed end, Fig 26(a). Incorporate flanges where possible in long shells to avoid distortion of the panel by releasing stresses during the trimming operation; ideally wall curvature, in plan, should be provided to counteract the effect introduced by the die radius (b). Use symmetrical cups

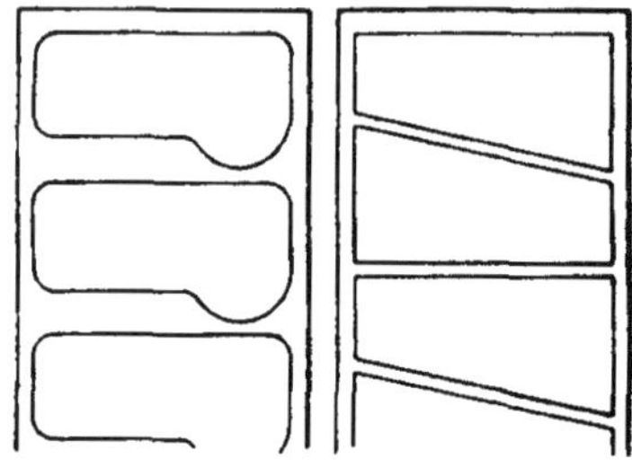

Fig 21: Blank nesting

Fig 22: Developed length

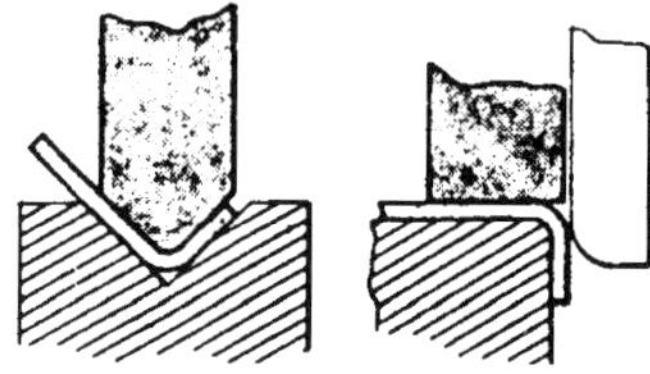

Fig 23: Bend radii minima

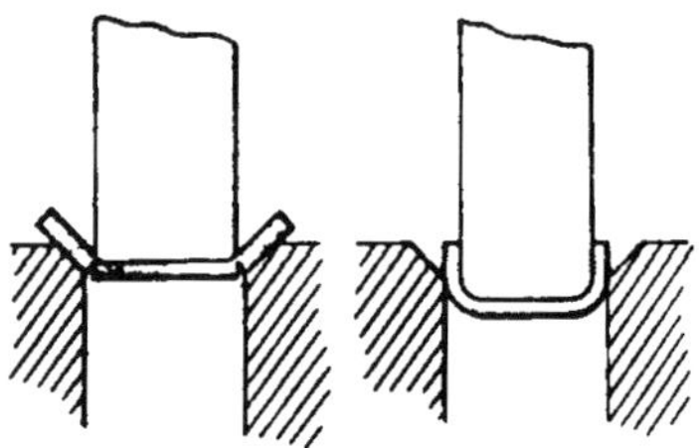

Fig 24: Draw radii minima

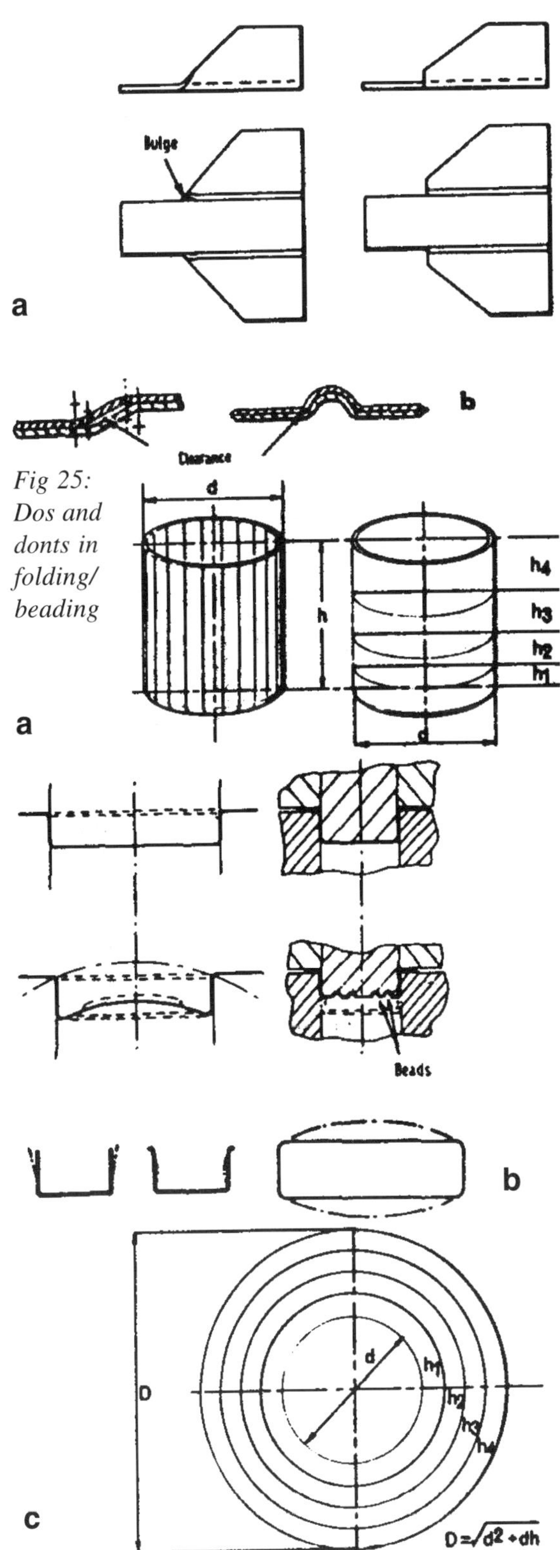

Fig 25: Dos and donts in folding/ beading

Fig 26: Dos and donts in drawing

where possible so that area of the final pressing is the same as that for the blank; cylindrical portions are then equal in area to annular ones, on the blank (c).

Pressworking steel behaviour Advanced methods are now being considered for avoiding the various pitfalls which can occur during the set-up process in preparing blank and die for a production run. The determination of correct retaining pressure on the side-flanges of a boxed-shaped deep-drawn part can be estimated by predictive techniques, Figs 27 and 28, without the expense of building prototype dies. Insufficient retaining pressure allows excessive material to be drawn in from the blank, which results in wrinkling. Conversely, too high a retaining pressure, together with too much friction on the upper die surface, can lead to excessive straining of the centre of the panel, resulting in spring-back. At British Steel's Welsh Laboratories the effect of surface texture on formability is being studied: surface texture provides the main control over friction in the press. A compromise has to be found between rough surfaces which retain lubricant, and avoids pressure welding of the strip to the tools, and providing a texture which is fine enough to present a smooth surface after painting. Rolls of strip are subject to electro-discharge and laser texturing techniques, seen as the best means of controlling strip texture to the optimum levels. A double-acting press has been installed for these experiments; it has two moving platens, one inside the other. Forces up to 500 tonnes on the inner platen are used to form the part while the outer platen can excert forces up to 300 tonnes for restraining the edges of the blank which can be of up to 2.74 x 1.52 metres projected area. Press loading is computer-controlled and time histories of loadings can be printed out for recording optimum load cycles. The experiments are also being used to determine degree and mode of strain required for imbuing curved panels with maximum dent resistance. The work is also showing how suitable positioned holes in the edges of a blank can be used to stress-relieve the centre section for avoiding splitting tendency after deep draw.

At the University of Warwick studies are being made on low-cost press-tooling which could have value for the production of panels for low-volume niche-market vehicles, Fig 29. Alternatives to cast iron dies under examination include zinc alloys, resins, ceramics and sprayed metals. Production runs of up to a million pressings are being mooted for zinc dies from alloys with much higher durability than that used in the past for prototyping dies. Resin systems under investigation include epoxy and polyurethane combinations with wear resistant fillers such as slate dust incorporated. The resin forms the surface layer of the die while the backing may be concrete, zinc alloy or cast iron. Even the more conventional cast iron dies can be reduced in price substantially by using cheaper grey iron in place of alloyed nodular iron and then chromium plating to produce surface hardnesses up to 900 VPN. These are particularly suited to coated steels as the chromium layer prevents the build up of zinc particles. In recent work reported by Roger Pearce[6], spring-back is examined in some detail. In the uniaxial bending case practical solutions include successive overbending until correct shape is obtained; bending with applied tension so that the neutral axis is moved out of the section depth of the sheet. The latter can be achieved by setting negative clearance between punch and die so that both surfaces are put in tension. When bending over a curved line, shrink or stretch flanging can occur, Fig 30, according to whether a concave or convex curve is involved.

Draw-beads are one of the ways of controlling the flow of metal between a die and blank-holder. They are positioned around the die-opening such that by subjecting the moving sheet to a series of bending and unbending cycles, sensitive control over metal flow is obtained. The bead acts like a brake which can give high restraining forces with relatively low blank-holding pressures. In Fig 31, the resulting strip sections are shown for various bead penetrations while Fig 32 shows increase in hardness of the sheet due to the bending and unbending process at an average bead penetration.

Computer programs are now available for simulating the effect of pressworking process variables. Matra Datavision have recently announced OPTRIS which can accept data from a variety of 3-D CAD systems. A fine finite-element mesh is used to describe the sheet metal blank and its deformation can be observed on-screen. It can predict thickness distribution and forecast wrinkles and tears. Volume vehicle builders have found that proving-out time for dies has been reduced by half, using the program. It can also analyse multi-stage drawing and thus determine the minimum number of redrawing operations required. Following one operation, the residual strain at each mesh node is stored for the simulation of the subsequent operation.

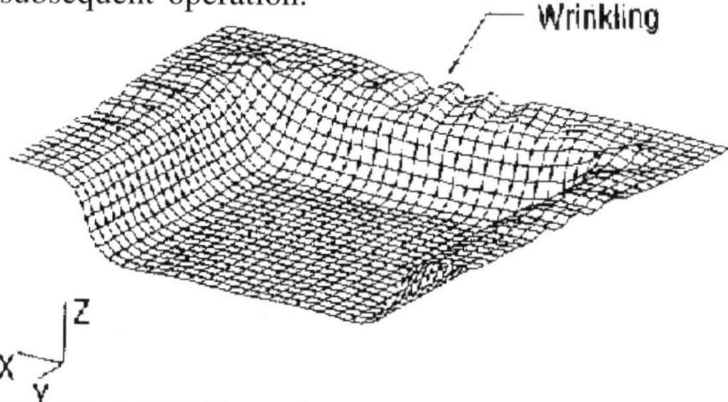

Fig 27: DYNA3D code used by HW Structures can predict the onset of wrinkling in an incorrect die set-up

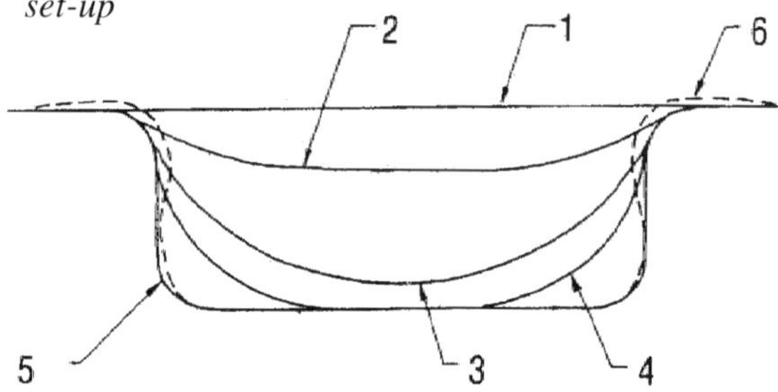

Fig 28: Spring-back prediction: 1 blank; 2-4 deformation under increasing pressure; 5 spring

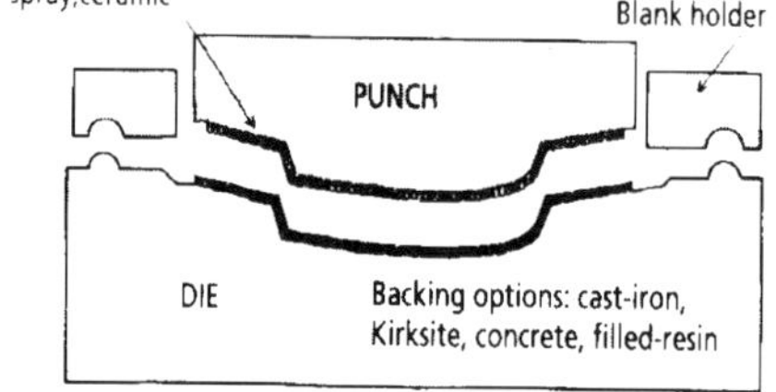

Fig 29: Low-cost tooling (Warwick University)

Fig 30: Bending along a curved line

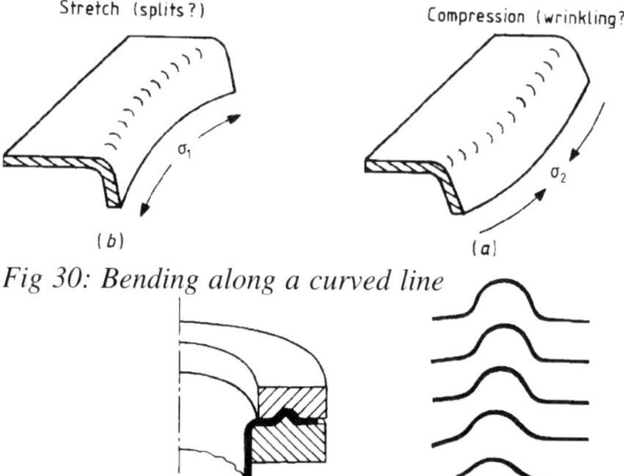

Fig 31: Effect of bead penetration on sheet contour

Deep-drawn presswork

While general engineering knowledge may suffice for designing components and panels for shallow presswork, deep drawing requires a much closer understanding of the production process if costly mistakes are to be avoided. Some of the ground rules are provided below.

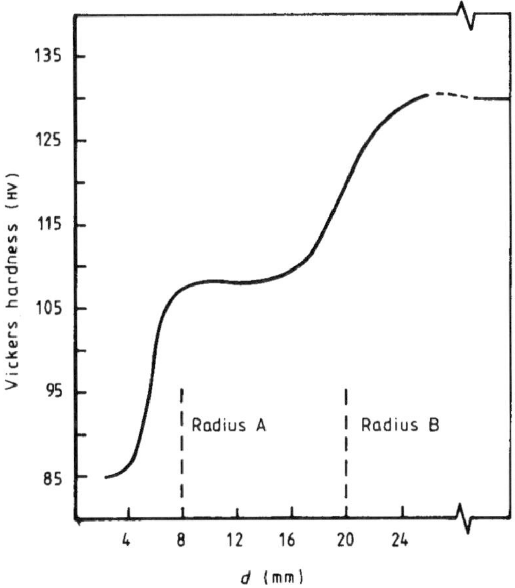

Fig 32: Change in properties as strip passes through draw-bead

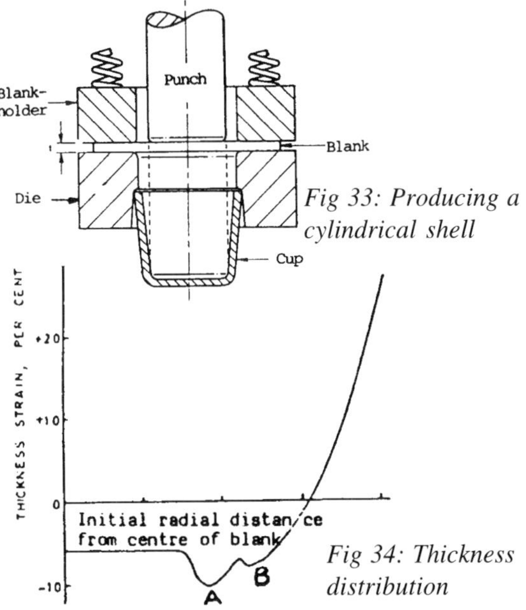

Fig 33: Producing a cylindrical shell

Fig 34: Thickness distribution

British Standard BS 4342 sees it as a process in which blank or workpiece (usually controlled by a pressure-plate or blankholder), is forced into and/or through a die by means of a punch to form a hollow component in which thickness is substantially the same as that of the original material. The process by which a component has its thickness reduced and length increased is known as ironing.

Deep drawn shells can be further processed by additional drawing operations (redrawing) or by reverse drawing, in which the punch travels in the opposite direction to the component with that used for the previous draw. The material undergoes compression in the direction of the periphery of the shell

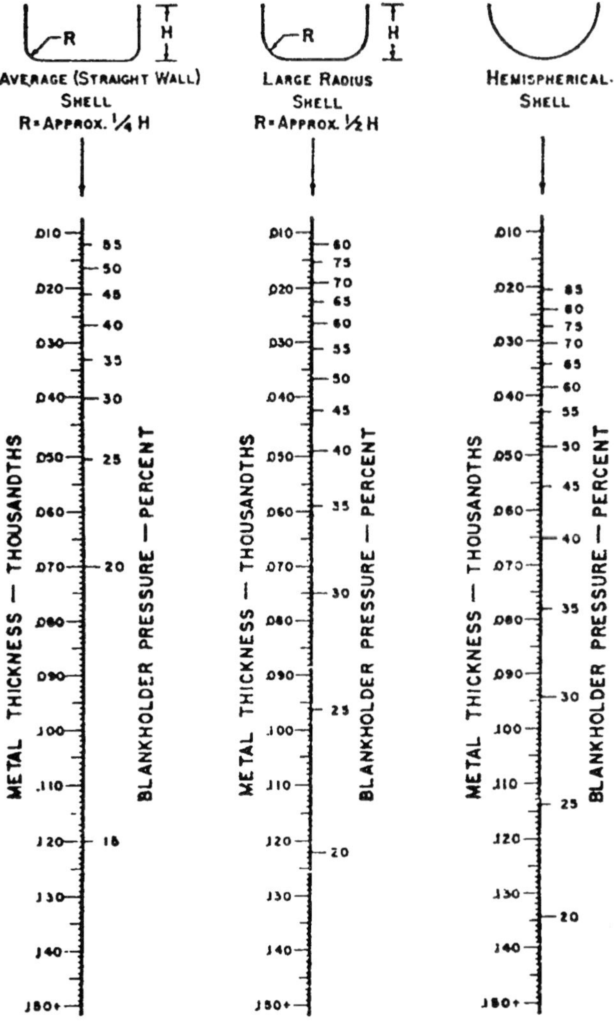

Fig 35: Parameter optimisation

— so that it tends to flow along its length and sheet thickness, virtually, does not change.

Fig 33 shows the set up for producing a cylindrical shell from a circular blank using a flat headed punch. The advancing punch causes an elemental ring of sheet to be bent round the punch head profile. Simultaneously the material in the flange is dragged radially inwards to the die opening where it is bent and unbent as it passes over the die edge radius. The circumference of each elemental ring shortens and sustains a compressive stress — as illustrated in the earlier article. Also a tensile radial stress is imposed as the flange is dragged inwards.

Fig 34 shows the typical thickness distribution which results across the sheet. In a correctly designed tooling set, failure caused by an oversized blank, or faulty material/lubrication properties, should occur by fracture at neck (A). Generally the limited drawing ratio (LDR) of maximum blank diameter to punch diameter, is equal to 2 — producing a cup with height slightly less than the punch diameter. Amount of draw can be expressed as percentage: *D - d/D x 100* where *D* is blank diameter and *d* punch diameter.

Blankholder force should be kept as low as possible commensurate with avoidance of wrinkles caused by the radial friction force between the blank and the tool. Avoidance of puckering is also necessary — caused by the circumferential compressive stress in unsupported material between punch and die where radial drawing is taking place. It is controlled by restricting the size of the profile radii of punch and die to decrease volume of unsupported material — also by increasing blankholding force. Fig 35 shows how, by expressing blankholder pressure as a percentage of draw pressure, that tables can be drawn up to optimise the different parameters for various punch shapes and material thicknesses.

There are also basic parameters of the pressing steel of interest to formability. These are work-hardening exponent, plastic strain ratio and strain rate hardening exponent. Circle grid strain analysis is a useful technique for seeing how the material deforms. The sheet to be formed is etched with a circle

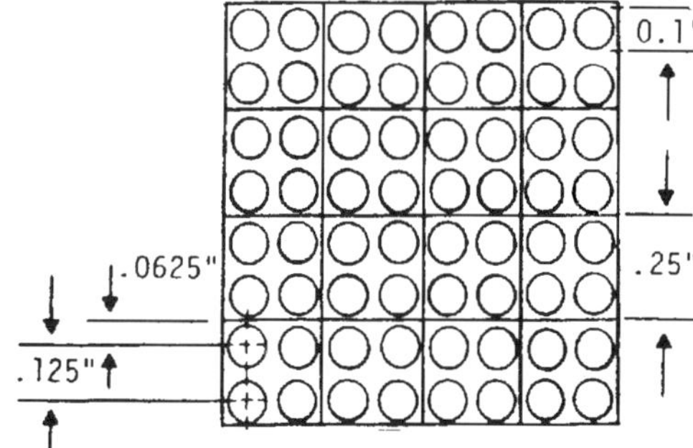

Fig 36: Circular grid technique

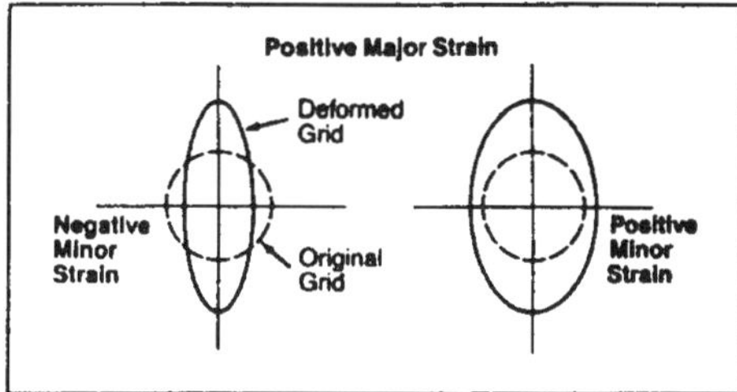

Fig 37: Major and minor axes

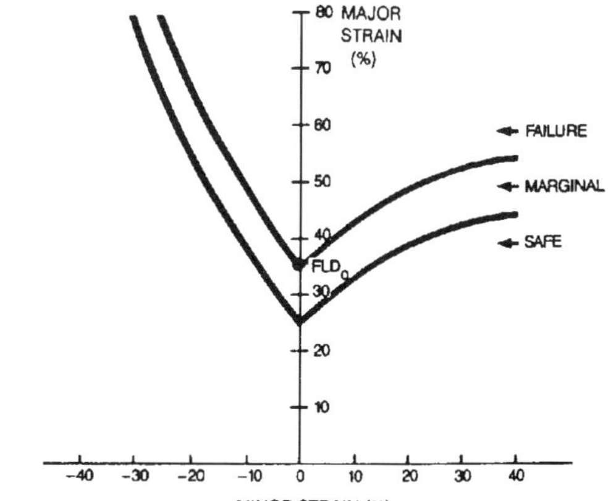

Fig 38: Forming limit diagram

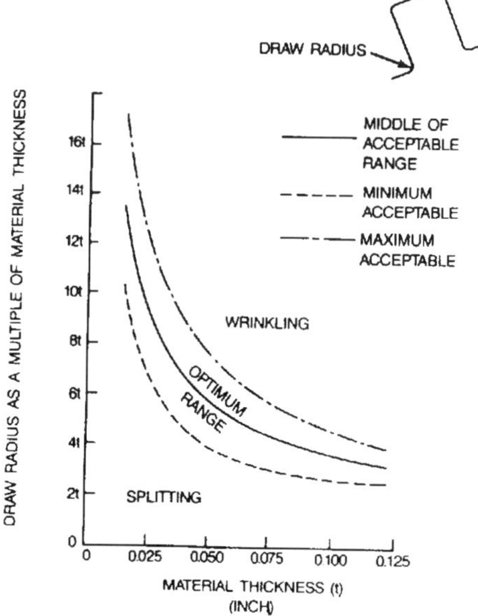

Fig 39: Draw radius vs thickness

grid, Fig 36. The amount of plastic strain at each surface can be measured at different stages in forming by observing major and minor axes of the resulting deformed ellipses, Fig 37. This is used to derive a forming limit diagram, Fig 38.

By intelligent use of such diagrams, variables like blankholder pressure, die clearance, lubrication, steel grade and thickness can be altered to optimise the forming operation. Draw radius relates to material thickness as shown in Fig 39. If a deeper cup is required, relative to the diameter of the cup, one or more redraw operations are necessary, Fig 40a. Redrawing is commonly employed to produce rectangular or irregular shapes. Fig 40b shows how a deep cylindrical shape is obtained by redrawing. Depth is increased at each stage, while the diameter is reduced.

Presswork automation and simulation

While, traditionally, presses would have been arranged in rows, and a workpiece taken from one press to another, for each stage of a deep drawing sequence. Automation has now been applied to the removal of the blank from a stack, the application of lubricant and the loading of the blank into the die. Also, presses have been linked, with the work transferred automatically from one to another for redrawing. Whereas the output of a manually operated press might have been 240 pieces/hour, a fully automated line of linked presses would be capable of twice this throughput.

Further increases in output have been obtained with the introduction of transfer presses in which the sequence of deep drawing operations is combined in one machine. A machine of this type produces some 720 pieces/hour and can range up to 8000 tonne capacity. Whereas, a decade ago, a production run might have been the quantity of pressings needed for three months, the current target could be no more than three weeks. This has led to improvements in the operational flexibility, involving computer-based controls. There is also a need for tooling that can be changed quickly by mounting the dies on moving bolsters. Roll-feeds are now commonly used to feed presses with strip or coil stock and computer control synchronizes feed rates with the requirements of the press. There is also a growing tendency to obtain

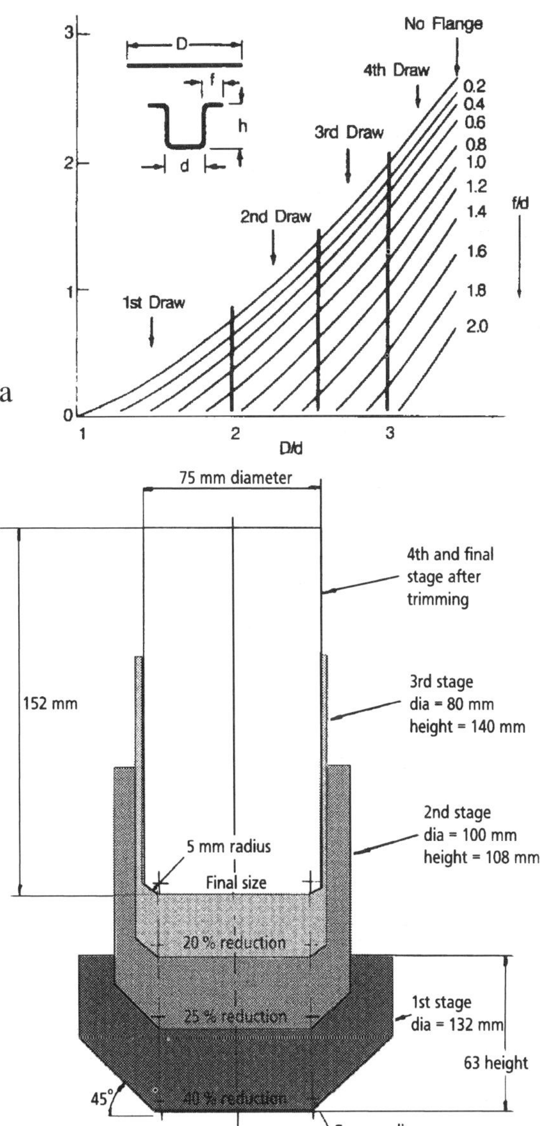

Fig 40: Redrawing operations: above, multiple redraw operations; below, how a cylindrical pressing is produced from a circular blank in four drawing operations[7]

sheet metal components from preferred suppliers who may provide assembled units and, in turn, source the pressings from subcontractors.

Presswork simulation While such parameters as work-hardening, contact area and friction area, that affect the drawing process, were decided by experienced pressworkers, computer-based simulation is increasingly the current norm. Here the effects of changes to process variables can be studied before the tool design is finalized. One approach to an aid of this kind is OPTRIS from Matra Datavision and developed by Dynamic Software. An explicit time integration method, as opposed to solving linear system of equations, was selected for OPTRIS in order to reduce demands on the central processor. Simulating the drawing of sheet metal, the software runs on Silicon Graphics and similar high performance workstations.

OPTRIS accepts data from EUCLID and other three-dimensional CAD systems. An extremely fine mesh describes the sheet metal blank. Both tooling and blank are displayed on a monitor screen, and the user is able to handle three-dimensional objects with ease. As well as predicting thickness distribution it can forecast faults such as wrinkles and tears. The program can also enable the minimum of redrawing operations to be determined and can be used to analyse multi-stage drawing. Following one operation, the residual strain at each mesh node is stored for the simulation of the subsequent operation. Traditional rule-of-thumb methods tend to introduce unnecessary operations out of caution so the elimination of one or more operations is economically valuable.

Drawing a car body floor is simulated by OPTRIS in Fig 41. At (a) the blank is depicted as a planar face and shown with the press tools. A cross section has been obtained by a dynamic cutting plane, one of the software features. (b) shows the pressing and tooling together. The dark portion corresponds to the blank holder. Moving tools are shown in lighter shades, while the fixed die is of a mid-shade. In this visualisation of the pressing (c), OPTRIS predicts an unsatisfactory product owing to the presence of wrinkles.

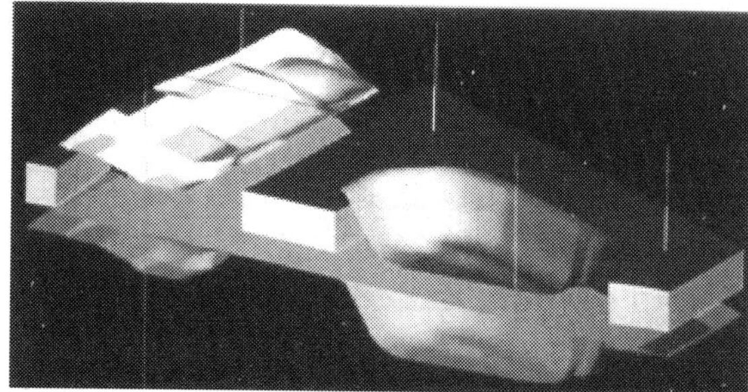

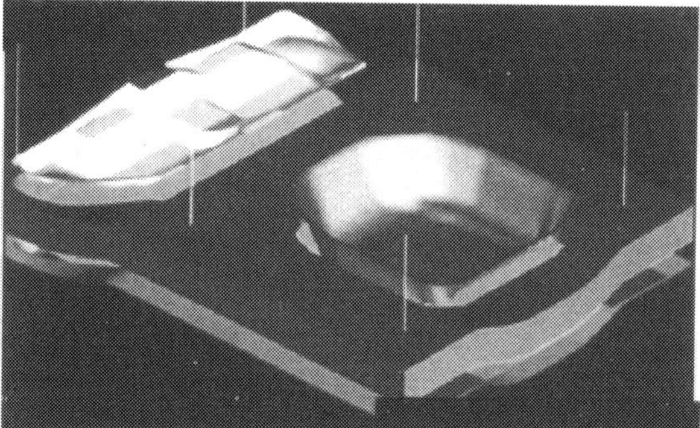

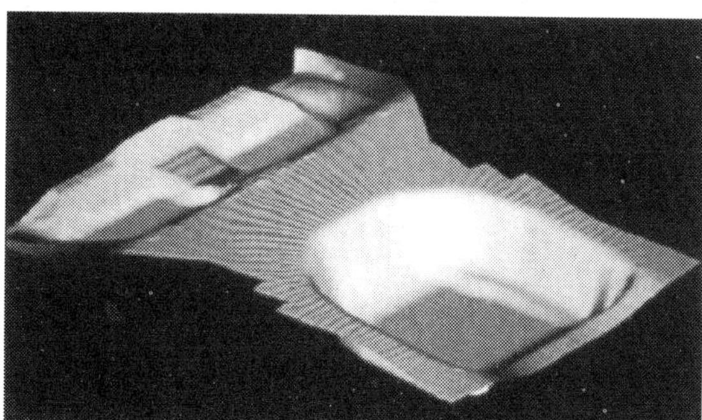

Fig 41: Simulation of car body panel forming: a, top, blank and press-tools; b, centre, tooling with pressing; c, bottom, wrinkles on final part

Hydroforming

Since many structural members of vehicles are box-sectioned, and made from pressings joined together, TI Vari-Form has developed a hydraulic forming process, Fig 42, which can be applied to produce this type of component from tube.

The welded round tube, in steel or aluminium alloy, is in straight lengths, typically between 1000-4000 mm. First, a tube is bent by conventional machinery to a configuration similar to that of the end product. Only one forming die is required for the process and a bent tube is placed in this die, which is then closed. Filled with liquid at high pressure, the tube is forced to adopt the shape of the die cavity. The material flows to its final form, and the objective is to obtain a constant wall thickness throughout the product. If the cavity is too large in places, there may be excessive wall thinning; also, wrinkling and buckling may occur where the cavity is too small.

Punches are mounted in the die to pierce holes during the forming operation. These holes can be of any size or shape: they can provide clearances, access to the interior and manufacturing locations. When

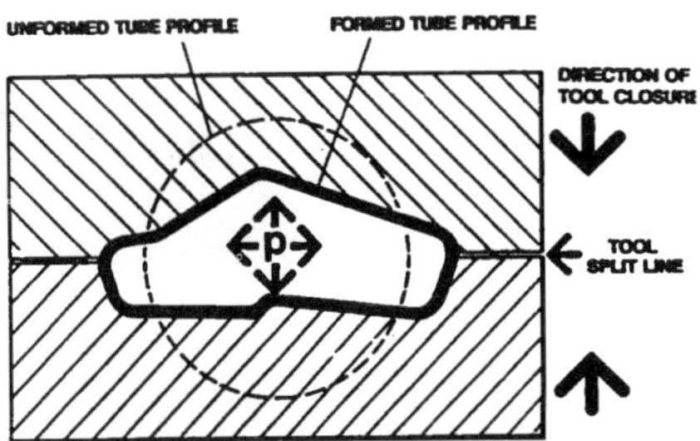

Fig 42: The hydroforming process

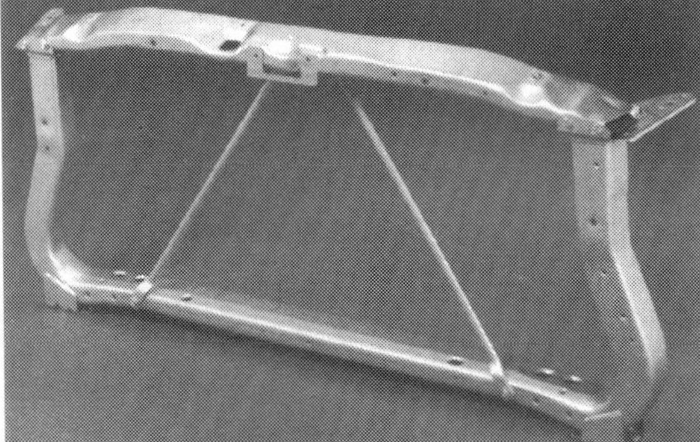
Fig 43: Hydroformed suspension sub-frame

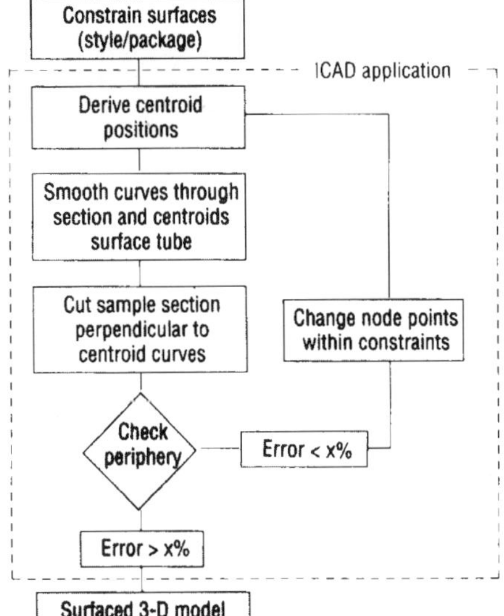

Fig 44: How ICAD replaces CAD steps

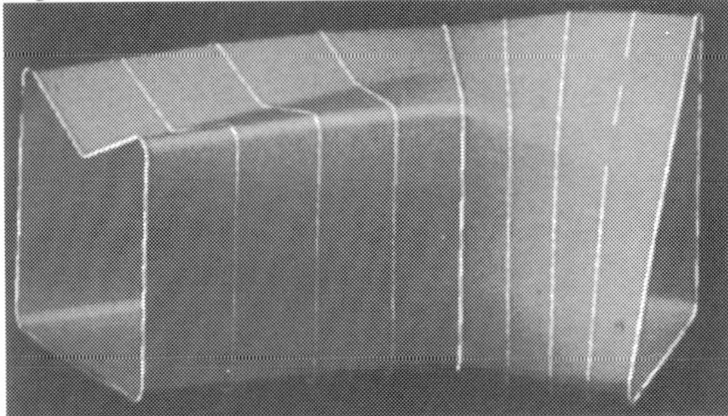
Fig 45: Print-out from tube design software

the product is required to accept a self-tapping screw, the punch can produce a hole with a collar.

Since pressings for conventional box-section members feature flanges, they are assembled by joining the flanges together, typically by intermittent MIG welding or spot welding. One reason for the weight savings achieved by Vari-Form is that flanges are unnecessary. In the case of a radiator support assembly, there were 16 components when the design was based on pressings A change to hydroformed tube reduced the number of components to eight, while weight fell from 15.1 to 11.5 kg.

One of a number of applications that car-maker Lotus has been studying for the process has been the production of a one-piece cant rail of a four door car with some of the surfaces either being externally visible or forming the frames against which the door seats need to close. This obviously demands accuracy and repeatability for it to be economic. Also, the cross sectional shape needs to change all the way along the component according the task each element is performing. Other typical components already being manufactured by hydroforming are the Mercedes S-class rear suspension sub-frame and the Audi 100 front sub-frame, Fig 43.

Until recently, designing the hydroform mould has been very difficult and relied on an empirical approach. As well as the need to maintain the section perimeter, there has also been a need to limit deformation as the component progresses from one cross-section shape to another. When Lotus started looking at the cant rail design, it took 16 weeks for the designer using a state of the art Computervision CADD-4X system to reach a satisfactory design. Now, with ICAD's (KBE) System processing the geometry initially built up with the CADD system, it takes just a few hours. The Lotus development of the ICAD package overcomes the problem, Fig 44, that designing for a complex hydroformed component is a highly iterative process. The nature of this process is best seen by describing some of the steps involved. This starts with an approximate line for the 'tube' or closed section component. Some of the surface geometries will be 'hard' (defined by forming an outer surface with another component) while others will be 'soft' (can be varied within limits to permit the cross-sectional perimeter to remain within the set limits), Fig 45. Various design constraints are also fed in such as minimum bend radii and maximum extension permitted in forming the material.

An approximate centre line for the tube component is also defined. This need not be accurate as it will be made more accurate as the process continues. A series of sections normal to this 'centre line' are produced. The spacings between sections will depend on the rate with which the cross section will change over length and the local accuracy required. In a typical component, 200 sections might be expected. These sections are then analysed to ensure that the rules relating to perimeter and material extension are observed. In addition, the centroid position for each section is established. A curve joining the centroid positions is described. The position and line of this will be different from the approximate centre line produced originally, and is likely to mean that many of the sections are no longer normal to it. In consequence, a series of new normal sections, at the same spacings as previous, are now produced. These are re-analysed, the centroid positions again calculated, and a third 'centre line' described.

Eventually the amount of difference between iterations is small enough to be ignored and the process can be considered complete. (For instance a four degree difference between one section plane and the next may be considered to within limits). In the experience of Lotus, 4-5 are sufficient. The resulting geometry, both of the 'hard' surfaces and of the 'soft' surfaces are then available to be fed back into the CAD system for inclusion in the eventual design.

A Lotus case study applied to the design of a saloon-car B-post is typically represented by Fig 46. The function-based design approach involved defining the member as part of the overall body structure and then as a member for supporting hinges locks and side impact forces. The overall structure is designed rationally from first concept in this function-based approach even from the point of choosing between triangulated space-frame, monocoque, platform or ladder frame construction for the structure. With regard to a more detailed study of the B-post, alternative sideframe options are considered, as in Fig 47, and a rational choice made. Joint and beam section properties are next identified for primary load path directions. The following step takes in manufacturing considerations in which tests can establish the advantages of different processes for the key elements as in Fig 48. Best design options are next assessed for predicted performance based on the criteria of structure performance, mass, manufacturability, cost and robustness of design. This structural scheme then allows detail design and packaging to proceed.

Key structural requirements for the B-post involved the load cases shown in Fig 49. Overall structure studies identified a requirement for high shear stiffness in the plane of the sideframe while 'sensitivity' studies showed stiffness of B-post to

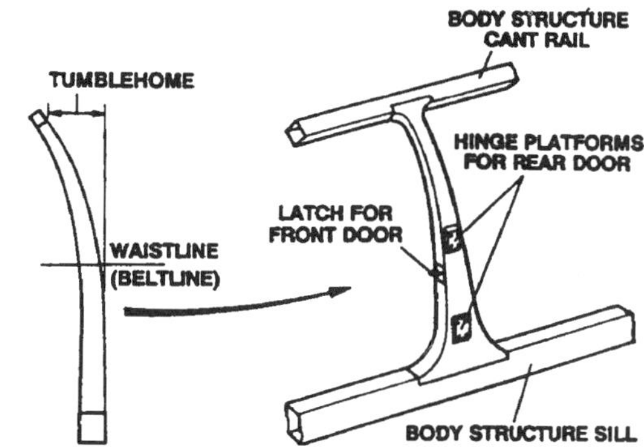

Fig 46: B-post under investigation

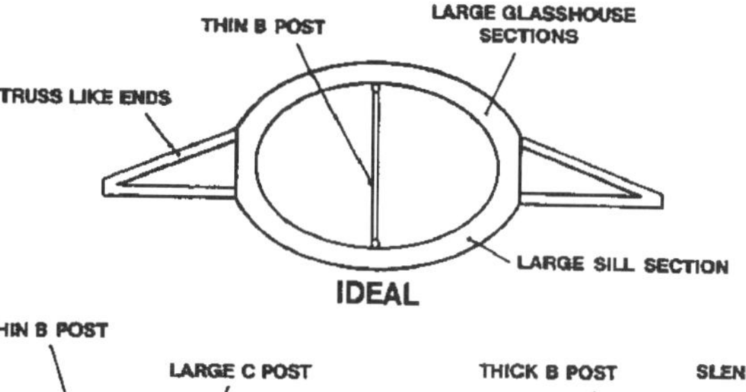

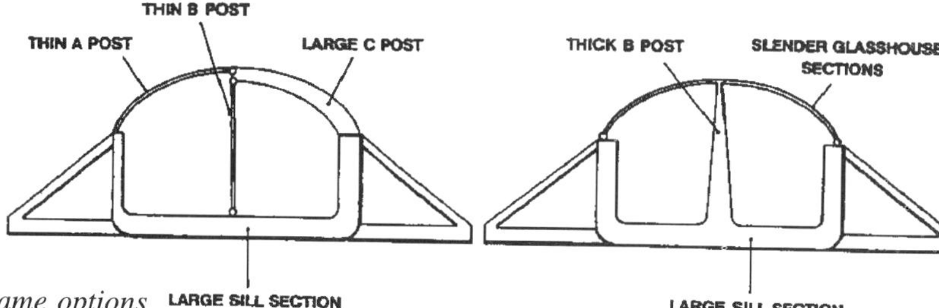

Fig 47: Alternative sideframe options

roof-rail joint about a transverse axis Ry to have the largest effect on overall torsional stiffness and the third largest effect on bending stiffness. In the design philosophies for the post summarised in Fig 48, the last gives particularly high robustness to end joint rotational stiffness while this parameter has very low value in the second. The first places emphasis on the packaging allowance above the waistline (Figs 50/ 51) and is a combination of the last and second and is

Fig 48: Aternative manufacturing processes for sill: from left, a extruded, b pressed or rolled, c hydroformed

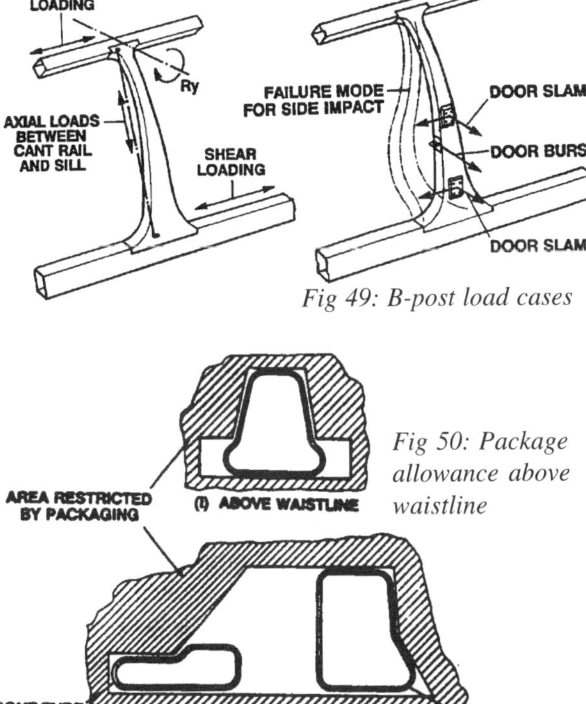

Fig 49: B-post load cases

Fig 50: Package allowance above waistline

Fig 51: Package allowance below waist-line; front tube is 70 mm and rear tube 90 mm diameter

well suited to the packaging requirements. Bending loads are limited to the short free-beam length above the waistline and the waistline-joint can be enhanced by making the upper beam and one lower leg in one piece. There is also a smooth transition of properties at the joint. Tumblehome of the upper part involves the vertical lower half changing into a curved profile to meet the cant rail, the structural offset resulting in torsion and bending loads as well as shear.

Packaging space defined the structural cross-sections for an analysis to determine section properties. These predicted 1.2 mm thick mild steel would be satisfactory for maximum stress in the static load cases. This is a key dimension for hydroformed tube as the fillet radii for structural packaging are dependent on it. Another analysis, of the manufacturing processes had already revealed the ability to form complex sections in curved paths to be crucial in maximising the structural package, particularly the ability to alter section shapes at the ends of the post to suit optimised joining with the upper and lower rails. Fig 52 shows the resulting structural solution under finite element analysis which demonstrated good continuity at the waistline joint but some manufacturing complexity which had to be reassessed and solved by laser-cutting of the formed tube to ease blending of the formed tubes.

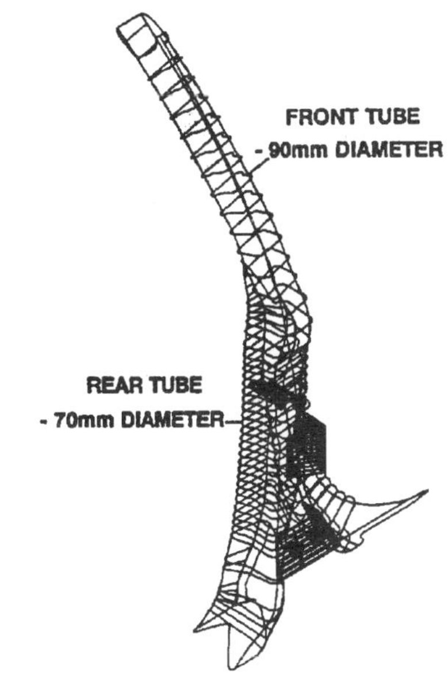

Fig 52: FE analysis

Panel trimming

Trimming of deep drawn pressings, without the expense of hard tools, is economically attractive for prototype panels and small quantity production. The availability of high-power continuous-wave YAG lasers is proving significant in this process, too.

Conventionally, CO_2 lasers equip the multi-axis cutting machines used for trimming. The beam from this type of laser is transmitted by a mirror system to the cutting head of the machine. Now, ShinMaywa has designed its RC series of machines to operate with YAG lasers transmitting through fibre optics. As a result, the machine is simpler and cheaper and the beam transmission system is unaffected by vibration.

Such machines are capable of cutting 1.2 mm thick steel sheet at speeds between 3 and 6 m/min, depending on whether the steel is coated. In configuration, machines are based on a cartesian robot and speeds in its linear axes are up to 11 metres/min. The cutting head moves about two rotational axes.

In Europe, the first machine to become operational is installed in the Coventry factory of Motor Panels. It replaces conventional trimming equipment, and is being used to profile body panels for the Aston Martin DB7, Fig 53. Laser profiling enables the company to reduce lead times and a feature of this installation is its twin pallet changer. Loading a panel on a pallet is performed away from the machine; at the same time, another panel can be trimmed on a second pallet in the cutting station.

Trimming with waterjets has the advantage of being applicable to a wide range of materials. ABB have introduced a unit for trimming vehicle components such as parcel shelves and door panels. This robotized waterjet cutting system is programmable, and an alternative to the use of dedicated punch tools Another approach to trimming car body panels is cutting with a fine plasma torch, Fig 54. Kuka is marketing robot-based systems that produce precise cuts whose quality is said to approach those made with lasers. However, fine plasma systems are less expensive than comparable laser systems.

To improve the accuracy of small holes cut with fine plasma, Kuka has developed a CNC 2-axis trepanning head which can be installed together with a height sensor on the IR 364/15 6-axis robot. This device produces holes within 50 mm x 50 mm, and as small as 5 mm diameter. It can be programmed to cut different shapes including circles, ovals, squares and slots. Kuka suggests that the small hole system would be suitable for installation at the end of a production line to work on assembled pressings. It could produce the holes needed in car bodies for aerials, bumpers, windscreen wipers and other such items.

Punch presses are still a most effective method of blanking and trimming, a good example being Al-ko trailer chassis sidemembers, Fig 55, produced on

Fig 53: Continuous pulse laser trimming of Aston Martin DB7 panel at Motor Panels

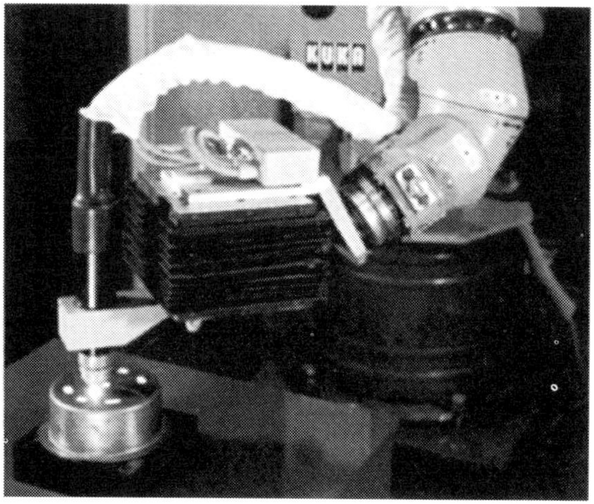

Fig 54: Fine plasma hole cutting by Kuka

Alloy	Condition	0.2% Proof Stress	UTS	Elongation
Supral 100	0	130	220	12
Supral 100*	T6	300	420	9
Supral 150*	0	120	195	9
Supral 150*	T6	270	390	9
5083 SPF	0	150	300	20
7475 SPF	T5	340	420	11
7475 SPF	T5**	360	450	8
7475 SPF	T6**	525†	574†	10
8090 SPF	T6**	350	450	4

Fig 56: Alloy properties compared
Alloy Condition 0.2% Proof Stress UTS Elongation

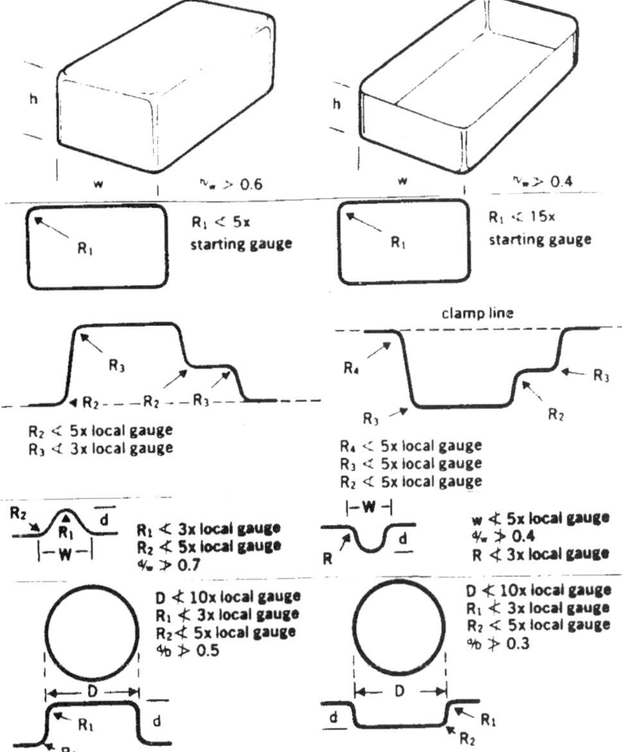

Fig 57: Corner radius recommendations

Trumpf 260 machine. Steel sheet up to 3 mm thick is punched and blanked and a Prime Medusa CAD program is used to nest the components on a strip for maximum economy in material usage. Steel sheet is positioned automatically beneath the 250 kN ram of the press and can be moved at up to 90 metres per minute for positioning. The magazine of the press holds up to 10 punch tools and these can be changed in 6.5 seconds. The press can also be used in nibbling mode to achieve apertures of any required shape and a series of slots is cut into the sheet which allow finish-components to be quickly separated.

Superplastic forming

A number of metal alloys can be produced in a superplastic state such that their strength drops to zero and elongations of some 1000% are possible. The products of two suppliers are reviewed here to show the potential for producing contoured panels economically in comparatively low volumes.

To achieve superplasticity the alloy must have a very fine grain size which must remain stable up to at least half the melting temperature of the material. The alloy is worked above this temperature threshold and carefully chosen strain rates are required in forming. A popular alloy for superplastic forming is 78% zinc and 22% aluminium which has a room-temperature tensile strength comparable with low-carbon steel. When raised to 250 C this alloy will become superplastic at a strain rate of 20 in (508 mm)/minute.

With properties in this state similar to thermoplastics the material can be formed by such methods

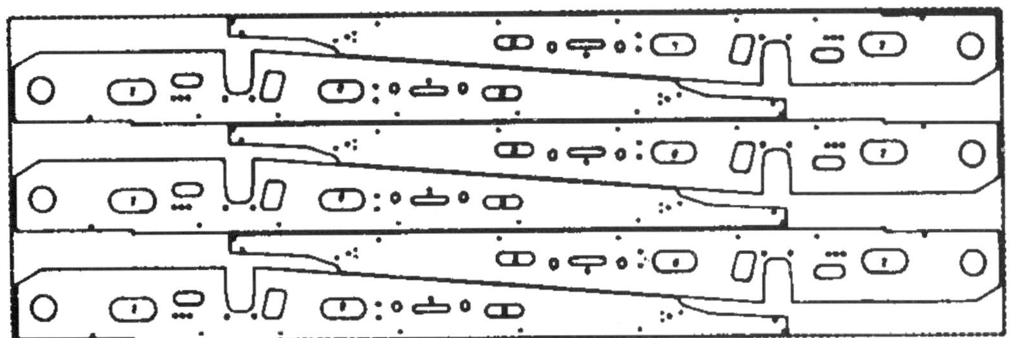

Fig 55: Nested Al-ko sidemember blanks

as vacuum-forming — using relatively low-cost tooling. Alongside the ability to undergo large elongations in tension is the ability to conform well in compression such that very fine detail can be reproduced in a forging process, for example.

One approach to achieving the superplastic state is closed-die forming in which slugs of alloy are forced through conical dies to reduce the grain size prior to forming. Forgings thus produced will not fracture in a brittle manner like die-castings and properties shown in Fig 56 are typical. The resulting alloy can be fabricated by conventional techniques such as spot-welding, brazing and soldering while special techniques are available for fusion-welding the material to steel.

Aston Martin was a pioneer user in the industry, using the Supral alloy for the Lagonda luxury car from TI Superform. With this material both male and female forming can be carried out over a single tool using 10 bar air pressure at around 500 C. Cycle time can be from 8 to 50 minutes depending on sheet thickness; sheets from 4 mm down to 0.8 mm can be handled.

Spring-back problems are virtually eliminated with the process and panels are reproducible to a tolerance of +/- 1mm over a length of 2.5 metres. There is a limit on depth of drawer, however, of 60% relative to panel width — with an overall limit of 400 mm using male forming and a draft angle of two degrees. Corner radii are critical, recommended values being given in Fig 57.

Superform Metals describe four superplastic forming techniques as in Fig 58. All operate at temperatures between 470 and 520 C. The company can provide a trimming service for formed panels, as well as the incorporation of machined cut-outs. Some interesting automotive applications include the sliding door for a Volvo ambulance shown in Fig 59 and a 4mm thick racing car bulkhead for March Engineering, Fig 60.

Another supplier is Custom Metalforms whose product is branded Super Plastic Zinc (SPZ), with properties as shown in Fig 61. Their thermoforming process is seen in Figs 62-63 . The four hydraulic rams give 100 tonnes clamp load capacity and will form sheets up to 1.2 x 0.65 metres. Dies or moulds are usually precision sand-cast in aluminium alloy or iron, then finish-machined and polished.

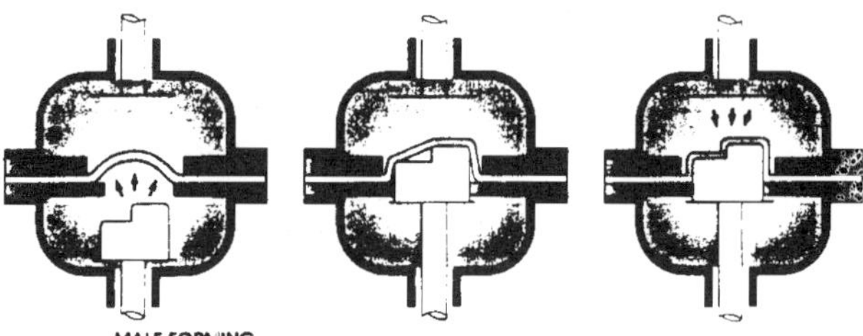

MALE FORMING
By combining air pressure with tool movement, deeper parts with more uniform wall thickness can be made

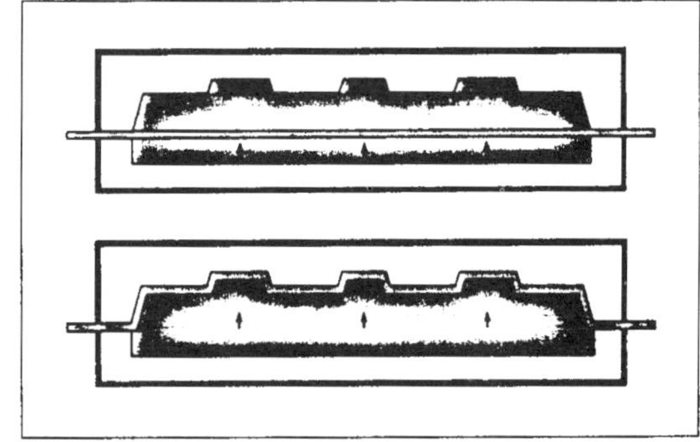

FEMALE FORMING
The superplastic sheet is stretched into the tool by applying air pressure

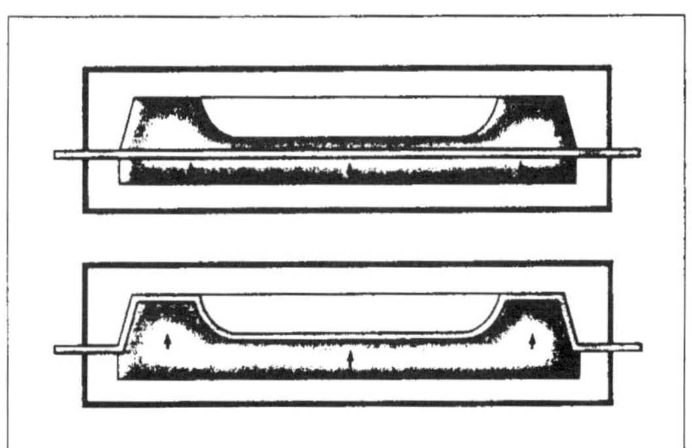

DRAPE FORMING
The air pressure forces the superplastic sheet into the cavity and over the male tool

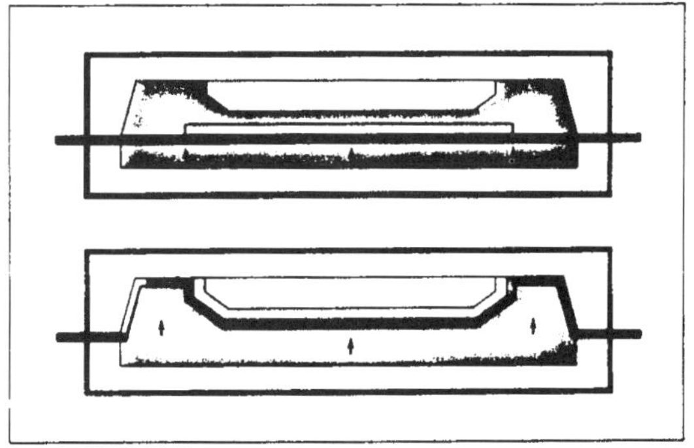

DIAPHRAGM FORMING
When design and performance criteria require the properties of a conventional sheet aluminium alloy, it can be formed over the tool using a superplastic alloy sheet diaphragm giving excellent control of thickness distribution

Fig 58: Alcan forming methods compared

Fig 59: Ambulance door

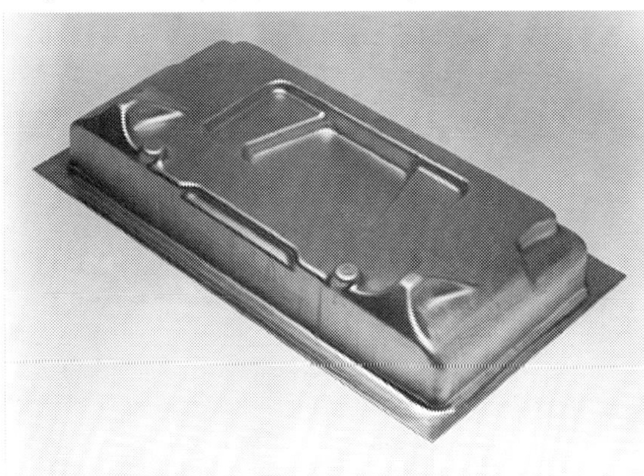

Fig 60: Racing car bulkhead

	Imperial units	SI units
Tensile strength (a)	64,000 psi	441 MPa
Elongation	10%	10%
Design strength (b)	7000 psi	50 MPa
Density	0.187 lb/in^3	5.3 gm/cm^3
Thermal conductivity	85 BTUft/hr/ft^2/°F	147 W/M/K
Specific heat	0.13 BTU/lb/°F	544 J/Kg/°C
Electrical conductivity	34%IACS	34%IACS
Coeff of Therm Exp	13.5 × 10^{-6}in/in/°F	25 × 10^{-6}°C
Melting range	790-930°F	420-500 °C
Electro magnetic shielding	Zinc offers protection against magnetic fields	-
Corrosion resistance	Zinc alloys have good corrosion resistance	-
Finishing	Can be plated, painted, buffed or left in the as-formed state	-

Fig 61: SP zinc properties

Preheat.
A suitably sized blank is heated between two platens.

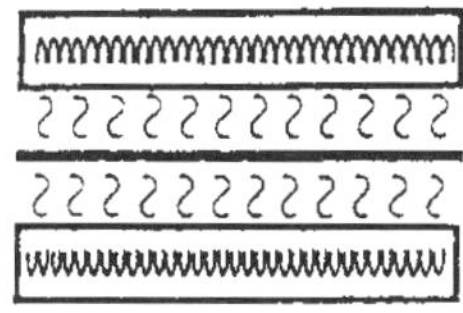

Clamp in Die.
The preheated blank is transferred to the die where it is clamped around the die opening, forming a pressure vessel.

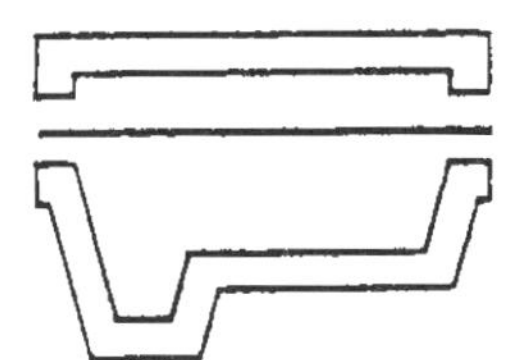

Apply Air Pressure.
Air pressure is applied to one side of the sheet, forcing it to move towards the die face.

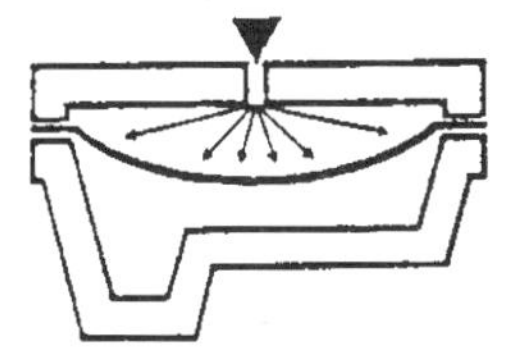

Form Part.
The sheet moves until it conforms totally to the die shape. The formed part is then removed from the die.

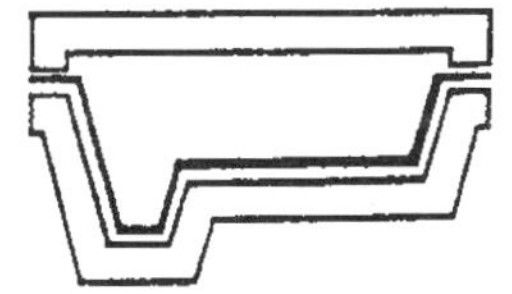

Trim and Finish.
After cooling, the formed part is deflanged and finished as required.

Figs 62-63: SPZ thermoforming

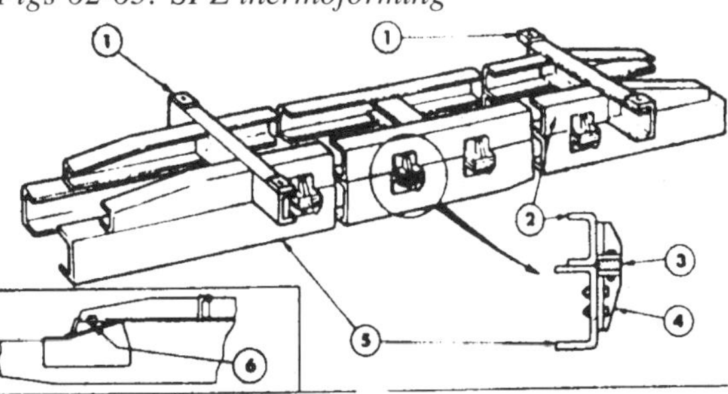

Fig 64: Chrysler truck fabricated by press brake and six press-braked parts of a Ford chassis (above)

Press-braking/roll-forming

Bending of metal sheet in one plane is carried out either by press-brake or roll-form in most chassis and bodybuilding situations. A number of successful specialist vehicles (Fig 64) have been designed for production in these techniques, saving the capital cost of draw-forming on conventional presses. Some guidelines are given here to optimise the use of the former equipment.

The bending of metal involves changing shape without foreshortening length or width so that the 'blank' area remains intact and no stretching takes place. Between the plastically deformed extreme fibres and the undeformed neutral axis are elastically deformed fibres which cause spring-back. This requires a degree of over-bend — necessary to reach an exact angle. There is also a shift of the neutral axis at the bend towards the compression side, of 0.3 to 0.5 times the metal thickness. An allowance has to be made on the blank to allow for the bend radius. For bend angle a and radius r, length allowance (normal to the bend) is $(a/180)\pi(r + 0.4t)$ when r is about three times the thickness t. For bending along a straight line the press-brake can be set to either 'air-bend' or 'bottom' the workpiece. The first case involves three point bending by the tip of the punch and the edges of the die while the second involves full contact by the total working surface of punch and die with the workpiece, Fig 65. Reasonably complex sections can be built up by progressive bend operations. Special tooling can also be used on press-brakes to obtain edge seaming and joggling — also tube forming. Load required to make a simple right-angle bend is a function of the die ratio (bottom tool vee opening/metal thickness) which is optimum at 8. Above this value, load is reduced; when high tensile materials are employed, however, the ratio itself is increased. The value of the load (tonnes) is obtained from $8L_f t/D_w M_f$ where L_f is load factor; D_w is die width (mm) and M_f the material factor (which is 1 for mild steel, 2 for aluminium and 0.67 for stainless steel).

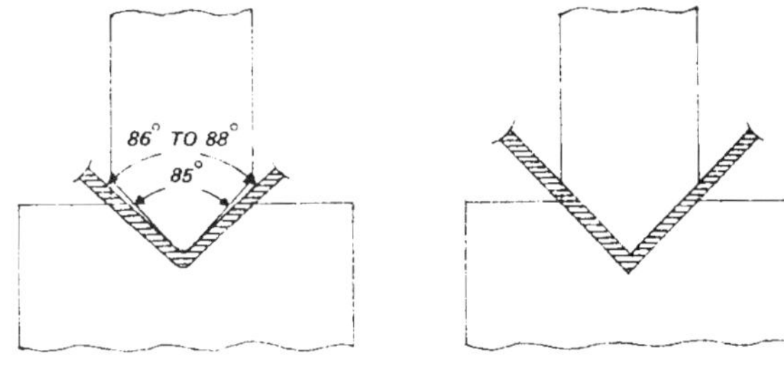

Fig 65: Air bending and bottoming

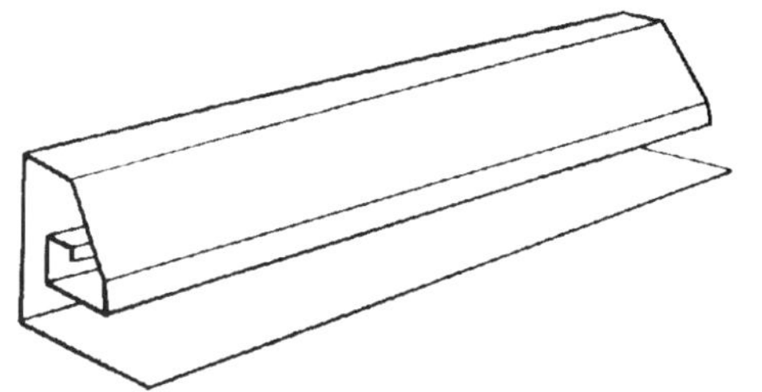

Fig 67: Part with five internal bends

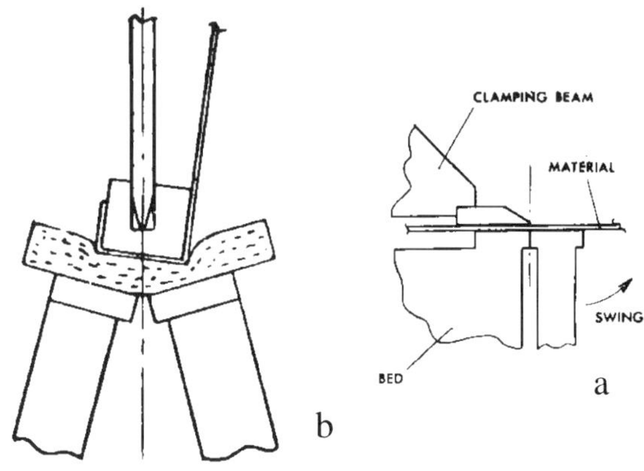

Fig 69: Folding machine and hydro-form compared

Metal Thickness	0·91	1·22	1·63	2·03	2·64	3·25	4·76	6·35
Load Factor, F	2	2·5	3·5	4·25	5·5	6·75	10	14
Metal Thickness	7·94	9·53	12·70	15·88	19·05	22·23	25·40	--
Load Factor, F	16·75	20	27·75	33·5	40·25	46·75	53·5	—

Fig 66: Load factors

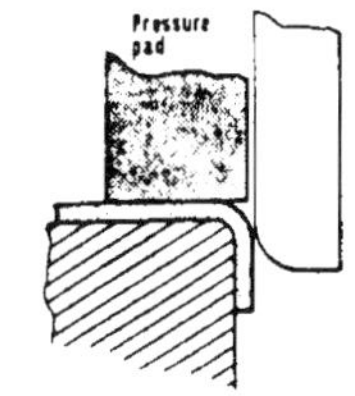

Fig 68: Simple right angle bend

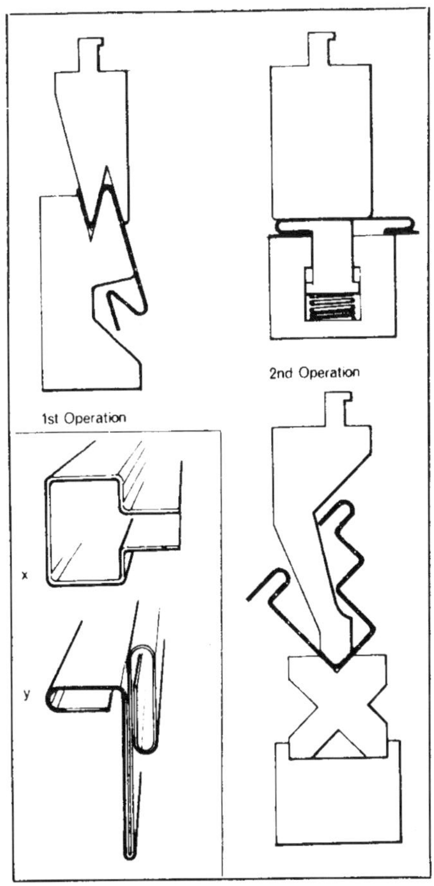

Fig 70: Roll-forming

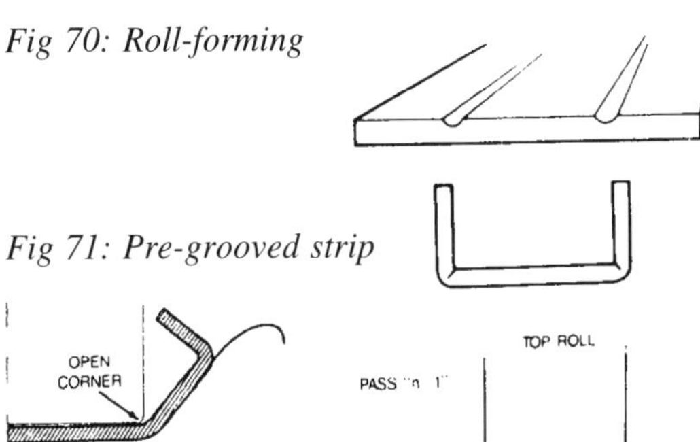

Fig 71: Pre-grooved strip

Fig 72: Secondary rolling

Load factors are shown in Fig 66 for different metal thicknesses.

A number of refinements have been developed for press-brakes which widen the scope of metalforming. The Belgium LVD company are reported to have supplied a machine to the US Caterpillar concern which has a six metre bed length. It has automatic thickness compensation built in and linear encoders each side of the ram to ensure parallelity and bending accuracy. Tool changing is also automatic and a plate support system is provided to prevent deformation during forming operations. LVD have also developed a CADCAM system which automatically calculates bending process parameters, after a part has been designed, and generates an NC program. Available on Promecam hydraulic pressbrakes is a CNC unit which provides synchronised back gauge and bending depth control. A typical product which can be made at one handling, with eight internal bends, is shown in Fig 67.

Trial-and-error is often the best procedure for obtaining optimum set-up in press-braking. Even on the moderately simple shape of Fig 68 any 'rule' for producing the right angle bends would be subject to exceptions. Minimum bend radius, for example, can vary from $t/2$ for soft material up to $7t$ for full temper steels. This applies, however, only if the bend is across the grain; should the bend be with the grain, these radii would cause cracking and would need raising by 50 to 100%. According to whether an open-vee or 'sidedrawing' tool is used, the length allowance for the bend could vary considerably.

An alternative to press-braking is hydro-forming which involves lower thrust forces to achieve bends. While a press-brake overcomes the restrictions of a simple folding machine, Fig 69a, the hydroform can operate within the same space freedom by with lower loads, Fig 69b. This is because the load is applied to the swinging wings, or folding beams, of the machine. The beams are located in trunnions which pivot on a fixed fulcrum pin, giving a sweeping action on the material, making it ideal for coated and polished workpieces.

Roll-forming

This technique is of course used to obtain higher speeds of bending of strip stock — but will usually involve longer set-up times in positioning rollers for a particular profile. The strip is pulled through and rolled into the desired section shape simultaneously. Roll speeds of about 20-70 metre /min are typical for mild steel of 0.15 to 10 mm thickness. Forming lines may incorporate pre-notching and/or pre-punching presses to achieve profiles with shaped holes and cut-outs; alternatively the rolls can be fed with pre-cut blanks. Beyond the rolls, cut-off presses can also be adapted for extra operations such as embossing. Fig 70, due to AISI, shows the step-by-step formation of a typical profile. Note that while the corner line travels in a straight line, the edge is constrained to a helical path which involves some stretching and recompressing. Care must be taken in design to avoid any deformation of the finish part due to locked-up strains which result.

As well as pre-judging the flow of material as it passes through the mill, the tool designer must also consider the shape of the finish section with respect to the lengths of unstiffened projections — also the orientation of the workpiece as it emerges from the mill. Spring-back effects must also be allowed for and minimum radii with respect to different materials have to be observed. It is sometimes possible to roll below these minima if the strip is pre-grooved as shown in Fig 71. When the shape of the section is such as to preclude access of the rolls to certain corners, techniques of secondary rolling, as shown by Fig 72, may be adopted to straighten back the section.

Complex sections will be rolled in successive passes as shown in Fig 73. Working tolerances quoted by experienced formers are +/- 0.010 in (0.25 mm) on the profile and a length *L* to straightness factor of *L*/300. Bend radii down to half sheet thickness are possible without grooving on hot-rolled grades of steel and zero radius for cold-rolled grades of deep-drawing quality to BS 1449. Steels up to BS 968 grade have been successfully cold-rolled when very high specific strength ratios have been required. Bumper bars are often roll-formed but a preform may be produced by press-forming, for wrap-around designs — the preform being subsequently rolled in a machine having one roll replaced by a form-block slipper. The straight preform is clamped to the slipper on its front profiled edge and a matching form roller applies reaction pressure against the form-block which is itself rotated to form, simultaneously, the bumper in cross-section/contour, Fig 74.

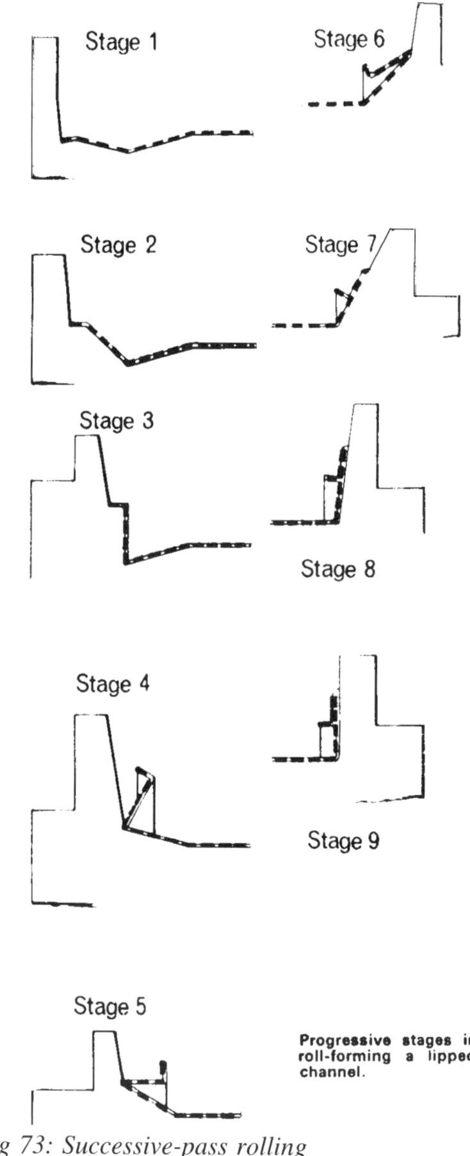

Fig 73: Successive-pass rolling

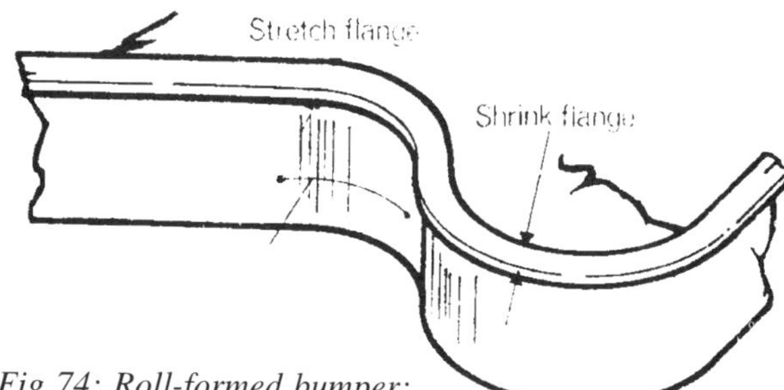

Fig 74: Roll-formed bumper:
Stress puckers will form in the bottom at the point arrowed unless control is exercised to make the metal exceed the plastic limit

Flex-forming

Sheet metal forming with flexible tools can offer potential benefits to short-run work, with the possibility of reducing die-cutting cost as only one half the die need be made. Recent developments of the flex-forming process now allow relatively complex shapes to be formed and high accuracies required of prototype production can be realised.

Either the male block or the cavity can be made from relatively inexpensive materials, the other tool half being a flexible rubber diaphragm. In the case of the system now being sold by ABB, pressurised hydraulic fluid actually forms the metal at the other side of the diaphragm and presses are available for forming pressures up to 2000 bar. The industrial realisation of the process has depended on the successful reaction of the very high thrust forces these pressures inflict on the press. This has been achieved by enshrouding structural parts in pre-stressed wire, so that they react with the forces without change in shape which would impair accuracy of forming. Fig 75 describes the process in which the blank (2) is placed on the rigid tool-half (3); the pair are then run into the press cylinder where the diaphragm (1) is located. Oil pumped into the fluid cell forms the blank by wrapping it around the rigid tool half, even into undercuts. Presses with trays up to 2.0 x 4.0 metres have been built.

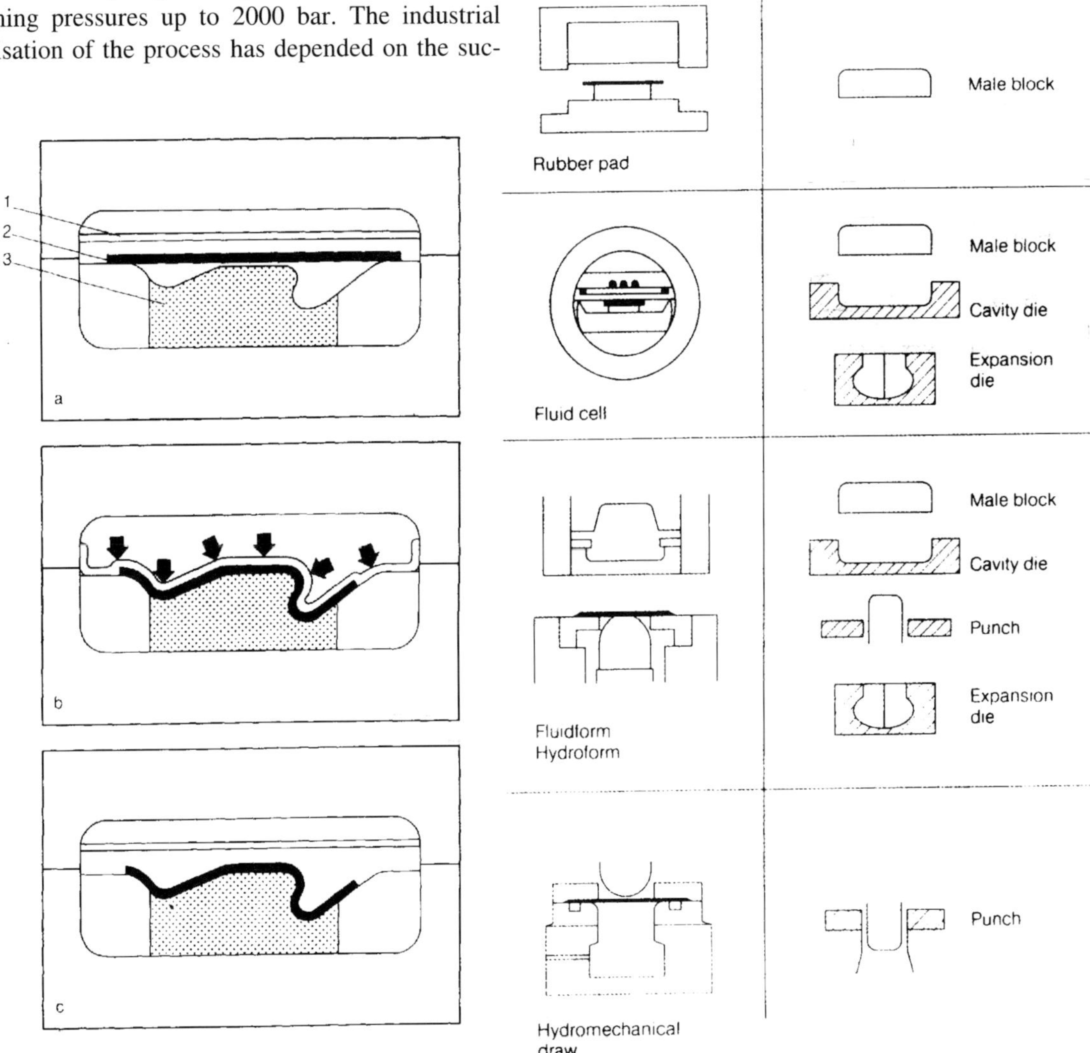

Fig 75: Stages in the flex-forming process

Fig 76: Fluid-forming comparisons

In prototyping work, sheets of different blank thicknesses can easily be tried out. There is also the advantage that in-built stresses arising from hand-forming are largely avoided and 90% cost reductions have been claimed as well as halving of lead times. Series production runs of several thousand units are also feasible if complexity of shape is not too great and provided a tougher tooling material for the rigid half is used. Choice of the tool material is also governed by the blank material and minimum forming radii.

Considerably greater depth of draw is achievable with the rubber diaphragm process than earlier fluid forming processes which used soft rubber pads. Sometimes more than one hit may be necessary for deep draw and, exceptionally, use of the Fluidform rather than Fluidcell process may be necessary. The Fluidform press has a telescopic ram system carrying a moving punch, the diaphragm being carried in the circular pressure unit. A similar process but without the diaphragm is hydromechanical drawing; Fig 76 shows the processes compared diagrammatically. Two different forming radii are involved in flex-forming, that by the rigid tool half and that by the diaphragm. The latter is known as the free-form radius and governs the pressure required. Mild pressing steels generally need pressures below 1400 bar. Pressure intensifiers are used to achieve these higher pressures over the 200 bar 'system' pressure. Cycle time of the operation is normally 1-3 minutes, depending on forming pressure — but productivity can be increased by using pallets pre-loaded with tools and blanks for simultaneous processing.

Fig 77 shows the rear fender of the BMW Z1 roadster made in galvanised 0.8 mm steel. It is formed in several steps, with intermediate trimming, and uses a pressure of 700 bar. The rigid tool half is made from Kirksite low melting point zinc alloy. The latest technique is to integrate the trimming of the sheet component. Here the rigid tool half has a cutting edge which trims the metal under the pressure of the diaphragm, a protective layer of rubber being inserted between diaphragm and blank, Fig 78.

Other potential benefits of fluid-forming processes, generally, are freedom from elongation of the blank when the part is open-ended and avoidance of surface scuffing. The latter is of importance in the pressing of coated steels or pre-printed panels.

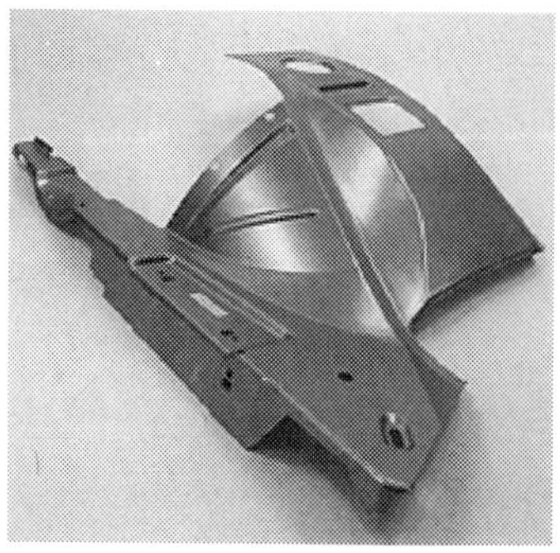

Fig 77: Rear fender and door pressings

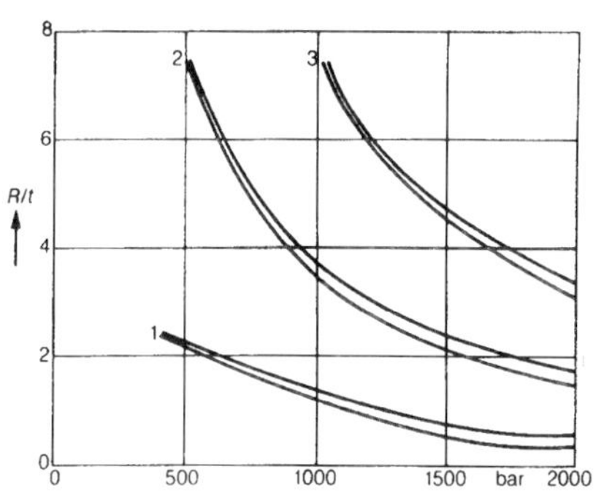

Ratio of radius R to the sheet metal gauge t for different materials as a function of the forming pressure p

a Cavity die
b Punch
1 Pure aluminium
2 Deep drawing steel
3 Stainless steel

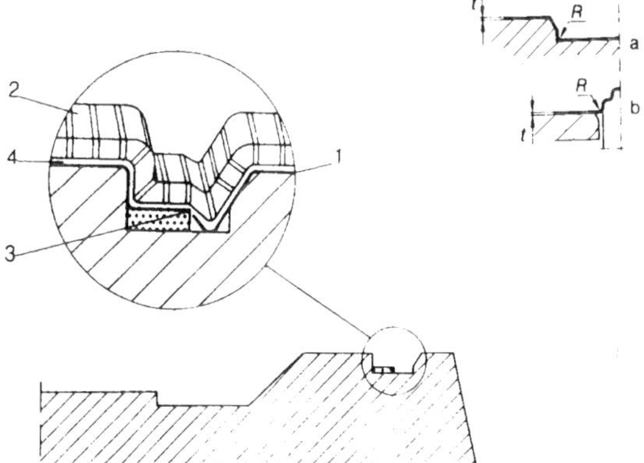

Fig 78: Integration of trimming

How trimming can be integrated in the flexforming process. The blank (1) is stretched over a sharp cutting edge (3) by the diaphragm (2) until the excess material is torn off. The diaphragm is protected by a layer of rubber (4).

Spinning and flow-forming

Very large circular forms and those requiring heavy metal gauges can be expensive to produce on press-tooling but relatively cheap by spinning. Here the basic parameters of the method are considered along-side the slightly more sophisticated process of flow-forming.

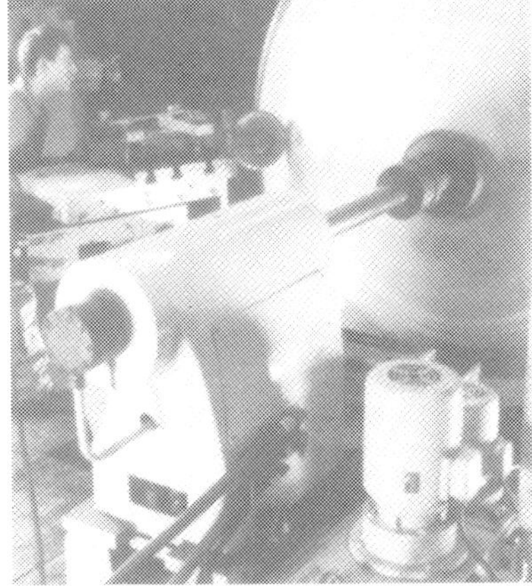

Fig 79: Tanker dished-end production

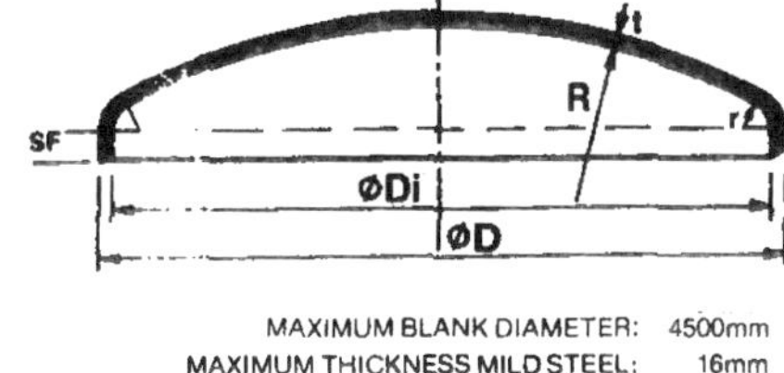

MAXIMUM BLANK DIAMETER: 4500mm
MAXIMUM THICKNESS MILD STEEL: 16mm
MAXIMUM THICKNESS STAINLESS STEEL: 12mm
MAXIMUM THICKNESS ALUMINIUM: 40mm

Spherical Radius	Corner Radius	Corner Radius
1700mm	35mm	0.5/8"
1900mm	38mm	0.3/4"
2100mm	40mm	1 1/2"
2300mm	45mm	4"
2600mm	50mm	6"
2900mm	115mm	8"
3200mm	150mm	
3600mm	200mm	
4000mm	225mm	
	250mm	
	300mm	
	325mm	
	350mm	
	400mm	
	450mm	

Fig 80: Dished end dimensions for a main supplier to the tanker market

Fig 81: Flow-forming

Both spinning, and the related flow-forming process, are used in the production of many other vehicle parts, from wheels to air-tanks. The advent of computer numerical control has enabled both low and high volume manufacture of spun parts. Tooling costs are relatively low compared with presswork and prototype production is a particularly attractive proposition, Fig 79; the range of sizes available from Metal Spinners (Newcastle) are seen in Fig 80.

Spinning is basically a technique of kneading sheet metal about a rotating form into a desired circular shape. The process can be carried out on a lathe with suitably large swing clearance and the form can be made of wood for low production quantities. The form is fastened to the headstock and the blank pressed against it by the tailstock. Blank and form are rotated at high speed and the operator, beginning at the centre, works the metal against the form with blunt ended hardwood or steel rods. Tolerances down to 1/8 th in (3.2 mm) can be held on work over 4 ft (1.22 m) diameter. Thicknesses are limited by the strength and work-hardening of the material.

In flow-forming the spinning blank is formed by shaped rolls. Wheel discs for drop centre wheels have been made for some time by the method and more recently discs for taper bead seat and tractor wheels are being produced. Rotational cold forming is another name given to the process and continuous reduction in wall thickness can be obtained from base to edge. Wheel discs thus produced have accurate diameter and section profile, good surface finish and low out-of-balance. Blank tolerance has no effect on finish-part tolerance.

The blanks are centred in front of the mandrel by an automatic feeder and clamped by a tailstock pressure pad. Two or three work rollers flow-form the blank, Fig 81. Rollers are either controlled by copying attachments which follow templates — or more recently by CNC. When finish thickness is reached, rollers are retracted, the work saddle drives to its limit position, the main spindle is braked and the disc stripped from the former. Saddle feed rates are particularly important. Since feed rate in a radial direction is governed by the axial feed rate and the copying profile, forming relationships change in the course of one pass. Radial feed must be matched to the degree of deformation and thus machines are provided with as many as four axial rates, switchable from a control panel.

Chapter 3: Body finishing and hardware manufacture

Finish and quality; anti-corrosion pre-treatment and coatings; finish painting; body sealants; coated steels, plastic coatings and pre-coated panels; glazing techniques; trim pad construction; body-hardware hot and cold forging; aluminium and ferrous casting; metal-matrix composites; structural timbers; metal injection moulding

Finish and quality

Quality is first perceived by the customer at first sight of the vehicle but, of course, is felt by him at all stages of the vehicle's operational life. The perception of poor quality in UK manufacture in the early post-war years, and beyond, some say has led to the decline in economic performance of indigenous manufacture and consequently its acquisition by former overseas rivals. This has led, albeit a bit late in the day, to a dramatic 'soul-searching' as to what are the key factors involved in bringing industry up to world-class quality levels. Both the first perceived surface finish, and the through-life operational quality, are of major consequence to customer acceptability.

Opinions vary widely as to the causes and their relation to the effects but Hill's[1] classic work pointed out the lack, at the time of writing, of production decision making in UK manufacturing industry which accompanied the then vastly increasing levels of inter-company competitiveness. In strategic terms manufacturing management still takes a subordinate role to marketing and finance, it is sometimes argued.

Quality is, of course, related to price and productivity. The familiar picture of gloom that surrounded UK international manufacturing competitiveness in 1984, 10 years after the first oil crisis and when the UK still had its own major contender in high-volume vehicle making, was explicit in productivity about a quarter of that of France/Germany and about one tenth that of USA/Japan. In the motor vehicle and component sector, ratio of UK imports to home demand rose 223 percent from 1973 to 1983. Political moves to inspire greater competitiveness by exposing UK firms to overseas competition has regrettably been unmatched by the necessary industrial response, Hill asserts. Not just labour, but total-factor productivity is the key to successful measurement of performance. In overall vehicle production unit terms, number of units per employee was over 3.6 times greater for Toyota in Japan that Ford in Europe, Hill revealed. This situation could not at that time, be accounted for by Ford, Hill maintained, until their delegation went to Japan to see for themselves.

Japanese industrial engineering consultant Shigeo Shingo was prominent in the design of production systems at Toyota and has described the set-up at that company in some detail[2]. He considers various zero defect inspection systems. 'Successive inspection', so called, during stages of a production process, involves the operator, at the second stage, inspecting the processed product from the first stage then carrying out himself, the second stage process — and so on by respective operators along the line of work, Fig 1. 'Self-inspection' is better, it is argued, in avoiding any proneness to 'make a compromise' or 'a misinspection unintentionally' than in 'Successive'. These two mistakes can be overcome by physical detection means that are fool-proof (*Poka Yoke* in Japan). Fig 2 shows an example for inspecting a 'handed' automotive part. A limit-switch on the press is used to avoid mis-positioning on the hole, corresponding to handing, by switching off power and sounding an alarm: this results in 100 per cent inspection. Source inspection it is explained, prevents defects by controlling causes at their origin and com-

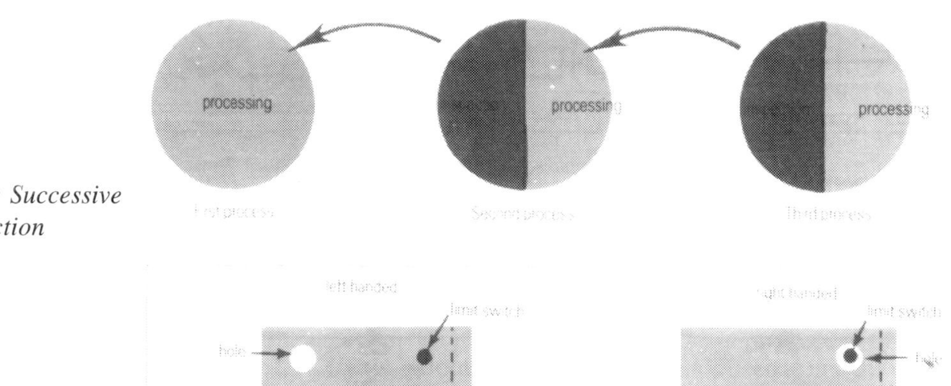

Fig 1: Successive inspection

Fig 2: Poka-Yoke inspection

prises vertical and horizontal types, the former tracing back the feed lines of the process, and associated control conditions, the latter finding conditions within the process itself that affect quality. For example, as checking the colour of a paint is a 'sense' rather than 'physical' inspection, 100 per cent reliability cannot be guaranteed. However by controlling paint colour density, quality and air-spray pressure, physically, discharge quality and hence colour tone can be controlled, the author asserts.

Examples are provided, in another of this author's works[3], which refer to the Kanto Auto Works at Yokosuka. In Fig 3, means for ensuring proper tightening of rear suspension damper bolts is shown. By adding microsensors and a reaction force detector to the two-axis nut-runner, rotation of the jig is prevented — which in turn stops the operation — if either torque value or number of parts tightened is incorrect. In Fig 4, parts mounted within an engine compartment are kept in correct orientation by means of limit switches. The assembly jig only clamps into position, for the work to begin, if the part specification detected by the limit-switches is confirmed.

Set-up time improvements The same author describes the SMED system in another work[4], referring to Toyota, where improvements were made to inter-press handling systems. The time taken to change the handling equipment, when the pressed parts changed from one run to another, had caused unacceptable delays. One technique involved mechanical

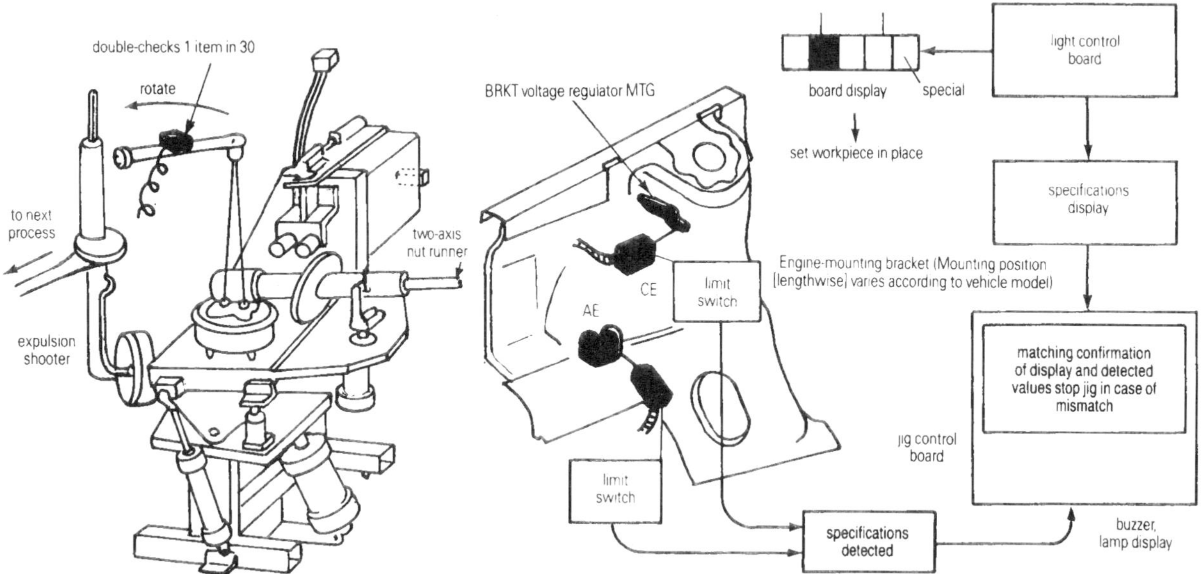

Fig 3: Set-up for fool-proof tightening of damper bolts

Fig 4: Ensuring correct parts orientation

fingers in lieu of conveyors (Fig 5a) to eliminate conveyor set-up time. The changing of dies also caused delays between pressing runs, reduced by using roller-tracks in lieu of a hoist-crane for die moving. Positioning of dies on the bolsters was simplified by locating stoppers on the moving bolsters and arranging corresponding cut outs in the die-bases (Fig 5b). A spring-action bobbing centre-key speeds the process (Fig 5c).

Overall shortening of production lead times at Toyota is considered by Monden[5] who describes a moving final assembly concept, similar to the classic 'Ford System' but in which a unit of a finished vehicle is produced in each cycle time and simultaneously an output at one stage being sent to the next process stage, known as Single-unit production and conveyance (*Ikko Nagare* in Japan). Toyota extend this from final assembly into machinery, pressing and fabrication areas so that a company-wide integrated single-unit flow system is obtained. Workplace layouts allow multi-process operation by multi-function workers; each worker controls several types of machines simultaneously. In gear making, one worker controls 16 machines, for example.

The author defines autonomation (automation with a human mind), Fig 6, to distinguish the process from 'traditional' automation in which no feedback exists for error detection. The concept is in fact extended to manual process lines so as to halt parts-feed when errors are detected. All machines/processes stop automatically when the required number of good parts have been produced, making JIT possible and increasing adaptability to changes in demand. The immediate calling of the operator's attention to defects is said to stimulate improvement activity and thus raise job satisfaction. Key to preventing defects via human judgement is that every worker has power to stop the line.

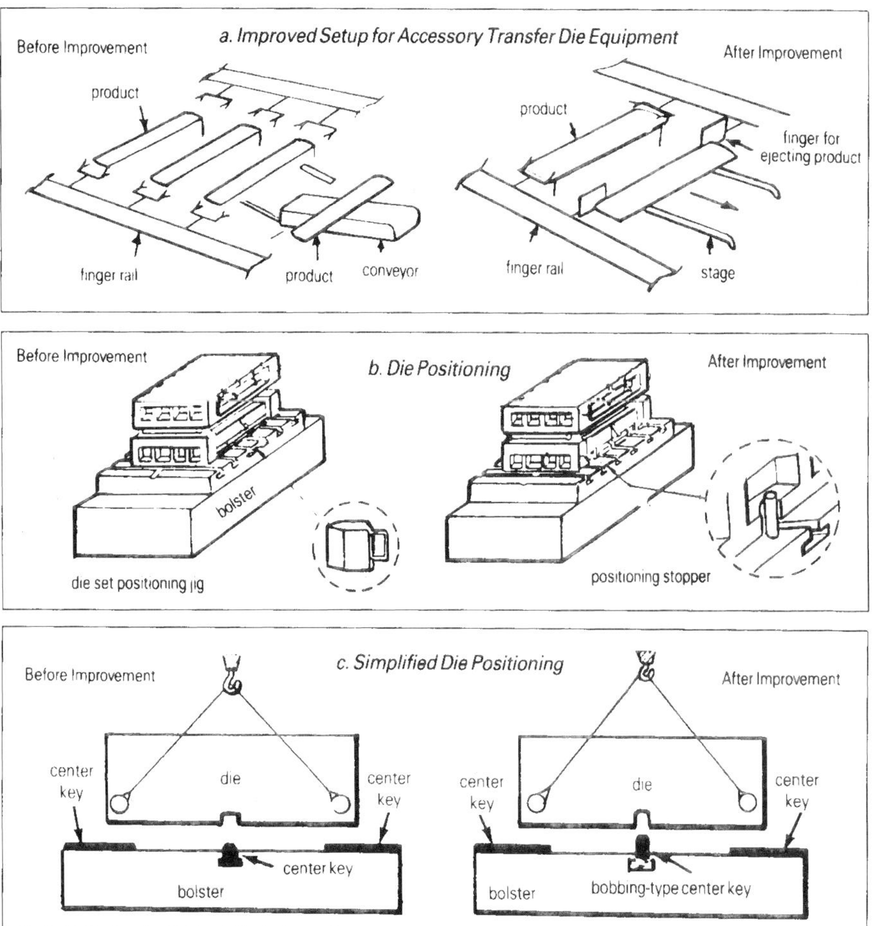

Fig 5: Speeding up presstool die changing

When described to US industrial engineers at conferences in 1981 many said their employers could not be trusted to this extent, saying 'their employees ...would stop the line simply to avoid work'. The opposite was true in Japan, the author explained, with morale being so high that 'they often fail to stop the line when they should...!'

This attitude aligns with a view of automation expressed in an article[6] describing an SERC research project which revealed UK production managers 'recoiled in horror' at the thought of emulating a Swedish semi-automated factory in which workers move between workplace and stores on scooters — to speed feeding of machines. The British thought their people would 'spend all day racing each other up and down the aisle instead of cutting metal'. This adds credence to views expressed by another journal on a contemporary Ford strike[7] that Ford workers are envious of 'practices that they believe to exist at Nissan's Sunderland Plant'. The management, too, is asking for them 'to accept many of the working arrangements... at Sunderland'.

To show that the cultural change, necessary to achieve management of high quality manufacturing, is not beyond the British, Peter Wicken's account[8] of the Nissan (UK) experience is the real evidence. The following excerpt from the book epitomises the total change in attitude necessary: 'Senior Japanese management do not do the job of the production people — as sometimes happens in British industry. What they do do is to spend significant periods on the shop floor — management by walking around. They give specific attention to the production process and concentrate both on quality issues and on ways of making the

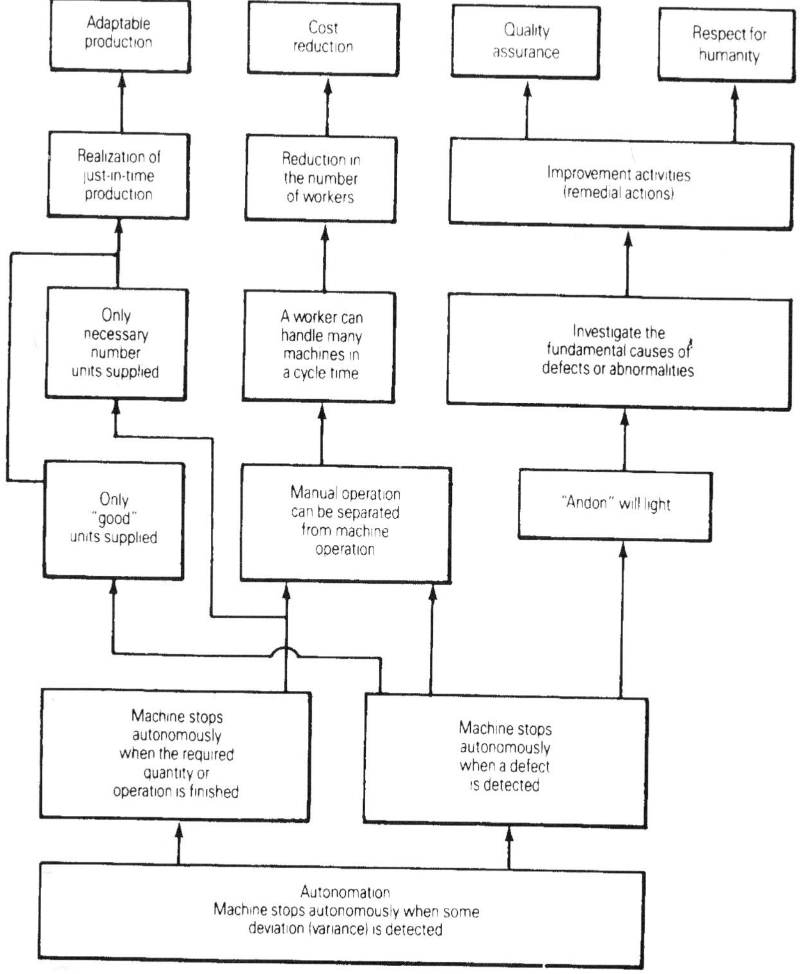

Fig 6: Autonomation concept

job easier. The attitude is that if the job is difficult quality will suffer and if quality suffers the job must be made easier. In a somewhat apocryphal story emanating from the GM-Toyota joint project in Freemont, California, a senior Japanese engineer, seeing an assembly line worker struggling with his job, spent an hour or so trying to determine how to put it right. After working out a successful solution he apologised to the operator because he, the engineer, had failed in his responsibility to make the production operation as easy as possible!'

Hale, Hoelsher and Kowal[9] describe the US approach to quality in some detail, exemplified by the Tennant Company who have reduced percentage sales cost of warranty, scrap and rework expense from 1.62 to 1.15% over a five year period. The total cost of quality (associated with not doing things right the first time: costs of prevention, appraisal and failure) fell from 17 per cent to seven per cent. The company initiated a zero defects policy and have identified five key factors for success: management commitment; employee involvement; co-operative, non-adversarial worker/management relationships; rewards for the people; time, energy and determination A much wider group of case studies including Tennant, and 25 others as well as Toyota Auto Body in California, are provided by Schonberger[10]. Toyota's approach is also examined in great detail in another Shigeo Shingo work[11]. This is undoubtedly the work which, through the authors 'sayings' during his distinguished career, provides a real insight into the philosophy behind quality improvement, Japanese style. Fig 7 shows the scientific thinking mechanism concept which is essence of the book.

British experience as seen by consultants Coopers & Lybrand finds the concept of variance not just in manufacturing but throughout the whole of a firm (Fig 8). Two classes of variation, due, respectively to 'assignable cause' and 'random variation' are identified. The former can be eliminated and the latter requires measurement and control, the authors argue. A more analytical approach to quality via risk analysis is provided by Singh and Kiangi[12] who argue that using computers to model situations can be a valuable tool in analysing, and subsequently reducing, risk. Simulating probabilities can replace hunch by

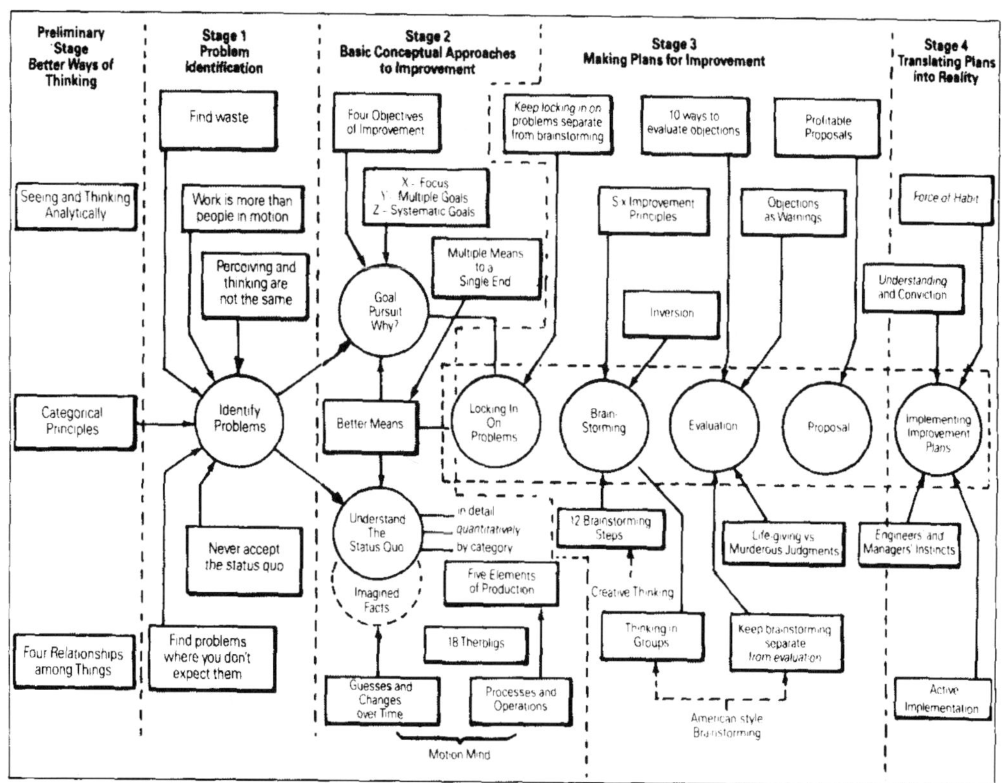

Fig 7: Scientific thinking mechanisms

careful analysis of the 'uncertainty profiles' relating to elements in a design sequence or manufacturing process. Sensitivity analysis is used to find the effect of controlling factors on such a process and reliability-functions can be defined. Several case studies are provided but none directly in automotive engineering.

Manton[13] argues that proven techniques are available to the engineer in pursuit of enhanced quality and customer satisfaction that go beyond the generally understood principles of statistical process control. Their adoption however, will not accomplish the goals of total quality if taken in isolation. Engineers share the responsibility for creating an environment conducive to continual improvement and the elimination of variability. Conformance to specification in both product development and manufacturing processes will not assure ongoing competitiveness and has to be replaced by an engineering process that links customer requirements and market opportunities to the potential manufacturing technology. It must also translate them into target parameters that will drive the entire product development, production, and supply process. The implications for the organisation of company functions are that barriers to co-ordinated teamwork must be removed and a 'common purpose' established across company functions for a planned approach to quality improvement through *prevention* and not inspection. Total quality philosophy attributed to Dr W Edwards Deming involves a fourteen point plan, Fig 8.

The author gives the example of a database generated in the design of automotive press tooling and created jointly by body engineers and tool designers, which reproduces with precision the styling department's intent, eliminating what used to be the free, although expert, interpretation by the toolmaker on whose skills and 'eye' the finished surface depended. Left-hand panels now mirror right-hand panels exactly, and quality of the finished product is enhanced and repeatability in body framing assured during the assembly process. Complex processes such as body seam sealing, through the deployment of robotics, achieve levels of consistency that are just not achievable any other way. While work patterns change and quality and productivity improve, further potential for product quality enhancement can be realized by virtue of the wealth of data generated and retained on the computer system.

Capabilities of processes are readily determined from recorded data and provide the means for further optimization. The product designer cannot ignore the power of these technologies, with processes defined in quantifiable terms. The engineer is drawn closer to the capabilities and potential of the processes and new technologies through the data now available to him. For example, utilizing laser vision technology, non-contact measurement of vehicle body assembly assists in defining the design requirements of, say, the windscreen glazing system. Working jointly with the process engineer, the product engineer responsible for glazing, possibly with the involvement of the body engineer, is able to conduct trial evaluations of alternative dimensioning and bonding processes in order to achieve the required performance consistency.

At another level, the entire design and production process is linked through the common database. High levels of investment are required if the data communication and central databases, protected from contamination by unauthorised personnel, are to be established. Commitment to the infrastructure has to

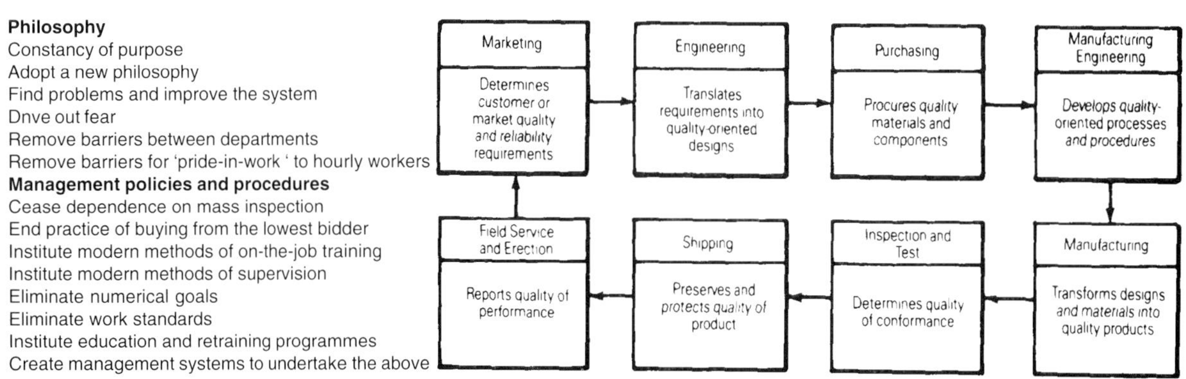

Fig 8: Deming's 14-point plan

be made ahead of full utilization, such that databases grow within the network as they are developed. The resulting 'database' is owned by all technical functions, having been brought into being through joint effort. A preferred approach is to think in terms of target values for the specified parameters to which processes will be controlled and any deviation will incur costs according to a relationship that takes into account both internal and external losses.

This concept is attributable to Taguchi and is the technology of process control said by the author to be replacing SPC and zero defect principles. Quality loss is defined as deviation from a target value and 'the cost to society' is all-embracing, as previously outlined. Taguchi defines off-line-, in addition to on-line-quality control, in relation to design for quality and importantly distinguishes between system functional design and parameter (target) design.

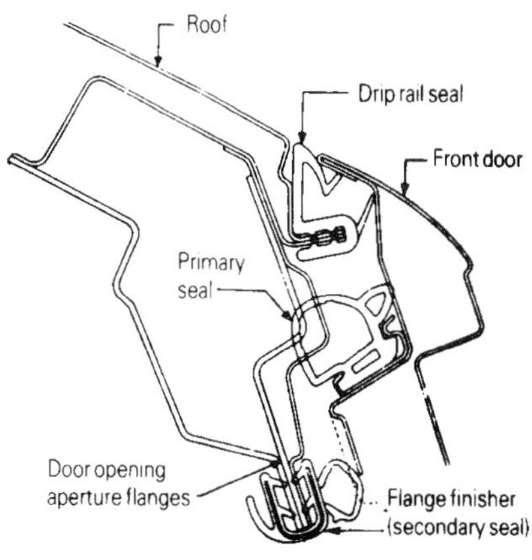

Fig 9: Automotive door-seal system

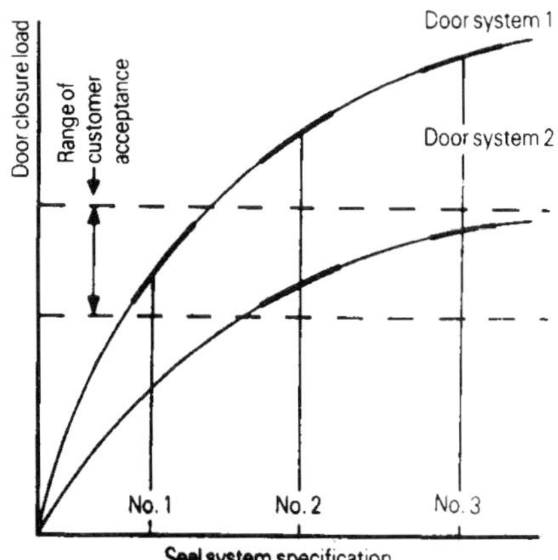

Fig 10: Performance criteria

Taguchi approaches the design of tolerance from the standpoint of 'permissible' costs using a parameter cost relationship. The latter concept although not difficult to grasp does not lend itself readily to practical implementation. Those who have attempted to specify allowable limits by this method have little concern over this as the precise costs turn out to be critical. All the efforts are, in preference, directed at reducing variability and achieving target values. This is the real power of the Taguchi method, the author explains.

The basic principle applied to the design stage for the section of the system of functionable design aims to reduce variability in the performance of the end product. If the specification of more expensive materials or the call for elaborate process controls are to be avoided, then an understanding of the causal factors of variability, in the end product or manufacturing processes proposed, must be such that the sensitivity of the design to assignable causes of variability can be minimised.

Figs 9 and 10 serve to illustrate this point. In this example, relating to the selection of door systems for automotive application, seal system number 1 works adequately in meeting the customer requirements for door shut load, but the variability is such that much of the acceptance range of shut load is taken up. Seal systems numbers 2 and 3 do not meet the requirements at all. With door system number 2, seal systems 2 and 3 now meet the requirement by number 3 has reduced variability, albeit at the higher end of the acceptable range. Perhaps by applying new materials, system number 2 would achieve reduced variability while retaining lower door-shut loads. But additional costs are associated with this.

Is the lower range of loads rather than the high range represented by seal system 3? Variability is more likely to cause customer complaint, particularly if differences are detected between say, front passenger and driver's door, than any specific concern regarding the general level of shut loads around the car, provided it is within the range of acceptability. Customer complaint results in refitting by the service agent and incurs cost and waste.

The Taguchi method leads the designer into

determining optimum parameters, having fully investigated the variability, or more specifically, the sensitivity of the system specification of optimum parameters for the constituent parts of the system. This is complex; in the door system example, the function of the seal system encompasses weathersealing, wind noise, and ease of door shut. Nevertheless off-line quality control as prescribed by Taguchi provides a procedure for parameter optimization that identifies the controlling factors that have most influences on these parameters.

In broad terms a series of experiments are conducted to test the effects of selected factors, nominated through consultation with all functions that possess experience of or a feel for the design and processes under review. Optimization of the controlling factors is arrived at through the evaluation of signal-to-noise ratios. Expressed another way, the method determines the contribution of controlling factors that reduce variability in the parameter, and so desensitizes the design to variations in those factors while optimizing their target values, Manton argues.

Putting these techniques into practice the engineer, through a minimal number of experiments according to a prescribed test plan, optimizes the target value for the customer related parameter of the system and also the constituent component parameters. Experiments are conducted on the processes themselves and therefore relate to actual conditions and provide the basis for continual improvement. The process variability is fully charted and the performance of the product evaluated in advance of series production. All functions participate in the experiments and arrive at an agreed specification that is seen to reflect the common goal of customer satisfaction . The purchaser or end user of the product is also the initiator of the process. Just how to interpret the needs of the end customer is the task facing the engineer responsible for the design specification and its realization. This is perhaps more difficult in the case of manufacturers reacting to the perceived requirements of a consumer market than it is for a specialist component or service company responding to the specific purchase requirements of another company or enterprise. A systematic approach to the investigation and analysis of the customer requirement is critical if this is to be the driving factor for all subsequent company activities.

Companies now emerging as world class, and market leaders, display new thinking in organizational terms but not of organizational/developmental theory. From striving to achieve higher levels of effectiveness across functional boundaries, there is a clear preference for project or product line structures. Emphasis is placed on collective responsibility via multi-functional teams with a single purpose, either by a product grouping or market sector. Responsibilities embrace the entire product programme from conceptual design through to maturity in the product's design and its subsequent replacement. Japanese organizations have been likened to a rugby team where the players advance up the field together, passing the ball. Compare this with the relay race that is more typical of Western companies where handoff from function to function leads to sequential activity. The resultant lead times for new product development, and hence response to changing conditions, are uncompetitive and interaction between functions, when setting a new direction, is not available to the players, the author maintains.

Quality at Nissan UK

According to Nissan director Peter Wickens, quality has many different meanings. To many, it simply means 'does it work — does everything stay together?' To others the basic specification is important — a painted car body totally hides the sheet metal and anti-corrosion treatment and, superficially, the car is assessed on the quality of its paintwork. But that paintwork can cover a high grade or low grade steel or a comprehensive or minimal anti-corrosion process. Japanese products are celebrated for the fact that they do not go wrong, that they last a long time — and that is all about specification. The Japanese car manufacturer sets a very high specification for all components and knows that it will be achieved. He does not inspect 'goods received', for over the years he will have built up a long-term relationship with a single supplier for each component, which gives him total confidence in the ability of that supplier to produce components of the required quality. 'We aim to build profitably the highest quality cars sold in Europe' is Nissan's philosophy statement to all employees.

All British personnel who visited Japan became convinced that NMUK could achieve Japanese attitudes to quality and reach, and possibly beat,

Fig 11: 'Neighbour checks' at Nissan ensure no defect is passed from one process to another

Japanese standards but there was no single solution. The British do not respond to slogans exhorting greater efforts to beat quality targets. Money spent on posters seeking to persuade people to improve quality is probably even more wasted than that spent on safety posters. You cannot achieve quality by sending people on courses and most of all you cannot mouth platitudes and then do everything to demonstrate that you do not mean what you say.

Much of the supervisor's morning meeting with his staff is about quality, how to resolve problems and continuously improve. Kaizen workshops have been introduced where employees with ideas for new tooling or other changes can go to try them out. All employees (not just production) are encouraged to study the quality of the cars in the vehicle evaluation bay, Fig 11.

In most British auto manufacturers a system of tag relief is operated whereby relief men are built into the group numbers and each individual is relieved in succession without the lines being stopped. In addition to the problems this creates for the foreman in trying to control relief time, it also makes it much more difficult to determine which operator was working on a particular task at a particular time. Thus, if a quality problem arises, it is not impossible to blame someone else. In Japanese plants a system of block relief is used, meaning that the production lines are halted for the break period. In addition to reducing manning levels, and making the foreman's job easier, it also makes it possible to identify the source of poor workmanship should this occur. When combined with the attitude that such problems need correction

Fig 12: At the Nissan plant a system of 'help-lamps' attach to a wire which runs along the assembly track. Operators can summon help with any quality problem which occurs

rather than punishment it is easy to determine the advantages of the Japanese system over the British.

Quality circles Without doubt, however, Western interest in the Japanese approach to quality is mainly centred in Quality Control Circles: in simple terms the thinking goes something like 'Our quality is poor, Japanese quality is good. The Japanese have Quality Control Circles — if we introduce Quality Control Circles our quality will improve'. Nothing could be further from the truth. QCCs work in Japan because they are part of the total philosophy which puts quality first and to which everyone is committed. When they are successful they emerge almost naturally as an extension of the normal way of doing the job. To believe that QCCs can simply be introduced into the UK and be successful, without other fundamental changes in attitude, is a delusion .

In Japan, QC activities begin in the production areas, Fig 12, particularly involving the foremen — and it was this process which developed into the QCC activities. It took however until 1962 for the movement to reach take-off point with the first Quality Control Conference for Foremen and this was followed in 1963 by the first All Japan QC Circle Conference.

It is interesting to note, however that in the country which enthusiastically embraced the QCC concept, the purpose has always been much wider than simply quality. The Nissan document, which explains the history and development of QCC in the company, lists the basic idea behind QCC activities 'carried out as a part of company-wide quality control activities': contribute to the improvement and development of the enterprise; respect humanity and build a happy bright workshop which is meaningful to work in; display human capabilities fully and eventually draw out infinite possibilities.

Deep down, however, when analysing the process the author and his colleagues would use in Nissan, the company were convinced of three things. First the programme had to be about something more than quality; second that it should be a natural extension of the way a team normally operated and third, the company did not want to establish an external bureaucratic structure which might destroy what was hoped would be achieved. NMUK was, however, also committed to the policy of team working and the development of the individuals within the team. In the words of a paper prepared within NMUK, it was felt 'essential to introduce Quality Circles as a means of improving individual and team development and the participation of staff in the general day to day running of their working areas. The key to success in achieving this development is in ensuring that Quality Circle activity is fully integrated into the job and is not seen as a separate activity'. It was therefore regarded that QC activity was just one aspect of a programme aimed at achieving employee commitment to continuous improvement in all areas.

However, to give substance to the activity and to ensure that Nissan did make progress it was thought sensible also to establish a more formal framework — particularly helpful when considering the larger issues. Thus, in addition to the informal activities, part of the everyday job, the company also set up a Steering Committee, very importantly under the chairmanship of the Director of Production, comprising representatives of all staff, to plan the more formal side. Because of the view that problem solving and continuous improvement are part of everyone's job Nissan made the decision not to give any financial reward for achievement— to do so would be to make it special. But recognition of good work is important and thus the company left it to the Steering Committee to decide what form this should take. (In the same way Nissan does not have a Suggestion Scheme. Formalising the process makes it something special. When continuous improvement is part of the job you do not need a bureaucratic, time consuming procedure to generate ideas.)

The title chosen for this activity by Nissan is 'kaizen team'. Kaizen workshops are areas of the plant where manufacturing staff can go to make or improve tooling or fabricate aids to ease the manufacturing process. It recognises that for some issues a formal process to resolve problems may be necessary and that that process will use analytical techniques which will have to be taught. It also means the full involvement of supervisors — in Nissan they did not need convincing, for having experienced the process in Japan, they needed little encouragement.

Because kaizen is part of the team responsibility, it is important that it becomes an integral part of a supervisor's responsibility (in fact not only production supervisors for the company determined that this activity would take place in all areas of the company) and that all supervisors be fully trained in the problem-solving techniques commonly associated with QC activity, Fig 13. The supervisor, the genuine leader of the group, then naturally becomes the leader, at least initially, of problem-solving and continuous improvement activities — although, as experience develops, other team members will be expected to assume leadership of problem-solving groups. The supervisor is the leader of the group. It is his responsibility in his section to ensure continuous improvement, and to appoint someone else as the formal leader of a kaizen team is artificial and to a certain extent diminishes his authority.

Commitment to teamwork Another specific reference in the Nissan philosophy statement is 'We recognise that all staff have a valued contribution to make as individuals but in addition believe that this contribution can be most effective within a team working environment...our aim is to build a company with which people can identify and to which they feel commitment'. In simple terms therefore teamworking means having all employees committed to the aims and objectives of the company — recognising that each individual has a valued contribution to make but that should be aimed for everyone working in the same direction.

Behavioural theorists have listed a number of factors important indeveloping commitment — which Nissan management found to be of considerable value: employees should understand the link between effort and performance; employees should have the competence and confidence to translate effort into performance; control systems should be introduced only when necessary; performance re-

Fig 13: There is a daily static evaluation at Nissan of a sample of production body paintwork and other items

quirements should be expressed in terms of hard but attainable goals; employees should participate in setting these goals, feedback should be regular, informative and easy to interpret; employees should be praised for good performance; rewards should be seen to be equitable; rewards should be tailored to individual requirements and preference; employees' psychological and physical well-being should be seen as important; jobs should be designed to maximise skill variety, task identity, task significance, autonomy, feedback and provide opportunities for learning and growth; finally, organisation and job changes should be brought about through consultation and discussion.

Nissan managers developed a philosophy statement aimed at building an effective company in which all are working towards the same aims and objective. Within the overall framework of wishing 'to achieve mutual trust and cooperation between all people within the company', the philosophy statement includes: 'we will develop and expand the contribution of all staff by strongly emphasising training and the expansion of everyone's capabilities; we seek to delegate and involve staff in discussion and decision making, particularly in those areas in which they can effectively contribute so that all may participate in the efficient running of NMUK; within the bounds of commercial confidentiality we would like everyone to know what is happening in our company, how we are performing and what we plan; we want information and views to flow freely upward, downward and across our company."

In Nissan if there is one aspect to be singled out as important in team building and commitment it is the five minute meeting at the start of the day. This is something the company have learned directly from Japan, it is successful in Nissan's plant in Smyrna, Tennessee and its importance in NMUK cannot be overestimated. In Japan, employees are prepared for work at their place of work at the start of the working day. But the place of work is not their position on the production line; it is at their meeting place, usually the rest area. It is here, in Japan, that the famous exercise period takes place — usually two minutes. To anyone who has pushed his or her way through the Tokyo rush hour or is about to start an eight hour shift on the production line, this exercise period can add little, physically, to the well-being of the individual. However, to the Japanese foremen the important point is that his team is doing something together. Even more important however is the fact that the exercise period is followed by a further few minutes in which the team talks together. Every day the foreman talks with his people. This talk is frequently about quality but it will also be used to discuss schedule changes, work redistribution or process changes. In Nissan the morning discussion takes place in the team's meeting area. It is typical in British industry to have what virtually amounts to a 'no-go' area — where foremen fear to tread. In other companies there is no such rest area at all. Whichever way round, it is rare for the foreman to be able to socialise with his workers. Often relief breaks will be on a staggered basis so that, whatever the facilities, the group as a whole can never get together. The Nissan solution has been to construct large meeting areas (about 20 feet by 15 feet), one for each supervisor and his team. These are spacious well-lit places where the supervisor has his desk, where there are lockers, tables and benches, blackboards, notice boards, recreational facilities and most importantly a boiler to make tea and coffee. In some the team has brought in a refrigerator, there are often photographs of the team members, others have comprehensive notice boards covering everything from press cuttings to spaces for team members to write up ideas for improving the job. Block relief is taken with, often, someone stopping a few minutes early to make the tea, thus on several occasions each day the supervisor and his team are able to be together as a group.

Eastern approach to quality

Opinion is divided as to whether quality is a cultural phenomenon, arising from deep-seated attitudes in different world civilisations or a temporal one due to different levels of industrial performance at different periods within a nation's economic cycle. A more radical analysis requires stepping outside the established literature of engineering science. A good approach to this has been made by Pirsig[14] who shows that quality can suffer from the dualistic lines of thought people take towards technology and uses the motorcycle to demonstrate how a machine can be described in many different ways.

The 'classic' and 'romantic' divide

This author explains how a motorcycle may be divided, for purposes of classic rational analysis, by means of its component assemblies and by means of its functions. If divided by means of its component assemblies, its most basic division is into a power assembly and a running assembly. The power assembly may be divided into the engine and the power delivery system. The engine consists of a housing containing 'powertrain', fuel-air system, ignition system, a feedback system and lubrication system. 'The powertrain' consists of cylinders, pistons, connecting rods, a crankshaft and a flywheel... and so on... the author suggests. 'That's a motorcycle divided according to its components. To know what the components are for, a division according to functions is necessary: a motorcycle may be divided into normal running functions and special, operator-control led functions. Normal running functions may be divided into functions during the intake cycle, functions during the compression cycle, functions during the power cycle and functions during the exhaust cycle... and so on. This description would cover the 'what' of the motorcycle in terms of components, and the 'how' of the engine in terms of functions. It would badly need a 'where' analysis in the form of an illustration, and also a 'why' analysis in the form of engineering principles that led to this particular conformation of parts.

The first thing to be observed about this description, he suggests, is so obvious you have to hold it down or it will drown out every other observation. This is that 'It is just duller than ditch water'... but some other things can be noticed that do not at first appear. The first is that the motorcycle, so described, is almost impossible to understand unless you already know how one works. The immediate surface impressions that are essential for primary understanding are gone. Only the underlying form is left. The second is that the observer is missing. The description doesn't say that to see the piston you must remove the cylinder head. 'You' aren't anywhere in the picture. Even the 'operator' is a kind of personality-less robot whose performance of a function on the machine is completely mechanical. There are no real subjects in this description. Only objects exist that are independent of any observer... it is argued. The third is that words 'good' and 'bad' and all their synonyms are completely absent. No value judgements have been expressed anywhere, only facts. The fourth is that there is a 'knife' moving here. A very deadly one; an intellectual scalpel so swift and so sharp you sometimes don't see it moving. You get the illusion that all those parts are just there and are being named as they exist. But they can be named quite differently and organized quite differently depending on how the knife moves. For example, the feedback mechanism which includes the camshaft and cam chain and tappets and distributor exists only because of an unusual cut of this analytic knife. If you were to go to a motorcycle-parts department and ask them for a feedback assembly they wouldn't know what the hell you were talking about. It is important to see the knife for what it is and not to be fooled into thinking that motorcycles or anything else are the way they are just because the knife happened to cut it up that way. It is important to concentrate on the knife itself.' In the book the author shows how an ability to use the knife creatively and effectively can result in solutions to the classic and romantic split.

The divide of logic

Two kinds of logic are used, inductive and deductive. Inductive inferences start with observations of the machine and arrive at general conclusions. 'For example, if the cycle goes over a bump and the engine misfires, and then goes over another bump and the engine misfires, and then goes over another bump and the engine misfires, and then goes over a long smooth stretch of road and there is no misfiring, and then goes over a fourth bump and the engine misfires again, one can logically conclude that the misfiring is caused by the bumps. That is induction: reasoning from particular experiences to general truths.

Deductive inferences do the reverse. They start with general knowledge and predict a specific observation. For example, if, from reading the hierarchy of facts about the machine, the mechanic knows the horn of the cycle is powered exclusively by electricity from the battery, then he can logically infer that if the battery is dead the horn will not work. That is deduction. Solution of problems too complicated for common sense to solve is achieved by long strings of mixed inductive and deductive inferences that weave back and forth between the observed machine and the mental hierarchy of the machine found in the manuals. The correct program for this interweaving is formalized as scientific method, which is broken down into six categories: (1) statement of the problem, (2) hypotheses as to the cause of the problem, (3) experiments designed to test each hypothesis, (4) predicted results of the experiments, (5) observed results of the experiments and (6) conclusions from the results of the experiments," the author argues.

Problems of 'how-to-do-it' handbooks

A creative artist may not be able to comprehend, or may be 'put-off' an engineering manual because of the lack of smoothness and continuity. 'He's unable to comprehend things when they appear in the ugly, chopped-up, grotesque sentence style common to engineering and technical writing. Science works with chunks and bits and pieces of things with the continuity presumed, and the artist works only with the continuities of things with the chunk and bits and pieces presumed. What he really wants is artistic continuity, something an engineer couldn't care less about. It hangs up, really, on the classic-romantic split, like everything else about technology.

Technology presumes there's just one right way to do things and there never is. And when you presume there's just one right way to do things, of course the instructions begin and end exclusively with the machine. But if you have to choose among an infinite number of ways to put it together then the relation of the machine to you, and the relation of the machine and you to the rest of the world, has to be considered, because the selection from among many choices, the art of the work is just as dependent upon your own mind and spirit as it is upon the material of the machine. That's why you need the peace of mind.'

'Stuckness': good and bad Pirsig asserts that rationality can be tremendously improved, expanded and made far more effective through the formal recognition of Quality in its operation. Before doing this, however, the author goes over some of the negative aspects of traditional maintenance to show just where the problems are. "The first is ... a mental *stuckness* that accompanies the physical stuckness of whatever it is you're working on: a screw sticks, for example, on a side cover assembly. You check the manual to see if there might be any special cause for this screw to come off ... hard, but all it says is 'remove side cover plate' in that wonderful terse technical style that never tells you what you want to know. There's no earlier procedure left undone that might (*otherwise*) cause the cover screws to stick.

If you are experienced you'd probably apply a penetrating liquid and an impact driver at this point. But suppose you're inexperienced and you attach a self-locking plier wrench to the shank of your screwdriver and really twist it hard, a procedure you've had success with in the past, but which this time succeeds only in tearing the slot of the screw. The book's no good to you now. Neither is scientific reason. You don't need any scientific experiments to find out what's wrong. It's obvious what's wrong. What you need is an hypothesis for how you're going to get the slotless screw out of there and scientific method doesn't provide any of these hypotheses.

The right facts, the ones we really need, are not only passive, they are damned elusive, and we're not going to just sit back and 'observe' them. We're going to have to be in there looking for them or we're going to be here a long time.

The difference between a good mechanic and a bad one, like the difference between a good mathematician and a bad one, is precisely this ability to select the good facts from the bad ones on the basis of quality. He has to *care*! This is an ability about which formal traditional scientific method has nothing to say. It's long past time to take a closer look at this qualitative preselection of facts which has seemed so scrupulously ignored by those who make so much of these facts after they are 'observed.' I think that it will be found that a formal acknowledgement of the role of Quality in the scientific process doesn't destroy the empirical vision at all. It expands it, strengthens it and brings it far closer to actual scientific practice. I think the basic fault that underlies the problem of stuckness is traditional rationality's insistence upon 'objectivity,' a doctrine that there is a divided reality of subject and object. For true science to take place these must be rigidly separate from each other. 'You are the mechanic. There is the motorcycle. You are forever apart from one another. You do this to it. You do that to it. These will be the results.'

This eternally dualistic subject/object way of approaching the motorcycle sounds right to us because we're used to it. But it's not right. It's always been an artificial interpretation superimposed on reality. It's never been reality itself. When this duality is completely accepted a certain non-divided relationship between the mechanic and motorcycle, a craftsmanlike feeling for the work, is destroyed. When traditional rationality divides the world into subjects and objects it shuts out Quality, and when you're really stuck it's Quality, not any subjects or objects, that tells you where you ought to go.

By returning our attention to Quality it is hoped that we can get technological work out of the non-caring subject-object dualism and back into craftsmanlike-self involved reality again, which will reveal to us the facts we need when we are stuck. Romantic reality is the cutting edge of experience. It's the leading edge of the train of knowledge that keeps the whole train on the track. Traditional knowledge is only the collective memory of where that leading edge has been. At the leading edge there are no subjects, no objects, only the track of Quality ahead. The good mechanic doesn't make this separation. One says of him that he is 'interested' in what he's doing, that he's 'involved' in his work. What produces this involvement is, at the cutting edge of consciousness, an absence of any sense of separateness of subject and object.'

The need for 'gumption'

The author insists that 'if you're going to repair a motorcycle, an adequate supply of gumption is the first and most important tool. If you haven't got that you might as well gather up all the other tools and put them away, because they won't do you any good. Since it's a result of the perception of Quality, a gumption trap, consequently, can be defined as anything that causes one to lose sight of Quality, and thus lose one's enthusiasm for what one is doing. There are two main types of gumption traps. The first type are those in which you're thrown off the Quality track by conditions that arise from external circumstances, and I can't see these 'setbacks.' The second type are traps in which you're thrown off the Quality track by conditions that are primarily within yourself; examples are intermittent faults. When intermittents recur, try to correlate them with other things the cycle is doing. Do the misfires, for example, occur only on bumps, only on turns, only on acceleration? Only on hot days?" These correlations are clues for cause-and-effect hypotheses, the author maintains.

Of the value traps, the most widespread and pernicious is value rigidity. This is an inability to revalue what one sees because of commitment to previous values. 'In motorcycle maintenance, you must rediscover what you do as you go. Rigid values makes this impossible. If you have a high evaluation of yourself then your ability to recognize new facts is weakened. Your ego isolates you from the Quality reality. When the facts show that you've just goofed, you're not as likely to admit it. When false information makes you look good, you're likely to believe it.

Anxiety, the next gumption trap, is sort of the opposite of ego. You're so sure you'll do everything wrong you're afraid to do anything at all. Often this, rather than 'laziness,' is the real reason you find it hard to get started. This gumption trap of anxiety, which results from over-motivation, can lead to all kinds of errors of excessive fussiness. Boredom is the next gumption trap that comes to mind. This is the opposite of anxiety and commonly goes with ego problems. Boredom means you're off the Quality track, you're not seeing things freshly, you've lost your 'beginner's mind'. Impatience is close to boredom but always results from one cause: an underestimation of the amount of time the job will take. You never really know what will come up and very few jobs get done as quickly as planned. Impatience is the first reaction against a setback and can soon turn to anger if you're not careful, Persig suggests.

Truth traps are concerned with data that are apprehended and handled by conventional dualistic logic and the scientific method. But, he points out, 'there's one trap that isn't — the truth trap of yes-no logic. Yes, and no ... this or that ... one or zero. On the basis of this elementary two-term discrimination, all human knowledge is built up. The demonstration of this is the computer memory which stores all its knowledge in the form of binary information. It contains ones and zeros, thats all. Because we're unaccustomed to it, we don't usually see that there's a third possible logical term equal to yes and no which is capable of expanding our understanding in an unrecognized direction. We don't even have a term for it, so I'll have to use the Japanese mu.

Mu means 'nothing'. Like 'Quality' it points outside the process of dualistic discrimination. Mu simply says, 'No class: not one, not zero, not yes, not no.' It states that the context of the question is such that aye or no answer is in error and should not be given. 'Unask the question' is what it says. One is told over and over again that computer circuits exhibit only two states, a voltage for 'one' and a voltage for 'zero.' That's silly! Any computer-electronics technician knows otherwise. Try to find a voltage representing one or zero when the power is off! The circuits are in a mu state. They aren't at one, they aren't at zero, they're in an indeterminate state that has no meaning in terms of ones or zeros. Readings of the voltmeter will show, in many cases, 'floating ground' characteristics, in which the technician isn't reading characteristics of the computer circuits at all but characteristics of the voltmeter itself. What's happened is that the power-off condition is part of a context larger than the context in which the one-zero states are considered universal . The question of one or zero has been 'unmasked'.

Or if he takes whatever dull job he's stuck with — and they are all, sooner or later, dull — and, just to keep himself amused, starts to look for options of Quality, and secretly pursues these options, just one for their own sake, thus making an art out of what he is doing, he's likely to discover that he becomes a much more interesting person and much less of an object to the people around him because his Quality decisions change him too. And not only the job and him, but others too because the Quality tends to fan out like waves.' This philosophy of Pirsig's could truly alter our approach to achieving quality in manufacture, as well as in maintenance.

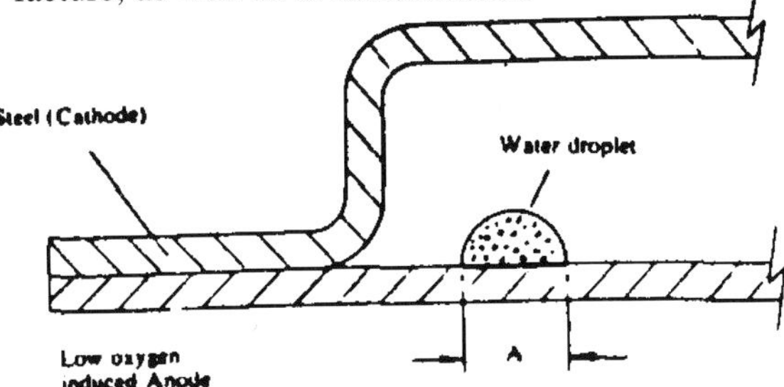

Fig 14: Cathodic corrosion

Body finishing techniques for corrosion prevention

Perceived quality lies primarily with the customer's satisfaction that vehicle bodywork withstands the rigours of the operational environment. An understanding of the basic mechanisms of corrosion is useful for the selection of the appropriate treatment. Not only steel, also aluminium alloys and even stainless steel need appropriate protection. Careful attention to joint design and 'ventilation' of box members reap valuable rewards in subsequent service of the vehicle.

Corrosion is generally understood to be a nature-cycle phenomenon of a metal slowly reverting to the state in which it occurs as a natural ore. The corrosion process is electro-chemical — as in a battery — with the dilute acid arising from road salt and atmospheric carbon/sulphur dioxides dissolved in water. As in a battery, the anode dissolves and the cathode collects positive ions from the electrolyte (acid). A corrosion 'cell' of this sort can be within a single drop of impure water (dilute acid) on the surface of the vehicle body. The area beneath the drop is starved of oxygen and becomes anodic while the surrounding area becomes a cathode, Fig 14.

Serious body corrosion is generally accepted to start from the inside on unprotected areas of metal and then work outwards to the paint film, which blisters. Two techniques used to protect the interior surfaces are 'wax' spraying and sacrificial protection. The latter involves using a coating such as zinc (more negative in the electro-chemical series than iron), on the surface of the steel, thus promoting cathodic action and protecting the steel by sacrificing the zinc.

Spraying with petroleum- or waxed-based inhibitors is a valuable technique — especially when

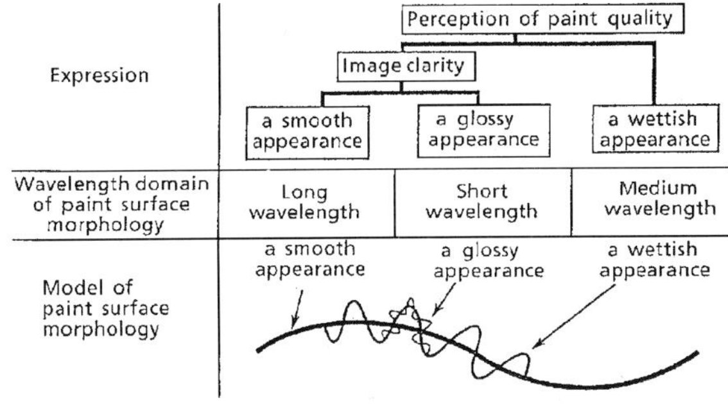

Fig 17: Elements forming perception of quality

bodywork has also received an electrocoat primer. The last process involves making the body an anode of a dc electric circuit and the paint tank the cathode. The specially formulated pigment blend is electro-conductive and tends to deposit evenly on all exposed surfaces.

Besides low-oxygen corrosion, the problem of crevice corrosion is prominent in vehicles with a multiplicity of spot welded flanges. Paint molecules can be simply too large to enter joint crevices and subsequent 'working' of the mating surfaces in service can promote the sucking in of moisture by capillary action. Another means of attack is by pitting corrosion caused by rupture of the paint film by stone chipping. Lastly, galvanic corrosion caused by metal-to-metal contact of metals wide apart in the electrochemical series, is another common enemy.

While aluminium alloys have good resistance to corrosive environments, due to the protective surface oxide film, this can be breached by acids formed from de-icing salts and subsequent pitting corrosion occurs. The alloys are also anodic to most metals and therefore good protection is required against galvanic corrosion. Stainless steels offer even better protection — again due to formation of an oxide film which itself requires some protection against local rupture.

Joint design for corrosion prevention Avoidance of both crevice and galvanic corrosion is necessary in joint design. Crevices should either be eliminated or shielded and insulation should be provided to separate dissimilar metals, with preferably the surface area of the cathodic metal being less than that of the anodic one. Fig 15 gives some 'avoid' and 'adopt' situations while Fig 16 gives pictorial advice on paint access and drainage conditions.

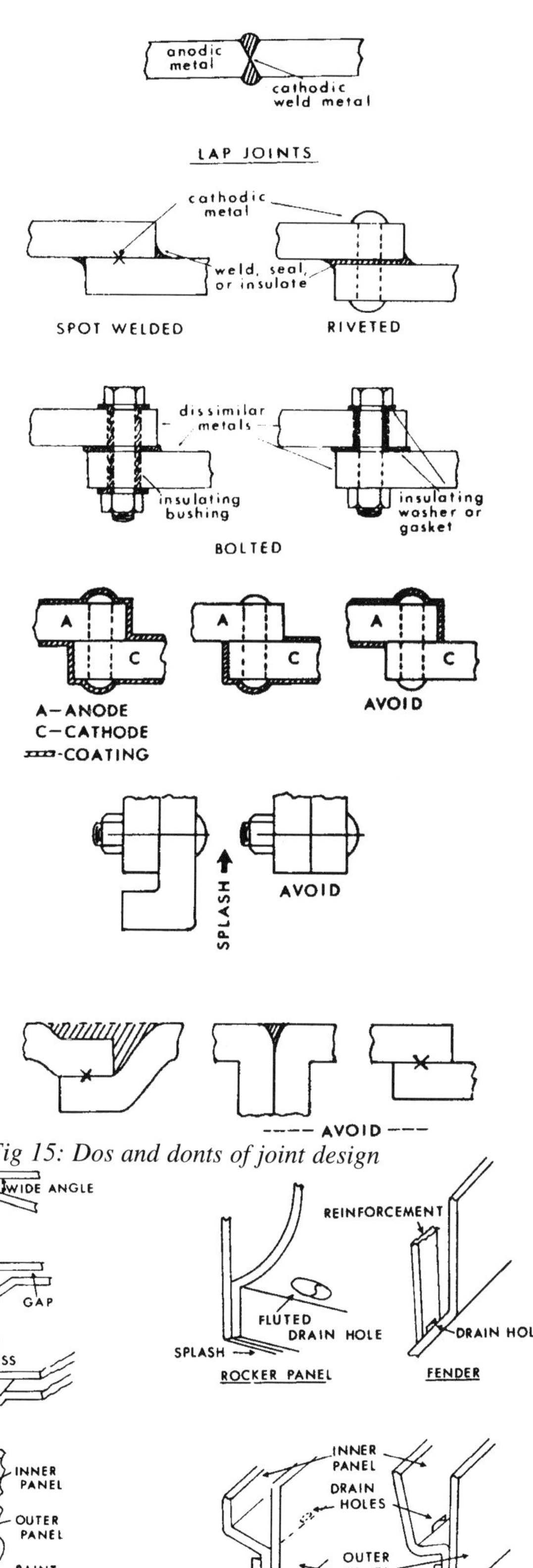

Fig 15: Dos and donts of joint design

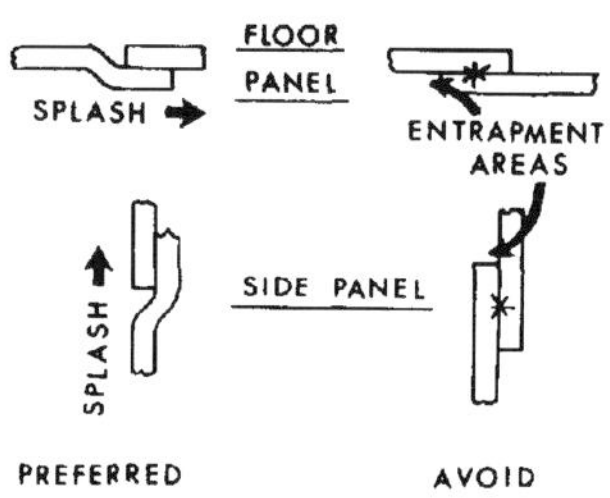

Fig 16: Draining paint from closed sections

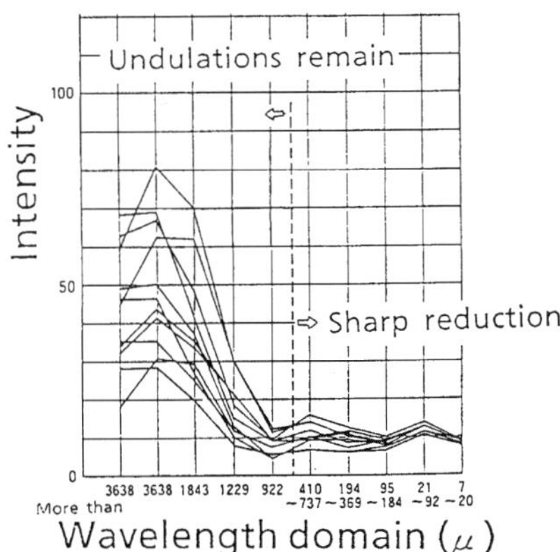

Fig 18: Frequency analysis of finish-painted steel sheet

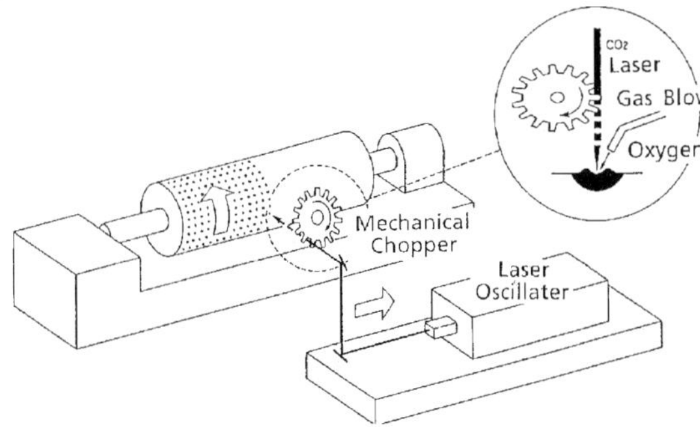

Fig 19: Roll-engraving by laser beam

Specimens	YP (N/mm^2)	TS (N/mm^2)	El (%)	n	r	SRa (μm)
Low Type of Laser-Texture						0.72
High Type of Laser-Texture	171	298	47	0.25	1.80	2.36
Shot-blast						1.17

Fig 20: Sheet specimens tested

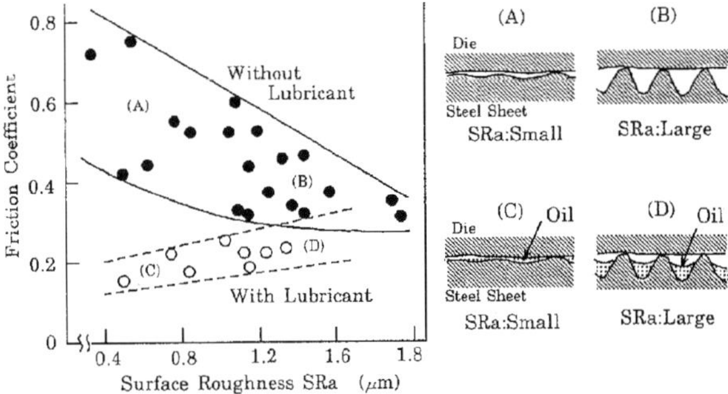

Fig 21: Friction and roughness relationship

Surface treatment to improve paint finish Work reported by Nissan[15] has shown the manufacturer to be exploiting laser-textured dull-steel in improving paint finish on its vehicles. Working with Kawasaki Steel, the company has overcome the problem of shot-blasted dull-steel, used to obtain suitable surface roughness for improving formability, by rolling the steel strip between laser-processed rollers.

The criteria shown in Fig 17 were used for assessing painted surfaces, the company having noted that glossiness and smoothness accounted for 70% of the subjective judgement of paint appearance. A fine dull surface finish is traditionally recognised as providing better image clarity but too fine a surface gives rise to galling and handling problems during forming. Frequency analysis of several samples of automotive sheet steel showed that sheet with surface undulations component were retained in the paint film surface as seen by Fig 18.

When conventional sand-blasted rollers are used, these undulations were difficult to eliminate because of the random pattern of craters produced in the roller surface by the process resulting in wavelengths over 800 microns. By using a laser beam to engrave the surface, using the process shown in Fig 19, a regular pattern of craters can be obtained, which resolves the problem by producing only short wavelength undulations.

To test the effect on formability the company used the specimens shown in the table of Fig 20 where *SRa* is the arithmetic mean deviation of the surface profile accounting for the area, Fig 21 showing its relationship to friction co-efficient. Trials with and without press lubricant showed that without it, galling is less likely when the area of contact between the die and the steel sheet is smaller; with lubricant, the latter is more effective when the peaks and valleys of the sheet surface are smaller. Fig 22 showed the results of deep-drawing tests in which the low-roughness type of laser-textured steel showed best formability, attributable to its lower friction. The practical case of an inner wheelbox panel, Fig 23, was also used to examine both formability and galling. Similar conclusions were reached. The effect of surface roughness on fracture behaviour was also examined on the basis of the reduction in panel thickness of the side wall corners where risk was greatest. The results in Fig 24a indicated that high-roughness type of laser-textured steel sheet allowed a large decrease in panel thickness even at a low blank-holding force, suggesting that this steel would be susceptible to fracture.

Galling tests were based on stamping 100 consecutive panels using a medium roughness laser-textured sheet and two other types, having rejected the high roughness one on grounds of fracture proneness. All three samples passed successfully. The effect of varying the amount of metal flow was also investigated, the results being shown in Fig 24b. Here the vertical axis indicates the length of the flange end *L1* along the long side of the panel. This length increased with increased number of stamping, making metal flow more difficult. Results indicated that steel with a higher *SRa* value are somewhat superior in respect of anti-galling performance.

Overall, the summary in Fig 25 showed that to improve paint image clarity a certain degree of surface roughness and crater density is required, better formability being obtained in the region of high crater density and low surface roughness. Anti-galling property is improved in the region of relatively rougher surface finish.

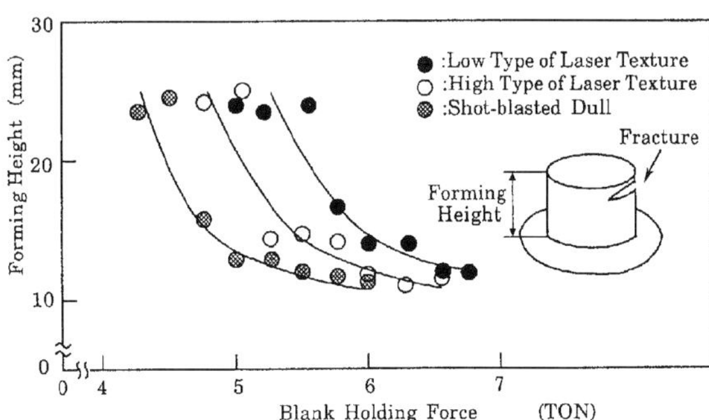

Fig 22: Deep-drawing test results

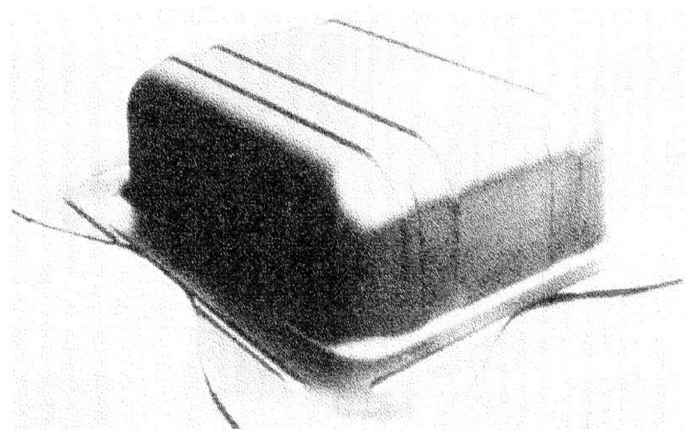

Fig 23: Inner wheelbox panel

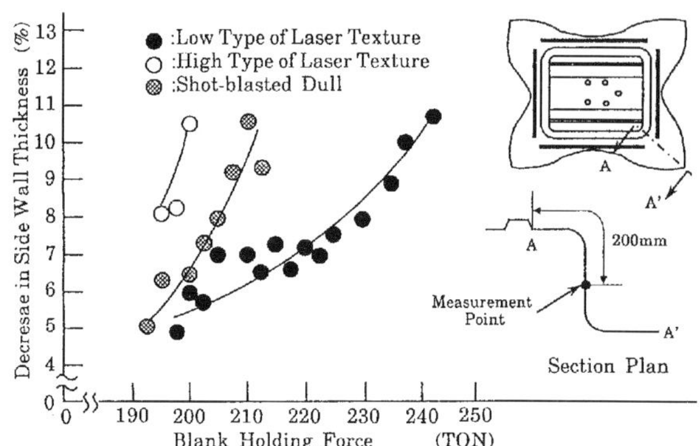

Fig 24a: Reduction in sidewall thickness

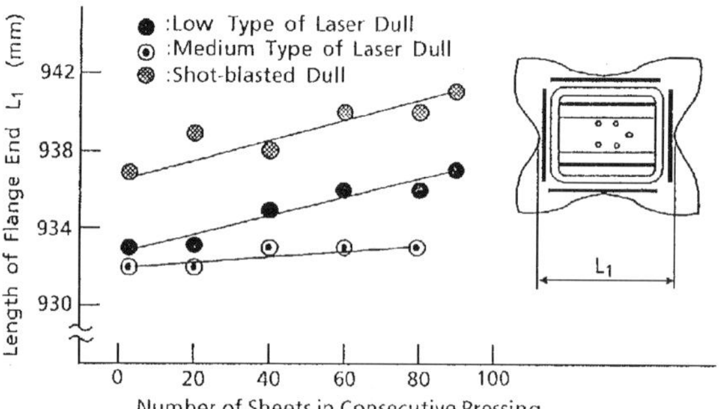

Fig 24b: Metal flow change in successive stamping

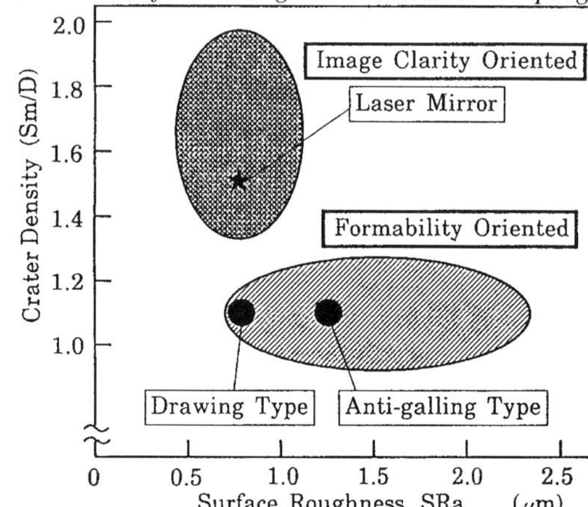

Fig 25: Application of laser-textured steel

Finish-painting techniques

Painting is an area in which the volume vehicle builder is likely to lead the specialist builder in new techniques and processes. Here, some recent developments could apply to both 'sides' of the industry and apply respectively to metal and plastic/GRP substrates. In recent years, emphasis has been on making metal bodies resist rusting out rather than making top coats more durable. This has led to phosphating plants, dipping tanks and electropaint.

Basic components of a paint system pigment, binder and solvent are worth considering separately for a proper understanding of comparative systems. Pigment provides both colour and opacity (hiding or covering power) but can also contribute to durability of both finish and priming coats. In the priming coat the pigment's main task is to resist corrosion and promote adhesion. Binders are the film formers which also contribute to the substrate bond as well as binding the pigment together. The solvent makes the mixture of pigment and binder thinner for ease of application and evaporates by 'air flash' at room temperatures or by heating up to 80 C panel temperatures for quick dry.

Broadly, paint types divide between cellulose synthetics, oil-based, low-bakes, two-pack materials and thermosetting acrylic enamels. Celluloses require multiple coats because of the low thickness of each film — also solvent retention can cause softness for some time after application. Oil and synthetic based paints rely on final film hardening (arising from oxidation of the paint) and re-coat periods can be critical for subsequent rectification procedures; however their high solid content gives them good covering power. Low-bakes involve a full cure of up to 60 minutes followed by a cooling down period during which oxidation makes the materials non-thermoplastic. The 'twin packs' require the paint, and a separate hardener (usually a toxic isocyanate), to mix together for the cure. An unusually hard film results which can be force-dried to allow the vehicle to enter service quickly.

The most common system in use by the volume builders is thermosetting acrylic enamel based on the chemistry of methylol acrylamide and hydroxylated acrylic monomers. The latter vary in chemistry according to their four main contributions of either providing: hardness, flexibility, cross-linkages or accelerated cure. The systems are suitable for the formulation of metallic finishes — due to the high order of gloss, polishing and weathering qualities. Melamine resins are now commonly used to accelerate the cure down to 30 minutes at a stoving temperature of 120-130 C. A key advantage of the system is the ability to reflow during 'repair in process' rectification procedures on the assembly lines of the major OEMs. Two such procedures are bake-sand-bake and bake-and-respray, both involving initial

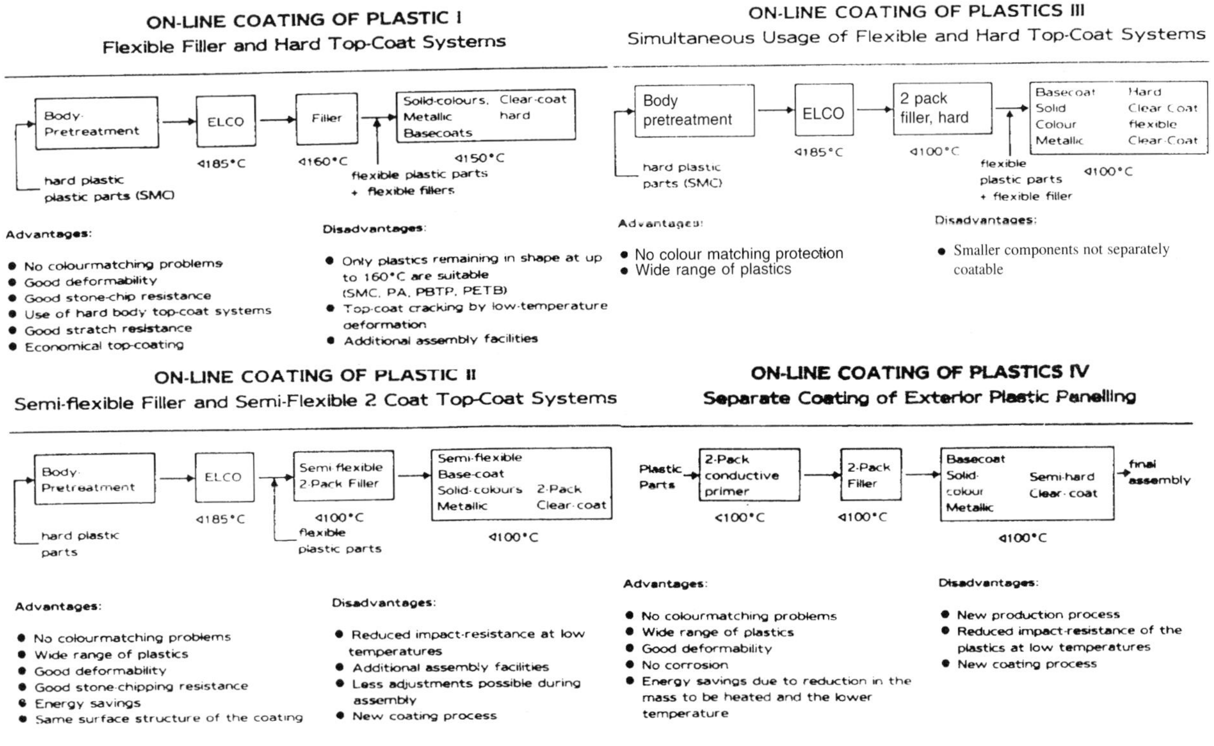

ON-LINE COATING OF PLASTIC I
Flexible Filler and Hard Top-Coat Systems

Body Pretreatment → ELCO → Filler → Solid-colours, Metallic Basecoats / Clear-coat hard

hard plastic plastic parts (SMC) ◁185°C ◁160°C ◁150°C flexible plastic parts + flexible fillers

Advantages:

- No colourmatching problems
- Good deformability
- Good stone-chip resistance
- Use of hard body top-coat systems
- Good stratch resistance
- Economical top-coating

Disadvantages:

- Only plastics remaining in shape at up to 160°C are suitable (SMC, PA, PBTP, PETB)
- Top-coat cracking by low-temperature deformation
- Additional assembly facilities

ON-LINE COATING OF PLASTICS III
Simultaneous Usage of Flexible and Hard Top-Coat Systems

Body pretreatment → ELCO → 2 pack filler, hard → Basecoat Solid Colour Metallic / Hard Clear Coat flexible Clear Coat

hard plastic parts (SMC) ◁185°C ◁100°C flexible plastic parts + flexible filler ◁100°C

Advantages:

- No colour matching protection
- Wide range of plastics

Disadvantages:

- Smaller components not separately coatable

ON-LINE COATING OF PLASTIC II
Semi-flexible Filler and Semi-Flexible 2 Coat Top-Coat Systems

Body Pretreatment → ELCO → Semi flexible 2-Pack Filler → Semi-flexible Base-coat Solid-colours Metallic / 2-Pack Clear-coat

hard plastic parts ◁185°C ◁100°C flexible plastic parts ◁100°C

Advantages:

- No colourmatching problems
- Wide range of plastics
- Good deformability
- Good stone-chipping resistance
- Energy savings
- Same surface structure of the coating

Disadvantages:

- Reduced impact-resistance at low temperatures
- Additional assembly facilities
- Less adjustments possible during assembly
- New coating process

ON-LINE COATING OF PLASTICS IV
Separate Coating of Exterior Plastic Panelling

Plastic Parts → 2-Pack conductive primer → 2-Pack Filler → Basecoat Solid colour Metallic / Semi-hard Clear coat → final assembly

<100°C ◁100°C ◁100°C

Advantages:

- No colourmatching problems
- Wide range of plastics
- Good deformability
- No corrosion
- Energy savings due to reduction in the mass to be heated and the lower temperature

Disadvantages:

- New production process
- Reduced impact-resistance of the plastics at low temperatures
- New coating process

Fig 26: Azco plastic painting process

bake at 70-100 C during which solvents evaporate but the material remains thermoplastic though hard enough to be rubbed down. After tak-wiping, in the first procedure, restoving takes place at 140 C during which reflow hides the sanding marks — while in the second these are resprayed over without the need for masking. Ford spec ESB.M32J.102A released in 1967 is widely accepted as a standard for reflow performance. Hydroxylated acrylic finishes provide an ideal carrier for the aluminium flakes used in metallic finishes and are superior to the alternative alkyds. Key to good reflow performance with metallics is that flakes are evenly dispersed initially.

Powder painting systems The first materials used in powder coatings were the epoxy resins, applied at high air pressures to achieve film thicknesses of 100-125 microns after 30 minutes of stoving at 200 C. Now a wide range of materials can be sprayed with film thicknesses down to 35 microns, involving a 10 minute cure at 160 C. Polyesters are now the common coatings and usually entail only a single coat system. Generally, mechanical properties are superior to wet painted coatings and a variety of machining and forming processes can be applied to the finish panel. Resin, pigment, hardener and additives are pre-mixed as solids and fed to an extruder, the barrel of which is heated to a temperature at which the mix melts. Cooled extrudate then passes through a kibbler (crushing machine) from which pellets are subsequently ground to a powder prior to spraying. Total cleanliness of the substrate is necessary, including the removal of oxide films in the case of aluminium alloys.

Future painting developments?

There is obviously considerable impetus to reduce the number of stages in any painting process for vehicle bodies. Increasing the thickness of electrocoat primer films from the current 17 to around 35 microns is one way of reducing the number of subsequent coats. A single coat E-coat primer is already in view — for application after phosphate pretreatment. Next stage will be to eliminate primer-surfacer films. Other possible developments include a twin-coat metallic finish involving flashing-off between base and clear coat as well as paints which are sympathetic to both metallic and plastic substrates.

The Netherlands company Akzo Coatings have been prominent in the development of paints for plastic substrates. Problems to be overcome in painting of plastic include surface etching by solvents, mechanical strength reduction, reduced low-temperature flexibility, differential thermal expansion and blister formation. Successful twin-pack flexible PUR coatings are available for plastic panels of nylon, polycarbonate, PUR-RIM, PBTP and polypropylene; there are also special systems for GRP and ABS. Akzo point out that flexible paints can never quite rival the finish of hard coatings used on metal substrates.

For a part like the ZMC tailgate of the Citroen BX car, a post-coating of 2-pack epoxy based conductive primer is used which is dried for 20 minutes at 140 C. The part is subsequently coated with a conventional filler and top-coat then baked at 140 C. Next stage in development will be an in-mould coating which is targeted to harden within 60 seconds. Fig 26 shows the process procedures under development by the company. A variation of the fourth is used for the BMW Z-1 Roadster-special. This involved the so-called Varioflex system which suited the different types of plastic used on different panels of the body.

Pre-coated metal sheet subsequently formed into body panels is a further line of development. Another is the use of paint-film 'prepregs' in lieu of sprayed paint. According to Bruce Johnson of 3M's Automotive Systems Laboratory, the system shown diagrammatically in Fig 27 has been produced with a conformable paint coating, reinforced with a flex-

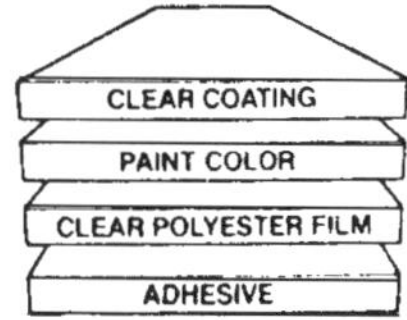

Fig 27: Paint film prepreg

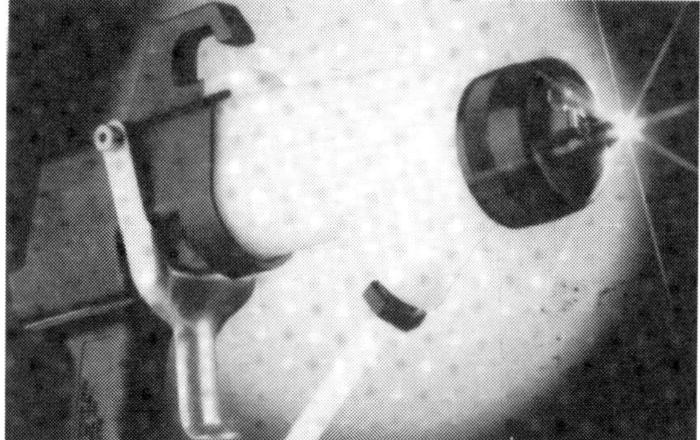

Fig 28: Electrostatic spray gun

ible layer and bonded with a hot-melt adhesive. Tests carried out by the company have shown that such dry films actually perform better than sprayed paint for certain attributes such as resistance to impact and road chippings. Application methods for dry films include vacuum thermoforming, for porous parts, and compression moulding for other 'solid' parts. For E-coated steel panels, bonding is carried out prior to forming — with the carrier film removed after the operation.

The DeVilbiss Company are pioneering the use of electrostatic paint application systems in the medium volume automotive industry and recently introduced the Aerobell rotary atomiser which operates on air-bearings at 60,000 rpm. The company have also introduced a manual spray gun with the benefit of electrostatic operation, Fig 28. The EFX-100 receives a low-voltage input which is converted to 20-80 kV for maximum transfer efficiency. The guns will apply solvent-based coatings using either air atomisation or air-assisted 'airless' systems and water-borne coatings using air atomisation.

Becker Industrial Coatings have introduced a new application process for polyurethane paint known as Vapour Injection Curing. It is suitable for coating three dimensional objects in either steel, plastic, aluminium or wood substrate. The atomising air carries a tertiary amine gas catalyst which mixes with the paint on release from the gun-head — to give accelerated curing speeds, Fig 29. A new family of PUR coatings has also been developed for the process including primers, pigmented and base/clearcoat finishes.

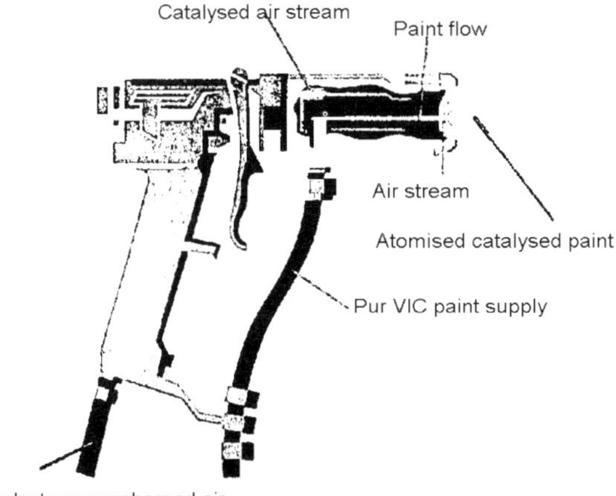

Fig 29: Vapour injection curing gun

Body sealants/protection coats

The demarcation between gap-filling adhesives and sealants proper is blurred. Compared to adhesives, sealants have less cohesive strength but unlike adhesives may be required to have such properties as non-sagging and ability to adhere to oiled substrates. The need to match formulation to specific duty is considered here.

In volume car and CV manufacture many sealants are putty mastics which can be pumped to different parts of the production lines for application by extrusion-guns, either manually or robot-operated. Lower volume specialist vehicles can benefit from particular sealants such as those in uncured tape form which can be cured either by heat or chemical activators.

Most sealants are, of course, of mastic consistency and are based on filled or unfilled elastomers — with thixotropic agents added to prevent sagging. The elastomeric binder is an important constituent as it largely determines mechanical and physical properties. Pigments are added for colourising, obviously, but their presence also increases density to reduce flow tendency. Conversely, solvents may be added to achieve a more pourable consistency.

Two basic groups of elastomer-sealants are the high and low molecular-weight categories. The low-weight materials are generally one-part gun-sealants. The higher weight materials are used to compound ribbons and tapes or are dispensed with two-part guns in which the elastomer changes after mixing from a syrup to a rubber-like solid.

The other major categorisation is between interior or exterior vehicle application. The latter, of course, often require compatibility with paint finishes while the former may be used for spot-welded joints on the floor pan and require special expansion characteristics to obtain a water-proofing function. Such expandable systems can be applied to the joint prior to welding, the effect of the weld-through being to expand the sealant material.

For windscreens, different systems are used for the bonded and non-bonded systems, necessarily; with the latter, the screen is mounted in a moulded weatherstrip and the sealant is required to provide a watertight joint, reduce glass movement and reform ruptured bonds when movement takes place. Polyisobutylene systems are used for this purpose whereas the bonded-in type employs a fast heat-

Fig 30 (right and opposite): Sealant applications and relative costs

Chemical Type	Base	Pretreatment	Remarks
Oleo-Resins	Drying and non-drying oils, resins and fillers.	Require primers to give adhesion to porous surfaces.	Best suited to lap joints, but can be tamped into butt joints. Available in strip form, or for knife or gun application
Rubber-Bitumen	Natural or synthetic rubber compounded with bitumen or pitch.	Movement accommodation of the order of 10%	Cold poured (set by solvent action) or hot poured.
Butyl	Butyl rubbers or degraded butyl rubbers, used alone or in combination with solvents, oils, extenders, etc.	Can be skin-forming, non-skinning, non-drying or curing (curing types cure by solvent evaporation).	Very wide range of compositions from soft, sticky compounds to hard rubber strips.
Polysulphide	Single or two-part mixtures based on polysulphide polymers. Two-part mixtures cure by chemical reaction fast.	Primers required on most porous and friable surfaces.	Wide range of compositions from soft elastic materials to hard rubbers. Available in a wide range of colours. Usual base of caulking compounds.
Silicone	Single-part chemically curing silicone elastomer.	Require primers on most porous and friable surfaces.	Remain fully elastic. Available in a range of colours
Polyurethane	Two-part self-curing mixtures based on polyurethane polymers.	Primers required on most surfaces for adequate adhesion.	Remain fully elastic.
Acrylic	Mixtures of acrylic polymers.		Plastic in nature.

curing polysulphide. An alternative to this is a pre-formed butyl self-adhesive tape applied to the glass before insertion. There are also systems using PUR sealants and an interesting alternative using a neoprene tape with embedded resistance wire. Exterior floor-pan seam sealants are required to be non-flow types which are impermeable and immune from cracking and blistering. For under-vehicle waterproofing, the sealant can be applied as a liquid and cured to an expanded solid in paint stoving ovens. For other exterior situations non-flow, anti-corrosive and self-levelling systems are available. Welded seams with small gap irregularities such as drip-rails can be sealed with vinyl plastisols prior to finish paint. Self-levelling ensures a smooth finish on which to apply the paint. The anti-corrosive systems are used on areas inaccessible for painting. *Tables in Fig 30 give sealants, applications and costs.*

Anticorrosion coatings

The interaction between coating systems and the basic mechanisms of corrosion are discussed here alongside precoated zinc and plastic systems as well as traditional pretreatments.While the best protection against corrosion lies in good body design , nevertheless there are difficulties inherent in such areas as interweld crevices (Fig 31) and the remoter regions of box section interiors which require supplementary coatings. Good protection is obtained by ventilation of the enclosed area, using wax injection and where possible specifying pre-coated metals.

Capillary penetration of wax into interweld crevices is crucial to overcome the lack of oxygen replenishment in these regions which promotes the development of corrosion cells. Electrocoat primers also may not reach such areas and there could be a tendency for road de-icing salts to remain in trapped liquid left behind after winter slush has dried by evaporation. Waxes are effective in repelling water which forms into globules on the surface of the wax.

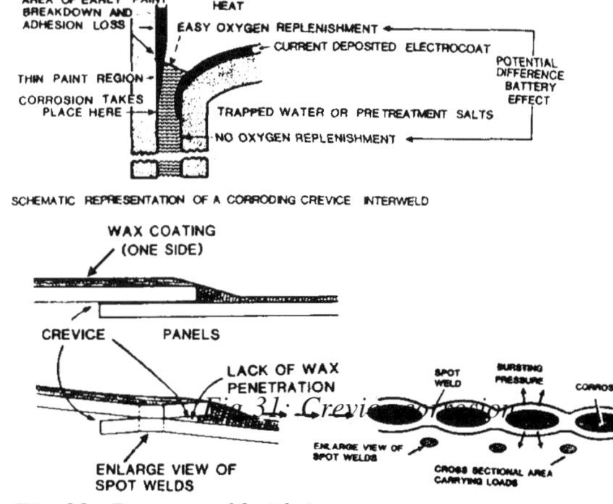

Fig 31: Crevi

Fig 32: Danger of bridging

	Material	Application(s)	Relative Cost
NON-HARDENING	Bituminous	Caulking, general purpose sealant	1 – 4
	Butyl	Metal-to-metal, metal-to-glass, caulking	5 – 10
	Acrylic	Caulking metalwork, fibre joints, gaskets	8 – 10
	Polybutane	General purpose construction caulking and sealers	3 – 7
HARDENING FLEXIBLE	Neoprene (solvent-release)	General purpose sealant; also for dissimilar metals	8 – 12
	Butyl (solvent-release)	Sealant for metal, glass, masonry, etc	5 – 9
	Acrylic	General purpose construction sealant, caulking and glazing	8 – 10
	Silicone (one part)	General purpose construction sealant, caulking and glazing	20 – 40
	Silicone (two part)	Sealant for heat shields, etc; porting and encapsulation	40 – 100
	Polysulphide (one part)	General purpose sealant, caulking, glazing	17 – 25
	Polysulphide (two part)	Caulking, porting and encapsulation	12 – 15
	Polysulphine/Bitumen	Flame and fuel resistant caulking medium, sealant for expansion joints	4 – 6
	Polyurethane (one part)	General purpose sealant, caulking, glazing	12 – 15
	Polyurethane (two part)	Caulking, porting and encapsulation	12 – 15
	Epoxy (two-part modified)	Caulking, porting and encapsulation	7 – 15
	Fluorinated hydrocarbon	Oil and heat resistant sealant	80 – 120
HARDENING RIGID	Oleo-Resin	Caulking, general purpose sealant	1 – 6
	Epoxy	Pipe sealing, porting, encapsulation of electrical compounds	7 – 14
	Polyester	Gaskets, pipe thread sealant, porting, moulding, encapsulation	6 – 12

Its viscosity must be chosen between the requirements of ability to penetrate the crevice on the one hand and provide a suitable thick surface film to prevent wicking of the water up the crevice on the other. Relatively high viscosities are necessary to maintain a thick film on a vertical surface but it must not be too high as to cause bridging of the crevices as shown by Fig 32. It is also important that the crevice does not draw up only the solvent suspending the wax and not the wax itself.

Electrolytic zinc-coat steels used in protection against corrosion are increasingly being applied. Fig 33 shows possible applications for British Steel's Nizec zinc-coated steel for one side and both-side coated cases. The CR code in the designation, signifying cold rolling, is usually followed by a number depicting the thickness of the zinc layer in microns. Zinc itself corrodes to sacrificially protect the steel and in so doing forms a colourless salt which itself protects the zinc. Since the zinc coating can be

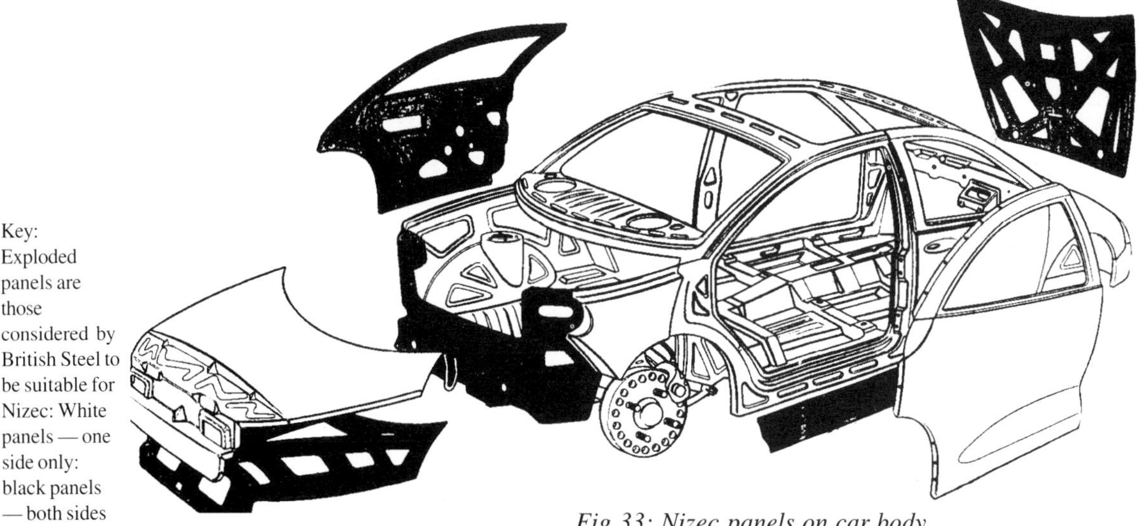

Key: Exploded panels are those considered by British Steel to be suitable for Nizec: White panels — one side only: black panels — both sides

Fig 33: Nizec panels on car body

ALKALINE CLEANERS		COLD EMULSION	ACID CLEANERS	
FERROUS	No. 3, 4, 5, 7, 8, 14, 16, 17, 18, 19, 20, 22, 23, 24, 26, 28, 30, 62	EMULSION CLEANER 'R'	DERAN, 'B' D & P.S.	Warm, derust & passivate
			DERAN 'C' & WALTERGEL P.5	Brush/swab on derust and passivate.
ALUMINIUM & ALLOYS	7, 10, 11B, 19, 22 26, 28, 62,	EMULSION CLEANER	D.P.1	Cold Combined degrease/pickle
			D.P.2 & D.P.3	Warm Combined degrease/Pickle
ZINC & ALLOYS MIXED METAL	6, 7, 14, 18, 19, 22, 26, 28	CONCENTRATE NO.1	P.S.4 & P.S.5	Heavy duty acid pickles
			P.S.7	Cold – Cuprous metals

Complete degreasing, cleaning, rust & scale removing processes. Pickling & passivating. Immersion soak, spray or brush/swab. Power washer cleaners.

PHOSPHATING PROCESSES approved to D.E.F 29A & BS 3189

'C.1'	Iron Class 1. Immersion		Fasbond A.S.	Zinc Class III Spray	
'MC'	Manganese Class 1. Immersion		D.W. & I.L.W.	Zinc Wire drawing – Immersion	
'E' SR & S.R.H	Zinc – Class II	Immersion Low Sludge & Scale	DXL & DXH	Zinc Cold Extrusion – Immersion	
Fasbond	Zinc	Paint Bonding	A.P.	Class IV Lightweight Iron – paint bonding. Tube drawing	Spray
A & F		Class II Immersion	A.P.3		Mixed –metal
'J'		Class III Immersion			
Cold		Class III Immersion ambient	A.P.4. A.P.5 A.P.6		Spray or Immersion

Phosphating processes for Corrosion protection Paint bonding Rubber bonding Tube & wire drawing Deep pressing & extrusion Chemical blacking Paint strippers.

CHEMICAL BLACKING PROCESSES · Walterblak & Walterblak 4

CHROMATING PROCESSES			
Cromcote	AL	Gold coatings for aluminium	Paint bonding & protection
	GL	Blue/Green coatings for aluminium	
	ZC	Zinc & Galvanised	
	ZH	Galvanised	
	ZF	Galvanised	
D.D.3 – Sealant additive for hot rinses			

CHEMICAL OXIDISING PROCESSES	
Walterisation 'L'	Aluminium, Paint bonding & corrosion resistance
SATIN ETCH PROCESS for ALUMINIUM	
WALTERBRYTE C.H. Aluminium & alloys	
COLD BRUSH ON TREATMENTS	
FOSCOTE R.S. – AL – GA For Steel – Aluminium – Galvanise	

Chromate conversion processes for:– Corrosion protection and paint bonding on:– aluminium, zinc, copper and their alloys.

Chemical oxidising & satin etches for aluminium & alloys.

DEWATERING FLUIDS	
D.F.F.	Film free – limited protection
D.F.F. 2	Thin film – good protection
D.H.	Black variant of above
BLACK D.H.	Dry resin film – good protection
D.M.	Thin mineral oil – good protection

SEALANTS	
V.S. 1 & 2	Temporary – for storage
SD6 & 9	Black – spirit stains
GM OIL	Mineral – corrosion inhibited
G.C. OIL	Mineral – exceptional protection
EMULSIFIABLE OIL No.2	Water/oil emulsion – dry finish excellent protective

Dewatering fluids – filming & non filming; lanolised, mineral or resinous. Corrosion inhibited oils and protectives. Shellac dyes & water stain One pack metal conditioners etching primers & lacquers.

Fig 34: Traditional pre-treatments

vapourised locally around spot welds there is a case for further wax injecting to protect the weld crevices, as before. The objective with zinc coating is to make metallic contact with the steel to promote a cathodic reaction in which both metals are equally exposed to the liquid environment. The mechanism is one of the electrons produced by the dissolving zinc travelling through the metallic contact and being changed into negative ions at the steel surface. These go back through the liquid environment to form a new zinc compound with the dissolved zinc ions.

In vehicles a problem occurs in that the high resistance of a thin film of water absorbed into the paint film reduces the flow of ionic current. However, a compromise is reached in that the very reduction of this current is the basis of the separate barrier mechanism of the paint for reducing corrosion. It is also necessary to overcome the difficulty of a less good adhesion of paint with galvanised steels. Finish painting systems are being underlaid with increasingly sophisticated primer coatings. Fig 34 shows the range of products and processes offered by Walterisation. Much of these treatment processes are for ensuring good adhesion of the finish coats — and indeed the primers and intermediates. Procedure for building up a good substrate is to abrade the surface with 180 'wet & dry'; wash down and dry then apply the phosphoric acid etch. After subsequent hot-water wash and blow dry a single coat of self-etch primer is applied. Next, three coats of filler are applied and baked for half an hour and this layer rubbed down with 600 'wet & dry'. Finally the surface is inspected and stopped up with stopper as necessary and a further filler/sealer coat applied as necessary after blocking down.

An interesting alternative to plastic coating of metals is the use of pre-painted metals in specialist body assembly. Several factors need to be considered in the choice of coating in the ability of such panels to withstand subsequent forming and fabrication. The key coating parameter is dry heat resistance being the resistance to film fracture after the film has been stretched and exposed to heat. Generally this varies inversely with the molecular weight of the coating polymer and amount of three dimensional cross-linking undergone in the cure cycle. Preferred primers for the coatings are the higher molecular weight epoxides, urethanes and linear polyesters. Among the special techniques used in welding pre-coated sheets are the use of weldable primers on the reverse side, indirect projection welding and the use of short bursts at very high energy levels.

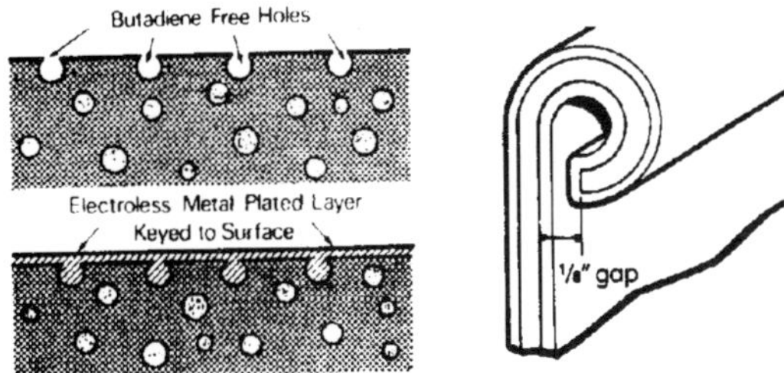

Fig 35a (left): Etching and coating; b(right): Rolled edges;

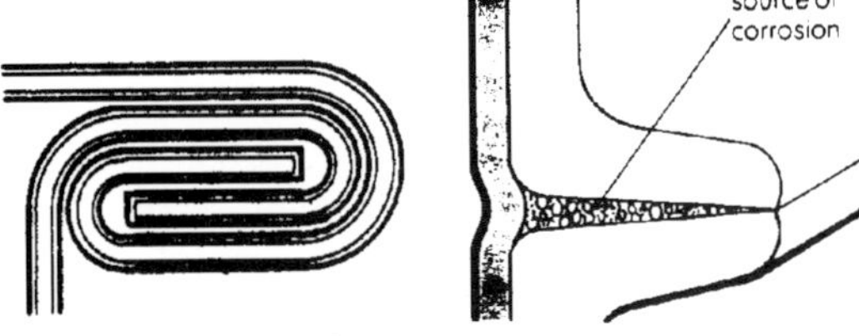

Fig 36a (left): Roll joint; b(right): Flange joint

Plastic coating

Plastic coatings and coatings for plastics generally involve the metal to polymer bond which needs care in design to avoid either peeling or corrosion. Careful adherence to rules of form and application will ensure the proper bond which is key to satisfactory in-service life. Most of the bright metal coating is applied to ABS plastics but higher temperature-tolerant plastics such as PPO can also give good results. In general, if a part has to flex, it should not be plated. However, selective plating may be employed and if, say, it is just a question of avoiding snap-fit lugs, then selective plating for the remainder of the part is quite feasible. The approach is normally to electroless coat the part with nickel and then regard it as metal for subsequent chromium plating. Uniform wall thickness in the order of 3 mm is preferred by platers; 5 mm is considered to be a maximum and where possible, abrupt change in thickness should be avoided. Where ribs are used, these too should be restricted in depth to around 50-60% of wall thickness. Because plating exaggerates the very smallest surface imperfection in the plastic, it is necessary to avoid any chance of sink-marks. Ribs should thus be well radiused and subject to normal draft of about 1 deg. In the part as a whole, external radii should be not less than 1mm and internal ones 0.5 mm minimum. A groove of some 1.5 mm deep should be used at the boundary of plated and unplated surfaces.

Plating thicknesses will of course depend on part application. Tests reported on ABS have shown that below 0.4 micron, microporsity can occur but above 0.8 micron, only crazed chromium films are considered to produce good quality. Adhesion strengths on ABS of up to 4-6 kgf/mm^2 can be expected but it is best to retain a temperature spread within 40-90 C. After cleaning the substrate prior to plating it is usual to etch the surface with acids which remove the butadiene, Fig 35a. Following the electroless nickel coating stage, applied by dipping into a catalytic reduction solution, very low current density is used in the electroplating so as to maintain the optimum thin coating of copper prior to the finish chromium layer. To avoid surface blistering at high temperatures, higher grade plastic substrates like PPO are chosen. A brand such as Noryl PN235 has heat deflection temperature of 112 C and water absorption at 23 C of about 0.08% in 24 hours.

Coating metals with plastic A number of dip-moulding and similar processes have been available for the plastic coating of metals but particularly strong coating with materials like polyamides are possible with plastic powder coating. ATO Chemicals' Rilsan technique is one such and interestingly the nylon is developed from caster oil rather than the mineral oil which is the source of most plastics. Parts to be coated must, however, be capable of sustaining the fusion temperatures involved. Coating film thickness is commonly 75 to 200 microns and fusion temperature between 200 and 225 C. An electrostatic projection coating process is used for parts too unwieldy for dip-coating — normally large thin-sheet fabrications. Powders are passed through a nozzle carrying a high electrical potential; particles become charged and attract to the metal substrate at zero potential.

In the alternative fluidised bed coating process, the part is oven-heated and then dipped into a fluidized bed of powder, akin to dip coating, for thicknesses from 325 to 375 microns. Really large items can be coated by a third process known as flame spraying which causes the powder to melt and flow. Again, radiused edges are the rule for good coating. Rolled edges, Fig 35b, should be left open with a gap of about 5mm. Roll or crush joints, Fig 36a, can be coated before rolling. By contrast, edge-welded flange joints, Fig 36b, cannot be used as it is virtually impossible for the powder to penetrate the entire joint.

Pre-coated panels

With heightening awareness of ecological matters, pre-coated (painted) steel and aluminium sheet is now of increasing interest to the motor industry as painting pollution problems can be dealt with earlier and perhaps more effectively by the coil-coater.

Coaters say because the bond between the metal substrate and the coating system is continuous, coil-coated metal can also withstand subsequent fabrication without damage to the constituent elements. Tests by the American Institute of the Steel Industry have shown that body panels can be formed from top-coated pre-painted sheet in existing metal forming presses using existing dies. Because the pre-coaters apply paints to both sides of the metal, corrosion problems during storage and pre-fabrication are minimised and panel galling is reduced by the absence of metallic contact with the dies. Dissimilar metal substrates can also be used without the usual risk levels of bi-metallic corrosion.

Prime candidates for the process are instrument panels and bumpers also cover-panels such as bonnet, boot and door. In general, roll-forming rather than press-braking is preferred for folding pre-coated sheets since there is less chance of coating damage. Then, in subsequent drawing stages, the main concern is to ensure that press-tools are designed to allow for the coating thickness. Edges can be protected with edge-trims or touch-up paints; rolled edges or returned flanges can also be incorporated, Fig 37.

A number of fastening systems have been developed for the material, and recommended by the European Coil Coating Association, which avoid the need for welding. Self-piercing rivets, self clinching nuts and self-drilling/self-tapping screws are the three principal contenders, Figs 38/9. Since they do not require pre-drilled holes they eliminate possible damage to the coating in machining. A variety of spring tension fasteners are also suitable — and, for

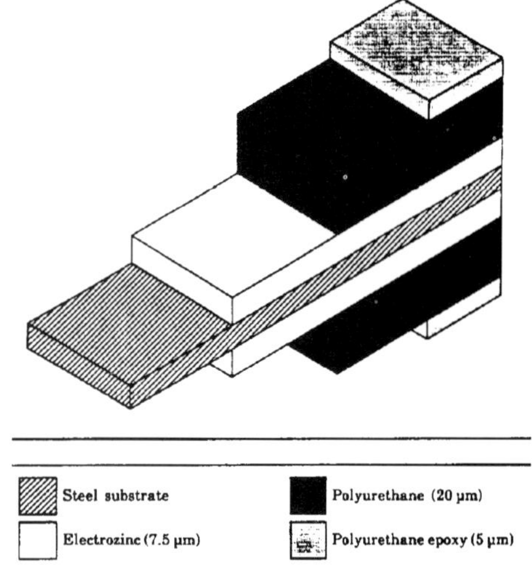

Fig 43: Flexgard T pre-coated sheet

Fig 37: Edge protection methods: clockwise from top left — folded; rolled; flanged; double closing strip; single closing strip

Fig 40: Clinched open section

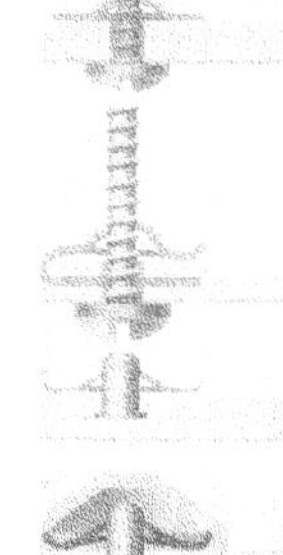

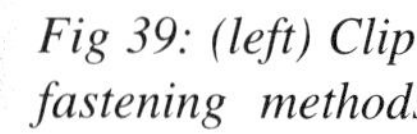

Fig 38: (far left) Insert fastening methods

Fig 39: (left) Clip fastening methods

Fig 41: Clinched closed sections

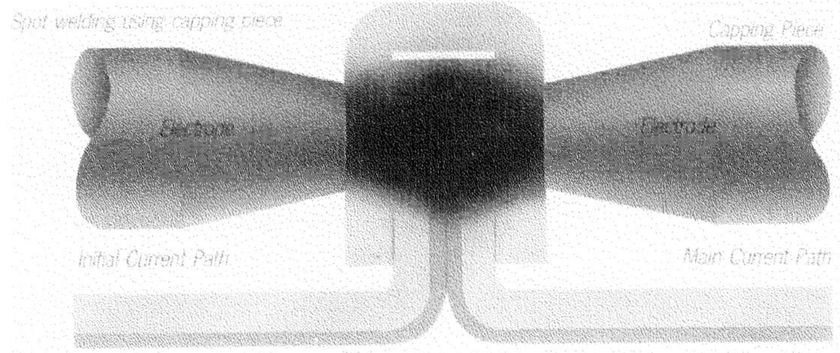

Fig 42: Bridge-conductor for spot-welding

PROPERTY	SPECIFICATION
% Gloss	25 - 45
Hardness	F - 2H Eagle Turquoise
Film Thickness Dry Wet	0.15 - 0.25 0.35 - 0.55
Cure Test	50+ Double MEK Rubs
Reverse Impac t	1.5 x Metal Thickness No pickoff
Flexibility	OT Bend : No pickoff

Fig 44: Flexgard T specifications

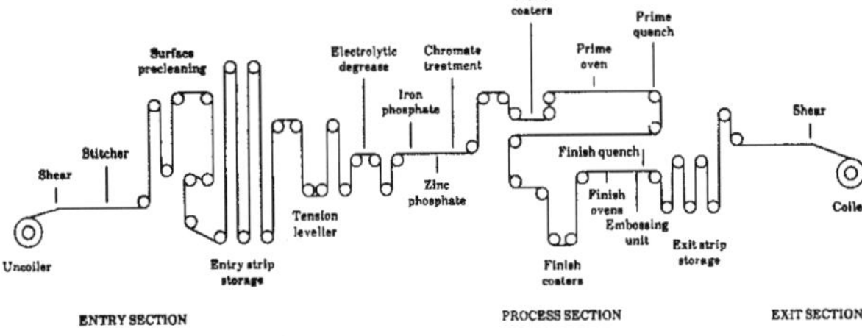

Fig 45: Production of Colorcoat

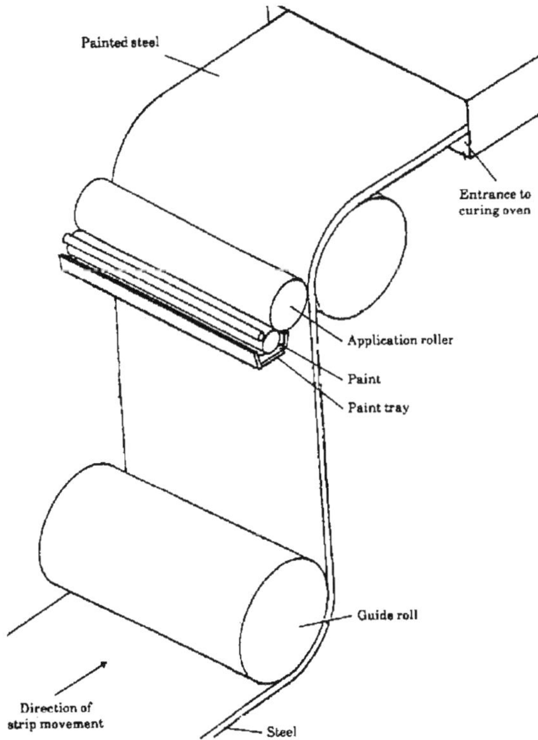

Fig 46: Rolling of Colorcoat

non-structural flanged joints, spring clips are a convenient fastener.

Clinched sections have also been developed to provide interlocking joints suitable for box-sections, Figs 40/1. Portable stitch-folding guns are also available which punch out and fold two tabs over one another in a single operation. While resistance-welding by conventional methods is not possible, due to the electrical properties of the coating, spot-welding of in-turned flanges can be carried out by placing a channel capping piece over the flanges to provide a low resistance contact, Fig 42. Capacitor-discharge stud welding can also be used to fix a stud to a hidden area of a component. A combination of short welding time and high current meets the requirements for welding without marking the components. Adhesive-bonding is feasible provided a correct choice of adhesive is made according to the particular organic coating.

British Steel have recently performed industry-standard testing of pre-primed sheet steel systems and have found superior crevice corrosion performance over an equivalent electropainted panel. The pre-prime system tested was Flexgard T developed by PPG, Fig 43. Its specification is seen in the table of Fig 44. Shown in Fig 45 is the Colorcoat line of British Steel at Shotton where both PUR-based primer and epoxy-PUR top-coat are applied wet to the pre-treated zinc-coated steel by roller coaters spanning the width of the sheet strip, Fig 46. After the primer is applied, the painted strip enters an oven to cure the organic film and expel excess solvent also to produce cross-linking of the polymers.

Glazing techniques

Glazing of vehicles has reached the interesting stage where glass is potentially a substantial contributor to the structural integrity of the body. Since the direct bonding of glass into the body apertures the material has made a contribution to stiffening the 'closure' frames by the addition of a diaphragm member.

Traditionally the key 'structural' application has been the windscreen — in which there have been recent trends to single and double curvatures — and for which Motor Vehicle Construction and Use Regulation 1017 requires the use of safety glass. This is usually laminated plate of 4.4 to 8 mm thickness and must provide distortion-free visibility even in areas of high curvature. General specification is provided in British Standard BS 857 and the latest type encompasses the best features of both toughened and laminated glass. It comprises high stressed inner glass ply bonded to a low-stressed outer ply by a highly penetration-resistant interlayer.

Variations are available from major suppliers Triplex — with clear or tinted glass, silk screened obscuration bands, graduated shade bands and fine wire heating elements. Toughened 'body' glass for side and rear windows (and to a limited extent for windscreens) is available in 3 to 6 mm thicknesses — again with a variety of silk-screened patterns also in Sundym and bronzed varieties. Another product is Hotscreen, the windscreen glass with a grid of very fine tungsten wires embedded in the laminate — and, similarly, Defogger backlights. This product typically involves 500 yards of 20 micron tungsten wire — dissipating 600 watts per square metre on a 12 volt system enabling a uniform defrost within two minutes. If higher voltages can be made available, gold and silver films can be used for demisting and defrosting which are more efficient again — with the potential to dissipate 1500 watts per square metre.

The trend in volume cars towards higher glazed areas has been accompanied by a parallel one of reducing glass thickness — so as to avoid nett increases in body weight. This trend is likely to apply to many other vehicle sectors and already 3mm toughened glasses and 5 mm laminates are being specified. There can be a cost penalty with thinner glass arising from its greater fragility at the manufacturing stage and the greater difficulty in controlling any curvature required. On the other hand thin glass can be used advantageously if curvature, coupled with rigid mounting, is present — as suggested in Fig 47 which also shows thickness effect on sound attenuation and heat transmission. It is even possible to bend and toughen glass, on a pilot scale, down to 2mm. Solar control systems under development include the use of photochromic and heat-rejecting glass — also special forms of double glazing, Fig 48. European Safety Regulations limit the amount by which 'tinted' glasses can be used — in the interests of maintaining optical clarity, Fig 49a. On front screens light intensity may not be reduced by more

Sound attenuation values

Typical car body shell	26 dB
5 mm toughened glass	26 dB
3 mm toughened glass	23 dB
5 mm laminated glass	28 dB

Resonance of glass plate, 1500 mm × 500 mm

Shape	Thickness (mm)	Edge support	Resonant frequency (Hz)
Flat	4	Rigid	45
Simple curve	4	Rigid	146
Simple curve	3	Rigid	143
Complex curve	4	Loose	15

Typical heat transmission factors ($\Delta T = 20°C$)

Road speed (km/h)	U for toughened glass ($W\ m^{-2}\ °C^{-1}$)	Glass thickness (mm)
7	5·6	5
	5·6	3
47	6·7	5
	6·8	3

Fig 47: Properties of glass

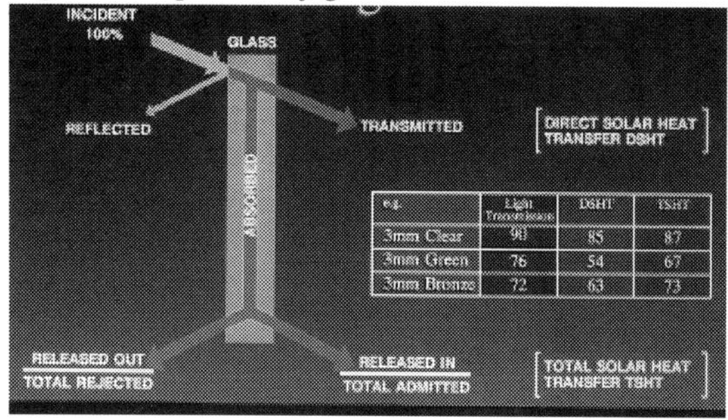

Fig 48: Optical effects of double glazing

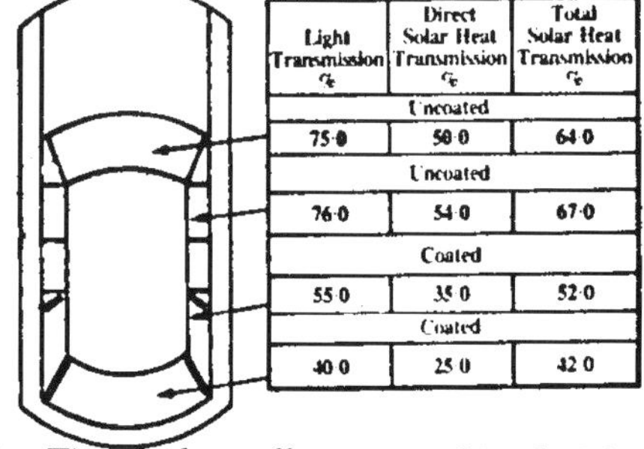

Light Transmission %	Direct Solar Heat Transmission %	Total Solar Heat Transmission %
Uncoated		
75·0	50·0	64·0
Uncoated		
76·0	54·0	67·0
Coated		
55·0	35·0	52·0
Coated		
40·0	25·0	42·0

Fig 49a: Tinted glass effect — combined with coating

than 25%. On side windows through which rear-view mirrors are viewed this figure becomes 30% and for the vehicle as a whole 45% with the rear window alone having an allowance of 60%. Even 'clear' glass can reduce intensity by as much as 10% and it has to be remembered that regulations also impose limits on maximum image distortion. However, while clear glass transmits about 84% of solar energy, it is possible to procure optically acceptable tinted glass which will reduce heat build-up in stationary vehicles by 'between a quarter and a half', say the makers. When glass becomes part of the vehicle structure, and subject to diaphragm stresses, there is risk of noise due to 'drumming'. As glass cannot easily be swaged without detriment to its optical performance alternative means have to be found to reduce this effect. One such is a form of double glazing comprising two sheets of optical glass, precisely matched in curvature, with an air gap between them, Fig 49b. In the case of roof lights, liquid crystal systems can be used with which light transmissibility can be changed subtantially by the operation of a switch, Fig 50. An alternative is an electrochromic system, Fig 51, in which an electric current generates lithium and hydrogen ions from the ionic conductor layer. These then flow to the electrochromic tungsten dioxide layer which changes its colour to a blue tint.

Encapsulation of glass is another trend, in which the glass, seals, trims and clips are pre-assembled as a module which is relatively easy to install. Encapsulation can also encompass flush glazing systems and indeed provide easier achievement of the tight tolerances involved. Dow have developed a technique using reaction injection moulding (RIM), Fig 52, involving relatively low pressures and temperatures that thus keep costs to a reasonable level. The process is even said to be capable of producing complete sliding-window units. The two reactive components are blended by high pressure impingement and then injected at low pressure into a closed mould. Reactive edge adhesive is applied to the glass which is then placed in the mould for a curing time of about 30 seconds, Fig 53.

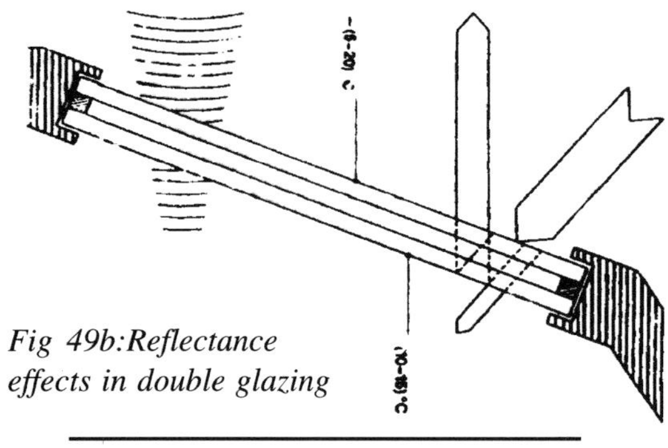
Fig 49b: Reflectance effects in double glazing

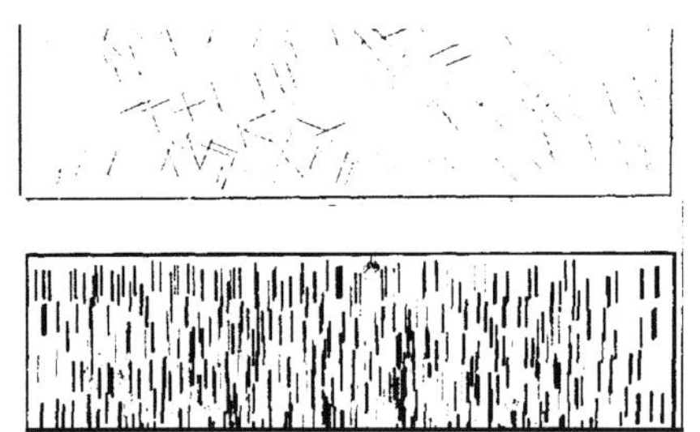
Fig 50: Switching of liquid crystals: with (above) and without (below) electric field

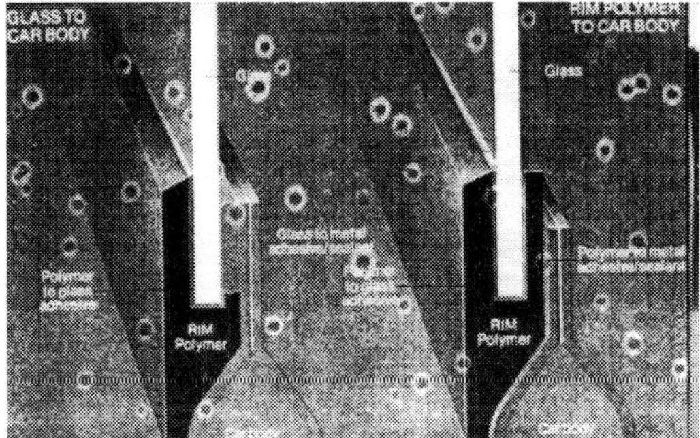

Fig 51: RIM encapsulation of glass

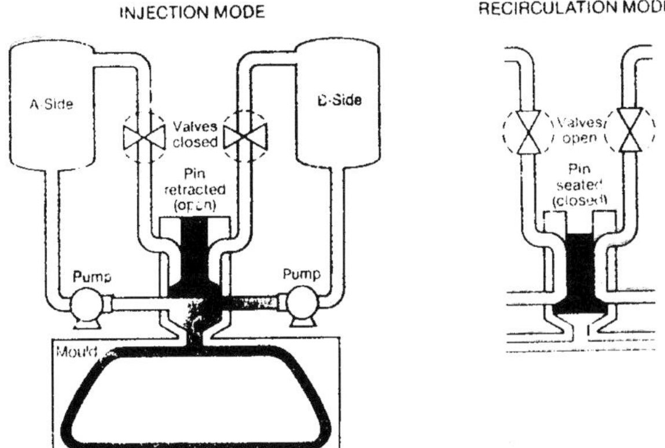

Fig 53: RIM encapsulation equipment

Fig 52: Electro-chromic glass

Hardware manufacture: hot-forging

While forging is a process for reducing a slug of metal as close as possible to its finish-dimensions, with minimal machining, thus being of considerable value in volume manufacture, it as also valuable in improving mechanical properties by altering grain-flow. This can have important consequences in specialist vehicles such as racing cars where extra manufacturing cost in low volume can be justified for increased performance.

The forging process can be used in the refinement of grain structure, directional control of 'flow-lines' as well as breaking up and redistributing unavoidable inclusions. In hot-forging, working of metal is carried out above recrystallization temperature but below the melting point, to exploit the reduction in reshaping force required compared with cold-working. Finishing temperature is usually arranged close to the upper limit of the cold-working range so that the finest possible grain size is ensured.

Sometimes cold-working is carried out as a post-process to finish closer to final size and enhance strength by work-hardening.

The ability to make the flow of metal, in forging, follow the contour of the part is the important advantage of the process as the grains and inclusions are altered at the same time. The part can also be given anisotropic properties by suitable control of the forging process — such that structural efficiency is considerably increased and a lighter and tougher part results. The grain-flow in a sample part can be checked by slicing the part, in the plane of interest, then polishing and etching the exposed surface. The resulting 'fibrous' appearance are due to the elongated inclusions which define the grain flow.

The US Forging Industry Handbook provides relationships between mechanical properties and grain flow, summarised in Fig 54. The benefits to a part such as a gear-wheel are particularly graphic when compared with machining the part out of solid-bar as might be the temptation for a one-off component. Even the basic up-setting of a slug of metal, from which to machine the part, would be helpful as the 'fibres' are bulged radially such that strong grain orientation exists at the base line of the teeth, where maximum bending stresses occurs. It would be better to form the blank as a section as shown in Fig 55; best

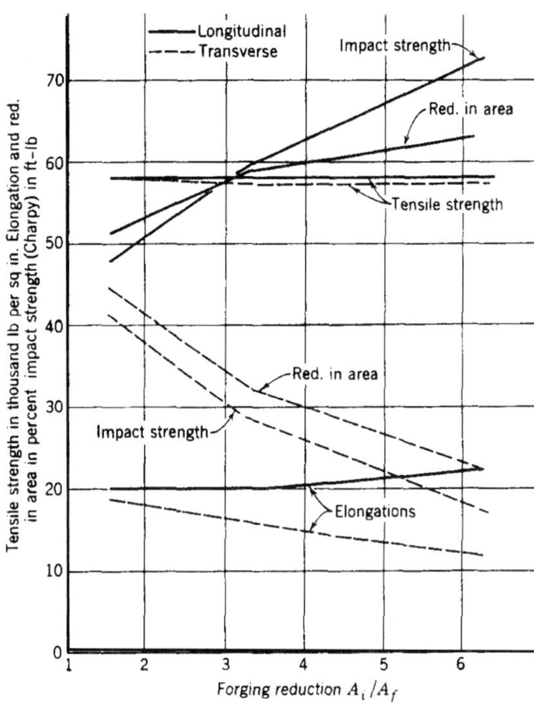

Fig 54: Grain flow effect on properties

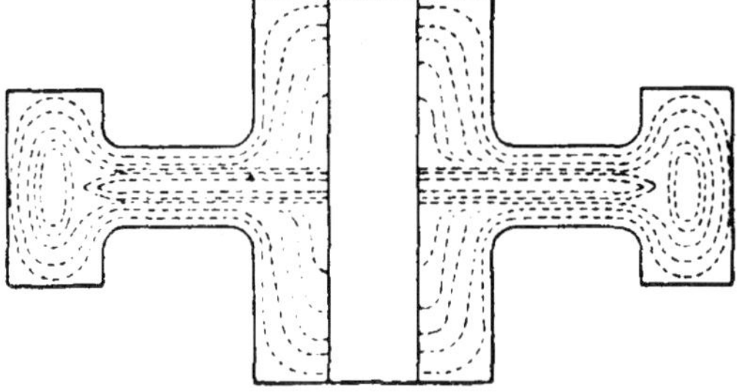

Fig 55: Blank form for better grain flow

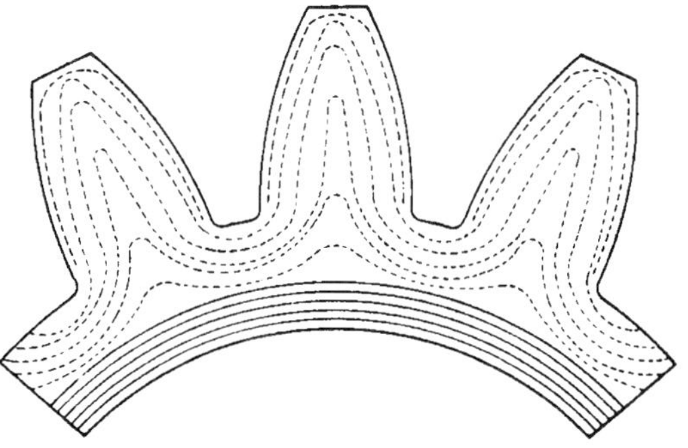

Fig 56: Blank form for best grain flow

of all would be to rough form the teeth as shown by the section in Fig 56. The degree of reduction in the forging of the blank from the billet is also significant as can be seen in Fig 57.

The same principles apply to a variety of stressed parts, the individual shape and performance requirement needing consideration by the designer. The flow of metal in a forging die is of considerable importance. It is essential, of course, to consult the forger on the exact metal flow pattern with respect to the surfaces of the die and the chilling action they produce. It may be necessary to carry out preforming operations to distribute the metal more gradually towards its desired shape. British Standard 4114 for forging was compiled on the basis of many international inputs and is a good guide for detail design. A useful paper on more general aspects of design has been produced by GKN[16]

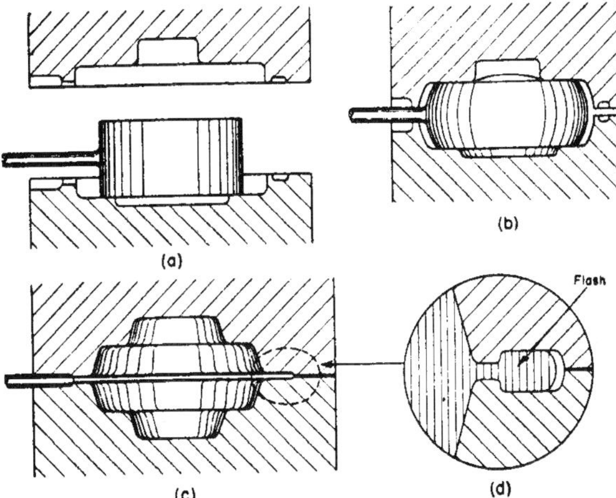

Fig 57: Blank reduction during forming

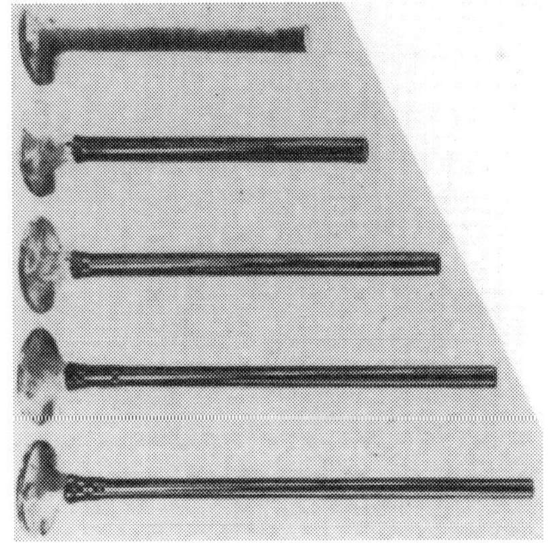
Fig 58: Stages in production of an axle shaft which starts as an upset forging left and goes through four stages of cold-extrusion.

Cold forging and extrusion

Both these are processes introduced to produce parts in high volume with very little loss of metal in swarf thus reducing costs by exploiting economies of scale and minimising use of material. However, the working of the material can also produce parts of better structural performance because grain flow is improved and surface hardness increased.

Drive shafts throughout the transmission line of the vehicle can all benefit from these processes and, weight-for-weight, produce parts with greater resilience and fatigue resistance, Fig 58. With more refined developments such as orbital cold forging, gear wheels, too, of superior performance can be manufactured. Even a low-volume vehicle could benefit by selecting shafting produced by cold-forging from the parts inventories of the high volume builders. Given that the right performance specification can be found it is more than likely that saving in weight and increase in durability will be the result.

Cold extrusion The increase in hardness and ultimate strength of steels before and after cold extrusion is shown in Fig 59 in which very substantial percentages can be noted. Work hardening during the

Steel AISI	Condition	Reduction, %	Hardness	Ultimate Tensile Strength, ksi (MPa)	Ultimate Yield Strength, ksi (MPa)
1016	Before extrusion		R_B 67	61.8 (426)	45.3 (312)
	After extrusion	50	R_B 93	109 (752)	105 (724)
		70	R_B 96	114 (786)	111 (765)
1045	Before extrusion		R_B 81	85.8 (592)	43.2 (298)
	After extrusion	50	R_B 96	146 (1007)	140 (965)
		65	R_B 100	154 (1062)	146 (1007)
1340	Before extrusion		R_B 80	79.6 (549)	46.9 (323)
	After extrusion	50	R_B 98	140 (965)	135 (931)
		65	R_C 28	148 (1020)	142 (979)
8620	Before extrusion		R_B 70	67 (462)	48.1 (332)
	After extrusion	50	R_B 98	122 (841)	118 (814)
		65	R_B 99	130 (896)	124 (856)

Fig 59: Steel properties with cold working

working of the material is such as to avoid the need for heat treatment after the part has been formed. Depending on the amount of deformation in the process, yield strengths can rise by up to 300%. Extruded parts are limited to the forming of symmetrical parts and a drive shaft can be cold-headed during processing to provide an integral flange; even internal splines can be formed during working of the shaft. With this combined upsetting of the ends, shafts up to two metres in length are possible, Fig 60. Methods employed for cold extrusion include backward, forward, radial, combination, impact and continuous extrusion, the first two being shown in Fig 61. With impact extrusion particularly, large reductions are possible because of the magnitude of the impact force. The generally accepted limits for maximum upset diameters are shown in Fig 62.

Advanced forging techniques

Orbital cold forging is carried out by the Floform company of Welshpool using machines from the Swiss Schmid company. In this process the cold metal slug is forced by hydraulic pressure between upper and lower dies, with the upper one orbiting around its vertical axis, Fig 63. The forming forces are thus concentrated on a small area of the part which is being rolled over its complete surface area. This reduces the load required to cold-forge a typical part by a factor of six and leads to the use of smaller machines. Tooling, too, is simpler and cheaper. More accurate form filling is also claimed. A 200 tonne machine at Flowform produces parts up to 85 mm in diameter. Tolerances in the die and punch capacities are quoted as +/- 0.03 and 0.06 repectively — with a length tolerance od +/- 0.2 mm. Finish formed bevel gear wheels can even be formed on the ends of shafts and relatively low runs of 200 parts per week are viable.

D
H
d
Free upset
$D \leqq 2$ to $2\,1/2d$
Carriage or step bolt
$D \leqq 3d$

Fig 62: Upset diameter limits

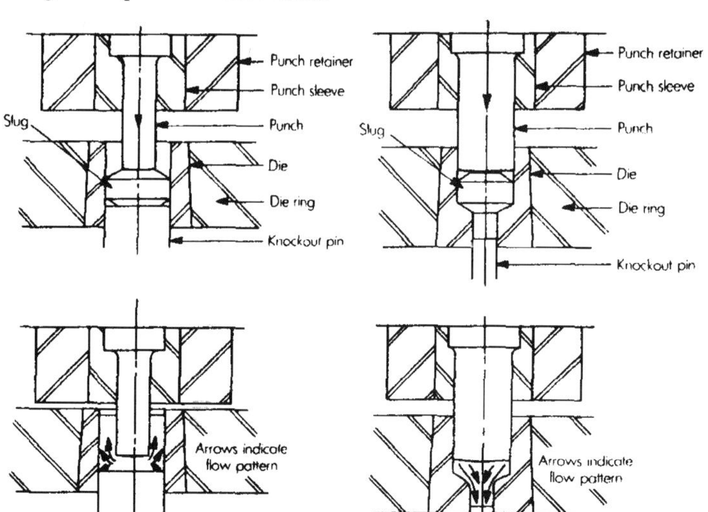

Fig 61: Forward (left) and backward (right) extrusion

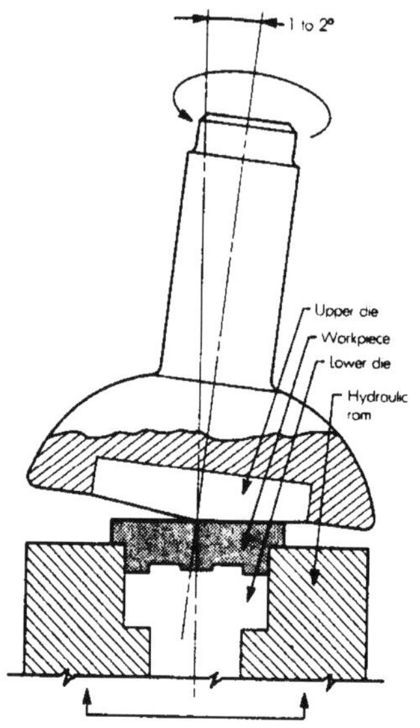

Fig 63: Orbital forging

Fig 60: Stages in the Dynaflow process of combined extrusion and forging:

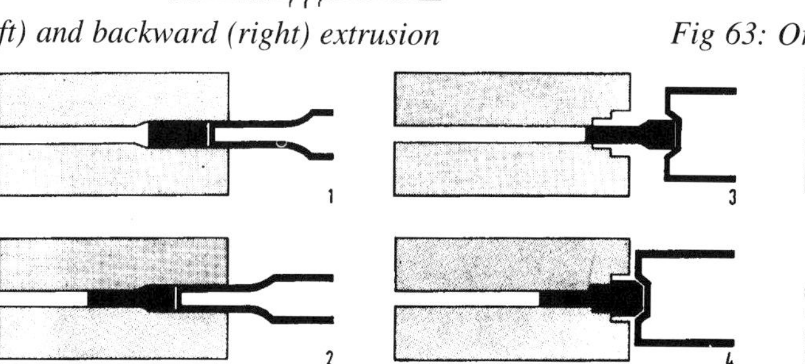

Dynaflow process: **1** and **2** represent forward extrusion, while **3**, **4** and **5** cover transfer to a second die and subsequent two-blow forging of the head

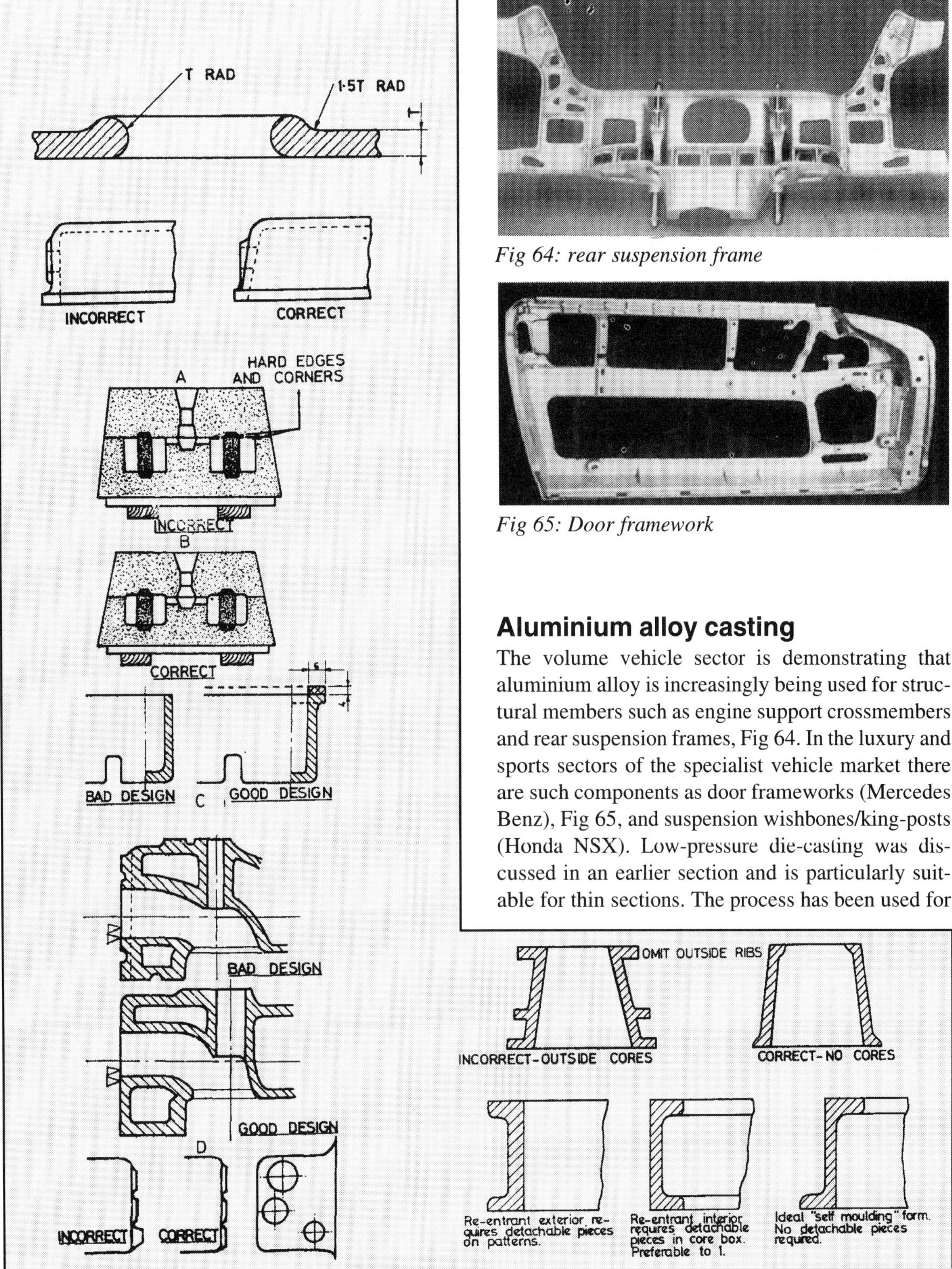

Fig 64: rear suspension frame

Fig 65: Door framework

Aluminium alloy casting

The volume vehicle sector is demonstrating that aluminium alloy is increasingly being used for structural members such as engine support crossmembers and rear suspension frames, Fig 64. In the luxury and sports sectors of the specialist vehicle market there are such components as door frameworks (Mercedes Benz), Fig 65, and suspension wishbones/king-posts (Honda NSX). Low-pressure die-casting was discussed in an earlier section and is particularly suitable for thin sections. The process has been used for

Fig 66: Do's and dont's of sand-casting

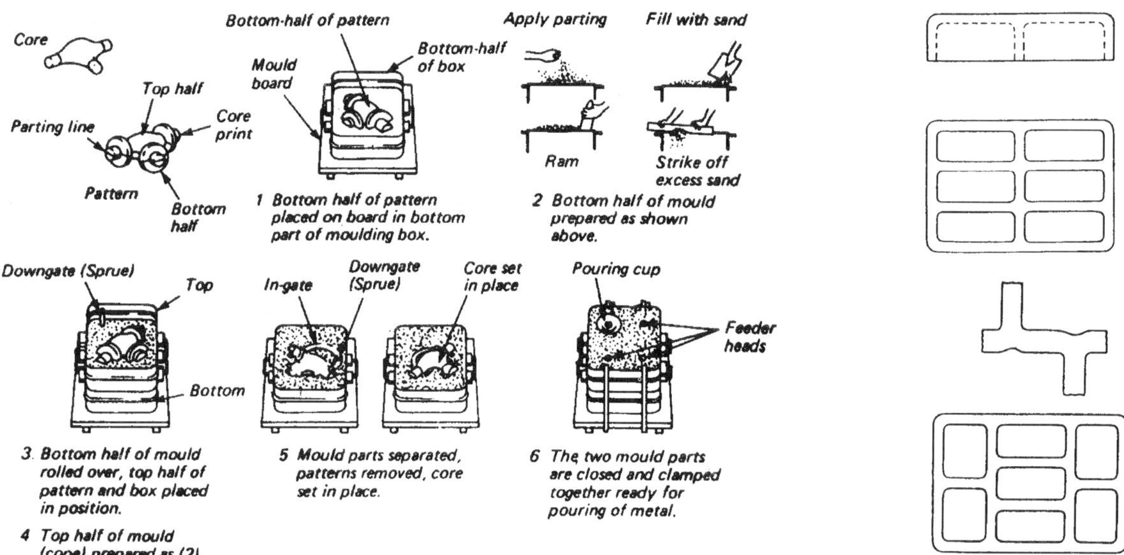

Fig 69: Sand moulding techniques

	Investment casting	*Premium sand casting*	*Conventional sand casting*
Advantages	• Small tolerances • Small wall thickness	• High mechanical properties for all wall thickness • Small wall thickness • Small tolerances	• Large dimensions • Low costs • Local high mechanical properties by local cooling
Disadvantages	• Lower mechanical properties	• Higher costs for castings and tooling	• Larger tolerances • Larger wall thickness
Part size (maximum overall dimensions)	1000 × 800 × 500 mm	3600 × 2000 × 1000 mm	5000 × 1500 × 1500 mm
Wall thickness, depends on: • outer dimensions • part geometry • alloy	T_{min} = 1.5 ± 0.15 mm	T_{min} = 1.8 ± 0.4 mm 0.2 (Local: 1.5 ± 0.2 mm)	T_{min} = 2.5 ± 0.5 mm (Local: 2 ± 0.5 mm)
Tolerances (D = considered dimension)	According to countries standard For example: VDG-P690 for Germany	T_{min} = ± (0.4 + 1.500/ 1000 mm	According to countries standard For example: DIN 1688 for Germany
Surface roughness	R_a = 3.2 μm	R_a = 3.2–6.4 μm	R_a = 6.4–12 μm

Fig 67: Aluminium casting overview

special road wheels and involves alloys such as BS LM 25 having elongation from 7-13%. Sand castings might better suit low volume work and some dos and donts are given in Fig 66.

An aluminium casting process overview is given in Fig 66, due to MBB. The alloys for casting (Fig 67) are chosen for good mould-filling capabilities. A357 is the standard casting alloy which contains silicone (G-AlSi7Mg0.6) is particularly favoured — with strength of 330 N/mm^2. For higher strength requirements, A201 is preferred which has a figure of 414 N/mm^2. The basic techniques of sand moulding are shown in Fig 68. The core, of course, is made separately by compacting bonded sand in a core box. Extensions shown on the core represent prints which accurately locate the core in the mould. A variety of additions convert the hand process shown to one of machine control in most cases. In the case of shell moulding, the sand grains are coated with resin which makes green sand flow as easily as dry sand. It is blown on to a heated metal pattern to form a resin bonded shell. Machines thus equipped are capable of higher outputs and the castings produced characterised by a smoother than normal sand-cast surface.

When finish casting is to be laid out for machining the drawing should show reference points. These should not be influenced by core-shift or movement of the drag. In general they should be as far apart as possible and on the same side of the parting line. The latter should be on one plane to facilitate parting. Pattern dimensions must differ from those on the finish-part drawing, to allow for shrinkage. All surfaces of the pattern also need a slight draft angle.

In designing ribbed sections, relationship of the rib section to the main section should be such as to permit a uniformly graded metal section. Fig 69 is an example where the design is altered by staggering the ribs to eliminate X-junctions and obtain the rib detail, top right. Three types of rib design are compared in Fig 70. The section at (b) is satisfactory but that at (a) poor because of high stress at the extreme edge; most preferred is that in (c) provided moulding procedure does not become unduly complicated.

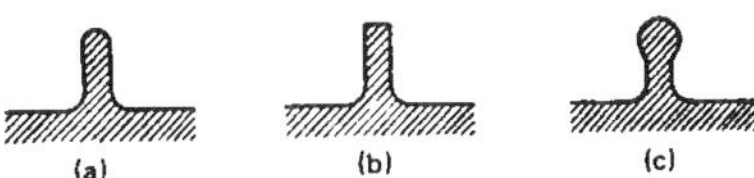

Fig 70: Rib configurations

BS1490 Reference No.	Minimum Ultimate		Typical Brinnel Hardness Number	Thermal Conductivity W/mK	Castability (Gravity Die)	Pressure-Tightness	Atmospheric Corrosion Resistance	Machinability	Anodic Film	Fusion Weldability	
	Bearing Stress MN/m^2	Shear Stress MN/m^2								Gas	Argon Arc
General purposes alloys											
LM4	320 560	150 —	85 115	133	Good	V. Good	Good	Good	Cloudy grey	Fair	Good
LM6	370	130	60	162	V. Good	V. Good	V. Good	Fair	Dark grey	Good	V. Good
LM8	320 370 430 560	130 150 170 220	60 70 80 110	179 171	Good	Good	Good	V. Good	Light grey	Good	V. Good
LM25	320 370 430 560	130 150 170 220	60 70 80 110	159	V. Good	Good	Good	V. Good	Light grey	Good	V. Gord
LM16	560 650 610	210 220 —	95 115 100	166	Good	Good	Good	V. Good	Light grey	Fair	V. Good

BS1490 Reference	Composition (Nominal) Per cent	Temper or Condition	Alloy (Alcan Ref)	Principal Specification and Requirements[1]			Other Related Spec'ns	Typical 0·2% Proof Stress MN/m^2
				U.T.S. MN/m^2	Minimum% Elongation on 5·65$\sqrt{So}$	U.T.S. Ton f/in^2		
General Purpose Alloys								
LM4	Al-Si5Cu3	M TF	117	160 280	2	10·0 18·0	L79	108 —
LM6	Al-Si12	M	160	185	7	12·0	4L33	90
LM8	Al-Si5Mg	M TE TB TF	B116	160 185 230 280	3 2 5 2	10·5 12·0 15·0 18·0	716A 722A 727A 735A	100 160 120 270
LM25	Al-Si7Mg	M TE TB7 TF	D135	160 190 230 280	3 2 5 2	10·4 12·3 14·9 18·0		100 160 120 270
LM16	Al-Si5Cu1-5Mg	TB TF TB7	C125[2]	280 325 310	12 4 10	18·0 21·0 20·0		150 300 201

Fig 68: Casting alloys

Advanced casting and die-casting techniques

Semi-solid metal (SSM) and squeeze casting processes introduced by Buhler offer potential for weight saving in automotive chassis, steering and body parts. Weight saving is also considerably enhanced in ferrous metals, using the process from Sintercast.

Heat-treatable lightweight components are seen as the promise for the new generation of casting machines from Buhler, some of which can handle composite materials. Although conventional die-casting of light alloys for automotive parts has been widespread, the lack of heat-treatability and inadequate mechanical strength has precluded the process from a number of applications.

The table in Fig 71 shows the potential weight saving in a front suspension system made possible by substitution of aluminium alloy for steel. These new casting processes make the economical production of such parts an increasingly realisable proposition, exploiting shot-control techniques. High injection power and precise velocity control also means that thin-walled weldable components are made possible with application in space-frame vehicle structures and the processes are threatening more expensive forgings for such applications.

In squeeze-casting, the handling of the liquid metal ensures there is non-turbulent, laminar die-filling followed by controlled freezing under high pressure. It is mainly used for thick-walled aluminium alloy parts. With SSM casting a special raw material is used having a microstructure which allows casting when the material is only 40% liquid. Very close to net shape casting is possible, reducing finish-machining to a minimum. Fig 72 shows mechanical properties of SSM castings compared with parts cast by more traditional rival processes.

SSM in detail Sometimes called Thixocasting, the process has three stages, commencing with the cutting of specially processed alloy billets into slugs. These are heated to a precise temperature at which they have both solid and liquid portions; finally the heat-softened slugs are placed into the shot sleeve of the die-casting machine. The special alloys are produced using direct-chill billet casting in combination with powerful electro-magnetic stirrers. This breaks up dendrites and transforms the fragments into globules which gives the alloy its special properties, Fig

Front suspension system Part name	Source: Alusuisse Buhler Weight in grams			
	Steel	Aluminum	Savings	
Upper control arm, front	750	250	500	67%
Upper control arm, rear	800	300	500	62%
Suspension arm	1850	700	1150	55%
Steering control arm	2100	1100	1000	48%
Support	190	120	70	37%
Suspension arm bearing	300	140	160	53%
Shock absorber bearing	185	130	55	30%
Bearing for steering control arm	370	280	90	25%
Knuckle	6950	3900	3050	44%
Total per side	13495	6920	6575	49%
Total per vehicle	26990	13840	13150	

Fig 71: Weight saving potential of aluminium alloy

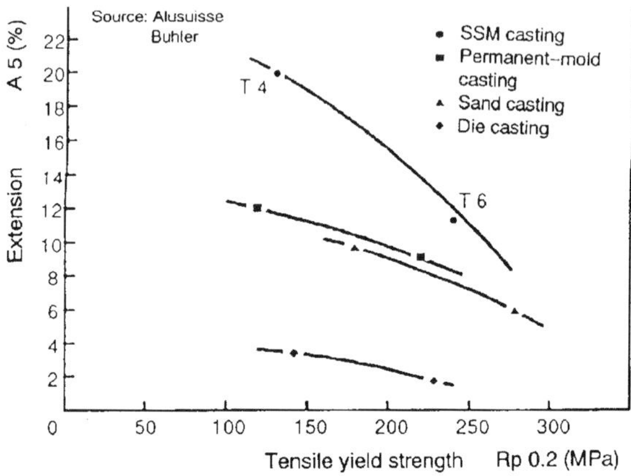

Fig 72: Properties of SSM castings compared with rivals

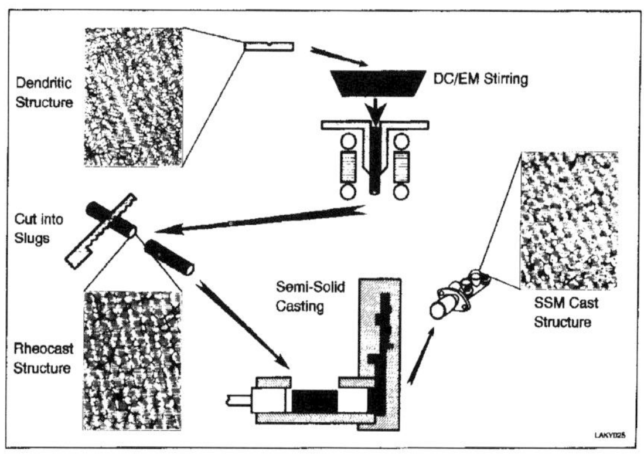

Fig 73: The SSM process

Fig 74: Potential for net-shape casting

	Ultimate Tensile Strength		Yield Strength (estimated)		Elongation
	MPa	ksi	MPa	ksi	%
T5 heat-treated 4 hours at 180°C	265	38	190	28	10
T6 heat-treated 4 hours at 520°C 4 hours at 160°C	309	45	245	36	11

Fig 75: Properties of an SSM cast shaft

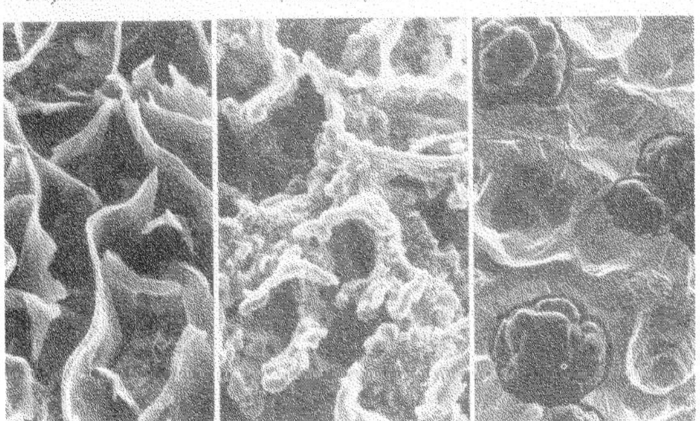

Fig 76: Microstructures compared

Fig 77: Sintercast cylinder block

73. The resulting SSM slugs retain their shape when heated but can flow like liquid, under pressure. These are currently available in type 356/357 AlSi7Mg.

Three to four metre long billets are bought from the SSM supplier and cut to length with sufficient accuracy to obtain correct shot weight. Since they retain shape during heating they can be mechanically handled into the shot sleeve to automate the production process. As the plunger compresses the metal its viscosity drops and the material flows like a plastic to fill the cavity with near perfect shape conformity and as the material is only 40% melted solidification, shrinkage is minimised and cycle times are reduced by up to 25%.

Typically a T5 heat-treatment is carried out, involving quenching straight from the die, followed by low temperature ageing (4 hours at 180 C). Fig 74 shows how considerably less machining is required on the final part compared with an equivalent permanent mould casting and the table in Fig 75 shows some typical properties obtained from the centre of a 20 mm cast shaft.

Sintercasting Compacted graphite iron is an iron matrix containing graphite particles, in a vermicular shape which provides a coral-like network, quite distinctive from grey flake iron or nodular iron, as seen in Fig 76. The continuity of graphite gives good thermal conductivity, vibration damping and machinability while the rounded edges give the high strength, stiffness and fatigue behaviour characteristic of nodular iron. Thus the material combines the benefits of grey and ductile irons and permits a highly castable tough material.

At elevated temperatures its fatigue strength is five times that of aluminium alloy which makes the material a significant contender for lightweight engine blocks and cylinder heads. Its good machinability also makes it an important contender for axle housings and the material is considerably cheaper than aluminium alloy.

Exploitation of this useful material has been made possible by the Sintercast process. Extremely accurate process control is the key and resort to titanium inclusions, used by earlier methods, is not necessary. Good control is based on Sintercast's claimed unique 'know-how' of the solidification behaviour of molten iron, a high-resolution measuring device, computer analysis techniques, which do not disturb production flow, and an automated correction of the melt composition just before casting.

Between 10-25% weight reductions have been achieved in engine blocks and it is claimed that CGI cylinder bores expand 20-30% less than with grey iron. Future 80 mpg eco-cars driven by direct injection diesels will be very strong contenders for the material as the company claim that light alloys will be unable to withstand 25:1 compression ratios and associated combustion temperatures of 630 C. Fig 77 shows a Sintercast block.

Metal-matrix composites

Fibrous or particulate reinforcements in a matrix of metal can be used to enhance physical and mechanical properties of the parent material — to the extent that Honda, for example, employed a selectively reinforced aluminium cylinder block in one of the recently introduced Prelude models. This eliminated the need for cast-iron cylinder liners and raised power-output from 135 to 140 bhp without increasing either engine size or weight.

With proprietary systems such as Duralcan, tubular components like propeller shafts, Fig 78, are being produced which promise radical improvements in specific strength and rigidity. Most MMC applications have centred around aluminium and usually involves the impregnation of a preform of ceramic fibre and the squeeze-casting process. Applications have included fibre-reinforced pistons, connecting rods and valve-train parts for engines. The fairly costly material is usually best exploited in rotating or reciprocating parts for reducing inertia as well as dead-weight of the vehicle.

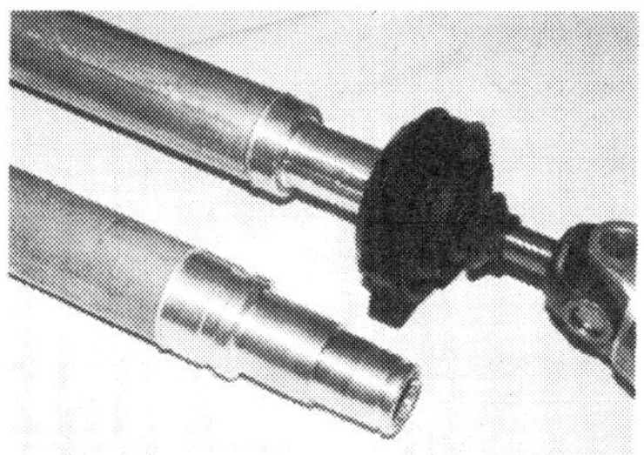

Fig 78: MMC drive-shaft

Volume Percent Al_2O_3	Ultimate Strength (ksi)	Yield Strength (ksi)	Tensile Elongation (%)	Elastic Modulus (Msi)
W6A.xxA				
0	45 (38)	40 (35)	20	10.0
10	51 (47)	43 (38)	10	11.8
15	53 (49)	47 (42)	6	12.9
20	54 (50)	51 (46)	4	14.1
W2A.xxA				
0	76 (68)	69 (60)	13	10.6
10	77 (72)	72 (66)	3	12.2
15	77 (72)	73 (67)	2	13.6
20	75 (70)	73 (67)	1	14.7

Fig 79: Tensile properties of two grades of Duralcan (maxima and minima)

Material	W6A.20A (6061/20 vol% Al_2O_3)
Condition	T8511
Nominal Tube Size	4.000" diam x 0.085" wall thickness
Diameter	±0.005"
Wall Thickness	±0.004"
Ovality	±0.006"
Eccentricity	±0.004"
Straightness	0.025" TIR maximum on a 6-ft length

Fig 80: Tolerances achieved by Duralcan drive-shaft

Infiltration of ceramic fibre preforms by the metal has been achieved with the Liquid Metal Forming (LMF) process. This involves liquid metal being displaced upwardly into a die, by gas pressure. The die is pre-evacuated and, after filling is complete, high pressures are applied to ensure pre-impregnation of the preform. Alternative approaches involve spray deposition of metal and reinforcement to produce semi-finished material which is then rolled or extruded to final size.

Production of near net shape MMCs has also been achieved by casting a slurry of particles in a metal alloy matrix. Brake disc rotors have been made by this process in Alcan's Duralcan material, an Al-SiC or Al-AL_2O_3 particulate MMC. Duralcan tube has now reached the stage of readiness for commercial exploitation; cylinder-liners and propeller shafts are seen as the other main automotive applications initially. This is now seen as a cost-effective means of achieving higher modulus, wear resistance and strength. The commonest materials, to date, contain some 20% alumina reinforcement.

In the 6061 aluminium alloy, addition of alumina increases yield and ultimate strengths as shown by the table in Fig 79, but decreases the elongation. The most noticeable property enhancement is in elastic modulus, however, which rises by 40% for 20% alumina addition. Particle distribution remains basically isotropic after extrusion and little directionality of properties with extrusion has been noted. There is some loss in toughness and an increase in density.

Because the critical speed for whirling of a propeller shaft can be expressed as:

$$(15\pi/L^2)[(E/\rho)g(R_o^2 + R_i^2)]^{1/2}$$

where L is length, R radius, r density and E elastic modulus; the only material characteristic affecting critical speed is E/r. For steel and aluminium alloy this ratio is similar but for Duralcan the specific stiffness is 32% greater, resulting in a 15% increase in critical speed. Alternatively, for the same critical speed, a 7% increase in shaft length or 13% reduction in shaft diameter is possible.

Achievable tolerances with Duralcan are shown in the table of Fig 80 and seen to be similar to those attainable with the matrix alloy. The MMC most suitable for prop-shaft tubes has the properties shown in the table of Fig 81a. In making up prop-shafts the tubes had forged aluminium alloy yokes welded to each end using 5356 alloy welding wire.

Property	Value
Yield Strength	45.7 ksi
Ultimate Tensile Strength	50.8 ksi
Tensile Elongation	2.1%
Elastic Modulus	14.2 Msi
Specific Stiffness	1.35 x 10^8 in.^{2}/s^2
Torque at Yield	
0.1% Offset	4700 ft·lb
JAEL*	4100 ft·lb
Maximum Torque	5100 ft·lb

Fig 81a: Properties of a large diameter Duralcan drive-shaft material

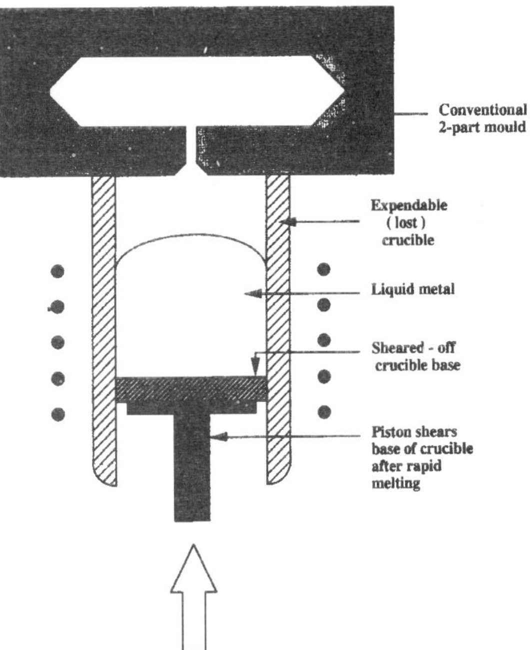

Fig 81b: Lost-crucible technique

In cast rather than extruded forms greater variety of shapes are, of course, attainable. One promising casting technique has been developed by Alcan at their Kingston, Ontario, laboratories. The 'lost-crucible' technique involves melting MMC as a slug in a single shot melting process then, when the material is at the correct temperature, a ram is pushed up the base of the crucible. This shears it off from the walls, causing the crucible base to act as a piston, with the walls acting as a cylinder, so as to displace the MMC into the mould. This avoids the problems of settling of the solid phase and reduces the chances of picking up hydrogen or oxide films, Fig 81b.

Structural timbers

The use of timber in structural and semi-structural roles was common in cars, trucks and buses — traditionally (Fig 82) — but lost ground with the increasing wage rates of skilled wood-workers and the relative cost reductions of metalworking processes. This is not to mention the increasing price of timber. There are now signs of a fresh interest in 'renewable' raw materials following official initiatives.

The European Commission has a 12 million ECU budget available over the next three years aimed at increasing availability of forest resources. There are also advances in resin-impregnated wood-fibre products which blur the boundaries with plastics-composite technology. The very high specific strength of timbers such as spruce was shown to advantage as skins of sandwich panels for the World War 2 Mosquito aircraft.

Notable automotive applications for other timbers include Frank Costin's historic Marcos structure (Fig 83). This famed vehicle structural engineer was later to use a timber monocoque for the structure of his own Costin Amigo car (Fig 84). The central portion of the Amigo structure comprised two wide and deep box-section side-booms and two transverse bulkheads, also of box section. They are connected underneath by a flat floor panel with a built-up non structural transmission tunnel along the middle. Inboard extensions of the side-booms project forward to form bearers for the engine and frontal space frame, of steel-tubular construction, which also carries the front suspension crossmember to close the torsion loop. Behind the rear bulkhead is an extension of stepped-box form with full depth side-webs to provide rear suspension anchorage points as well as enclosing luggage and spare-wheel housing.

Gaboon plywood of top-grade non-marine specification was employed having 0.8, 1.5, or 3 mm thickness. Parana pine of 12.5 and 25 mm square section was used for the joints — at the time cheaper than the Sitka spruce used in the original Marcos structure. Local reinforcement at anchorage points is particularly easily achieved by glueing on backing blocks. Adhesive used throughout was Ciba Geigy's Aerolite 300. Bonded-in diaphragms helped to stabilise the side booms and the triple-panel bulkheads are stabilised by longitudinal diaphragms. Transverse diaphragms are used in the bearer portions to take

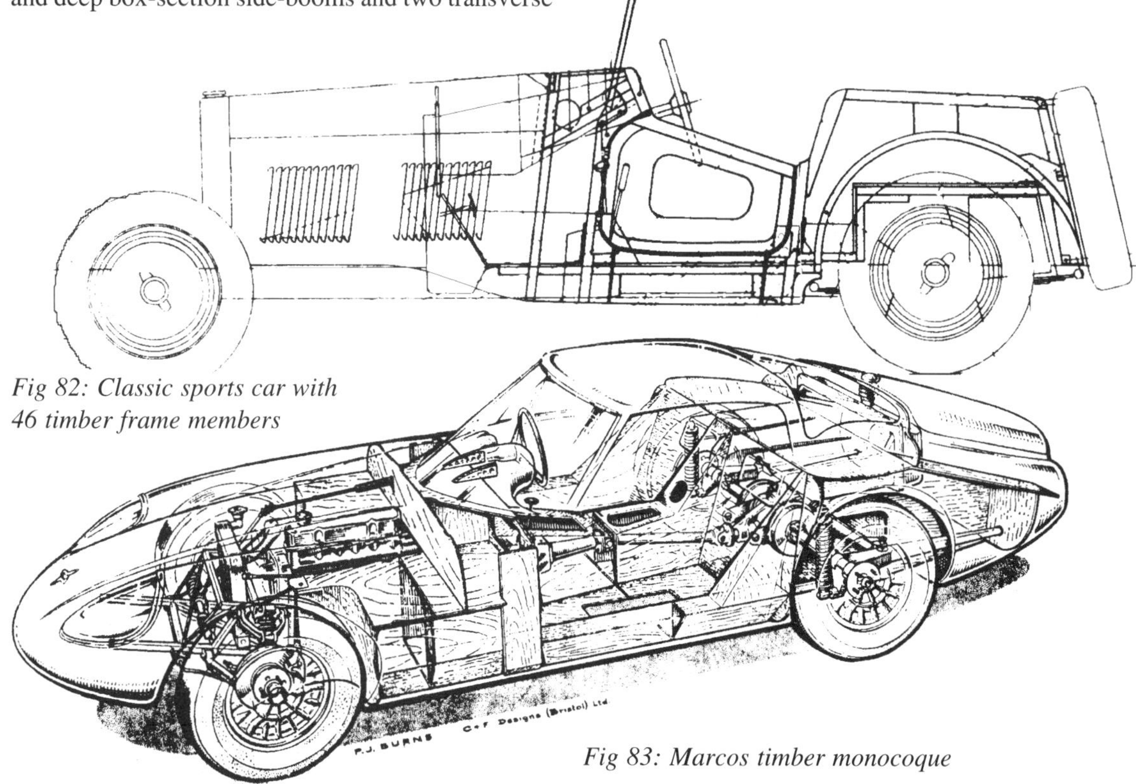

Fig 82: Classic sports car with 46 timber frame members

Fig 83: Marcos timber monocoque

shear loads from the engine mountings. For protection, the entire structure was painted with a two-part polyurethane primer and topcoat; also the top surfaces of the plywood was given a nylon sheathing to prevent abrasion and the engine bay treated with fire-resistant paint. Kerb weight of the car was 1468 lb (670 kg) and torsional stiffness over 3500 lbf ft (4745 Nm)/deg. In commercial vehicles, timber has of course played a structural role in decking and dropsides of open-platform trucks as well as in the frameworks of bus and box bodies. It also has some appeal for closure frameworks on refrigerated vehicles because of its relatively low thermal conductivity compared with metal equivalents. Structural plywoods are also widely used for decking — 12.5 mm thick material supported on 355 mm spaced crossbearers typically replacing planked decking of 35mm thick solid timber. Trim panels for cab doors are an application for moulded woodfibre and finisher panels for cab rear walls can involve 3D mouldings made by preforming rather than profiling hardboard sheet. Usually a mixture of 75% mechanical woodpulp and 25% long-fibre kraft pulp is blended in water with small quantities of resin and other additives.

The mixture is preformed on a screen to the shape of the finished product and then pressed between matching dies at 260 C with a pressure of 200 lb/in^2 (1.4 MN/m^2). One die is slotted and covered with a perforated screen to allow the water and steam to escape — thus only one side is smooth. Density varies according to resin impregnation and wall thicknesses usually vary between 0.06 to 0.12 inch (1.5-3 mm). Lateral pressure resistance for a 2.5 mm thick panel is typically 350 lbf/in^2 (2.4 MN/m^2).

Structural properties of timbers

Woods are classified as soft or hard according to whether they come from cone-bearing or broad-

Fig 84: Costin Amigo monocoque detail

Standard name	Botanical species	Density	Maximum bending strength: Equivalent fibre stress at maximum load	Stiffness: Apparent modulus of elasticity	Energy consumed to total fracture: Total work	Resistance to suddenly applied loads: Maximum height of drop of 50 lb hammer	Resistance to suddenly applied loads: Toughness. Energy consumed	Resistance to suddenly applied loads: Izod notched bar test. Energy consumed	Maximum compressive strength parallel to grain
Softwoods		lb/ft^3	lb/in^2	1000 lb/in^2	in lb/in^3	in	in lb	ft lb	lb/in^2
Fir, Douglas	*Pseudotsuga taxifolia (douglasii)*	37	14 800	2260	28·2	34	—	—	8480
Hemlock, Western	*Tsuga heterophylla*	30	11 100	1690	17·6	26	—	—	6390
Pine, Parana	*Arancaria Augustifolia*	33	14 200	1510	15·3	26	—	—	7970
Redwood (Baltic)	*Pinus sylvestris*	30	11 200	1420	—	24	70	—	6360
Spruce, European (Whitewood)	*Picea abies*	23	9700	1140	12·4	19	—	—	5110
Spruce, Sitka	*Picea sitchensis*	27	10 900	1750	22·8	26	90	6·4	5800
Hardwoods		lb/ft^3	lb/in^2	1000 lb/in^2	in lb/in^3	in	in lb	ft lb	lb/in^2
Agba	*Gossweilerodendron balsamiferum*	30	10 200	1070	12·9	21	74	6·5	5310
Ash	*Fraxinus excelsior*	42	15 100	1860	41·8	43	180	—	6990
Balsa	*Ochroma spp. Heartwood*	12	2800	470	—	—	—	1·5	1680
Balsa	*Ochroma spp. Sapwood*	17	5300	710	—	—	—	4·0	3730
Beech	*Fagus sylvatica*	43	16 200	1950	27·4	45	151	10·6	7870
Birch, Canadian yellow	*Betula lutea*	43	17 200	2270	44·6	61	—	—	8880
Dahoma	*Pipta deniasdrum Africanum*	43	15 800	1620	37·6	36	—	—	8520
Danta	*Cistanthera papaverifera*	47	16 100	1790	26·3	39	—	—	9080
Elm	*Ulmus procera*	32	9300	1090	14·6	23	68	4·5	4740
Gurjun	*Dipterocarpus spp.*	46	15 900	2260	28·0	37	—	—	8330
Idigbo	*Terminalia ivorensis*	29	12 000	1400	10·8	19	78	5·8	6320
Keruing	*Dipterocarpus cornutus*	50	18 800	3110	24·2	39	154	13·3	10 580
Makoré	*Mimusops heckelii*	44	12 900	1420	—	—	—	—	6220
Mansonia	*Mansonia altissima*	40	16 800	1690	30·1	45	169	10·2	8190
Oak	*Quercus robur, Q.petraea*	43	13 300	1560	19·1	33	131	8·7	7210
Obeche	*Triplochiton scleroxylon*	24	8400	900	11·5	21	59	5·0	4190
Opepe	*Sarcocephalus diderrichii*	47	16 500	2070	21·3	28	104	8·3	10 020
Red Muranti	*Shorea dasyphylla*	30	12 100	1620	19·9	30	—	—	6990
Sapele	*Entandrophragma cylindricum*	40	9500	1450	—	23	102	7·4	8320

Fig 85: Properties of soft and hard woods

leafed trees and does not always mean hard or soft in the physical sense. The values in Tables 1 and 2 of Fig 85 refer to air-dried timbers (12% moisture content) as for some green timbers (50% moisture content) strengths are markedly lower. Those shown are ultimate strengths which should be appropriately factored for their working design values. Grain effects are obviously most important in predicting design performances. This is to some extent less in recently introduced plywoods having 45 deg angle between grain directions of adjacent laminations. Modulus of elasticity of wood parallel to the grain is 15 to 20 times higher than at right angles to it and in plywoods, bending stress is by no means proportional to distance from the neutral axis (Fig 86).

Bending moment is related to stress by $M = KfI/y$ where f is ultimate stress, I the second moment of area of ply having grain parallel to the span and y the distance from the neutral axis to the outermost face of the included plies. K is normally taken as 1.5 for 3-plies (grain of outer plies parallel to span) and 0.85 for others. A nomograph for solution of this expression is shown as Fig 87.

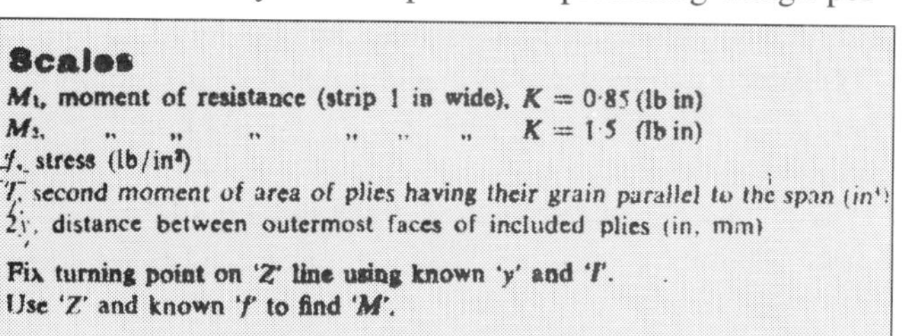
Scales

M_1, moment of resistance (strip 1 in wide), $K = 0{\cdot}85$ (lb in)
M_2, " " " " " " " $K = 1{\cdot}5$ (lb in)
f, stress (lb/in²)
I, second moment of area of plies having their grain parallel to the span (in⁴)
$2y$, distance between outermost faces of included plies (in, mm)

Fix turning point on 'Z' line using known 'y' and 'I'.
Use 'Z' and known 'f' to find 'M'.

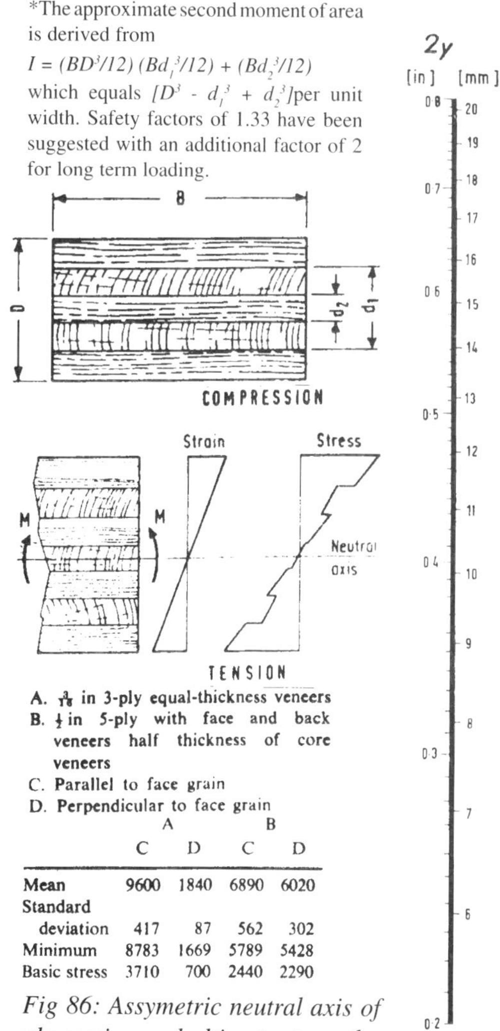

*The approximate second moment of area is derived from $I = (BD^3/12)(Bd_1^3/12) + (Bd_2^3/12)$ which equals $[D^3 - d_1^3 + d_2^3]$ per unit width. Safety factors of 1.33 have been suggested with an additional factor of 2 for long term loading.

A. $\frac{3}{16}$ in 3-ply equal-thickness veneers
B. $\frac{1}{4}$ in 5-ply with face and back veneers half thickness of core veneers
C. Parallel to face grain
D. Perpendicular to face grain

	A		B	
	C	D	C	D
Mean	9600	1840	6890	6020
Standard deviation	417	87	562	302
Minimum	8783	1669	5789	5428
Basic stress	3710	700	2440	2290

Fig 86: Assymetric neutral axis of ply section and ultimate strengths

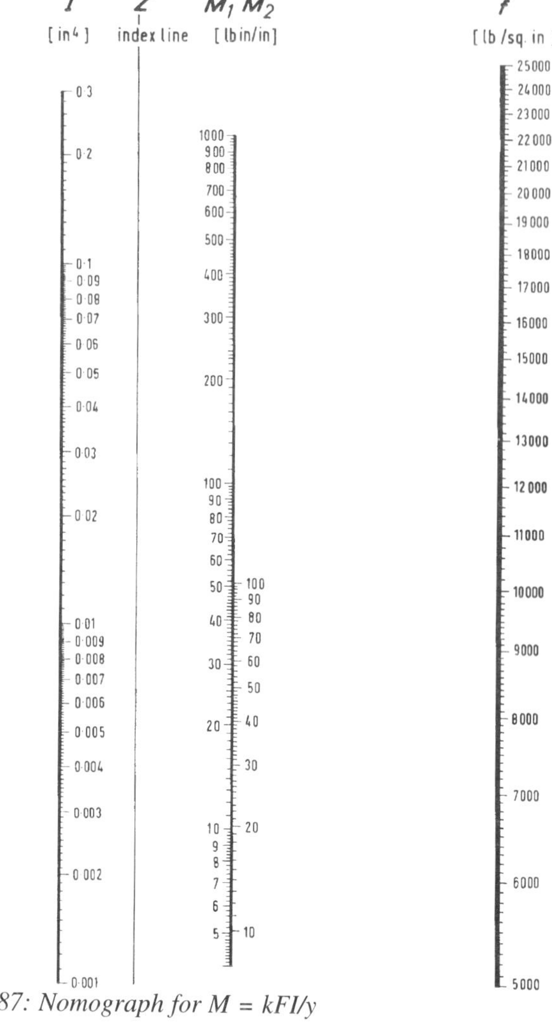

Fig 87: Nomograph for M = kFI/y

Metal injection moulding

For this process of forming metals as though they were plastics, the Israeli company Metalor Ltd has developed the capability to produce parts by MIM with high surface finish and dimensional accuracy of 0.5% which has alluded the technique since its introduction in the mid 1980s.

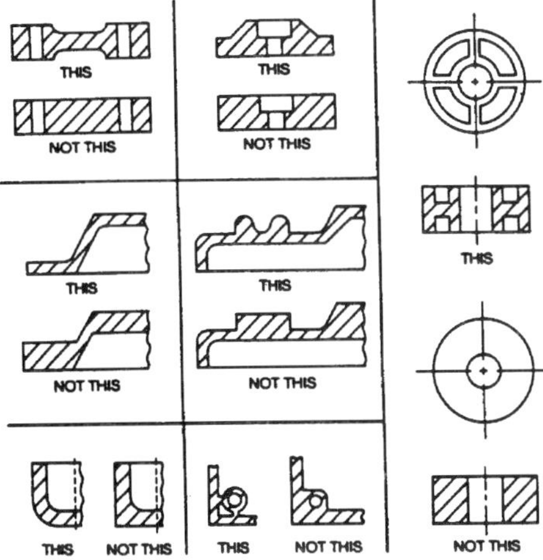

Fig 88: Uniform wall thickness and coring

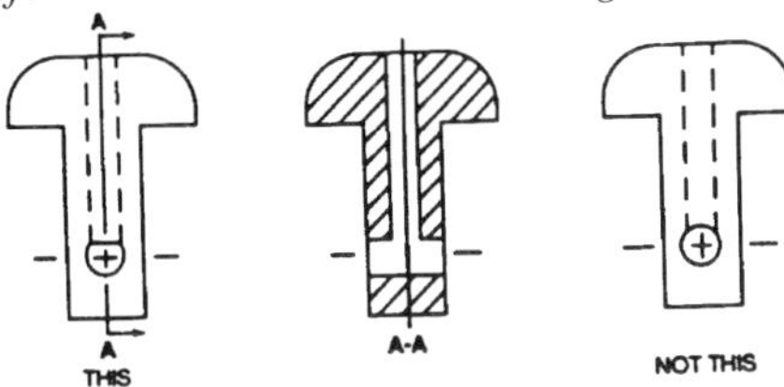

Fig 89: Holes perpendicular to one another

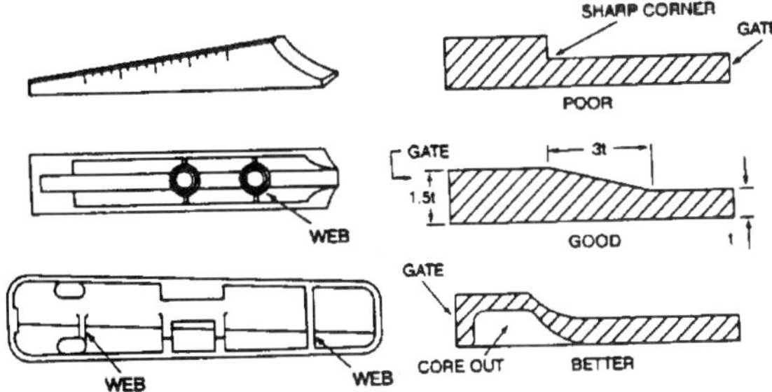

Fig 90: Webbing Fig 91: Wall thickness transition

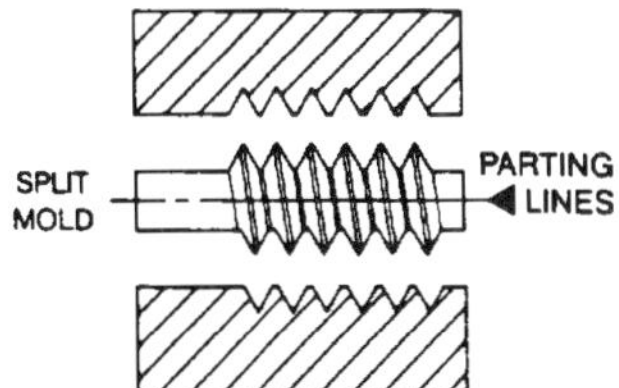

Fig 92: Moulding external threads without side core

As well as the higher accuracies the process can now achieve densities of up to 97% of the theoretical solid density of the base material and stronger components than are possible with sintering, are now said to be available. Generally, any shape that can be plastics injection moulded can now be produced in metal and production runs of less than 2000 are claimed to be economically feasible. Better control of the moulding, and subsequent drying, than is available with sintering is the secret. With shrinkage of some 18%, between initial green moulding and finished product, tooling design is paramount and currently the technique is best suited to parts of 13-150 grammes, with a size envelope of 50 x 50 x 50 mm.

Part design follows the recommendations for plastics injection moulding, with uniform wall thickness being critical to the preservation of accuracy during shrinkage. Coring can be used to attain this uniform thickness, as seen in Fig 88. This can also reduce cost by reducing material volume and cutting processing times. Often cores can be achieved by adding holes formed by pins protruding into the mould cavity. Blind holes formed by pins supported only at one end should be limited in depth to twice the pin diameter otherwise deflection might result from flow of feedstock.

In the special case, Fig 89, of holes perpendicular to one another, making one of the holes in a D-section, will allow the tooling to function better, be stronger and minimise flashing. Fig 90 shows the use of ribs to make uniform wall thickness easier to attain, but warpage, sink marks and stress concentrations must be guarded against. Where wall uniformity is impossible a gradual transition between thicknesses is desirable as seen in Fig 91. The mould should be gated at the heavier section to ensure proper packing of the feedstock. Draft angle for mould ejection can be as low as 0.5 to 2 degrees while fillers and corner radii should be from 0.4 to 0.8 mm.

Threads can be moulded into a part and the least expensive way is to locate the parting line on the centre of the thread as in Fig 92. Dimensional tolerance of +/- 3 % can normally be achieved and one third this value for very small parts. Surface finish is 0.8 microns which is better than most investment cast parts.

Powder metallurgy

According to Bloom[17], although sintering is a versatile and flexible process, there are some rules to be observed in order to get the best and most economical results, as seen in Fig 93. Unlike molten metals or thermoplastics, powdered metals do not flow. Pressing in one direction is the only thing that can be relied upon. Steps should be avoided; re-entrant angles, threads, grooves, etc, in directions other than that of the closure of the die are impossible without great complication and expense. Component length/diameter ratio should not exceed 3: 1, otherwise there may be a density gradient through the component. Holes in the direction of pressing are made by core rods incorporated in the tool set and should be so dimensioned that the cores are as short and thick as possible to avoid high stresses. Where outside radii or chamfers are to be made, a land must be provided so that the tools do not need sharp edges, which would wear badly during use. Different metals require different compaction pressures and so do different applications. These pressures are usually between 10 and 50 tonf/in^2. With the higher pressures, press capacity may be a limiting factor. There is bound to be at least 10% porosity, and where it is important to keep this within certain limits the matter should be cleared with the supplier. Porosity may also be exploited, by impregnation with oil up to about 40%.

In the production of powdered metals the base metal is melted in an electric induction furnace and released through a nozzle in the base of the melting vessel by gravity. High-pressure jets of water, air or other gas then act upon the stream of molten metal and break it down into a powder having the desired shape characteristics. After filtration and drying, the powder is milled, screened, blended, refined or heat treated, according to its ultimate use.

Various mesh sizes from 18 to - 400 are used, depending on many factors but chiefly on the composition. Carbonyl nickel powder is one of the finest, at about 5 microns. The processing method determines the fineness of the powder an important property. In the sintering technique, the powder is first compacted into the correct shape and then carried through the furnace where the actual sintering takes place. The property of cohering in the pre-sintered condition, called 'green strength', is usually lacking in very fine powders. Particle size has a bearing, too, on the degree of compaction needed, which is inconveniently high with very small particles.

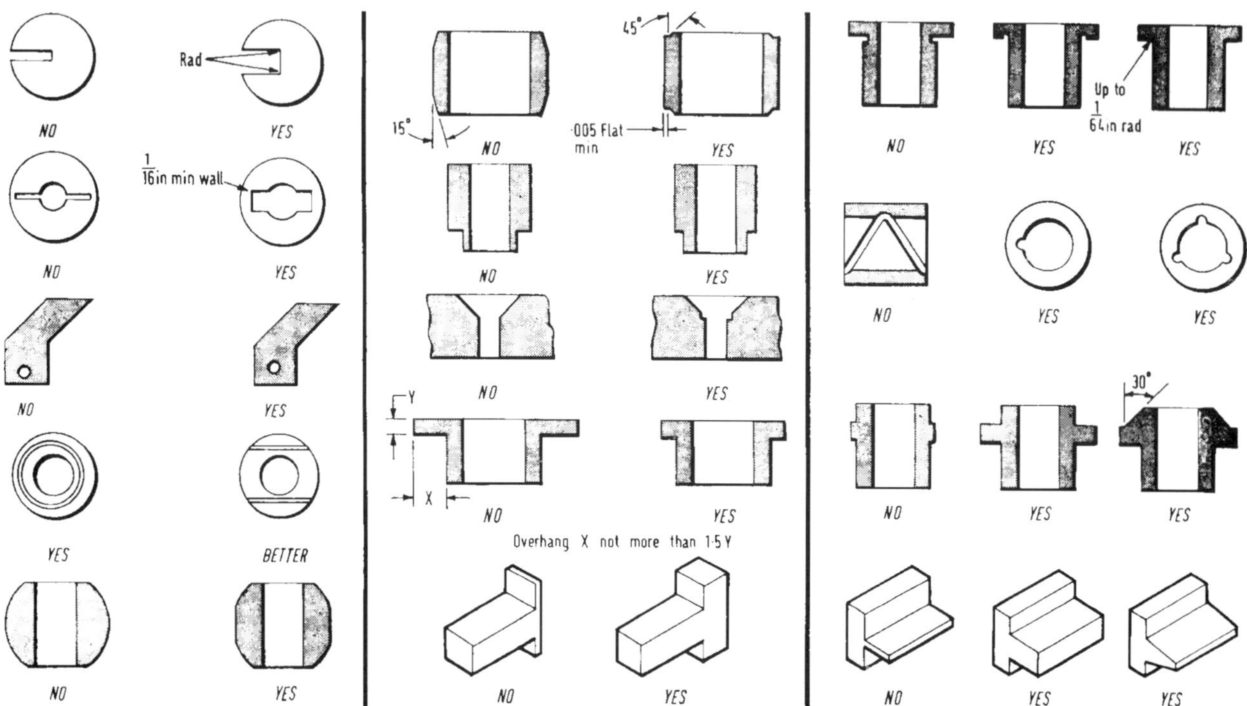

Fig 93: Dos and donts in designing for sintering

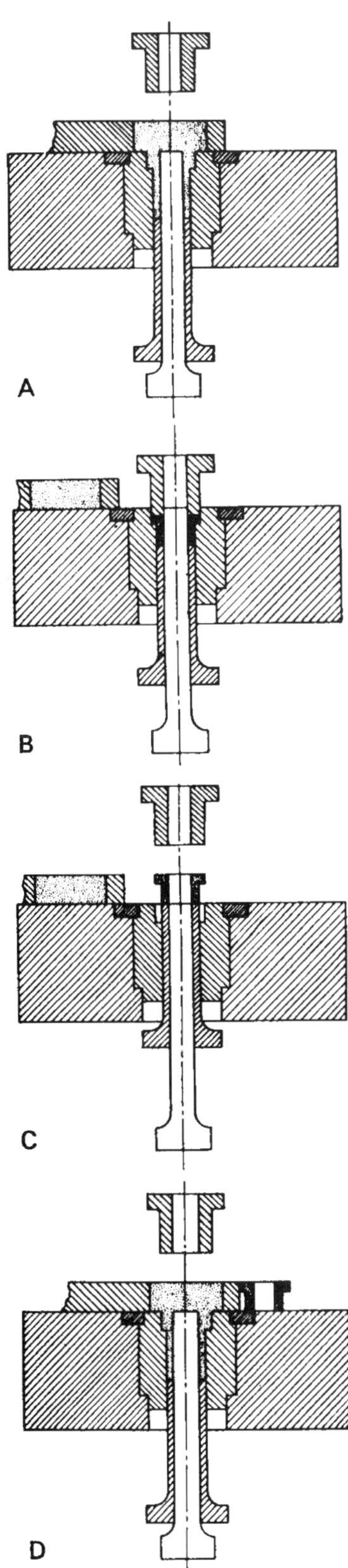

Fig 94: Firth Cleveland sintering process

In designing a component for sintering it is most important that sudden changes of section are avoided, since metal powders will not flow around corners. The process also has its own shrinkage characteristics which vary from metal to metal. A 5% nickel steel, for instance, might shrink some 0.020 in/in during production and a nickel-molybdenum steel only 0.005 in/in. To allow for this in designing the die, the composition of the powder mix must be settled before the production tools are designed, if necessary by tests conducted with a prototype component machined from a sintered billet. The sintered part can have tolerances approaching +/- 0.001 in. Sintered components are inclined to be more brittle than many, however.

With a sintered component there is no problem in specifying a work-hardening base metal since machining is unnecessary; subsequent work hardening (especially of surfaces such as teeth or splines) is exactly what is wanted. Therefore, by suitable choice of material both machining and hardening operations can be eliminated. Initial hardness of a work-hardening sintered part can be up to 190 Vickers, rising to 450 and higher as a surface rubs against another part in service. Firth Cleveland's sintering process is shown in Fig 94, illustrating four stages in the compacting of a flanged bush. The tool consists of a die held in the press table, movable top and bottom punches, and a fixed core rod positioned within the die cavity. At (a) the upper punch has been raised and the bottom punch lowered to enable the feeder shoe to fill the die cavity with powder. The amount of powder fed into the die is a predetermined ratio of between two and three times the volume of the finished part. At (b) the shoe moves away and both top and bottom punches move inwards to compact the powder into the required size and shape. In view (c) the top punch has moved away and the part is ejected by the bottom punch. At (d) the part is pushed aside by the feeder shoe which refills the die cavity and the cycle recommences.

Surface cold densification and isostatic pressing According to Langdon[18] the first of these two developing sintering techniques could be used to improve the performance of PM components in applications subject to high surface contact stresses. In the USA, this approach had already introduced PM products into bearing race applications: the same route might be used soon for transmission gear products. For enhancing shape capability with sintering, cold isostatic pressing has potential, says the author, particularly in eliminating density gradients, and consequent length/wall thickness ratio limitations, because of the low frictional forces involved.

However, the US suppliers say the means of enhancing shape capability with the most potential for automotive applications were those most closely aligned to conventional PM forming technology, Fig 95. One of these was split-die compaction, Fig 96. Another was the rotation of dies and punches during compaction, to allow the forming of helical gears. A third was the assembly of a number of simple shapes to produce a complex component: the composite camshaft was an example of this.

Fatigue performance According to the German Fraunhofer Institut, many engineers still compared and evaluated the fatigue behaviour of different alloys by using data obtained only with unnotched specimens, usually under alternating bending. It was, for example, commonly said that, owing to porosity, the fatigue behaviour of sintered steel was inferior to the behaviour of wrought steel. For this reason, some design engineers might disqualify sintered steels, it was really necessary to compare results obtained with notched specimens, as most components had geometrical discontinuities which acted as stress raisers. In the case of notches higher than two, which were usual in design, and for densities around and higher than 7 g/cm^3, similar endurance limits for wrought steels, cast nodular iron and different sintered steels had been obtained.

Pre-alloying Lindskog Consulting consider pre-alloyed powders preferable to systems involving

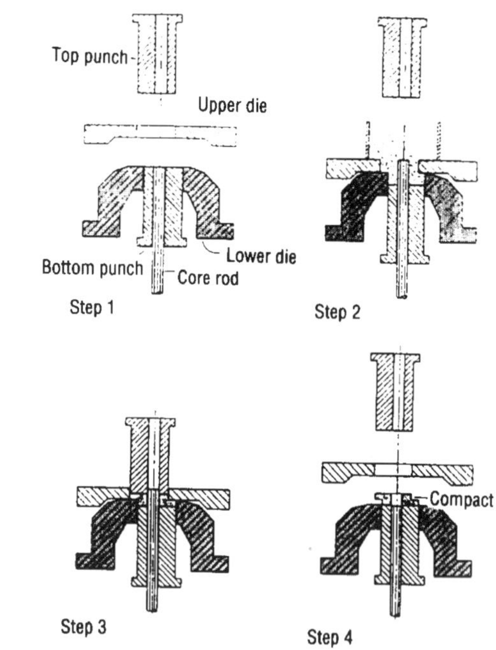

Fig 96: Bound Brook split die compaction method

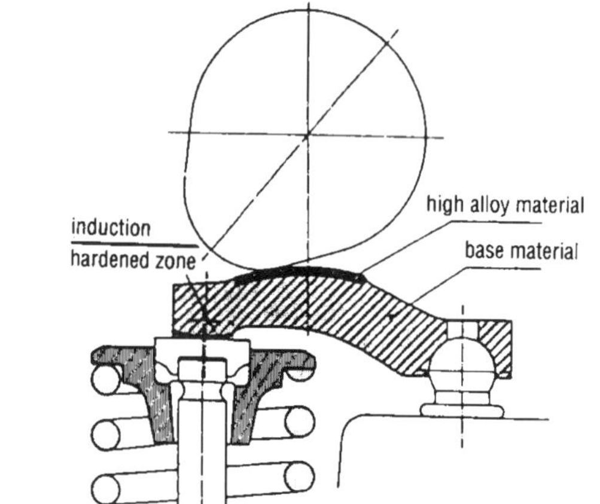

Fig 97: Miba rocker arm

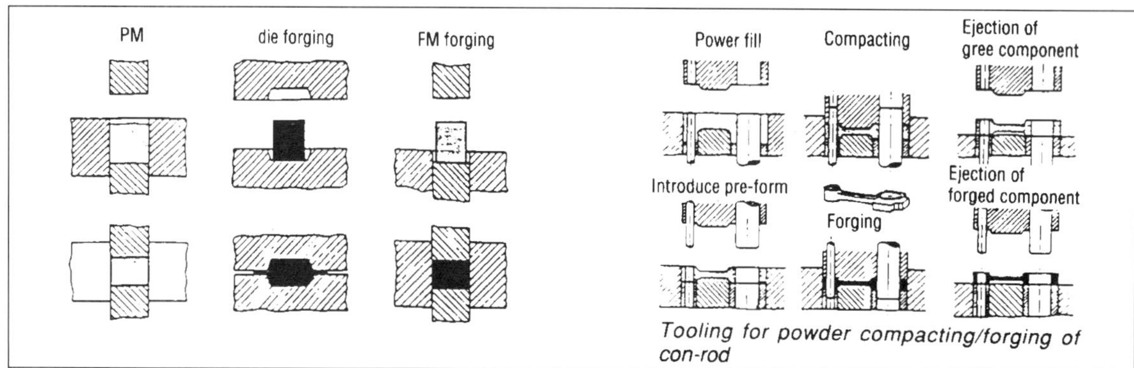

Fig 95: Tooling requirements compared

direct reduction of iron ore whenever it was important to have a completely homogeneous distribution of the alloying elements, as in PM stainless steels. Their drawback was that the alloying additions raised the strength not only of the sintered material but also of the powder particles, and hence reduced the compressibility of the powder. Alternatively, different types of powder could be bound together by either sintering or gluing, when the hardening effect of the alloying elements on the powder particles would be avoided. When bonding was achieved by sintering, the powder was termed partially pre-alloyed.

Haganas AB had developed a partially prealloyed powder called Distaloy for sintered components in manual transmissions. The requirements included a high density and higher strength than used in earlier sintered-parts as well as controlled dimensional change upon sintering. Dimensions could not be corrected by sizing owing to the hardness in the sintered condition. Distaloy contained nickel, copper and molybdenum finely dispersed and adhering to pure iron particles. While the latter particles kept their high inherent compressibility, the alloying substances were prevented from demixing. Synchronizer hubs and rings were now possible, by a simplified process in which double pressing, double sintering and sizing could be eliminated. Tensile strengths up to 1000 MPa in the as-sintered condition and, after heat treatment, up to 1700 MPa were attainable with Distaloy which produced heat-treatable components. The molybdenum addition permitted through hardening; case-hardening by carburizing or carbonitriding was also feasible.

High strength alloys PM high speed steels were now being used either alone or in combination with cheaper materials. The rocker arm tip was commonly made of sintered high speed steel and in a recent development by MIBA Sintermetall AG, the whole rocker arm was made as a PM bimetal component, the bulk of which was made from a partially prealloyed powder with copper, nickel, molybdenum and carbon: only the wear surface in contact with the cam was made from a 2 mm thick highly wear resistant molybdenum-alloyed PM steel. The surface facing the valve rod was, in this case, made wear resistant by induction hardening, Fig 97.

Metal injection-moulding is the subject of interest for the manufacture of PM automotive components according to T&N Technology. For the PM process, the metal powder is mixed with a binder to produce a plastic substance that can be injected into a split mould. The binder is removed in a subsequent heating operation, after which the moulding is sintered. With fine metal particles, metal injection moulding produces components of high density. However, shrinkage in the debinding process is high, and must be predicted precisely if accuracy is to be achieved.

Chapter 4: Polymer systems

Innovation in polymers; designing in GRP and high-strength composites; thermoplastics; designing in polypropylene; ABS and the styrenes; thermo-forming of plastics; load-bearing plastics; reinforced thermoplastics and engineering plastic alloys; semi-rigid PUR foams and sandwich panel construction; metal/polymer composites; reaction injection moulding; resin transfer moulding

Innovation in polymers

An in-depth business intelligence report[1] on automotive polymers, published early in 1996, forecast that by 2003 average weight of plastics per car was set to increase by 15% from its already substantial base. On the assumption of a global rise in car production of 44%, the authors expect a 50% growth in plastics and composites usage. Greatest expected increases are for external and structural components. Fig 1 shows a comparison of material usage expected, with potential for polypropylene rising to 2.36 million tonnes on a world annual basis by 2003.

In the more established vehicle interior polymer application area, there is likely to be a continued trend towards parts integration into single mouldings as discussed in detail in Chapter 1 (page 52). Delphi Automotive have focused considerable attention to this approach. Reductions in cost, weight and assembly time of a vehicle door hardware system are all made possible by their recently announced Super Plug system. This is the name given to a single moulding in which several components are integrated. Developed by GE Plastics and Delphi Interior and Lighting Systems, it is the result of a five-year collaborative project. Delphi adopted gas-injection moulding to achieve the objectives of the project. This is a process which obtains low weight, stiffness and dimensional stability in injection mouldings. GE Plastics developed a Xenoy PC-polyester with a 30% blend a glass fibre for Super Plug.

What makes the approach, Fig 2, truly remarkable is the number of components that are produced as one. The total is 61 for the prototype. They include the window regulator backplate, window regulator cam, window guidance channels, window motor attachments, inside remote handle support, armrest supports, wire harness attachments, speaker mounts and various mechanical fasteners. Current studies by Delphi indicate that, compared with conventional design, the potential reduction in number parts is 75%. Some benefits of this approach to door hardware systems are obvious. Assembly, for example, takes less time owing to the reduced number of components. Also, assembly lines can be shorter than normal, and less assembly tooling is needed. The lightweight Super Plug is easy to handle, and many ergonomically difficult operations in the assembly of metal door components are avoided. Cost savings are said to be from 5% to 10%. In the case of the prototype, the weight reduction for each door was 1.5 kg. Quality is also improved, and the door hardware can be delivered as a pre-tested system to the assembly line. With fewer component interfaces, a reduction in noises such as squeaks and rattles may be expected. The frame material of Super Plug absorbs motor and gear noise, too. And, compared with an assembly of metal components, there is, of course, improved resistance to corrosion.

Despite the advantages already obtained, GE Plastics believes the prototype system to be a halfway stage. The company looks forward to panels being incorporated in Super Plug. These could be door inner panels, perhaps, or the outer skins.

Gas injection-moulding Hollow plastics components are produced by an injection moulding machine when the Cinpres process is applied. Here, an inert gas is injected into the molten polymer. It follows the path of least resistance, where the polymer's viscosity is low, forming a hollow core. Weight savings of 50% can thus be obtained for thick section mouldings. During the moulding process, the timing, pressure and speed of gas injection have to be accurately controlled. Fig 3 shows the process sequence for producing the grab handle for a car interior. Gas injection moulding avoids the need to equip moulds with long, cantilevered cores. It is also an alternative to structural foam moulding: with structural foam, the surface of the moulding may be blemished by escaping gas bubbles.

Gas injection moulding can produce hollow channels through the ribs that are commonly used to reinforce thin moulded structures. The pressure of the gas forces the polymer against the mould surface during solidification, and prevents the formation sink marks. As a result, an external panel can be strengthened by ribs without the usual sink marks being present on the exposed surface. An ability to combine thick and thin sections improves freedom for the designers of mouldings.

Polymers for light-weighting

The other substantial driver for automotive polymer applications is weight reduction of the total vehicle. This has been the subject of a quite revolutionary European programme of co-operative research. Mosaic is an acronym of Material Optimization for a Structural Automotive Innovative Concept. Conceived and managed by Renault, it was given Eureka status and brought together several European producers of raw material. Research was focused on experimenting with different methods to lighten the structure of vehicles. The findings were expressed in two prototype structures based on the Renault Clio. One of these was made of steel, and achieved a weight saving of 10%. The second was a hybrid structure comprising aluminium and composites for which the potential weight saving was up to 30%. Clio specification standards were met by the composite front end, a structure extending ahead of the firewall, Fig 4. Firmness, endurance and vibration-deadening levels were described as excellent. The composite structure behaves differently from metal on impact: whereas metal side members bend, composites absorb energy through crumpling. In a head-on crash test at 50 km/h, the cabin remained intact but caving-in was slightly higher (7%) than on a production Clio.

Developed by DSM to meet the demands of the front structure in the Mosaic project, Shimoco HMC, or high modulus compound, is derived from SMC (sheet moulding compound) technology. HMC offers increased toughness without loss of strength and improved resistance to corrosion by under-bonnet fluids. These properties are due to the use of Atlac 810, a vinyl ester thermosetting resin optimum blend for Mosaic contained 45% of glass fibres, 25 mm long. Other components of the blend were mineral fillers and additives such as curing agent and pigment. An innovative design concept based on HMC, reduced the number of components from 80 originally in the Clio to nine with HMC. Among requirements were that wall thickness should be as constant as possible, while ribs and undercuts were to be avoided. Conventional glass-fibre reinforced composites have overall strength properties equal to, if not better than, steel, but their modulus is between 10 and 20 times lower and with lower elastic elongation.

	N America	Europe	Asia	ROW	**TOTAL**
PE	126	127	165	49	**467**
PU	267	592	223	144	**1226**
PP	336	969	825	196	**2326**
ABS	212	209	50	87	**558**
PVC	127	285	116	68	**596**
Eng	446	556	343	246	**1591**
THset	127	159	31	59	**376**
TOTAL	**1641**	**2897**	**1753**	**849**	**7140**

Fig 1: Estimated global usage of plastics in cars for 2003 in thousands of tonnes

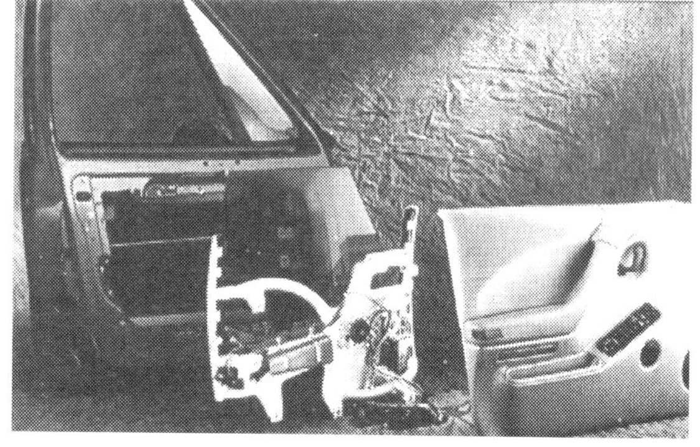

Fig 2: Delphi Automotive Super Plug system for polymer integration on an inner panel of a door assembly

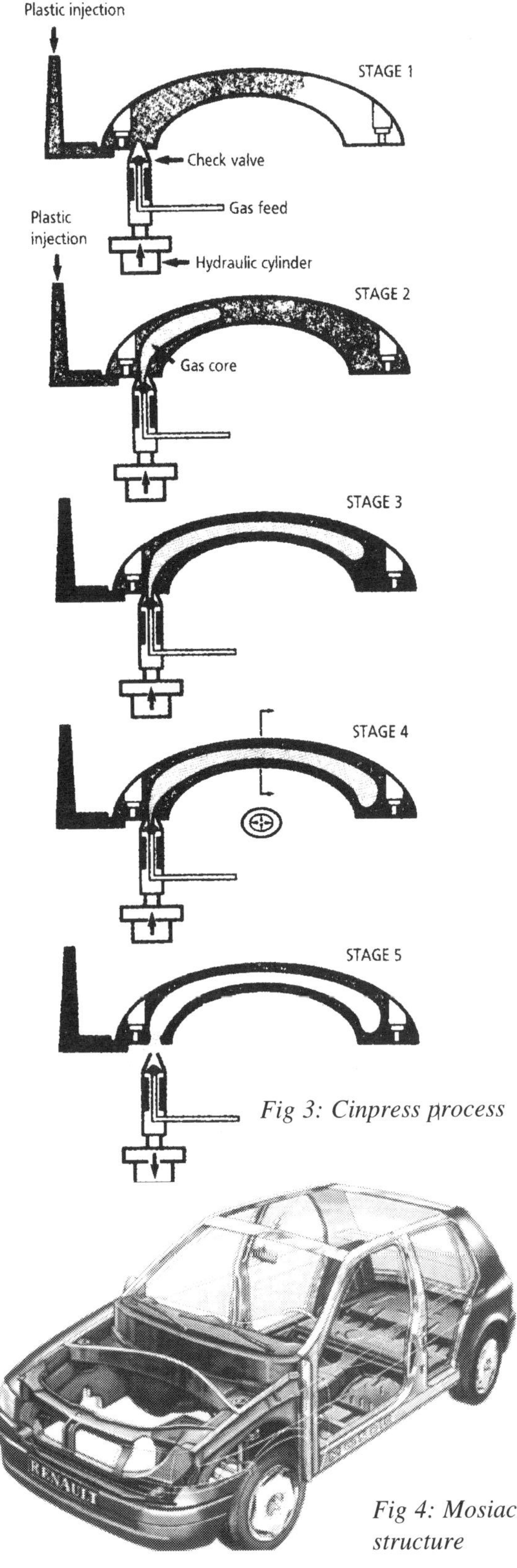

Fig 3: Cinpress process

Fig 4: Mosiac structure

The HMC was developed to meet the specific demands of a car structure. These included high elastic elongation, modulus and strength together with good fatigue resistance. A high degree of elongation without cracking and good fatigue endurance through excellent glass-fibre adhesion has been obtained from HMC. Glass content was reduced gradually from 50% to assist flow in the moulding process. Recycling has also been investigated, and retrievable HMC components can be ground down and used as filler in new HMC or SMC up to a level of 25%.

Commercial vehicle applications According to the world's largest truck maker, Daimler-Benz, an innovative approach to the application of composites could see the replacement of a number of steel parts. Carbon fibre composites, the company argues, outperform conventional aluminium alloys by a factor of three in strength and rigidity. They also have the ability to absorb considerably more impact energy, weight for weight, than metals. However that usually lack malleability and fail suddenly, without warning, as impact loads increase to the limit. D-B is looking at traditional textile technologies for optimizing the structural performance of composite reinforcements. Modified looms and braiding machines are being used to integrate the reinforcement profile into complex three dimensional parts. Sandwich structures are also being produced by the technique, involving pile-threaded cores. D-B asserts that CFRPs will sink considerably in price as demand rises but the company is not yet satisfied with the surface characteristics of such systems. Random fibre reinforced plastics are, however, already in use in the rear headrests and dash-panel of the Mercedes Station Wagon. For the company's Unimog SPV range a carbon fibre prototype cab has been developed by the Dornier aviation division. It weighs 60-70 kg against 230 kg for the steel counterpart. Resin transfer moulding techniques are used for impregnating the fibre with epoxy resin, production time being about 20 minutes. The cab is also much quieter than the steel equivalent, reportedly.

Van bodies and tractor shells Innovation is also evident in CV body construction. Sandwich panels for the Volumaker body, Fig 5a, are produced by a modified resin transfer moulding process. To obtain long life, the moulds are made of stainless steel. They are packed with continuous-filament glass and PVC foam. In addition to stiffening the panels, the foam

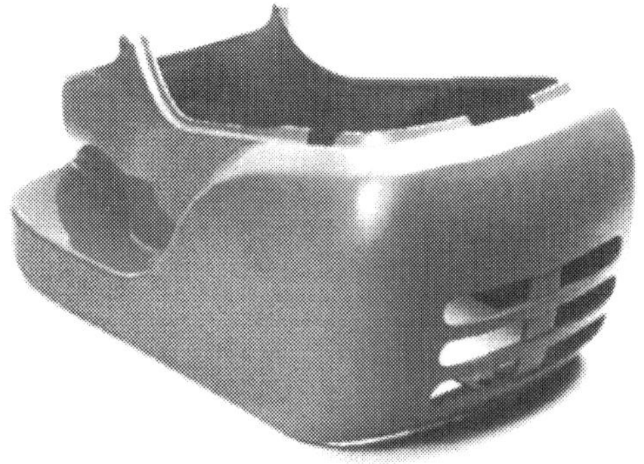

Fig 5: Polymer innovation in CV bodies and tractor shells

confers impact resistance. Colour is applied in the form of a gel coat. When the mould is closed, resin is injected and drawn into the mould by vacuum. Heating the mould obtains a high-gloss finish on the exterior skin.

Holden Hydroman, with Telenor of France, have developed Telene polymers for innovative application to farm-tractor body shells, Fig 5b. The polymers are thermoset olefins formulated with DCPD (dicyclopentadiene) - based comonomers. Owing to the relatively low viscosity of these resins, large and complex structures can be produced by the RIM process. Internal mould pressures are low in comparison with those that occur in injection moulding, so RIM reduces the initial capital expenditure in moulds. The polymers are said to have an excellent combination of properties including stiffness and impact resistance.

Designing in GRP and high-strength composites

A specialist technology is building up around GRP and allied composites and is beginning to take over as a standard construction technique for specialist sports cars and low-volume production truck cabs.

In the broad classification of particle, laminate and fibre reinforced composites, GRP falls mainly into the last category. Reinforcement can, however, range between almost particle-shaped chopped-strands to woven mats akin to those used in laminates. The resin generally used is polyester but phenolics and epoxies are two of the alternatives. The glass reinforcement can be extremely strong, depending on conditions; for example, a single filament drawn to about 12 microns has an ultimate tensile stress of 3.5 GN/m^2.

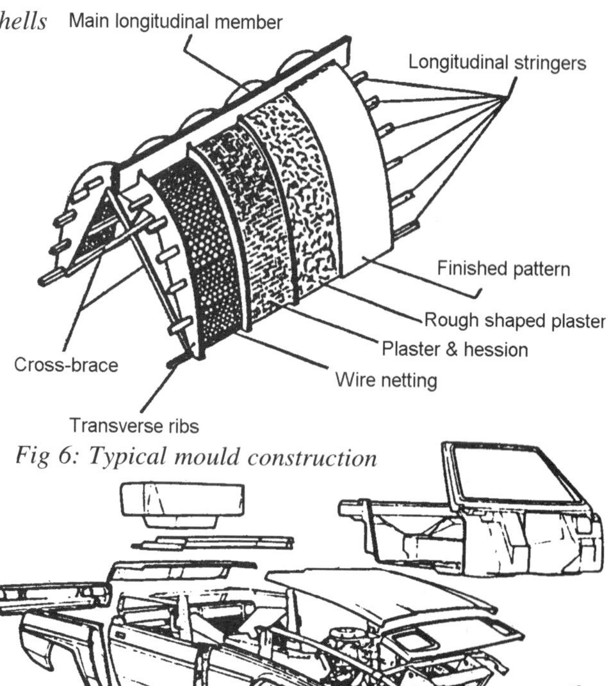

Fig 6: Typical mould construction

Fig 7: Reliant Scimitar mouldings

Fig 8: Making a rotating preform

Material	Glass fibre by weight %	Flexural strength lbf/in² x 10³	Flexural strength kN/m²	Flexural modulus lbf/in² x 10³	Flexural modulus GN/m²	Tensile at Yield lbf/in² x 10³	Tensile at Yield kN/m²	Tensile modulus lbf/in² x 10³	Tensile modulus GN/m²	Ult. Tensile elongation %	Izod Impact ft. lbf/in notch	Flammability	Continuous heat resistance °F	Continuous heat resistance °C
SMC	15-30	18-30	124-207	14-20	9.6-13.7	8-20	55-138	16-25	11-17	0.3-1.5	8-22	Slow	300-400	150-200
DMC (BMC)	15-35	10-20	69-137	14-20	9.6-13.7	4-10	28-69	16-25	11-17	0.3-0.5	2-10	to	300-400	150-200
Preform/mat	25-50	10-40	69-276	13-18	8.9-12.4	26-30	172-207	9-20	6-14	1-2	10-20	non-burning	150-400	65-200
Cold press moulding polyester	20-30	22-37	152-255	13-19	8.9-13.1	12-20	82-138	—	—	1-2	9-12	Non-burning	150-400	65-200
Spray-up polyester	30-50	16-28	110-193	10-20	6.9-13.7	9-18	62-124	8-18	5.5-12.4	1.0-1.2	4-12	Burn to SE	150-350	65-176
Moulding compound (phenolic)	5-25	18-24	124-165	30	20.6	7-14	48-117	26-29	17.9-20	0.25-0.6	1-6	SE	325-350	165-176
Nylon	6-60	7-50	48-345	2-26	1.3-18	13-33	90-228	2-20	1.3-13.7	2-10	0.8-4.5	Slow to SE	300-400	150-200
ABS	20-40	23-26	158-179	9.2-15	6.3-10.3	11-16	76-110	6-10	4.1- 6.9	3-3.4	1-2.4	Slow	200-230	93-110
Polypropylene	20-40	7-11	48-76	3.5-8.2	2.4- 5.6	5.5-10.5	38-72	4.5-9	3.1- 6.2	1-3	1-4	0.85-0.75 in/min	300-320	150-160
Thermoplastic polyester	20-35	19-29	131-200	8.7-15	6.0-10.3	14-19	96-131	13-15.5	9-10.7	1-5	1.0-2.7	Slow	275-375	135-190

Fig 9: GRP properties compared

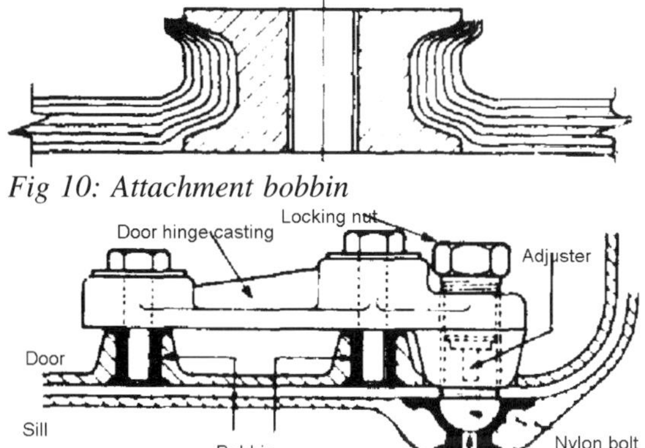

Fig 10: Attachment bobbin

Fig 11: Mechanical attachment

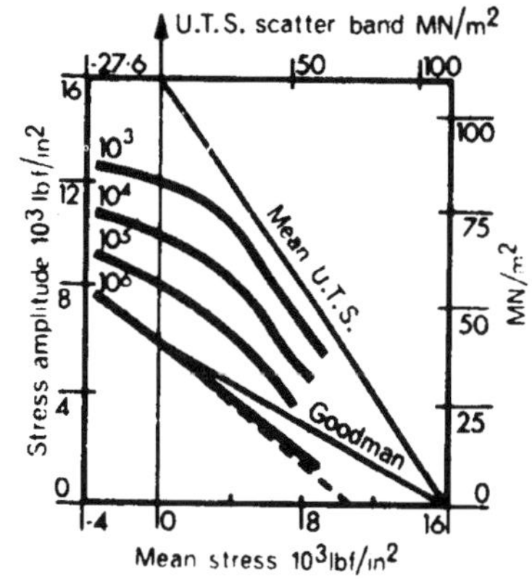

Fig 12: Goodman fatigue diagram

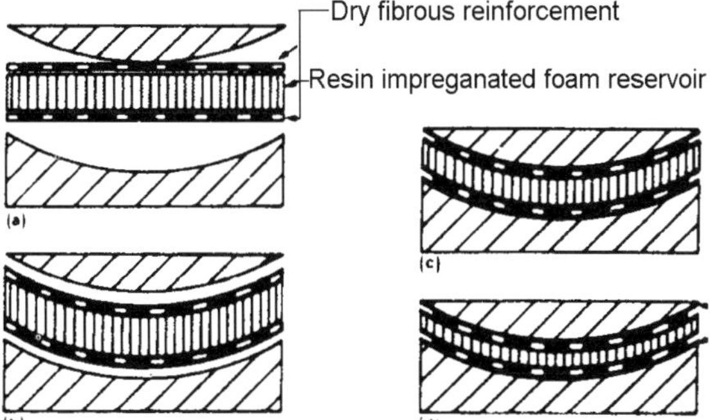

Fig 13: Reservoir moulding

The unsaturated (crystic) polyesters can be cured from liquid to solid state either with heat plus catalyst or, at room temperature, using an accelerator plus catalyst. A wide variety of techniques can be used to speed up the curing time but the present state of the art precludes GRP from very high volume manufacture (above 200,000 parts per annum). The main attraction, to low volume manufacturers is the relatively low tooling cost. A typical mould construction is shown in Fig 6, from Scott Bader's "Crystic Polyester Handbook".

A variety of methods are used for producing the composite. Traditional hand layup (bucket and brush) is still used where large complex shapes are involved; Reliant are keen exponents and use the method on the inner body moulding of their 2-seater Scimitar car, Fig 7. Its bonnet skin is made by the VARI process, perfected by Lotus, while the boot skin is a cold pressing using matched dies. Matched dies are also used to produce the body shells of the Chevrolet Corvette, at a rate of 90 units per day, working at pressures of 125,400 lbf/in² (850 kN/m²) but with dies heated to 215-300 F (102-149 C). Curing of the resins under these conditions takes 25 minutes. The glass reinforcement is preformed by spraying chopped fibres and resin binders on to rotating preform screens, Fig 8. The preform is stoved for 23 minutes before placing on the mould; a total of 89 mouldings is used in each body.

GRP material properties

A typical GRP mix for heavy duty automotive use is glass fibre 40.0, resin 40.0, monomer 0.41, filler 16.5, lubricant 0.08 and preform binder 2.0 percent. Mechanical properties for such a mix are tensile strength 24,000 lbf/in² (165 kN/m²), elastic modulus 1.4 x 10^6 lbf/in² (9.6 MN/m²) and shear strength 16,000 lbf/in² (110 kN/m²). Fig 9, due to Owens/

Corning Fibreglass, provides a more complete description of properties giving an indication of the effects of different resin systems. Failure in a plastic material occurs usually at the flexural, or cross-breaking, strength limit of the material with no definable yield point. Experience shows that minor impacts cause only a spring-back deflection while limiting stresses cause fracture but no permanent set. With hollow sections, however, tension rather than compression failure is likely to be the criterion.

Thin walls are generally desirable in press moulding for reducing cure times. Thickness should be in the range $0.3 < t < 0.25$ in ($7.6 < t < 6.35$ mm) for sheet. Design approach is to integrate parts into as small a number of individual mouldings as possible so as to save fabrication cost and avoid stress concentration. Double curvature should be used, where possible, to increase stiffness with well-radiused curves preferred to assist fibre distribution and air removal. As much taper as possible should be allowed in deep-draw mouldings and undercuts avoided to assist removal from mould. Local strength and stiffness can be enhanced by moulding in ribs, hollow sections and lightweight cores; as stress concentration is not relieved by yielding, it is best to use large area washers and metal inserts to spread shear and bending loads at attachment points. Figs 10 and 11 show mechanical attachment systems used on the Lotus Elan body. Diecast oval bobbins set in prior to moulding have high resistance to twist and tension. The door hinge attachment comprises a diecast plate into which is screwed a bolt with a nylon ball at the end. The balls engage cup bobbins laminated into the adjacent bodywork. Recommended safety factors in design are:

Short term static	2
Long term static	2
Variable unidirectional	4
Repeated	5
Reversed alternating	5
Impact loads	10

Fatigue strength has been found to lie between 0.20 and 0.35 times the mean static strength under fully reversed stress conditions at one million cycles. Fig 12 shows a Goodman diagram in which it is seen that experimental mean stress amplitude falls below the Goodman line for structural metals.

GRP special construction techniques

Reservoir moulding is a method of producing GRP sandwich panels. Here a reservoir of flexible open-cell foam is impregnated with a resin and sandwiched between two layers of dry fibre reinforcement. The three layer sandwich is placed between dies under a pressure of just 12 kg/cm^2. The foam acts as a sponge which, during moulding, squeezes the resin into the

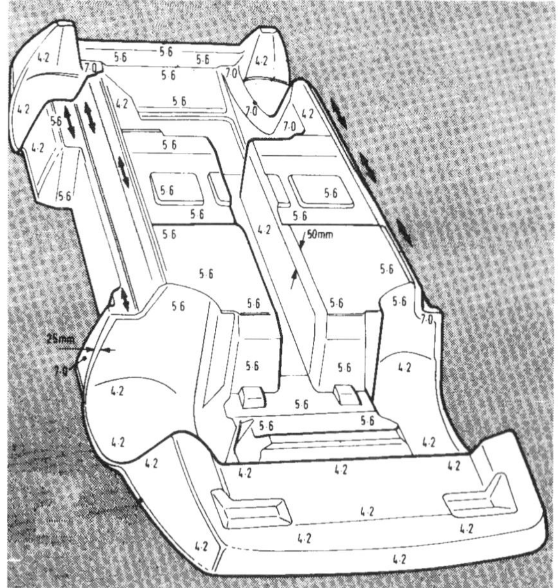

Fig 14: Weight of glass in moulding

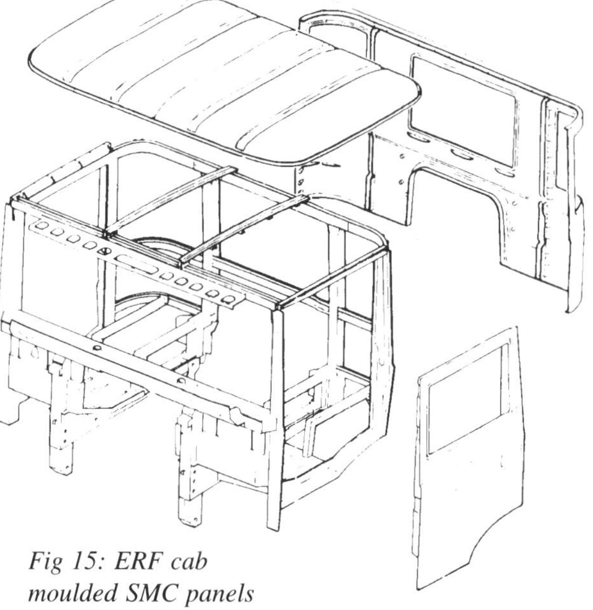

Fig 15: ERF cab moulded SMC panels

fibre, Fig 13. Sandwich stiffness can be altered by varying compression pressure and the tooling can be simplified by using one rigid and one flexible die-face such as a liquid-filled bag. The VARI (vacuum assisted resin injection) process was pioneered by Lotus for car body shells but is now available for a variety of licensed manufacture applications, including truck cabs. Here the part is made between matched mould surfaces after laying up of the reinforcement by hand. Vacuum is then applied to the space between the mould faces and resin automatically drawn in to the cavity. Preformed foam cores can also be placed in the mould to achieve localised box sections within the main part. It is also now possible to preform the glass reinforcement to speed the lay-up process. Another recent development is a technique for making metal faced moulds which can be heated to further shorten the cure cycle. Fig 14 shows the bottom half of a Lotus car structure, made by VARI, with numbered panels indicating the weight of glass (in lb/ft^2) used in each area of the moulding.

Sheet moulding compound (SMC) The ability to produce sheets of glass fibre reinforcement impregnated with a catalysed layer of polyester, within protective layers of polyethylene film has led to sheet moulding compounds (SMC). These simplify the process of matched die moulding to give relatively high production rates. Low profile resin systems can be employed for exterior body skin panels requiring high surface smoothness. Four types of SMC reinforcement are common: short chopped glass rovings in a random pattern for equal strength in all directions; endless rovings arranged parallel to chopped fibres for unidirectional strength; medium length rovings arranged parallel but staggered to obtain better mould flow than in the last case; finally a wound formation of crossed roving tapes laid at 20 deg again to effect directional strength. Fig 15 shows the larger panels of ERF's truck cab which are fitted over a steel frame.

Non-glass composites Structural performance comparable with metals can be had with the addition of materials such as para-aramids and carbon fibres. These materials are already commonplace in racing car structures and offer possibilities for such parts as slim-line windscreen pillars on road cars and trucks. The Kevlar brand of para-aramid, by Du Pont, exists in two states: firstly as a 0.12 mm thick endless fibre with cut length varying from 6 to 100 mm, secondly in the form of 'pulp'. It is said to be 30 per cent stronger than glass. The company also make a so-called meta-aramid, Nomex, which is a 'paper' that can be formed into lightweight honeycomb cores of sandwich panels. Courtaulds Carbon Fibre Division have tabulated the properties of high strength composites, compared with metals, as in Fig 16, which includes S and E-glass composites as a datum.

High-strength composites Thermosetting composites are now being specified for 'mechanical' as well as 'structural' parts. The former require close attention to form design for properly exploiting the special properties of both the reinforcing materials and the resin matrix. The National Engineering Laboratory has been prominent in the UK development of advanced composites, co-ordinating research work of a consortium of materials, component and vehicle manufacturing companies. An important outcome has been the development of a composite suspension system for light vehicles. It is based on a hot press moulded arm incorporating continuous glass reinforcement. It has been made using filament winging and resin transfer moulding techniques and follows on NEL work to co-ordinate the production of GRP springs and suspension arms.

Volvo researchers on the team came up with the twin-leaf design, Fig 17, which integrated sus-

	Tensile strength (GPa)	Tensile modulus (GPa)	Specific gravity	Specific strength	Specific modulus
A-S/epoxy	1.59	113	1.5	1.06	75
XA-S/epoxy	1.90	128	1.5	1.27	85
HM-S/epoxy	1.65	190	1.6	1.03	119
S-glass/epoxy	1.79	55	7.0	0.90	27
E-glass/epoxy	1.0	82	2.0	0.50	21
Aramid/epoxy	1.29	83	1.39	1.00	60
Steel	1.0	210	7.8	0.13	27
Aluminium L65	0.47	76	2.8	0.17	26
Titanium DTD 5173	0.96	110	4,5	0.21	25

Fig 16: High-strength composite properties

pension arms and springs in one piece. This comprised a crossmember connecting the twin-leaf trailing arm/springs. Additional constraints on the 'springs' were located at the forward end where the trailing arms attach to the vehicle and at an intermediate point restricting the arm's vertical motion. Epoxy resins were used as the matrix for the glass. Reaction blocks between the leaves were given shear flexibility by layers of composite laminated with elastomeric material; the blocks are bonded in position such that there are no sliding surfaces in the whole design.

NEL have listed the different fabrication processes for composites as shown in Fig 18. Reinforcing materials now include glass, aramid and carbon fibres — also a number of other mineral fibres and mats including cloth and paper. A number of different weaves are used with the woven fibre reinforcement configurations. These include plain-weave, with warp and weft woven alternately; satin-weave in which the weft passes beneath three warp yarns and over one — and vice-versa; twill-weave in which groups of fibres are plain woven to produce a diagonal pattern; finally uni-directional weave which signifies fibres lain together in parallel strands for maximum directional effect in physical properties.

The common thermosetting-resin systems are polyester, vinylester, epoxy, phenolic and bismalyamide. Because thermosets will not revert to their liquid state when heated, their temperature tolerance is generally higher than for thermoplastics. Compressing resin and fibres together sufficiently to get a reliable bond between them has led to systems of curing mouldings under pressure such as vacuum-bag and autoclave. Designing in press-moulded GRP involves a similar approach to form-design in unreinforced plastics. Maximum parts integration into a single moulding; increase stiffness by using double curvature and well radiused curves (this also assists fibre distribution and air removal); as much taper as possible to facilitate mould removal; use of ribs, hollow sections and lightweight cores for localised stiffness; use of large area washers and metal inserts to spread shear and bearing loads at attachment points (as stress concentrations are not relieved by yielding as with ductile metals): these are the main ground rules. Fig 19 gives design rules in pictorial form.

Thermoplastic materials/composites

Separate, related, features deal with grp composites and moulding design for ease of production; here we examine the properties of unreinforced thermoplastics and some of the application opportunities. These materials span end-uses from soft trim through to mechanical linkages and the choice of polymer system is crucial to a proper exploitation of properties.

Classification of thermoplastics is best rationalised by going back to the beginnings of synthetic materials and tracing the family heritage of some of the present key contenders amongst the existing family of useful polymers. The 1860s saw some of the early successes with synthetic materials based on the

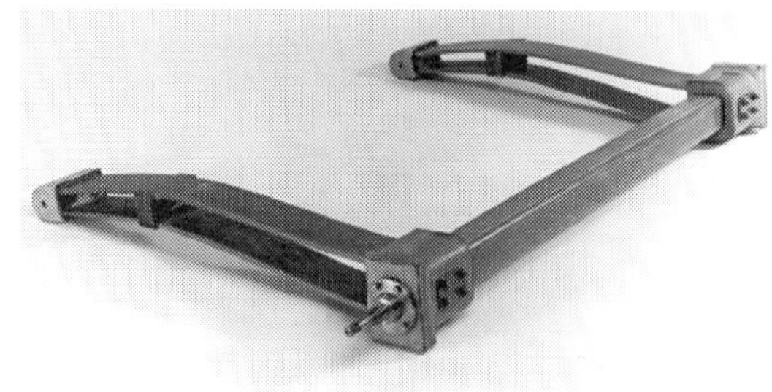

Fig 17: Volvo twin-leaf composite spring

Hand lay-up	**Layers of reinforcing mat or fabric are cut to fit, laid in the mould, and saturated with resin mix. The product has only one finished side. Boat hulls are often manufactured with this process.**
Spray-up	**The resin/reinforcement mixture is sprayed into a mould of relatively complex shape. Used for objects like tubs and showers.**
Pultrusion	**As rods and tubes are pulled through a moulding machine, resin is applied. This process is used when high strength along the length of the finished product is desired. Fishing poles and construction beams are fabricated with this process.**
Compression moulding	**If volumejustifies tooling and set-up cost, this is the most efficient way to produce large, contoured, high-strength components. The resin/reinforcement mixture is applied between matched male and female parts and subjected to high heat and pressure.**
Injection moulding	**Resin and reinforcement are forced by a screw or plunger into a matched metal mould where it takes shape and cools. This process is suitable for manufacturing a large quantity of small parts quickly. Decorative knobs and handles, pulley wheels, and electrical junction boxes are examples of injection moulded products.**
Filament winding	**Fibre is would around a suitably shaped mandrel in a predetermined pattern. Resin impregnation can occur during or after winding. This process produces products with exceptional strength. Underground storage tanks and pipes are the most common filament-wound products.**

Fig 18: Fabrication processes

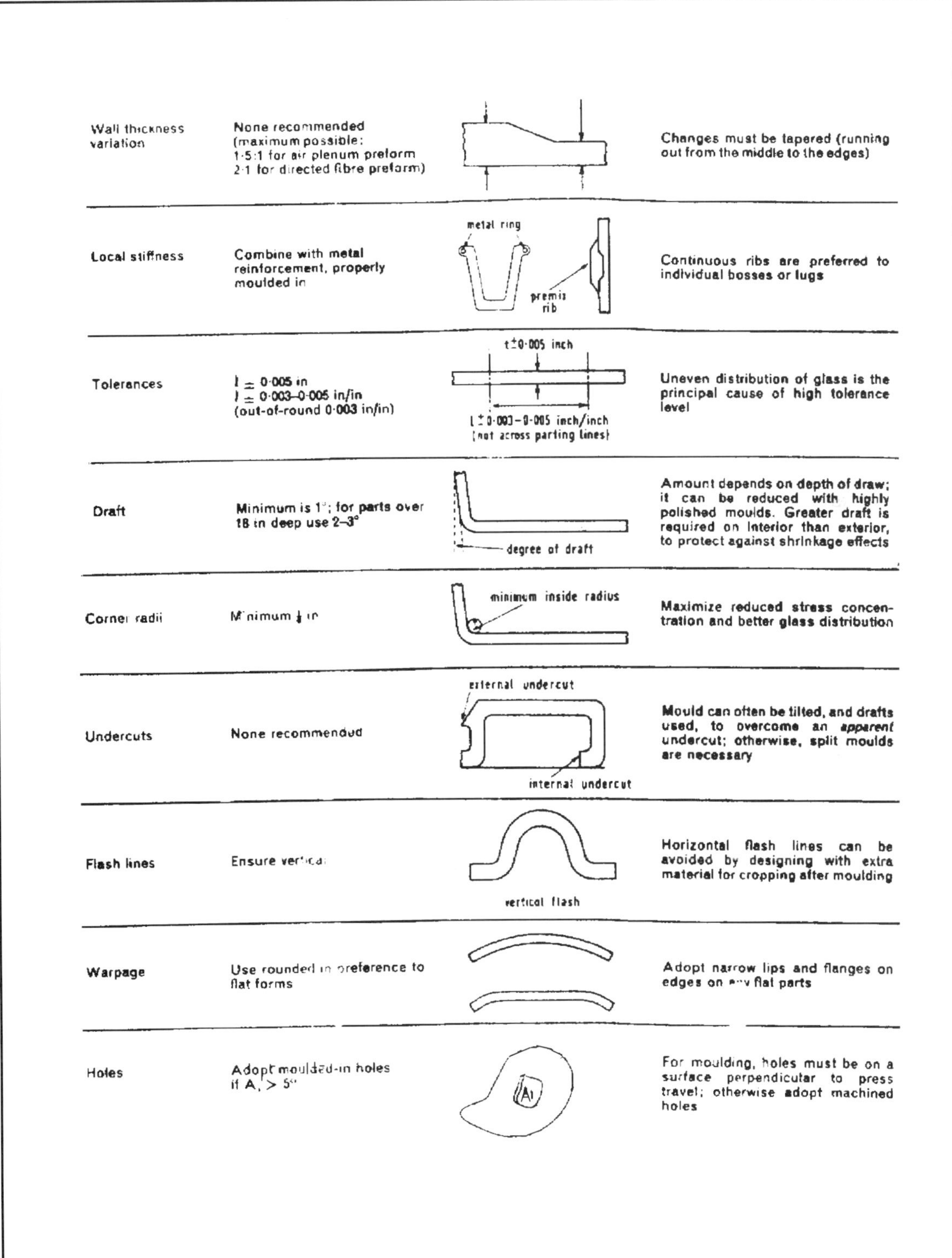

Wall thickness variation	None recommended (maximum possible: 1·5:1 for air plenum preform 2·1 for directed fibre preform)		Changes must be tapered (running out from the middle to the edges)
Local stiffness	Combine with metal reinforcement, properly moulded in	metal ring; premix rib	Continuous ribs are preferred to individual bosses or lugs
Tolerances	t ± 0·005 in l ± 0·003–0·005 in/in (out-of-round 0·003 in/in)	t ± 0·005 inch; l ± 0·003–0·005 inch/inch (not across parting lines)	Uneven distribution of glass is the principal cause of high tolerance level
Draft	Minimum is 1°; for parts over 18 in deep use 2–3°	degree of draft	Amount depends on depth of draw; it can be reduced with highly polished moulds. Greater draft is required on interior than exterior, to protect against shrinkage effects
Corner radii	Minimum ¼ in	minimum inside radius	Maximize reduced stress concentration and better glass distribution
Undercuts	None recommended	external undercut; internal undercut	Mould can often be tilted, and drafts used, to overcome an *apparent* undercut; otherwise, split moulds are necessary
Flash lines	Ensure vertical	vertical flash	Horizontal flash lines can be avoided by designing with extra material for cropping after moulding
Warpage	Use rounded in preference to flat forms		Adopt narrow lips and flanges on edges on any flat parts
Holes	Adopt moulded-in holes if $A_1 > 5°$	A_1	For moulding, holes must be on a surface perpendicular to press travel; otherwise adopt machined holes

Fig 19: Design rules

co-valent bond with carbon of such elements as hydrogen, oxygen, nitrogen and chlorine. The large molecule 'polymers' (meaning 'many bricks'), having extended chains of elements, gave rise to solid substances with unusually strong forces holding the elements together. These forces were modified with plasticisers to achieve varying degrees of flexibility and, with those materials known as thermoplastics, by applied heat which causes melting rather than decomposition. Operating extremes of rigid thermoplastics are defined by low and high temperatures leading to either 'glass-like' or 'rubber-like' states. As well as plasticisers, fillers are added which reduce cost for a given bulk; also stabilisers to improve weathering, colourisers and, to improve mould flow, lubricants.

PVC exists in many forms according to the amount of plasticiser added; cracking after long use usually signifies loss of plasticiser. Polythene exists in both crystalline and amorphous states representing two markedly different densities; as the material is difficult to keep stable under ultraviolet light, it is used mainly in hidden situations such as cable insulation. Polypropylene and acetal resins are similar in derivation but very much tougher. Polyamides, like nylon, come from the separate phenol resin derivative, with the ability to align crystals in the direction of a common fibre giving rise to relatively high strengths though when untreated, water absorption is also high. Another resin derivative is polyester leading to the styrene polymers such as ABS. Higher strength polymers come from the epoxy resin branch. Non crystaline plastics, with lightly cross-linked polymer chains, are the synthetic 'rubbers'; as well as the flexible forms, certain polyurathane rubbers can be cast as solids.

Engineering thermoplastics

As well as injection-moulding, these can be thermoformed, extruded and blow-moulded and are often reinforced. Projected price reduction, relative to hot-rolled steel and aluminium, suggests economic substitution by the early 1990s with likely exploitation in all classes of special vehicle bodywork. In Europe the main, multi-application, contenders in engineering plastics are considered by the British Plastics Federation to be ABS, polyamides, polyacetals, polycarbonates, polyphenylene oxides and reinforced polypropylenes.

Polypropylenes and polyamides can be modified in many ways to improve tensile strength, creep resistance, thermal expansion characteristic, low-temperature impact strength and high-temperature stiffness. Chemical modification includes addition of ethylene as a co-monomer to polypropylene and rubber to polyamides — both to obtain greater impact strength. Particulate fillers, such as glass and mineral beads, can be used to improve heat distortion temperature, rigidity, dimensional stability and creep resistance. Thirdly, the use of glass and other fibres (usually shorter than those used with thermosetting resins in GRP composites) can improve heat ageing as well as most of the characteristics listed above — but with the disadvantage of causing some directionality of strength and rigidity.

Generally the amorphous thermoplastics — PC, PPO, and ABS — are valued for their low shrinkage and good impact strength but fatigue and chemical resistance characteristics may not suit certain operating environments also reinforcement is said to do little for these polymers. By contrast, crystalline thermoplastics such as PA and PP have a

Material	Flexural modulus GPa	Notched Izod impact strength J/m	Heat distortion temperature °C
Polypropylene homopolymer	1.7	45	65
Polypropylene + 40% calcium carbonate	2.8	53	75
Polypropylene + 40% talc	4.2	35	97
Polypropylene + 30% coupled glass fibre	6.5	100	148
Development grades PXC 51370	3.3-5.3	300-230	144-152

Fig 20: Polypropylene co-polymer properties

To meet motor industry requirements for adhesion, water-soak and stone-chip resistance:
1 Apply adhesion-promoting primer and dry in hot air from 70-90 C.
2. Apply surfacer coat for exterior parts, typically an alkyd stoved at 120-140 C.
3. Apply solid colour top-coat, typically a polyester system stoved at 135 C.

Fig 21: Painting procedure for polypropylene

lower temperature tolerance — but this can be corrected by reinforcement. PP has been used as a fascia panel material, with both mineral and rubber additions — in a polymer termed olefinic thermoplastic elastomer (OTE). Glass reinforced versions can be used for such applications as bumpers and spoilers. Further specialist polypropylene co-polymers have properties with potential for exterior body panel application, Fig 20. A suitable painting procedure for such panels is shown in Fig 21. A stampable sheet form of the material is particularly attractive for general structural and enclosure work — with properties as shown in Fig 22. Here a continuous glass mat is used for the reinforcement, giving good strength but not particularly smooth surface finish. Complicated ribs and bosses can readily be formed into finish structures and moulding times are from 20-50 seconds using cheap cast metal tooling.

Stampable sheet for bodywork

Understanding of plastics fabrication as a 'flow' process, in which there is only one step to a completely finished part, is crucial to successful design. As the material flows into the heated mould cavity, it permits varied section thicknesses, bosses, ribbing and complex contours. As the finish plastic has an 'elastic' modulus about 1/30th that of steel it will normally be required to have a thickness three times that which is effective for a design in steel. The ability to use ribbing to obtain flexural rigidity does, however, mean that reduced thicknesses are possible. Fig 23, due to the US company Allied Fibres & Plastics, shows the effectiveness of different rib types. The technique can also be used to stiffen brackets by the addition of a flange as shown in Fig 24. Also corrugated walls can be a useful form of stiffening, Fig 25.

At points of attachment, either spreader-washers, for through bolts, or metal inserts, for tapped holes, should be used. Technique for calculating stresses for a press-fit insert are shown in Fig 26.

A common stampable thermoplastic is Radlite GMT which is 30-40 per cent glass-reinforced polypropylene; this is suitable for non-painted semi-structural applications such as aerodynamic undertrays. Typical properties are shown in Fig 27

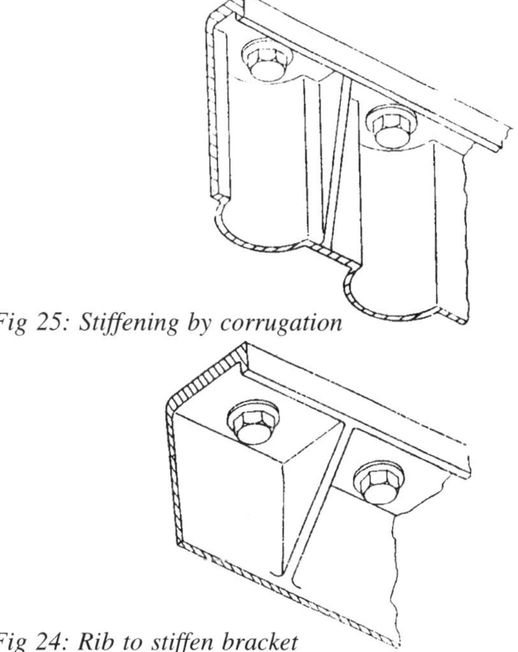

Fig 25: Stiffening by corrugation

Fig 24: Rib to stiffen bracket

Property	HW60GR30	'Azdel'
Density/(g/cm^3)	1.12	1.19
Flexural modulus/GPa	6.5	5.5
Tensile yield stress/MPa	86.2	75
Heat distortion temperature/°C (18.6 kg/cm^2)	148	155
Notched Izod impact strength/(J/m)	100	535
Coefficient of thermal expansion/(10^{-6} °C^{-1})	3	2.7

Fig 22: Properties of Azdel stampable pp sheet

Case	Shape	Change	Moment of inertia	Increase in I ΔI	Increase in weight ΔW	Ratio I/wt $\Delta I/\Delta w$
1		Base 2" x 1/4"	0.0026			
2		Double height	0.0208	700%	100%	7
3		Add 1/8" wx 1/4" H Rib	0.0048	85%	6.25%	14
4		Add 1/4" wx 1/4" H Rib	0.0064	146%	12.5%	12
5		Add 1/8" wx 1/2" H Rib	0.0118	354%	12.5%	28
6		Add 1/4" wx 1/2" H Rib	0.0194	646%	25%	26

Fig 23: Ribbing configurations

and modulus distribution in a flow-formed part in Fig 28. Effect of mould flow is to produce glass orientation which can give directionality of strength and stiffness. Usually the 'flats' receive the greatest moulding pressure and thus have the best material properties. Side walls are moulded at lower pressure which results in lower strength. If the flow path is restricted by a long thin area, the resin can 'outrun' the glass resulting in further loss of strength. The moulding process involves cutting sheet into specific blank sizes, preheating the blanks in an infrared oven and placing them in a presstool. Pressure is applied for 30-45 seconds.

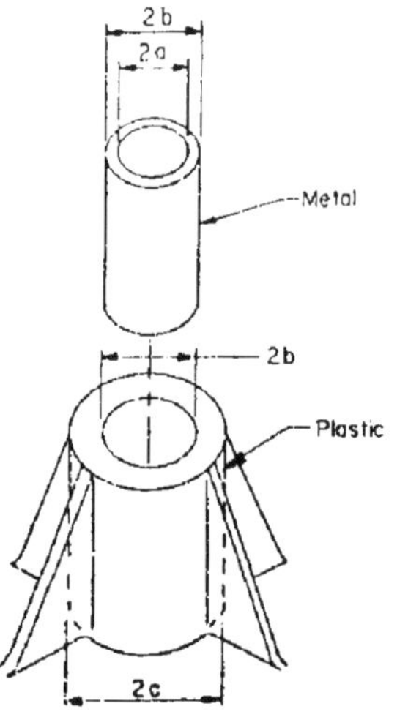

$$\Delta = pb\left[\frac{1}{E_1}\left(\frac{b^2+c^2}{c^2-b^2}+\nu\right)+\frac{1}{E_2}\left(\frac{b^2+a^2}{b^2-a^2}-\nu\right)\right] = pbK$$

$$K = \frac{1}{E_1}\left(\frac{b^2+c^2}{c^2-b^2}+\nu\right)+\frac{1}{E_2}\left(\frac{b^2+a^2}{b^2-a^2}-\nu\right)$$

$$P = \frac{\Delta}{bK}$$

Δ = Sum of radial deformations

$E_{1,2}$ = Moduli of Elasticity of Outside & Inside Cylinders

P = Shrink Fit Pressure

S_D = Allowable Design Stress

ν = Poisson's Ratio

$$\max S_{hoop} = P\frac{b^2+c^2}{c^2-b^2}$$

$$S_D > S_{hoop}$$

Fig 26: Press-fit insert stress calculation

ABS and the styrene plastics

For semi-structural applications at relatively low cost ABS is a useful plastic. Related styrene plastics such as SAN also have significant automotive application as does the sheet and laminate forms of ABS

The styrene plastics involve four main products of commercial interest, styrene homopolymer, butadiene-styrene (BS) copolymer, styrene-acrylonitrile (SAN) copolymer and the familiar ABS copolymers. Styrene homopolymer has limited use at ambient, or above, temperatures but toughened polystyrene has structural uses, particularly in expanded form, as a self-skinning foam. SAN is best known for such applications as household cups but has a structural application in an 'alloyed' form to be discussed below. ABS is one such alloy, different grades of which depend on relative proportions of nitrile rubber, butadiene, styrene and SAN. Alloying with PVC and polycarbonate also yields useful properties. The material is, of course, widely used for interior body

Composition by weight			
polypropylene:	75	60	per c
glass fibre, 13mm	25	40	per c
Density (consolidated)	1.08	1.22	g/cc
Flexural Properties			
modulus	5500	6500	MPa
peak strength	130	165	MPa
Heat Distortion Temperature (1.8 MPa)	154	159	°C
Impact Strength			
(notched) Izod	400	650	J/m
(Notched) Charpy	45	70	KJ/m²
Falling Dart (3.5mm thickness)	17	20	J

Fig 27: Properties of Radlite

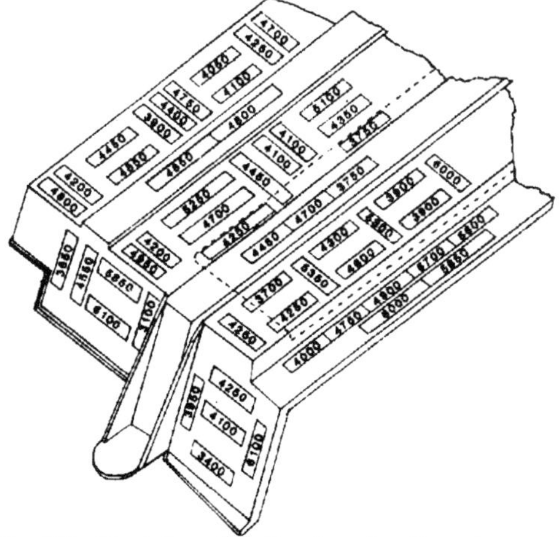

Fig 28: Modulus distribution for flow-formed part

As the butadiene content is reduced,

these properties diminish	*these properties increase*
Impact resistance	Tensile strength
Elongation	Creep resistance
Deflexion under load	Rigidity
Flex life	Flexural strength
Opacity	Gloss
	Abrasion resistance
	Chemical resistance
	Surface hardness
	Translucency

Fig 29: Varying butadiene content

	styrene acrylonitrile copolymer	Lowest Type 1	Intermediate Type 2	High Type 3	Highest Type 4	Heat resistant
Impact, Izod, notched ⅛ in × ½ in bar (ft lb/in)						
@ 73°F	0·40	1·1	2·0	3·2	4·3	2·0
@ 0°F	—	0·8	1·4	1·7	2·0	—
@ − 40°F	0·40	0·6	0·8	1·0	1·4	0·5–1·0
Elongation (%)	2·5	10–15	15–20	20–25	25–40	5–10
Tensile strength at yield (lb/in²)	11 500	8000	7000	6500	6000	8500
Flexural strength at yield (lb/in²)	15 500	13 000	11 500	10 000	9000	12 000
Flexural modulus (lb/in² × 10⁻⁵)	5·1	4·3	3·8	3·5	3·2	3·5–4·0
Rockwell hardness (R scale)	122	113	110	95	90	112
Heat distortion temp. (°F), 264 lb/in² load	205	190–196	186–193	185–191	183–190	225

Fig 30: Physical properties: ABS derivatives of SAN

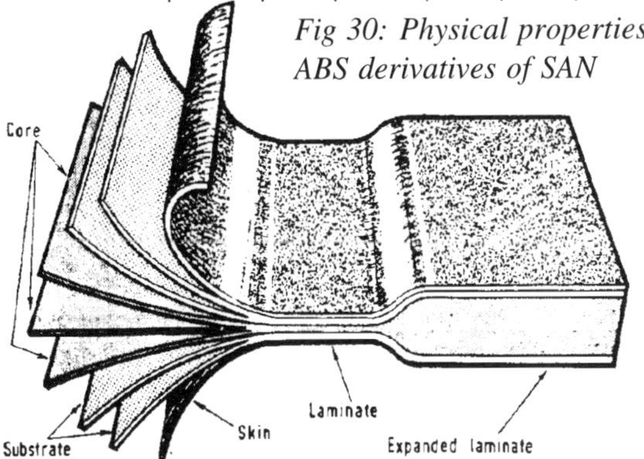

Fig 31: Expanded laminate

hardware and is ideal for ducting and utility trim-panels, under the general heading of 'light-engineering' plastics. General grades suit applications where heat requirements are not excessive. ABS alloys can range from low-impact high-rigidity grades to very high impact ones having considerable flexibility. The effect of varying the butadiene content is seen in the table of Fig 29; its complete removal from ABS leaves SAN. This last material possesses unusually high rigidity and is transparent so has found automotive application in cover panels for dashboard instruments. Typical physical properties in quantitative terms are in Fig 30. The distinguishing properties corresponding to most of the ABS copolymers are easy fabrication, good chemical resistance and outstanding dimensional stability within temperature limitations. Durability and creep resistance under load are both good by plastics standards.

The temperature-dependence of mechanical properties requires that special steps need to be taken in design calculation. Sheet forms of ABS can be calendered to obtain a range of surface textures, including simulated leather frequently used in automotive trim panels. Transparent SAN can also be used in conjunction with vacuum-metalizing to obtain pleasing display panels. Electro-plating of ABS is also a valuable surface treatment property — unsurpassed by most other thermoplastics. Details of plating ABS were given in the earlier section on coatings for plastics in Chapter 3, as was the thermoforming of ABS. An interesting variant on the sheet form is expanded laminate, Fig 31, which can also be thermoformed and is rigid enough for vehicle body shell construction. The performance of Royalex grade, from US Rubber, is seen in Fig 32; this brand has been used in cab production for the White Truck Corporation and in a variant of the Ford Cobra as a body shell material.

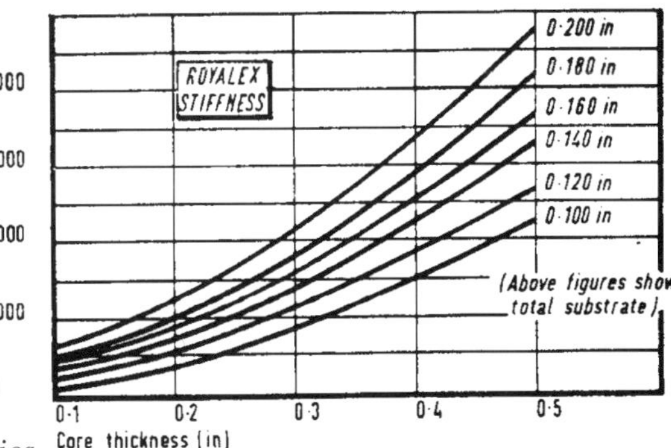

Fig 32: Stiffness characteristics

Thermoforming plastic sheet

Vacuum and pressure forming are used for shaping thermoplastic sheet around a former and as low-cost processes, are attractive for small runs of relatively low stress-bearing materials. Electric car bodies have been formed in materials such as polypropylene co-polymer and acrylic. More common automotive uses, however, are for cab and body, Fig 33, interior parts in ABS and its co-polymers.

ABS is particularly suitable for vacuum-forming, in which a heated and softened sheet is drawn into a mould by creating a vacuum between it and the mould surface. Moulds can be of wood, sheet-metal or filled epoxy resins. The limitation of vacuum-forming is the excessive drawing down at sheet corners which can lead to undue thinning of the material. This has led to drape-forming which involves insertion of the mould into the softened sheet before the application of vacuum. Two further variations are air-assist and plug-assist forming. The former employs partial inflation of the heated sheet by air pressure and the latter has a plug in the form of a mould lowered into the softened sheet prior to application of the final vacuum.

Large formings beyond the scope of injection mouldings can be produced. In pressure, rather than vacuum, forming, considerable accuracy of detail can also be obtained. However, it is not generally possible to obtain abrupt changes in wall thickness as is the case with injection moulding. Since internal stresses in the material are consequently low, with thermoforming, a stronger thin-walled item can generally be obtained. Generally, a minimum draft angle of 2-3 degrees applies to male moulds but female ones are more tolerant because of the tendency of the material to shrink away from the mould; and draft angles down to one degree are possible.

Shallow female moulds can even tolerate parallel sides. Generous radii are required, however, in order to minimise internal stresses. Absolute values depend upon wall thickness; see Fig 34a. In Fig 34b is seen how the inside and outside surfaces of thermoformed parts must essentially be parallel in contrast to those of injection mouldings. It can also be difficult to obtain sharp detail on the side opposite the mould when thick walls are being formed, Fig 35. Accurate consistency of wall thickness across the part certainly cannot be relied upon although this can to some extent be controlled by zoned heating and cooling. As a guide, parts formed on male moulds have thick bases and thin sides while those formed on female ones have the opposite.

Fig 36a, due to West[2], shows the air-assist technique used to improve wall thickness distribution. Vents are best drilled in areas of the mould where the material is last to form. Materials suppliers should be consulted on the appropriate hole size for

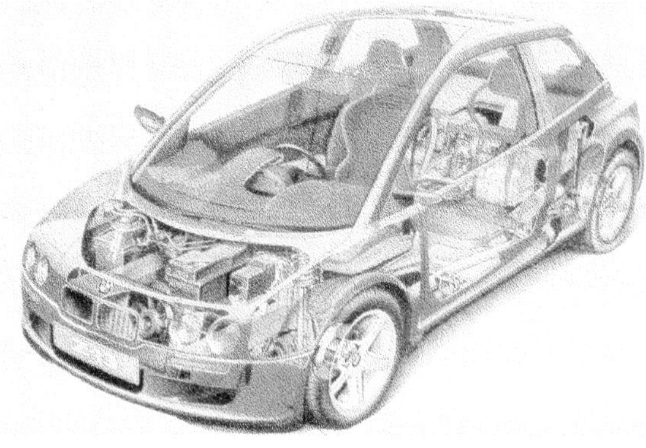

Fig 33: BMW Z-13 with thermoformed plastic body

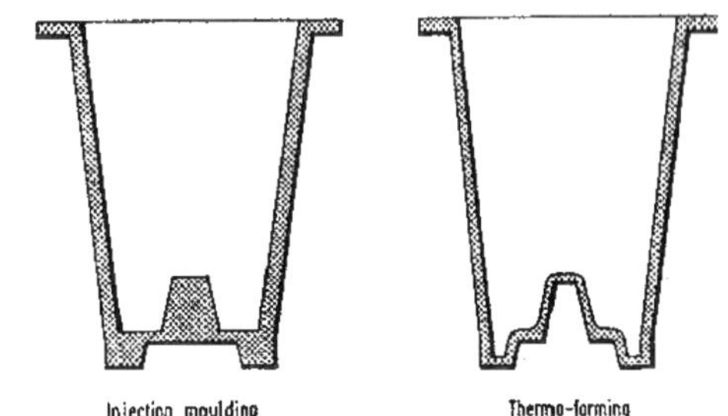

Fig 34a: Radius dependence on wall thickness

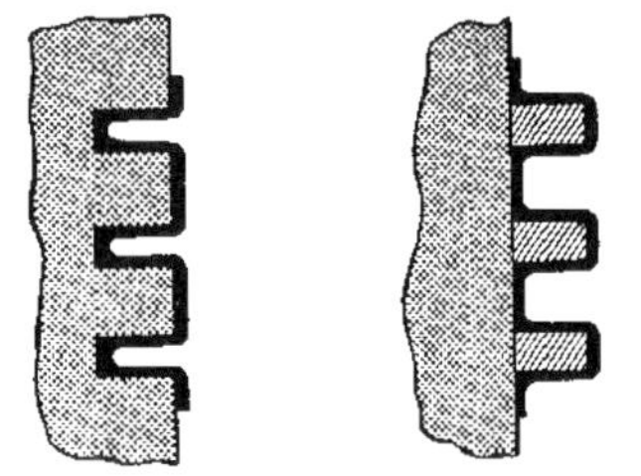

Fig 34b: Parallel wall requirement

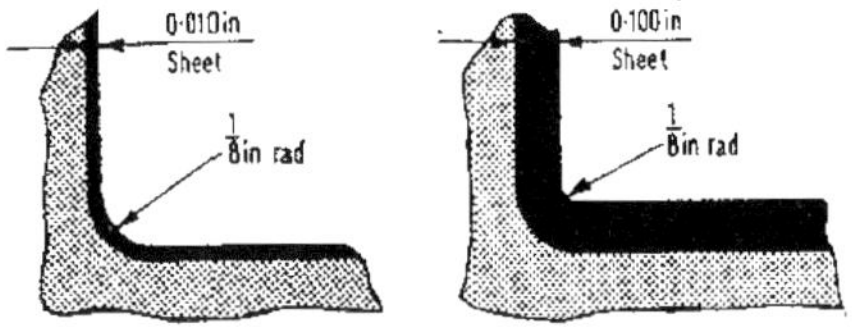

Fig 35: Sharp wall detail

their material, which will minimise witness marks. Sometimes it is possible to use a slot as seen in Fig 36b, in lieu of holes, depending on the shape of the part. Borg Warner have produced a useful guide to vac-forming ABS sheet to help determine draw ratio (Ratio of surface area S to area of blank within the clamp frame s). For a rectangular shaped part, length l, width w, and height h the ratio is obtained from:

$$S/s = [lw + 2lh + 2wh + 2(1.15l \times 0.1w) + 2w \times 0.75l]/1.15l \times 1.2w]$$

Sheet blank starting thickness T can be found from $T = CRp$ where $C>1$ for $R>3$ and is $=1.3$ for negative forming, 1.2 for positive forming and 1.1 for air-assist forming. Tolerances on finish-parts should allow for 0.7-0.8%, for negative, and 0.3-0.5% shrinkages for positive moulds respectively. Heat cycle time varies with sheet gauge, surface finish, colour and heat source. In the case of ABS sheet with double-side heating to 165 C surface temperature, gas fired machines typically require from 20 to 310 seconds for sheet gauges from 1 to 10 mm. These should be factored by 1.4 for gauges up to 7 mm with single-sided heating systems. Cooling channels should be provided in the moulds to reduce component temperature down to 65 C prior to withdrawal from the mould. Surface finish on the mould should be within 5-10 microns for optimum air evacuation and recommended vacuum line cross-sectional area is 2000 mm^2/m^2 of mould area and total area of vacuum holes is equal to 1.2 times that of the vacuum line.

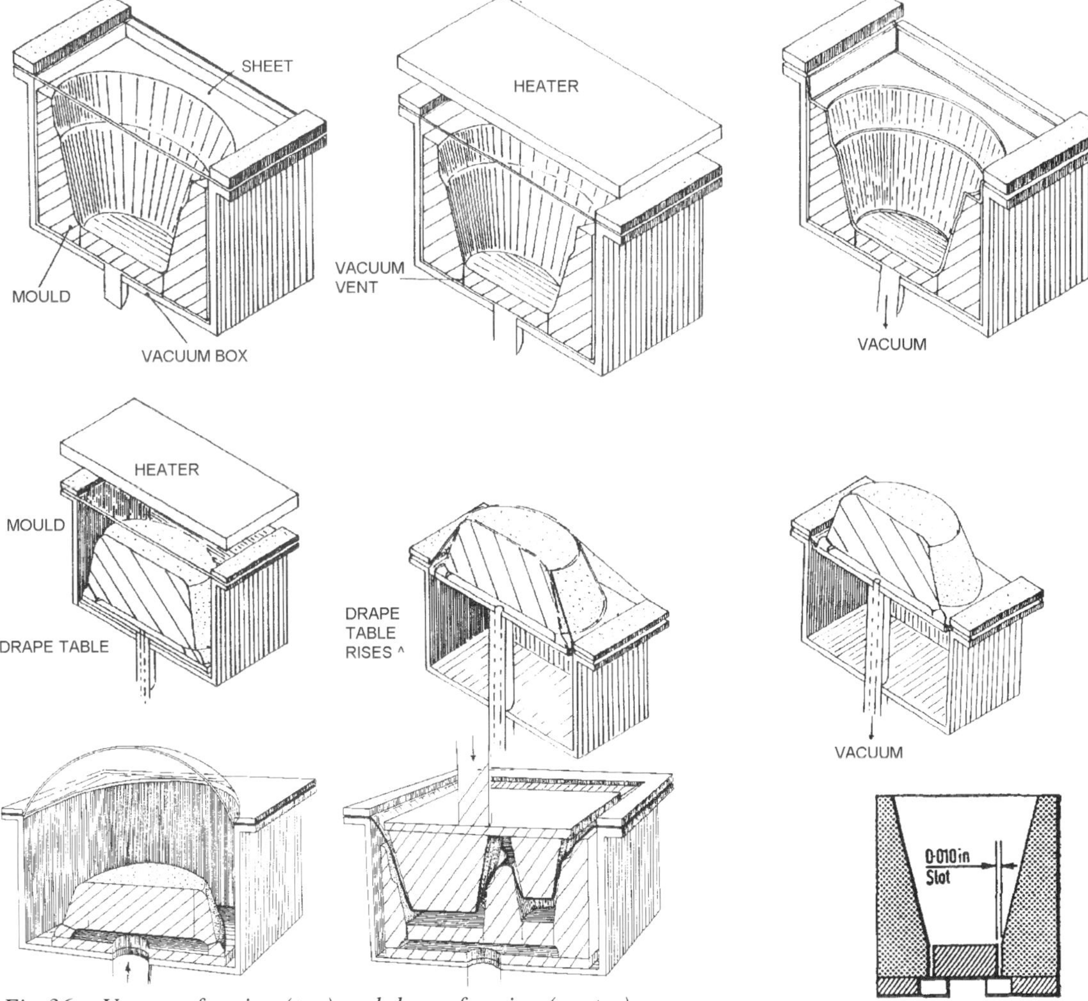

Fig 36a: Vacuum-forming (top) and drape forming (centre) compared with air-assist technique (bottom)

Fig 36b: Use of slot

Load-bearing plastics

Design techniques for load-bearing plastics inevitably differ from those applying to elastic materials and continuous attention must be paid to creep properties when predicting stresses and deflections. This section gives some rules for predicting stresses and deflections of loaded parts.

Plastic material specifications must state service temperatures and duration of loads when using the so-called pseudo-elastic design method. The value of 'elastic' modulus used must be relevant to these service conditions — and because, for most plastics, modulus value is unlikely to be above a few GN/m^2 at the very best, high inertia cross sections are the rule for minimising deflections.

Creep is defined as the strain, which is time dependent, resulting from applied load — and associated stress relaxation is the stress, which is time dependent, resulting from the applied deformation. Given a defined duration of loading, a (secant) creep modulus may be calculated as the ratio of the applied stress to the applied strain at the loading time and temperature of interest. Modulus is generally strain-dependent and in carrying out material tests it is useful to represent creep data by cross-plotting to obtain isochronous stress-strain curves for different times under load.

Design approach is to first identify maximum service temperature and duration of load; next to calculate maximum stress applying to the particular component; then to read the strain from the appropriate creep data and calculate the value of creep modulus; finally to use this value in elasticity calculations to predict deformations as stress relaxation in the component — arriving at the final dimensions by an iterative process.

Considerable increases in bending stiffness per unit weight can be obtained by exploiting the section-inertia increases which are so easily achieved with plastics parts. Deep thin webs or wide flanges must however must be examined against the possibility of premature buckling failure.

The following worked example applies to an extruded T-section strip in acetal co-polymer, 1 in wide x 1 in deep (2.5x2.5 mm), with general wall thickness of 0.2 in (5 mm). It forms a beam fixed at both ends and is assumed to have to support its own weight at a temperature of 20 C for 10 years. Simple calculations show the section inertia (second mo-

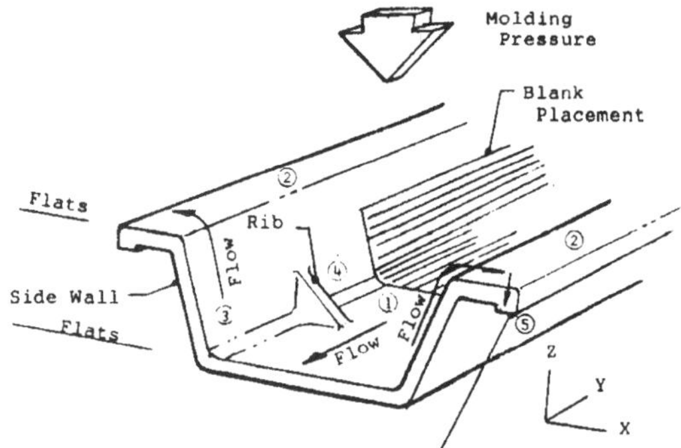

Fig37: GMT structural section

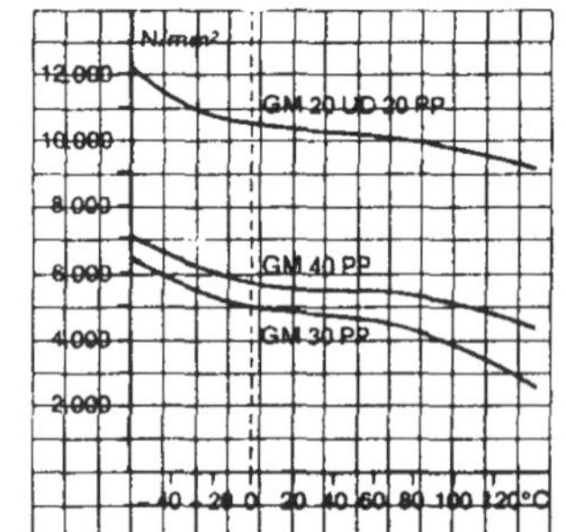

Fig 38: Flexural modulus vs temperature

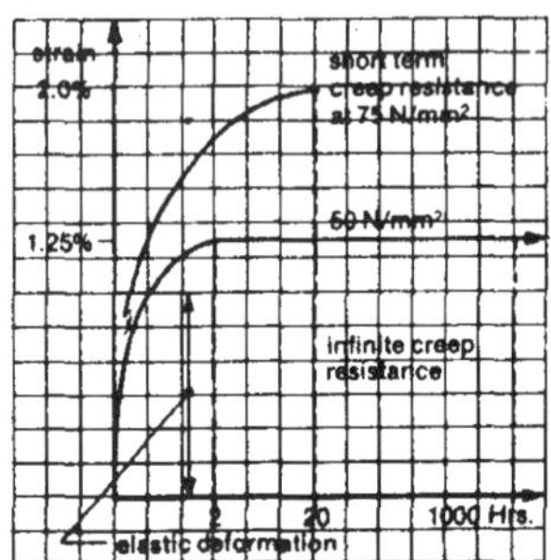

Fig 39: Creep resistance of GM 40 PP

Fig 40: Chrysler seat structure

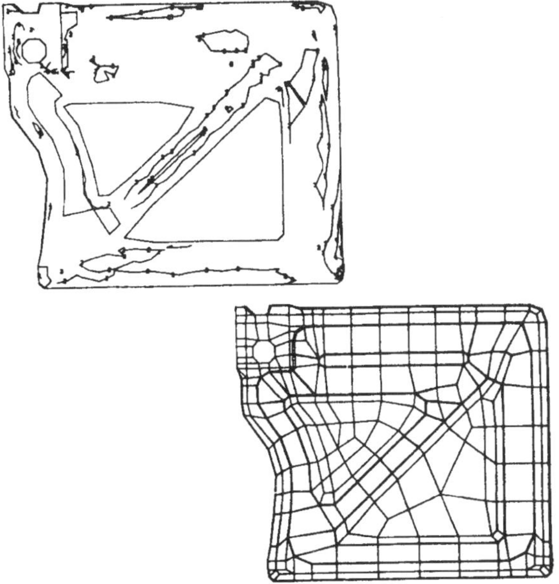

Fig 41: Von Mises stress plot

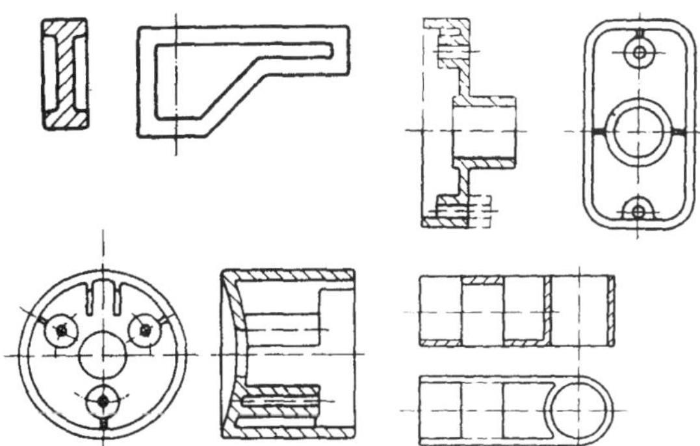

Fig 42: Rib design

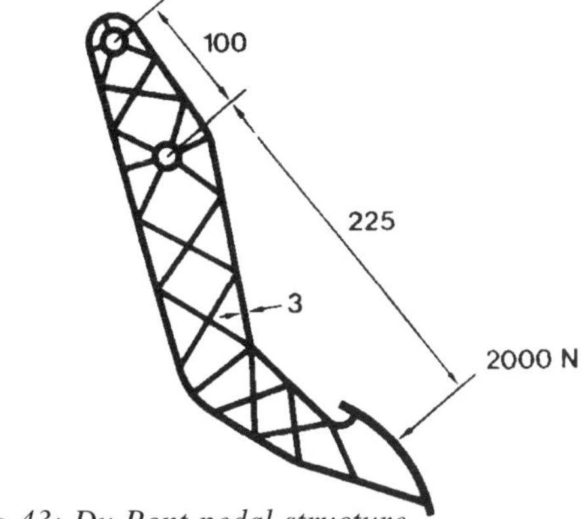

Fig 43: Du Pont pedal structure

ment of inertia) to be 0.0313 in^4 (from imperial tables), cross-section area 0.36 in^2 (2.32 cm^2) and density 0.51 lb/in^3 (14 g/cm^3). Assuming a strain of less than one per cent; estimated tensile creep modulus, from tables, is 1.1 x 10^5 lbf/in^2 (758.5 MN/m^2) at 1% strain, 10 yr, 20 C. The appropriate bending deflection formula then gives a maximum central deflection of 0.18 in after 10 years.

It is important in design to note that notch-sensitivity to impact failure is particularly prone with some plastics. Different grades of a generic class of materials can also have quite different impact properties. Stress concentrations must be reduced by the use of generous blend radii at corners and elsewhere. When load is removed from a thermoplastic, strain is recoverable provided the material has not yielded. Initially the recovery rate is high but it decreases with time. So for periodic loading conditions, the strain at the end of each loading period increases with time at a lower rate than if the load were maintained constant.

For a component like a seat structure, the load imposed by the occupant is rarely constant and effects such as braking and acceleration must be taken into account. The latter increases deflection response and vice versa. It has been found by experiment that strain response to increase in load decreases with number of times the event has occurred.

The form of the component affects its mechanical properties. In a member such as that in Fig 37, the 'flats' receive the greatest moulding pressure and have the best material properties; the sidewalls are moulded at lower pressure and so have lower strength. Typical modulus and strength figures for various elements of the GMT section shown in the diagram should be ascertained in its assessment. Fig 38 shows the material properties vs temperature for GMT while Fig 39 shows creep resistance of the unreinforced polypropylene which is the base material.

In seat design, deflection is the usual limiting criterion and tests by Chrysler have shown, for example, that a seat may reach limiting US Federal Motor Vehicle Safety Standards deflections while experiencing only 30 MPa stress levels. However, fatigue stress levels are important, it is argued; under a 800 N rearward load, the seat should not experience more than 50 MPa to obtain 'infinite' life. If the seat design can be modelled for FE analysis, a Von Mises stress plot should be predicted (this is a combination of 3-dimensional tensile and shear stresses that represent a failure criterion) — as shown in Fig 41 for the split back-folding rear car seat of Fig 40.

An example of rib design to minimise deflections is shown in Fig 42. In the pedal member designed by Du Pont in Fig 43, test deflection criterion for the part was taken to be a maximum of 10 mm at full load which is considered as 200 N for a throttle pedal, 500 N for a clutch and 2000 N for a brake pedal. The way different elastic moduli effect the result, for four of Du Pont's thermoplastics, is shown in Fig 44. For its Delrin brand co-polymer the company have released time-dependent stress-strain curves as shown in Fig 45. Some designers favour calculating long-term creep modulus from secant modulus E_c for a given deformation h; a value is suggested of $(1/2)E_c/0.1h$ for 1% strain. This is factored by 0.7 per 70 F to allow for thermal loading. Additionally, safety factors are used to approximate the static load condition as 1.5-1.9 for amorphous and 1.2-1.8 for semi-crystaline thermoplastics. VW designers suggested the approximate load factors of Fig 46 for moderately stressed shaped plastic parts. This shows how little remains of the original elastic modulus after long-term loading at 212 F.

For their Delrin resins, Du Pont recommend designing where possible with uniform wall thickness to avoid internal stresses caused by moulding. The transition from one wall thickness to another should also be gradual. Ribs should be specified with caution otherwise they will distort the part after moulding. Ribs and strengthening members should be 1/2 - 3/4 as thick as the walls they reinforce and deep ribs may require 1/4 - 1/2 degrees of taper for easy ejection from the mould. By drawing a circle at the intersection of the rib and wall, section thicknesses can be compared, as shown in Fig 47. Rib thickness combined with blend radius could easily overstep the wall thickness limit criterion; see Fig 48 for creep resistance.

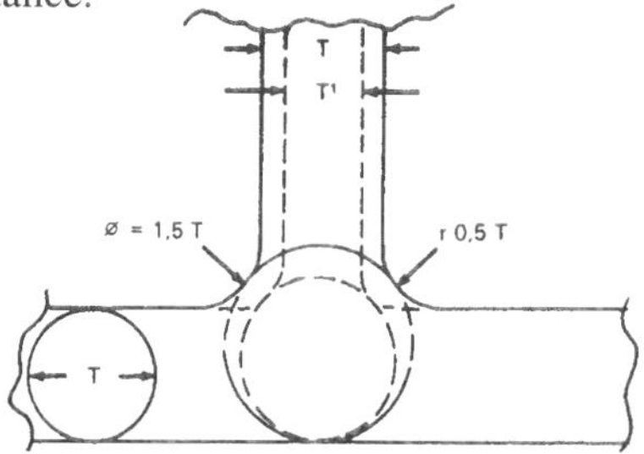

Fig 47: Finding true section thickness

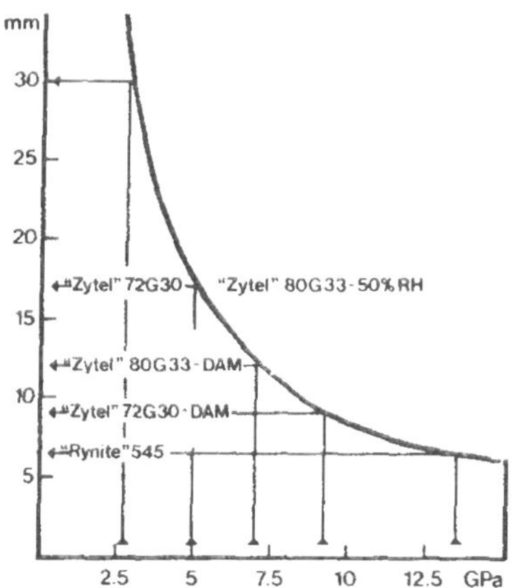

Fig 44: Modulus variation with grade

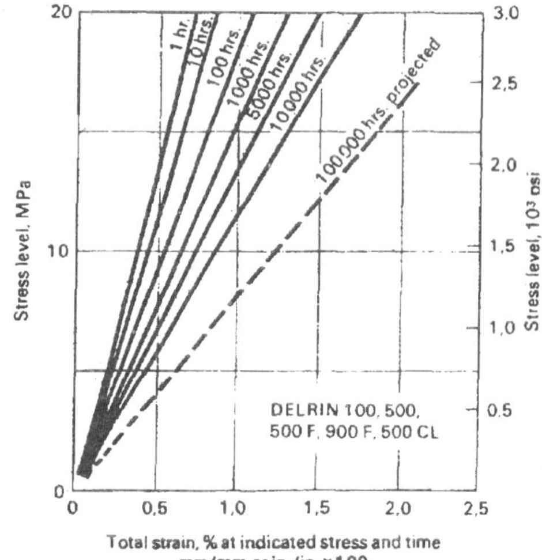

Fig 45: Time-dependent stress-strain curves

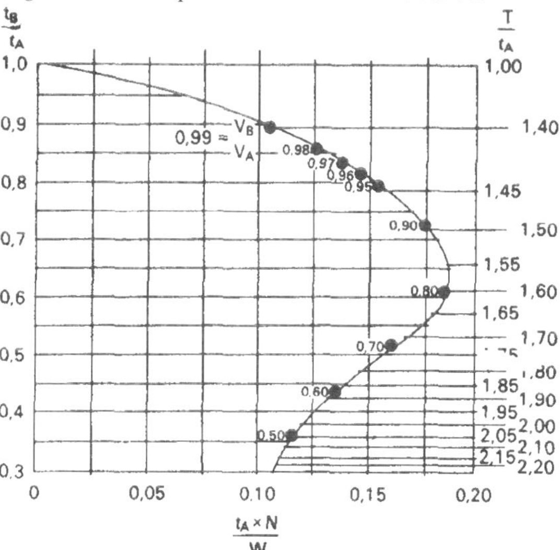

Fig 46: Load factors

Designing in polypropylene

This important semi-structural thermoplastic is widely favoured for the balance between its mechanical properties and ease of moulding. ICI were early in the introduction of the material into automotive applications and have built up a useful bank of design and application data, some of which is reproduced here.

ICI produce some 40 homopolymers and co-polymers of polypropylene under the Propathene brand — including filled, reinforced and grades which can be plated. Common to all are the properties of good temperature and chemical resistance also suitable flow properties for easy processing. Mechanically they are mostly semi-structural with low weight for a given strength; they also possess good fatigue and environmental stress cracking resistance.

The relatively low cost of the material has also ensured polypropylene's widespread use for automotive applications. Homopolymers are suitable for less structurally demanding interior trim applications but for areas like foot-well kick panels, requiring greater toughness at low temperatures, general-purpose co-polymers are appropriate. For fascia crash-pad work, tougher high-grade co-polymers are needed.

Grade GYM 202 of 10,000 kP/cm^2 flexural modulus is used, for example, in the full-width one-piece fascia panel for a cross-country vehicle. This is a large thin-section panel requiring the combination of easy-flow mould filling and toughness in service. GWM 101 of 14,000 kP/cm^2 flexural modulus has been used for 'organ-type' accelerator pedals with integral hinge of the flexing plastic. CV and tractor cab sound-insulating panels are often made of foamed polypropylene. Reinforced grades are also used in foamed form for such applications as tractor seat pans and rear parcel shelves in cars.

Commonly used for radiator fan shrouds, heater casings and ducts are the talc-filled grades such as 22T20/40H, of 35,000 kP/cm^2 flexural odulus, because of their high rigidity and heat distortion temperature. In chassis and exterior applications generally, 20% glass filled grades are popular for such applications as headlamp bodies while greater toughness is available in GWM 203, with 350 J/m Izod impact strength at 23 C, for bumper inserts and wheel-arch liners. PXC 4717 is a grade specially developed for plating, the pre-treatment process achieving a plating peel strength of 4.5 - 7 kg/cm.

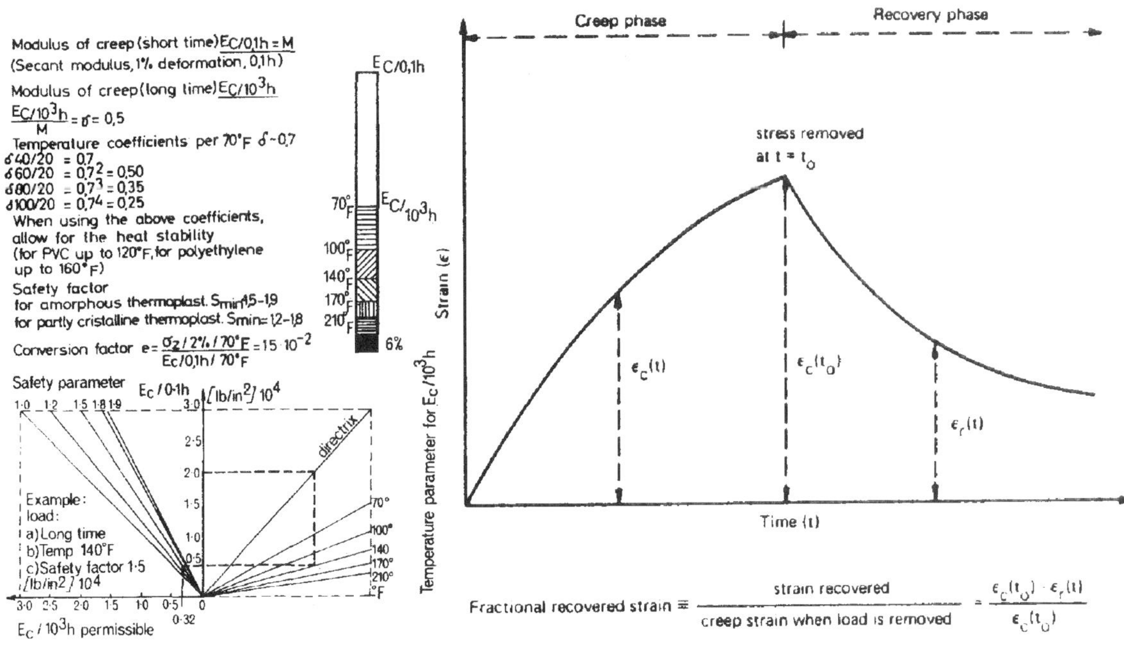

Fig 48: Nomogram for creep resistance

$$\text{Reduced time} \equiv \frac{\text{recovery time}}{\text{duration of creep period}} = \frac{t - t_0}{t_0}$$

Fig 49: Creep and relaxation of a thermoplastic

Creep of plastics

Deformation, fracture after dynamic fatigue and resistance to impact are the important mechanical properties. ICI supply data based on both creep, Fig 49, and stress-relaxation tests. The first involves observation of strain as a function of time when the specimen is held at a constant load while the second involves measuring stress as a function of time while maintaining constant strain.

Creep data may be represented as strain vs log time, Fig 50. These can be presented in different ways, to suit particular requirements, by drawing sections at constant times or constant strains. A further derivation of possible use is tensile creep modulus vs log time.

ICI point out that Propathene will recover after long periods of creep and this property can be exploited in components subject to intermittent loading. However, the recovery, like creep itself, is time-dependent and the data an be presented most concisely if new co-ordinates are defined, Fig 51. General recommendation by the company, for design, is to use the criterion of maximum strain to signify 'failure'. Limits are 1% where welded joints are used, 2% for injection-moulded components and 3% as an absolute upper limit.

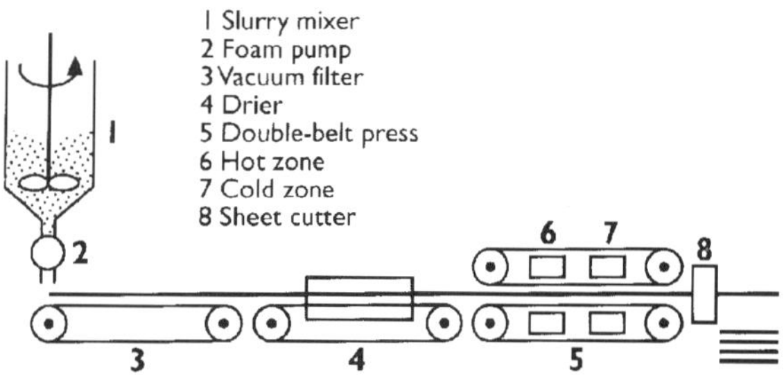

Fig 53: Slurry process for GMT

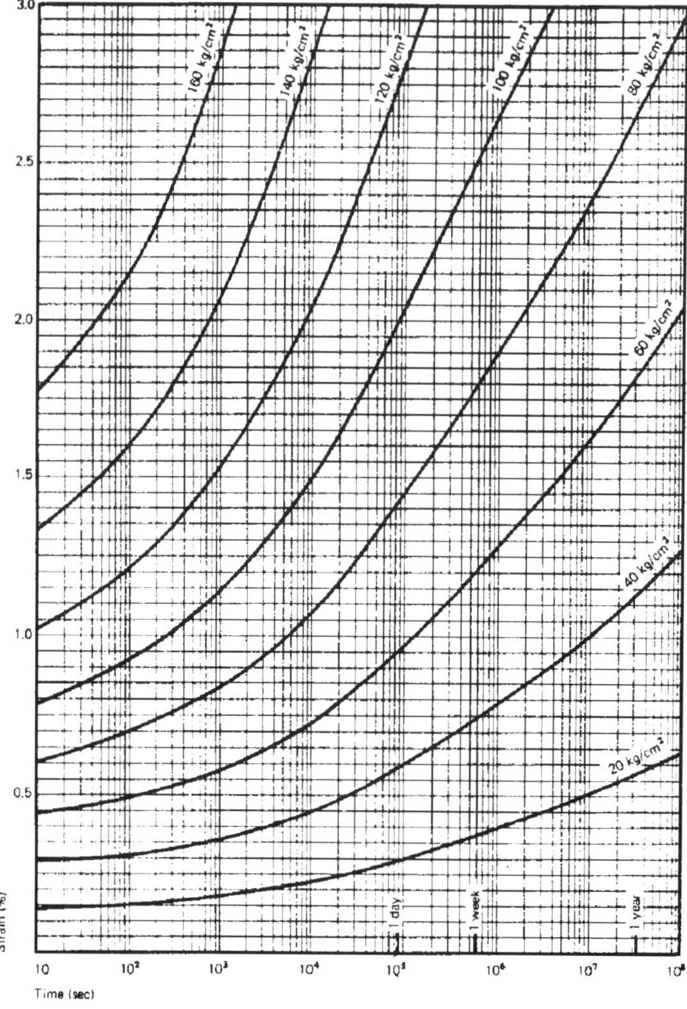

Fig 50: Strain vs log time

Reinforced thermoplastics

Apart from structural RIM, a number of other reinforced thermoplastics, including those based on convenenientl handlable prepregs, are available as a ready means of providing structural panels for medium volume levels of production.

An early entrant into the market for reinforced plastic structural materials was ICI whose Sandwich-Moulding process of the early 1970s was successfully used to produce structures of relatively high specific stiffness. It involved a new injection moulding technique, for the time, which allowed the production of 'solid' skins and foamed cores in one operation — but skins were not exclusively glass-reinforced. As well as the greater flexural rigidity of thermoplastic panels produced by the process, surface quality, too, was improved over 'solid' panels by virtue of absence of sink marks at the intersection of the panel with reinforcing ribs or wall normal to the surface. Also the skin materials did not have to be the same as that used for the core material, with foaming agent. An early prototype part made in the process was a bonnet panel which weighed just 10 lb (4.5 kg) and had a thickness of 0.4 in (10 mm). A conventional steel equivalent at that time was three times less stiff in bending and weighed 22 lb (10 kg) while a solid plastic one, of the same material as the sandwich skins, required a thickness of 0.21 in (5 mm) yet still did not have sufficient torsional stiffnesss.

Relatively high volumes had to be used because tool costs were even 10% higher than conventional injection moulds. Two polymer formulations

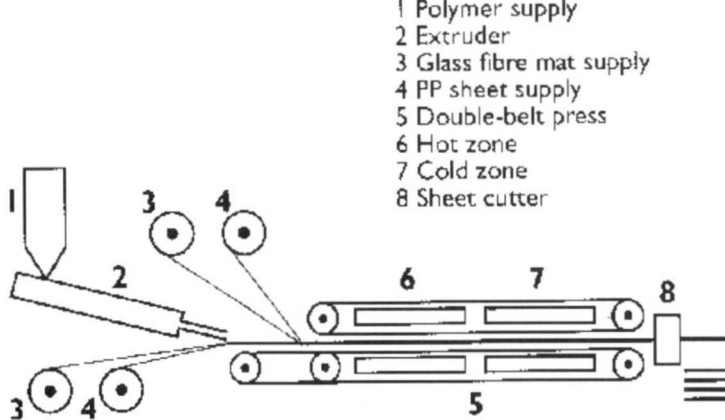

Fig 54: Separate process for random fibre mats

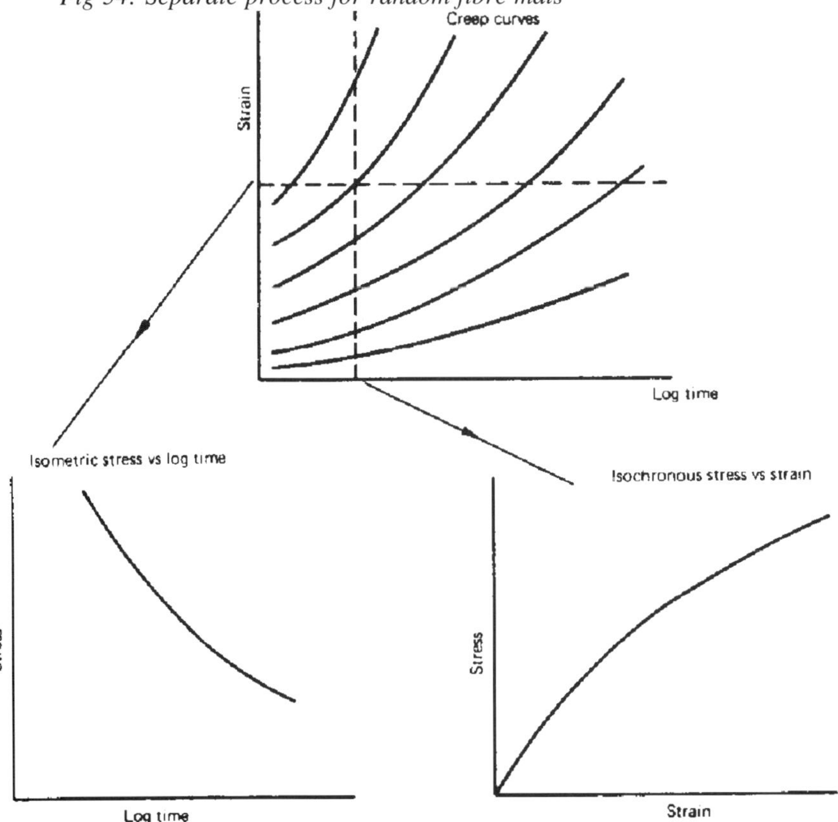

Fig 51: Defining new co-ordinates

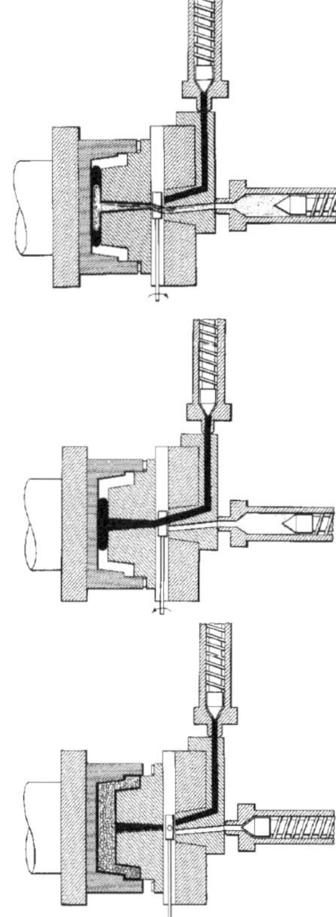

Fig 52: Stages in ICI sandwich moulding process

were injected from separate injection units, one after another, into a mould via a common sprue. A short time after filling, the mould was opened to allow the foamed sandwich to expand, Fig 52.

In designing structures made in the process ICI found that calculations used for solid plastics could be employed but with flexural modulus divided by the expansion ratio in the moulding process, given by final thickness of moulding/ unexpanded thickness. Where different skin and core polymers were used, the two moduli were summed and then divided by twice the expansion ratio to arrive at an equivalent modulus figure. For a 25% coupled-glass filled polypropylene skinned panel of 0.5 m^2 area at 20 C subject to short-period loading, approximate mid-point deflection is 0.1 mm per kg uniformly distributed. At 1.8:1 expansion rato thickness required would be 10.4 mm and weight 2.8 kg assuming an expanded polypropylene core and an equivalent elastic modulus of 1.58 GN/m^2 .

GMT systems

The general case for thermoplastic matrix composites has been well made. A DTI pamphlet[3] argues that the limited shelf-life of GRP thermoset prepregs, due to chemical instability, advanced the use of glass mat thermoplastics (GMTs) and advanced thermoplastic composites (ATCs). The latter resulted from a need, initially by aerospace users, for tougher materials with fibre volume fractions above 50% using a greater proportion of aligned filaments. While polypropylene was once the predominant material, usage has widened to include nylons, PBT/PET, polycarbonate and polyphenylene sulphide.

Generally chopped fibres are processed in a wet slurry process, akin to paper-making during which chopped fibre, polymer powder and processing aids cause fibre bundles which form filaments. The slurry is pumped on to a vacuum filter belt and the water removed, the resulting non-woven web of mixed fibres and polymers is then dried, Fig 53. This process is particularly well suited to high fibre loadings (60-70% by weight). For random fibre mats, these are impregnated with molten polymer by sandwiching the extrudate between two layers of mat as in Fig 54.

GMT is usually itself processed either by stamping or high speed compression moulding, the former being used for simpler-shaped parts. Stamping involves placing a near net shape pre-heated blank in a

	ISO standard	PP-GMT*	PET-GMT**	PA-GMT**
Glass content	1172	30	30	30
Tensile strength	527	80	100	120
Elongation at break	527	--	2	2.5
Modulus of elasticity	527	--	9000	6500
Charpy Impact 23 °C	179/2D	45	60	70
HDT-A	75	140	245	215
Relative density	R1183	1.13	1.60	1.37

Fig 55: Akzo GMT properties

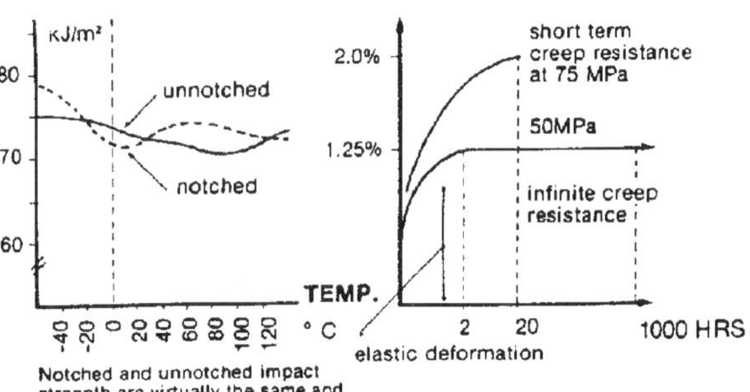

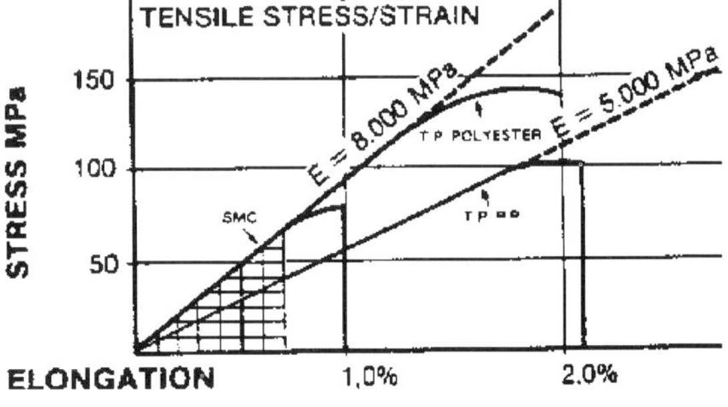

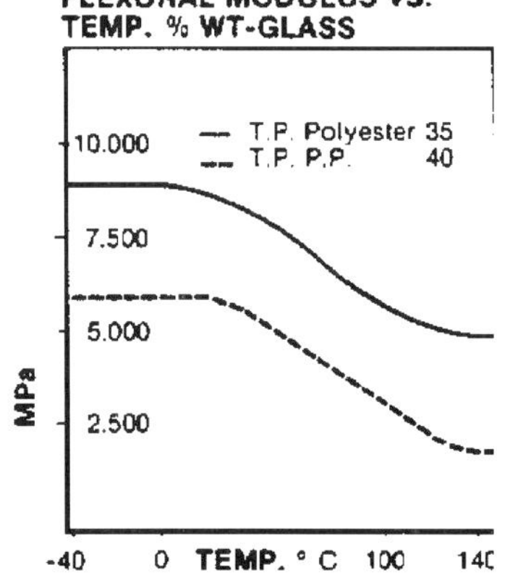

Figs 56,57,58: Performance comparisons with polyesters (top left: energy; top-right: strain)

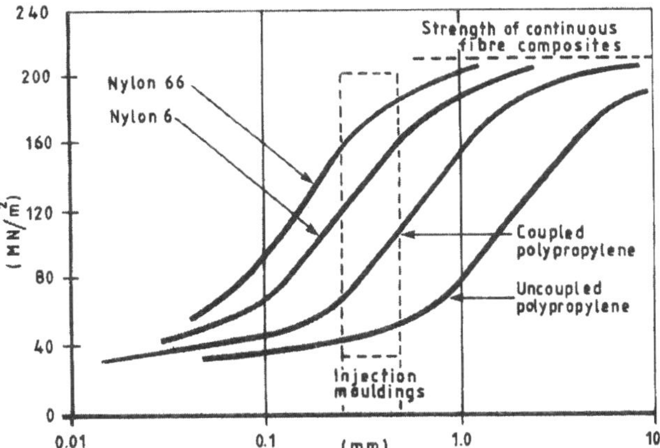

Fig 59: Nylon and polypropylene composite tensile strengths compared, for given fibre length

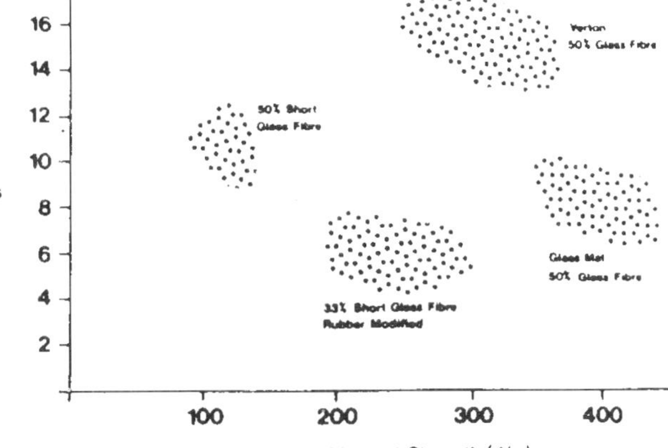

Fig 60: Properties of long-fibre Verton reinforced thermoplastics compared with other GRPs

lower temperature moulding press and pressing, in a cycle time from 15-45 seconds, sufficiently high pressure producing good surface finish. High speed flow-moulding of larger, more complex parts can allow variable wall thicknesses to be obtained and flow can provide the part with directional properties.

The table in Fig 55 summarises selected properties of three GMT types available from Akzo. With stamping of panels in these materials surface finish would be inappropriate for exterior applications, so main uses are in such areas as boot floors and hidden seat supports. Because the product cools in the mould, wall thickness is limited to a maximum of 5 mm and to prevent stress concentrations minimum fillet radius of 0.1 mm should apply. To ensure good filling, ribs of no more than 3-5 times wall thickness should be used. Forms of reinforcement possible include short fibres, 0.2-2 mm, random mat with fibres up to 10 mm, uni-directional tapes and woven or knitted fabrics when compression moulding is used. Figs 56, 57 and 58 show performance comparisons with polyester systems.

Competing resin matrices for composites

Shorter cycle times are among the advantages likely from longer, and continuous, fibre reinforcement techniques available for thermoplastics, over thermosetting counterparts, in developments reported by ICI[4]. Fig 59 shows increase in tensile strength obtainable for polypropylene and nylon by improving fibre-to-matrix chemical bonding and increasing fibre length up to 10 mm for so-called second-generation

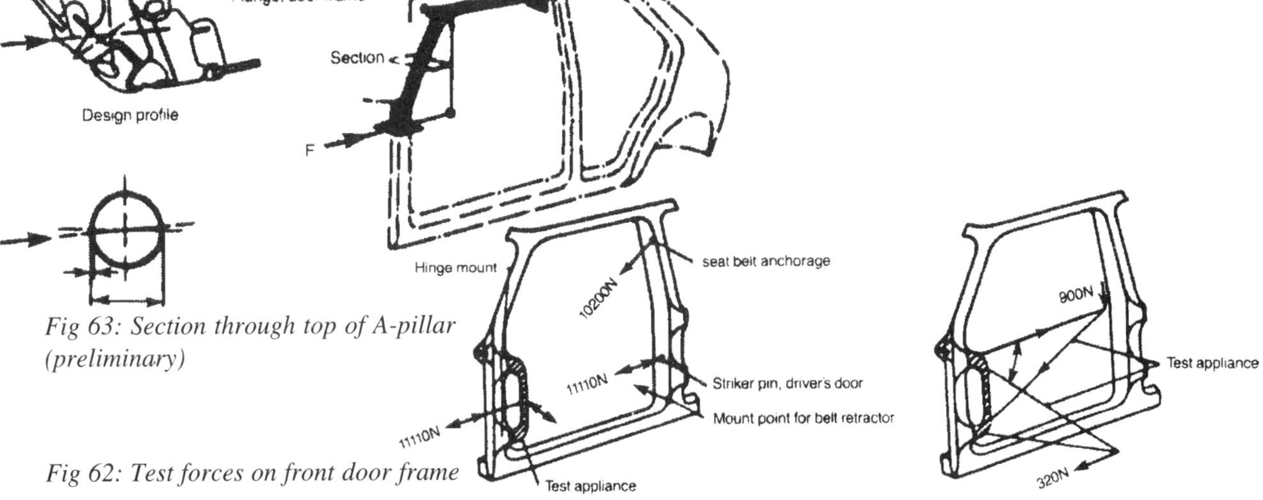

Fig 63: Section through top of A-pillar (preliminary)

Fig 62: Test forces on front door frame

systems. Objective is to preserve the fibre length in the as-moulded part — minimising the shearing action of the moulding machines — and to produce prepregs for sheet forming. Fig 60 compares the properties of second generation composites with conventional injection moulding compounds in respect of room-temperature stiffness and impact strength, in each case double the short-fibre values. This puts the materials in line for body-panel and underbonnet applications, say ICI.

Cycle times of 30-40 seconds show a distinct advantage over thermosets. Seat shells, load floors and battery trays are good applications. Tapes of 'third-generation' material have also been developed with high volume (55 per cent) fractions of continuous unidirectional fibres being used to impart high stiffness in particular directions (Fig 61) . Sections of the material can also be used for in-mould reinforcement and pre-preg thicknesses from a 'few hundred microns to several millimetres' are available. Thermoformability of the materials makes possible compression-moulding, filament winding, roll forming and rubber hydroforming. Most interest has been shown in 50 per cent carbon-fibre/PEEK composites having flexural modulus of 120 GN/m^2 retained up to 143C (50 GN/m^2 at 300C).

SMCs reasserting their claims High strength composites have been proposed for car passenger-compartment frameworks by VW[5]. Prior to considering the complete framework, initial studies were carried out on the front door surround-frames; these which showed 20 per cent weight reduction, against steel, could be obtained without loss of rigidity. Normal test forces subjected to a steel frame are shown in Fig 62 and used to test the composite frame, made from 60 per cent (by weight) glass reinforced SMC with two-thirds continuous filament of one third random cut fibres. With carefully designed fibre orientation 30 GN/m^2 elastic modules can be obtained (against 210 for steel).

Section modulus at the top of the A-pillar is a

	Fibre	Unfilled	30% Short fibre	65% Continuous Long fibre
PET	Glass	60	140	800
Nylon 66	Glass	85	180	650
PES	Glass	90	145	500
PEEK	Glass	91	150	780
PEEK	Carbon	91	215	1100

Fig 61: Effect of continuous fibre reinforcement on flexural strength in MN/m^2

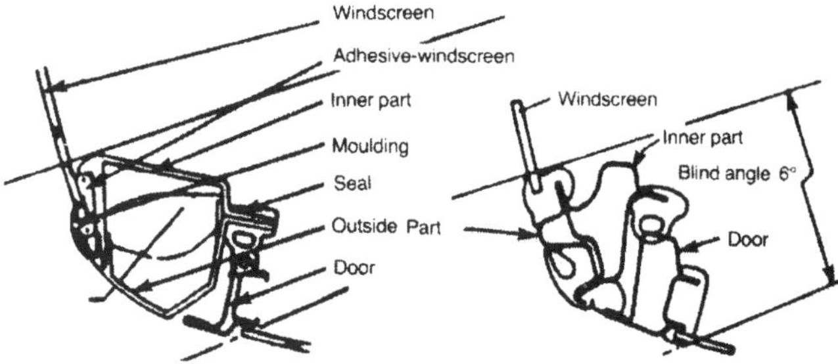

Fig 63: Section through top of A-pillar

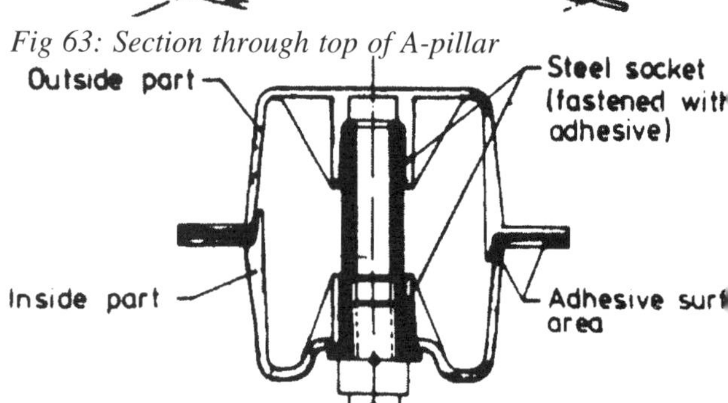

Fig 64: Feeding seat belt forces into the B-pillar

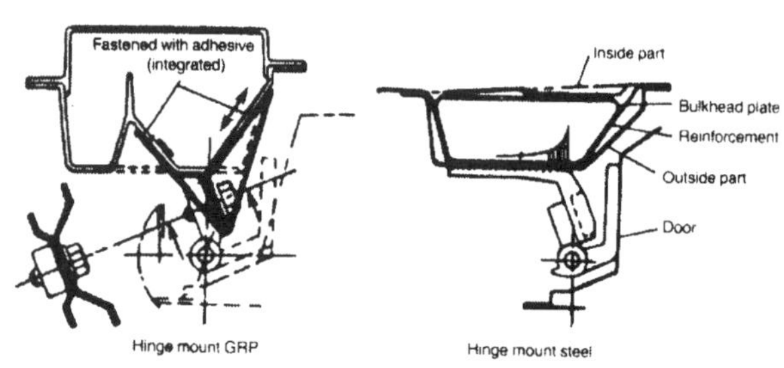

Fig 65: Hinge mounts compared: steel and synthetics

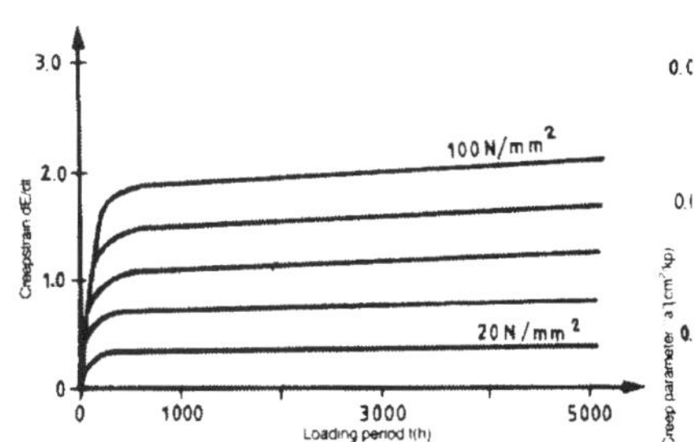

Fig 66: Creep strain of GRP showing exponential and linear regions

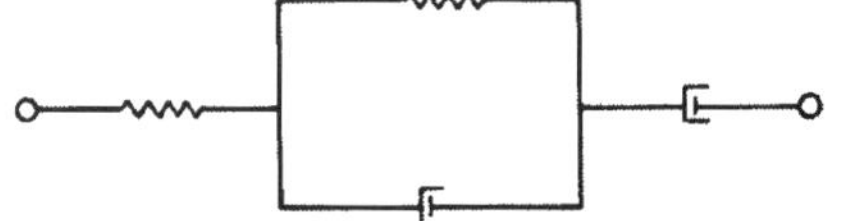

Fig 67: 4-parameter fluid model of viscoelestic plastic

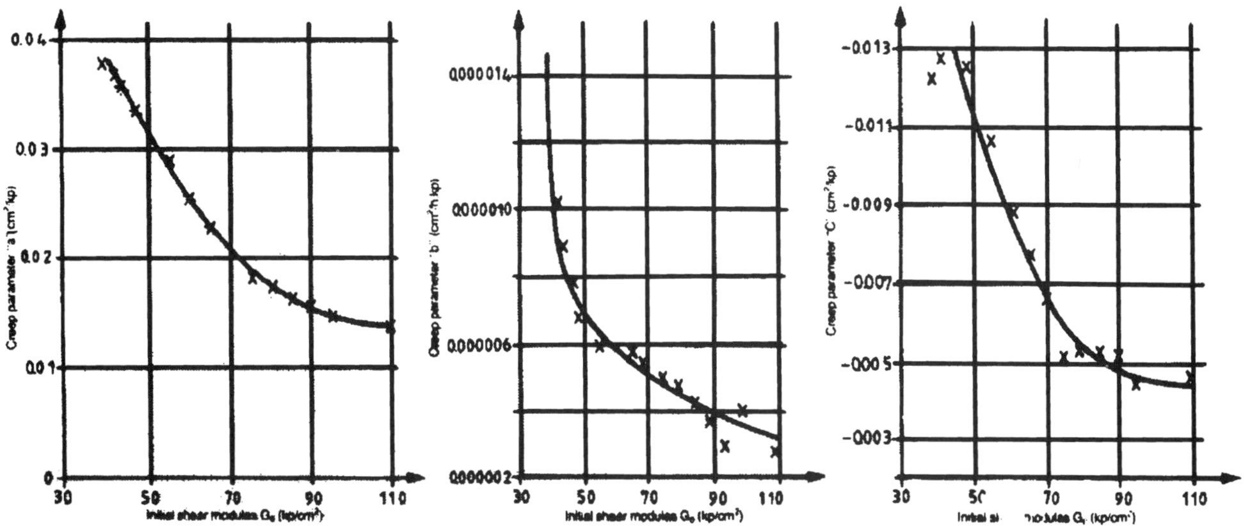

Fig 68: PUR foam creep parameter dependency on initial shear modulus of a, b, c respectively (left to right)

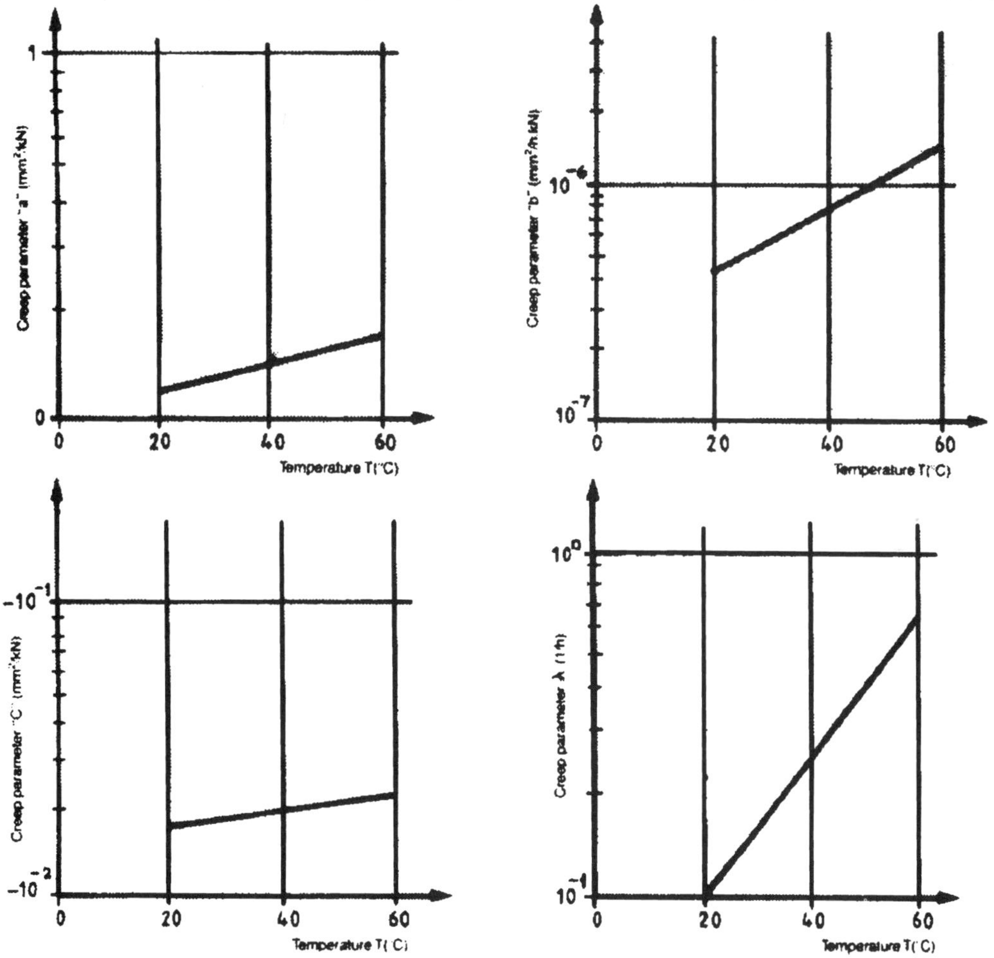

Fig 69: GRP creep parameter dependency of a, b, c and l (left to right)

critical parameter and use of a tube shape in composite, against irregular pressings in steel, is shown in Fig 62 to be an important factor. At the top joint with the cant rail, composite construction allows the use of corner webs to stabilise the section against surface element yielding. Eventually the pillar section in Fig 63 was chosen against the traditional steel shape shown for comparison. Another critical area is the seat-belt anchorage on the B-pillar and for the composite construction the special design in Fig 64 was adopted, dividing load between inner and outer shells of the frame. Redesign is also necessary at the hinge mountings (Fig 65).

Composite sandwich panels VW[6] has also addressed the vital topic of designing against the fundamental weakness of plastics: duration of loading causing creep and relaxation. The research analysed GRP-PUR-GRP sandwich panels in particular, introducing new visco-elastic FEM foundations describing material properties in the quasi-linear stress regions between -30 to +80C under long term loadings. Thin GRP skins have minimal bending stiffness themselves and so carry predominantly normal stresses in overall bending of the sandwich; the core material reacts almost the entire shear force — in the flat panel case — as well as stabilising the skins against compressive buckling. Maximum normal stress in the skins can only be reached if shear stress in the core, and its adhesive bond to the skin, is not exceeded. In the case of profiled parts, the author points out that the skin's own bending stiffness is no longer negligible so shear stress in the core is reduced and allows higher short-time structural loading. Under all load conditions the profiled structure shows better time-dependent mechanical behaviour than the flat one.

Linear and slight non-linear behaviour of visco-elastic materials — with initial exponential creeping followed by constant linear creeping — under a sustained load (Fig 66) can be described by a fluid model as in Fig 67 in which springs represent: stress = strain x elastic modulus and dampers: stress = shear force x rate of change of strain. Time dependent creep y and relaxation $\tau(t)$ describing the model under random loading are:

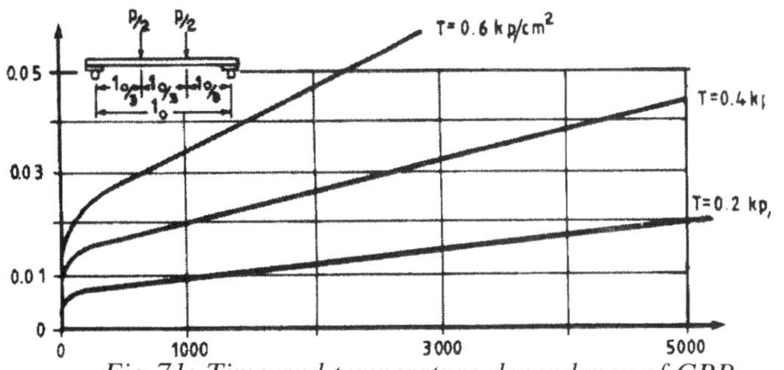

Fig 71: Time and temperature dependency of GRP elastic modulus; deformation against a base of loading period

$$y(t) = \tau_o I(t) + \int_o^t + I(t-t')(d\tau'/dt')dt'$$

$$\tau(t) = y_o I(t) + \int_o^\tau + Y(t-t')(dy'/dt')dt'$$

where τ_o and y_o are initial stress and deformation values; $I(t)$ is deformation per unit stress; $Y(t)$ stress per unit deformation (relaxation modulus):

$$I(t) = a + bt + ce^{-\lambda t}$$

where creep parameters a, b, c and λ takes values from characteristics in Fig 68 and Fig 69. Temperature dependent shear modulus:

$$G_T = G_o/\{1 + (K_T^+/K_T^-)\Delta T\}$$

Subscript *o* means initial value at 20 C while K_T^+ and K_T^- are O.019 and 0.0033 respectively for rigid PUR foam from +20 to +80 C and -30 to +20 C; ΔT is temperature difference. Creep parameters of rigid PUR foam and GRP shown in Fig 70 and 71 have been obtained by analysing test results at constant shear and tensile stress respectively at various temperatures. The value has been found to be constant at $0.0117h^{-1}$. The author has verified these expressions in the quasi-linear stress regions, by experimental tests. Results demonstrated that despite the familiar creep of plastics under sustained load, the materials possess good recovery capacities due, the author says, to their exponential and linear division of creep functions.

FEM analysis of a composite car body floor (Fig 72) with 1.5 mm GRP skin and 12 mm thick rigid PUR foam core was carried out to examine these load cases: temperature-dependent creep under bending by four 100 kg passengers; static simulation of front impact at 14 *g* with and without 20 *g* seat belt forces and dynamic loading induced by the sub-idle engine shake mode. The analysis showed that compared with a steel floor the composite panel showed higher stiffness even under the constant load case of 1.5 days at 60 C.

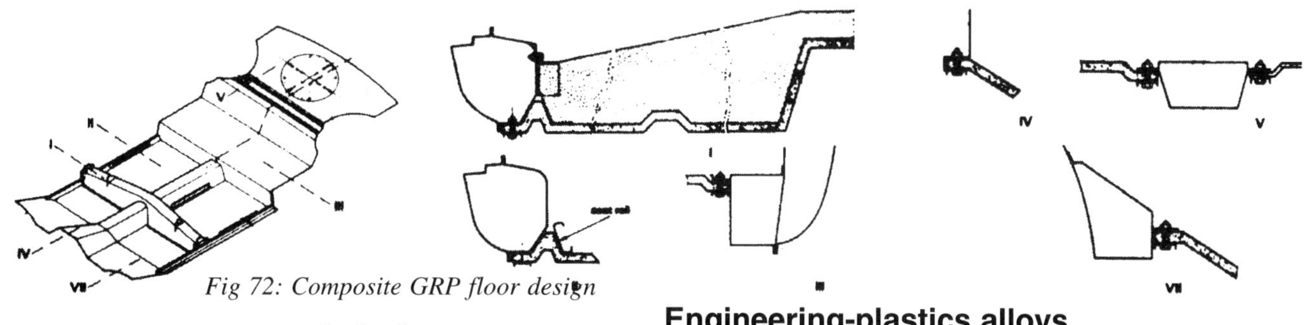

Fig 72: Composite GRP floor design

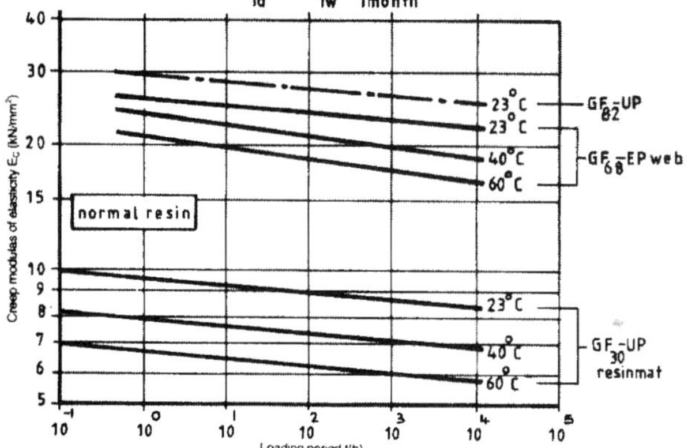

Fig 70: Creep of rigid PUR foam under constant shear

Engineering-plastics alloys

As well as the familiar plastic monomer categories, a series of co-polymers has been developed by the material suppliers; these can provide very useful combinations of properties . Here the products of one company, which specialises in this area, are examined.

A plastics materials supplier which has been particularly involved with plastics alloys is G E Plastics. Among this supplier's principal engineering co-polymers are Lexan polycarbonate, Noryl polyphenylene oxide (PPO), the Valox terephthalates, The Xenoy blends of semi-crystalline plastics, Noryl GTX PPO alloy and Ultem polyetherimides.

NORYL* GTX Typical properties*	SI UNITS	TEST METHODS DIN	ASTM	900	901	910	810	820
Mechanical								
Tensile strength at yield	N/mm²	53455	D 638	60	54	60	—	—
at break	N/mm²	53455	D 638	57	56	57	90	120
Elongation at yield	%	53455	D 638	5-7	5-7	5-7	—	—
at break	%	53455	D 638	100	50	100	15	10
Tensile modulus	N/mm²	53457	D 638	2150	2150	2150	4500	6000
Flexural strength	N/mm²	53452	D 790	100	90	100	130	170
Flexural modulus	N/mm²	53457	D 790	2100	2000	2100	3000	4000
Charpy notched impact								
23°C	kJ/m²	53453	—	13	12	13	6	7
−20°C	kJ/m²	—	—	9	8	9	4	5
Izod notched impact								
23°C	J/m		D 256	240	175	250	80	80
−20°C	J/m			190	125	190	60	70
Hardness, H 358/30	N/mm²	53456	—	115	85	115	124	132
Thermal								
Oxygen Index**	%	—	D 2863	< 20	< 20	< 20	26	26
Flame retardancy*** (1.6 mm)		—	UL Stand 94	94 HB	94 HB	94 HB	94 HB	94 HB
Heat Resistance:								
Vicat, method B	°C	—	D 1525	> 220	215	> 220	> 240	> 240
Vicat, B/120	°C	53460	—	195	170	195	210	225
DTUL, 1,82 N/mm²	°C	—	D 648	—	—	—	175	205
Mould Shrinkage	%	—	D 1299	1.2-1.6	1.2-1.6	1.2-1.6	0.5-0.7	0.4-0.6
Thermal Conductivity	W/m°C	VDE 0304/1	C 177	0.23	0.23	0.23	0.24	0.25
Coefficient of linear thermal expansion	m/m°C	VDE 0304/1	D 696	9×10^{-5}	9×10^{-5}	9×10^{-5}	$4\text{-}5 \times 10^{-5}$	$3\text{-}4 \times 10^{-5}$
Physical								
Specific gravity	—	53479	D 792	1.10	1.10	1.10	1.16	1.23
Water 23°C/50%RH/24h	mg	53495	—	29	28	29	31	30
absorption 23°C/50%RH/24h	%	—	D 570	0.4	0.3	0.4	0.5	0.4
Equilibrium, 23°C/50%RH	%	—	D 570	0.9	0.9	0.9	1.1	1.0
Saturation, in water 23°C	%	—	D 570	3.5	3.4	3.5	3.6	3.5

Fig 73: Noryl GTX alloys

The alloy blends are a series of products rather than one particular percentage mix of polymers. The Valox family, for example, involves polybutylene, polyethylene and polycyclohexylene in three separate co-polymers PBT, PET, and PCT. All combine good mechanical, thermal and di-electric properties and so find application in electrical applications such as heavy-duty connectors. There are a variety of glass-filled grades, too, some of which suit electric motor frames.

Ultem polyetheimide is inherently flame retardant and combines the physical properties of the more exotic plastics with the processing ability of normal engineering plastics. Supec polyphenylene sulphide (PPS)-based resins are high-performance crystalline polymers which combine short-term high heat resistance with long-term thermal stability, plus chemical resistance, which allows under-the-bonnet application. The Zenoy series was specifically developed for the auto-industry and is generally noted for mechanical strength, chemical resistance and dimensional stability combinations. Particular examples have high resistance to petrol and lube-oils and have high impact strength down to -40C; some grades (CL200 series) are off-line paintable, too.

Noryl GTX has both the inherent advantages of the PPE monomer as well as the chemical resistance of a semi-crystalline material. The result is a material with heat performance of over 190 C Vicat B which makes on-line painting possible. Its low-density also gives it a weight advantage over a number of glass-filled thermoplastics on a strength-for-strength basis. Noryl GTX is however, itself available in glass-reinforced form (GTX 800) and a 20% glass-filled form (GTX 820) has a flexural modulus of 4000 MPa while GTX 830 goes up to 7000 MPa. The chart in Fig 73 covers the properties for whole range. The unfilled GTX 901 is particularly notable for being able to withstand braking temperatures in the wheel trim application. The material requires a three-zone screw for plasticising in injection moulding, Fig 74. The unfilled GTX 901 is notable particularly for being able to withstand braking temperatures — useful in the wheel trim application.

Case-study in plastics alloys

G E Plastics' Scope concept car had its tailgate moulded from this material to demonstrate a structural application for a part which was to be painted on-line. Both interior and exterior skins, Fig 75, are moulded in the material. The complete part was designed with 25 degrees of rake for the rear window so as to preserve lamina air-flow across it So as to avoid loss in rear vision with this arrangement the vertical section of the tailgate above the rear bumper is fitted with a transparent panel of Lexan. The part is also notable for having the rear-lamp bodies, hinge and strut anchorages integral with the inner shell.

After assembly, the part is primed with a two-component conductive primer prior to mounting on the body shell for painting on the electrostatically applied top-coat paint line, the paint being baked at 150 C. The assembly operation involves both structural adhesive bonding and heat-staking, the pins for the latter being positioned through holes in the window flange and under the license plate —so that the glass sealant and lifting grip cover the heatstake points. The stakes serve the purpose of locating the two shells in place until the adhesive has cured. The electrical connections are also heat-staked, being metal strips which attach to the ribs of the interior shell. A connector block combined with the lock-pin transmits electrical power into the door. These pins are also heat-staked and a soft buffer moulding hides the joints from view.

Fig 74: Three-zone screw

N 3 2 1 H

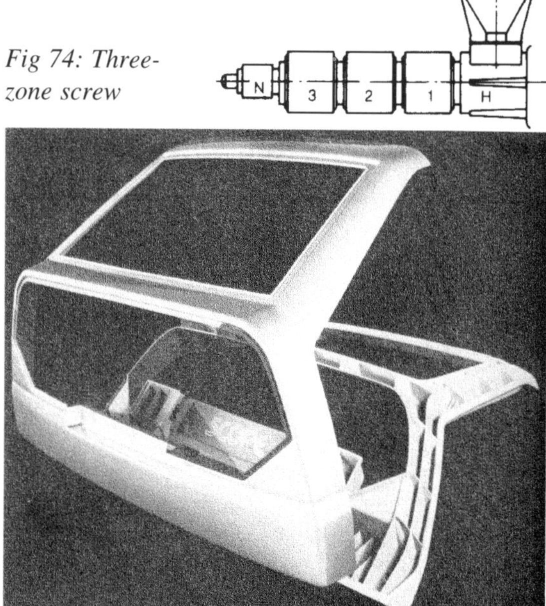

Fig 75: Scope tailgate in GTX

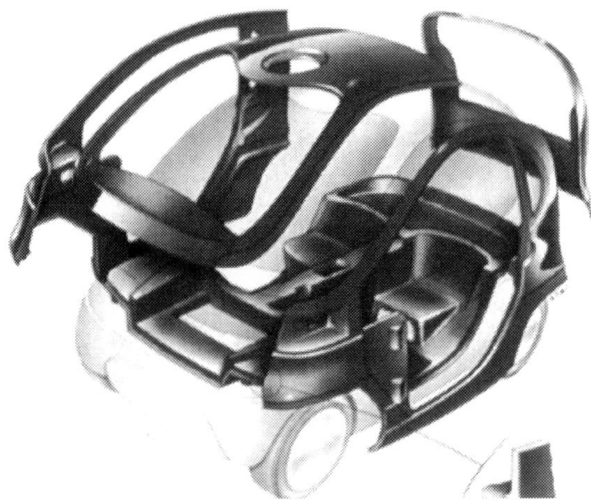

Fig 76: Tulip concept car and structure

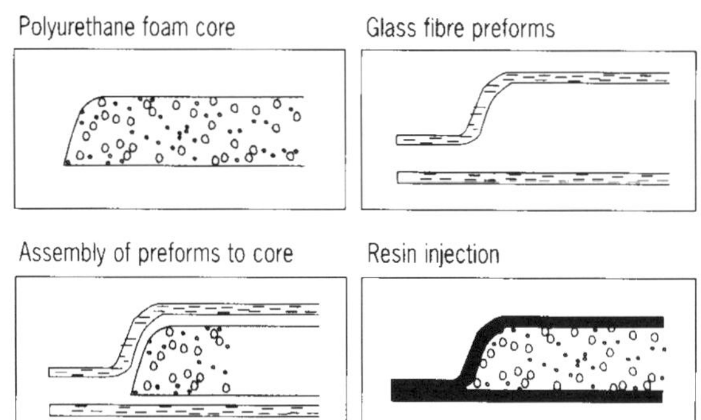

Fig 77: Sandwich configuration

High-strength composites for passenger cars

PSA Research Department, in partnership with the SORA Composites Group developed the innovative Tulip concept-car which is composed of just five primary elements bonded together.

These five parts constitute both the exterior and interior of the vehicle bodywork, Fig 76. In effect, the seats, dashboard, centre console and so on, form an integral part of the structure which has significant benefits in terms of the rigidity of the vehicle. Tests have also shown that its capacity to absorb impact is superior to that of steel, with progressive deformation. NVH and thermal insulation characteristics are also enhanced by the use of the technology, it is claimed.

The process involves resin transfer moulding (RTM) and each of the five parts comprises a rigid polyurethane foam core of 110 kg/m^3 density. Glass fibre mat is pre-formed and wrapped around the foam core before being placed into the low pressure injection tool. To ensure accurate location, the glass mat is retained in the part line of the tool. Polyester resin is then injected into the tool which impregnates the glass fibre and completes the sandwich construction, Fig 77. With a 40% by weight ratio of glass reinforcement to resin, the resultant assembly weighs approximately 30% less than equivalent steel structure. The material used also contributes to the safety of the vehicle, both for its occupants and to pedestrians. The energy absorbing characteristics of these panels have been shown to be 87% higher than for standard steel parts, say the company.

The tooling investment is kept to a minimum due to the low moulding pressures involved in the RTM process. To complement the precise resin injection system, the tools have a compression chamber in place of the traditional vents. The tools are also designed to maintain close temperature control across the entire surface. Typically +/-2 C is achievable to ensure consistent polymerisation of the resin and the use of chromed steel or highly polished nickel shell tools allows tor parts with class 'A' surface finish to be moulded. DSM had a role in the project to optimize the resin system to suit the Sotira RTM process. Mould flow analysis tests have therefore been performed with the long term aim of reaching body panel production rates of 200 per day by end of 1997.

Composites in impact

Impact absorbent properties of automotive composites came under the spotlight in a report of the Defence Research Agency[7]. Technology transfer from aerospace has been used to quantify the performance of high strength composite sandwich structures of the type used in racing car construction, with a view to application in military and other vehicle categories.

A key property is impact energy absorption by nose-cones and other vulnerable parts of the structure. High energy levels can be absorbed by the ability to trigger numerous failure modes simultaneously such as fibre fracture, delamination, matrix cracking and fibre pull-out. Hitherto it has been possible to model structural performance of composites using finite element techniques prior to failure, but not after.

The DRA study involved sandwich structure cores of unit cell arrangement, aimed at considering possible scaling effects when cells are stacked, also the effects of altering ply stacking sequence within the web core, its inclined angle, and the presence of a flexible polyurethane foam. The DISPLAY III/ NISA II FE package was used in the investigation; two core configurations were examined: triangular, for in-plane loading performance, and rhomboid for loading normal to the sandwich panel skins. Skin ply angle was a constant 0 deg while web ply angles of 0/ 90 deg (cross-ply) and +/- 45 deg (angle-ply) were tested experimentally and by FE simulation. Predictions tor the initiation of buckling, and elastic energy, are seen in Fig 78, loading being uniformly distributed normally to the top skin.

For the experimental tests, two-ply laminates were made in an autoclave on moulds from woven carbon-fibre/resin prepreg, web core and skins being subsequently bonded. Foam rated at 1/2 J/mm energy absorption was injected after the sandwich panel was bonded. A falling-weight impact machine was used to dynamically crush the specimens. Typical force-time plots for the impacts are seen in Fig 79, one for through-plane and the other for edgewise loading.

Fig 80 shows load to cause buckling in through plane impact which compares favourably with the previously shown predicted results. Fig 81 demonstrates the useful amount of energy absorbed. Figs 82 and 83 are similar results for edgewise loading. These showed that the rhomboid structure showed better performance due to its shorter web length, and although the foam did not contribute to energy absorption, it did help web stability. Effects of web stacking sequence can also be judged.

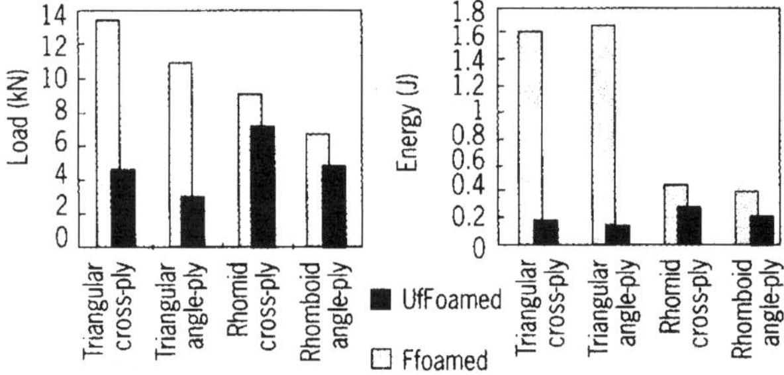

Fig 78: FE prediction of buckling load (left) and elastic energy at buckling (right)

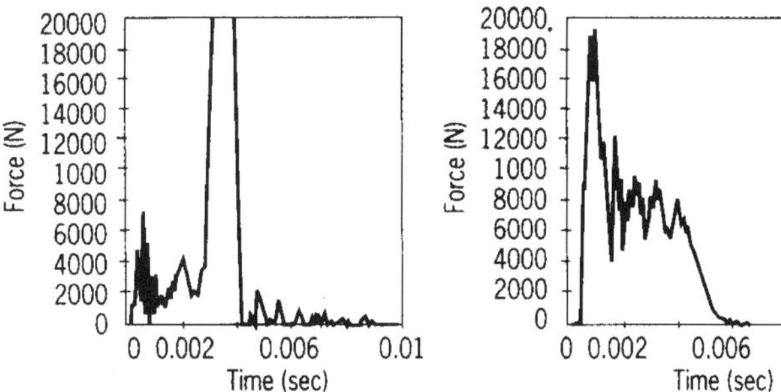

Fig 79: Force/time plots for triangular angle-ply web sandwich without foam, through, plane, left, and for rhomboid angle-ply with foam, edgewise loading, right

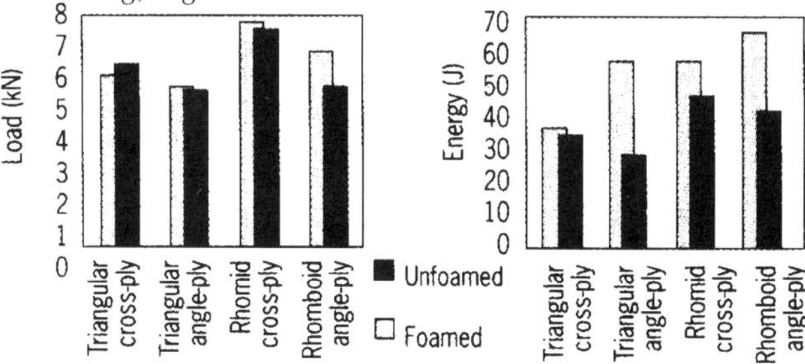

Fig 80/81: Through-plane crush-initiation-load/ energy absorbed

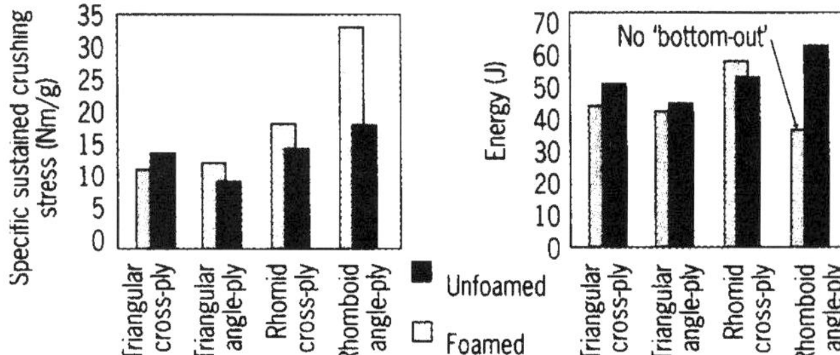

Fig 82/83: Edgewise impact crushing-stress/ energy absorbed

High performance composite applications

Two recently published case studies show how thermoplastic and thermoset systems still compete strongly with each other for high-performance underbonnet automotive applications, with selection influenced by strength, rigidity, warpage resistance, temperature stability and material damping.

The spread of structural plastics applications is illustrated by Solvay Automotive[8] in a brake pedal housing in aluminium alloy redesigned for manufacture in thermoplastic, Fig 84. This is part of a modular sub-assembly comprising brake pedal mounted to a metal pivot pin located in the top of the housing. A 5.67 kg vacuum servo unit is bolted to the front face of the housing and connected to the pedal with a bracket. A pedal force of 1.5 kN had to be withstood and reacted by four bolts through the base plate, connecting the vehicle body structure.

Finite element analysis defined a suitable wall thickness for the thermoplastic part as 4 mm, complemented by a minimum radius of 5 mm on all features in the load path. The decision was taken to exclude ribbing from stressed areas with the objective of allowing the material to flex within its up to yield range. Mouldflow injection moulding computer simulation was used to define initial process settings and confirm even mould filling with absence of gassing problems. The package also allowed its warping prediction facility to predict the degree of distortion arising from anisotropic shrinkage due mainly to glass filling in the nylon resin (PA66 — GF30%). Shrinkage was estimated at 1.1% across, and 0.16% with, the material flow. Injection point was at the top of the part as viewed in Fig 85 and venting at the bottom, as viewed. The predicted distortions of the base plane gave 2.75 and 3.15 mm for front and rear fixing positions respectively. Applying Taguchi methods in design, and using a mould counter-crowned on its base by 2.0 mm, these distortions were reduced to 0.5 and 1.8 mm respectively, considered allowable for the stress level they induced into the component on assembly.

In the initial FE analysis a 30 % overload was applied, 1.9 kN, to the pedal and the model revised to ensure maximum stress of no more than 100 MPa within the part (a safety margin within the 130 MPa allowable). The model was further revised to maintain stress below 75 MPa and a predicted failure of the part at an externally applied pedal load of 2.8 kN.

Because damping of reinforced thermoplastics is reduced with increase in glass reinforcement or heat-resistant fillers, researchers at Mitsubishi[9] have developed special formulations of SMC to increase its vibration damping properties for such applications as engine rocker box covers. The key to the

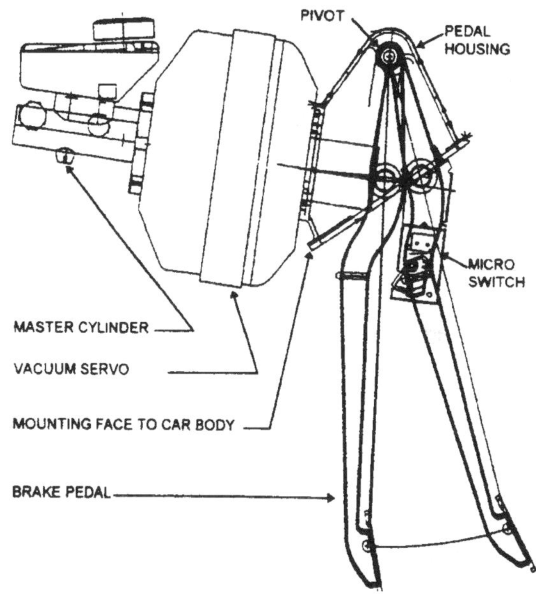

Fig 84: Brake pedal and housing assembly

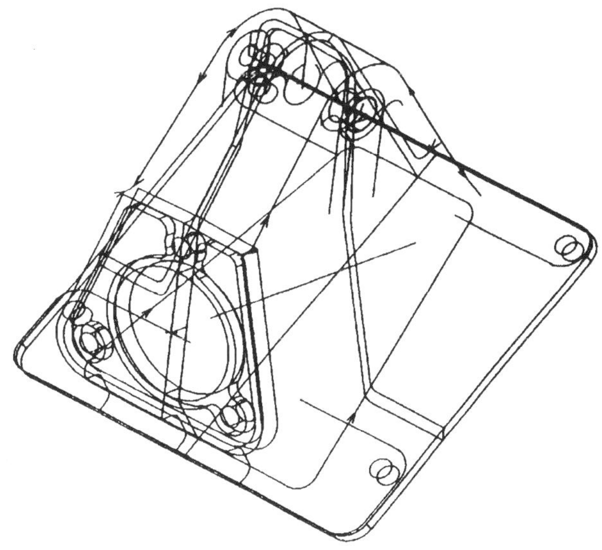

Fig 85: The housing moulding

development is shifting the glass transition temperature *Tg* of the material to nearer the normal operating temperature of the part. This is achieved with a combination of three polymers, vinyl esters VE having high *Tg*, unsaturated polyester UP having low *Tg* and polyisocyanate used as a thickening agent. With an optimized formulation it was possible to demonstrate a 1.0 dB noise reduction for a rocker cover compared with one made from aluminium alloy.

Compared with nylon, SMC has the inbuilt advantage of high strength, high heat resistance, comparatively low material cost and accuracy of moulding for large parts. A loss factor of no more than 0.05 was the vibration damping target for the material at an operating temperature of 60-100 C. The loss factor due to friction between the engine and the component is considered to be 0.01-0.03. Flexural strength of 100 MPa and flexural modulus of 5 GPa, minima at 90 C, were also set as material targets, to obtain adequate oil sealing at the interface. Fig 86 shows the blend ratio in relation to *Tg* as well as the flexural modulus at 90 C. The 5:5 (type 16X) and 7:3 (type 15X) blend ratios are seen to have suitable modulus properties and the vibration damping properties for these two types is seen in Fig 87 against those for conventional SMCs. The 16X was eventually chosen for its improved temperature dependence of flexural strength and modulus properties.

Fig 88 shows the structure of the rocker-cover and Fig 89 the vibration levels obtained in test. The levels are obtained by determining a transfer function by the hammering impact test shown inset. The results confirmed that the resonance peaks were increasingly lower as the temperature rises from 20 to 100C.

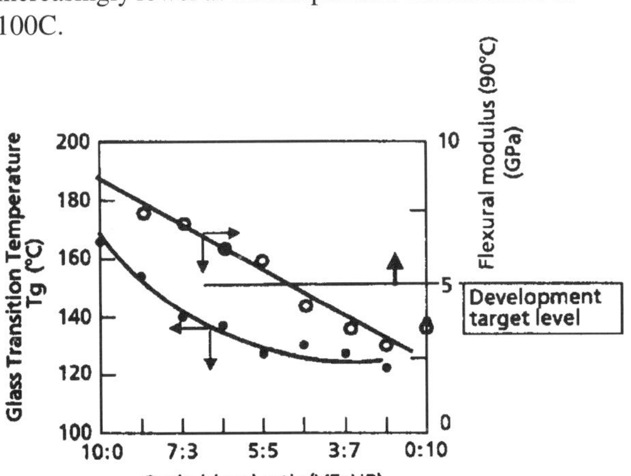

Fig 86: Tg and flexural modulus at 90 C

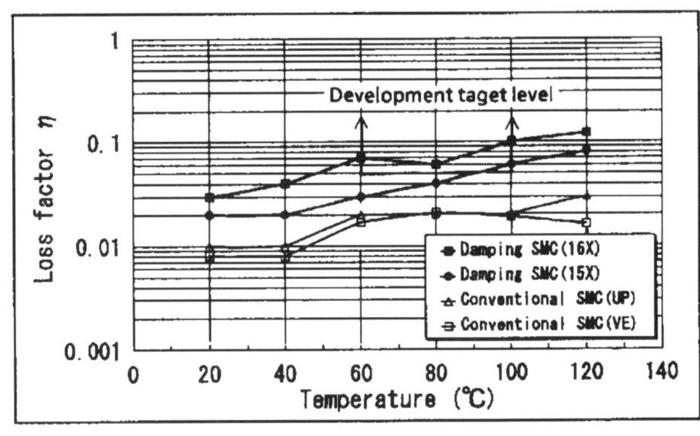

Fig 87: Vibration damping property

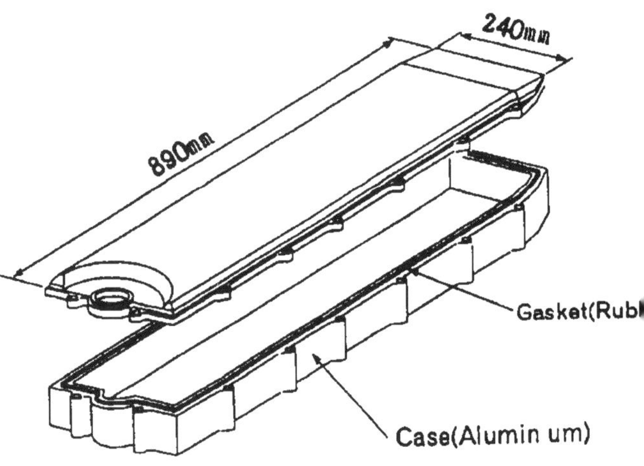

Fig 88: The rocker-cover structure

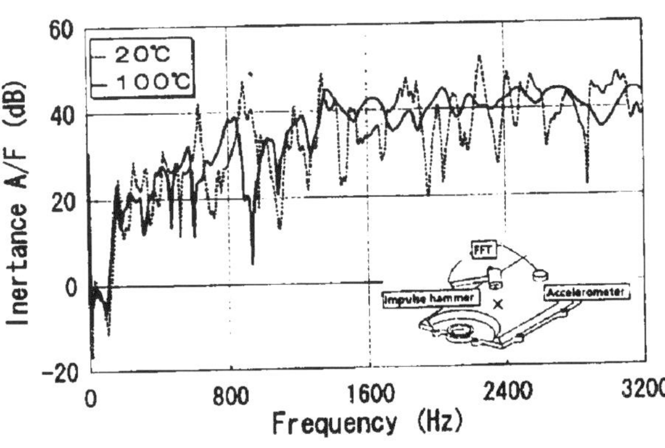

Fig 89: Hammering impact test results

Composites for volume production

Results of the four year Mosiac project, involving Renault and six major European materials suppliers, demonstrated the possibilities for composites in high-volume production. An alternative Clio car structure based on aluminium and polymer composites, achieved weight saving 25% over the standard vehicle. Designed for volume production, considerable parts consolidation has been achieved — from 400 required of a steel body shell to 40 for the weight-saving prototype.

The Mosaic Clio has a superstructure of extruded aluminium alloy framework with aluminium roof and quarter panels; dash and floor pan are in GRP. Alternative aluminium alloy and composite solutions have been used for the front-end. The front-ends meet legal crashworthiness requirements but the composite one involves a 7% larger crush displacement at 50 mph frontal impact than the production Clio. There is also a 20-30% cost penalty for this sub-assembly in composites and a 40-60% penalty in aluminium alloy.

The composite 'chassis', comprising front-end, front body panels, floor pan and rear wheel housings, consists of only 13 parts in composites. In the case of the aluminium frame, extrusions are channelled to carry wiring and grooved for skin panel fastening. A variant of SMC was developed by DSM for the composite front-end; this an extended form of high-modulus material known as HMC. This needs to be moulded with a near constant wall thickness but with relatively simple tooling. There also BMC (bulk moulding compound) cones between the bumpers and their support beams for energy absorption. The HMC material is claimed to have almost elastic elongation as well as superior stiffness to GRP. Resin is a polyester/vinylester hybrid which is also said to imbue good fatigue resistance — important in that the vehicle's suspension connects direct to the front-end structure. A 45% glass content is used and mechanical strength is said to be 40% above SMC.

HMC panels were made in matched-die compression moulds which produced part-to-part dimensional variation of less than 0.5 mm. The front-end panel extends from the bumper back to the windscreen and dash-panel mounting. It includes the firewall and two side-panels to which the suspension is mounted. Nine parts are involved compared with over 80 in the steel structure. Structural bonding is used to join the nine parts in an industrialised process taking just 50 seconds. Rivets on the glue line are needed to meet crashworthiness requirements and allow adhesive curing at room temperature. The floor-pan was developed by Montedison using Unifilo glass mat into which resin is injected on cannon machines. The part is then moved to a clamping station, total cycle time being 3 minutes. The part weighs 13 kg and has 50% glass content. Flexural modulus is 13 000 MPa.

Natural and knitted fibres for composites On the initiative of Silsoe Research Institute and the UK Biocomposites centre an interesting project has been undertaken which may one day lead to synthetic fibres being replaced by natural ones. Cellulose fibre is obtained from linseed, grown increasingly for its vegetable oil content. By chemical treatment of the surfaces and then using the material to reinforce cellulose based polymers, from waste products, potentially this could result in engineering plastics as biodegradable as required. Each hectare of linseed produces two tonnes of seed and two tonnes of straw, presently a waste product and containing 20% potentially long fibres. Specific tensile stress of the 25 mm long fibre cells is 1.05 GPa (glass is 1 GPa) while specific modulus of 40 GPa compares with 28 for glass. With processing machinery developed thus far, a profit could be made at a raw fibre sale price of £200-250/tonne compared with £1200-1500 for polyester.

Use of unidirectional continuous fibres for composites reinforcement has not been exploited for complex shaped parts due to the time-consuming hand lay-up required. Prepreg mats such as GMT

* Fibre orientation is defined spatially by stitch geometry
* Knit structure coherency guarantees fibre orientation during large plastic deformations
* Weft knits can be drawn in wale and course directions by up to 150 %
* Drapability of preforms allows unham pered shaping
* Spatial interweaving of stitches hinders formation of laminates
* Deep drawing is economical

Fig 90: Knitted fibre properties

permit glass contents of no more than 30% so interest has been shown in a recent technique for knitting fibres. Their properties, shown in the table of Fig 90, allow high-strength freely shaped parts to be made, using automated shaping techniques.

The Swiss Federal Institute of Technology have derived relationships between these properties and related production processes using knitted carbon fibre reinforced PMA and PEEK resins. A knitting technique has been developed which reduces fibre stress during the stitch build-up phase. The weft knit has a symmetric structure shown in Fig 91 whereas in a densified composite, Fig 92, two stacking sequences were adopted. The cell like structure which results has allowed the Institute to model the structural performance using six different elastic constants. Densification to give fibre content up to 60% is possible using a knit of small area weight but large stitch size.

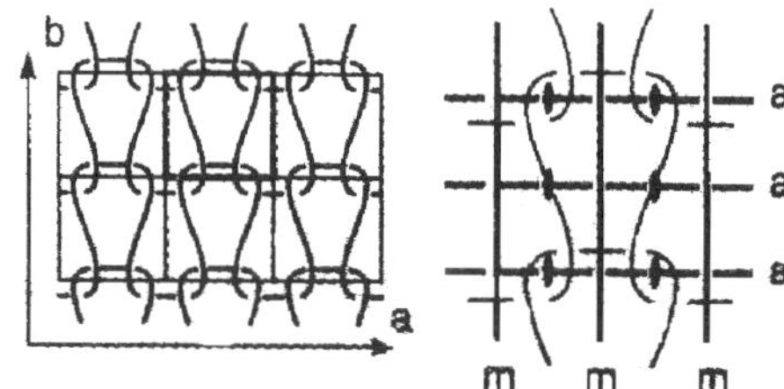

Fig 91: Unit cell (left) of single weft knit with symmetric dimensions (right)

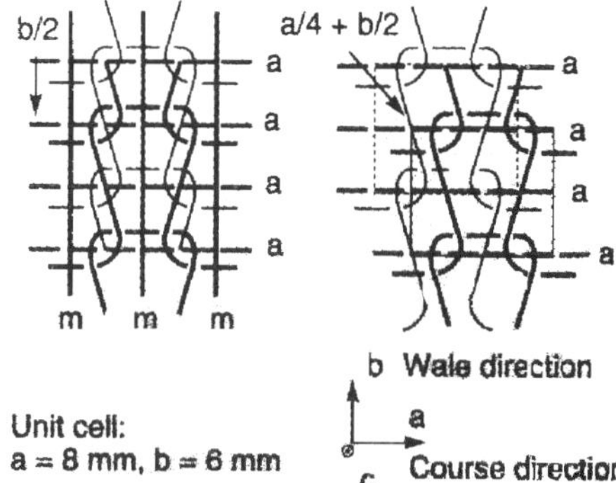

Fig 92: Symmetry in weft knits and possible stacking arrangements

Polymer/metal composites

Besides the familiar sandwich panel, other attempts which resemble latticed girder structures have been used to combine metals and polymers for maximum structural efficiency

A structural composite system has been described by Bayer[10] combining plastic and metal, Fig 93, the technology involving the terms 'insert' and 'outsert'. An example of the insert technique is injection-moulding around threaded inserts or metal parts used for transmitting a force. The outsert method refers to the combination of a metal billet with, for example, moulded plastic studs and bearings. Another variant is the joint technique, in which metal tubes or profiles are joined together or fitted with connecting pieces by means of injection-moulding; brake pedals have been made this way. The plastic/metal composite differs from previous composites, in that the plastic is used to stabilise thin-wall metal shells, thereby creating a composite component with completely new properties.

The principle involves bonding metal and plastic in an injection-moulding process, to produce a complex, load-bearing component. Plastic + injection-moulding means economical manufacture and good integration whereas steel + deep-drawing means mass production and stiffness. The combination of

Fig 93: Different versions of Plasti-metal

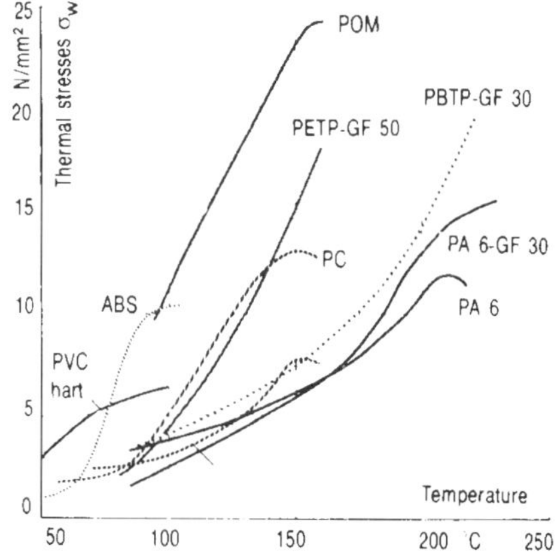

Fig 94: Thermal stresses after single heating/cooling

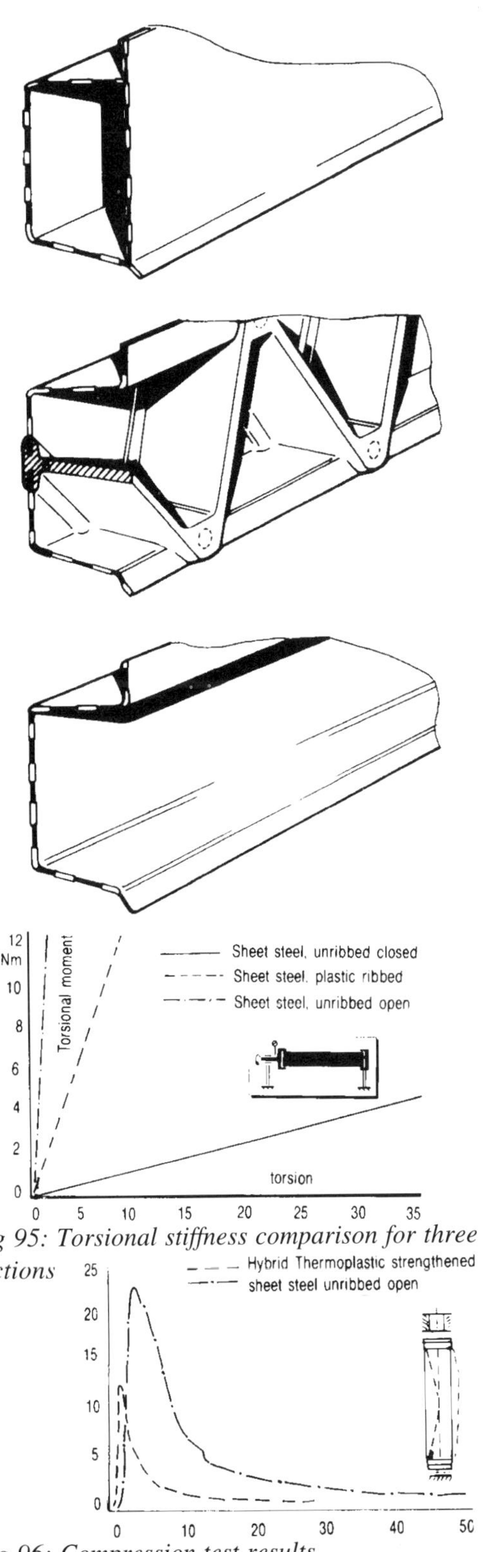

Fig 95: Torsional stiffness comparison for three sections

Fig 96: Compression test results

the two materials and processes results in the high-volume production of a complex component: place a deep-drawn and perforated piece of steel in an injection-moulding die and inject the plastic around this part.

Thermal stress investigations have been carried out in order to measure the contraction stresses as the melt cooled in the die, so that the ejection forces required could be determined. To this end, a test rig was constructed, in which test-pieces (samples), rigidly tensioned, could be heated in an oil-bath to temperatures in the thermo-elastic range. The samples were prevented from buckling by side-supports. At first, compressive stress built up during heating. This gradually returned to zero as the temperature increased and the plastic relaxed. During cooling, tensile stress developed in the sample as the plastic contracted. Fig 94 shows the various thermal stresses (tensile stresses) after the test-pieces had cooled down related to differing heating temperatures.

These stresses also occur in a plastic/ metal composite when the plastic is interlocked with the metal by both force and shape. For this reason, it is preferable to use semi-crystalline plastics, such as PA or PBT, which can reduce the stresses by relaxation. Glass-fibre reinforced plastics are best, since the reduced contraction (moulding shrinkage) also leads to lower thermal stresses. The relaxation of these stresses is related to time and temperature. At low temperatures (-30 C) the stresses will increase in accordance with the thermal linear expansion or contraction. With a coefficient of linear expansion of 40×10^6 and temperature difference of 50 C, the increase in expansion will be around 0.2%. Basic trials were carried out using test pieces of varying design and different materials; these were: PA 6-GF with primed sheet steel + bonding agent; PBT mod. with sheet steel/ribbed; PA 6-GF 30 with perforated steel/ribbed; PBT mod. ribbed

Fig 95 shows different torsional stiffnesses of three test-pieces. By ribbing the plastic/metal composite profile, the torsional stiffness of an open metal profile can be increased by a factor of 12 (geometrically). Further improvement can be obtained by using different ribbing designs and plastics. However, the torsional stiffness of the closed steel profile can not be achieved economically by the composite. The load-bearing capacity of such composite profiles with regard to compression in a longitudinal direction is shown in Fig 96. By ribbing the metal profile, premature buckling could be prevented and the load-bearing capacity increased by about 80%. The bending strength of such composite profiles could also be significantly increased. Even thin-wall closed metal profiles fail, through buckling, before bending failure of the composite profile (Fig 97).

Component for a car door Bayer AG assumed responsibility for two sub-projects under the Eureka CARMAT programme which entailed the development of an underbody and car side door based on a Citroen AX design. The supporting structure is not just a load-bearing structure, but also an assembly module. Consequently, the window-winding and door-closing mechanisms, the window frame and the panel components all have to be taken into consideration. Once the geometry of the metal body had been established, Bayer had to design the plastic ribbing. First a paper model was made to establish the crude structure of the ribbing and the flat bonding areas for the plastic 'coring'.

Since the geometry of part had been drawn up using CAD the same geometrical data could be used for the FEM calculations. Fig 98 shows the support-

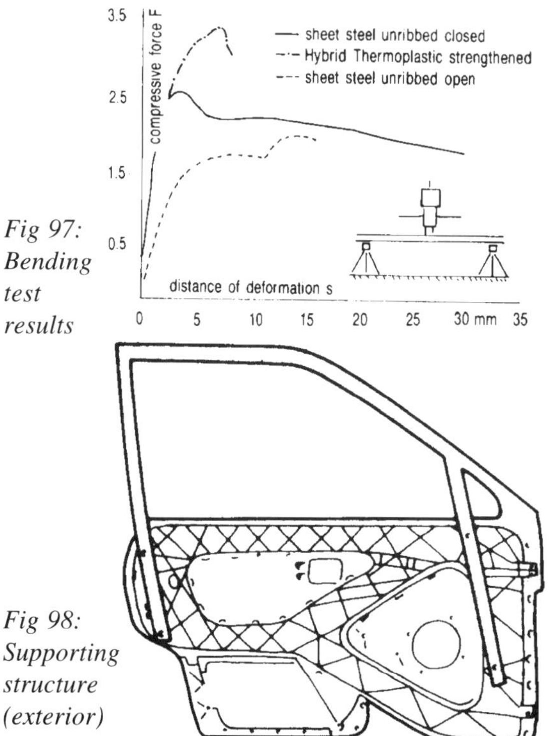

Fig 97: Bending test results

Fig 98: Supporting structure (exterior)

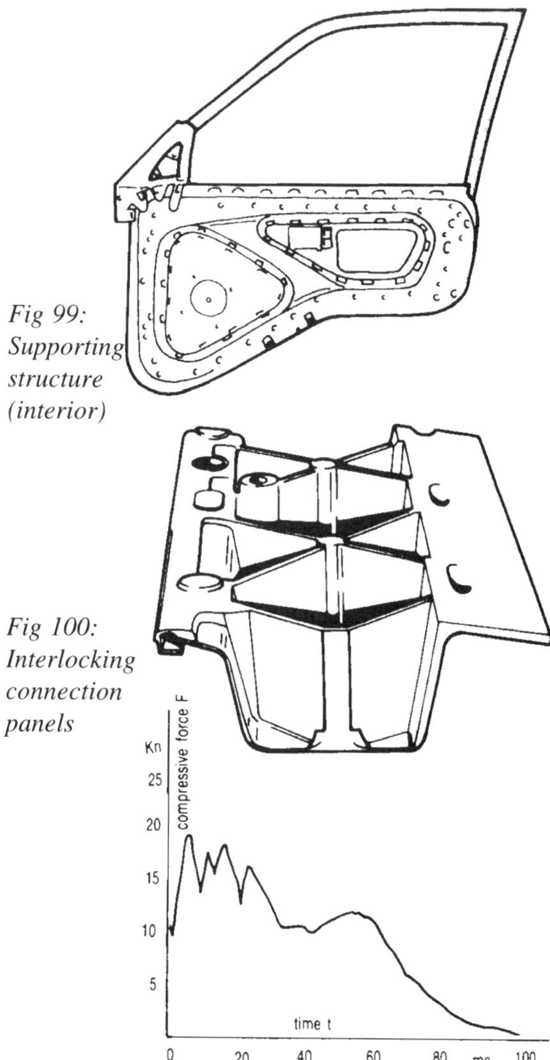

Fig 99: Supporting structure (interior)

Fig 100: Interlocking connection panels

Fig 101: Pendulum test results

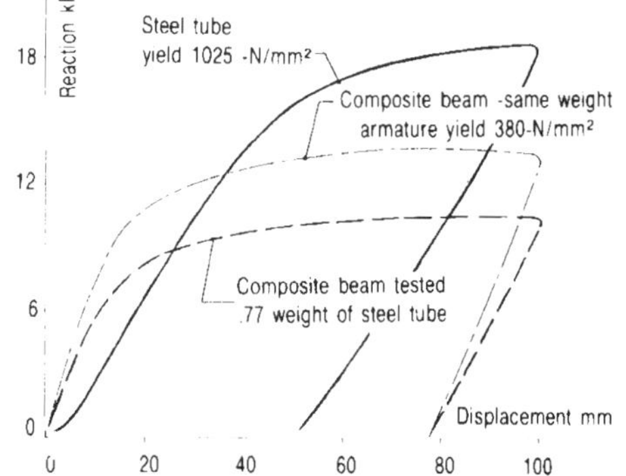

Fig 102: Three point bending test results for side impact beam

ing structure from the outside with loudspeaker housing, map-pocket and the housing for the door latch and an interior view is seen in Fig 99. Attachment points, including that for the mirror base, can be seen. Bonding areas can be seen in Fig 100 which shows the die head connection on the closed side of the metal profile. The plastic melt enters the countersunk openings in the metal, forming a die head between the wall of the recess in the injection-moulding die and the metal part. Effectively the die head connection takes place directly in the injection-moulding die; no additional operation is required. Fig 101 shows the force-time curve from the pendulum test on the part (pendulum weight: 780 kg, pendulum speed: 8 km/ hr). The door exhibits a high degree of resilience, since a relatively constant force is exerted over a long period, followed by a sharp drop in force in a very short time.

Stabilised core metal/polymer composite (SCC)

Vehicle structural innovator, Gordon Wardill developed light-weight panels, initially for bus structures, based on self-skinning polyurethane moulded around a metal reinforcement structure. Fig 102 shows test results for a side-beam.

The Stabilized Core Composite (SCC) has since been applied to car doors in experimental work with PSA group. One of the key features of the SCC process is the fact that the encapsulated pieces are joined mainly by the bonding action of the urethane, which is injected under pressure during the RIM process. Apart from making stiffer joints, this means that it is possible to manufacture relatively large body sub-assemblies in a 'one shot' injection process without the use of metal welding techniques and fixtures and with the use of very low cost tooling, Fig 103.

In a feasibility study recently compiled for PSA, prototype SCC doors were compared with all-steel originals. All stiffnesses were equalled or improved upon. The gains in window frame stiffness were particularly notable. This was due to the ability of SCC to increase joint stiffness without the cost and weight penalties usually involved. An overall improvement in strength and energy absorption was experienced. Tests for axial crush, as experienced in vehicle frontal impact, showed that the SCC door was three times stronger than the original. Inherently high lateral strength means that the conventional side intrusion beam is not required. The SCC door was 1.5 kg lighter than the original. Inherent in the design is the fact that a plastic composite or metal outer skin can be mechanically fastened to the load bearing inner panel as a final operation, thus enabling a robotised 'plant-on' operation to be performed for all mechanical and glazing items, in an estimated time of 90 seconds. Thus a considerable reduction in assembly costs can also be achieved.

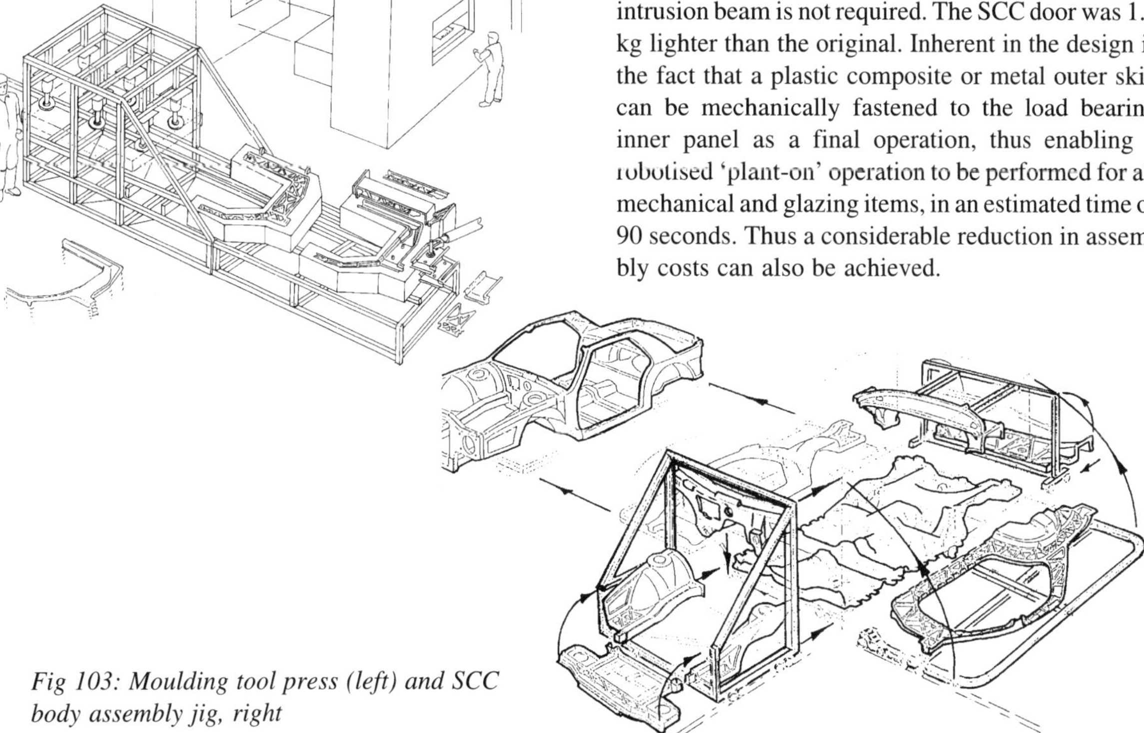

Fig 103: Moulding tool press (left) and SCC body assembly jig, right

Hawtal Whiting (HW) and SCC's originator explained[11] that the technique had exhibited results of cantilever bending tests on a section of a seat-back beam, comparing the SCC beam with an equivalent weight thin wall box beam (steel pressing). The all steel beam exhibited a 'classical' thin walled beam failure mode. A discrete compressive buckle hinge formed after some 20 mm of end displacement and, thereafter, a rapid loss of bending resistance was experienced. In the case of the SCC beam, a tensile failure occurred after some 260 mm of end displacement. The SCC absorbed 170% more energy than the all steel equivalent.

To study SCC in passenger-car body structures, manufacture and testing of a car door side impact beam was undertaken. Three designs of beam were investigated: a thin walled pressed steel beam using the maximum allowable depth (50 mm). The material yield strength required was 380 N/mm^2; a thick walled steel tube beam having an outside diameter of 35 mm and a wall thickness of 4 mm (material yield strength required was 1025 N/mm^2); a beam manufactured in SCC which also used the maximum depth available of 50 mm (core yield strength was 380 N/mm^2)

In the results of thesimple three point bending tests in Fig 102 it can be seen that the thin walled beam exhibited a high bending stiffness over the initial phase of displacement (useful in side impact). It was of no real use because of its 'classic' early buckling related collapse. The thick walled steel tube displayed a high ultimate load, due to lack of local buckling. This feature was not very useful because this peak resistance occurred at a point where door tensile resistance was adequate. Prior to this, in the linear bending phase, the tube showed poor linear resistance, due to the geometrical effects of its small diameter. The SCC beam did not have as high an ultimate bending strength as the steel tube but had a much higher bending stiffness over the first 35 mm of displacement. As a result, overall, the SCC beam absorbed 25% more energy than the steel tube in the useful phase of displacement.

In a test on an SCC beam door to the specifications of FMVSS 214, the very high initial stiffness apparently caused a premature failure at the beam/steel structure interface. As a result, it just met the requirements but did not realise its ultimate potential. It was concluded that it would have been better, for

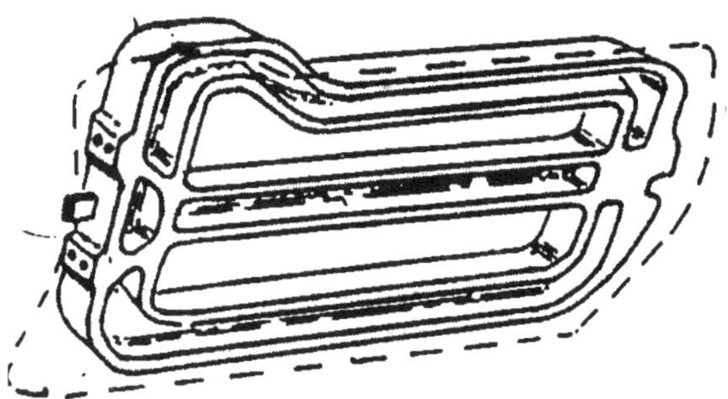

Fig 104: SCC door structure

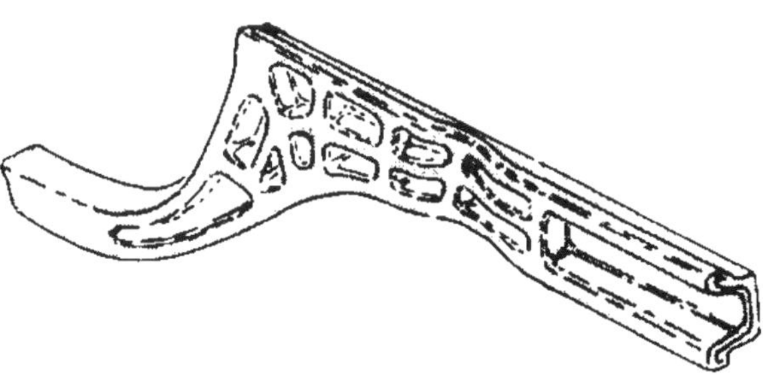

Fig 105: SCC energy-absorbing front sidemember

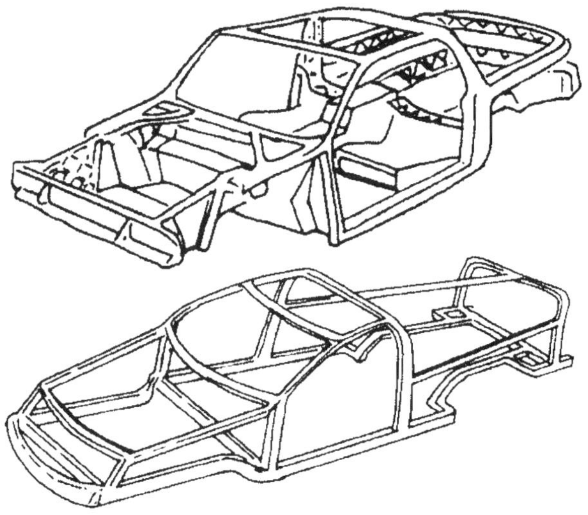

Figs 106/107: Body shell studies

this reason alone, to develop a door structure which was entirely in SCC, Fig 104. Also considered was the possibility of using an SCC front side member in order to improve the collapse behaviour during frontal impact (Fig 105). Complete body shells were also considered, the results of engineering and manufacturing studies being shown in Figs 106 and 107.

Detailed layouts, FE calculations, manufacturing development studies and, eventually, tests were effected on a crucially loaded area of the car — the front door pillar to rocker joint (Fig 108). In this case, a vehicle was chosen which had previously been designed by HW using the conventional pressed steel panel approach. Prototypes of this vehicle had already been manufactured and tested and thus a known steel joint was being compared with an SCC equivalent.

The composite version had to adhere to one essential manufacturing requirement. The favourable manufacturing economics are at their most advantageous when components, such as the sideframe, are manufactured in a 'one shot' injection moulding process. In order to accomplish this, 'open' sections had to be employed. Load-displacement tests were carried out on a rig specially developed for joint testing. Results were obtained for both linear pre-failure displacements and post-failure large displacements. Results of the joint linear FE and test studies are summarised in the table of Fig 109

The SCC joint gave an increase in specific stiffness of 37% under fore-and-aft loading and an increase of 31% under lateral loading. The latter implies good torsional stiffness in the rocker area of the joint. The large displacement test results are shown in Fig 110. The steel joint displays typical thin wall structural behaviour. Failure was due to extensive areas of compressive buckling which lay on a diagonal line across the joint. The SCC joint achieved a load which was 44% higher than that for steel. Tensile failure occurred at the composite/rig interface. The joint itself remained intact up to this point (5 degree rotation). The composite had absorbed 10% more energy than the steel. This rotation is about as much as can be safely taken by the A pillar.

Behaviour of SCC under load If a beam having an open channel section is considered and a bending moment is applied about a transverse axis, then areas of the free edges are subjected to compressive forces.

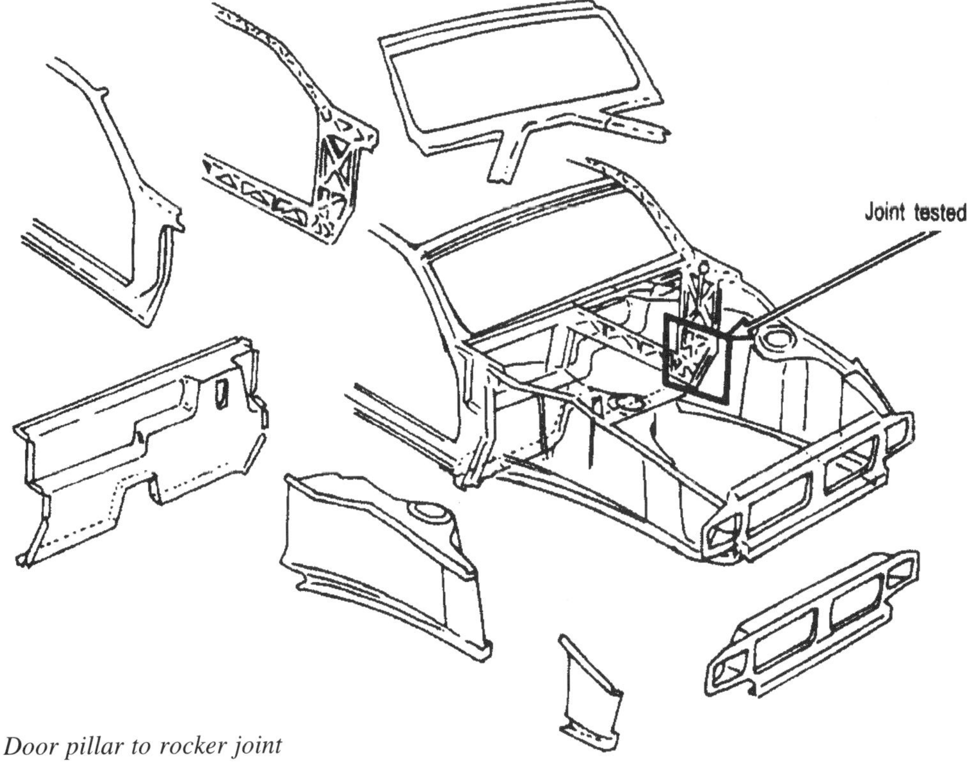

Fig 108: Door pillar to rocker joint

In the 'all steel' case, with wall thickness typical of automotive structures, a likely form of failure will be some form of compressive instability, such as local buckling. The stress at which this occurs is likely to be, at the most, the yield stress of the material. Very soon after the buckle has been initiated, the load carrying capability of the beam usually drops to a low level, due to a discrete 'hinge' being formed.

One way of delaying the onset of buckling is to stiffen the material by increasing the thickness. This would, of course, increase the weight of the beam. However, if one halves the thickness of the metal and supports it on either side by means of rigid cellular self-skinning urethane giving a total wall thickness of 10 x sheet thickness then the weight of the section remains unchanged. In this case, the modulus of elasticity of the urethane is likely to be one 250th of that of steel. Since the buckling load of the free edge of the flange varies approximately as the thickness cubed, then, ignoring secondary effects, SCC/steel buckling load ratio = 4. The net effect, in the case of a beam, is to trade-off a fraction of the tensile strength for a four-fold increase in bending strength. It should be noted that for maximum efficiency, the core must be supported on both sides.

Advantages can be realised in closed section beams and vehicle joints in general, with derived effects which result in improvements to the torsional stiffness of beams: important for those which have a large influence on body torsional stiffness — rocker assemblies for example. Bending stiffness of longitudinal beam flanges affects body torsional stiffness. SCC flanges are stiffer in this respect than steel and therefore an improvement in beam torsional stiffness is possible.

Two other important beneficial effects have become apparent. It is well known that many joints display a reduction in stiffness, due to quite large changes in overall shape. These changes can be prevented by means of internal diaphragms placed across sections of the joint. It has always been difficult if not impossible to economically achieve this in the conventional steel structures, due to welding and alignment problems during assembly. With SCC, however, it is no more difficult to include internal ribs than it is not to.

Secondly, some of the flexibility of steel joints is due to the effect of inter-spot-weld buckling. This is totally prevented with SCC because, during the RIM injection process, perfect bonded joints are produced between the simple armatures that are used — and virtually no welding need be employed in the construction, say, of a car body.

One major drawback of SCC is the difficulty in designing components, as truly three-dimensional structures are involved The effects of three dimensional volumes of rigid cellular urethane have to be considered. The initial definition and parametric optimization procedures for SCC are much more time consuming than those for the sheet metal approach.

Steel	3.75 kg	Mass
SCC joint	3.67	
Steel	419 N-mm/rad	Rotational
SCC joint	565	stiffness
Steel	112 N-mm/rad/kg	Specific
SCC joint	154	stiffness
Steel	19 N-mm/rad/kg	Lateral
SCC joint	25	stiffness

Fig 109: Joint performance comparison

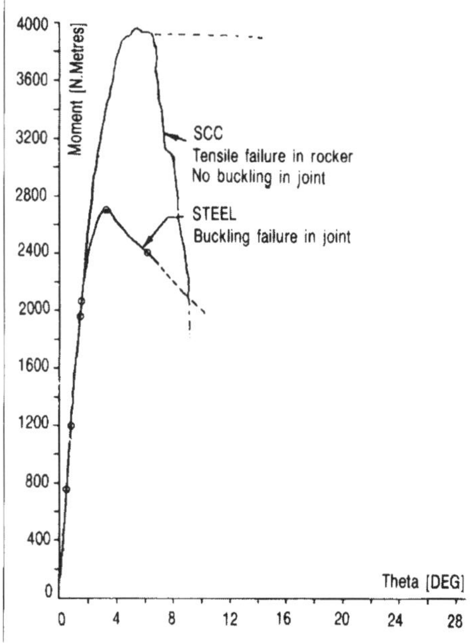

Fig 110: SCC joint large-displacement test results

Semi-rigid polyurethane foam

With this polymeric material, which is in a separate category from most others used in automotive work, a variety of techniques can be used to obtain compounds of different physical properties. Here the semi-rigid foam variant is considered in some detail.

Urethane polymers are strictly speaking 'rubbers' and are difficult to classify in 'thermoplastic' and 'thermosetting' terms. They are made from combining polyesters or polyethers with di-isocyanates and are unusual in that their physical properties do not depend on compounding materials. In fact they cross-link and undergo chain extension to produce a wide variety of useful materials including hard and soft foams, elastomers and rigid-panels, when suitably reinforced. These reinforced reaction injection moulded panels are to be considered separately in a future article as are the flexible foams. The semi rigid foams considered here are of interest in that they can be foamed in situ or cast in slabs — even with a so-called 'self-skinning' high-density outer surface. Urethane polymers have good tensile strength, load-bearing capacity, elongation potential and both good hardness and abrasion resistance.

	Material type	Density kg/m³	K	r	Temperature variation
1	Urethane	31.5	312	0.02	None
2	Urethane	75.5	730	0.06	- 20%
3	Isocyanurate	50	385	0.06	- 20%
4	Isocyanurate	53.5	374	0.05	- 20%
5	Phenolic	43.5	67	0.08	None
6	Phenolic	50.5	327	0.03	None

(Units for K are such that $K\varepsilon^r$ is in kN/m^2.)

	Material type	Density kg/m³	r	Temperature variation
7	Urethane	95	0.11	- 70%
8	Urethane	85 - 89	0.16	- 90%
9	Urethane	48,66	0.18	?
10	Urethane	120	0.09	?
11	Urethane	90	0.15	?
12	PVC nitrile	161	0.18	- 90%
13	Ethylene	154	0.09	?
14	Ethylene	37	0.04	?
15	Scotfoam	52	0.30	?

Fig 111: Strength increase with density

Energy absorption is a key property for the material's application in bumpers and interior crash-padding. The material is also used as an acoustic insulator in combination with solid PVC or EPDM sheet, the latter providing a dense layer to reduce sound transmission while the PUR foam absorbs the sound energy. The material is also combined with carpeting to form shaped floor mats which are convenient for assembly purposes and avoid joints and gaps in the floor-covering. Ability to use low-cost moulds and relatively simple production equipment makes the material valuable in its more structural role in specialist vehicles. Complete vehicle front-ends can be made — also such aerodynamic aids as spoilers and air-dams moulded at modest cost. For items such as fascias and crash pads, PUR may be foamed into a skin of PVS or ABS, previously shaped by vacuum-forming, drape-forming or slush-moulding — exploiting the material's valuable adhesive properties. Self-supporting liner panels can be produced in the so-called Press Foam process with which fabric skins can be incorporated. The PUR is foamed in a flat tray then clamped in a press before curing is

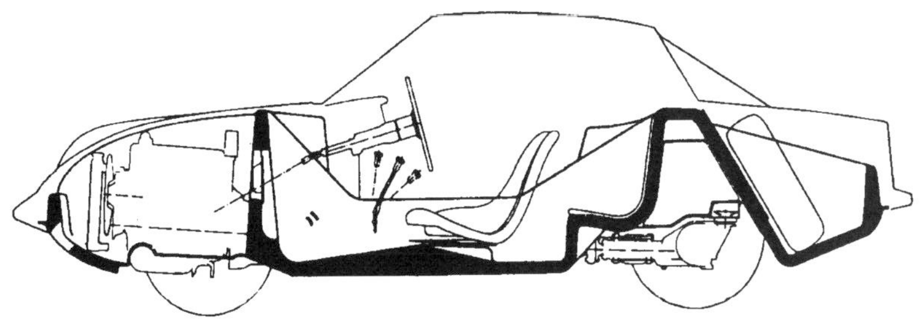

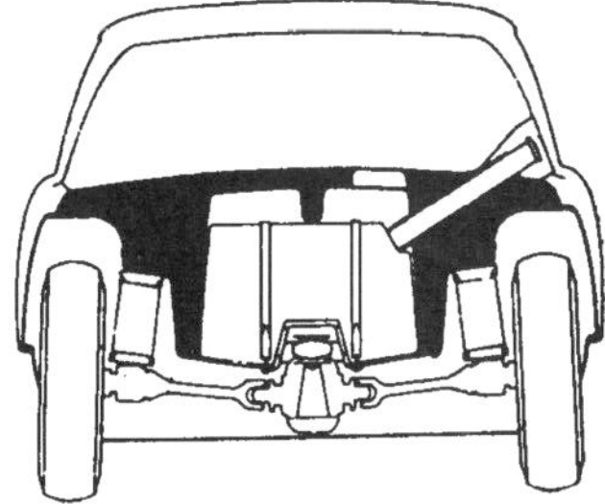

Fig 112: Bayer experimental sandwich construction car

completed. Differential compression of such partly cured foams enables its density to be varied so that strength and hardness can be varied across a panel.

One of the most interesting structural uses for PUR was proposed by Bayer who built a complete car structure using a rigid foam within GRP sandwich skins, Fig 111. Another important application is in the core of refrigerated vehicle bodies where its closed cellular structure and low thermal conductivity play crucial roles. Again structural strength can be varied according to density as seen in Fig 112. Three in-situ foaming techniques are available, pour-in-place, spray-in-place and froth-in-place. The first allows bodies to be built with minimum capital expenditure. All that is required for a typical hand mixing kit is protection equipment, contact thermometer, heater unit, power-drill and propeller-type stirrer. The foam typically exerts a pressure of 2-3 lb/in^2 (0.14-2.0 bar) as it rises, after mixing and pouring, so the cavity must be adequately jigged to avoid any panel distortions.

Where jigging is too difficult to achieve, spray-in-place permits panel insulation on one side of a panel to be applied and it is of course possible to trim the expanded surface and bond on the second panel to obtain a sandwich structure. For spraying, a system based on methylene diphenyl di-isocyanate (MDI) is preferred to the longer established toluene di-isocynate (TDI). Frothing-in-place involves the addition of a frothing agent, under pressure, into the mix to give rapid expansion prior to the foaming action. Jigging costs can be reduced by this method and greater heights can be filled from a single 'pour'.

The assessment of physical properties of foams for crash-padding effectiveness can be carried out on test machines which apply a fixed rate of strain to the foam specimens. The compressive stress on the foam is related to strain by *Ker* where *K* and *r* are material constants, Fig 113. Typical behaviour for semi-rigid foams is shown in Fig 114. For rigid foams, yield occurs at approximately 10 per cent strain, beyond which the cell structure collapses. Semi-rigid foams, however, show good recovery at these strains, by contrast, and can survive up to 60 per cent strain levels. The semi-rigid foams are seen to be more strain-rate dependent than the rigid ones. The above apply to flat test-loading surfaces and the effect of a cylindrical indenter is seen in Fig 115 superimposed on the 'flat' case.

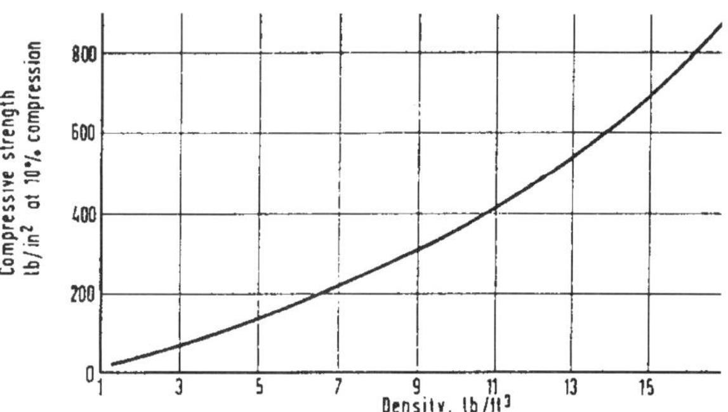

Fig 113: Stress/strain relationship

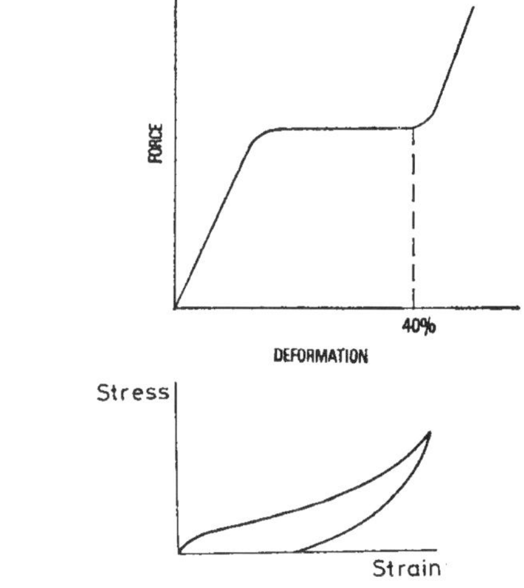

Fig 114: Semi-rigid foam characteristics

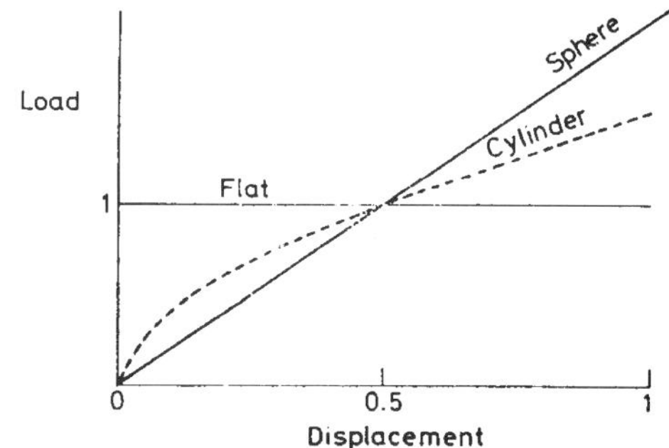

Fig 115: Semi-rigid foam testing

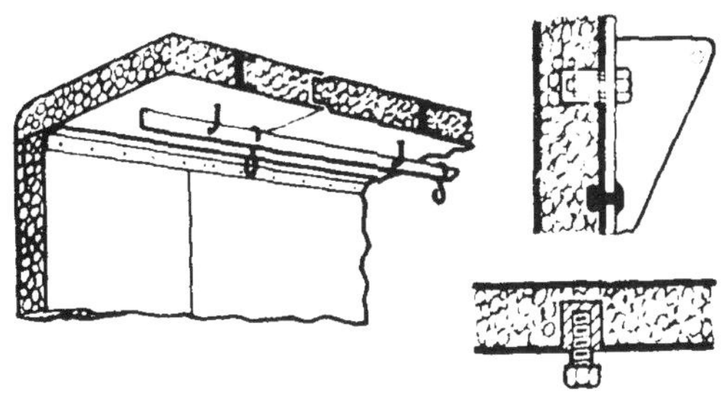

Fig 116: Plugs and inserts

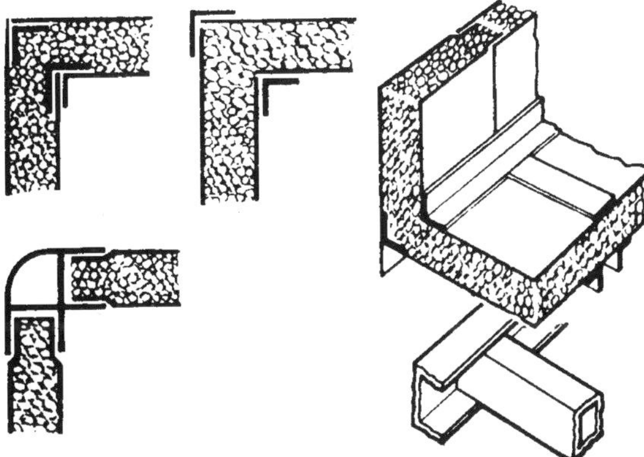

Fig 117: Corner joint techniques

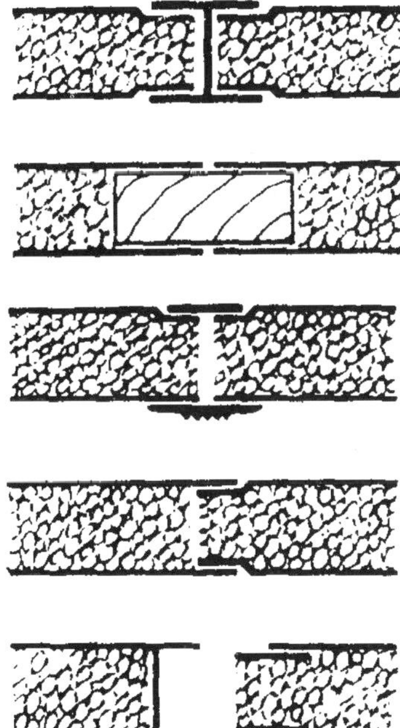

Fig 118: Reinforcement techniques

Sandwich panel construction

The need to stabilise thin sheet structures against buckling and wrinkling, under either compressive or shear loading, has led to the use of sandwich construction. Cost and complication in assemblies of pillars and rails, to support large area panels, is overcome by using light-weight core materials sandwiched between thin skins of higher strength facings.

Besides the clear structural advantages of sandwich panels, those of heat and sound insulation, also the ability to achieve good surface smoothness over wide areas, add to the appeal of this form of construction in many special vehicle categories. Recent automated build techniques have further widened their scope and developments of plastic foam technology has enabled extremely light-weight core materials to be used of high structural efficiency.

The key parameters of the panel which have to be related to the external loads and the maximum stress capabilities of the face and core materials, as discussed in a later chapter. Any lateral loading of the panels (at right-angles to the faces) must also be considered in relation to the reaction and support points of the structure using bending and stability theory to be discussed later.

Bonding of the core to the skins is a critical factor of structural integrity. Foamed in-situ core systems are usually self adhesive to the skins; otherwise complete coverage of the surface must be ensured if the continuity of shear load path between core and skin, on which the strength and rigidity of the panel relies, is to be guaranteed. A common method of applying pressure between skins and core, during bonding, is the vacuum bag which allows atmospheric pressure to be evenly applied across the surfaces.

Concentrated loads, at attachment points on the panel, must be fed through to the core without inducing peeling loads between it and the skin. A variety of plugs and inserts can be bought for this purpose, Fig 116. Corner joints will normally require reinforcement of the type shown in Fig 117, again to protect against peeling loads. Butt-jointing of panels needs to be as strong as the main area of the panel and should involve the use of gap-filling adhesives. Some reinforcing techniques are shown as Fig 118.

Foam and honeycomb cored panels

One technique of automating the application of bonding resin between core and faces was perfected by Duramin Engineering in the 1960s. Shown in Fig 119, the system depends on cutting transverse irrigation grooves into the surface of the slab-stock expanded polyurethane core, (A), as well as inlet and outlet manifolds, (B) and (C). The core is sandwiched by glass-cloth laminates, (E), held between aluminium alloy mould faces (F) backed by timber stiffeners (Z). A colourised gelcoat is shown at (G) and (H) shows extrusions used to close the open ends of the mould. An encapsulating vacuum bag (I) is sealed at (J). Air is evacuated from the mould by an exhauster (K) and resin drawn in through container (L). The Lotus VARI process, described in a separate chapter, has also some similarities as well as the ability to handle curved panels and incorporate a wide variety of core shapes.

Proprietary foam-cored sandwich panels can also be purchased from a number of sources. Some properties of Schmitz Ferroplast panels are shown in Fig 120. Another popular core material is honeycomb cell aluminium or plastic available in Aeroweb sandwich panels from CibaGeigy of Duxford in Cambridgeshire. Such cores are favoured for demanding structural duties such as racing car body construction. Typical fastening, edging and cornering methods for such panels are shown in Fig 121.

Aeroweb arimid honeycombs are also available in hexagonal or 'overexpanded' rectangular format, the latter (Fig 112) particularly suiting curved panels. A variety of cell sizes and 'paper' thicknesses can be specified to give differing strengths and densities. The material must also be designed against the various possible failure modes shown in Fig 113.

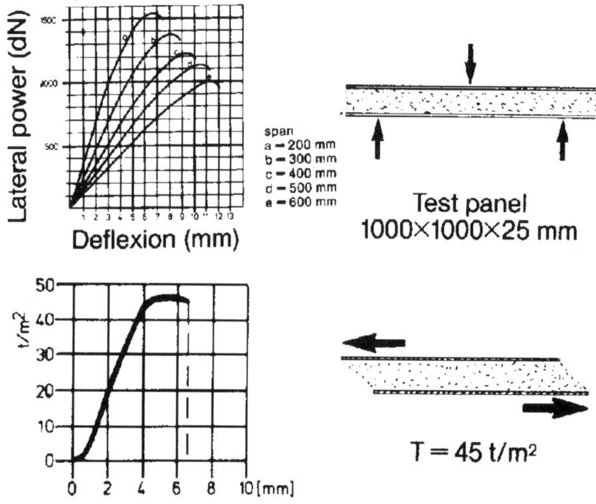

Fig 120: Ferroplast panel properties

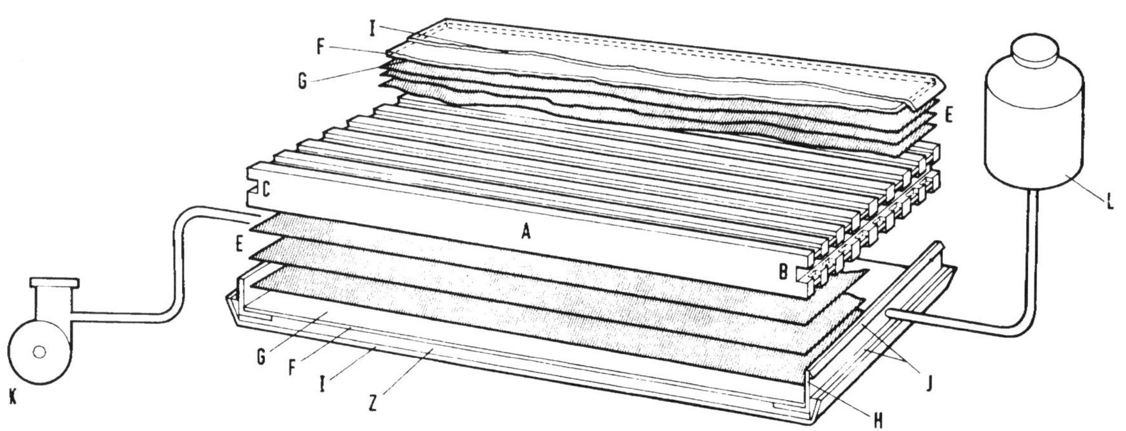

Fig 119: Automated resin-injection

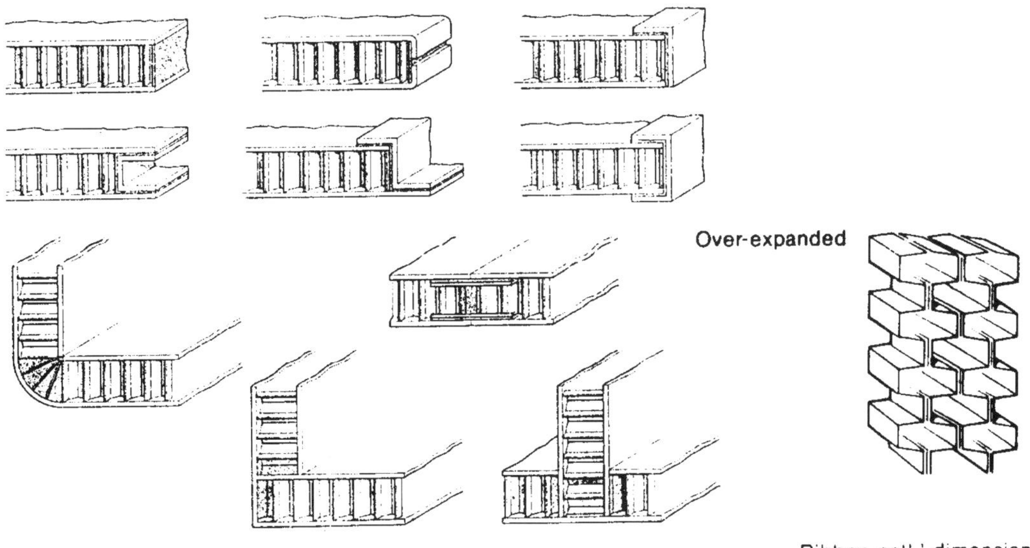

Fig 121: Aeroweb jointing techniques

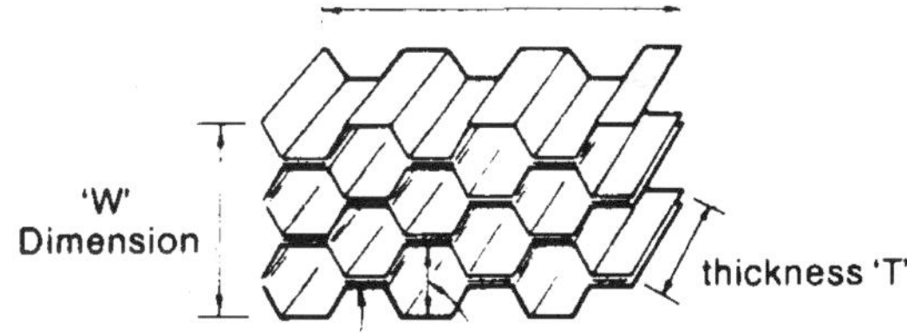

Fig 122: Aeroweb variations

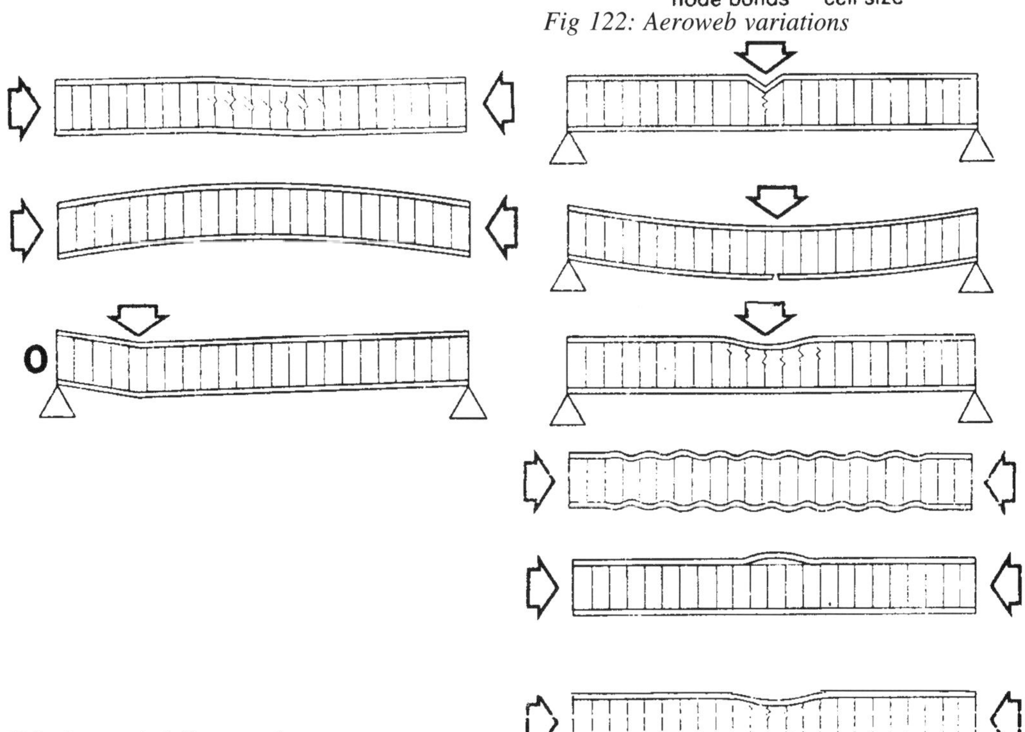

Fig 123: Aeroweb failure modes

Reaction injection moulding

Rapidly becoming to thermoplastics and urethanes what GRP-moulding is to thermosets, RIM is now challenging GRP for composite body panels in relatively low volumes. The process is modest in tool costs and produces panels of high stiffness when the recently developed polyurea systems are employed.

By using polymerisation in the mould, the RIM technique is quite different from other plastic moulding methods and can be used for producing quite complex parts and panels without unduly high tooling investment —since mould pressure is low. Two or more components flow into the mixing chamber, Fig 124, at relatively high pressure (100-200 bar) and then are expanded into the mould at much lower pressure. The streams impinge at high speed to obtain thorough mixing and intiate polymerisation as they flow into the mould cavity at a pressure of about 100 bar. Low viscosity during mould filling is one of the key attractions of the process as a relatively small metering machine can make large parts. The low viscosity also simplifies reinforcement with, for example, the possibility of using continuous-fibre mat placed in the mould.

Some 90% of RIM production is in polyurethanes and urea-urethanes, the latter being uniquely suited to the process as they do not melt-flow like normal thermoplastics and therefore conventional injecton-moulding is not possible. ICI have developed a family of polyureas for body panel applications with unusually good processability and physical properties. Gel times of two seconds are possible and mould temperatures are less than 93.5 C. Overall cycle time is about 1.5 minutes and further development promises less than one minute, Fig 125. Filler packages are also becoming available which allow part surface finish comparable with steel; moisture stability is high compared with competing thermoplastics and the materials can tolerate temperatures of 190.5 C. Figs 126 and 127 show show heat sag properties, of the so-called 9002 material, which are seen to be dependent on post-curing and mould temperature. Fig 128 shows typical properties for a formulation that would suit body panels but others are available which raise the elastic modulus as high as 200,000 psi. The company also produces medium-modulus systems for fascias and bumpers, properties of two such being shown in Fig 129.

As well as increasing strength and rigidity of panels, the addition of glass-fibre or other reinforcements considerably improves the compatibility of

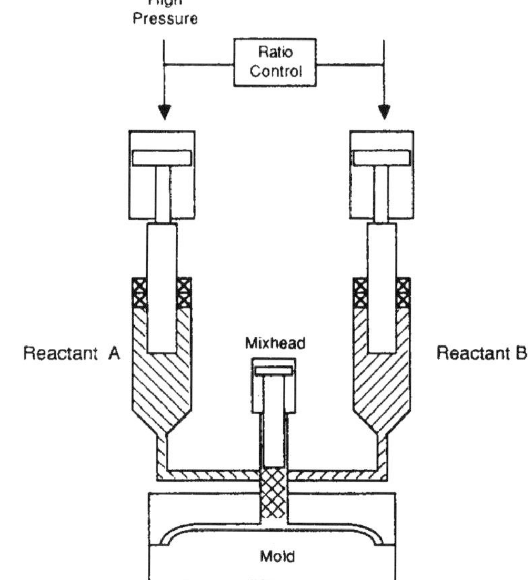

Fig 124: RIM machine

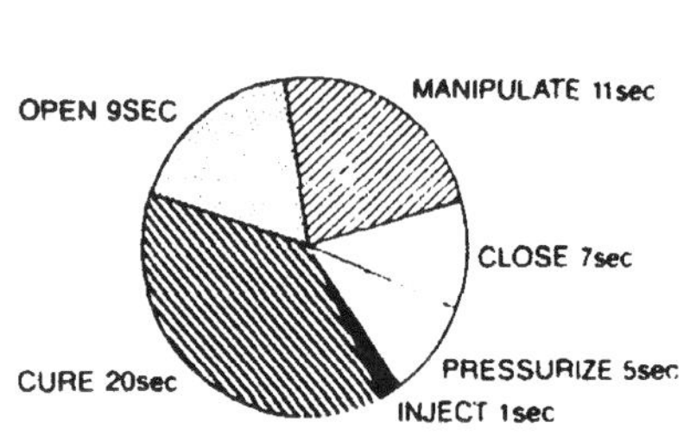

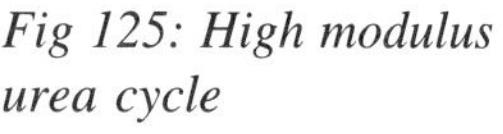

Fig 125: High modulus urea cycle

Fig 126: Post-cure effect on heat sag

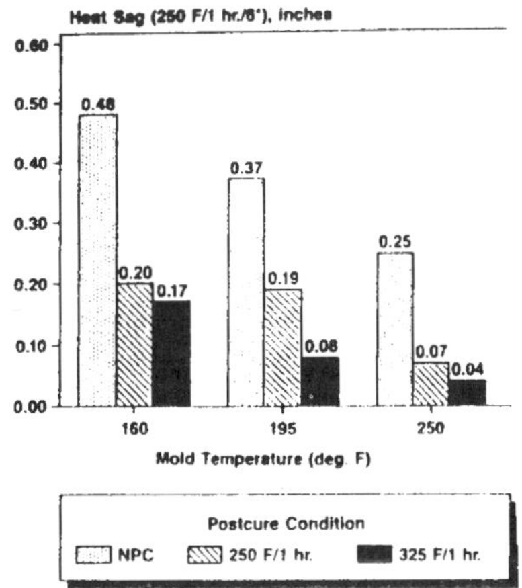

Fig 127: Mould temperature effect on heat sag

Typical properties

Specific gravity	1.10
Flexural modulus, psi	
at 73°F	144 000
at −20°F	220 000
at 158°F	95 000
Modulus ratio −20°F/158°F	2.31
Tensile strength, psi	5100
Elongation at break, %	95
Gardner impact, ft-lb at −20°F	10.1
Mould temperature, °F	160
Component temperature, °F	110

Fig 128: Body-panel RIM formulation

Milled Glass	0	15
Specific gravity	1.02	1.12
Flexural modulus, psi		
at 73°F	35 000	56 000
at −20°F	80 200	125 000
at 158°F	23 500	45 000
Modulus ratio, −20F/150°F	3.40	2.78
Tensile strength, psi	3250	2400
Elongation at break, %	180	100
Tear strength, Die C, lbs/in	430	480
Heat sag, 6" overhand, inches		
250°F/1hr	0.28	0.18
Mould temperature, °F/1hr	250	250

Fig 129: Fascia/bumper formulation

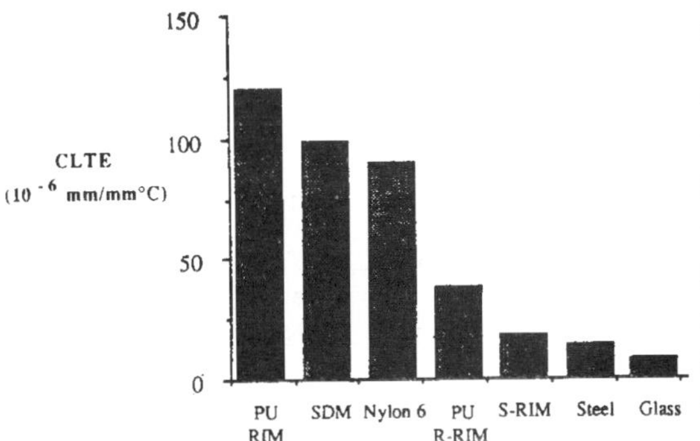

Fig 130: Comparison of processes

thermal expansion co-efficient with such materials as steel and aluminium. Since polymers typically have co-efficients some 10 times greater than steel, a metre-long part hung on to a steel body could change in length by 1 cm between summer and winter temperatures. The way in which this can be modified by reinforcement is clear from the comparisons of Fig 130. In this figure CLTE is co-efficient of linear thermal expansion; SDM is styrene dimethyacrylate; RRIM is reinforced RIM which in this case is 10% by volume 1.6 mm chopped glass fibres; S-RIM involves a styrene dimethacrylate injected into 10% by volume continuous fibre-glass mat.

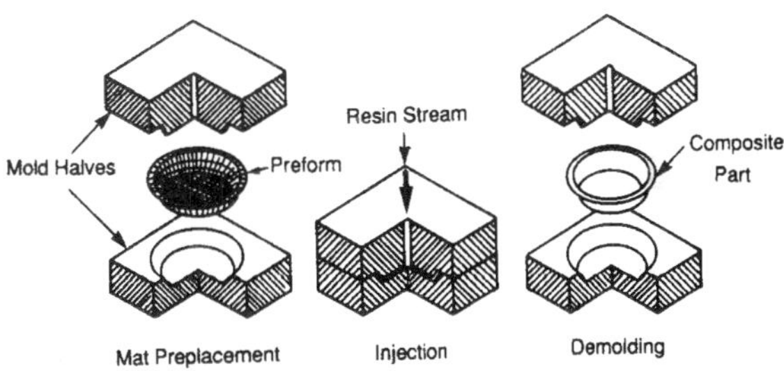

Fig 131: Steps in the S-RIM process

	RTM	S-RIM
Equipment Cost	\$30,000	\$500,000
Flow Rate (kg/min)	2.3	55
Mixing	static mixers	impingement
Mold Pressure (MPa)	0.3	2.4
Void Content (vol%)	0.1 - 0.5	0.5 - 2.0
Mold Materials	epoxy	steel
Mold Temperature [b] (°C)	25 - 40	95
Component Viscosities (MPa·s)	100-550	< 200
Cycle Time (min)	10-60	2-6

Fig 132: S-RIM and RTM compared

	Isocyanurate [a]	Urethane [b]	Acrylamate [c]	Epoxy [d]
Random Glass Mat (wt%) (3 mm thick part)	38	44.8	40	40
Specific Gravity	1.54	1.53	1.46	--
Void (vol%)	1.5	1.5	--	--
E_f (MPa at 25°C)	8,100	9,600	8,700	9,200
Tensile Strength (MPa)	150	150	125	160
Elongation (%)	7.3	2.0	2.1	1.2
Izod Impact (J/m)	510	660	790	~800
Heat Distortion (°C)	184	189	240	>200
Thermal Expansion (m/m°C) x 10^{-6}	~20		27	18

Fig 133: S-RIM-composite properties

S-RIM is the process in which long-fire mat is placed in the mould and reactive monomers injected on to it, Fig 131, the process being akin to resin-transfer-moulding but with high-pressure impingement mixing to accelerate the reaction, Fig 132. Typical properties of S-RIM composites are shown in Fig 133. A recent development is so-called low-density S-RIM which is particularly attractive for trim-panel applications; its comparative properties are seen in Fig 134.

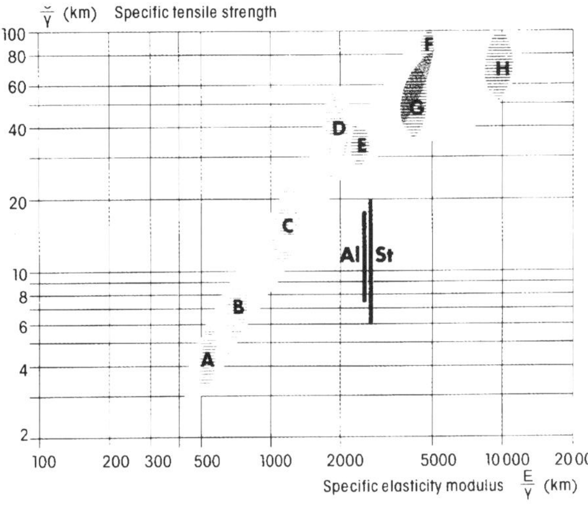

A - SMC
B - Random Fibre (RTM)
C - Glass Cloth (RTM)
D - UD Glass (RTM)
E - Aramid Cloth (RTM)
F - UD Aramid (RTM)
G - Carbon Fibre Cloth (RTM)
H - UD Carbon Fibre (RTM)
Al - Aluminium
St - Steel

Fig 137: Properties of RTM reinforcement systems

Resin transfer moulding

In resin transfer moulding (RTM) a low viscosity fluid is injected into a mould containing pre-placed fibre reinforcement. RTM has been used to mould large non-automotive FRP parts as an alternative to open moulding because it can provide two good surfaces and make parts with sandwich cores.

For relatively short model runs (10-20 000), RTM is a lower capital cost process compared with SMC compression moulding. The stages of the process are shown schematically in Fig 135. First the glass reinforcement or preform is placed in the mould. Once the mould is closed the resin is injected with no or little movement of the glass. After mould-filling the part is left curing in the mould until it is dimensionally stable so that it can be demoulded without losing its shape. The fact that the reinforcement is pre-placed in the mould gives the process the potential for making parts with better surface and mechanical properties than SMC where the fibre orientation is usually less controlled and favourable because of its flow.

During the filling stage, the resin being injected usually contains filler and can exhibit a viscosity with some shear thinning behaviour. However, for typical RTM compositions this effect is small and as a first approximation can be neglected.

Property	Units	LD-SRIM	Woodfiber	ABS	PP
Specific gravity	g/cc	0.55	0.85	1.05	1.04
Filler content	%	20 (glass mat)	-	-	20 (talc)
Flexural modulus	MPa	2370	3790	2000	2090
Tensile strength	MPa	41	13	36	25
Dart impact	J	5.7	3.0	11	13
HDT @ 1.8 Mpa	°C	95	90	83	130

Property	Requirement	Result
Thickness, mm	per drawing	2.72
Specific gravity	per drawing	0.55
Warpage, %	4 max.	0.40
Dimensional stability, % Δ	0.6 max.	0.40
Tensile modulus, MPa	12 min.	41
Flexural modulus, MPa	1750 min.	2370
Flexural strength, MPa	25 min.	67
Impact resistance, J	0.9 min.	no cracking
Flammibility, mm/min	101.6 min.	self extinguishing
Odor	no objectional	none
Stain resistance	no staining	non-staining

Fig 134: LD-SRIM properties

Property	Unit	Value
Epoxy Resin		100 P.B.W.
Hardener System		20 P.B.W.
Mixed Viscosity at 25°C		7.0 - 10.0 Poise
Tensile Strength	MPa	87.1
Tensile Modulus	GPa	2.92
Elongation at Break	%	6.9
Flexural Strength	MPa	140.8
Flexural Modulus	GPa	2.97
Compression Strength (Yield)	MPa	121.7
Heat Distortion Temperature		153°C
Tg		164°C

Fig 138: Properties of resin systems

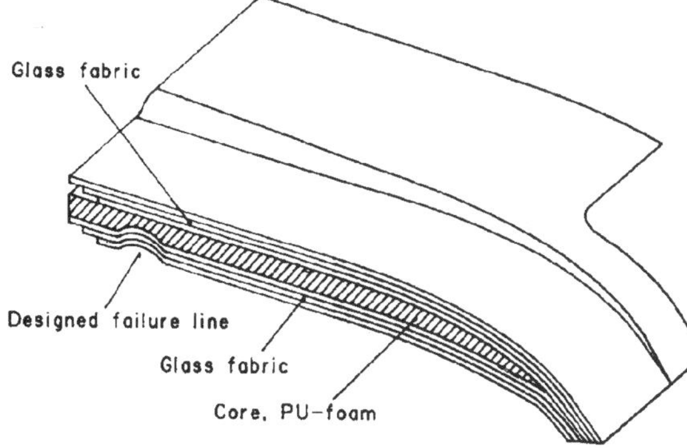

Fig 139: Designed failure line in bonnet panel

According to the Seger & Hoffmann subsidiary of Dow Chemical, the areas in which development effort has been concentrated are preform technology, essential for industrial processing; obtaining a class 'A' surface from the mould; modifications of resin systems to improve flow and fibre wet-out; resin metering, mixing and injection procedures; optimisation of the 'fibre package' for flow, mechanical and visual properties. To increase the efficiency of the moulding stage it is best to perform as many operations as convenient outside the RTM mould; the optimal placement of fibre into the RTM mould can be time-consuming particularly if the shape is relatively complex.

The preform procedure should satisfy the following requirements: produce a precisely formed shape that can be easily placed in the RTM tool; be capable of handling fibres which may need accurate orientation; produce a preform having sufficient integrity to withstand industrial handling. Methods available for preforming complex shapes include spraying fibres and binders on to a perforated mould or the use of mats and/or fabrics pre-treated with thermoplastic binder which can be shaped by pressure when heated. The former process tends to be slow and labour-intensive but has better control of the preform shape and generally has low material wastage. A disadvantage is, however, the inability to accurately orient fibres. A particular problem when preforming cloths, which have to be drape-formed, is that fibre orientation and density change at sharply changing shapes in the tool. This means that the properties of the composite at these points can be difficult to predict.

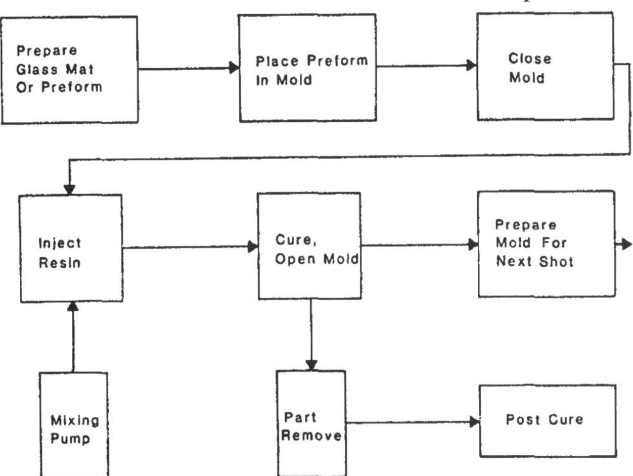

Fig 135: Stages in RTM process

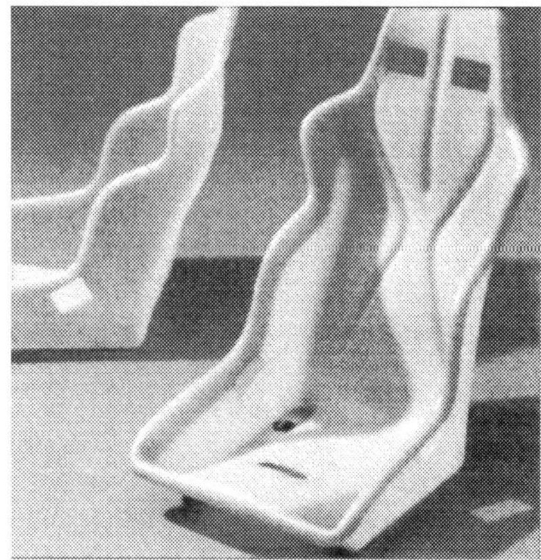

Fig 136: Seat shell made by Dow

The seat shell of a sports seat is one example successfully produced by the company. The part had previously been produced by hand lay-up, producing shells of inconsistent quality and surface finish. The part is complex with a deep draw, multi-curvature shape and multi plane shut-off. The method of preforming chosen used continuous filament random glass fibre mats (Fig 136). The first-attempt preforms and mouldings had been made with a single plane split-line with the associated problems of waste. Optimisation of the preform blank, preform procedures and RTM mould modifications resulted in a 30 - 40% saving in material during preforming compared to the first stage mouldings.

The preforming of fibre reinforcement for flat mouldings such as a bonnet may not be absolutely necessary. The handling of the fibre mats, however, can be difficult as they are rarely self-supporting. A bonnet can be designed as a slim sandwich structure instead of inner and outer mouldings. The rigid foam core thus provides a convenient method of supporting the fibre reinforcement during transportation of the fibre to the RTM mould.

An RTM bonnet moulding has to satisfy both the normal horizontal body panel mechanical requirements and have a class 'A' surface finish from the mould. Many fibre types and cloth constructions can be used in the RTM process as seen in the property diagram (Fig 137). Because of its overall price versus performance, E-glass is expected to be the most widely used reinforcement material for automotive applications.

Resin injection

Epoxy resin systems, with their low volume shrinkage of between 1-3% together with their good mechanical properties, are ideally suited to class 'A' surface body panel applications. Painting at temperatures up to 150 C can also be met with specifically formulated epoxy resin systems. The table in Fig 138 gives the most important mechanical properties of a typical resin system developed for RTM. Faster curing systems based on vinyl ester resins can be used for RTM structural moulding where surface finish is less critical. Vinyl ester resin-based systems are capable of giving mould closed times of under 3.5 mins. High speed mixing and dispensing technologies developed for the polyurethane industry are also likely to be adapted for the RTM process.

Metal inserts to spread loads at high stress areas, such as hinge attachment points, can be incorporatcd into the RTM moulding. The shape of the polyurethane core moulding can provide a method of locating and transporting these inserts during the preform process. A particular design feature of the bonnet is the designed failure line (Fig 139). During high speed impacts the bonnet fails along this line.

As with most composite processes, consolidation of parts is the key to successful and economic exploitation; Dow cite the apron structure in Fig 140 as a good example.

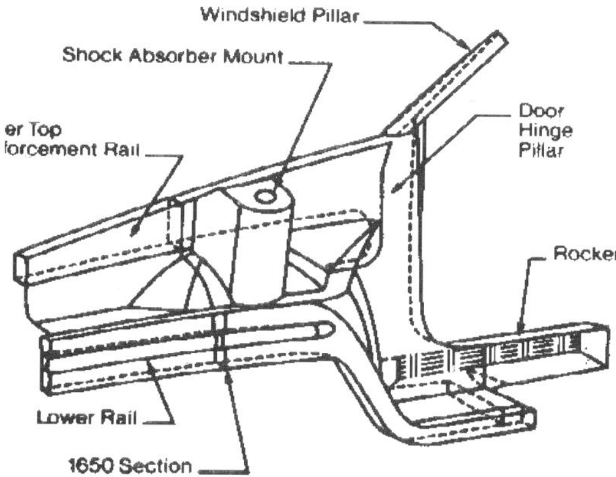

Fig 140: Parts consolidation in apron structure

Fig 141a: Rotational moulding applications

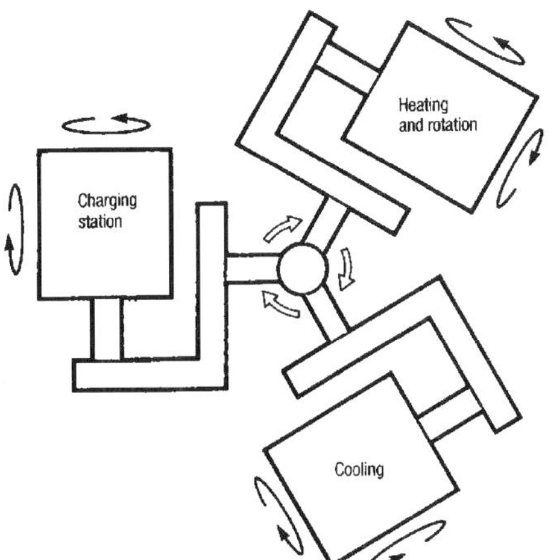

Fig 141b: Process configuration

	INSIDE RADIUS	OUTSIDE RADIUS
	(in inches)	
POLYETHYLENE		
Ideal	.500	.250
Commercial	.250	.125
Minimum	.125	.060
POLYVINYLCHLORIDE		
Ideal	.375	.250
Commercial	.250	.125
Minimum	.125	.080
NYLON		
Ideal	.760	.500
Commercial	.375	.375
Minimum	.187	.187
POLYCARBONATE		
Ideal	500	.750
Commercial	.375	.375
Minimum	.125	.250

Fig 143:Recommended draft angles.

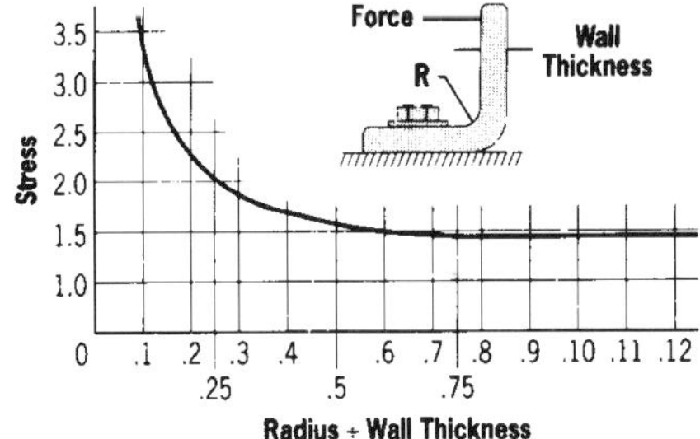

Fig 144: Corner radii

	OUTSIDE SURFACES		INSIDE SURFACES	
	Minimum	Better	Minimum	Better
POLYETHYLENE	0°	1°	1°	2°
PVC	0°	1.5°	1°	3°
NYLON	1°	1.5°	1.5°	3°
POLYCARBONATE	1.5°	2°	2°	4°
POLYCARBONATE	1°	1.5°	1.5°	3°

Fig 145: Recommended radii

Rotational moulding

Hollow sectioned end-products ranging from on-board fuel tanks to tanker bodies up to 9000 litre capacity, for small trailers, are among the potential applications of rotationally-moulded plastics, Fig 141a, with the basic process configuration being seen in Fig 151b.

T&D Rotomoulding[12] have shown that advances in the process have made it a cost-competitive alternative to blow-moulding, vacuum-forming and GRP moulding. The ability of the thermoplastic material to free-flow within the cavity of the mould as opposed to being forced, as in blow- or injection-moulding, is said to be an important benefit. The ideal shape for the process is a part which has smooth blends between adjacent contours of its surface.

The inner wall of the part will be free-formed, as the outside is defined by the mould, and chosen wall thickness is critical to the the design. A size of 3 mm is said to provide good compromise between cycle time, ease of processing, strength and cost but ranges form 1-20mm are not uncommon. Tolerance of 20% on this dimension would be typical. By arranging for differential temperatures over the surface of the mould, a degree of differential wall thickness can be obtained.

The steps in Fig 142 are a guide to optimum design. Flatness tolerance on typical PE or PVC parts would be around 0.5 mm per 50 mm, reducible to 0.3 mm for nylon and polycarbonate. Generally it is best to minimise the use of flat surfaces on the part. Adequate distance must be allowed between parallel walls, *A*, to allow powdered plastic to contact all cavity surfaces. A minimum of five times wall thickness is recommended. Likewise corner angles must be of adequate size, *B*. PE, PVC and nylon parts can be moulded with angles down to 45 degrees but harder flow materials require a higher minimum. PE, PC and PVC can go down to 30 degrees and nylon has been taken down to 20 degrees. There must also be an adequate radius where two walls meet.

Stiffening ribs are a valuable feature of rotomoulded parts and generally small multiple ribs are preferred to large single ones. The ribs must themselves be hollow and normally depth *Y,* and width *X*, *C*, should be 4 and 5 times wall thickness respectively. Rounded ribs,*D*, can also be used but sidewalls *Z* need to have draft sufficient for easy mould release; individual material specifications should be referred to. Use of 'kiss-off' ribbing, *E*, is also a possibility where inner and outer walls can be joined at intervals. The contact area is normally 1.75 times wall thickness. It is also a useful place to introduce holes into the part.

Liberal use of draft angle throughout the design will reduce the mal-effect of high mould-release forces. As the part cools it shrinks and the external surfaces draw away from the cavity, making mould removal easy; however, the internal surfaces shrink on to the mould and removal is difficult without generous draft, *F*. Different materials have different draft-angle requirements, Fig 143. Where textured surfaces are used, higher draft angles will be required.

Inreasing the size of corner radii distributes corner stress over a wider area and improves the moulding of corners. Stress raisers occur when the inside corner radius is less than 25% wall thickness and corner strength enhancement occurs when the value is over 75%, *G*, Fig 144. Inside and outside of corners should be radiused to minimise thickness reduction effects at the corners, *H*. Recommended radii for different plastics are seen in Fig 145.

A vent hole is required somewhere on the part, to equalise pressure inside and ouside during moulding, unless the part is almost spherical in shape. An inward or outward projection parallel to the parting line of the mould is an 'undercut' which must be deformed in withdrawing the part from the mould. Most moulds are of two or three piece construction so as to accommodate undercuts, side-cored holes or inserts, *I*.

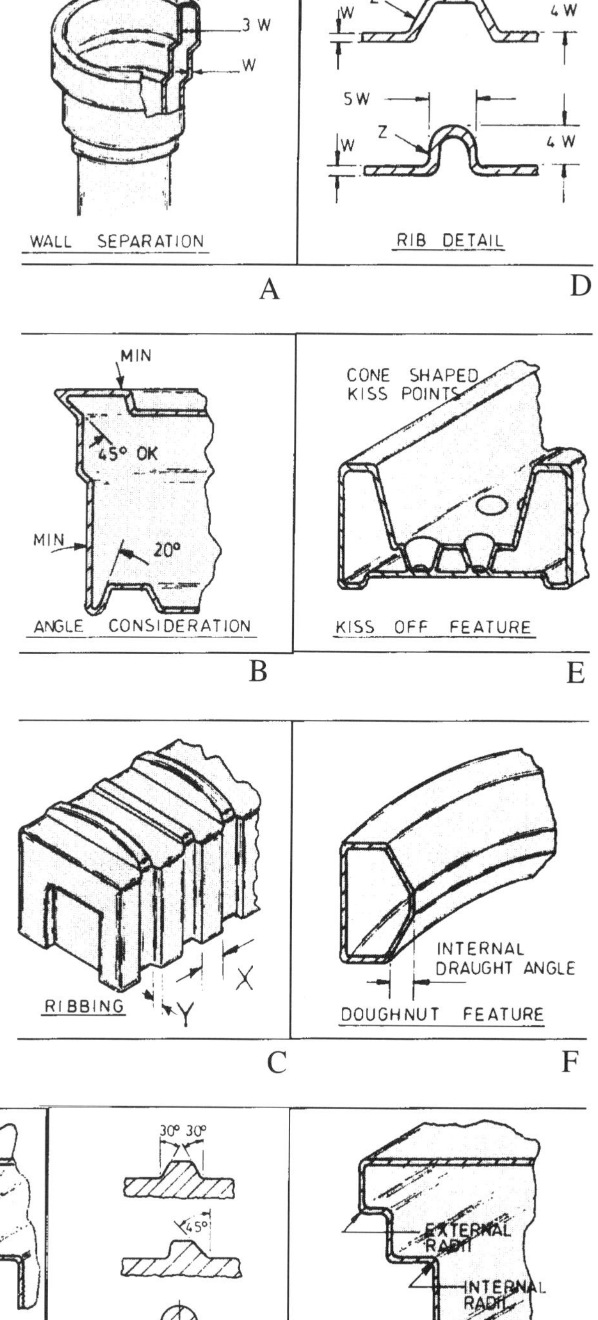

Fig 142: Optimising the design

Trim- pad construction

For pouring foam into thermoformed skins, when making up such parts as crashpads and spoilers, polyether-based PUR is usually employed. Either TDI or MDI foaming can be used, the latter allowing a one-shot process.

Density can be varied by altering the amount of blowing agent used; the higher the density the easier it is to obtain a well finished moulding. Resilience, too, can be altered by compounding techniques. Fig 146 shows tables relating four typical formulations for semi-rigid foam and indicating how water and triethanolamine content affect density and hardness.

	A	B	C	D
Propylan 555	100	100	100	100
Triethanolamine	5	5	10	10
Water	2·5	3·5	2·5	3·5
Propamine A	0·8	0·7	0·8	0·7
Crude MDI	61·0	76·8	75·3	90·5

	A	B	C	D
Core Density lb/ft³	4·9	4·0	6·0	5·2
Compressive Strength ... ... (Load in lb/in² to 50% compression)	3·1	2·7	6·6	4·3
Compression Set ... ... (50% compression for 22 hrs @ 70°C)	8·1	15·7	10·1	17·4
Tensile Strength lb/in²	18·9	15·6	25·2	16·7
Elongation	55	50	43	40

Fig 146: Semi-rigid foam formulations

Property	Value
Density kg/m³	48–160
Tensile Strength kN/m² (BS3379)	138–414
Tear Strength N/m ASTM-D 1564	157–525
Compression Set 75%	25–35
(BS3379) 90%	80–90
Elongation at Break % (BS3379)	50–90

Fig 147:Crash-pad foam characteristics

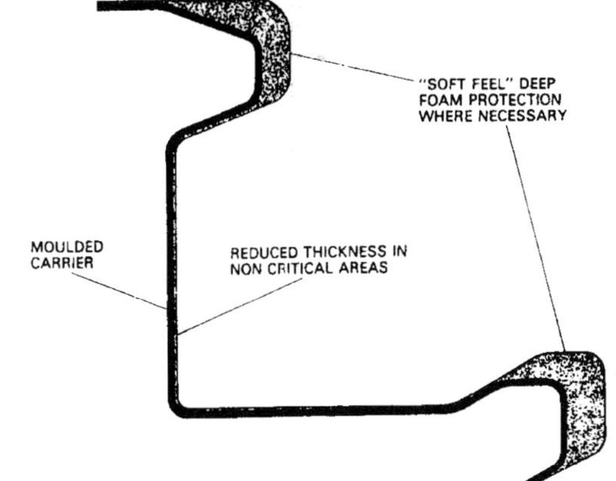

Fig 148: Microsorb technique

Cycle time involves several phases: the cream time takes from 2-60 seconds and allows the unreacted chemicals to reach the extremities of the mould; then rise time is that for the liquid of 1.15 gm/cm³ to foam into a density of around 0.07 gm/cm³; cure time is the period in which polymerisation takes place. Catylists are used to vary the relative timings.

Self-skinning PUR foams now form a substantial part of those used for automotive trim panels. These high-modulus foams can be produced by either the prepolymer or one-shot processes. The former involve reacting excess TDI with polyol then completing the foam-reaction with addition of water and catalysts. With the one-shot system, the individual chemicals are mixed together at the time of moulding. While such processes can effect cure without heat, cycle times can be shortened by passing through an oven at 38-71 C prior to demoulding. Moulds of glass or epoxy reinforced resin can be used, or for higher volumes, cast aluminium alloy.

Typical characteristics for a crash pad foam are shown in Fig 147. Covers may be vac-formed ABS or PVC, blow-moulded or blow-formed PVC — also rotationally cast as slush-moulded PVC. A variety of support structures or inserts can be used including metals, plastics or moulded woodfibre. Particular care is needed with deep drawn parts to avoid excessive thinning of the core or loss of surface grain on the skins.

Flooring systems from PUR foam are made by spray forming a PVC skin into an electro-form, the skin, after curing, being filled with semi-rigid foam then backed with a thin PVC sheet. In the case of door trim panels, a typically blow-formed PVC sheet of about 0.3 mm might be used; the pressure developed during foaming then forms the PVC blank against a contoured mould surface to eliminate the expense of the vac-forming operation and permitting the use of a less-expensive cover material. Fibre-board or similar inserts are required to ensure thermal stability of the moulded pad.

Self skinning foams involve a cellular core whose structure gradually changes towards the outside of the moulding to form a closed integral outer skin. Hardness can vary from that of a flexible cushion to a rigid form — in densities from 100 to 800 kg/m³. Such foams cure without external heat normally but post-curing can be applied to obtain a particular physical characteristic. Again a wide variety of mould materials can be used, with or without grained surface. An important consideration is the need to completely vent the air, to avoid surface imperfections. A vac-formed polystyrene sheet liner can be used for this purpose, which has the additional attraction of dispensing with a release agent. Acrylic and PUR mould coatings permit colour keying with other body parts. A comparatively recent advance by Marleyfoam is the Microsorb technique of obtaining soft-touch surfaces for such parts as crashpads, Fig 148.

Chapter 5: Metallic materials

Steel re-challenging 'light-alloys'; reducing weight with improved steel structures; intelligent steel selection; HS steels; aluminium alloy sheet and extrusions; austenitic and ferritic stainless steels; magnesium alloys; copper and shape-memory alloys; bonded steel/composite laminates

Steel re-challenging 'light-alloys'

Market and production pressures are forcing vehicle manufacturers to seek new high performance materials. With the increasing challenge of 'light alloys'. British Steel Welsh Laboratories has responded to this need by developing a range of high strength products which allow both body and structural steels to be manufactured from thinner strip, leading to reduced weight and improved fuel efficiency.

Compared with competitive light-alloys and polymer composites, sheet steels offer wide ranging combinations of strength and formability at relatively low cost. Vehicle assemblies fabricated from complex panel shapes, which are still the rule in current designs, exhibit high rigidity, energy absorption and predictable surface characteristics — essential for obtaining a consistent paint finish. Sheet steel must also meet stringent requirements of formability, weldability, consistency in both physical property and dimension.

In order that steels can be formed consistently into complex shapes they must exhibit high levels of plastic strain ratio in areas of the pressing biassed towards deep drawing. In other parts of the body, steel may be required to stretch rather than draw. To achieve this, an adequately high work hardening index, or *n* value, is a crucial element. Skin panels place heavy demands on formability because of the higher degree of automation on press lines. Pressings need to maintain their shape after the forming operation, therefore minimal spring-back and high shape flexibility are essential.

The weldability of steels and the safety of welded joints are important considerations; increased use of robotic assembly lines, has meant that pressed parts need to possess uniform consistency of shape to facilitate trouble free assembly. Other emerging assembly techniques include laser welding, adhesive bonding and improved mechanical connections. Adhesive bonding is not currently used for structurally critical applications, but used in conjunction with mechanical fixings, the combination providing the advantages of improved structural stiffness and reduced component weight.

A major advantage of strip steel for vehicle panel applications is the surface quality that can be achieved, resulting in a high distinction of image rating. Welsh Laboratories has carried out research into the measurement and control of surface characteristics to ensure that the sheet steel surface has a specific texture that is consistent and reproducible. This has resulted in roll preparation techniques designed to produce strip surface textures that retain lubricants to improve forming performance and facilitate good adhesion of autobody primer and paint combinations.

High-strength and coated steels Development of a range of high strength steels which offer an effective balance of strength and formability offer the potential for vehicle weight reduction. In some cases, high strength steels have been utilised to compensate for reduced gauge; in others to offer enhanced component strength. Reduction in formability has been overcome with bake-hardenable steels, which can be readily formed into complex shaped body panels the strength of which increases during paint baking.

Organic coated steels listed in Table 1 fall into three general classes: thin organic coatings, thick organic coatings, and pre-primed steel. Thin organic coatings act as lubricants during complex forming operations and offer protection against corrosion, since the coatings cover those areas of metal that are usually inaccessible to conventional electropriming processes. Ultra thin, one micron thick organic coatings can be satisfactorily used without impeding weldability. Thick organic coatings are usually prepared by applying zinc rich paint to a metallic coated steel substrate which improves corrosion and chip resistance to the fully finished panels. Thick organic coatings possess limited weldability and special care should be taken when producing good quality joints.

Preprime strip steel is presently being developed. The steel substrate, typically electro zinc coated steel, has a two coat system applied on a conventional coil coating line. The resulting product is intended to bypass the pre-treatment, electropriming stages and provides at low cost a product which exhibits good corrosion and improved stone chip resistance combined with a high quality surface finish aiding the ultimate image clarity of the painted part. Preprimed steels which presently cannot be welded are currently being developed for the spares aftermarket where components can be stored in conditions preventing corrosion prior to use.

Metallic coated steels involve zinc or zinc alloy coatings, applied to a cold reduced steel substrate. They confer corrosion protection by acting as a barrier against the ingress of moisture and, additionally, the zinc provides sacrificial protection if the surface layer is damaged or mechanically perforated. The life of a zinc coating can be extended or the coating mass reduced by incorporating alloying such as nickel. Zinc alloy coatings for body panels are harder and provide improved weldability when compared to pure zinc with the same coating mass. Terne-coated steel, widely used for petrol tanks, has been improved by the use of a nickel flash process initiated by Welsh Laboratories. Ternex, offers a level of consistency and ease of handling that allows high press rates and fast assembly, thus helping to cut unit production costs.

Hot dip coatings are produced by a continuous coil coating process involving the use of strip accumulators and on line welding and sharing, thereby avoiding the need to stop the line whilst coils are changed. The feedstock used on hot dip coating lines are cold reduced coils that are first annealed. Hot dip zinc coated steels are presently used primarily for underbody and non-visible parts. However, advances in production technologies and surface treatments have resulted in high quality surfaces that are smooth and free of visible spangle.

Iron-zinc alloy coatings are produced by induction heating the coated strip as it emerges from the zinc bath; these are used throughout Europe for internal parts. Extensive development work is currently taking place to ensure that the quality require-

TABLE 1. ORGANIC COATED AUTOMOTIVE STEELS

SUBSTRATE	*ORGANIC COATING*	*TRADE NAMES*	*APPLICATION*
Cold reduced steel	Zinc rich paint (13 microns)	Zincrometal	Bonnet
Electrozinc	Zinc rich paint (10 microns)	Bonazinc Welcote/Zincroplex	Bonnet
Zinc-nickel	Clear organic resin (1 to 10 microns)	Durasteel Welcote N	Door inner
Hot dip zinc	Zinc rich paint (20 microns)	Bonazinc	Bonnet
Terne	Zinc rich paint (one side) aluminium rich paint (other side)	Dorrlform	Petrol tanks
Electrozinc	Two coat primer	Preprime	Body panels

TABLE 2 HOT DIP METALLIC COATED AUTOMOTIVE STEELS

COATING	*TRADE NAME*	*MAIN CLAIM*	*TYPICAL APPLICATIONS*
Zinc	Galvatite	Standard hot dip	Wheel arches floor pan
Iron-zinc alloy	IZ Galvanneal (7-1 2% Fe)	Good weldability and wet adhesion	Floor pan Wheel arches Body panels
Zinc-aluminium alloy (55 % Al)	Galvalume	Corrosion and heat resistant	Silencers and heat reflector
Aluminium	Aludip	Heat resistant	Exhaust systems
Tin-lead (15% tin)	Terne	Good formability and weldability	Petrol tanks & radiator straps
Tin-lead-nickel flash (7-10% Sn)	Ternex Nickel-terne	Good formability and weldability	Petrol tanks & radiator nickel straps

TABLE 3 ELECTROPLATED AUTOMOTIVE STEELS

COATING	*TRADE NAME*	*MAIN CLAIM*	*TYPICAL APPLICATION*
Zinc	Zintec (E2)	Standard product	Exterior panels
Zinc-chrome-chrome-oxide	Zincrox	Good corrosion resistance	As for electrozinc
Zinc-iron	EZA Excell Zinc	Corrosion resistance with good finish	As for electrozinc
Zinc-nickel (11-15% Nl)	Nizec EZN	Higher weight coatings with good weldability & formability	As for electrozinc
Tin-lead (4% Sn)	Electro-terne	Good finish corrosion resistance	Petrol tanks

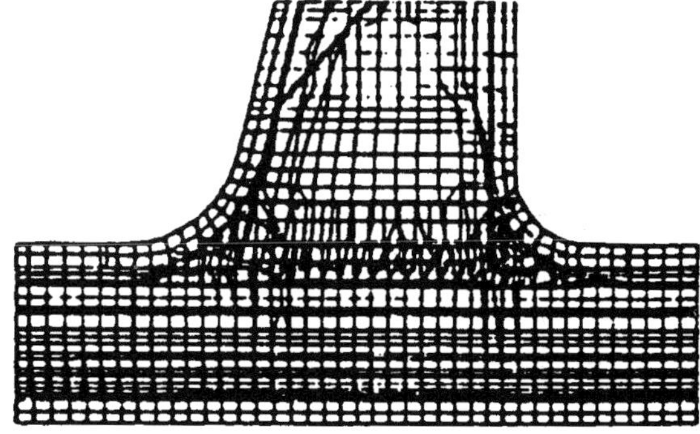

ments of iron-zinc alloys for surface-critical full-finish body components are met. Examples of hot dip metallic coated steels and typical automotive applications are given in Table 2.

Fig 2a: Application of pre-painted steels

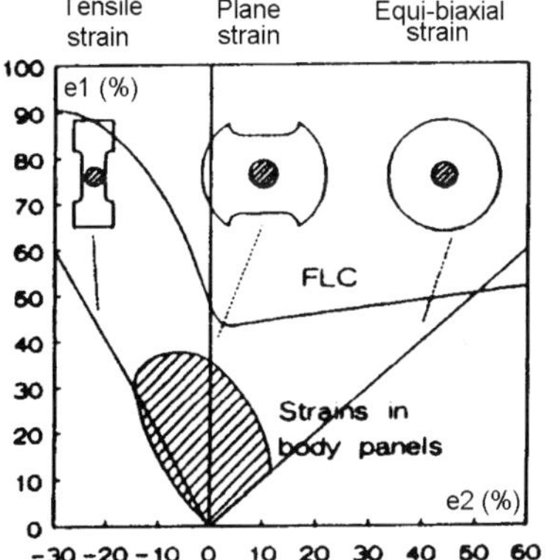

Fig 2b: Forming-limit diagram for pre-painted steel

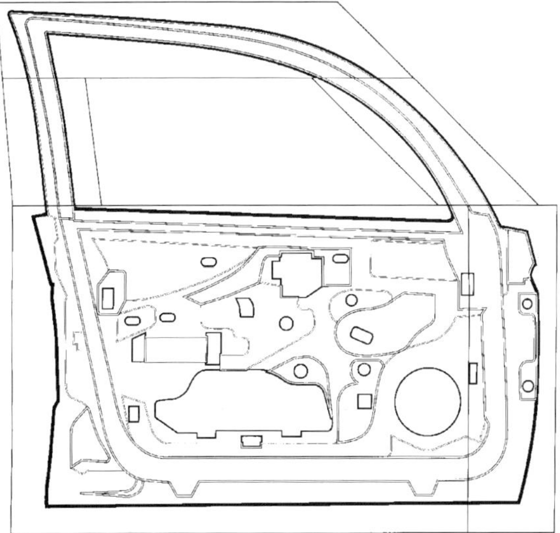

Electroplated steels are cold-reduced pre-annealed steels, electrolytically coated with a uniform deposit of zinc or zinc metal alloy. As with the hot dip processes, electroplating has been improved to produce higher quality steels. Recent developments include the production of single sided material for the automotive industry, improved cell designs which simplify the production of thicker coatings, and the use of insoluble anodes to improve overall production efficiency. Types and automotive applications of electroplated steels are given in Table 3.

Weight saving with HS steels High-strength steels featured strongly in work undertaken recently by British Steel aimed at achieving lighter, more fuel-efficient vehicles. Phase one of this work, which included input from International Automotive Design, highlighted the potential for weight savings using high-strength steels combined with new manufacturing technologies involving adhesives and weld-bonding. Phase two concentrated on structural integrity and manufacturing feasibility, using finite element and design sensitivity analysis, based on a current medium-sized saloon. The work demonstrated that a 20% or more weight reduction of the car body could be achieved. The American Iron and Steel Institute (AISI), Ford Motor Company and Porsche Engineering Services have also collaborated on a project to optimize car body design, reduce weight and improve stiffness and structural efficiency. The iterative analysis yielded a body structure some 15% stiffer and nearly 20% lighter than the existing base saloon car. The savings, of the order of 140 lbs in vehicle weight with a cost saving of over £25 per unit, would increase the cost advantage of weight reduction with steels compared with more expensive alternative materials. Part of the British Steel weight saving project involved determining the effect of reducing spot-weld pitch from a nominal 50 mm. The effect on the torsional bending stiffness of varying each of the body joint stiffnesses by an arbitrary unit amount in both the fore/aft and the inboard/outboard directions was determined. Stiffness of the B-pillar to sill joint was analysed, Fig 1, with a number of spot-weld pitches and the information used to factor the performance of other joints.

Pre-painted steels The Dutch Hoogovens Group have developed prepainted deep-drawing steels in both weldable and non-weldable forms. Shown in Fig 2a are inner panels seen as likely applications for the material. Other applications are seen in closed box sections which are otherwise inaccessible to the deposition of catalytic primer. Developed in conjunction with Akzo, the weldable product can replace the degreasing, pretreatment and priming stages in body-in-white manufacture. The non-weldable product, developed in conjunction with PPG can replace all the stages prior to top-coat application. Both are based on paint systems on top of either hot-dip or electro galvanised steel. A 12 micron thick polyester film with a conductive pigment (Ferrophos) is used with the weldable material, the non-weldable one using a 25 micron epoxy-polyurethane coating.

Forming limit diagram for Hoogovens prepainted sheet, Fig 2b, has the x-axis displaying main strain and the y-axis the strain in the plane of the sheet, perpendicular to the main strain. Material is deformed in three ways: uniaxial strain e1 positive, e2 negative; plane strain e1 positive, e2 = 0; equibiaxial strain e1 = e2 = positive. Shown inset are the test specimens used. A forming limit curve is constructed when the points corresponding with the highest deformation of the material can take in each direction are connected. The hatched area shows deformations that occur in automotive forming.

Laser-welded tailored-blanks By using laser butt welding to make up a single tailored blank (*a detailed description is given in Chapter 1*) from a patchwork of separate blanks, a part like this door, Fig 3, can be press-formed with a single die. This technique allows the selective use of steel types and thicknesses to meet the differing performance requirements of various areas of the part, and significantly reduces the scrap resulting from punching openings out of a single sheet. The boxed lines show the patchwork of separate blanks that constitute the tailored blank: the curved areas of the door indicate where those blanks have become part of the single door-pressing. Laser welded blanks are also being used instead of traditional one piece blanks to produce a number of formed and deep drawn automotive components such as body side rings, centre pillars, frame members and damper towers. Automobile manufacturers are looking to this technology to derive cost savings and vehicle weight reduction; reduced scrap in the blanking or cut-off process; complete elimination of extra reinforcing components; reduced use of higher cost materials and fewer manufacturing steps for sub-assemblies.

HSS application trends British Steel's HyPress high-strength steel range made its name originally for commercial vehicle heavy stampings while the 1980s saw widespread extension into cars, commencing with products such as medium gauge suspension sub-frames. The more recent spread into lighter gauge parts with tighter corner radii has been the subject of further metallurgical and structural development. Changes in tooling have been made to accomplish substantial formability even with much reduced elongation values of the higher yield grades of HyPress. By appropriate tooling configurations spring-back can be eliminated or even negative values obtained (Fig 4).

For critical pressings such as the wheel centres of car wheels niobium grades of the material have been developed to obtain desired conditions in the critical bell-mouthed stud-hole region. A special hole expansion test procedure (Fig 5) has been developed for the material. A whirling mass fatigue test has also been developed which reveals creditable life properties, typically three times greater than mild steel. Choice of hot or cold-rolled HyPress is a function of price, the HR material being typically £100/tonne cheaper but with maximum width capability rising from 445 to 530 mm.

Interest in higher strength steels for vehicle bodies was spurred by the quest for lower weight by gauge-reductions from 0.9 down to 0.70 mm being a general requirement. This bought the response of rephosphorised steels from the strip mills but the additional cold work process causes such steels to have a somewhat higher price. Interestingly, too, HyPress yields are still some 10 per cent higher than re-phos steels although the latter provides a 30 per cent gain over mild steel. With these thin gauge materials care is needed not only to enhance rigidity but also to avoid 'oil-canning' on large flat closure type panels (where a shaped profile might not be aesthetically acceptable) and vibrational drumming, by the use of internal stiffeners and damping pads. Bake-hardening and dual-phase steels which were first developed in Japan arose mainly as a follow-on from the installation of continuous annealing lines, initially purchased as a means of ensuring closer control of batch-to-batch quality of all sheet steels, British Steel maintain. Given the annealing process, the move to dual-phase was a secondary step towards obtaining a range of high-strength steel. However dual-phase sheets are more costly than mild steel. Aftermarket panel replacements are supplied in the primed condition and, whilst this ensures the panels are often at bake-hardened strength level, it does have implications on repair costs.

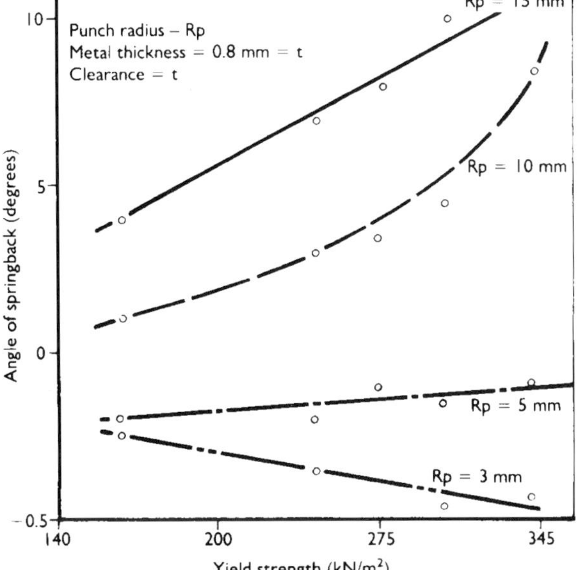

Fig 4: Spring-back characteristics

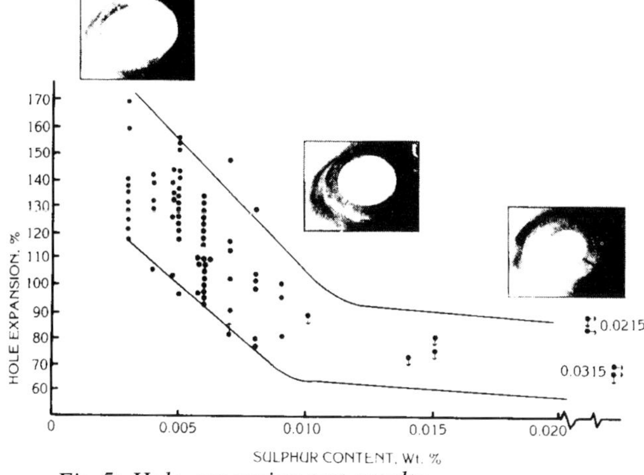

Fig 5: Hole-expansion test results

Changing chemical composition (per cent) of carbon steels

	C	Mn	N2	Al	Ti
Static	0,040/0,060	0,250	0,0055	0,03–0,08	–
Continuous	0,025 (ELC)	0,200	0,0035	0,02–0,05	–
	0,005 (ULC)	0,150	0,003	0,02–0,05	0,06

Changing chemical composition (per cent) of H S Steels

E 280 Steel	C	Mn	P	Al	Nb	
Batch annealing	0,05	0,300	0,015	0,015	0,040	0,025
Cont. annealing	0,05	0,300	0,015	0,015	0,040	–

Fig 6: Annealing affect on steel properties

Continuous annealing Belgium company Cockerill-Sambre point out that continuous annealing, already known for galvanized sheets and tinplate, is now employed for cold-rolled thin sheets. Here the evolution of temperatures both during heating and cooling, is much more rapid than in a static annealing cycle. The treatment of a coil of steel (Fig 7) takes a few minutes as opposed to several days required for static annealing. This development has had a major influence on the kinetics of the crystalline transformation and precipitation phenomena produced in the steel during the procedure, the company say.

A speed of cooling adapted to both mild steels and high strength steels can be achieved by cooling in two steps. The first is achieved between the annealing temperature and an intermediary temperature of about 400 C by quenching in boiling water. The second, between this intermediary temperature and the final temperature (200 to 100 C), is done: either directly with misting jets in the case of high strength steels; or with gas jets (jet cooling) after over-ageing for mild steels. The use of boiling water has given its name to this procedure, Howaq being an acronym of Hot Water Quenching. The first factor which can be affected is the chemical composition of the steels. The content of manganese, nitrogen, aluminium, and particularly carbon must be diminished with respect to that admitted for steels annealed in a conventional furnace; see Fig 6 for data on extra and ultra low carbon steels.

The second critical factor is the rate of cold reduction; it must be increased to lower the recrystallization threshold and to increase the drawing capacity. Reduction rates of 70 per cent and more are envisaged. One way of obtaining them is to supply the rolling lubricant directly and no longer only in emulsion. Finally maintenance at a temperature of around 400 C for a certain time depending on the level of the characteristics envisaged, is called over-ageing. The object of this treatment is to permit the atoms of carbon and nitrogen in super-saturation in the ferrite to emerge and precipitate, particularly at the edges of the grains. In this way, it counters their deleterious effect on the ageing. The processing of high strength steel is opposite to that of the mild steels; what had to be avoided there is now desirable. In the first place, the chemical composition must be altered as in Fig 6. The quantities added are less than in the classic formulas for batch annealing. This

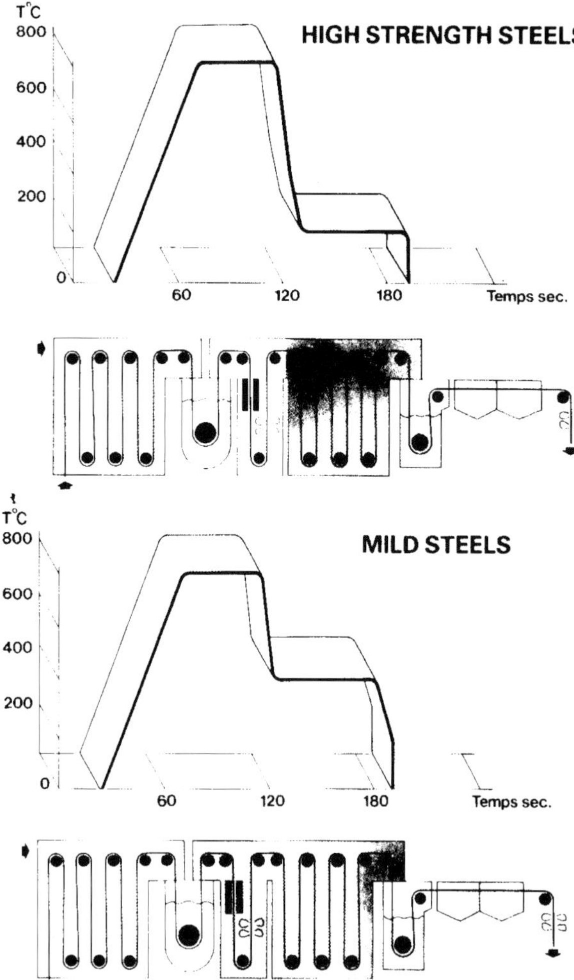

Fig 7: Continuous annealing process

improves the homogeneity of the properties and the welding capacity. Finally, the annealing cycle is characterized by a low annealing temperature and the absence of over-ageing treatment: this is replaced by a passage of a few seconds at 200 C to precipitate partially the carbon in supersaturation in the ferrite because of rapid cooling and to avoid the softening effect of over-ageing.

By means of a climatic chamber and an automatic camera to take photographs at predetermined times, it has been demonstrated that for a Howaq sheet, relative to steels annealed in classic furnaces with tight coils; the time until the appearance of the first rust spot is multiplied by 3 or 4; the number of spots per surface unit is divided by 5 or 6; the speed of appearance of surface oxidation is divided by 15 or 16, say Cockerill-Sambre.

Reducing weight with improved steel structures

The steel industry has also reacted to the initiatives of the aluminium and polymer composites industries, in reducing vehicle structure weight, by international combined-company research programmes into structural optimization.

A number of the claims made by the aluminium suppliers are based on improved construction techniques, particularly weld-bonding, which can equally be applied to steel bodies; furthermore, considerable scepticism is being shown to claims of superior crashworthiness by aluminium alloys. Often these claims are made for unrealistically low impact speeds. At real-life impact speeds the positive strain-rate sensitivity of steel shows to the advantage of steel — as aluminium alloy is unaffected by the strain-rate of impact deformation, the steel makers argue.

There has already been a halving of panel gauges since World War 2 and sheet thicknesses of 0.65 mm are now typical for mild pressing steel; this can be reduced further of high-strength steel panels which already account for some 10-15% of the body-in-white weight for a car structure. A British Steel research initiative has involved a study carried out by IAD on a Kia body shell. The study took place in separate phases, new technology, new materials and a combination of the two. An FE analysis of the Kia shell showed that selective reduction of panel thicknesses for the less stressed panels, and the substitution of HS steels for others showed the first benefits, commensurate with necessary shell stiffness, strength, durability and manufacturability. A variety of new steels were also examined, including cold-rolled, bake-hardened, dual-phase, rephosphorised, micro-alloyed and transformation induced plasticity steels.

The particular application of top-hat closed box sections on the front quarter panels allowed analysis of the effect of changing parameters in the expression relating critical buckling stress on impact:

$$= [R(t/b^2/\beta(1 - v^2)]^{0.43}\sigma_y^{0.57}$$

where R is a function of section aspect ratio, t is sheet thickness, b a function of thickness to section width ratio, v is Poisson's ratio and σ_y is yield stress.

Doubling the yield strength can raise the critical buckling stress by 50%. Alternatively, tripling the yield stress allows a reduction in panel gauge of over 50%, something which would be difficult to achieve economically with an aluminium alloy on a volume-produced vehicle, presumably. When using high strength steels, too, a number of manufacturing considerations need attention. Increased 'crown' is required in pressings to overcome higher 'spring' characteristics. Strategic character lines over the full length of the panel would be required to hide the effect of surface deformation caused by welding or hemming close to skin surfaces. Increased corner and bend radii are required and closed corners requiring high stretch should be avoided. Appropriate offsets would also be required around depressions, to eliminate potential surface deformation.

Spot-weld pitch has a minimum value depending on tendency for weld current to shunt through on adjacent welds although this can be counteracted by the use of larger weld transformers. There would also be a reduction in electrode life arising from the need to increase welding speeds, with closer spot pitch, to match existing overall BIW production rates. However, as well as bringing small gains in BIW torsional stiffness, close pitch will improve the ability to withstand fatigue loading, British Steel maintain. The company also stress that steel has a positive endurance limit, normally corresponding to 45% of tensile strength whereas aluminium alloys have no such limit at all.

The importance of weld-bonding future steel structures, for greater structural efficiency, has also become clear from the studies — and over 8% gains in torsional stiffness have been shown or an equivalent percentage weight saving. While automated adhesive application equipment was readily available, shop-floor quality audit facilities needed careful scrutiny before the widespread adoption of weld-bonding. But the company stress that design for adhesives at the concept stage would vastly improve the chances of successful exploitation.

Honeycomb cores were another future possibility — for steel-skinned sandwich panels. In commercial vehicles there was scope for exploitation, too, of foam-cored sandwich panels which had been extensively developed for the building industry. Ferroplast panels used by Schmitz Cargobull were an example. But for cars, some type of formable sandwich was necessary unless the industry was prepared to radically change its approach to structural design. Limited trials have been carried out already on the use of rigidized sheet as a backing for the skin panel for such applications as bonnet covers. While such techniques could equally be applied to aluminium alloy panels, the company stressed that all aluminium panels had to be 'handled with kid gloves' in the press shop, and assembly areas, so that speed of throughput would bound to be affected in the volume production situation. Some care, too, was also necessary in the handling of zinc-coated steel panels and the use of chrome plated press tools was required to avoid surface scuffing. Higher currents through the welding electrodes were also required, to avoid high wear rates.

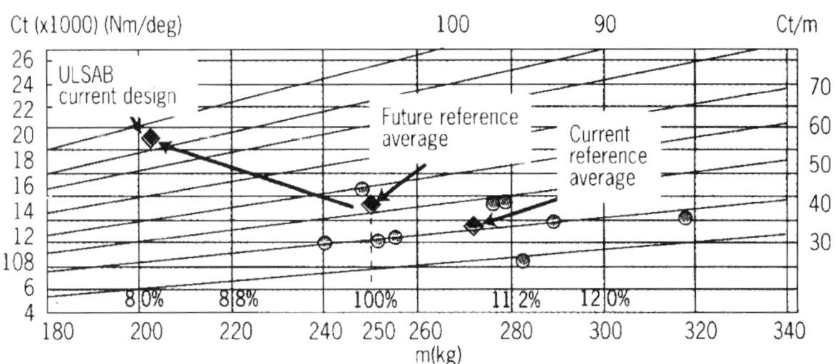

Fig 8: Benchmark torsional stiffnesses

The steel consortium approach to light-weighting

According to Ashley[1] the body shell represents about 20% of vehicle mass so it is not surprising that a light-weighting programme has been undertaken by a world-wide consortium of 32 steel producers representing 15 countries on five continents. They commissioned Porsche Engineering Services to design a lightweight steel body incorporating current standards of structure rigidity, crash-worthiness and manufacturability. The result of the programme is what is known as the ultra light steel auto body (ULSAB).

The basic package was defined by benchmarking and targeting a group of existing vehicles, which were the Ford Taurus, the Mazda 929, the Toyota Cressida, the Honda Accord, the Chevrolet Lumina, the Acura Legend, the Lexus LS400, the Mercedes E series and the BMW 5 series. The resulting body in white developed under the programme presented a dimensional statistical average of these vehicles normalised to the same wheel base.

One of the first benchmarks was the stiffness of the car body structure and measurement was made of the current stiffness for the quoted vehicles' body structures with glass in situ, Fig 8. The current average torsional rigidity was 11902 Nm/deg and the target for the ULSAB was set at 13000 Nm/deg. In a similar way targets were set for bending rigidity, the first body in white bending and torsion modes and the body mass. The composite average body in white of the bench mark vehicles was 271 kg and the target for the ULSAB was 200 kg.

The design method used was based completely on computer modelling and a number of beam models were created for initial analysis. A shell half-model was used for detailed design assessment and optimization, consisting of more than 37000 elements and over 36000 nodes. Numerous options were reviewed to provide an optimized structure and out of these options three basic construction methods were investigated.

Simulated crash testing performed on the design were a 35-mile per hour frontal impact, a 35-mile per hour rear-moving barrier (FMVS301), a EEVC side impact (Europe) and a roof crush (FMVSS 216). In all these cases a simulated engine, suspension and other necessary componentry was included in the model, and it was ensured that the model met the criteria with a significant margin.

Although it was found possible to make significant improvements in the unibody-type structure, the mass of the optimized shell still did not meet the programme target of a weight reduction down to 200 kg. The actual body mass achieved at this stage was 229 kg. In parallel, therefore, another design study looked at the use of hydroforming, where hydraulic pressure is used in tubular sections in order to provide an optimized shape. By maximum use of hydroforming, it was found possible to end up with similar results to that of the unibody structure, namely 229 kg. A number of other proposals were also considered to reduce mass, such as structural fuel tank, structural seats, rigid engine cradle, motor compartment brace and cast nodes at the A-pillar joint. Some of these proposals added mass and rigidity, but it was found that the calculated efficiency of the complete structure had decreased and hence the weight targets were not being further met.

It was therefore decided to stay with more traditional concepts, but to take known manufacturing and design technology to its limits by combining the unibody structure with hydroforming and maximum use of structural adhesives. The other key technologies were laser welding, tailor-welded steel blanks, and roll forming. The process was computer interactive all along, optimizing sections, improving joints and using different steel qualities and material thicknesses at appropriate points in the design to create the optimum structure.

Hydroformed side roof rail A number of parts that would traditionally be created by sheet metal stampings were replaced by one-piece hydroformed tubes. This required tight control during design of section perimeters and transitions and some unique joint designs. To attach other body in white components to the tube requires a one-sided attachment method and in this case laser welding was chosen. Fig 9 shows the hydroformed side roof rail. Particular attention was paid to the connection from the A, B and C pillars and to the integration of the rear shock towers. Fig 10 shows how the rear shock tower is integrated into the rear rail in order to provide optimal load transfer into the structure.

Structural bonding To improve the rigidity of the ULSAB, the body side inner assembly is weld-bonded to the body side outer assembly. Combined use of welds and high-temperature adhesive provides continuous bonding. Fig 11 shows the two principal application areas for bonding. For structural efficiency reasons, a non-traditional approach to the cowl to A-pillar joint was taken by running the lower cowl section through the A-pillar inner and outer creating a hole, Fig 12. This produced a large rigidity gain for the structure. It worked because the centres

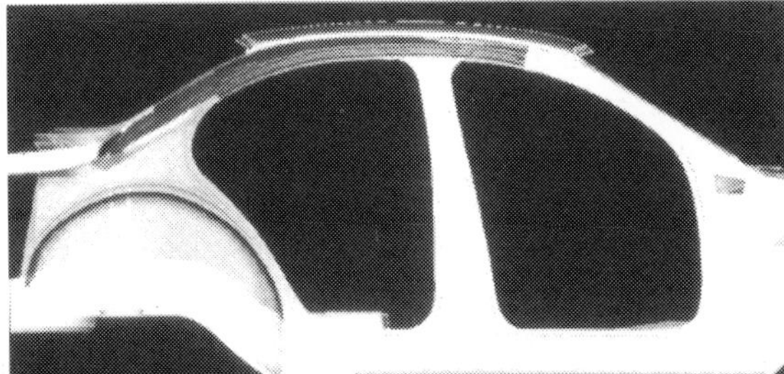
Fig 9: Hydroformed side-roof rail

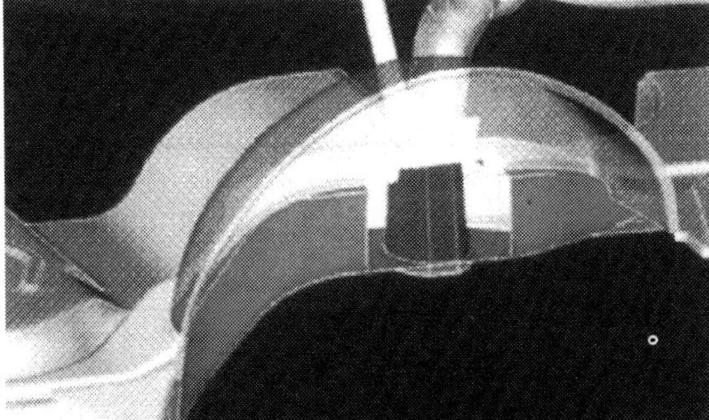
Fig 10: Integration of rear shock-absorber tower

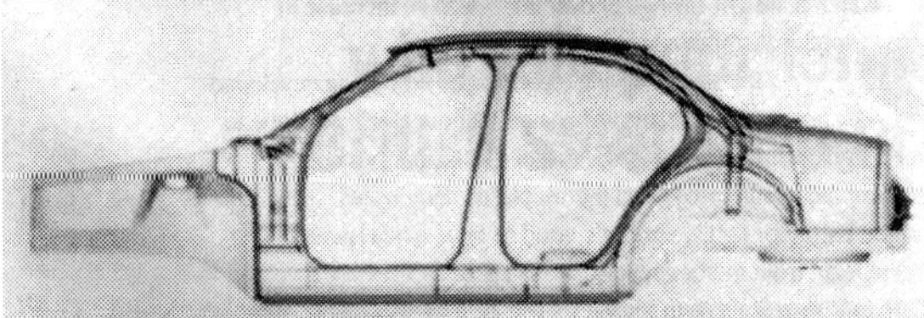
Fig 11: Principal application areas for bonding

Fig 14: Final body structure

of some traditional joints become a structurally dead region that can be removed. The front-area/shock-tower region is integrated into the skirt, which is itself laser-welded to the fender support rail, Fig 13. The wheel house is spot-welded to the front rail welding flange and, in the shock tower area, to the front rails lower flange, hence forming a well-integrated structure.

Tailor-welded blanks and special steels In order to reduce the part-count by eliminating reinforcement, also to reduce sub assembly welding and to increase the structural efficiency, a number of tailor welded blank parts are used in the design. These include the front and rear rails, the rocker, the A and B-pillars, and the wheel house fronts and inner wing panels. About 67 per cent of the structure of the ULSAB is made from special steels, either dual-phase or hard-baked dependent upon the difficulty of forming and the strength requirements. This compares with a low figure of 3-4 per cent typical of many manufacturers and approximately 35 per cent used by BMW and Toyota.

Corrosion protection and dent resistance At one time to build a vehicle out of lighter gauged steel would have problems with corrosion resistance of the structure. It is now becoming common practice to coat virtually the whole of the body structure and certainly everywhere where there is a corrosion problem. If this action is accepted as part of the normal cost of the structure, then the corrosion life of the ULSAB will be similar to that of existing vehicles. At the same time, the dent-resistance of the structure is preserved by the use of high tensile steels, particularly hard-baked steels on the external body structure. Even though there may be some reduction in gauge, the actual dent resistance is superior to that of normal mild steels.

The final body structure as a surfaced concept model is shown in Fig 14. All elements were designed for manufacture using current best practice forming processes. The assembly sequences and weld schedules were laid out to be suitable for volume production. The continuous laser welding was specifically designed to be accomplished in one station to minimise investment costs. The final body weight achieved was 205 kg and the body shell meets crash-worthiness requirements as described above, as well as all the body torsional and bending stiffness requirements. The principal limitation was in meeting the body bending specification; the torsional stiffness target was exceeded by almost 50 per cent. The table in Fig 15 gives a performance summary for the completed structure. Finally, a cost comparison exercise was carried out by going through the complete manufacturing process for the ULSAB, together with its bench-marking comparators. The ULSAB shows a significant reduction in piece-count from 195 parts to 169 parts, together with a reduction in mass from 270 to 205 kg. There is some increase in cost due to the more highly specified steel, but the overall effect was to bring down the estimated cost of the body structure from US $1116 to $962. It is also possible to make some comparison with the Ford AIV project where a Ford Taurus body shell was redesigned in aluminium alloy. That project achieved a body weight reduction from 270 kg to 146 kg whilst increasing the body torsional stiffness to around 17000 Nm/deg. This increased weight reduction was achieved at a significant increase in cost. It would thus appear that steel can achieve by redesign half the potential weight reduction through the use of aluminium at a cost benefit compared with existing practice.

Design highlights announced for the latest phase of the project include front rails, a hydroformed side roof rail, spare tyre tub and body side outer. The front rails are a hexagonal section (Fig 16a) compared to a more conventional 'hat' section which is more stable in deformation and thus is capable of absorbing more energy in a crash. For mass efficiency, the rails are made of high strength steels in a three piece tailor welded blank. The hydroformed side roof rail (Fig 16b), made from a thin high strength steel, creates an essential load path, reduces the total number of parts and maximises section size, allowing for both mass and cost savings. The spare tyre tub (Fig 16c) is made using a steel sandwich material which affords mass savings with no loss in performance and can now be pre-assembled as a module with the spare tyre and tools. The body side outer is the largest part in the ULSAB structure and represents a large percentage of the mass. The design eliminates reinforcements in the body side assemblies. The cost and mass of the component are reduced because of this consolidation of parts.

The European steel industry now expects half of all steels used by 2005 will be high performance grades. Certain models such as the Porsche Boxster (30 per cent), BMW 5 series (40 per cent) and Volvo 900 series (40 per cent) already use a high proportion of high strength steels. Volvo's new C70 cabriolet open-top sports model uses ultra high strength steel in key structural components to ensure an impressive level of crash resistance. Since the roof is normally a

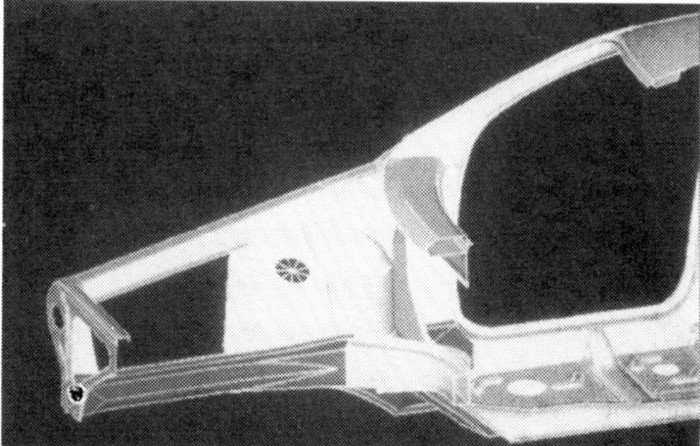

Fig 12: Running the lower cowl through the A-pillar

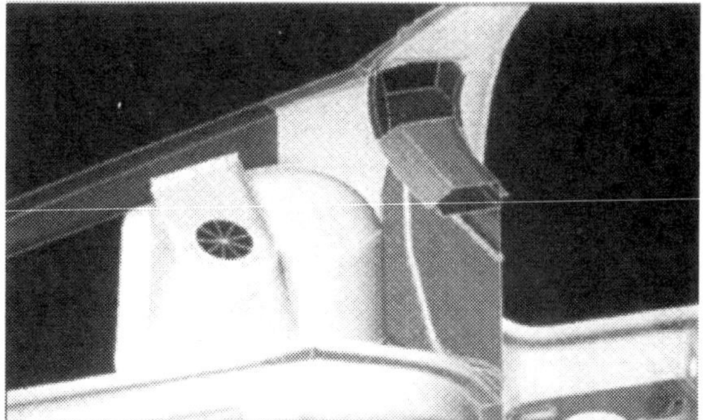

Fig 13: Integration of front shock-absorber tower

Criterion	Reference	ULSAB	Target	Difference.	%
Mass (kg)	271	205	200	-66	24
Torsional Rigidity (Nm/deg)	11531	19056	13000	+7.525	35
Bending rigidity (Nm/m)	11902	12529	12200	+627	5
First BIW Mode (Hz)	38	51	40	+13	34
Parts Count	195	169	-	-26	-13
Cost (USS)	1116	962	-	-154	-14
Price per Kilo, S.	4.12	4.69	-	0.57	13.8

Fig 15: Results of ULSAB project

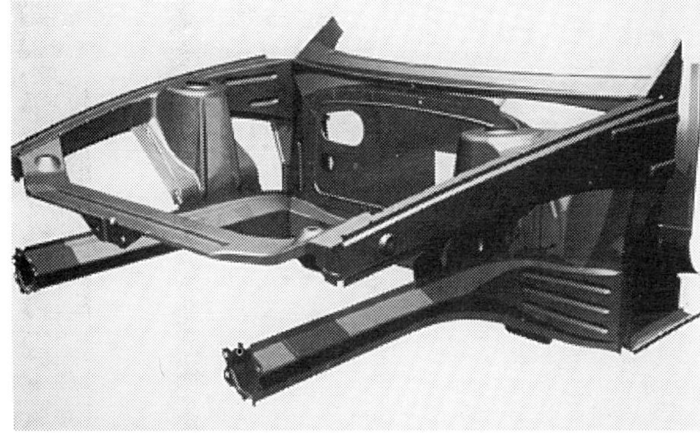

a

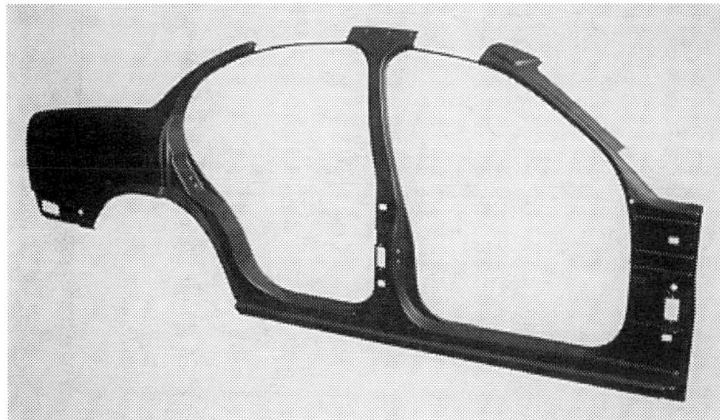

b

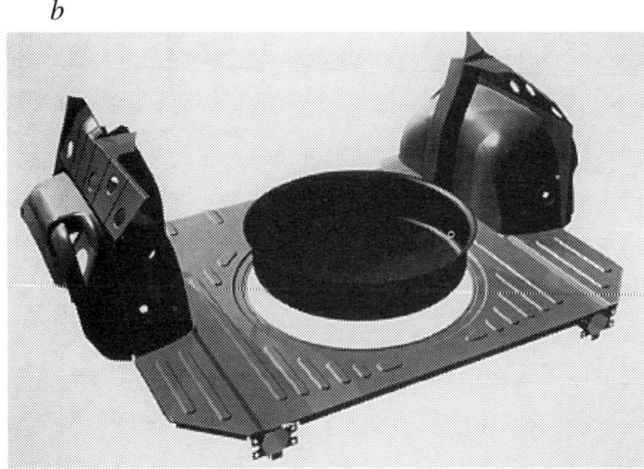

c

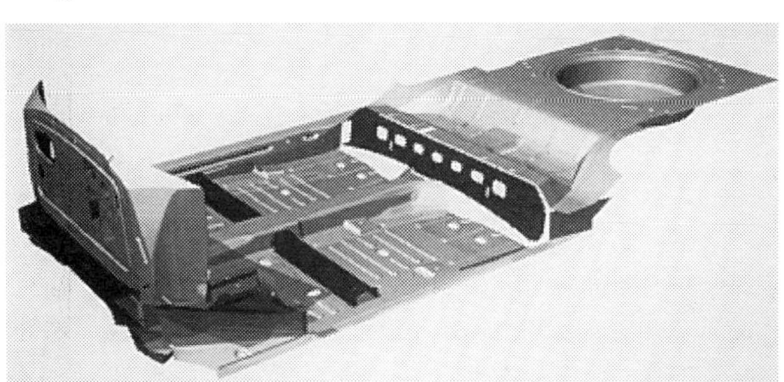

Fig 16: Key elements of the ULSAB Phase II body-in-white

vital part of the structure, other solutions had to be found. Part of the solution was to use ultra high strength boron steel in the windscreen structure to create a roll bar. Increasing the capacity for the ULSAB structure to absorb more crash energy has been achieved, for example, by using structural one-piece components that use steel of varying thickness and grades: tailor welded blanks. This applies to nearly half the USLAB project body and underlines the simultaneous engineering approach of combining design and manufacturing initiatives.

Latest ULSAB project in detail Up to 90 per cent of the new body frame comprises high strength steels which are engineered to maximise ductility. The ULSAB consortium predicts that by the year 2005, over half of all automotive sheet steels will be the new higher strength grades. Half the ULSAB mass now consists of laser welded tailored blanks, using steels of various thicknesses, strength levels and coatings, allowing the ULSAB structure to absorb high levels of crash energy in a lighter frame. Tailored steel components crumple at predefined points starting within the thinner gauge with low stress levels to thicker gauge at high stress levels.

The rear rail inner uses a tailored blank incorporating three different thicknesses: 1 mm, 1.3 mm and 1.6 mm. The body side outer uses a one-piece tailored blank, satisfying the different structural requirements of such a large panel. This component blank is produced from five different steels with thicknesses ranging from 0.7 to 1.7mm and the yield strength from 210 to 350 MPa. The raw material for the hydroformed side roof rail is a welded tube of 1 mm thick and an outside diameter of 96mm. These dimensions can present both manufacturing and forming challenges. In this case, the project partners were able to prototype the side roof rail without difficulty using forming simulation techniques.

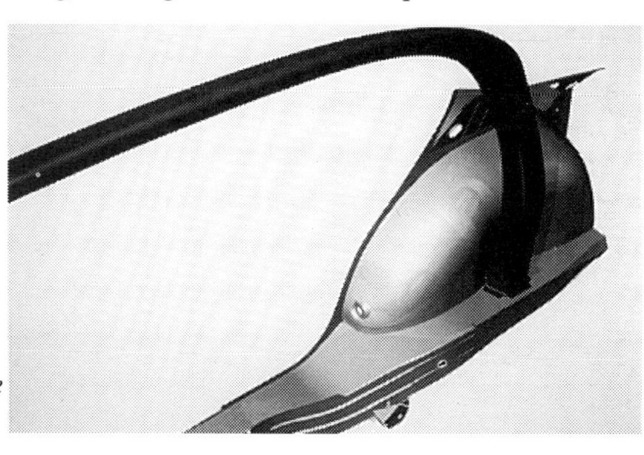

Intelligent steel selection

Steel has been proven to be one of the most successful structural materials and its high resilience has proved it be an effective competitor to the so-called 'light-alloys' now that reliable anti-corrosion treatments have been developed. Because there are so many varieties of the material, according to alloying elements and heat-treatment, the selection of the appropriate grade is often a daunting task to the non material-scientist.

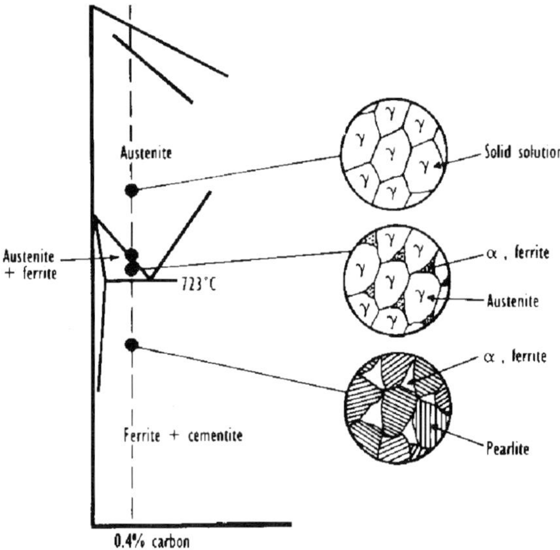

Fig 17: Slow cooling of a 0.4% carbon steel

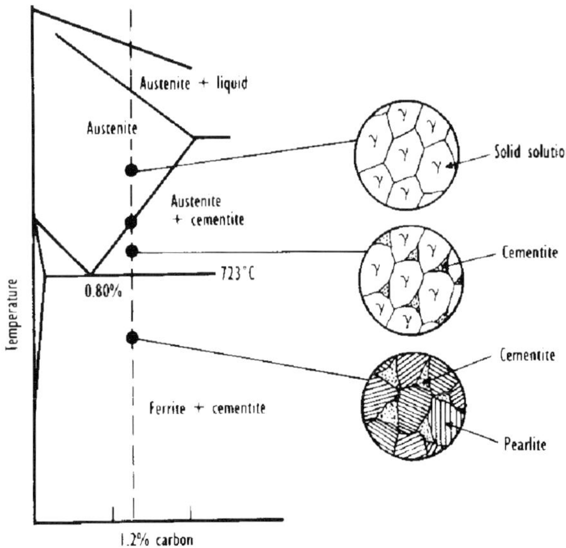

Fig 18: Slow cooling of a 1.2% carbon steel

Steel is classified according to a variety of crystal structures, dependent on the concentration of each alloying element; it is also affected by the way in which the material is cooled from furnace conditions and upon any cold-work carried out. The Materials Information Service of the UK Dept of Trade and Industry list the principle ones as ferrite, austenite, cementite, pearlite, martensite and bainite. Ferrite is the crystal arrangement for pure iron, which is found as part of the structure of most steels, and can absorb carbides by diffusion in the solid-state. Austenite is the single phase solid solution in which all steels exist at sufficiently high temperatures and with some alloys it is present at room temperature.

The crystal arrangement for these two principle forms is body-centred cubic and face-centred cubic respectively. Figs 17 and 18 show the effect of slow-cooling of 0.4 and 1.2% carbon steels respectively while Fig 19 shows the way in which chill rate can be varied to produce the other forms; Fig 20 shows the iron to iron-carbide systems available with different heat-treatments.

Work hardening is the resistance to continuing plastic flow which deforms the crystal structure — usually by microscopic shearing along definite planes. The extent of work-hardening depends on the number of intersections between crystal orientations at which slipping can take place. Most alloy steels have a complex structure with many such intersections and so are particularly susceptible to cold-work — which can usually improve strength and hardness at the expense of ductility — and the effects can only be cancelled by annealing.

Carbon and alloy steels

With up to 0.05% carbon, steels are ductile and similar in property to iron; they cannot be heat-treated. This is also the case for 0.05-0.2% steels but their hardness can be increased by surface carburising and there are no problems with heat-affected-zones in welding. Because ductility is high, machinability is poor unless high cutting speeds are employed or elements added to form free-machining inclusions.

Medium-carbon steels, of 0.2-0.5% C can be modified by heat-treatment and work-hardening; then the hardening effects in the heat-affected-zones of welds must be accounted for. Such steels are plentifully produced so are relatively cheap and their medium-strengths are useful in some structural applications. An intermediate range of 0.5-0.8% C are even more susceptible to heat-treatment and work-hardening; they can also be flame-hardened - and can be cold-worked to give yield strengths of up to 2000 MPa — but impact strength is poor. With the high-carbon steels, over 0.8% C, cold-working is not possible and these steels fracture at very low elongation. Machinability is good when normalised to reduce surface hardness.

For alloy steels, key features of the effects of alloying elements are as follows. Chromium improves high-temperature properties and corrosion resistance. Nickel lowers critical heat treatment temperatures and toughens steel; it also imbues good low-temperature performance without detriment to impact strength; vanadium generally improves mechanical properties due to its grain-refining effect. Molybdenum also gives a fine-grained structure with consequent improvement in strength. Tungsten is valued in steels requiring hardness and high-temperature stability. Boron can considerably increase hardness of low to medium carbon steels, while maintaining good weldability. Copper enhances corrosion resistance while aluminium is useful for nitriding and grain refining. Manganese lowers heat-treatment temperature and can provide high-carbon steels that are tough and workable. High chromium steels also offer good corrosion resistance and when a steel has more than 18% chromium it becomes a stainless steel whose properties are discussed later.

High Strength Steels

As with aluminium alloys, so with steel: appropriate alloy selection for a given set of design and loading conditions can result in substantially enhanced strength without altering structural rigidity. With the recent development of High Strength Steels (HSS) in sheet form, the opportunities for higher specific strength panels and pressings present themselves.

British Steel supply the motor industry with high strength steel in narrow and wide strip form. For volume cars in the UK something like 10% (200 kg per car) of the steel is HyPress High Strength Low Alloy (HSLA). The material first made its name for CV chassis stampings and a variety of grades correspond to the available fatigue lives shown in Fig 21.

Somewhat higher priced rephophorised steels are also available where even higher strengths and thinner gauges are required. In the case of HSLA, relatively low carbon content ensures reasonable ductility. According to grade, HSLA yield strengths can range from 275 to 550 MN/m^2 and elongation from 12 to 24%. Elastic modulus is virtually the same

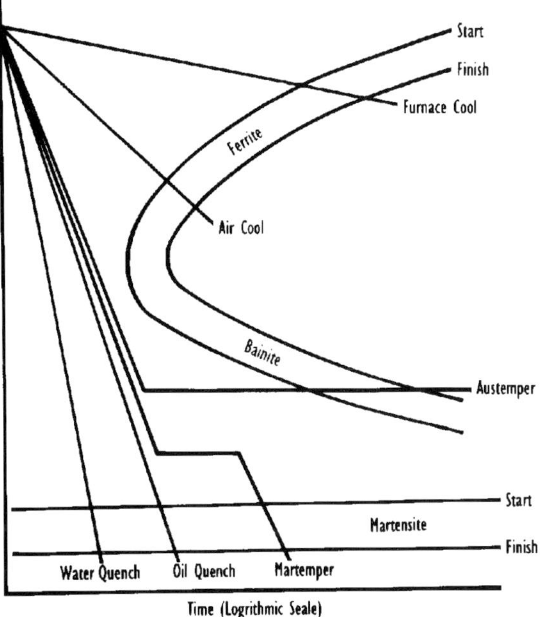

Fig 19: Time (x-axis)-temperature (y-axis) transformation diagram

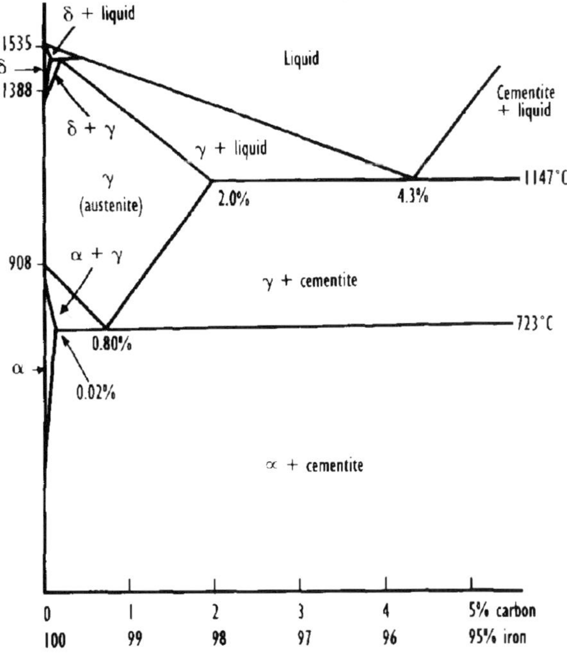

Fig 20: Iron carbide system (y-axis: temperature C)

as mild steel and therefore strength-critical rather than deflection-critical panels are the most promising candidates. On a direct-substitution basis it has been suggested that the materials will offer sheet thickness reductions of 15-35 per cent, for similar panel configuration and loading, at a cost penalty of about 20 per cent. Some prefer the niobium rather than titanium grades, for ease of welding, correct choice of alloy allowing the use of the same spot welding plant as for mild steel. There was the further advantage, in HS steels other than HSLA, of bake-hardening which can take place in paint stoving ovens, with a potential yield strength enhancement of 80 N/mm^2 or so. In the UK, most truck chassis members were made from hot-rolled HSLA in gauges ranging from 3 mm on crossmembers to 8 mm on sidemembers. Fig 22 provides a comparison of performance for different steel grades.

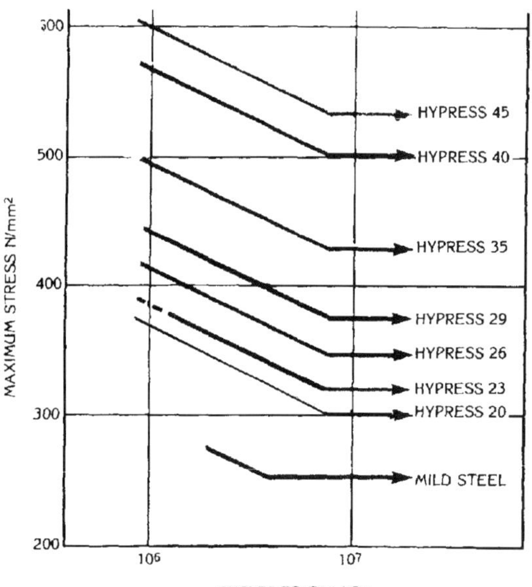

Fig 21: Fatigue curves for Hypress

Grade	Minimum Yield Strength ksi (MPa)	Minimum Tensile Strength* ksi (MPa)	Elongation (percent) Minimum in 2 in. (50 mm) Hot Rolled	Cold Rolled	Galvanized
35	35 (241)	50 (345)	23	23	23
40	40 (276)	55 (379)	21	20	20
45	45 (310)	60 (414)	19	18	18
50	50 (345)	65 (448)	17	16	16
55	55 (379)	70 (483)	15	N/A	N/A
60	60 (414)	75 (517)	N/A	10	10
70	70 (483)	80 (552)	N/A	8	8
80	80 (552)	-	N/A	4	4
120	120 (827)	-	N/A	2	N/A
140	140 (965)	-	N/A	-	N/A

Fig 22: HSS performance comparisons

In so-called Dual-Phase steels, strength is enhanced by a transformation hardening due to successive heating and then quenching. The resulting microstructure exhibits two phases, ferrite and martensite. The former is relatively soft while the latter exhibits strengths up to 1250 N/mm^2. Proportion of the two phases is controlled by the temperature from which the steel is quenched — and by the rate of quenching. Continuous annealing lines now used in the most modern steel plants have water quenching systems which allow rates up to 1000 C per second to be achieved.

Pressworking techniques The degree of panel thickness reduction is somewhat limited by the stretch marks which can occur during pressing. Compared with the press settings for mild steel, there should be modification of the gap between the mating dies to avoid either wrinkling due to uneven strain or buckling due to radial compression. In the case of parts requiring deep drawing, some modification of detail design will probably be necessary compared with that used for mild steel and the number of stages in the forming process often has to be increased.

Work hardening is also an important element in strength enhancement for some HS steels. Standard tension tests on specimens can be used to determine work-hardening exponent n and plastic strain ratio r as the strain rate hardening exponent m. Typical values, due to the American Iron and Steel Institute are given in Fig 23. The n value is related to the steepness of the stress/strain curve in the plastic deformation region and co-relates with elongation and ratio of tensile strength to yield-strength. If the metal strain hardens during the forming process, initial straining will cause the deformation zone to increase in strength — until deformation stops in that

	Yield Strength ksi (MPa)	Tensile Strength ksi (MPa)	Percent Elongation in 2 in. (50 mm)	n value	r_m value	HRB hardness
Commercial Quality	40 (276)	51 (351)	34	0.18	1.1	60
Drawing Quality	37 (255)	50 (345)	37	0.19	1.1	47
Drawing Quality Aluminum Killed	35 (241)	49 (338)	39	0.19	1.1	49
Deep Drawing Quality Aluminum Killed	30 (207)	46 (317)	41	0.20	1.5	45
Interstitial Free	25 (172)	43 (296)	42	0.21	1.8	40

Fig 23: Work hardening exponents

region and spreads to another. The higher the strain hardening the more uniformly will deformation be spread throughout the pressing.

The r value is a measure of the resistance to thinning as the metal is drawn. The weakest zone of a typical 'cup' pressing is at its base; in order for the base to deform and elongate, the thickness must be small. By specifying a material that resists thinning, a more uniform stress distribution will result. The m value works in the same way as the n value in delaying the start of local 'necking'. While high n extends the range of uniform deformation, high m delays the onset of the local neck.

Aluminium alloy

Recent increases in the price of aluminium alloys prompts the intelligent use of the material to fully optimise its value. Rather than making substitution of materials in existing designs, a redesign for the new material is always worthwhile if time permits.

In many areas of special vehicle building, like container van bodies, aluminium alloy is of course the rule rather than the exception. In other areas , such as luxury and high-performance cars, aluminium alloys were used historically, were next displaced by steel and now compete for varying shares of the market with steels and composites. Small cars, of course, were first produced with aluminium alloy and then, in the volume sector particularly, made in steel.

Series	Principal Alloying Elements
1000	Aluminium of 99.0% minimum purity
2000	Copper
3000	Manganese
4000	Silicon
5000	Magnesium
6000	Magnesium + Silicon
7000	Zinc
8000	Lithium and others
9000	Unallocated

Fig 24: BS numbering system

Melting point	660°C
Boiling point	2480°C
Thermal conductivity (0-100°C)	0.57 cal/cm/s/°C
Temperature coefficient of linear expansion (0-100°C)	23.5×10^{-6} per °C
Electrical resistivity at 20°C	2.69μ Ω cm.
Temperature coefficient of resistance (0-100°C)	4.2×10^{-3} per °C
Density at 20°C	2.69 g/cm^3
Modulus of elasticity	68.3 GPa
Modulus of Torsion	25.5 GPa
Poisson's ratio	0.34

Fig 25: Physical properties of 'pure' aluminium

E - Excellent G - Good
V - Very Good. N - Not recommended

Alloy	TIG MIG	Resistance	Brazing	Soldering
1080A	V	G	V	V
1050	V	V	V	V
1200	V	V	V	V
2014A	N	E	N	N
3003	V	E	V	V
3103	V	E	V	V
3105	V	V	G	G
5005	E	E	G	G
5083	E	E	N	N
5154A	E	E	N	N
5251	V	E	N	N
5454	E	E	N	N
6061	V	V	V	G
6082	V	V	G	G
7010	V	V	N	N
7020	V	V	N	N
7075	N	V	N	N

Fig 27: Jointing data

Description	U.K. Designation	Min. tensile strength MPa 1200 alloy*
Annealed, fully soft	O	70
As manufactured	M	—
Quarter-hard	H2	95
Half-hard	H4	110
Three-quarters hard	H6	125
Fully-hard	H8	140

Fig 26: Strength enhancement

	Specific gravity	Density, kg/m³	Melting point °C	Linear coefficient expansion, 10^{-6}/°C	Specific heat joules/kg/°C	Modulus of elasticity G.P.A.	Torsional modulus G.P.A.	Poisson's ratio
Aluminium	2.69	2699	580-660	24.0	0.96	69	25	34
Mild Steel	7.85	8097	1900	12.6	0.42	207	84	30
Copper	8.93	9200	1083	16.5	0.38	120	45	33
Magnesium	1.75	1803	565	29.0	1.02	44	17	33
Timber (Ash)	0.60	618	—	4.5	—	10	—	—
Glass	2.64	2520	1500	8.4	—	69	—	24
A.B.S.	1.05	1082	125	90.0	—	1.4-2.7	—	—
P.V.C.	1.40	1442	95	70.0	1.40	2.8	—	—
Nylon	1.14	1174	125	98.0	1.67	3.6	—	—

Fig 28: Material properties

It is still a surprise that so few small cars with steel bodies rival the unladen weights of separate chassised, and aluminium bodied, of the 1920s such as the Austin 7 Chummy. With the increasing use of plastics skin panels on small specialist sports cars there is some sign of aluminium alloys competing with steel for the construction of the base structures.

Fig 24 shows the British Standard numbering system for designating aluminium alloys while Fig 25 gives some of the physical properties of 99.99% pure aluminium. Most of the alloys have elastic modulus one third that of steel and weight about three times that of steel — hence the danger of direct substitution for a particular design. The strength of aluminium alloys, depending on composition and heat-treatment, can virtually match a number of steel alloys; however, selection must be made with care so as not to unduly compromise some other property or pay too high a price.

Two distinct groups of alloys are those that can and those that cannot be heat-treated. Those which cannot can have their strength enhanced by varying degrees of cold-working. This is expressed in the alloy by the letter H and a number from 1 to 8 indicating strength increase. There will normally be progressive loss of ductility with cold working and if this quality is at a premium then the O-designated alloy should be selected — signifying the fully annealed condition which removes any cold-working effect. The non heat-treatable series are 1000, 3000, 4000 and 5000 while Fig 26 shows strength enhancement of the heat-treatable ones. Heat-treatment has two basic stages; heating at prescribed high temperature followed by rapid cooling is solution heat treatment; heating at a prescribed low temperature for a prescribed time is known as ageing. The former increases the metal's ductility and the latter causes hardening; forming of the metal is thus sometimes carried out between the two stages. Alloys suitable for these processes are 2000, 6000, 7000 and 8000; while designations TB, TE and TF refer to solution treated and naturally aged; artificially aged and the combination, respectively.

The Aluminium Federation provide much data, such as that in Fig 27, on the joining of aluminium alloys. Because all of the alloys have a tenacious oxide film covering, provision has to be made for breaking through this prior to welding. Because the metal has much higher conductivity than steel, higher heat inputs for fusion welding are required; its high reflectivity also means that a different appearance from steel is exhibited in the solid to molten transition phase. Because heat applied in soldering, brazing or welding has a softening effect in the region of the join, work-hardened alloys will be locally annealed — an aspect which must be allowed for in design. Material properties for aluminium alloy ap-

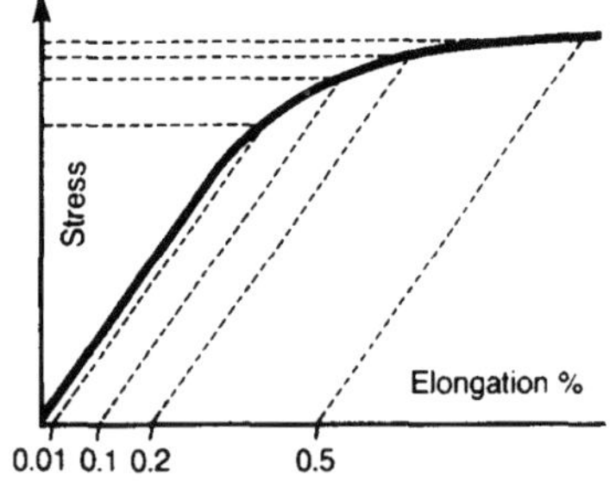

Fig 29: Typical stress/strain curve

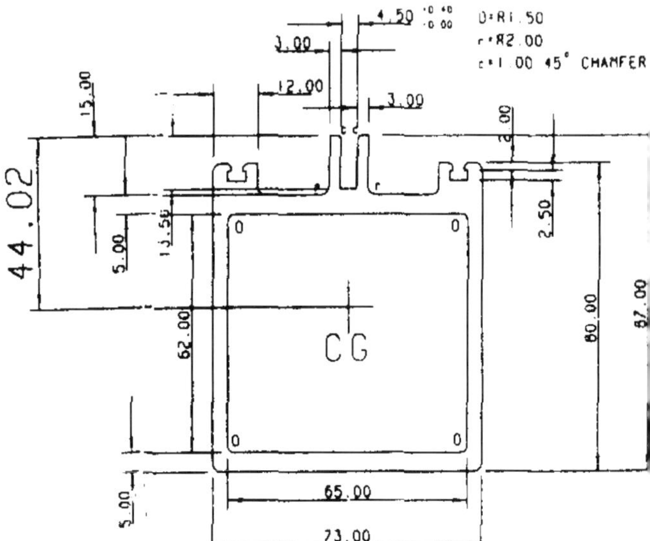

Fig 30: Calculating section properties

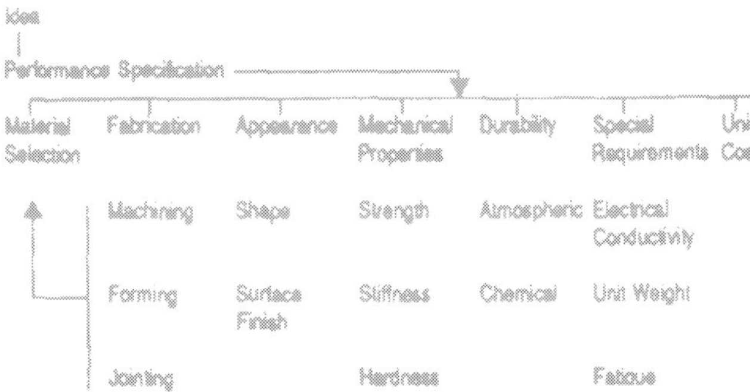

Fig 31: Design plan for aluminium alloy build

pear in Fig 28 alongside those for rival materials. Recently developed aluminium-lithium alloys have a lower than normal density and strengths and stiffnesses compare well with more expensive copper and zinc alloys with aluminium. It is important in design to realise that aluminium alloys have about twice the expansion co-efficient of steel — also that, unlike steel, the material does not have a clearly defined yield point, leading to the 0.2% strain proof stress being used in design, Fig 29. For applications like truck bodies the natural 'mill' finish of aluminium has been found acceptable. Upgrades on this are provided by anodic oxidation (anodising), liquid or powder organic coatings — for which BS 1615, BS4842 and BS 6496 apply respectively.

A recent change from riveting to adhesive bonding of aluminium alloy panels in vehicle construction has been reported for mini-bus production. Quieter operation of the vehicle is said to result and the continuous join line has resulted in a lighter gauge of framing. This and other design and fabrication problems is covered in the Shapemakers Extrusion Design Package, available from the Aluminium Federation. It comprises a comprehensive manual and a number of computer diskettes containing CAD programs for many design applications. Accent is on extrusions which are of course widely used in specialist vehicle body building; the software allows rapid calculation of section properties for complex extrusions — with considerable benefit to structural analysis, Fig 30. The 2000 and 6000 alloy series are described for the extrusion application and a flow chart for assessing design ideas is suggested as Fig 31. Extrusion allows the convenient addition of bulbs and root buttresses, Fig 32, to improve section properties and can provide even for integral hinges, Fig 33, for specialist application. Built-in mechanical fasteners, Fig 34, are another possibility. Alloys 2014 and 2024 have been used for truck chassis frames; one of the key factors in choosing high-performance alloys is the fatigue to ultimate tensile strength ratio depicted in Fig 35.

In this illustration, Al-Zn-Mg and Al-Mg-Si alloys are compared for fatigue property which like toughness and stress corrosion resistance reduces with increased ultimate strength. Code of practice CP118 from the BS I is particularly valuable for the structural use of the material.

Fig 32: Section bulbs

Fig 33: Extruded hinges

Fig 34: Fastener fixtures

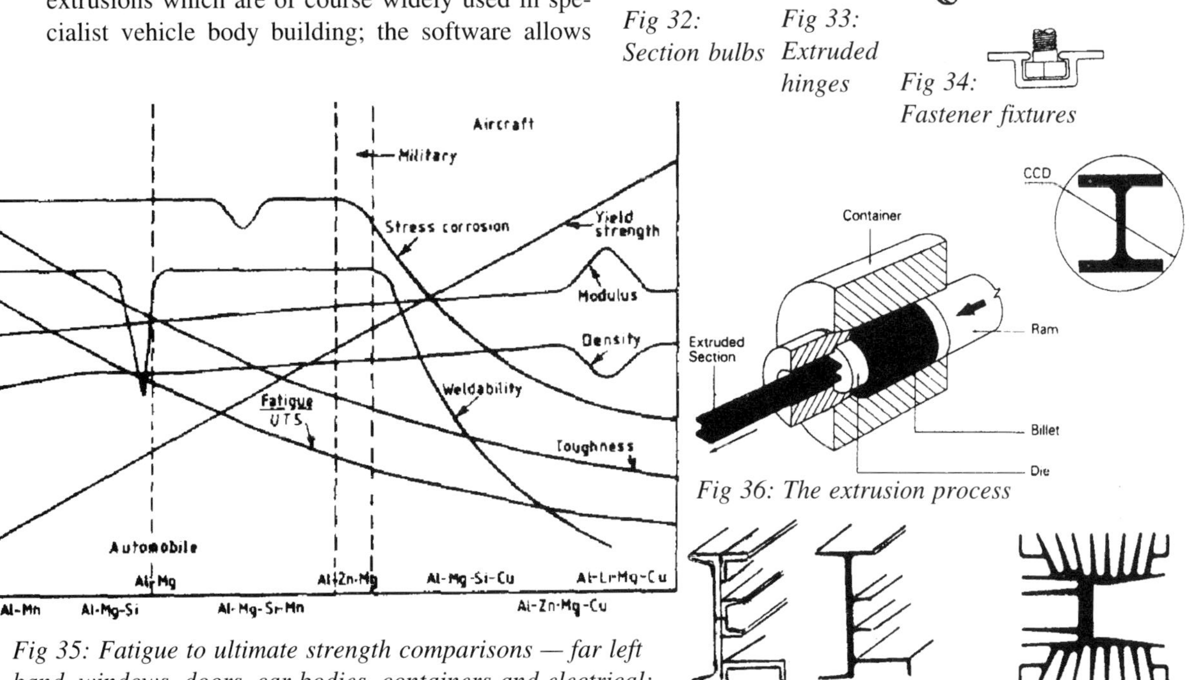

Fig 35: Fatigue to ultimate strength comparisons — far left band, windows, doors, car bodies, containers and electrical; near left band, road transport structures and tank ends

Fig 36: The extrusion process

Fig 37: Section intricacy with extrusion

Aluminium-alloy extrusions

Careful fabrication control together with imaginative structural design can bring about pleasing body shapes, for a variety of end uses, with the help of aluminium alloy sections. Given here are some design guidelines on how to exploit the imaginative sections available from the aluminium suppliers.

Use of aluminium extrusions for vehicle body structures —particularly in commercial vehicles — has revolutionised design in recent decades and brought to the smallest of workshops the ability to fabricate clean-lined vehicles with accuracy. Lengthy fitting and finishing operations are largely avoided by exploiting imaginative sections created by the aluminium companies for solving common assembly problems. To produce a complex extruded cross-section, a pre-heated ingot in its plastic state is forced through a steel die cut out to the section shape, Fig 36. For hollow sections, a mandrel is attached to a 'bridge', the extruded stream being split at the bridge and re-united before emerging from the die. Section advantages include intricacy, fine-tolerance and the ability to replace several parts, Fig 37. The ability to produce sections previously only achievable by roll-forming and machining are also great benefits.

Section shapes should be chosen where possible to have their various elements as near as possible of equal thickness; this is to ensure consistency of metal flow and hence resultant physical properties. Sections of high overall dimension, or circumscribing circle diameter (CCD), are more expensive to produce with high accuracy because of the difficulty of maintaining tolerance — arising from the tendency for slower metal flow at the periphery. Accuracy in compact sections is normally good enough to provide successful mating components — such as those in the designs of British Alcan in Fig 38. These also show how flush fitting surfaces can be obtained as well as sliding, 'drawer', effects. The elasticity of aluminium alloy also allows for the use of clip-fits, Fig 39, either permanent or releasable. Tolerances for straightness, twist and concavity/convexity are shown in Fig 40 alongside those for angular warpage of section elements. Tables are also available from aluminium suppliers for tolerances of extruded round tubing in terms of eccentricity caused by variation of mean wall thickness.

Aluminium alloy extrusions in platform and box van bodies are now familiar technology. Platform bodies have traditionally been at the more expensive end of the chassis/body combinations. Fig 41 shows typical crossbearer and longitudinal runner sections. The rear bearer may incorporate a shroud plate, as shown; usually bearers are solid-riveted or

A 0.5% max of B (for individual surfaces)

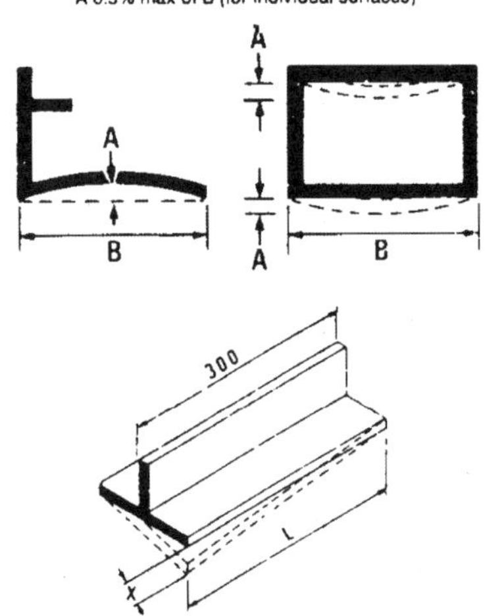

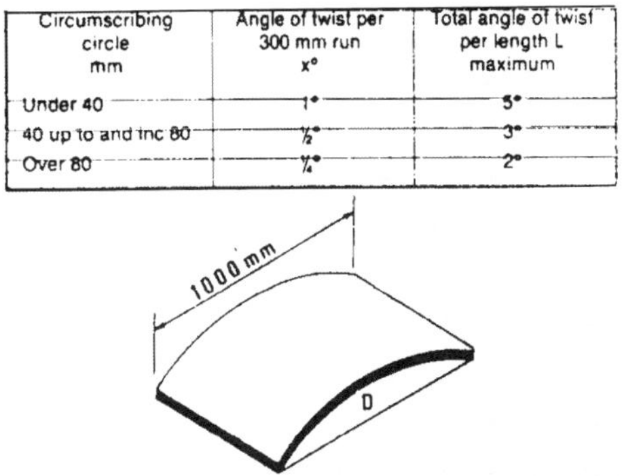

Circumscribing circle mm	Angle of twist per 300 mm run x°	Total angle of twist per length L maximum
Under 40	1°	5°
40 up to and inc 80	½°	3°
Over 80	¼°	2°

For diameter, or sections within circumscribing circle mm	Conditions	Departure from straightness over any length of 1000 mm D mm
Up to 100	All	1.5
Over 100	M	2.0
	All others	2.5

Fig 38: Secton variation with extrusion: bow, top: twist, centre; straightness, bottom

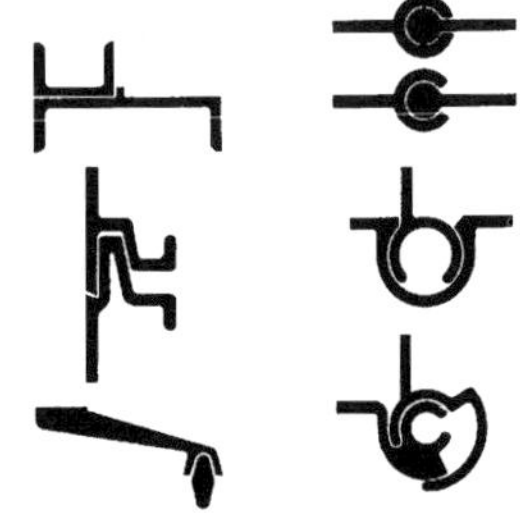

Fig 39: Clip-fit designs

bolted to longitudinals using angle brackets, as seen in Fig 42a. Side boards are increasingly made from extrusions with plank sections shown in Fig 42b; the conventional box van body uses a variety of sections including the sill and corner pillars, as in Fig 43.

Extrusions are also used with timber in so-called composite construction bodywork as seen in the cant rail section of Fig 44 used to connect plywood sidewall and roof panels. Usually the top corners of box vans are castings which enclose the cant rail extrusions. The kind of sill section used will depend on the requirement for a rub-rail. Extruded floor panel sections are also now commonplace and Fig 43 also shows recommended spacing and carrying capacities. It is also necessary to check the carrying capacity of the 'planks' in respect of the wheel loadings from hand- and power- operated pallet trucks. Individual planks may have to withstand concentrated loads of 1650 lb (750 kg) under these conditions. Timber inserts are also used where there is a requirement for securing loads by wood battening. Fig 45 shows a side rave section used to form the edge of platform floor. A wide variety of jointing recommendations for vehicle body extrusions is in the aluminium supplier's handbooks. It is important to realise, however, that many of these are not applicable to integral structures — requiring greater attention to the load paths between the bulkhead rings and the longitudinal 'stringers' of the structure.

Straightness tolerances for extruded bar, regular sections and extruded round tube

Sections within circumscribing circle	Condition	Departure from straightness (S) over any selected length of 1000mm
Up to 100mm	All	1·5mm
Over 100mm	As extruded (M)	2·0mm
	All other	2·5mm

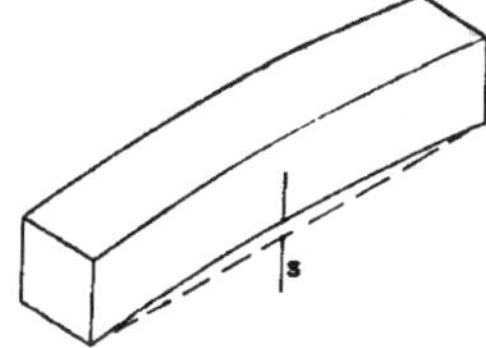

Twist tolerances for extruded solid and hollow sections

Diameter of circumscribing circle	Angle of twist (t) per 300mm run	Total angle of twist per length
Under 40mm	1°	5°
40 up to and inc. 80mm	½°	3°
For diameters over 80mm:		
Lengths up to 8000mm.	¼°	2°
Lengths over 8000mm.	¼°	3°

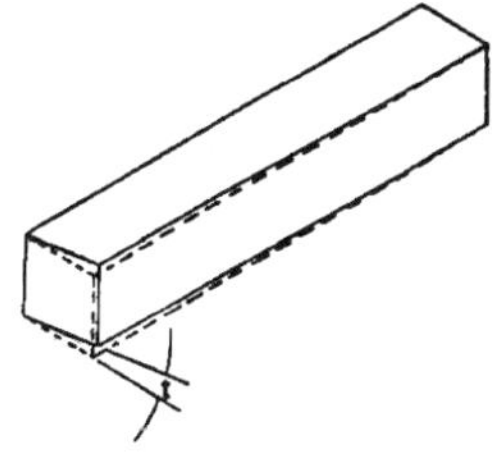

Concavity and convexity tolerances

Over the width (A) of the section the maximum tolerance on concavity and convexity (B) shall be 0·05mm per 10mm of width

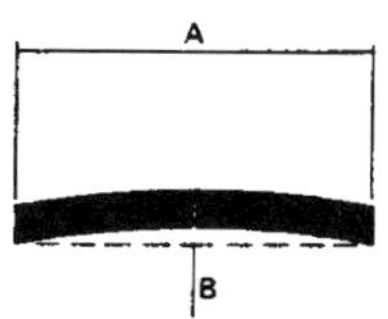

Normal angular tolerances

Nominal thickness of thinnest leg	Allowable deviation from angle specified (θ)
Up to and inc. 1·6mm	±2°
Over 1·6mm, up to and inc. 5·0mm	±1½°
Over 5·0mm	±1°

Angles are measured at the extremities of the section.

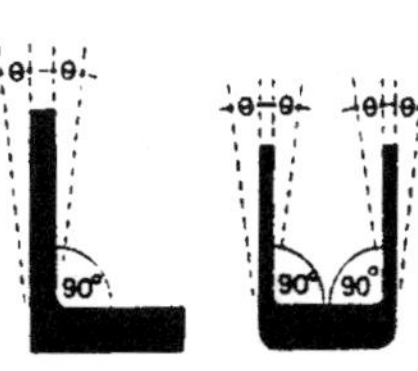

Fig 40: Tolerance requirements

Recently revealed properties for a range of structural extrusions recently introduced by Alcoa are shown in Fig 46. These high strength thin walled sections have predictable energy absorption characteristics and therefore can be used in car spaceframes such as that from the company shown in Fig 47. This kind of construction is said to result in 30-40% weight savings compared with steel unibody structures.

The Shapemakers organisation set up by the extrusion suppliers have published a comprehensive handbook, and associated software disks, of 2D computer-aided-design procedures, known as the Extrusion Design Manual. On a desk-top computer, section characteristics can be quickly obtained in terms of area, periphery, weight per metre, also section second moment of area. Among the sample calculation procedures provided is the following one for unloading ramps for a vehicle, Fig 48. The requirement was for the ramps to have maximum weight of

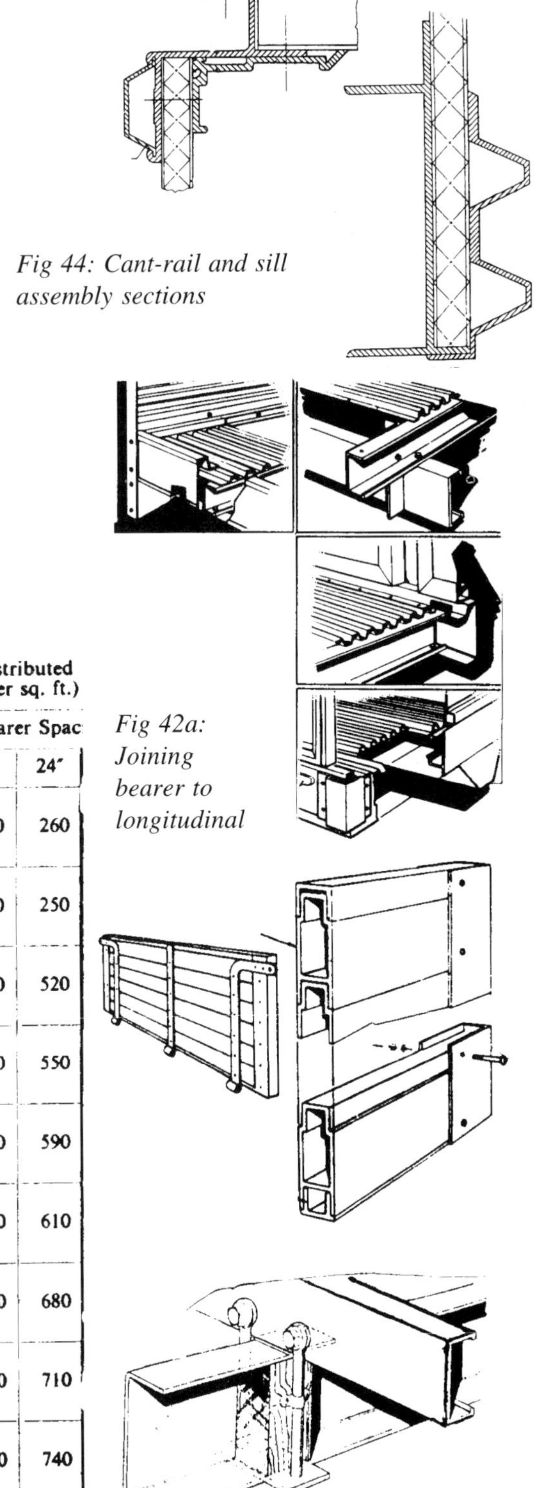

Fig 44: Cant-rail and sill assembly sections

Fig 42a: Joining bearer to longitudinal

Fig 42b: Drop-side and chassis clamp details

Weight as laid (lb. per sq. ft.)	Maximum Distributed (lb. per sq. ft.)			
	Cross-bearer Spac			
	15"	18"	21"	24"
1·60	670	470	340	260
1·86	650	450	330	250
2·99	1,330	920	680	520
2·52	1,420	980	720	550
2·48	1,520	1,050	770	590
2·42	1,580	1,100	800	610
3·36	1,750	1,210	890	680
3·27	1,830	1,270	930	710
2·02	1,900	1,320	970	740

Fig 41: Crossbearer and runner sections

50 kg each, span of 2.5 metres, operating angle up to 30 degrees and maximum wheel loading from pallet truck of 2 tonnes shared by four tyres of maximum width 200 mm. Specified channel section is 254 x 88(web) x 14(flange) mm.

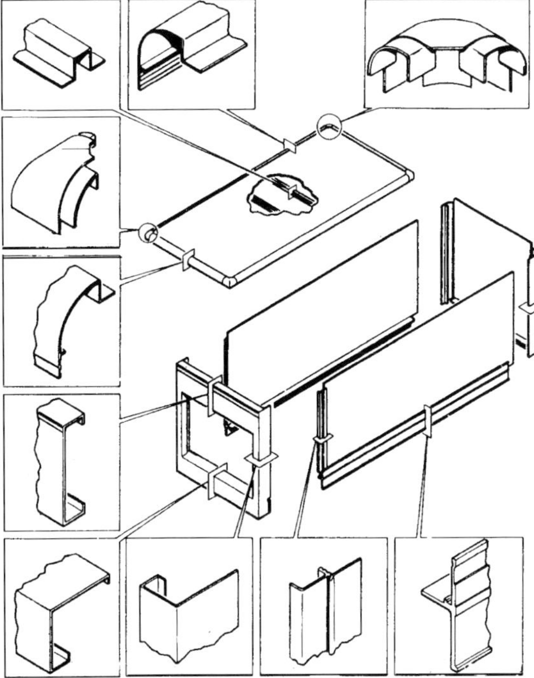

Fig 43: British Alcan CV body extrusion examples

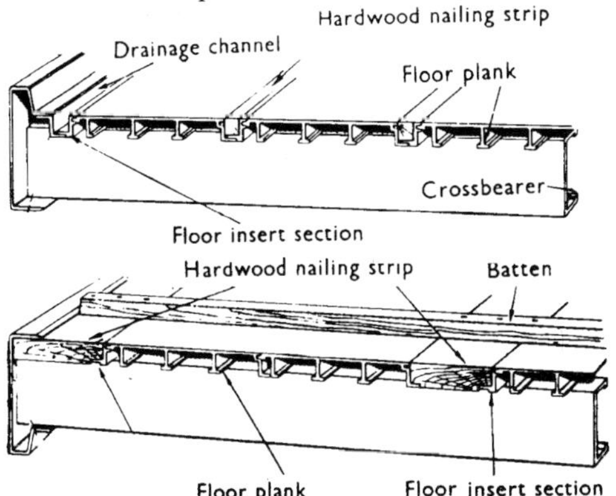

Fig 45: Floor-plank and side rave sections

Yield Strength, T4, ksi	16	(110 MPa)
Tensile Strength, T4 ksi	34	(235 MPa)
Percent Elongation, T4	21	
Yield Strength, T6, ksi	39	(269 MPa)
Tensile Strength, T6, ksi	44	(304 MPa)
Percent Elongation, T6	11	

Fig 46: ALCOA extrusion properties

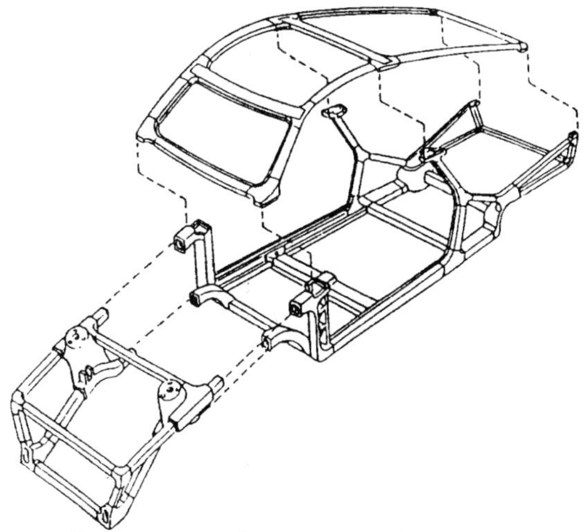

Fig 47: Extrusions for car space frame

Span L

Longitudinal restraint

Slope θ in degrees

Section properties:

Area		5030 mm²
Modulus	Zxx	54620 mm³
Inertia	Ixx	3459100 mm⁴
Radius of Gyration	r	26.2 mm
Weight/metre		13.39 kg/m
Alloy		6082 T6

Loading. As the vehicle is unloaded it moves out of the horizontal with a considerable shift in its neutral axis and the loading on the first set of wheels increasing. This will be a feature of the individual vehicle. For the purposes of this calculation it is assumed to be 10%, hence -

$$\text{Maximum individual wheel load} = \frac{19640\text{N (2 tonnes)}}{4} \times \frac{110}{100} = 5400\text{N}$$

Bending Stresses. The ramp acts as a simply supported beam and with normal wheelbased vehicles will have a central load as the worst condition. (Load Case 2.)

$$M = \frac{WL}{4} = \frac{5400\text{N} \times 2500\text{mm}}{4}$$

Maximum bending moment = 3375000 Nmm

$$\text{Maximum Stress} = \frac{3375000}{54620} = f_{bc} = 61.8\text{N/mm}^2$$

Allowable Stress Levels. See Table 3.2 (From British Standards CP118)

6082 T6 alloy

Bending P_{bc} = 154N/mm²

Deflection

$$\delta = \frac{WL^3}{48EI} \qquad \text{For 6082 E} = 68{,}900\text{N/mm}^2$$

$$\delta = \frac{5400 \times 2500^3}{48 \times 68900 \times 3459100}$$

$$\delta = 7.45\text{mm}$$

The deflection/span factor = $\frac{\text{Span}}{336}$

which is well inside the recommended value of $\frac{\text{Span}}{200}$

Fig 48: Sample calculation for loading ramps

Car space-frame in aluminium alloy

For the car space frame shown in Fig 47 which became the Audi ASF concept, Fig 49, both the extrusions and the die-castings were supplied by Alcoa in alloys specially developed to give the desired characteristics. The extrusions are not of normal commercial quality but are produced in a new heat-cured AlSi alloy; as they are situated in locations prone to impact damage. Curved extrusions like the windscreen crossmember were formed to the required shape before delivery to Audi, with an extrusion-bending processes which can handle complex shapes and tight radii, including the ability to bend parts in two planes simultaneously.

The die-castings are produced in an alloy which is easy to weld, extremely ductile and thus readily deformable in the event of a crash. They are made by the Vacural process, in which the die is evacuated before the molten metal is injected to ensure homogeneous composition with low gas content, then heat treated. The sheet material is heat cured, being pressed in the softer and more easily fabricated T4 state and then hardened to design strength, after assembly, when the bodyshell is heated as it passes through the paint line. Care is necessary in designing the dies for the skin panel stampings, to cater for the different elasticity and elongation limits of aluminium, also the flanges on many panels have to allow for the different spring-back characteristics of the metal.

Attachment of all the extrusions to the node castings is by continuous shielded arc welding, automated wherever possible. Jointing technique for the skin panels is punch riveting. This process gives a 30 per cent greater strength than spot welding, requires less energy and achieves consistently high quality. It is used for 68 per cent of all single-point joints, the remainder being spot welded or clinched. Typical applications are the flanges around the windscreen and rear window, the door frames and the B-pillars. Clinching, in which two panels are joined by beading over the mating edges, is employed where a lower static strength is acceptable, for the stiffeners on the bonnet for example. Epoxy adhesive bonding is utilised for the bonnet, doors and boot lid, and a modified epoxy for bonding the floor panel and suspension strut turrets. Details are given in Figs 50, 51 and 52.

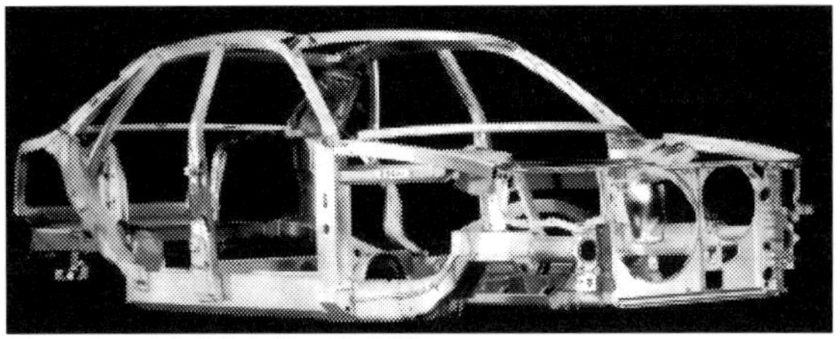

Fig 49: ALCOA/Audi ASF structure

A computer-simulation with about 38000 elements was created for the finite-element offset-crash model, using a 50 per cent overlap and an impact speed of 55 km/h. In the case of the lateral impact model for evaluation in accordance with FMVSS

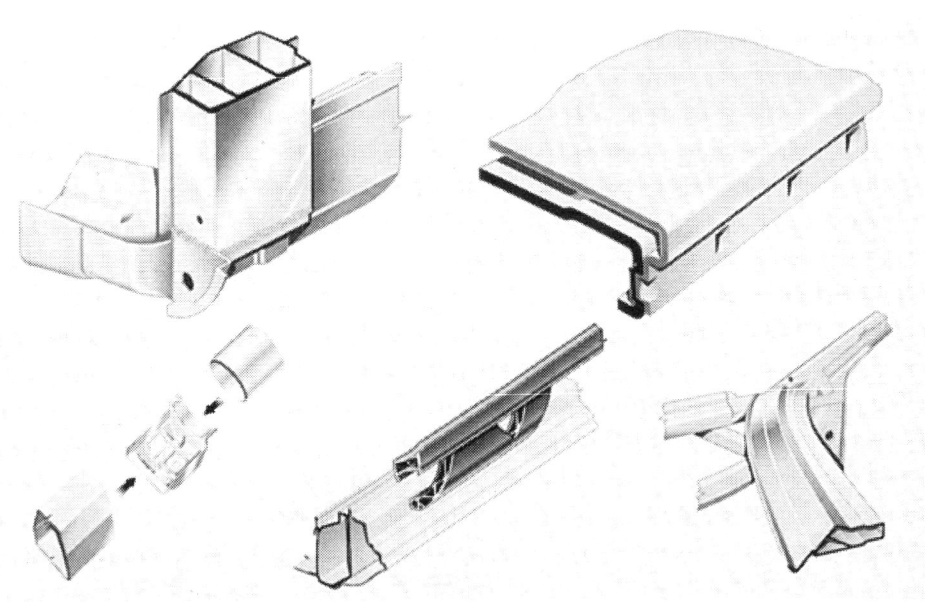

Fig 50: A-pillar extrusion, lower section, top-left; clinched joint, top-right; node-casting for subframe mount, bottom left; side-impact barrier, bottom centre; windscreen header rail, bottom right.

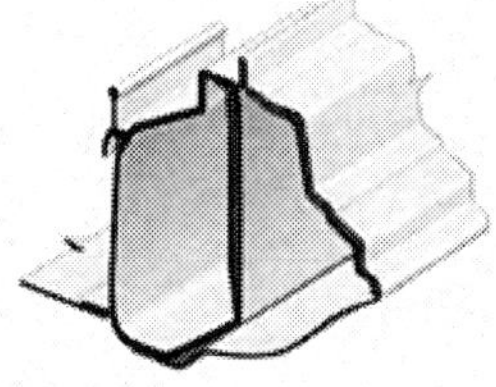
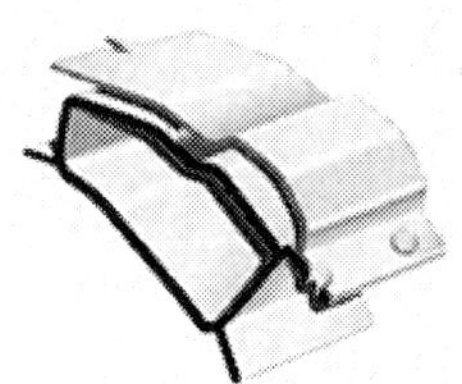
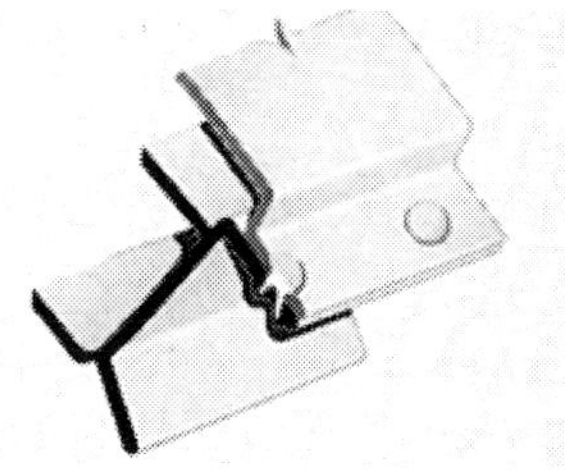

Fig 51: Door sill cross-section, left; upper side rail, centre; punch-riveted joint, right.

214, there were some 23000 elements. The front-end crash model for the '35 mph' test entailed 19,000 elements, whereas the rear-end crash model necessitated 26,000 elements.

Primary advantage of the ASF design is the welght reduction of up to 40 per cent. Static torsional stifiness is improved too, by about 40 per cent in comparison with a steel body. For instance, the figure of 23.8 kNm/deg for the D2 aluminium body compares very favourably with the 16.9 kNm/deg of the Audi 100 C4. There is a 30% reduction in the number of component parts in comparison with a steel body. A further benefit is the flexibility afforded by the space-frame concept, enabling different model derivatives of a car to be developed more quickly and cheaply.

The ASF meets the current US lateral impact standards, as well as the 55 km/h offset crash standard. In general, the body has been designed for rigidity, leading to increases in wall thickness by an average factor of 1.7 to 1.8 in relation to a sheet steel body. These increases result in a reduced overall stress levels, giving reserves of strength and an operating life of 300000 km, ALCOA claim.

An important feature of the design is the safety cage for passengers, which requires much larger forces to cause plastic deformation in the structure than is the case with current steel bodies. In order to obtain the safest possible survival space and the lowest degree of damage severity in an accident, the deformation areas have been given graduated degrees of rigidity: the resistance of the impact absorbers is lower than the deformation resistance of the front section of each longitudinal chassis member, which is lower than the resistance of the rear section of the chassis member, and this in turn is lower than the resistance of the cell as a whole.

As aluminium sections have a much higher energy absorption per unit of mass than steel equivalents, when used as longitudinal chassis members, they can provide the same energy absorption as steel members but with less than half the weight, the suppliers maintain. For example. the energy absorption capability of a crumpling 1.2 mm thick square steel beam is equalled by that of an aluminium beam of the same geometry with a wall thickness of 1.8 mm, but with a weight saving of 50 per cent. Even better is a tubular aluminium section, as this produces a much closer series of folds during crumpling and thus needs more deformation work per unit of mass. For lateral protection, a new system has been adopted comprising door impact members with latticed reinforcements and an overlap between the doors and the pillars and sills. Fig 53 shows how special consideration has been given to accident damage repair.

Aluminium applied to a conventional shell structure

Direct substitution of aluminium alloy for steel panels is the philosophy behind the recent Alcan/Ford joint venture project on turn-of-the century lightweight vehicles. Part of the agreement between the companies is a price stability agreement which will short-circuit the traditional fluctuations as the metal

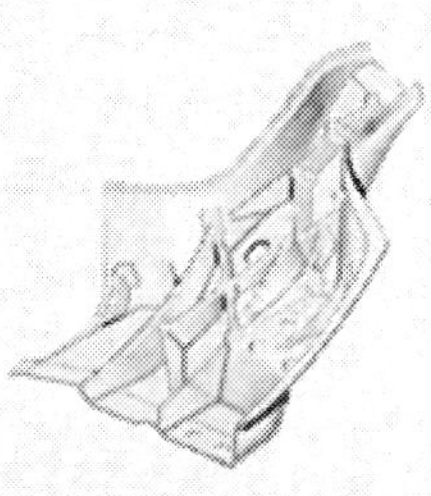
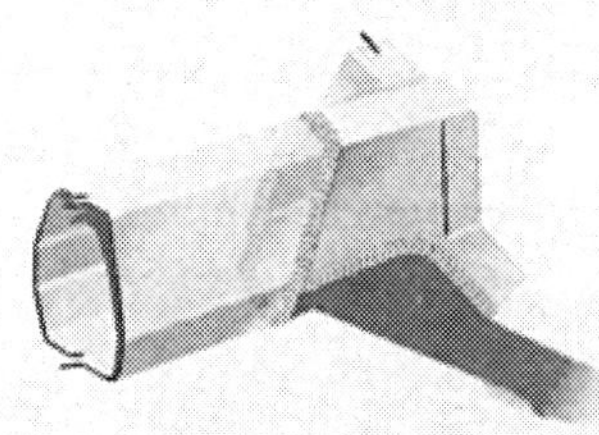

Fig 52: Node casting for chassis member, left; MIG-welded joint, centre; node casting for suspension strut turret, right.

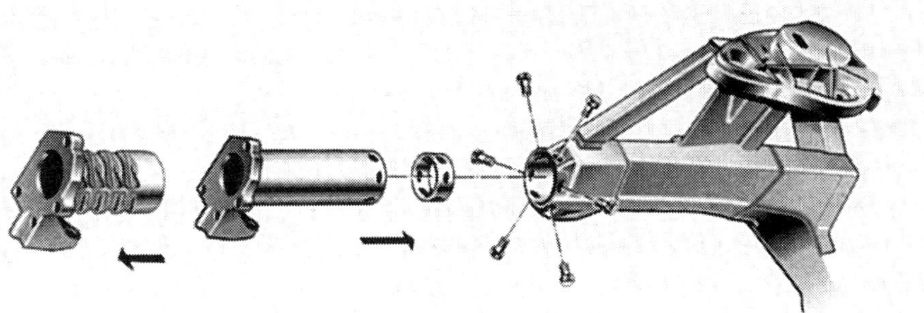

Fig 53: Bolt-on repair method for chassis-members

is traded on the London metal exchange. Alcan will contract to supply Ford over a fixed term at a contracted price. Little compromise is being made towards altering the basic body-structure design to better exploit the material's intrinsic properties, so much a feature of Alcan's designs described in Chapter 1. Ford clearly wants to play safe on the proven shell of the Taurus/Sable car, Fig 54, and has merely counteracted the material's one third elastic modulus of steel by stepping up the gauge of panels by an average of 50 per cent. By this means torsional stiffness has been increased by some 50 per cent although bending stiffness has been reduced. Higher twist resistance is explained mostly by the use of Ciba-Geigy adhesives in a weld-bonding technique, for connecting panels, which substantially improves joint-efficiency and coincidentally reveals the poor performance of conventional spot-welding used on traditional car bodies.

Further improvements are yet to be made, however; currently the pilot-build shells are being shown with the swages in the wheel-arch panels, for example, replaced by equivalent bowler-hat shape sectioned weld-bonded flanged stiffeners, because the material will not tolerate additional stretch in forming swages into an already deep-drawn panel. This will be avoided in future die and panel design. Alcan's established AVT (aluminium vehicle technology) process is used developed earlier for such projects as the experimental Rover Metros, also built by substituting aluminium for steel panels. A key element of this is the development of an adhesive which will not wash out under the influence of solvents used in subsequent painting.

The resulting Taurus shell has successfully met the 35 mph frontal impact requirements with a structure which is 46 per cent lighter, using 5754 alloy which has particularly good pressing properties. Closure panels are made in 6111 alloy. Interestingly Ford already claim to have the greatest experience of pressing aluminium closure panels of any car-maker. Ford have invested a further $35 million in achieving the necessary expertise in pressing and fabricating the panels of the main shell structure while Alcan considers it has spent some $100 million on AVT work and continues to do so at a rate of $10 million/year. The Ford AIV concept has been extended to a fleet of 40 vehicles based on the Sable. Each vehicle weighs 318 kg less than its steel counterpart but a further secondary weight saving of 135 kg is in prospect. Fig 55 explains this using North American units. Natural frequency of the shell is put at 30 Hz against 25 for the steel shell. The shell alloy has a 3 per cent magnesium content and the weld-bonding technique is said to improve its fatigue performance. The XD4600 Araldite adhesive meets toughness and impact criteria set by Ford, at -40 C, and creep resistance requirements at 100 C.

Ford claim that crash resistance of the aluminium alloy shell is superior to a heavier steel one with lower metal gauge but presumably this would very much depend on the design of the shell. The company assert that the demands of volume production limit the scope for redesign, however. With approximately equivalent design to a steel shell the AIV one is 182 kg lighter and in testing it has survived 50 000 miles 'rough-road' durability operation, without 'aluminium-related' failures.

Although the pressworking techniques will be very similar to those for steel panels produced in volume, some hint of a change in spot-welding technique was revealed, possibly involving the use of

Fig 54: Ford Sable -based AIV in 5754 aluminium alloy

DC rather than AC equipment. There may be a possibility of introducing some extruded sections into the structure, Ford suggested, and there would need to be substantial re-equipping of service stations for making economical repairs to damaged aluminium alloy panels. There was also mention of eddy current separation techniques being used to distinguish between cast and wrought aluminium to improve the prospects of recycling the material at the end of the life of the vehicle.

A contrasting approach to aluminum alloy car body structures, to that of both ALCOA and Alcan, is shown in Fig 56 to be that of Hydro Aluminium who have produced a space frame using welded joints for the extruded profiles making up the main structural elements. The concept has been used in the Pininfarina Ethos and BMW Z13/E1 experimental cars.

Fig 56: Hydro Aluminium space frame

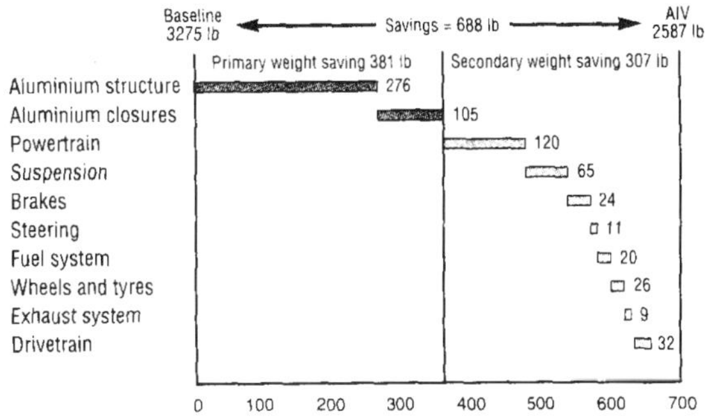

Fig 55: Prospected weight savings

Condition	Tensile strength (tonf/in²)	0·2% Proof stress (tonf/in²)	Elongation (%)	Hardness (H.V.)	Bend test 180°
Softened	35	14	35	200	—
Quarter hard	55	33	25	220	1t
Half hard	65	47	10	270	2t
Three quarter hard	75	58	7	320	4t
Hard	80	60	5	360	6t

Fig 57: 301 stainless steel

Grade Warm Worked	Tensile Strength	1·0% Proof Stress	Elongation
304 (18/10 chromium-nickel)	40 tonf/in²	24	30%
316 (18/10/3 chromium-nickel-molybdenum)	40 tonf/in²	28	20%
Nitrogen Bearing Grade			
304 (18/10 chromium-nickel)	38/51 tonf/in²	20·5 tonf/in²	35%
316 (18/10/3 chromium-nickel-molybdenum)	41/51 tonf/in²	22 tonf/in²	35%

Fig 58: Effect of warm working

Stainless steels

High-alloy corrosion resistant steels (otherwise known as 'stainless') comprise the ferritic, martensitic, precipitation- hardening and austenitic groups, the last of which has a potential for automotive structures depending on the degree of refinement of the design as well as the prevailing relative price of nickel.

Ferritic stainless steels, or chrome-irons as they are sometimes called, are considered in a later section. Here the higher grade austenitic stainless steels are examined — which can be successfully exploited if used in very thin sections and stressed to higher predicted levels when weight saving is crucially important. While strengths in the annealed condition are not very high, good ductility and work hardening ability means that it can be used in the cold-worked condition and still retain a good deal of formability. Fig 57 shows properties of 301 in various standard work-hardened conditions. Even the half hard state offers a proof stress of 50 tons/in^2 combined with a very useful elongation of 10 per cent.When used in the form of plate, for road-tanker shells, say, 0.2 per cent nitrogen can be added to the steel melt to beneficial effect. Up to 50 per cent increase in proof stress can be obtained above the normal annealed state — without any strength reduction in subsequent heat treatment of the plate. Another possibility is to 'warm-work' the otherwise untreated material - with benefits as shown in Fig 58.

Typical of tanker construction in stainless steel is the example shown in Fig 59 which also uses a special lightweight tandem bogie suspension. An example of stainless steel cladding section is seen in Fig 60, developed by a vehicle builder in conjunction with Firth Vickers from their 301 17/7 material of 28 gauge. Section properties are shown in Fig 61.

Types 304 and 321 are the normal British Standard austenitic stainless steels. The 321 grade is stabilised with titanium to provide complete immunity from intergranular corrosion and improve high temperature oxidation resistance in such applications as exhaust systems. An 18% Cr 9%Ni austenitic alloy is also made for components requiring even higher corrosion and oxidation resistance; austenitic grades are non-magnetic. The British Standard designation for stainless steels is found in BSI publication PD6290 — the three digits of the grades mentioned above are used for definition. All 300 steels are austenitic and

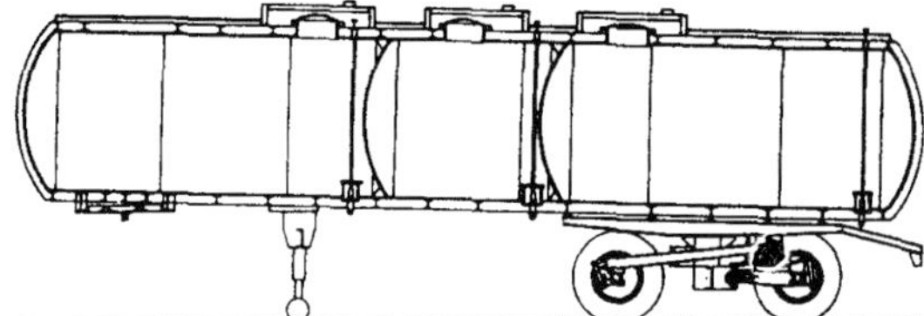

Fig 59: Frameless tanker in stainless steel

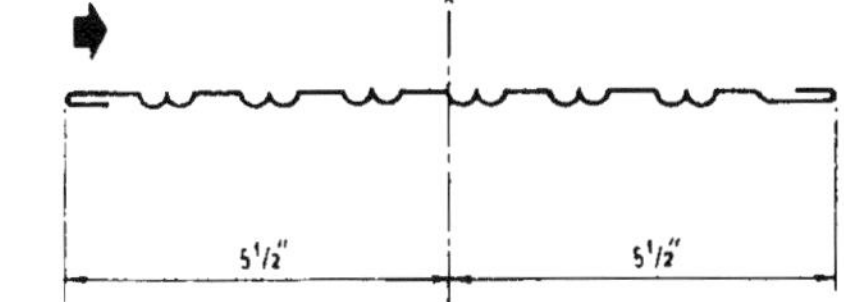

Fig 60: Cladding section

Properties of cladding section

Weight	0·63 lb/ft^2
Section area	0·1776 in^2
Inertia—about NA 'X-X'	1·958 in^4
Section modulus—about 'X-X'	0·356 in^3
Position of NA	Central in 11 in widths
Yield stress	30 ton/in^2
Uniform load capacity per span:	
48·4 tons	1 ft
5·3 tons	3 ft
1·3 tons	6 ft

Fig 61: Section properties

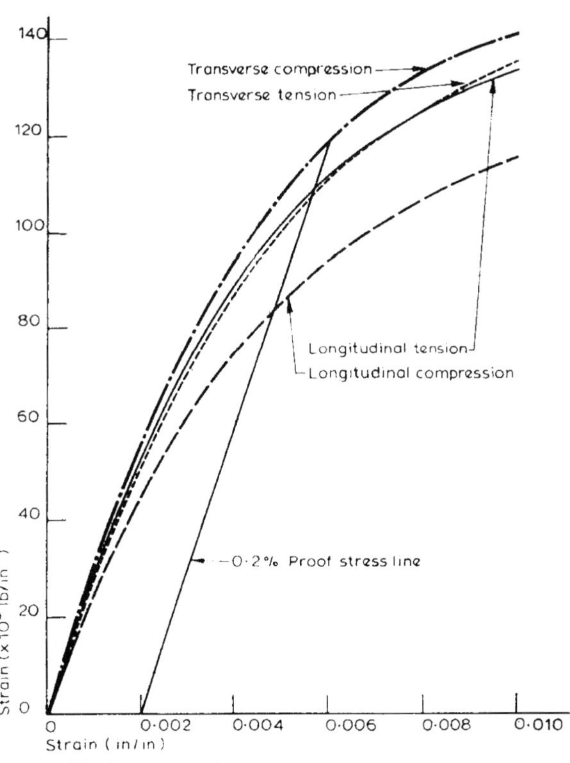

Fig 63: Stress-strain curves

the full range with suffixes is shown in Fig 61. Plate, sheet and strip are covered in BS 1449, billets and forgings in BS 970 and wire in BS 1554.

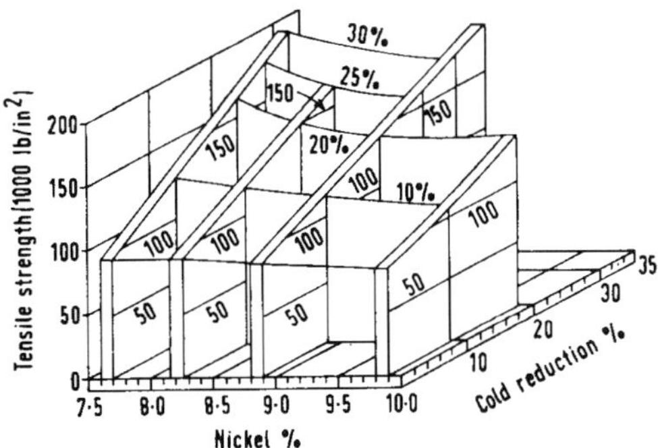

Fig 64: Effect of nickel concentration

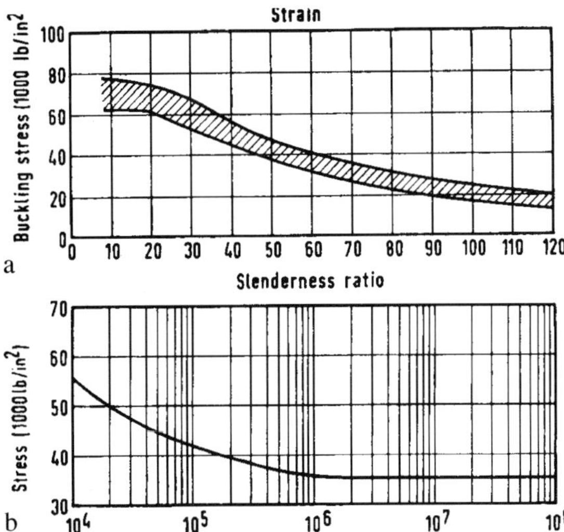

Fig 65: Critical buckling stress for 301 (above) and fatigue curve for 301 (below)

The stress/strain relationship of austenitic steels is unlike those of ordinary steels, Fig 62, there being no yield point. Working stress is usually taken as two-thirds of the 0.2% proof stress. It should also be noted that some anisotropy is found after cold working, the compressive properties in the direction of rolling being not strengthened as effectively as the longitudinal tensile or transverse tensile and compressive properties. Another peculiarity of stainless is that increase in compresssive stress level is accompanied by a falling off in elastic modulus. Struts should therefore be designed on the basis of tangent modulus obtained from the stress/strain curve at the operational stress. Fig 63 lists the wide range of surface finishes that sheet and plate forms of the material are made available from the rolling mills. Effect of nickel concentration is seen in Fig 64, due to Sandvik

According to International Nickel, the austenitic group is the most extensively used of all stainless steels because of good corrosion and heat resisting properties and its excellent weldability. If sufficient nickel is added to the chromium steels, the austenite is stabilised so that it is retained at room temperature, particularly advantageous as far as welding is concerned because it eliminates the need for any pre or post heat treatment, which ensures weld joints of good strength and ductility.

The material also has advantages under buckling, shear and fatigue loading. Fig 65a shows the behaviour of annealed 301 as a function of slenderness ratio while Fig 65b shows a typical *SN* curve. Unlike the ferritic steels, these materials are not particularly sensitive to notches.

BS Designation	En Steel replaced	Class*	Principal constituents (percentages)	Forms available†	Remarks
409S17	New grade	F/M	Cr 12, C 0.09 + Ti	1, 2	Low cost weldable type
420S37	56C	M	Cr 13, C 0.24	3	Not suitable for welding, hardenable.
430S15	60	F	Cr 17, C 0.10	1, 3	Not suitable for welding
301S21	New grade	A	Cr 17, Ni 7	1, 2	Work hardens most rapidly of austenitic grades
302S25	58A	A	Cr 18, Ni 9, C 0.12	1, 2, 3	Less suitable for welding than 304S16
303S21	58M	A	Cr 18, Ni 9 + S	3	'Free machining' grade
304S12	New grade	A	Cr 18, Ni 10, C 0.03	1, 2, 3	Low carbon variety more suitable for welding than 304S16
304S16	58E	A	Cr 18, Ni 9, C 0.07	1, 2, 3	
305S19	New grade	A	Cr 18, Ni 11, C 0.10	1	Work hardens least rapidly of austenitic grades
316S16	58J	A	Cr 17, Ni 11, Mo 2.5	1, 2, 3	Resists most common chemical loads
321S20	58B & C	A	Cr 18, Ni 9 + Ti	1, 2, 3	Titanium stabilized for welding in thicker section and high temperature service.

Fig 62: Range of 300 grade steels

Ferritic-stainless/galvanised steels

While nickel-rich austenitic stainless steels used in such applications as frameless tank shell construction are important in the context of both high structural strength and high resistance to corrosion, less costly ferritic types, such as Cromweld 3CR12, are now finding increasing application in special vehicle building. Meanwhile an increasingly used alternative to hot-dip galvanised steel is seen in the electrolytically zinc-coated cold-rolled steels such as British Steel's Zintec. The properties and potential of the two materials are compared here.

Weldability is one of the key attributes of Cromweld 3CR12, the relatively low cost stainless steel now used in a variety of structural monocoques which might once have been the province of the higher grade alloys of aluminium. By removing much of the anti-corrosion pre-treatment costs associated with mild pressing steel, its use becomes economically viable — particularly if effort is made to use the material in a structurally efficient way such that its high modulus, relative to aluminium alloy, can be exploited. In this context, the term 'stainless' is seen by its suppliers as applying to steels containing at least 10.5% chromium.

Best known in cutlery applications, early stainless steels had 12 per cent chromium content — but many other compositions have since developed. Main classification is between non-magnetic chrome-nickel steels and the magnetic straight chrome steels. However, there is a main sub-division of the latter category between martensitic and ferritic types. Of the last two, the former are heat-treatable but the latter not and, between the two, comes the comparitively recent addition of 3CR12 whose price, as unity, is compared with other materials in Fig 66.

Because the so-called chrome irons (ferritics) suffered from poor weldability, they became domi-

Material	Sheet (£/Kg)
Mild steel	0.4
3CR12	1.0
Ferritic stainless steel	
430 (17Cr)	1.5
Austenitic stainless steel	
304 (18Cr-10Ni)	1.8
316 (17Cr-12Ni-2.5Mo)	2.2
Duplex stainless steel	
2205 (22Cr-5.5Ni-3Mo-0.2N)	2.6
High alloy steel	
310 (25Cr-20Ni)	3.3
Ti-clad steel	7.5
Alloy 400 (Ni-32Cu)	15
Titanium (pure grade 2)	17
Alloy C276 (Ni-16Mo-15Cr-6Fe-4W)	25
Titanium (0,2Pd grade 7)	33
PTFE	38
Alloy 625 (Ni-21Cr-9Mo-2.5Fe)	40
Zirconium	45
Tantalum	250

Fig 66: Metal price comparison

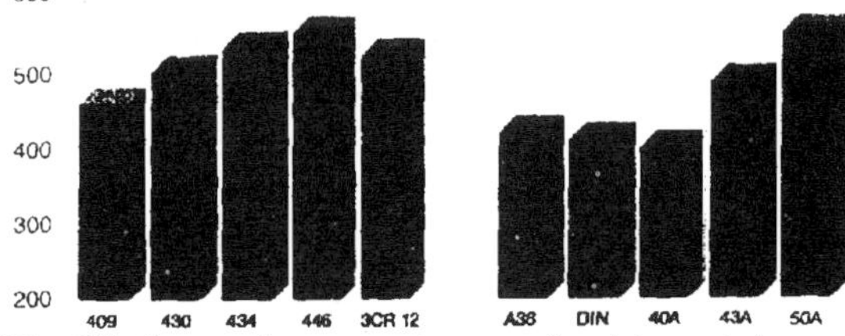

Fig 67: Strength comparison — ferritic stainless, left; structural steels, right

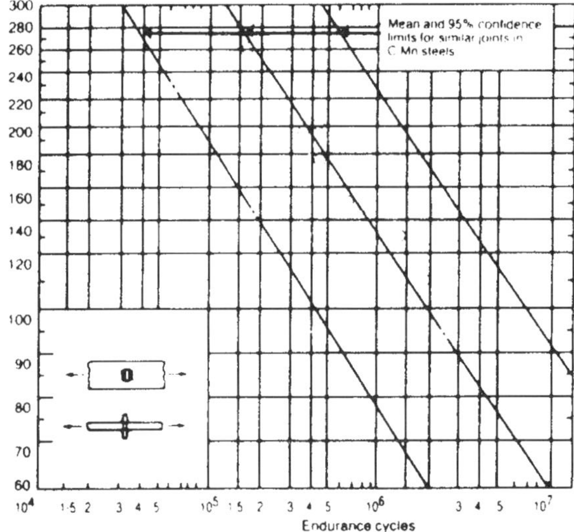

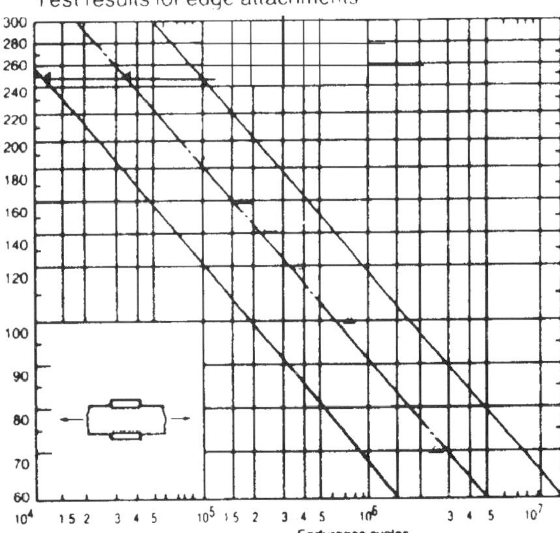

Fig 68: Notch toughness comparison

nant in very thin gauge applications such as specialist exhaust systems and trim — while the cutlery industry continued with martensitics. 'Utility stainless' is seen as the category for 3CR12, whose strength properties are shown as Fig 67 alongside competitors. Specialist vehicles incorporating these materials have shown that the ferritics represent about 10 to 20 per cent of the unladen weight; so that total vehicle price is only partially geared to price of the 'stainless'. The material can be welded in thicknesses up to 30 mm and it is available in 11 thicknesses of plate down to 3 mm. Lengths from two to six metres are available in widths from 1 to 1.5 m. Available sheet thicknesses are 4.0 down to 0.8 mm.

Notch toughness is as in Fig 68 while fatigue properties are revealed in the two sets of S-N curves shown in Fig 69 for different weld geometries. Fig 70 shows corrosion resistance performance compared with other materials — giving rate per year in microns for two, five and 10 year periods. Unlike the austenitics, 2CR12 is said to be immune from stress corrosion.

Electrolytic zinc-coated steel The substrate for BSC's Zintec material conforms to BS 1449: Part1:1983 and is available in various qualities for different depths of draw — as well as a high strength variant called Tenform. After coating, the material can be formed and fabricated in the same way as uncoated steel of equivalent properties. The coating, of course, offers 'sacrificial' protection yet does not impair welding and soldering. As well as Tenform high strength steel the material is also available with substrates of rephospherised steel. So-called Tenform PK has a forming limit diagram as shown at Fig 71 for a 0.7 mm sheet of 275 N/mm^2 yield stress material (B) compared with 180 N/mm^2 yield mild steel (A). Increases in press loads can be calculated by assuming a drawing force proportional to 0.5 x (yield + tensile strengths). Recommended resistance spot welding conditions for two Zintec sheet of 0.8 and 0.4 mm are shown in Fig 72a alongside typical spot-weld shear strengths in Fig 72 b.

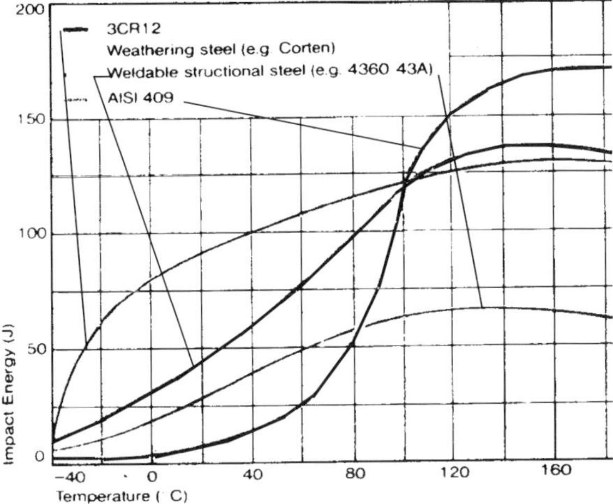

Fig 69: S-N curves for different weld geometries

Corrosion rate (in micrometres/year) after 2 years, 5 years and 10 years at various sites.

Period	2	5	10
Metal			
Mild steel	10,9	5,84	4,32
'Cor-Ten' A	6,10	3,56	2,29
Zinc	0,203	0,330	0,330
Copper	0,838	0,813	0,559
Aluminium			
3S	0,038	0,028	0,028
M57S	0,028	0,038	0,033
50S	0,051	0,056	0,028
B51S	0,046	0,028	0,033
Stainless steel			
430	<0.0025	<0.0025	<0.0025
304	<0,0025	<0,0025	<0,0025
316	<0,0025	<0,0025	<0,0025
3CR12 (Pickled)	0,007		

Fig 70: Corrosion resistance

Fig 71: Forming limit diagrams

Single Sheet Thickness mm	Weld Diameter mm	Shear Strength kN
0.7	4.2	4.0
0.9	4.7	5.5
1.2	5.5	8.0
1.6	6.3	11.0

a

Single Sheet Thickness Over mm	Single Sheet Thickness Up to and including mm	Electrode tip dia mm	Welding Conditions Force kN	Weld Time cycles	Weld Time s	Current kA
0.4	0.6	4	1.5	8	0.16	6.0
0.6	0.8	4	1.5	10	0.20	7.0
0.8	1.0	5	2.4	12	0.24	8.0
1.0	1.2	5	2.4	15	0.30	9.0
1.2	1.6	6	3.4	18	0.36	10.5
1.6	2.0	7	4.6	22	0.44	12.0

b

Fig 72: Spot-welding conditions and shear values

Magnesium alloys

While magnesium alloys have been used extensively in automotive applications, the historical impetus has come from the need for engine/gearbox weight reduction in such rear-engine vehicles as the VW Beetle car, AEC buses and a number of current Formula One racers. With more sophisticated alloying techniques and careful design analysis the material is finding important new structural roles.

It is still not widely realised that the cost per cubic inch of magnesium alloys is lower than for aluminium and steel; also it is highly competitive with zinc and plastics composites when specific strength is taken into account. As the material is abundant in sea water it seems likely that magnesium's future prospects are good, particularly when used as a metal matrix for fibre-reinforced composites. The material has also been used in CV body and cab construction in the USA, historically, and there are good prospects again in these areas for wrought products provided the properties are properly exploited in design. In the specialist vehicle sector, where specific performance requirements are available for the end-use, particularly good benefits are promised.

Supplying companies like Magnesium Electron, Norsk Hydro and Dow have been good providers of design and specificational data for optimising end-uses of the material. Considerable expertise has been gained in the demanding application of road wheels which have been a successful application. In high performance vehicles the 1.8 gm/cm^3 figure for

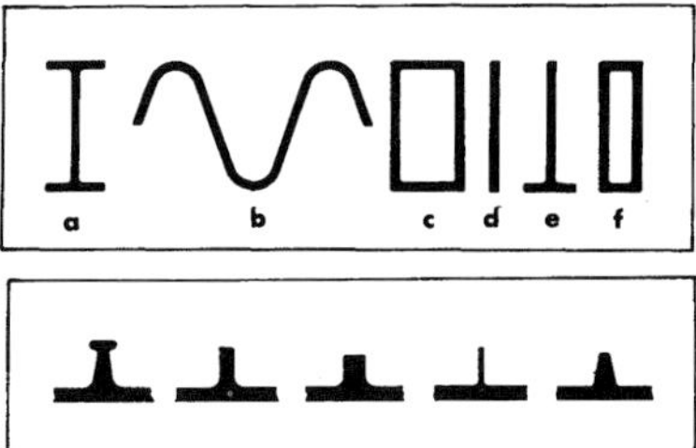

Fig 73: Section profiles

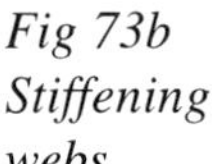

Fig 73b Stiffening webs

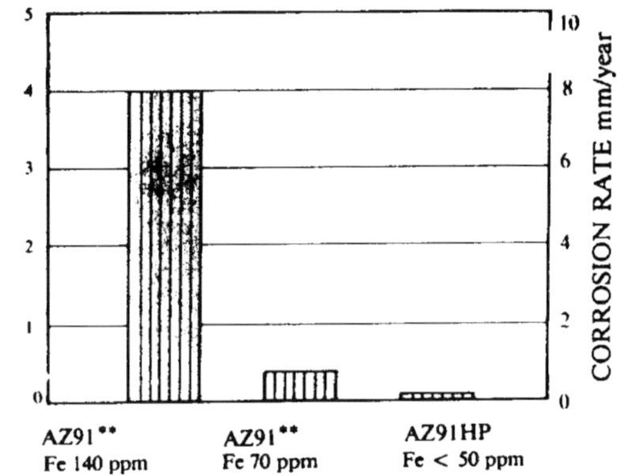

Fig 76: Corrosion rates compared

Alloy	Paint system*	Exposure time hrs	Scribe corrosion Creepage mm	in	Stone chip resistance Rating No.**
AZ91	A	240	7	0,25	—
AZ91	B	240	3	0,12	4
AZ91HP	A	240	0	0	—
AZ91HP	B	1000	0,5	0,02	1-2
AZ91HP	C	1000	0	0	1

Fig 77: Performance improvement with coating

	MAGNESIUM		ALUMINIUM		ZINC			REINFORCED THERMOPLASTICS		
	AZ91D Die Cast	ZE63A-T6 Sand Cast	LM24 Die Cast	LM25TF Chill Cast	BS1004A Die Cast	ZA8 Chill Cast	ZA27 Sand Cast	Lexan 3413	Torion 4203L	Zytel 70G33L
Yield Strength (MPa)	160	190	150	220	–	208	255	–	–	–
Tensile Strength (MPa)	230	300	320	280	283	240	320	130	191	124
Compressive Yield Strength (MPa)	160	195	–	185	414	210	255	120	172	414
Elongation (%)	3	10	1-3	2-5	10	1-2	8-11	3-5	15	14
Hardness (H_B)	63	60-85	80-85	90-110	82	85-90	90-110	–	–	–
Shear Strength (MPa)	140	–	185	205	214	241	225	72	–	86
Impact Strength (J)	2.0	0.5	3	8	58	–	58	100	–	133
Fatigue Strength (MPa)	97	115-125	138	95	47.6	51.8	103.5	40	–	–
Elastic Modulus (GPa)	45	45	71	71	–	85.5	80	8.6	0.7	–
Density (g/cc)	1.81	1.87	2.79	2.68	6.60	6.3	5.0	1.43	1.38	1.38
Specific Heat (J/kg°k)	1050	960	963	963	419	435	525	0.27	–	–
Thermal Expansion Coefficient ($\times 10^{-6}$/°k)	26	26.5	21.8	22.0	27.4	23.2	26.0	22	30	23
Thermal Conductivity (W/M°k)	72	109	96.2	151	113	115	122.5	0.21	3.1	–
Electrical Conductivity (% IACS)	11.2	30.9	24	39	27.0	27.7	29.7	–	–	–

Fig 75: Alloy properties

magnesium is a clear advantage over the 2.7-2.8 figure for aluminium alloys. The elastic modulus is 6.4 x 10^6 lb/in^2 (44GN/m^2) compared with 10 x 10^6 (69) for aluminium alloy so that deep, high modulus, sections are the rule in structural design. Resilience of the material is high and some alloys have particularly good damping capacity. Both these factors make the material a promising choice for impact-absorbing structures.

Typically, a magnesium alloy body panel would be twice as thick as a steel one but less than half its weight. The thickness will give benefits both in vibration reduction and resistance to denting by minor impacts. Fig 73a shows some extruded sections. Those at (a), (b) and (c) are preferred while (d), (e) and (f) should be avoided because of high edge stresses and difficulty in production. When using stiffening webs built in to the walls of castings, Fig 73b, those with smooth blend radii are preferred; thin webs and sharp corners should be avoided.

Tungsten arc welding can be used for all magnesium alloys; gas welding fluxes tend to be corrosive to the material. When spot-welding is used, spots should be placed at intervals of eight to16 times sheet thickness and be a distance four to10 times thickness from the edge of the sheet. While normally the surface carries a protective oxide film, contact with other metals should be avoided in damp conditions to avoid electrolytic action. Fire hazard only applies when the material is in very thin sections or finely divided chips. Even then they cannot usually be ignited below their melting point of 600 C.

In the design of flanged joints drawn up by bolts, flange faces should be generous in area and bolts should be spaced closely. Where bolts will be removed frequently, ferrous metal inserts are recommended. Rivets used with the material should be aluminium alloy with five to seven per cent magnesium and diameters should be three times sheet thickness, at most. Minimum pitching of rivets should be three times the rivet diameter and minimum edge distance twice the diameter.

Fig 75 shows a comparison of mechanical and physical properties of magnesium alloys with aluminium and zinc based casting alloys and reinforced thermoplastics — due to Magnesium Services. For three popular alloy grades, corrosion rates are compared in Fig 75 based on the ASTM B117-73 salt spray test. When finished with organic coatings, the performance increases to that of Fig 76. The company point out that alloy purity has no significant influence on galvanic corrosion and give the advice in Fig 77 for avoiding problems.

Trapping liquids at places where dissimilar metals are joined should be avoided and aluminium alloy washers from the 5000 or 6000 series should be used with thickness or diameter large enough to provide the clearances shown. When joining magnesium to other metals, fasteners should enter from the non-magnesium side, Fig 78, and aluminium shims should be used at the joint face. The use of shielding, as shown, should also be adopted.

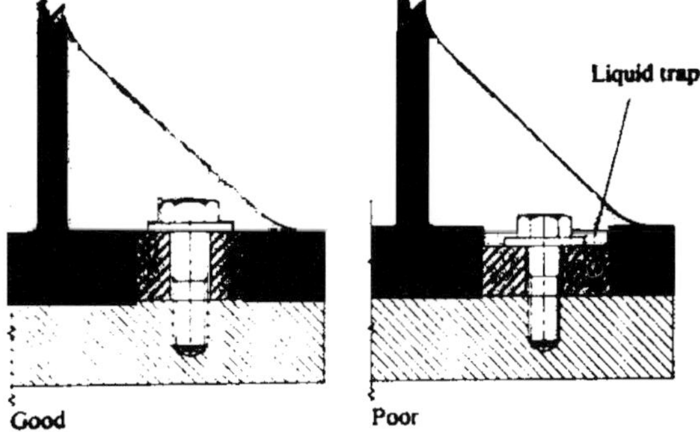

Fig 77: Avoiding trapped liquids

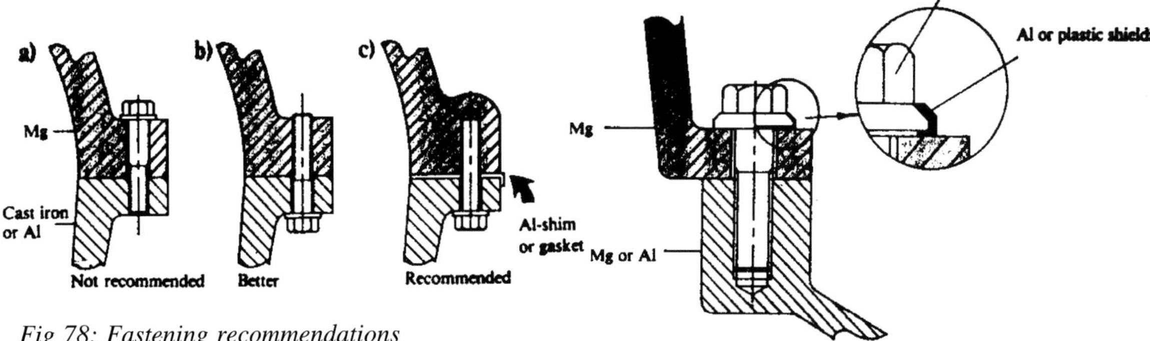

Fig 78: Fastening recommendations

Structural application of magnesium

Fiat engineers have described[3] an interesting structural application of magnesium alloy, in which a single-piece carrier beam under a dashboard which replaced an 18-part spot-welded assembly.

The successful exploitation of the material in this application arose from overcoming some of the traditional disadvantages of the material. High raw-material cost was overcome by a combination of factors, weight saving through optimum structural design; component integration and assembly cost reduction by optimum design for manufacture; and storage cost reduction through value analysis. A robust design in terms of section thicknesses and blend radii minimised the influence of the relatively low modulus on structural performance of the part which supports the dashboard, passenger-side air-bag, electronic system controllers, steering column and heater-radiator matrix.

The part was designed on Computervision's CADDS and CAE Engineering's IDEAS suite, as pre- and post- processor; NASTRAN and ABAQUS finite-element codes were also used. Success in crash performance prediction was proven in impact tests, only minor changes in attachment to the body-in-white shell being necessary. The three different fastening techniques finally adopted, Fig 79, included (a) screw rivets, (b) self-tapping screws and (c) double-screws.

The pressure die-casting production process necessitated the use of an open-section member, Fig 80, in place of the fabricated box-section of the original spot-welded steel assembly. A multi-web reticular structure, Fig 81, was the result; it achieved 50 per cent weight reduction; 80 per cent increase in XY bending stiffness; 30 per cent increase in YZ bending stiffness and 50 per cent increase in torsional stiffness, Despite the open-section. An AM60B alloy was used and on the basis of a 70000/year production run, using the cold-chamber die-casting process (Fig 82), 2-5 per cent cost saving was achieved over the steel assembly it replaced.

Fig 81: Multi-web section

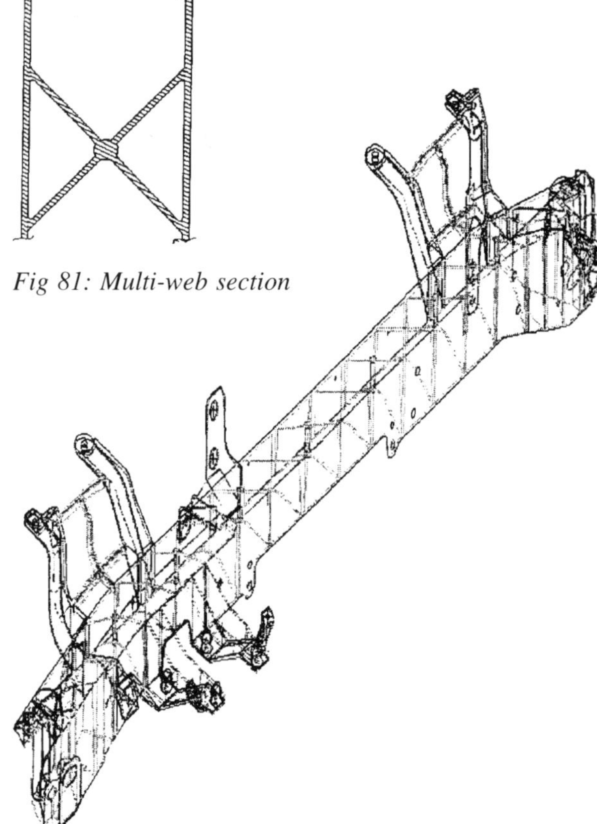

Fig 80: Open-section carrier beam

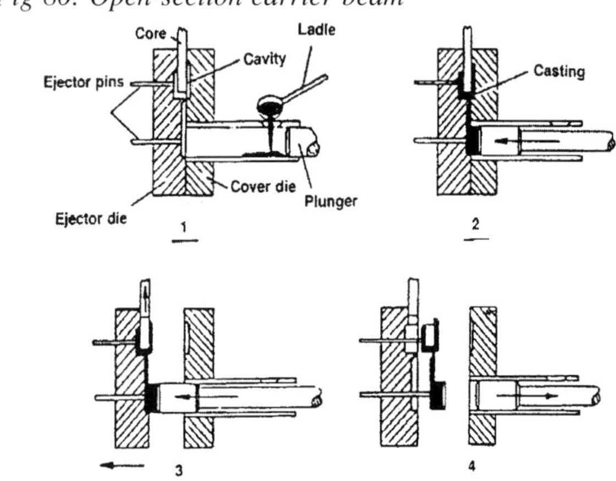

Fig 82: Die-casting process

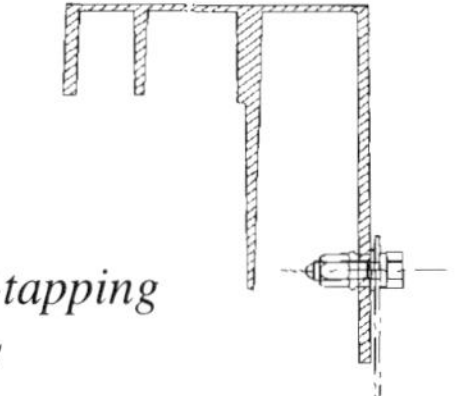

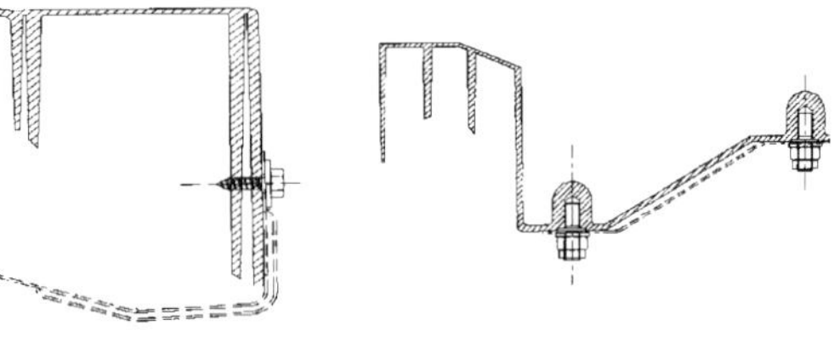

Fig 79: Screw rivets, self-tapping screws and double-screws

Copper alloys

Corrosion resistance, ductility, and, when suitably alloyed, fast machinability, characterise copper and its derivatives. The material is also important for its high electrical and thermal conductivity and even has valuable spring properties when alloyed with berylium. Brasses are, of course, one of the commonest copper alloys and in earlier times the material was important decoratively in automotive application.

Copper's usage in some notable architectural applications such as the window framing on Coventry Cathedral's famous curtain wall of glass prompts the question whether designers should reconsider the material for prestige vehicles such as luxury coaches and limousines. Widely used in the heat-exchanger cores of radiators, the material could have this role usefully extended to the walls of vehicle air-intakes if the long-heralded concept of surface-radiators comes about.

The material is already making a come-back for brake piping, particularly in passenger service vehicles, where very long service lives are required for this safety-critical application. There is serious consideration being given, too, on top-market cars as longer warranty periods are demanded — and, as quality expectations rise, even on more modest vehi-

Metal	BS No	Specification Designation	Metal Removal Rate cm³/min
Brass	2874	CZ121 Pb4	133
Aluminium	1474	FMA (2011)	80
		HE30 (6082)	44
Mild Steel	970	EN1A	36
Stainless Steel	970	304	6

Fig 83: Metal removal rates

Designation	Description	2870 Sheet Strip Foil	2871 Tube			2872 Forgings & Forging Stock	2873 Wire	2874 Rods & Sections	2875 Plate
			Pt 1	Pt 2	Pt 3				
CZ 101	90/10 Brass	•		(x)			[*]		
CZ 102	85/15 Brass	•		(x)			•		
CZ 103	80/20 Brass	•		(x)			•	(x)	
CZ 104	Leaded 80/20 Brass					(*)		•	
CZ 105	70/30 Arsenical Brass	(x)					(x)		•
CZ 106	70/30 Brass	•					•	(x)	•
CZ 107	2/1 Brass	•					•		
CZ 108	Common Brass	•		•			•		
CZ 109	Lead Free 60/40 Brass	(x)				•	(*)	•	(x)
CZ 110	Aluminium Brass	•		•	•				•
CZ 111	Admiralty Brass				•				
CZ 112	Naval Brass	•				•		•	•
CZ 113	Naval Brass, Special Mixture	Deleted—see CZ 133							
CZ 114	High Tensile Brass (low aluminium)					•		•	
CZ 115	High Tensile Brass (soldering quality)					•		•	
CZ 116	High Tensile Brass					•		•	
CZ 118	Leaded Brass 64% Copper 1% Lead	•							
CZ 119	Leaded Brass 62% Copper 2% Lead	•		•			•		
CZ 120	Leaded Brass 59% Copper 2% Lead	•							
CZ 121 3 Pb	Leaded Brass 57% Copper 3% Lead					•	(*)	•	
CZ 121 4 Pb	Leaded Brass 57% Copper 4% Lead					•	(*)	•	
CZ 122	Leaded Brass 57% Copper 2% Lead					•	(*)	•	
CZ 123	60/40 Brass (Low Lead)	•				(*)			•
CZ 124	Leaded Brass 62% Copper 3% Lead							•	
CZ 125	Cap Copper	•							
CZ 126	Special 70/30 Arsenical Brass	(x)		•	•				
CZ 127	Aluminium/Nickel/Silicon Brass			•		(x)		(x)	
CZ 128	Leaded Brass 60% Copper 2% Lead					•		•	
CZ 129	Leaded Brass 60% Copper 1% Lead					•		•	
CZ 130	Leaded Brass for Sections							•	
CZ 131	Leaded Brass 62% Copper 2% Lead							•	
CZ 132	Dezincification Resistant Brass					•		•	
CZ 133	Naval Brass (uninhibited)							•	
CZ 134	Naval Brass (high leaded)							•	
CZ 135	High Tensile Brass with Silicon					(*)		•	
CZ 136	Manganese Brass					(*)		•	
CZ 137	Leaded Brass 60% Copper. 0.5% Lead					(*)		•	
CZ 138							(*)		

Fig 84 (right and next page): Copper alloys and properties

cles. The brake-pipe application is more likely to be a copper-nickel alloy than a copper-zinc one and its protagonists argue that the long reign of tin/lead coated mild steel may be nearing its end due its limited life expectancy in severe operating conditions. Brasses containing around 37 per cent zinc are the so-called duplex alloys with excellent ductility at hot working temperatures used in forging and extrusion. Those most commonly cold-worked as sheet are the 30% zinc brasses. Free machining alloys are those which additionally contain 3 per cent lead to aid the chip-braking of the swarf which makes the high speed production of such items as pipe fittings so profitable. Fig 83 shows the remarkably high metal removal rate compared with competitive materials while Fig 84 provides the main categories and properties. Such specialist vehicles as fire-tenders and gully-emptiers are also big users of brass, arising mainly from its corrosion resistant properties. The availability of brass in precise preformed shapes such as extrusions, hot-stampings and die-castings eliminates much of the machining required by some of its less costly competitors. The alloying of copper with tin, to produce bronze, is important; when containing around 5 per cent tin and 0.3 per cent phosphorous the so-called phosphor-bronze becomes useful as a spring material — particularly when electrical conductivity (while being non-magnetic) is required at the same time. With higher than 0.3 per cent phosphorous, a different phase alloy is formed which has good bearing properties and finds uses in the bed-plates of turntables, for example. With 2 per cent berylium added to copper an even tougher alternative to phosphor-bronze occurs with better wear resistance. Heavily loaded bushes or springs with almost complete freedom from hysteresis can be made from the metal.

Composition															
Copper %		Tin %		Lead %		Aluminium %		Arsenic %		Iron %		Others %	Zinc %	Total Impurities %	Nearest ISO Designation
Min	Max	Min	Max	Min	Max	Min	Max	Min	Max	Min	Max			Max	
89.0	91.0				0.05						0.10		REM	0.40	CuZn10
84.0	86.0				0.05						0.10		REM	0.40	CuZn15
79.0	81.0				0.05						0.10		REM	0.40	CuZn20
79.0	81.0			0.1	1.0								REM	0.60	CuZn20Fe
70.0	73.0				0.075			0.02	0.06		0.06		REM	0.60	CuZn30As
68.5	71.5				0.05						0.05		REM	0.30	CuZn30
64.0	67.0				0.10						0.10		REM	0.40	CuZn33
62.0	65.0				0.30						0.20		REM	0.50 (Excl Pb)	CuZn37
59.0	62.0				0.10								REM	0.30 (Excl Pb)	CuZn40
76.0	78.0				0.07	1.80	2.30	0.02	0.06		0.06		REM	0.30	CuZn20Al2
70.0	73.0	1.0	1.5		0.07			0.02	0.06		0.06		REM	0.30	CuZn28Sn1
61.0	63.5	1.0	1.4										REM	0.7	CuZn38Sn1
57.5	60.5	0.60	1.25										REM	0.75	
56.0	59.0	0.2	1.0	0.5	1.5		1.5			0.5	1.2	0.3-2.0Mn 0.02 Max Sb	REM	0.50 (Excl Al)	CuZn39AlFeMn
56.0	59.0	0.6	1.1	0.5	1.5		0.2			0.5	1.2	0.3-2.0 Mn	REM	0.50	CuZn39AlFeMn
64.0	68.0					4.0	5.0			0.5	1.2	0.3-2.0 Mn	REM	0.50	
63.0	66.0			0.75	1.5								REM	0.30	CuZn35Pb1
61.0	64.0			1.0	2.5								REM	0.30	CuZn37Pb2
58.0	60.0			1.5	2.5								REM	0.30	CuZn39Pb1
56.5	58.5			2.5	3.5								REM	0.7	CuZn39Pb3
56.5	58.5			3.5	4.5								REM	0.7	CuZn38Pb4
56.5	58.5			1.5	2.5								REM	0.7	CuZn40Pb2
59.0	62.0			0.3	0.8								REM	0.30	CuZn40Pb
60.0	63.0			2.5	3.7						0.35		REM	0.50 (Excl Fe)	CuZn36Pb3
95.0	98.0				0.02						0.05		REM	0.25	CuZn5
69.0	71.0				0.07			0.02	0.06		0.06		REM	0.30	CuZn30As
81.0	86.0		0.10		0.05						0.25	0.80-1.40 Ni 0.80-1.30 Si	REM	0.50 (Excl Sn, Pb, Fe & Mn)	CuZn14AlNiSi
58.5	61.0			1.5	2.5								REM	0.50	CuZn38Pb2
58.5	61.0			0.8	1.5								REM	0.50	CuZn39Pb1
55.5	57.5			2.5	3.5		0.5						REM	0.75 (Excl Al)	CuZn39Pb3
61.0	63.0			1.0	2.5								REM	0.50	CuZn37Pb2
REM			0.2	1.7	2.8			0.08	0.15				35.0/37.0	0.5	CuZn36Pb2As
59.0	62.0	0.50	1.0		0.20						0.10		REM	0.4	CuZn39Sn
59.0	62.0	0.50	1.0	1.3	2.2						0.10		REM	0.2	CuZn37Pb2Sn
57.0	60.0		0.3		0.8	1.0	2.0				0.5	1.5-3.5 Mn 0.3-1.3 Si 0.2 Ni	REM	0.5 (Excl Sn Pb Mn and Ni)	CuZn37Mn3Pb2Sn
56.0	59.0				3.0							0.5-1.5 Mn	REM	0.7 (Excl Pb)	CuZn40Pb2Mn
58.5	61.0			0.3	0.8						0.2		REM	0.5	CuZn40Pb
80		0.5											REM		CuZn20Sn

Shape-memory alloys

Work on memory-metal actuators for automotive applications from Southampton University has been reported[5]. Low-cost SMA springs are said to be now available which alter in length rapidly as their temperature is changed and produce a larger force than bimetallic devices.

SMA springs can be thought of as bi-stable mechanical devices that witch quite rapidly between two temperature-controlled states. Possible automotive applications include engine cooling, fuel-injection and lubrication systems. A prototype actuator has been built at Southampton for control of a radiator blind.

With an SMA the deflection under load is elastic and reversible up to the elastic limit, like an ordinary metal, but beyond the limit the permanent plastic deformation can be removed by heating, Fig 85. The spring element is given its two-state characteristic by cycling a helical SMA shape through its transition temperature a number of times while the component is held by an external load. The effect of this is to build up a small amount of plastic deformation in the element, after which two-way memory behaviour appears. For a spring with open-coil shape above the transition temperature and a closed-coil one below it, reversion from one shape to the other is achieved by setting the temperature.

In a batch of springs made from 1.6 mm SMA wire, having 10 mm diameter and 10 mm length, measurements after warming in a water bath were recorded as in Fig 86, showing a doubling in length between 80 and 95 C. Here the transition temperature was 90 C but a different one could be obtained by varying the alloy composition. When cold the SMA springs are maintained compressed by a conventional bias spring. Rate for the hot SMA spring was noted as 0.9 N/mm.

For the blind actuator, six springs are used in parallel to obtain sufficient actuation force; these locate in pockets on a disc which is co-axially mounted with the plate holding the two bias springs and has an actuator rod protruding from it, and passing through the bias spring reaction plate. If a 20N load is applied to the actuator a 5-6 mm displacement, on heating, is developed. Fig 87 shows a load/displacement characteristic for a 4-spring actuator with bias springs alongside a theoretical prediction of performance.

To design an actuator for a given stroke, the stiffness, length and number of bias and SMA springs must be calculated. L_t, Fig 88, is determined from: $L_{CM} + L_{OB} - P$. When the actuator is in low-temperature state the SMA springs shorten until their coils are in contact, independent of any bias spring force. In the high temperature condition, there must be force between bias and SMA springs; forces produced by respective springs are $F_M = [L_{OM} - (L_{CM} + x)]M$ and $F_B = (x + P)B + W$. Since these forces are equal they can be equated to give: $W = M(L_{OM} - L_{CM}) - xC - PB$

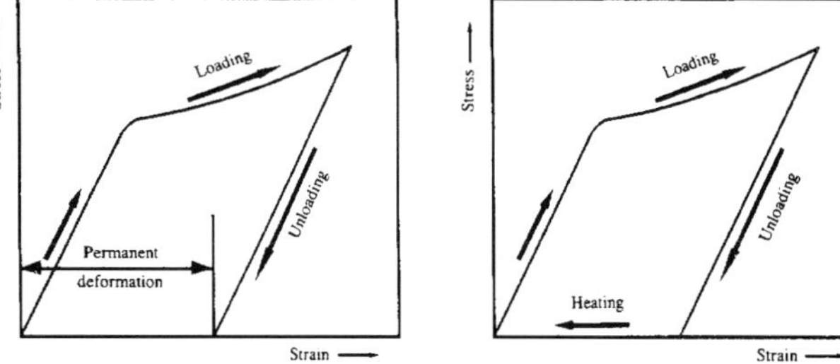

Fig 85: Thermomechanical properties of SMA

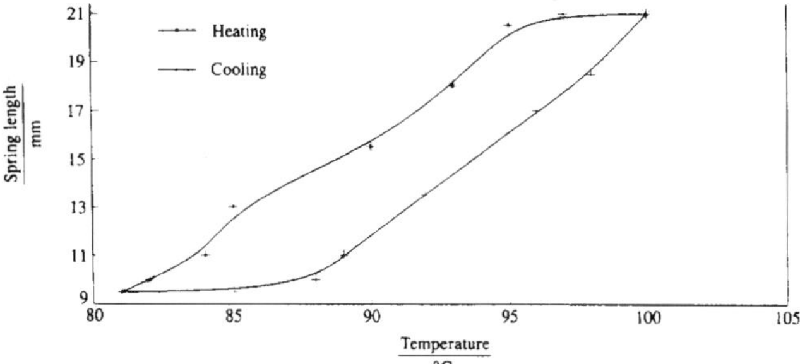

Fig 86: SMA spring hysteresis

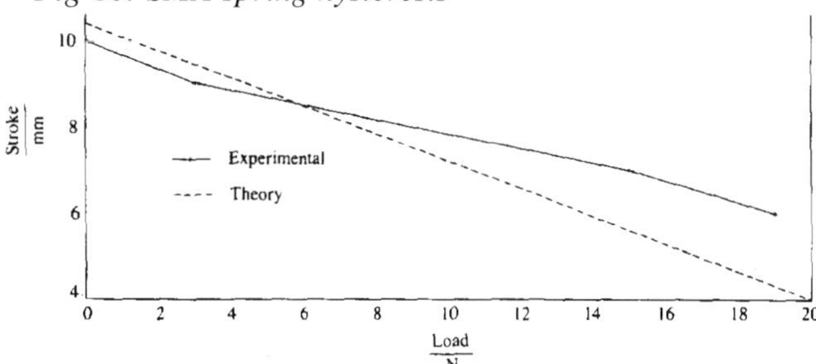

Fig 87: Characteristic of prototype actuator

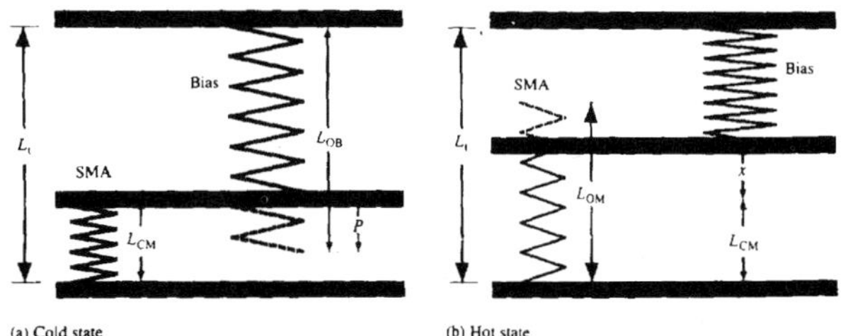

Fig 88: Notation for actuator dimensions

Bonded steel/elastomer systems

The French steel corporation Sollac has launched a rubber cored sandwich laminate which is showing automotive potential. Other suppliers offer interesting variants on this theme to give the possibility of formable thin sandwich panels.

Resilient sandwich

In recent years, the French steel concern Sollac[6] has marketed Solonconfort vibration damping steel sheet (VDSS) under a licence agreement with Nippon Steel. This is a thin sandwich panel of steel skins with a viscoelastic polymer core. Skin thicknesses range from 0.35-1.0 mm; core thickness is 45 microns and the maximum strip width available is 1650 mm.

Three different thermosetting polymers can be chosen for the core according to operating temperatures for the application, below 50 C, from 40-90 C or 70-120 C. Heat is developed in the core during damping of the vibration in the panel, by its shear deformation as the panel flexes, Fig 89. Skin materials available include cold-rolled steel, electro- or hot-dip- galvanised sheet.

Damping capacity of the sheet is defined by a loss factor expressing the ratio of dissipated energy to elastically stored energy, which is a function of temperature and frequency, Fig 90. The figure applies to a cantilever bar having a loss factor some 200 times greater than solid steel. The authors report substantial reductions in car rear-compartment noise levels when the material is used for rear wheel arch construction. In certain frequency ranges reductions up to 10 dB can be obtained.

Compromise has to be found between the depth of draw after forming the panel, and the consequent reduction in damping efficiency, with the chosen polymer, mechanical strength and bonding strength to the skin panels. A forming limit curve for Solonconfort is shown as Fig 91; while this is similar to a solid steel sheet, allowable strain is less. The core contains metallic particles which allows panels to be spot-welded without the use of a by-pass circuit. Using the material does not necessarily compromise weight saving, the authors explain. In the structural dash panel of the Mazda 929 car, the OEM claimed a 5dB lower vibration level while weight saving of 18 per cent was achieved.

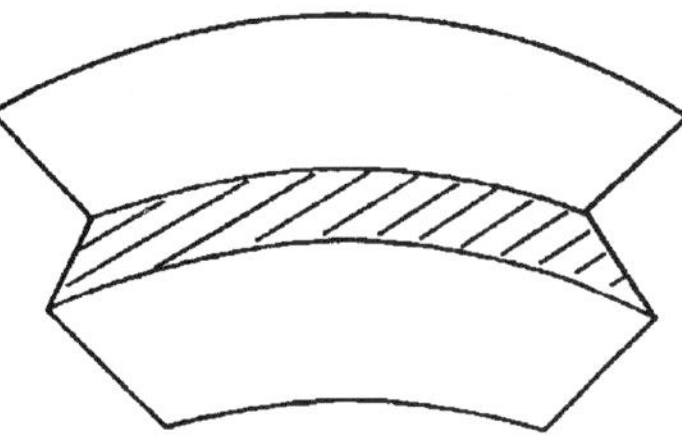

Fig 89: Shear deformation of core

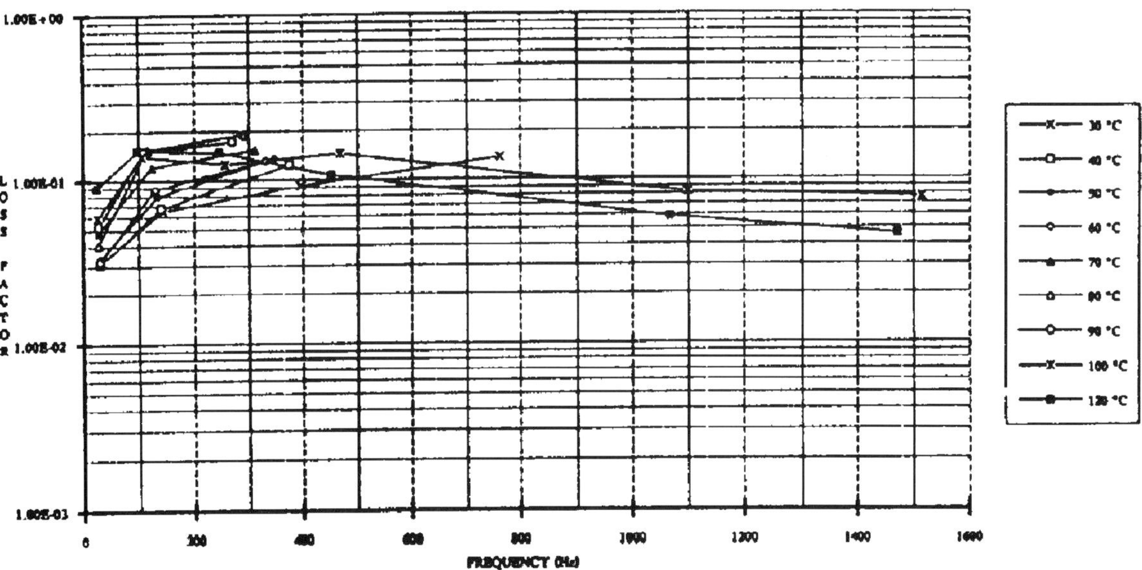

Fig 90: Damping loss factor

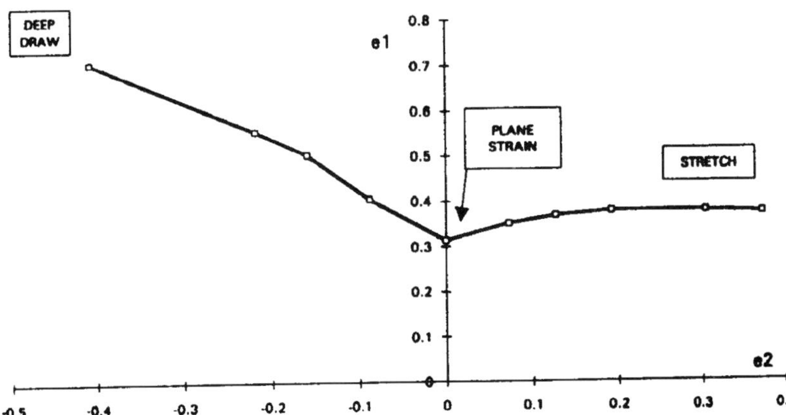

Fig 91: Forming limit diagram

Semi-rigid sandwich

The sandwich material Hylite (aluminium/plastic/aluminium) developed by Hoogovens Groep is claimed to be the lightest bodywork material outside exotic composites. It consists of two layers of 0.2 mm aluminium and a core of 0.8 mm polymer material. For equal flexural rigidity it is said to be 65 per cent lighter than steel, 50 per cent lighter than plostics and 30 per cent lighter than aluminium sheet. It can be deep-drawn on existing presses and is form-stable up to 150 C, which is of importance for painting. When mass produced, Hylite is more expensive than steel but is in the same range as aluminium and cheaper than plastic. The additional cost for one kilo of weight saving amounts to approximately one pound Sterling. An early application was the concept convertible based on Citroen XM shown in Fig 92. It also uses a combination of steel and aluminium extrusions developed by the company, and is claimed to save both weight and investment costs. The main problem in engineering a convertible is in ensuring the rigidity of the body. The XMple concept is claimed to overcome this by using a structured floor section made from hollow, thin-walled aluminium extrusion sections. The central part of the steel floor is replaced by aluminium, while the front and rear ends and the sills remain in steel. The hollow floor section on which the platform is based is 300 mm wide by 50 mm deep, with cross-ribs on the inside. An enclosed tunnel section, two side sections and two cross sections complete the structure which is welded together by robot. The complete floor is claimed to weigh only 50 kg, and the steel is joined to the aluminium floor which adhesive and monobolts. Rigidity is claimed to be similar to the original XM car.

Fig 92: XMple concept car

The plastic core material in Hylite is specified in a way that allows the shear modulus to retain an acceptable value over the operating range from -30 to +85 C; compared with sandwich cores made from viscoelastic materials, sound deadening properties are preserved throughout the operation range, the makers claim. The company have ascertained that a skin yield stress of 130 kN/m^2 minimum is required and have chosen an appropriate aluminium alloy grade, the 5182 designation containing 4.5 per cent magnesium. The PP core chosen is tolerant of paint stoving temperatures. This combination is also said to have satisfactory draw-ability with standard press-tools modified to account for the lower tear strength compared with solid aluminium alloy. Forming radii also must not be less than 5 mm and slightly curved drawbeads are needed to increase stretch level. Thus far the product is tolerant of pressing rates up to 50 mm/sec. A warm bending technique has also now been developed to enable radii down to 2 mm being achieved on the outer skin for flanging the edges of parts such as bonnet panels. For recycling the product a technique of cooling parts to -1OO C, using liquid nitrogen, is proposed at which temperature plastic and aluminium can be separated using a hammer mill, the materials having sheared apart by differential thermal expansion effect.

Chapter 6: Structural design: applying classic methods

Constraints on design; structural design basics; thin-walled structures; framework analysis; beam elements; stiff-jointed frames; torsion of box-beams; column and sheet stability; designing sheet steel structures; sandwich panel analysis

While advanced finite element techniques of computerised structural analysis are widely used for predicting crashworthiness of new body shell designs, it is also useful to apply traditional structural techniques in order to get a better feel for the design in the early concept stages.

Constraints on design

Expensive capital equipment used in forming and fabricating vehicle body panels has always to be in the mind of the vehicle body designer; legal and safety constraints on design must be understood, too. He or she must also be aware of how the structural concept for a new vehicle affects the load paths through its principal elements. In addition, vehicle design is becoming increasingly an exercise in packaging; trucks, vans, buses, cars and specialist vehicles all have to be designed within strict spatial envelopes. Power units, transmissions, suspensions, and support systems of all kinds have to be found a place among the occupants and payload goods. These have to be transported in comfort, safety, and often at very high speeds, and be allowed wide apertures for entering/dismounting or loading/unloading.

Rather than the structural requirements of the vehicle dictating the layout of the propulsion, transmission, and suspension systems, according to the optimum reaction of the load inputs involved, the tendency in recent years has been to leave structural considerations to later in the design process. In the early days of horseless traction, of course, it was different. When one looked at a vehicle one would nearly always see a stout chassis frame, not just wonder whether or not there might be one somewhere within. But then, of course, even the car chassis, with its sub-systems, was manufactured in one plant and sent to another for its coachbuilt superstructure, so clearly the technique and the design method were different. With the increasing importance of crashworthiness considerations, however, there is now a much earlier attention paid once more to the structure.

The other main purposes of the vehicle structure are, of course, to link up the mounting points for the vehicle's rear suspension and final drive, front suspension and steering, engine and gearbox, tank for fuel, and seats for occupants. It requires a rigidity to maintain accurate handling, lightness to reduce inertia and rolling drag, and toughness to sustain punishing fatigue loads from road, power unit, and driver. Most trucks still retain chassis frames; in buses they are being 'developed out' and in cars they have virtually disappeared. Even in racing and sports cars unitary construction has become the rule rather than the exception.

In this chapter we shall look at the basic scantlings of the structure and the purpose of the various elements, the way structures are developing, and application of the classical analytical methods.

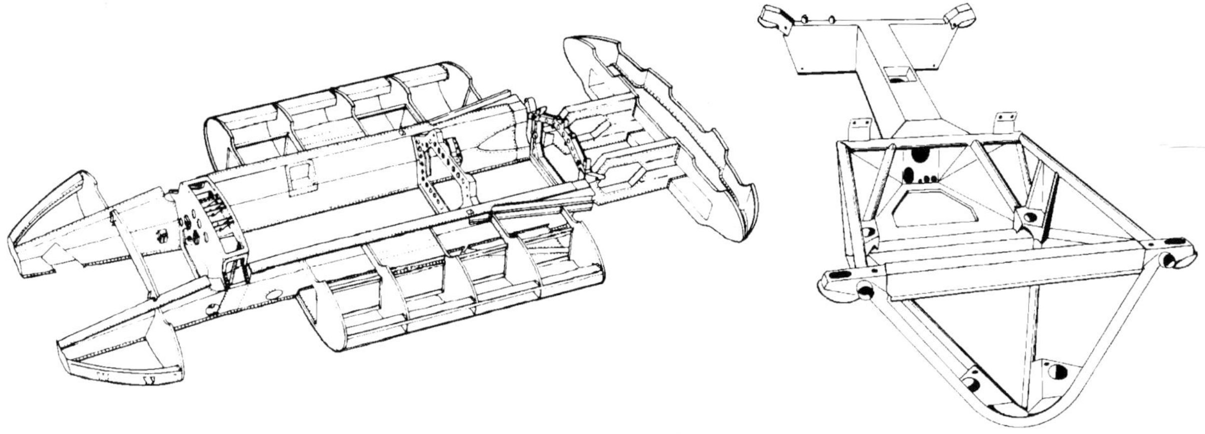

Fig 1: Sigma racing car structure

Fig 2: Lotus backbone chassis structure

Approach for the specialist builder

For the specialist builder the initial approach to 'shell' structure design can commence with a drawing layout aimed at optimum positioning of the main structural elements. The designer will need a minimum of four drawings to one fifth scale to depict front, side, and rear elevations, and a plan view. Since thin-walled shell construction involves large hollow closed members, additional cross sections might also be useful.

The first step is to draw in front and rear roll centre heights, from a suspension analysis, and the hub support members at either end. This enables the suspension linkage to be laid out and the vertical wheel travel determined. Next draw in front/rear axle centre lines, ground line, and undertray line to give adequate ground clearance at full bump. Reach diagrams for the drivers and occupants will help to fix pedal and hand control positions, from ergonomic theory, and the space left for packaging engine, transmission, radiator, and fuel tank can then be analysed. Structure design starts with the major transverse members or bulkheads, generally one each at front and rear axles and two in between for fascia and seat back, positioned according to whether the car is front or rear-engined. On a front-engined sports car the front bay is normally occupied by the power unit, the central bay by the occupants' legs, and the rear bay by the seats; on a rear-engined car, feet and legs are in the front bay, seats are in the centre, and rear bay is occupied by the engine/transmission.

Bulkheads through which driver's legs and engine/transmission protrude must have large cut-outs; these therefore need careful design to ensure that they are structurally stable. All the transverse bulkheads must be part of a structure which is rigid enough for minimum bending and torsional deflection. Large regular sectioned torque boxes are required in forming the longitudinals which must be deep enough for adequate bending rigidity. In the design shown in Fig 1 for the Pininfarina Sigma racing car structure, the two rearmost bulkheads are joined by deep box beams, themselves cut out in the webs to allow for movement of the drive shafts. The front three bulkheads are joined by more widely spaced box beams, in turn stabilised by outrigged half oval section members forming the fuel tanks and the streamline fairings for the wheels. In the critical area of the engine/transmission mountings additional members link the two support bulkheads, seen tapering in plan view. Beside it is a contemporary road car understructure in the 'high performance' category, Fig 2. The detail design of mountings on the bulkheads and longerons for engine/transmission, and all the others involving concentrated loads, is of crucial importance. Thin walled shell structure design is only successful if the full section proportions are exploited in resisting the loads, so that design of web stiffeners, stabilising rings, and spreaders need attention.

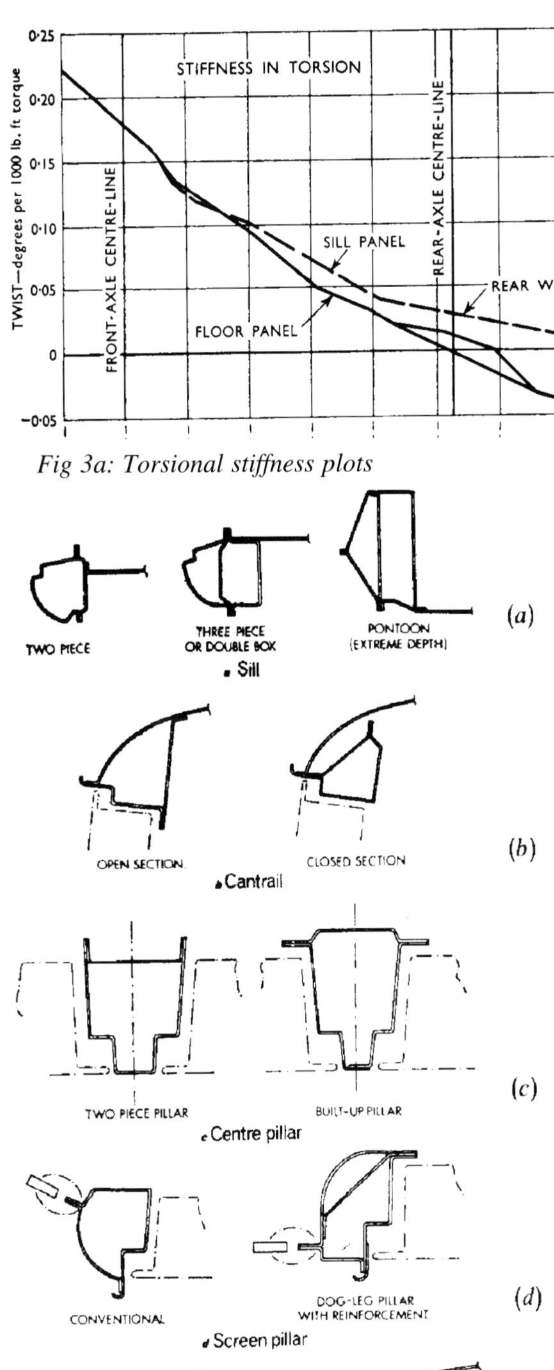

Fig 3a: Torsional stiffness plots

Fig 4: Main elements of the structure

Deflection as a design criterion

Colin Campbell[1] describes the limitations of the traditional chassis, likening it to a good bridge structure. He points out that whereas a good bridge is designed to sit on solid piers, a good chassis has to sit on a road that flutters and wobbles underneath—like the waves of a choppy sea. When diagonally opposite wheels lift simultaneously on an uneven surface severe racking strains are imposed and a considerable twist can take place. Torsional deflections of the somewhat flexible coach-built bodies and disturbance of the steering geometry could result in a rough ride. Further advances in vehicle development were to be an even greater threat to the traditional design of chassis frame. As car buyers, and later bus riders, demanded more comfort, softer suspensions had to be considered. These required much larger vertical wheel deflections than were possible with the beam-axled front (steering) suspensions of the early vehicles. Large vertical wheel motions with beam-axled suspensions would have meant severe changes in wheel inclination and the resultant development of high gyroscopic torques. Only by providing independent suspension of the front (steered) wheels could the high wheel displacements be achieved, which, with minimal camber and steer angle change, were the prerequisites of a soft ride. But with the discarding of the beam axle and single piece track rod of the coupled front suspension a new rigidity was required of the vehicle structure to which the independent front suspension linkages were anchored.

It is usual to express torsional stiffness in lbf ft/deg (Nm/deg) of twist between front and rear 'axles'. In a typical family saloon 4500 Ibf ft/deg (6100 Nm/deg) should be a minimum with 5000-5500 (6500-7500 Nm/deg) preferred. It is thus apparent that criterion for vehicle structures is normally for stiffness rather than strength, the structural analyses being of deflections rather than stresses. Mid span bending deflections for a car should not exceed 0.05 in (1.27 mm) and door aperture deformation should not exceed 0.05 in (1.27 mm) for a 1500 Ib (680 kg) mid span load.

Car structure classification

The main categories of structure are:

Integral: in which the main loads are carried by the members making up the structure. In practice this means that all the bodywork, except doors and hatches,

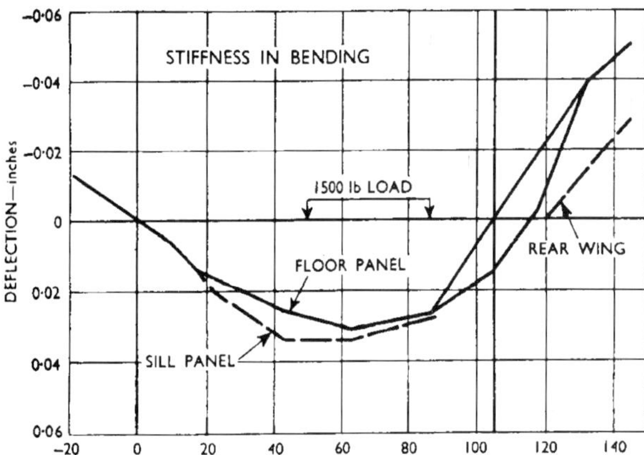

Fig 3b: Bending stiffness plots

is load carrying.

Semi-integral: in which there is a separate chassis frame but the remainder of the structure contributes substantially to the overall stiffness.

'Punt' structures: in which the floor is a box section with adequate torsional stiffness so that open-sports or lightweight GT bodywork can be fitted.

For the small family saloon, Platt[2] suggested that the ascendancy of the unitary (integral) body in Europe was due to the following:

1. More effective use of material resulting in weight saving or higher strength and rigidity for given weight.
2. Some saving in factory cost at the penalty of higher tooling cost.
3. 'All of a piece' construction giving freedom from destructive relative movements between chassis structure and body shell.
4. Well suited to the design of cars with low floors.

Platt points out that tests of torsional stiffness of complete shells do not merely provide absolute values of stiffness. By placing numerous dial gauges along the length of the structure, graphs can be obtained such as that shown in Fig 3 which would reveal any undesirable discontinuities in stiffness along the body.

The discontinuities would reveal possible failure points in road service under the action of cyclical racking strains as the vehicle negotiates uneven surfaces. Clearly the reinforcing at such points and repeating the stiffness test in search of negligible discontinuity would lead to a more durable structure.

The main objectives in structural design are to make the best use of material by arranging for each member to support as near as possible its maximum load potential and to make the structure direct and continuous by providing an unbroken path from point of application to point of reaction. For the specialist builder the advantages of mild pressing steel enjoyed by the mass producers are insignificant on a one-off special or short series build. By careful examination of the specific strength and specific stiffness of different materials, an optimum choice can be made, governed as well by the fabrication facilities available. In considering material in sheet form the flexural stiffness is a sounder criterion than the direct strain.

Typical metal thicknesses used in private car construction are 0.036 in (0.915 mm) for large area panels such as roofs, quarters, doors, and floors; these can reduce down to between 0.030 and 0.026 in (0.765 and 0.66 mm) where curved contours add stiffness and 'tightness'. In major structural parts such as cross members, sills, rails, and pillars, thicknesses of 0.039 and 0.049 in (1-1.25 mm) are common with 0.064 in (1.625 mm) used for local reinforcements.

Panels are normally spot welded together, typical strengths for given panel thickness being shown in Table 1. Pitch for spot welds normally varies between 1 and 2 in (25.4 and 50.8 mm) according to strength and sealing requirements; minimum pitch is

Table 1: Spot-weld strengths for given panel thicknesses

SWG	*Thickness (in) (mm)*	*Spot weld dia. (in) (mm)*	*Average shear failing load (ton) (tonne)*
20	0·036 (0·915)	0·18 (4·575)	0·63 (0·64)
18	0·048 (1·22)	0·22 (5·6)	0·80 (0·813)
16	0·064 (1·62)	0·25 (6·35)	0·88 (0·895)

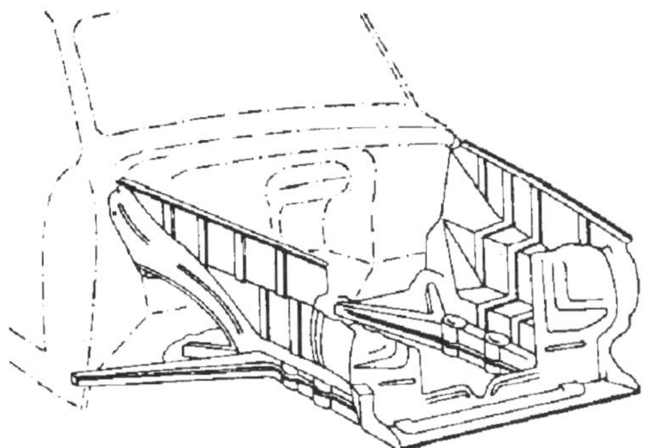

Fig 5: GM Holden front-end

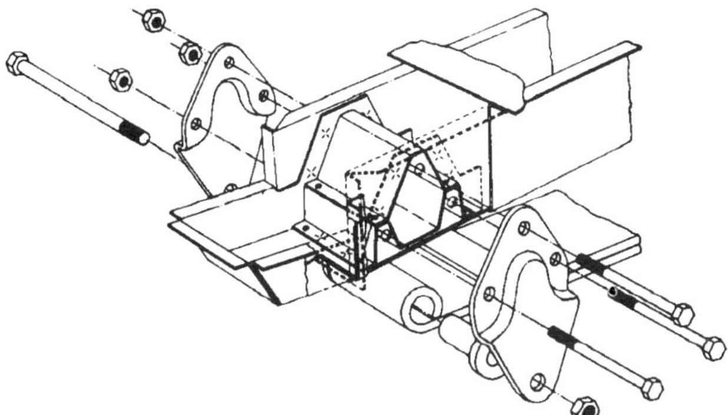

Fig 6: Structural attachment to box-member

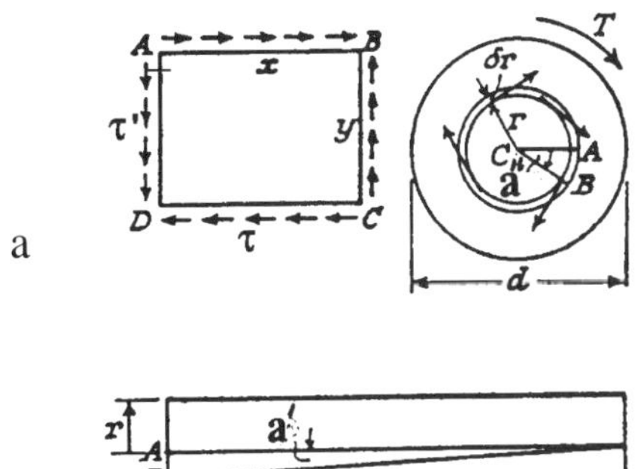

Fig 7: Shear in flat panel and circular section

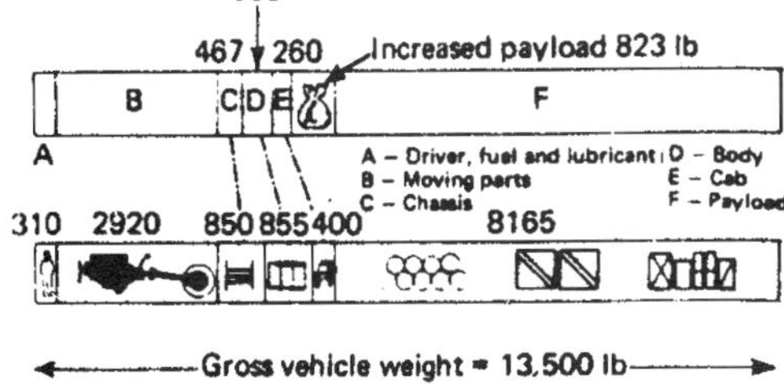

Fig 8: Economic rationale for structural efficiency

three times the spot-weld diameter.

In looking at structural configurations in integral structured cars Platt summarises the main elements of the structure (Fig 4) as an assembly of rails, sills, and pillars with practical sections as shown. He sites the GM Holden (Fig 5) to demonstrate the triangulation in elevation and plan designed to provide rigid connection between front suspension and bulkhead. This design had proved successful even on gruelling roads in the Australian outback.

Attachment of mechanical units to an integral body needs special attention to the spreading of load in avoiding local distortion and failure, also in limiting undesirable relative movement. The rear spring front hanger bracket is an important area in conventional leaf sprung front-engine/rear-drive layouts. This mounting has to transmit braking and drive loads horizontally as well as vertical loads due to weight and braking/driving torque reaction, and side loads due to cornering. The example illustrated in Fig. 6 shows that the boxed (longitudinal) body member has an internal stiffener and side plates with through bolts.

Structural design basics

Solid elastic materials such as steel and, to a degree, aluminium alloy or even plastics heavily reinforced with elastic materials, can be readily examined in respect of internal loads and deflections. This is on the assumption that stress σ (internal load per unit area) is proportional to strain ε (change in size per original size).

It is useful to think in terms of SI units because the gravity force on one kilogram load is 9.91 (approximately 10) and most units in the system are expressed in multiples of tens, hundreds, thousands and so on. Internal load is assumed distributed equally across the section of the solid body under strain in simple, linear, stress analysis and the ratio stress/strain is the elastic modulus, E, 70 kN/mm^2 for aluminium.

As well as direct stress, tensile and compressive, shear stresses often need to be considered. Here a shear modulus, G = stress/strain, applies. Note that any element under shear stress τ, Fig 7, is subject to a complementary shear stress, τ' — arising from the force balance on the element, to maintain equilibrium, as:

$\tau_{xz} y = \tau'_{yz} x$

This is important in materials like timber which are weak in shear along the grain.

A structural element under twist is subject to shear strain. As well as strain on the transverse sections of the element, there will be complementary shear straining of the longitudinal planes. In Fig 7b, twist α over length l can be equated to the shear angle α'' and radius r of the symmetrical transverse section, and the strain expressed as stress in $\tau/r = G\alpha$. The torque T on the element can be equated to the sum of those of each tangential stress, and the expression for the polar moment of inertia, J, substituted, to obtain $T/J = \tau/r$ showing that, for a given torque, shear stress is proportional to the radius and the torsional stiffness is:

$$T/\alpha'' = GJ/l$$

Structural efficiency

Immediately after World War 2, it was realised in the USA that lessons learned from aircraft design could be applied to ground vehicles. In a 1945 SAE paper presented by Mac Short[3], a particularly far-sighted view was presented. Fig 8 reproduced here demonstrated potential savings made possible by using structural design techniques. The example applied to a truck described in US units of measurement. He showed that by careful use of aluminium alloy, 30 per cent weight saving could be achieved in the cab and 35 per cent in the body and that 15 hours of stress analysis would be required per pound weight of structure saved, a figure that has been subsequently slashed by the emergence of computers.

Part of the suggested approach is to draw up load envelopes as shown in Fig 9 for torsion, bending and shock loads. Another key factor is use of careful design at the points of load application, to avoid structural discontinuity, by spreading concentrated loads as much as possible. Once stress intensities have been kept low by these means, the other main design objective, with monocoque structures, is to avoid buckling of the thin sheet materials. The immediate attraction of the lower density materials, for such purposes, is shown in Fig 10. This illustration merely reflects the effect of increased thickness on a plane panel under end load; even greater buckling resistance is, of course, obtainable with framework-stiffened panels and sandwich construction.

It is a good discipline to consider the purpose of the vehicle structure on starting a new design: to link up the mounting points for suspension and drive systems with the support platform for passengers and/or payload. Rigidity of the structure is required to maintain accurate handling; lightness to reduce inertia and rolling drag; toughness to provide fatigue resistance against vibration and shock loads from the ground surface.

When diagonally opposite wheels lift simultaneously considerable torsional loads can be applied to the vehicle structure. With the trend to softer suspensions, involving wheel displacements that could cause steer fight if careful geometric control is not maintained, the rigid disposition of suspension anchorages is paramount. In Fig 11, for spring stiffness c (N/mm) and road displacement d, load transmitted to the structure is cd. If torsional stiffness is C, torsional deflection of the chassis $D = cd/C$ (in mm). For spring rate 15 N/mm, a torsional stiffness of over 2000 N/mm will be required to limit structure deflection at the suspension anchorage to 1 mm when negotiating 100 mm road bumps.

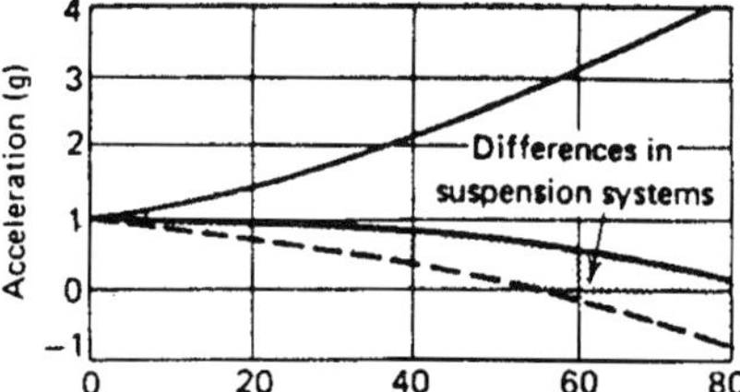

Fig 9: Load envelopes per mph of road speed

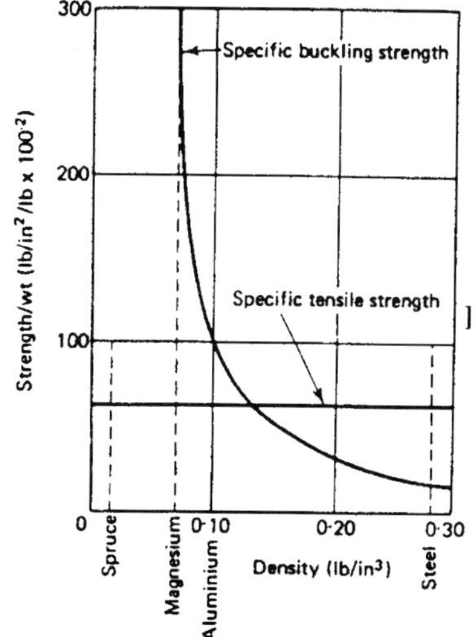

Fig 10: Structural efficiency of lightweight materials

The use of hollow closed section members for withstanding torsional loads can be visualised by considering the effect of slitting the tube along its length and twisting it. Some two thirds loss of stiffness results from the ability of the free edges to slide with respect to one another. For the rectangular section closed tube shown in Fig 12, the angle of twist in radians per unit length is given by:

$$T(b' + d') / 2Gb'^2 d'^2 t$$

and average shear stress by:

$$T/2tb'd'$$

There will be a local increase in the stress levels at the internal corners according to the graph shown in Fig 13. In many cases the structural member is loaded by a combination of stresses: direct stresses, direct stresses due to bending and shear stresses. Normal practice is to add direct to bending stresses; add bending stresses in planes at right angles to each other and then add shear stresses to the above. This results in principal stresses, in tension/compression and shear respectively:

$$(\sigma/2) +/- [(\sigma/2)^2 + \tau^2]^{1/2}$$

$$[(\sigma/2)^2 + \tau^2]^{1/2}$$

Finite element method Division of the structure into small parts (finite elements) allows analysis by the FE method. The SSS technique, introduced in the next chapter, involves a form of FE idealisation but, as explained there, the relatively 'course mesh' division into large parts allows a good appreciation of the behaviour of the main elements of the structure so important in design. Fine mesh division into small parts, either planar or solid, allows a more accurate prediction of the structural stress or deflection pattern.

The method is an extension of matrix structural design techniques also to be considered in a later section. In its simplest form, plane triangular elements are used and stress strain relationships, described above, used to obtain equations which are solved by a computer for structural displacements and stresses. Fig 14 shows some panel elements used and Fig 15, an early representation of a car body made from both panel and beam elements. The numbers show the node points of the model where the elements interconnect. Typically six equations, corresponding to three perpendicular force and torque balances at these points would define the internal loading of the structure.

For 'solid' shapes and multi-curved surfaces, more complicated elements, including three-dimensional ones, can be used. These involve a greater number of equations to be solved simultaneously. The use of matrices to obtain computer solutions is really seen to advantage in such cases. Proprietary computer programs for such purposes are, of course available. One such, under the acronym of PAFEC (Program for Automatic Finite Element Calculation) is available from the well known company of that name founded at the University Science Park, Nottingham.

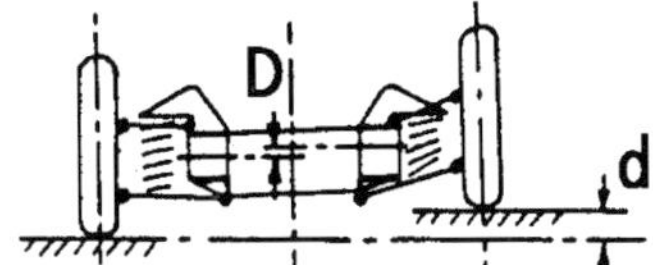

Fig 11: Twist due to frame and suspension deflection

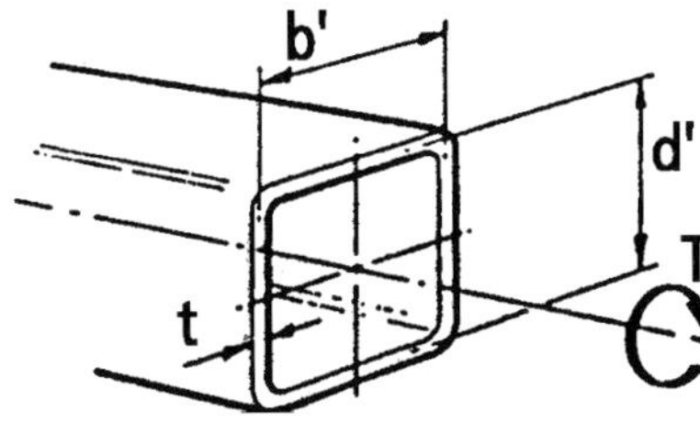

Fig 12: Twist in rectangular section tube

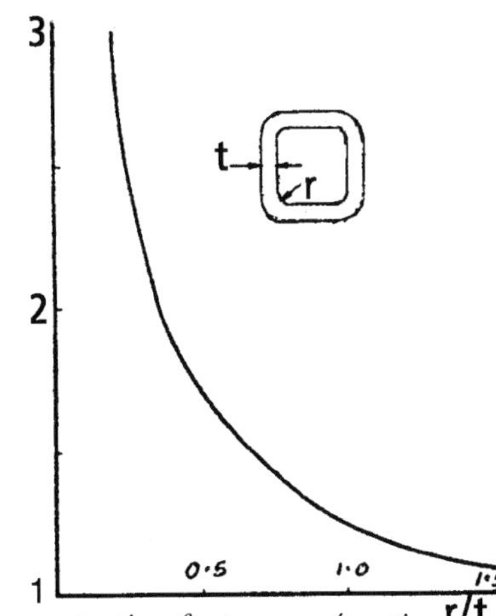

Fig 13: Stress concentration factor vs r/t ratio

Thin-wall structural analysis

The many simplifying assumptions that can apply to panelled structures make them surprisingly easy to calculate for loads and deflections. In applying the basic rules of structural mechanics, the designer soon develops 'structural feel' which is invaluable in visualising new concepts. Structural engineers have developed the design technique of 'thin-walled-structures' from the early theories of aircraft stress analysis and achieved surprisingly accurate results.

The first section of the chapter introduced one of the approaches to analysing thin-wall structures, the representation of one surface of a load bearing structure by a rectangular panel enclosed by end-load carrying bars at its edges. There are a number of other simplifications which can help turn real life structures into ones which are readily solved by uncomplicated calculation techniques. As well as assuming panels only carry shear and tensile forces in their own plane, it can also be taken that all loads can be ignored to which the thin wall is flexible or would collapse by buckling. Flanged edges of panels, and intermediate stiffening swages, can be represented by end-load carrying area added to the cross-sections of those used to idealise the basic surface panel.

While the application of simple stress/strain relationships to any structural design problem in a vehicle body might suggest the use of very fine gauges of metal, the practical limitations of sheet handling and accidental damage caused by highly

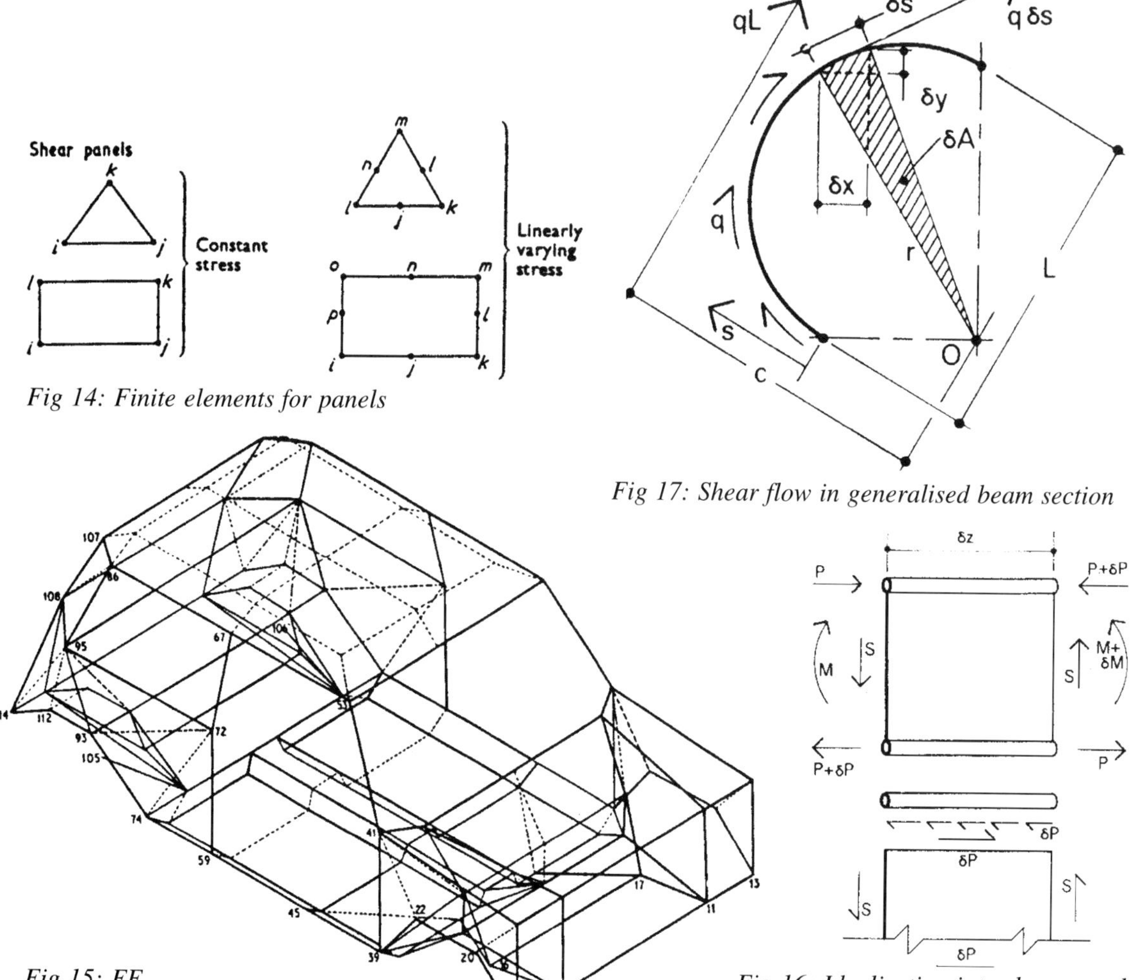

Fig 14: Finite elements for panels

Fig 15: FE representation of car structure

Fig 16: Idealisation into shear panels and end-load bars

Fig 17: Shear flow in generalised beam section

concentrated loads normal to the surface set minimum gauge thicknesses. In the case of volume-built cars and vans, the general body material employed is deep-drawing mild steel to EN 2A/1 with an ultimate tensile stress of 45,000 lbf/in^2 (310 x 10^6 N/mm^2) and elastic modulus of 30 x 10^6 lbf/in^2 (210 x 10^9 N/mm^2), 20 gauge being the common minimum thickness for general purpose. Many specialist vehicles use special materials and therefore thickness is much more variable; so too are the structural design criteria for the wide variety of possible special-purpose bodies.

Beam structural elements and bending theory are to be considered in a later chapter; that theory is not required for short thin walled beams as the principal panel loads are shear ones. Thus a small element of a thin-walled beam can be considered as in Fig 16 for design purposes. The magnitude of 'pure' shear stress in the panel, $t = S/ht$ for thickness t and shear force S. It is useful for calculation to express the shear in 'flow' terms: shear flow = force per unit length, so that the shear flow, $q = S/h = tt$. Under the action of the bending moment M longitudinal stress in the bar $s = P/A$ where A is the cross-section area of the edge bar or 'boom' in thin-wall parlance. Therefore $s = M/DA$ where D is the distance between booms. The moment of inertia, or second moment of area, of the whole beam cross section $I = sAy^2$ where y is the distance of the boom area from the bending axis (the centre, for a symmetrical section beam). At any distance L along the beam, from the point of application of S, as $M = SL$, boom end load $= qL$ and it is assumed, for short (elemental) lengths of the beam, elemental force $dP = dM/D = Sdz/D$ so that $dP/dz = S/D = q$. This implies that the shear flow in the 'web' of the beam is equal to the rate of change of end load in the boom.

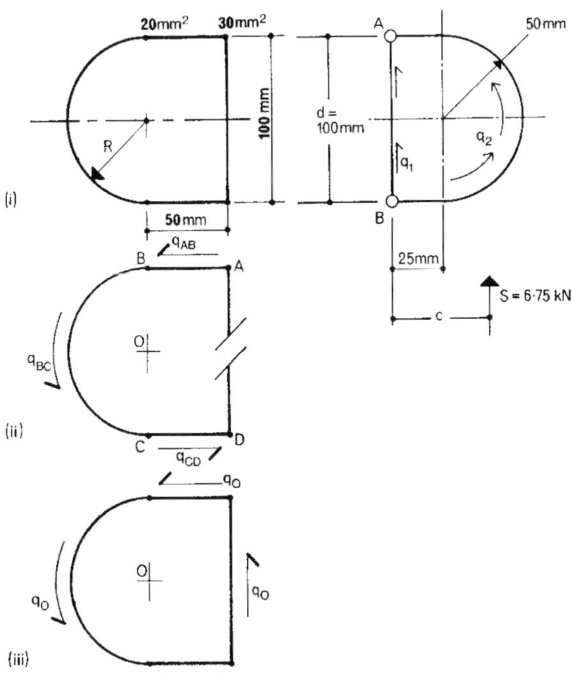

Fig 18: Car or van sill section idealisation

The assumption of constant shear flow in the web is also made for a curved web and an important theorem by Peery[4] showed that for a beam section such as that shown in Fig 17, twice the area enclosed by the curved web $[A]$ divided by 'chord' of the curve (100mm in this case) gives the distance c - moment arm of the resultant of the shear flow in the curved web. A car sill beam could well be idealised into the section shown in Fig 18. Generally, in closed section thin wall beams, the idealisation is made to enclose the shear panels between booms - the shear flows in the panels being in equilibrium with the externally applied load. This example is solved by using the expression $S/D = q$, derived above; with the two shear panels, in this case, $q_1 D + q_2 D = S$ so that $q_1 = (S/D) - q_2$ and in taking moments about boom B, $Sc = 2q_2 [A]$ giving:

$$q_1 = (S/D) - Sc/2[A] \text{ and } q_2 = Sc/2[A]$$

Then by substituting the numerical value of $[A]$, and choosing thicknesses of 3 mm and 1 mm for straight and curved webs, these two equations can be solved simultaneously for the condition of equal shear stress in the webs to find the value $c = 33.33$ mm which meets this condition. Thus, if the beam was to be loaded within its span, this would be the most efficient position across the section to fix the load attachment.

Another example to consider is a rectangular section beam having, say, four angle corner flanges of area A_f, depth d, breadth b and corresponding panel thicknesses t_d and t_b. This would be idealised by a four-boom box beam with boom section areas:

$A_f + (bt_b/2) + (dt_d/6)$

Calculation technique starts by making an imaginary cut to make the closed section momentarily into an open one by removing an element of the section - then to find 'open-section' shear flows. Next an opposite sense shear flow q_o is introduced on all elements, equal to the 'resultant shear flow' q^*, on the cut element. Final stage is to balance moments of the combined shear flows against the moment arm of the externally applied load. The q_o flows are assumed to set up torque $T = q_o.2[A]$.

Here it is useful to introduce an extended form of the expressions for torque and twist, given in the later section on torsion, for solid and thick walled sections. Akin to the Peery expression above, it can be shown for a closed thin-wall section member that twist per unit length:

$$\alpha''\{q/2[A]G\}\sigma(s/t)$$

and torque, $T = q2[A]$, for angle of twist a'', shear modulus G and s/t summed for elements around the section such that in the generalised box tube of Fig 19,

$$a''/l = [T/4(bd_2)G](b/t_1 + d/t_2 + b/t_3 + d/t_4)$$

Reverting to the box beam of the preceding paragraph, it is now possible to use these expressions to obtain $\sigma q^*(s/t) = 0$ when equating the twist per unit length to zero (as this symmetrically sectioned beam is only in bending and not torsion).

A triangular section beam can be evaluated in the same way; Fig 20, with equal boom areas 1,2,3 = a, is an example for which shear panels have width h and thicknesses t_l and t for vertical and angled panels respectively. To find the position across the section at which an external load will cause no twist of the beam (known as the 'shear centre'), the right hand panel is removed to first obtain open-section shear flows; but by taking moments about (2), q_{12} and q_{23} are isolated and q_o is balanced by the moment of the external force S at the shear centre, with offset e. By putting $S = I$,

$$(\alpha h^2/2)e = q_o h(3)^{1/2}/2.$$

Then by putting $a''l = 0$, as open-section shear flows $q_{12} = q_{23} = ah/2$, the value of e becomes:

$$(3)^{1/2}ht_l/(2t_l + t).$$

At a joint between box members such as the important rear-quarter to sill junction of a car, Fig 21a, consideration should be given to load transfer at the corner. In figures taken from the design of the Austin A30 structure by TK Garrett[5], bending moment of 51 000 lbf in (5.76 kNm) induced end loads of 33 000 lbf (146.7 kN) in the top and bottom flanges of the 5 in (127mm) sill. The 18.5 MN/m^2 safe working stress steel used in this application dictated an area of 0.123 in^2 (0.795 cm^2) edge material - a 3.4 in (86.5mm) wide flange of 20 gauge sheet. The webs of sill and quarter panels overlap as shown by the dotted lines to provide the necessary transfer of the 1900 lbf (8.45 kN) shear force applying in this example. It is also important to maintain the cross-section shape of thin-walled box-beams, both along their length and, crucially, at the corner joint where a diagonal diaphragm (bulkhead) might be required.

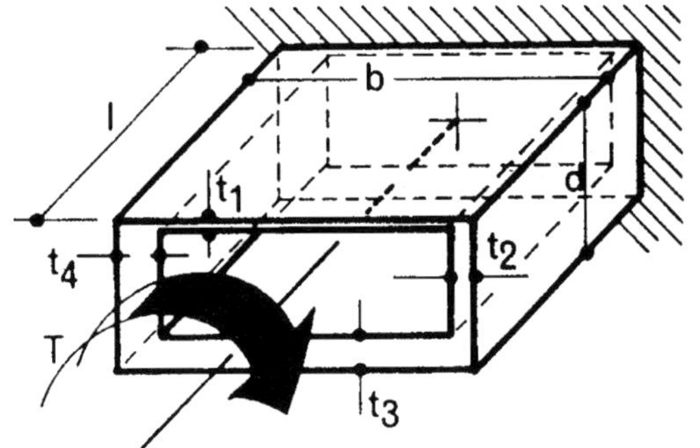

Fig 19: Generalised box-tube section

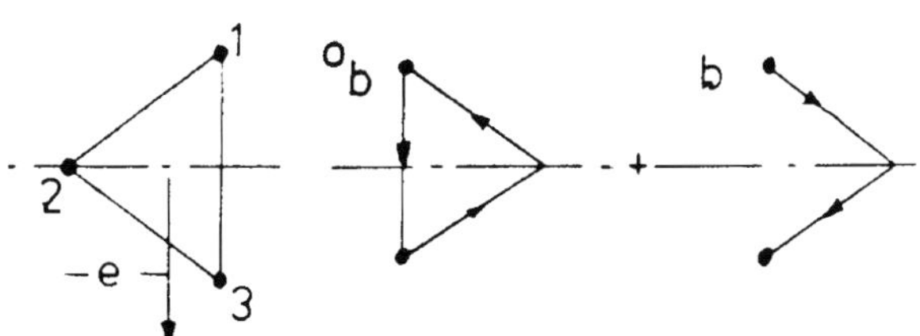

Fig 20: Triangular section box beam

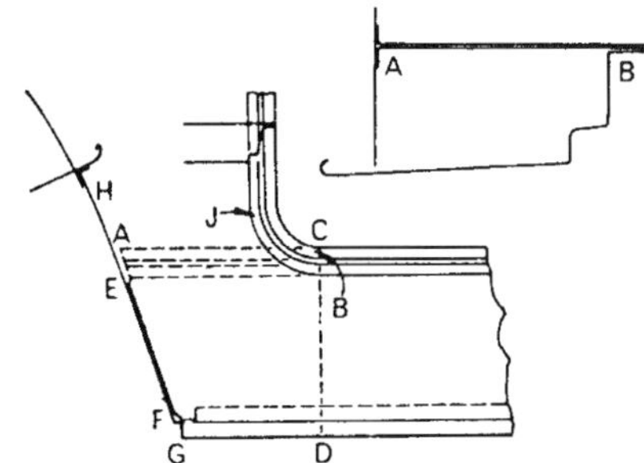

Fig 21a: Sill to quarter panel junction

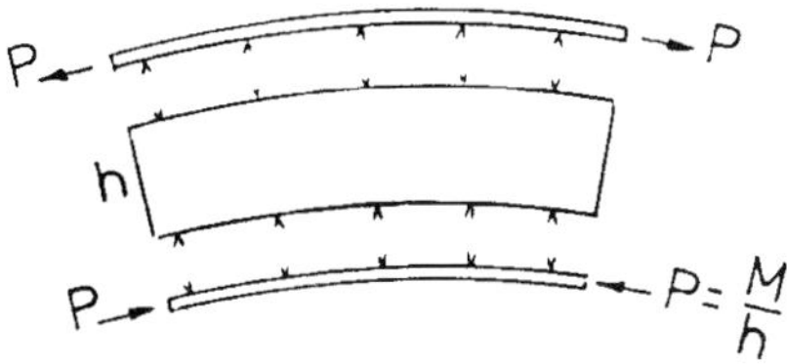

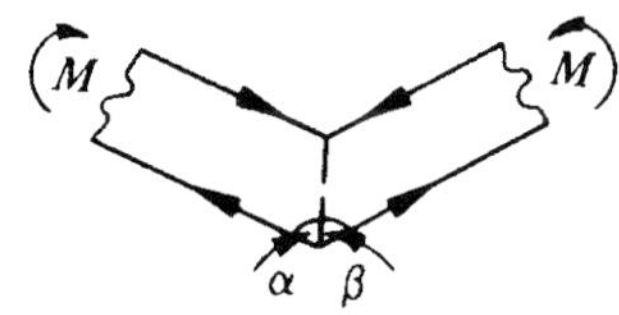

Fig 21b: Curved and angled corner joints

In the case of a curved corner joint between box beams, Fig 21b, the web is subject to direct stress *M/hRt* where *R* is the radius of curvature and *t* the web thickness. For the lightly angled joint shown in the same figure, a 'kink-strut' may be required at the junction to maintain the cross section against collapse. Load in the strut will be *M/d(cos a + cos b)*. If the beam has particularly deep webs, it may be necessary to provide vertical stiffeners which help stabilise shear panels against buckling. One technique is to swage the panel rather than fasten on additional parts. General rules for swaging are that all swages should be straight and not intersect; that they should run along the shortest distance between supported edges of the panel. In calculation they can be represented by end-load carrying bars.

A useful description of the twist of open-section thin walled members, for use with the expression $T/\alpha'' = GJ/l$, mentioned above, refers again to the generalised section in Fig 17. Polar moment of inertia $J = \sigma s t^3/3$ leads to the values for particular shapes shown in Fig 22 alongside some *J* values for representative closed sections.

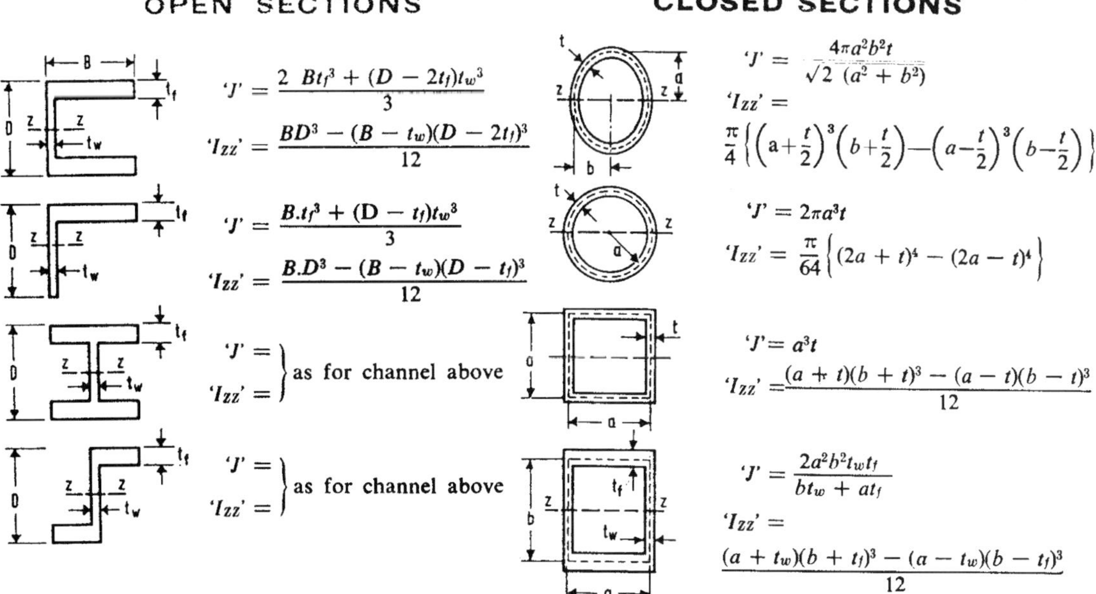

Fig 22: Section characteristics

Contemporary thin-wall analysis

Much of the theory covered in the last section was developed in the early works on thin-walled structures written in the formative years of the aircraft industry. Many of these are now out of print but a moderately priced contemporary book has now been published[6] for civil engineers which gives further exposure to such theory as shear flow in thin-walled box-beams. An example reproduced here, if scaled down in size by a factor of 10, would be the order of magnitude of a car sill section, Fig 23.

Here a singly symmetrical trapezoidal cross-section is considered and a method exemplified of finding the distance of the shear centre, through which a vertically applied load would cause no distortion of the sill, from the larger left hand face of the sill. The shear flow is given by:

$$q_s = -(S_y/I_x) \int_s^o tyds + q_{s,0}$$

where the second moment of area of the section about the x-axis is given by:

$$I_x = (12x600^2/12) + (8x300^2/12) + 2[\int_0^{800} 10\{150 + (150/800)s\}^2 \, ds]$$

The q_b shear flow distribution is obtained by 'cutting' the section at 0 and since y = S_A

$$q_{b.OB} = -(S_y/I_x) \int_o^{Sa} 8S_A dS_A = -(S_y/I_x)4S_A^2$$

From this expression values of $q_{b.B}$, $b_{b.BA}$, $q_{b.AD}$ and by symmetry, the shear flows in all the elements of the section can be found. And since shear centre position is defined by:

$$q_{s,0} = \int(q_b/t)ds/\int(ds/t)$$

$$\int(ds/t) = (600/12) + (2x800/10) + (300/8) = 247.5$$

by combining these and substituting for $q_{b.OB}$, $q_{b.AD}$ and $q_{b.AB}$ gives $q_{S.0} = (S_y/I_x)x1.04x10^6$. Taking moments about the mid-point of AD:

$$S_y x_S = 2[\int_o^{150} 786 q_{OB} dS_A + \int_o^{800} 294 q_{BA} dS_b]$$

By equating shear flows to eliminate them from this equation and then integrating it to eliminate S gives x_S = 282 mm.

For thin-walled open sections such as the complex extrusion shown in Fig 24, Megson provides an example for obtaining the torsion constant J and determining the maximum shear stress produced by torque T.

For thin walled sections generally:

$$J = (1/3) \Sigma st^3$$

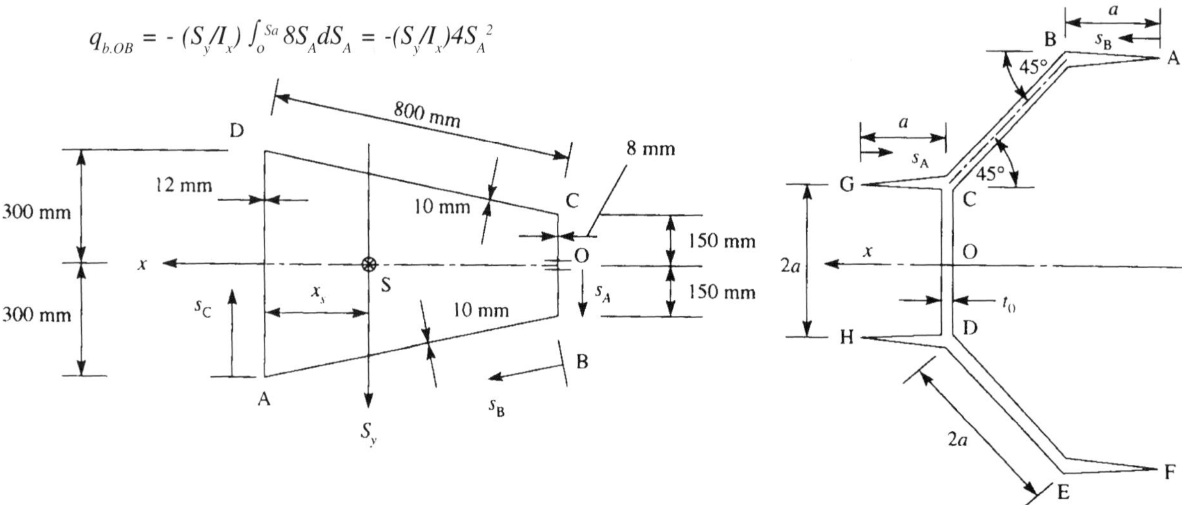

Fig 23: Analysis of a closed thin wall section in bending

Fig 24: Analysis of a thin-walled open section in torsion

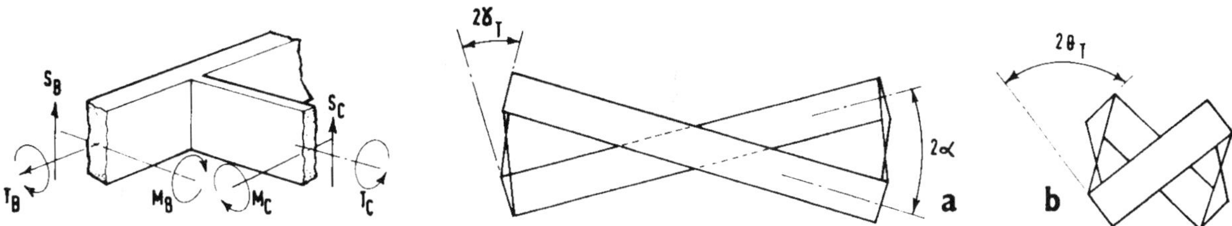

Fig 25: Chassis crossmember/sidemember behaviour

and when t varies with s,

$$J = (1/3) \int_{section} t^3 ds$$

so in this case:

$$J = (2at_o^{\,3}/3) + 2(1/3)\int_o^a (S_A t_o/a)^3 dS_A + 2(1./3)\int_o^{3a} (S_B t_o/3a)^3 dS_B$$

This gives $J = 4at_o^{\,3}/3$ and maximum shear stress

$$\tau_{max} = +/- Gt(d\theta/dz) = +/- Tt/J = 3Tt_o/4at_o^{\,2}$$

In an earlier article[7], Megson argued the case for calculating the torsional stiffness of semi-trailer chassis in order to provide adequate resistance to an offset payload causing an overturning accident on corners. Since overall torsional stiffness of the chassis could be approximated to the sum of the torsional stiffnesses of its members, he compared the stiffness of a closed tube:

$$GJ = G.4a^2b^2t/[2(a+b)]$$

with an identical open one, slit along one edge as:

$$GJ = G.2(a+b)t^3/3$$

By choosing a = 6 in, b = 2 in and t = 0.1 in, he showed that the closed section was 680 times as stiff, torsionally, as the open one. He based the chassis torsional stiffness assumption by considering the forces applying to a sidemember to crossmember joint as in Fig 25. Because both members have comparatively high bending strength and low torsional strength, he argued, the displacements due to bending and shear were negligible in comparison with the twist due to torsion and that the displaced shape as shown could be assumed. Thus the axes of the members remain straight, sections remain plane and each component is loaded as though by pure torque only.

He carried out experiments on a model frame, Fig 26, to verify these assumptions, Fig 27.

Fig 26: Model testing

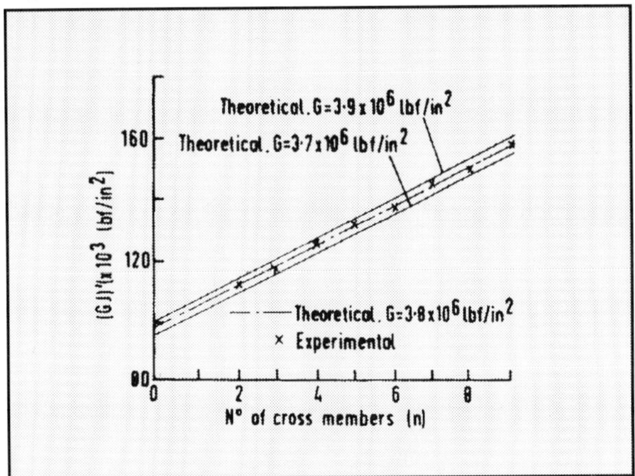

Fig 27: Experimental verification

Framework analysis

Both triangulated pinjointed frames and rectangular stiff-jointed frameworks are used in real and idealised body structures. The pillars and rails of car and van bodies are the key frame elements of private and commercial vehicles. In many specialist sports vehicles tubular space frames, which can be analysed by framework methods, form the structures which support the GRP shells.

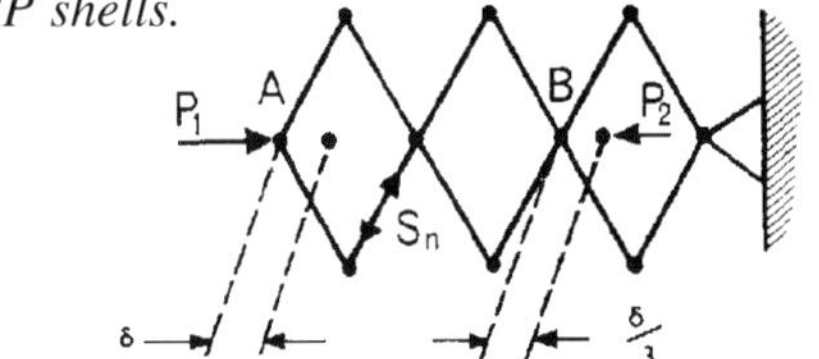

Fig 28: Linkage mechanism

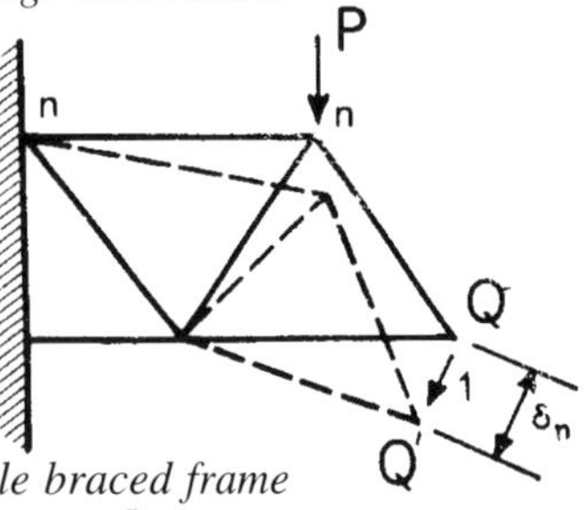

Fig 29: Stable braced frame

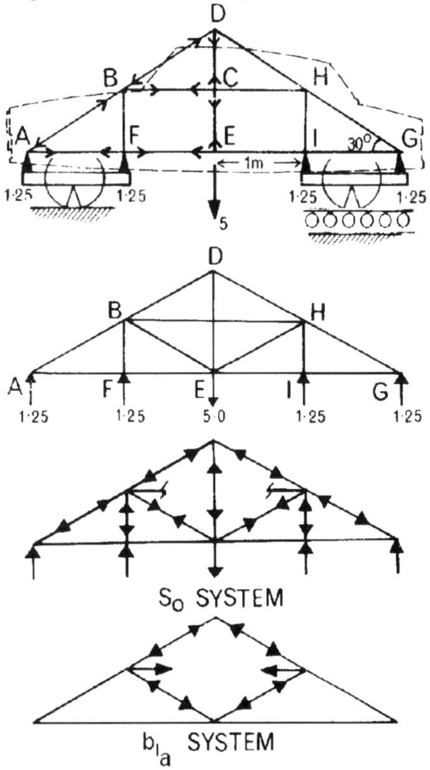

Fig 30: Car sideframe representation

Framework methods derived from civil and aeronautical engineering are valuable analytical tools which solve these structures and form the basis of other methods used in solving integral and monocoque ones. Earliest attempts to predict stresses and deflections, within the first all-welded steel car bodies, were based on the assumption that the body sides behaved like truss frames. The action of vertical loading from engine and occupant seat mounts was examined against the reaction at the axle spring hangers. These analyses were subsequently refined to take in the effects of stiff joint conditions at the A,B and C post connections with sole bar and cant rail. This last approach, however, requires the application of bending theory to be considered in a later section.

Frameworks of discrete links can be analysed by methods first derived for framed bridges, roof

	S_0	b_1	l	$S_0 b_1$ l/AE
(2X) AF	$+1{\cdot}25\sqrt{3}$ (X2)	$+{\cdot}25\sqrt{3}$ (X2)	1	·94 = 1·84
(2X) AB	− 2·5 (X2)	− ·5 (X2)	$\frac{2}{\sqrt{3}}$	1·44 = 2·88
(2X) BF	− 1·25 (X2)	− ·25 (X2)	$\frac{1}{\sqrt{3}}$	0·18 = 0·36
(2X) BC	$+2{\cdot}5\frac{\sqrt{3}}{2}$(X2)	$+\frac{5\sqrt{3}}{2}$(X2)	1	0·94 = 1·84
(1X) DC	+ 5	+ 1	$\frac{1}{\sqrt{3}}$	2·88 = 2·88
(1X) CE	+ 5	+ 1	$\frac{1}{\sqrt{3}}$	2·88 = 2·88
(2X) BD	− 5 (X2)	− 1 (X2)	$\frac{2}{\sqrt{3}}$	5·78 = 11·56
(2X) FE	$+1{\cdot}25\sqrt{3}$ (X2)	$+{\cdot}25\sqrt{3}$ (X2)	1	0·74 = 1·46
				26·54 /AE = S_E

Member	S_0	b_{1a}	l	$S_0 b_{1a}$l	b_{1a}l	$S = S_0 + b_{1a}X$	b_{1b}	$S_0 b_{1b}$l
(2X) AB	− 2·5	0	$\frac{2}{\sqrt{3}}$	0	0	− 2·5	− 0·5	1·46
(2X) AF	$+1{\cdot}25\sqrt{3}$	0	1	0	0	$+1{\cdot}25\sqrt{3}$	$+\frac{\sqrt{3}}{4}$	0·94
(2X) BF	− 1·25	0	$\frac{1}{\sqrt{3}}$	0	0	− 1·25	$-\frac{1}{4}$	0·18
(2X) FE	$+1{\cdot}25\sqrt{3}$	0	1	0	0	$+1{\cdot}25\sqrt{3}$	$+\frac{\sqrt{3}}{4}$	0·94
(2X) BD	− 3·75	$-\sqrt{\frac{2}{3}}$	$\frac{2}{\sqrt{3}}$	10	1·54	−2·975	−0·595	2·05
(2X) BE	+ 1·25	$-\sqrt{\frac{2}{3}}$	$\frac{2}{\sqrt{3}}$	−3·33	1·54	+ 2·025	+ 0·505	1·18
BH	0	+ 1	2	0	4	− 1·485	− 0·297	0·43
DE	3·75	$+\sqrt{\frac{3}{3}}$	$\frac{2}{\sqrt{3}}$	5	0·77	+ 2·975	+ 0·595	1·03

$X_{BH} = -\frac{\Sigma S_o b_1 \, l/AE}{\Sigma b_{1a}^2 \, l/AE} = -1{\cdot}458$ kN

$S_E = \Sigma S_o b_{1b} \, l/AE = \frac{16{\cdot}4}{AE}$

Fig 31: Member force tabulation

trusses and like structures. Such frames are often hinged linkage mechanisms, cross-braced (triangulated) — to form rigid structures. The method is thus best understood by an initial understanding of an unbraced mechanism, Fig 28. The concept of 'work', taken from basic mechanics theory, is used to make the rule that in a system in equilibrium (force balance) under a set of loads, the work done by the various applied loads is equal to the work done by their resultant. Internal (virtual) work is also done by the strained links of the mechanism and those who have operated a 'lazytong' riveting tool will know of the load magnification seen in the figure where:

$$P_1 s = P_2 s/3$$

which on applying the rule above gives $P_2 = 3P_1$. Internal forces S do virtual work U_i which can be equated to the external work $U_e = Ps$.

For a stable framework that has been braced, Fig 29, a two-stage visualisation of the load system is helpful. The first is the equilibrium of bars and joints, ignoring the effects of deformation, while the second is from the forces developed when compatibility of strain is considered at the joints. When trying to find displacement of any point in a frame, stage-one forces can be considered as fictitious (virtual) and an imaginary unit load be assumed to act (for convenience of calculation) at the point where the displacement is to be found. In stage two, the real event gives rise to displacements obtained from the work equation:

$$U_e = U_i \text{ or } Sn_1.dn(o)=\Sigma\, \sigma.1.\varepsilon_0$$

for member force S, deformation d, stress s and strain e in the nth member under virtual unit 1 and real original o loads. When member cross section A and length l are considered, then:

$$1.s = \Sigma(n)\, \sigma_1 .\varepsilon_o/Al$$

and for the framework shown, if only member n is extended:

$$1.S_n = b_{1n} dl_o$$

where b_1 is member force due to the unit load. And when all members deform,

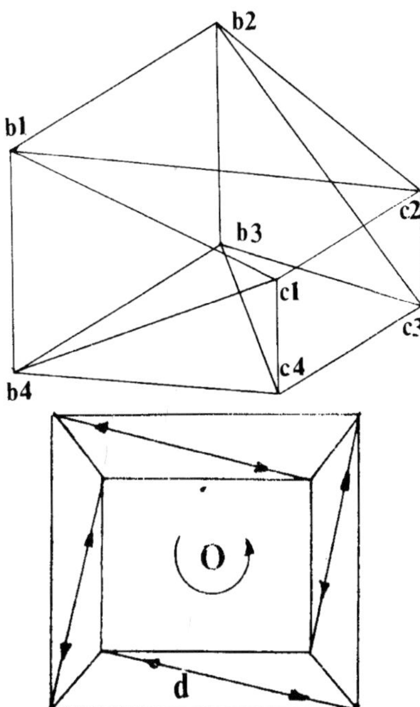

Fig 32: Single bay of space truss

$$S = \Sigma S_o.b_1/(l/AE)$$

since $e_o.l = s_o.l/E = S_o.l/AE$ where E is the elastic modulus.

To show how the method is applied, a somewhat oversimplified car side-frame is considered in Fig 30. A table is made up, Fig 31, for the member forces with unit load being applied at E where the vertical displacement downwards is to be found. Note that the frame is symmetrical so that only forces in half the frame members need be calculated and doubled up as appropriate.

The procedure is to start with the external reaction points A and F and use trigonometry to resolve the proportion of the reaction force which passes through to the member. As all joints are assumed to be pinned, the structure gains its rigidity by end-loading of the members only with a geometry giving a triangulated bracing effect. Members at right angles to the external forces at the joint receive no load and those in line receive equal and opposite load at each side of the joint. Load directions at the joint (see arrows) pair up with those at opposite ends of the member to define whether a tensile or compressive force exists. Away from the external load points, at B, say, resolving vertically to AD gives:

$$BF\cos 30 = BC\sin 30$$

therefore internal force BC (tensile, positive) is:

$1.25(3)^{1/2}$.

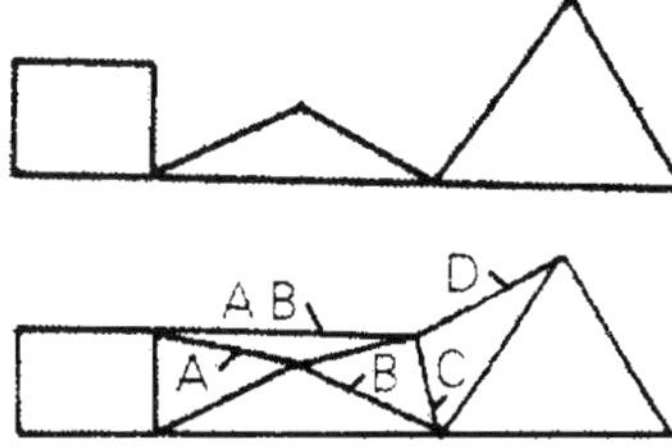

Fig 34a: Cockpit bay

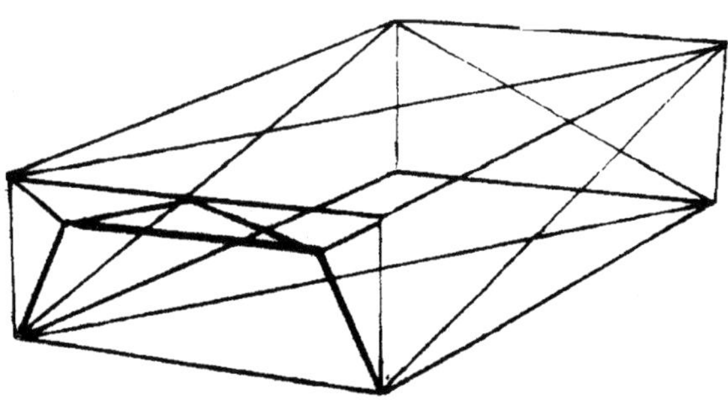

Fig 34b: Front-engine bay

Space frames

The last example was for a plane frame (two dimensional frame); for a three dimensional frame, resolving is a little more difficult and a method known as tension coefficients is applied. This is based on the fact that proportionality exists between both resolved components of length and of force. Tension coefficient:

$$t = S/L = F_x/X = F_y/Y = F_z/Z$$

for member force S, in a length l, projecting force vectors and length components, F_x, F_y, F_z and X,Y,Z on perpendicular coordinate axes.

Fig 32 shows one bay of a space truss, that might be an extension used to support an engine/gearbox unit behind a monocoque bodyshell structure. For analysis, the frame is assumed to have rigid plates at $b_{1,2,3,4}$ and $c_{1,2,3,4}$ which offer no force reaction perpendicular to their own planes (zero axial warping constraint). When considering the torsional load case, for the bay, b_1c_1 and b_4c_4 member forces are zero by resolving at b_1 and c_4, observing zero axial constraint of the truss. Remaining members (the envelope) have equal force components F_z at the bulkheads and tension coefficients for all of them are F_z/Z. This common value can be determined by projecting envelope member forces on to the bulkhead planes, as shown. By taking moments about any axis O, for any one member (c_3b_4, say, with projected length d), contribution to torque reaction is = rtd (half the area of the triangle formed by joining O to c_3 and b_4). Thus total torque is $2tA$, by summation.

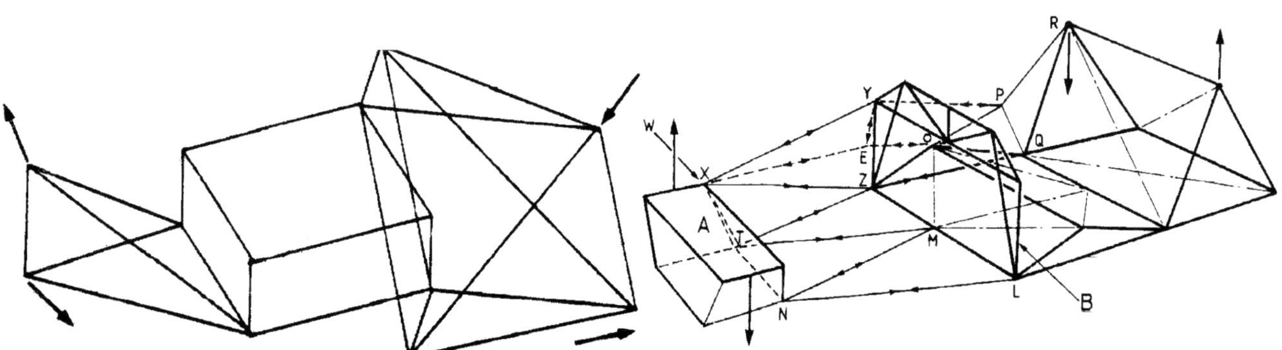

Fig 35a: Triangulated frame

Fig 33: Determinate space frame

Costin and Phipps[8] show how loads in a complete carspace frame are calculated similarly. Frank Costin, too, is a notable pioneer of space frame analysis; he conjectured that any two torsionally stable systems pin-jointed together, at any two node points, would result in torsional continuity between bays, which like that in Fig 33, is statically determinate. This led Costin to approach the design of a cockpit bay as in Fig 34a; tubes A,B,C,D are added to provide bending continuity, also. The front-engine bay is represented by a simple box with triangulated bracing, Fig 34b, and the complete vehicle frame emerges in Fig 35 (the Costin Amiga).

Statically indeterminate frames

The great majority of vehicle structures are indeterminate, due, for example, to stiffness at member joints assumed for the simplified analysis, above, to be pinned. Such frames need to be idealised as determinate structures or analysed by an extension to the above method. Real-life frames usually involve additional members which are redundant from the standpoint of statical determinacy but might still perform some useful additional purpose.

Slightly better idealisation of the car sideframe shown in the top view of Fig 30 would be that of the frame in the view beneath it in which the stiff-jointed door openings are represented by cross-braced pinned frames rather than linkage mechanisms. While the crossbraces are redundant from the standpoint of the frame retaining its shape under moderate load, and from the standpoint of calculation, they are more truly representative of the anti-lozenging stiffness of the openings.

The technique of analysing single-redundancy frames is to cut any redundant member and then calculate member loads for the remaining, determinate, frame. Here BH is cut to give the S_o and b_1 force systems shown below and a third force system b_{1b} determined for finding the required central deflection after a force equal to that in the cut redundant member is inserted. This is done again by applying unit load vertically downwards at E then obtaining b_{1b} forces by dividing S forces (from stage-one) by the real external load of 5 kN at E, giving member forces for unit external load. The table in the lower part of Fig 31 shows the central deflection to be reduced by a factor of 26.54/16.9 = 1.5 by the bracing which simulates the stiffness of the aperture frame.

Finding the degree of redundancy of a framework can be difficult when only considering the number of possible extra load paths in the structure. Another approach is to consider that equilibrium equations alone are insufficient to find the forces at any member connection, or external load input, point. This has led to the expression for number of redundancies = $3f + 2h + r + 3m - 3j$ where f is the number of fixed supports, h the number of hinged supports, r number of roller supports, m the number of members and j the number of joints. This reduces to $m + r - 2j$ for a two-dimensional (plane) frame. Of the three equilibrium conditions of horizontal and vertical force, and moment, balance, only the first two apply at a pinned joint, though pinned joints can be given extra degrees of freedom as shown in Fig 36.

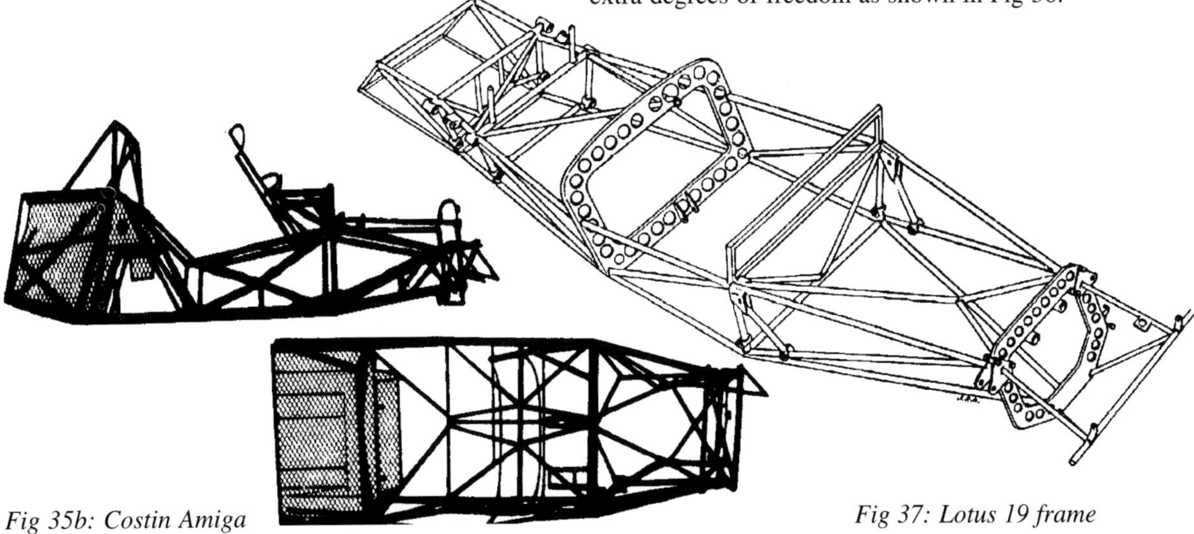

Fig 35b: Costin Amiga

Fig 37: Lotus 19 frame

Stable determinate frames are made by adding triangles formed by two extra members and one joint as shown. Addition of more or fewer members involves either a mechanism or a redundant frame in which some, or all, members can only deflect with accompanying deformation of the others. Note that a determinate frame has members which can deform independently. Determinate space frames are made by adding pyramids of three extra members. The Lotus 19 is an example, Fig 37 — before the adoption of the later backbone tube and space frame powertrain support.

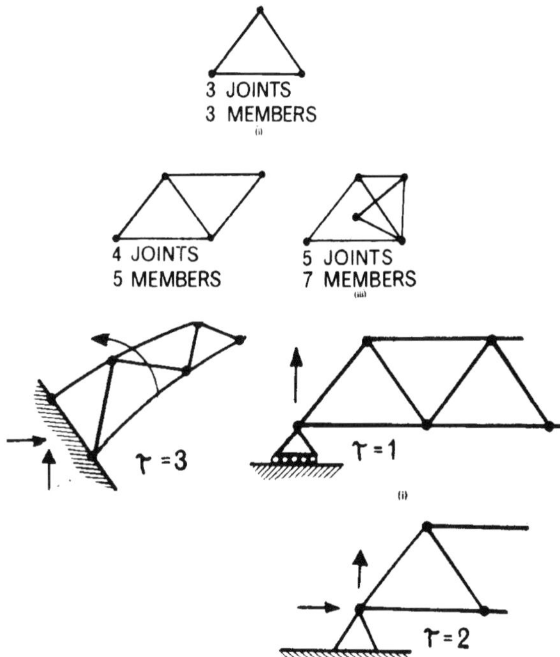

Fig 36: Number of redundancies

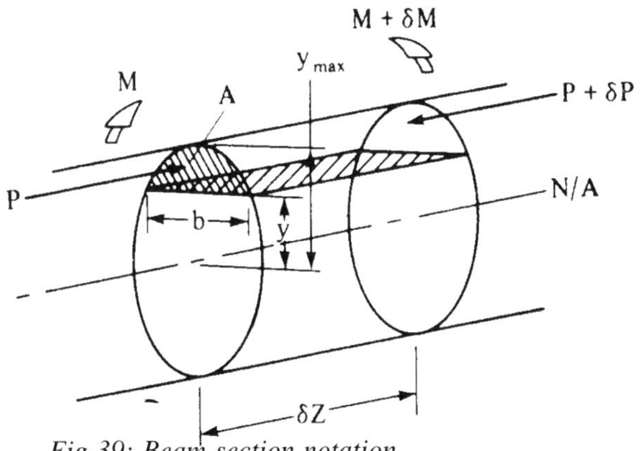

Fig 39: Beam section notation

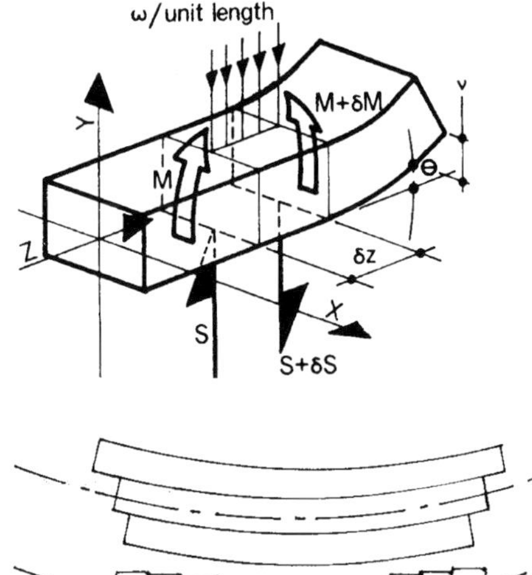

Fig 38: Beam calculation notation

Beam elements

Consideration of thin-walled structures in an earlier section introduced beam-elements under the combined action of shear forces and bending torques (or moments). For the relatively short and deep beams there considered, external shear forces caused the predominant internal loading of the beam. Here we consider the relatively longer and shallower beam for which the predominant internal loading is bending stress caused by external bending moments.

Familiar structural elements on vehicles, subject to bending rather than shear stresses, in flexure, are chassis sidemembers on trucks and integral body crossbearers on cars — which carry seating loads reacted at their end-connections with the body sidepanels and rails.

Fig 38 is useful in establishing a sign convention for the analysis of beams so that positive and negative values of forces follow a constant rule for different load cases. Shear force S is taken as positive clockwise; bending-moment M is positive upwards towards the compressed top fibres of the beam (clockwise left, anticlockwise right); load is negative applied downwards on the compressed edge of the beam and changes in slope θ and translational-deflection v of the beam are both positive in the positive direction of upward loading on the beam. The balance of forces over a short slice of the beam, length dz, along with basic stress/strain relationships established earlier, leads to the equation of flexure, for a

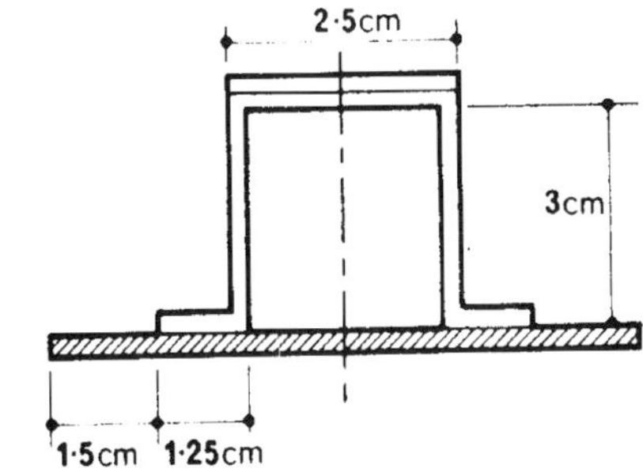

Fig 40: Fabricated box-beam

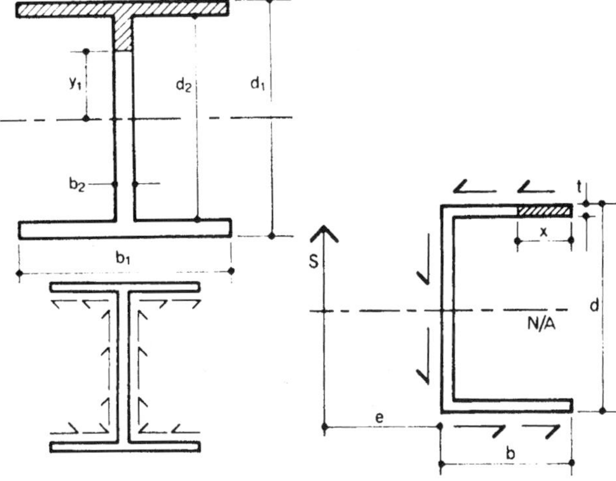

Fig 41: Thin-wall I-beam

Fig 42: Thin-wall channel

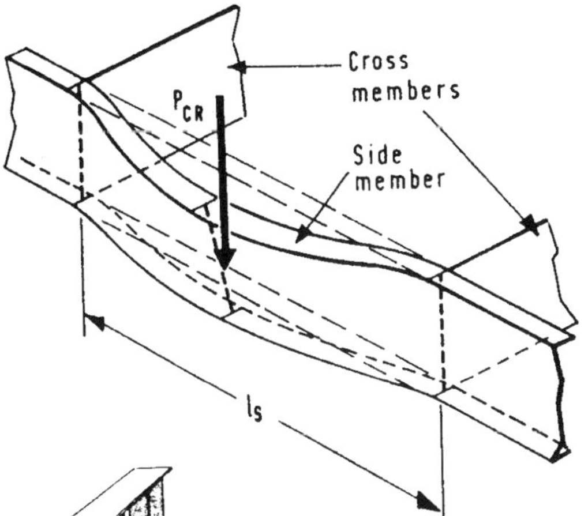

Fig 43: Lateral buckling

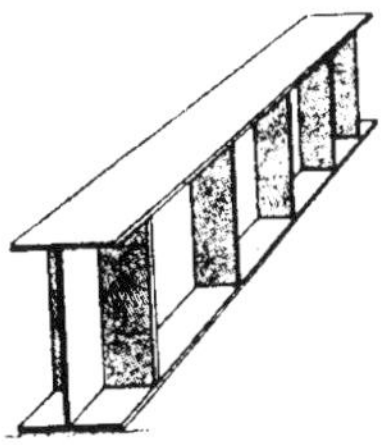

Fig 44: Web stiffeners

beam bent to radius of curvature R, used in calculation: $\sigma/y = M/I = E/R$ for bending stress σ at y distance over the cross-section of the beam from the 'neutral-axis' and I section second moment of area — where, at the centre of symmetry of the section, bending stresses are zero between increasing compression, above, and increasing tension, below, for the sort of loading depicted in Fig 39. Inset in Fig 38 is an imaginary picture of the beam element to help visualise the concept of 'complementary shear' by considering the beam to be comprised of vertical and horizontal slipplanes. To calculate the value of these complementary vertical and horizontal shear stresses an integration needs to be performed, for irregular shaped sections, based on further relationships between external and internal forces. Moving from left to right along the elemental length dz in Fig 39, the external load w per unit length equals the rate of change of shear force (first derivative, dS/dz) which in turn equals the second derivative of bending moment (and in turn the fourth derivative of distance y times the flexural stiffness, EI) where the elastic modulus is E. This leads to the formula for shear stress in beams:

$$\tau = SAY/Ib$$

where Y is the distance from the neutral axis to the centroid of the shaded area of the cross section seen in Fig 39. A use for this formula is in determining the required pitch of rivets (or spot welds) on the closing plate which might be boxing in a beam of otherwise open cross section (Fig 40). Table 2 shows a logical way to set out the calculation and the distance of the neutral axis from the lower face of the bottom reinforcing plate can be found from $\sigma[Ay]/\sigma[A]$ in this case = 2.64/2.15 = 1.23 cm. The section second moment of area for the total section is found from $I = I_g + Ah^2$ where h is the distance of the centroid of each element of the section from the neutral axis and I_g the element's inertia about its own axis. Here I = 3.877 cm^4 and the distance of the neutral axis from the farthest stressed layer of the beam = 3.3-1.23 = 2.07 cm. As the centroid distance of the reinforcing plate from the neutral axis is 1.18 cm, then for a given rivet shear load capacity, required pitch can be determined. For a bending moment of 300 Nm, this would work out to be a value of 2.05 cm in this particular worked example.

Table 2: Section modulus calculation

	Area A (cm^2)	Centroid distance y (cm)	Ay (cm^3)	I_g (cm^4)	distance to N/A, h (cm)	Ah^2 (cm^4)
Top 'flanges'	0·5	3·2	1·60	$(2{\cdot}5 \times 0{\cdot}2^3)/12 = 0{\cdot}00167$	1·97	1·94
Webs	0·6	1·6	0·96	$(0{\cdot}2 \times 3^3)/12 = 0{\cdot}45$	0·37	0·082
Bottom flanges	0·25	0·15	0·04	$(2{\cdot}5 \times 0{\cdot}1^3)/12 = 0{\cdot}00021$	1·08	0·293
Reinf. plate	0·8	0·05	0·04	$(8 \times 0{\cdot}1^3)/12 = 0{\cdot}00067$	1·10	1·11
	$\sum = 2{\cdot}15$		$\sum = 2{\cdot}64$	$\sum = 0{\cdot}4525$		$\sum = 3{\cdot}425$

Table 3: Values of gamma

$I_s^2 . G.J/E\,I_w$	0·4	4	8	16	32	64	96	128	200	400	
for P_{CR} concentrated at mid-span	466	154	114	86·4	69·2	—	54·5	52·4	49·8	47·4	*Load applied at section centroid*
for P_{CR} distributed evenly between lateral supports	673	221	164	126	—	101	79·5	76·4	72·8	—	*Load applied at section centroid*
	587	194	145	112	—	91·5	73·9	71·6	69·0	—	*Load applied at top flange*

Table 4: Permissible stress for main alloys — tsi (N/mm^2)

Alloy	Condition	Form	Thickness in (mm) from	Thickness in (mm) to	Axial p_t p_c	Bending p_{bt} p_{bc}	Shear p_q	Bearing p_b	λ_s
H30	WP	Extrusions, solid and hollow	¼ (6·4)	3[c] (76)	9·5 (147)	10·5 (162)	5·7 (88)	14·4 (222)	59
		Sheet	—	¼ (6·4)	9·0 (139)	9·9 (153)	5·4 (83)	13·7 (212)	61
		Drawn tube	—	1/16 (1·6)	9·1 (141)	10·1 (156)	5·5 (85)	14·4 (222)	
			1/16 (1·6)	—	8·6 (133)	9·6 (148)	5·2 (80)		64
		Plate	¼ (6·4)	1 (25)	8·5 (131)	9·5 (147)	5·1 (79)	13·7 (212)	
N8	M	Extrusions, solid and hollow	—	—	5·4 (83)	6·3 (97)	3·2 (49)	13·0 (201)	96
		Plate	¼ (6·4)	1 (25)					
	O	Extrusions, solid	—	—	5·3 (82) 5·0 (77)	6·1 (94) 5·9 (91)		12·2 (188)	101
		Plate	¼ (6·4)	1 (25)	5·4 (83) 5·1 (79)	6·3 (97) 6·0 (93)		13·0 (201)	
H9	WP	Extrusions, solid and hollow	—	6[e] (152)	5·7 (88)	6·3 (97)	3·4 (53)	9·0 (139)	81
	P	Extrusions, solid and hollow	—	⅛ (3·2)	5·2 (80)	5·7 (88)	3·1 (48)	8·3 (128)	86
			⅛ (3·2)	½ (13)	4·2 (65)	4·7 (73)	2·5 (39)	7·6 (117)	102

Table 5: C1 and C2 values for webs

Alloy	C_1		C_2	
H30-WP sheet	65		120	
H30-WP plate	67		124	
N8-O, N8-M plate	84		156	
Others	$150/\sqrt{p_q}$	p_q in tonf/in²	$280/\sqrt{p_q}$	p_q in tonf/in²
	$590/\sqrt{p_q}$	p_q in N/mm²	$1100/\sqrt{p_q}$	p_q in N/mm²

The analysis of the beam section in separate elements is particularly valuable in calculating thin walled beam performance. Such sections have shear stresses varying widely across the section — in contrast to a slender solid-section beam, the shear stress across which can often be assumed constant in calculation. For the I-section thin-wall beam of Fig 41, complementary shear flow acts horizontally in the flanges, as well as vertically in the web, which runs out to zero at the edges. Thus for thin-wall section beams, the assumption of constant shear stress is made only for the 'vertical webs' normal to the neutral plane of bending. For sections which are non-symmetrical such as the channel of Fig 42 the sum of the shear stresses in individual elements has a resultant at distance *e* from the web which passes through the shear-centre of the beam.

To avoid twisting of the beam, induced by bending loads, the external load should be applied through the shear centre, the value of *e* being obtained from $e = td^2b^2/4I$. To ensure best performance from thin-wall section beams, means must be provided to ensure that cross-sectional shape is maintained — if necessary by employing lateral stiffeners in the form of bulkheads or ribs. There is otherwise a tendency for certain sections to buckle in bending,

Fig 43. Another problem to avoid is that of fatigue cracking of compression flanges on open section beams; this can be caused by local buckling of flanges which are too wide to be stable in compression. This tendency can be reduced by the use of web stiffeners as shown in Fig 44. There is also a possibility of overall lateral buckling of flanges if too narrow a flange is chosen. A formula due to Timoshenko can be used to calculate optimum spacing of crossmembers to avoid such problems in which 'warping rigidity', rather than bending rigidity, is expressed as the product of elastic modulus and warping constant, $I_w = t_f h^2 b^3$ for use in the formula $P_{cr} = g\,(EI_w GJ)^{1/2}$ where γ takes values from Table 3; b and h are section height and width and t_f is flange thickness.

A common condition in vehicles is the support of deck flooring by cross-bearers. If the decking is made up of a number of discrete beams, or planks, and an elastic crossbearer is supporting them at mid-span, the deflection of the planks is given by:

$$(5wl^4/384EI) - (Rl^3/48EI)$$

from which can be found the radius of curvature of the deflected combination $R = (5wl/8) - (48EIy/l3)$ and the centre deflection of the crossbearer is:
$(w/a)[1 - 2(\cos bl/2)(\cosh bl/2)/(\cos bl + \cosh bl)]$ where $a = 48EI/l^3$ and $b = (a/4EI)^{1/4}$

Beam design methods

The design of beam systems in aluminium alloy is covered in British Standards[7]. This states that the direct flange stress in bending shall not exceed the permissible value given in Table 3. The average shear stress in the webs should not exceed that in Fig 45 for d/t not exceeding $C1$ in Table 5. It is also necessary to guard against failure in lateral buckling in those cases where the external loads might give rise to this situation — that is a tendency to twisting and sideways deflection. Thin-walled beams should not be stressed beyond compressive limits shown in Fig 46 where the lamda value (λ) of the y-axis is calculated from $k(t_f/t_2)^{1/2}$ for I-sections and channels or $k(t_f/b)^{1/2}$ for solid and hollow rectangular sections where k is found from Fig 47, t_2 is flange thickness and b is section width.

Intermediate transverse web-stiffeners must be provided throughout the length of a beam where the d/t ratio exceeds the value of $C1$ in Table 4 where t is web thickness and d is web depth. Stiffeners must be designed so that second moment of area of each, about the face of the web, is not less than $1.3d^3\,t^3/s^2$ where s is stiffener spacing. Normally it will not be required to have stiffeners at spacings greater than $1.5d$. Where the ratio d/t exceeds $C2$ in Table 5, longitudinal stiffeners, Fig 48, will be required as well as transverse ones. Fig 49 gives values of h/d such that panels above and below the stiffener are of equal strength. Stiffeners should be so designed that second moment of area about the face of the web is not less than $4dt^3$. Outstanding leg-widths of both transverse and longitudinal stiffeners must not exceed 12 times plate thickness.

Curved and cranked beams

The Unit Load method of structural analysis is particularly suitable for more complex shape real-life beams under a variety of load conditions. The method can be understood by imagining a small elastic

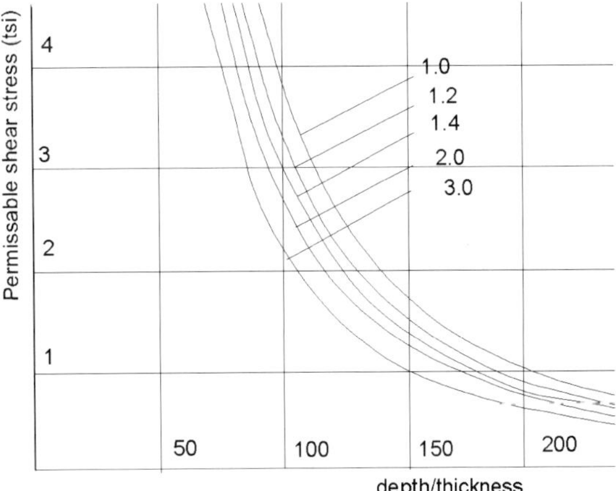

Fig 45: Permissable shear stress in webs - tsi

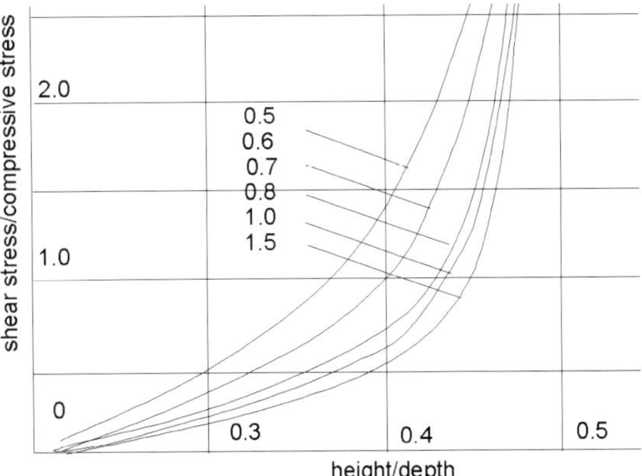

Fig 46: Compressive stress limits

element in a curved beam which is otherwise considered to be rigid. An imaginary load *w*, at *P* in Fig 50, is then considered to cause vertical displacement *D*; the beam deflects Δ under the influence of the bending moment and it can be shown that $\Delta = \int Mm/EI.ds$ where *m* is bending moment due to imaginary unit load at P and *M* due to the real external load system. An example might be a car floor transverse crossbearer having a 'tunnel' portion for exhaust-system and prop-shaft clearance, Fig 51. To find the downward deflection at B, the imaginary unit load is applied at that point and it develops reactions of 2/3 at A and 1/3 at D. The bending moments for the real and imaginary loads are shown in Table 6 which also illustrates the subsequent calculation to obtain the deflection due to the driver and seat represented by distributed load *w* over length *l*.

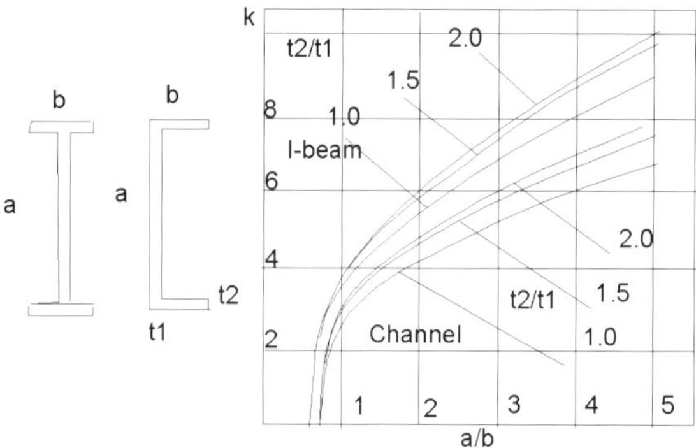

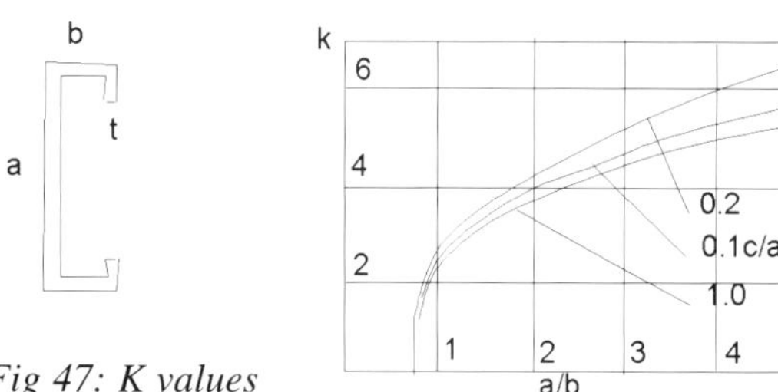

Fig 47: K values

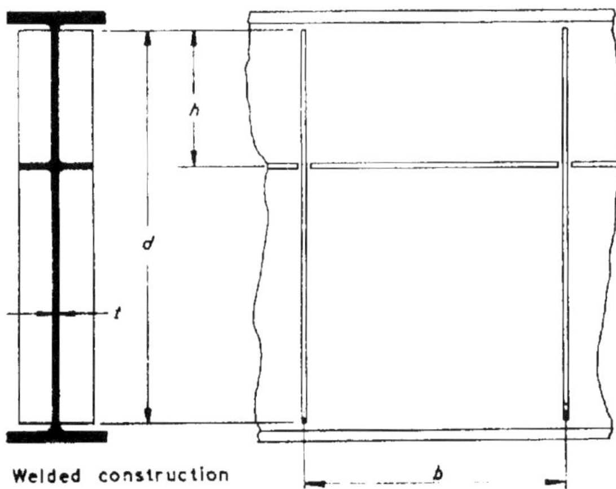

Fig 48: Longitudinal web stiffeners

Continuous beams

Where a beam passes over one or more mid-span supports, a 'redundant' structure exists which is 'statically indeterminate'. A more refined analysis therefore has to be undertaken known as the Moment Distribution method. This uses a hypothetical cut to determine distribution of bending moment along the beam. For the beam and load shown in Fig 52, firstly each support is hypothetically locked with a fixing which makes slope of the beam equal to zero at these points. The so-called fixing endmoments (FEMs) are then calculated at these points and their difference in value for adjacent bays noted (the sign convention states that all FEMs are assumed positive). These moments are then progressively relaxed by balancing moments which revert the 'fixed' supports back to pinned ones. The relaxing moments are apportioned according to the relative stiffness of the 'bays' each side of the support, distribution factors for this purpose being proportional to elastic modulus and section second moment of area but inversely proportional to bay length:

$$K_1 = (EI/l)1/[(EI/l))_1 + (EI/l)_2]$$

and $K_2 = 1 - K_1$

As the supports are 'relaxed', a carry-over moment is transferred to the next support, either side,

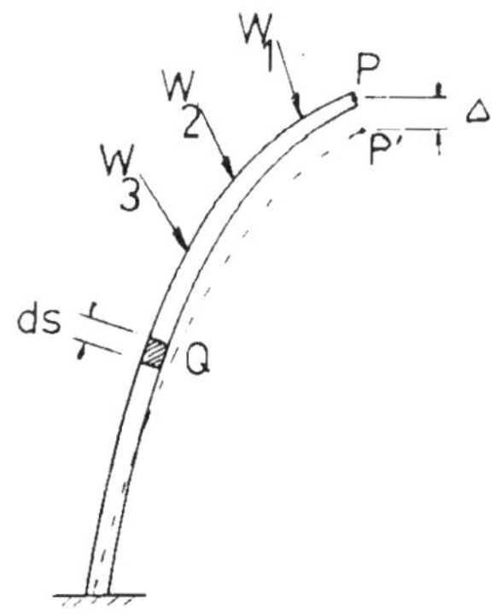

Fig 50: Unit Load method

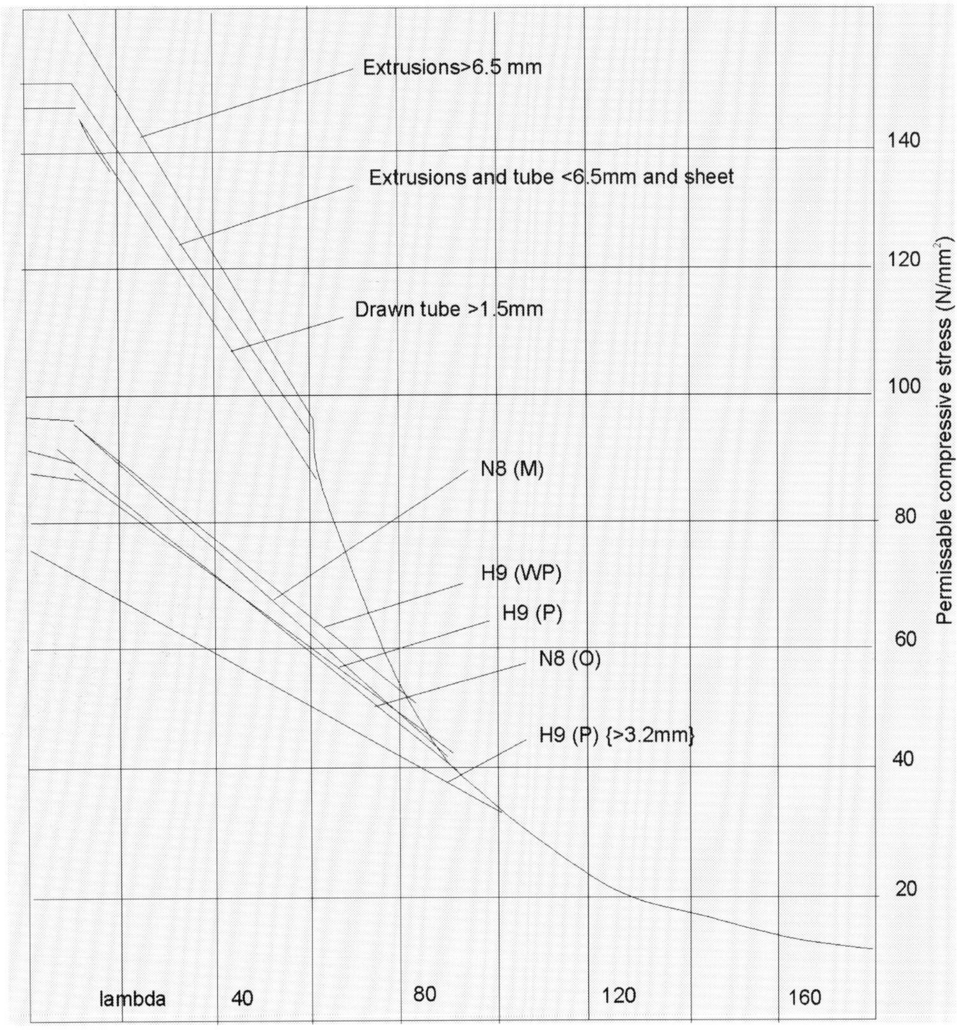

Fig 49: Values of h/d

according to the relation *Ma = (1/2)Mb*. The sign convention makes the moment 'differences' into an algebraic sum so that the negative sign becomes positive. In the figure, the bending moment diagrams for the two bays in the example, after 'fixing' the intermediate support, are shown beneath the loading diagram; these indicate the FEM imbalance. The relaxing operation in stage-3 is equivalent to the successive imposition of BM diagrams such as in (iv) — with the final BM diagram appearing in (v).

An automotive example of the use of the method is seen in Fig 53 to represent the chassis sidemember of a medium-weight truck having an applied load system representing front and rear engine/transmission mounts and payload acting through

Table 7: Sidemember calculation stages, from + 0.98 at B

B	C		D		E	
1	0.555	0.445	0.378	0.622	1	D.F
-0.35 -0.63	+0.35 +2.69	-5.2 +2.15	+5.2 -1.48	1.3 2.43	+1.3 +0.5	-1.8 F.E.M
+1.34 -1.34	-0.32 +0.59	-0.74 +0.47	+1.07 -0.50	+0.25 0.82	-1.22 +1.22	
+0.30 0.30	-0.67 +0.51	-0.25 +0.41	+0.24 0.32	+0.61 0.63	-0.41 +0.41	
+0.25 -0.25	-0.15 +0.17	-0.16 +0.14	+0.21 -0.16	+0.21 0.26	-0.27 +0.27	
+0.09 -0.09	0.12 +0.11	-0.08 +0.08	+0.07 -0.07	+0.13 0.12	-0.13 +0.13	
0.98	+3.16	-3.16	+4.26	-4.26	+1.8	1.8 kNm

bodymount points. Stepping of the frame approximates to the change in section modulus due to tapering and the distributed load pattern in (ii) approximates the real one with additional conversion to SI units. Pinned supports represent the ends of the laminated suspension springs and section second moment of area for the beam in mm^4 x 10^6 are for each bay as follows: AB 8.5, BC 12.5, CD 20, DE 16.5 and EF 12.5. Handbook values of the FEMs for the two bays are shown at (iii) and the analysis is carried out as in Table 7, from which the following distribution factors are obtained:

$$K_{cl}=(1.25/1.25)/(1.25/1.25+2.0/1.25)=0.555$$
$$K_{dl}= (2.0/1.25)/(2.0/2.5+1.65/1.25)=0.378$$

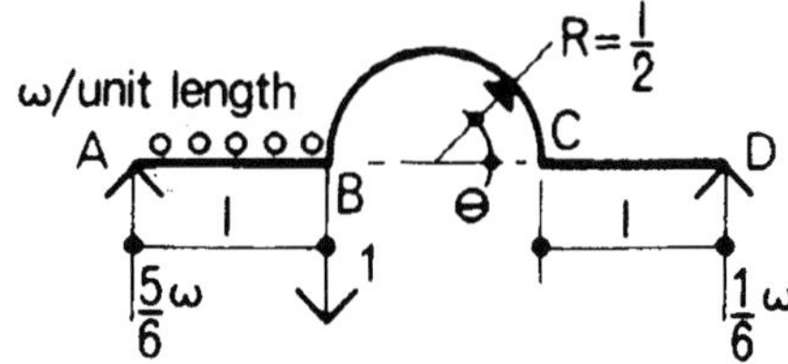

Fig 51: Seat support bearer

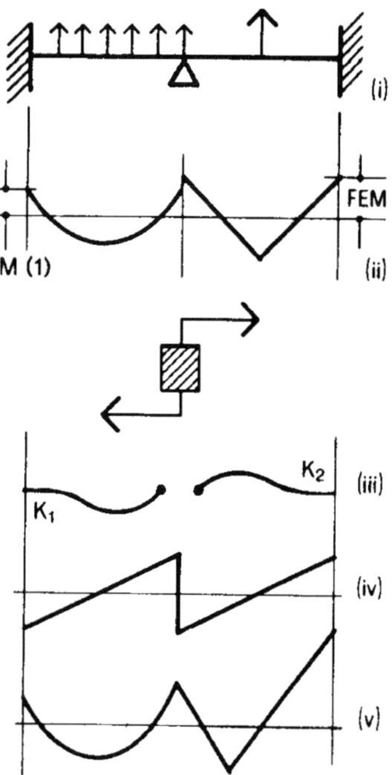

Fig 52: Moment distribution method

Table 6: Unit-load calculation

	M_0	m_1
AB	$\frac{5wl}{6}s - \frac{ws^2}{2}$	$\frac{2}{3}s$
BC	$\frac{wl}{6}\left(l+\frac{l}{2}-\frac{l}{2}\cos\theta\right)$	$\frac{1}{3}\left(l+\frac{l}{2}-\frac{l}{2}\cos\theta\right) = \frac{1}{6}\left(3-\cos\theta\right)$
	$= \frac{wl^2}{12}(3-\cos\theta)$	
CD	$wls/6$	$s/3$

So

$$\delta_B = w\int_o^l \left(\frac{5}{9}ls^2 - \frac{1}{3}s^3\right) ds +$$

$$\frac{wl^4}{144}\int_o^{\pi}\left(3^2 - 6\cos\theta + \cos^2\theta\right) d\theta + \int_o^l \frac{wl}{18}s^2\, ds$$

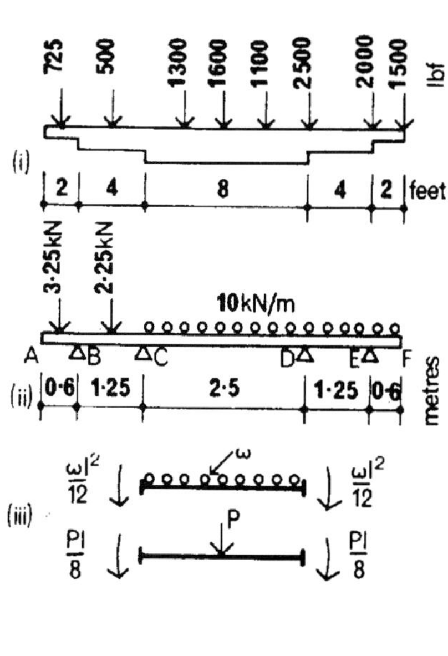

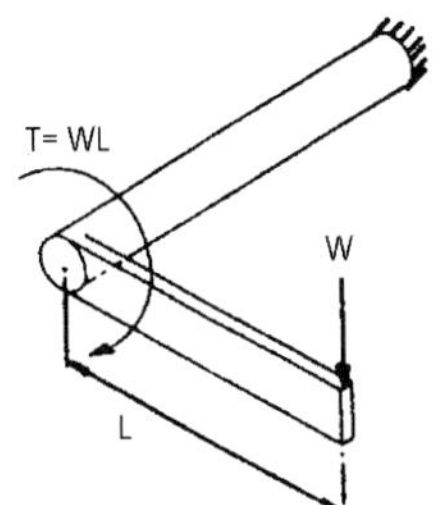

Fig 54: Element subject to twisting and bending

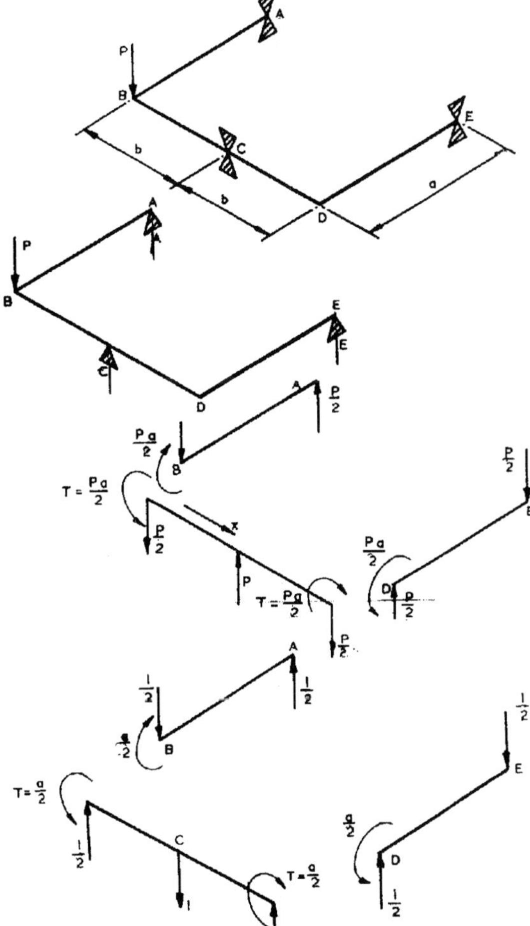

Fig 55: Portal frame under combined loading

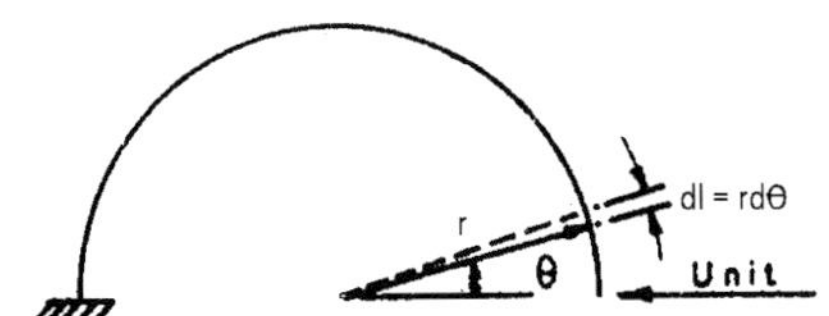

Stiff-jointed frames

A variety of structural elements of the vehicle body and chassis comprise stiff-jointed frames. These may be either simple rectangles or made up of arched and cranked members. All can be readily analysed by the Unit Load technique introduced in the previous sections.

The methods of calculation used for curved and cranked beams were also described above and can be adapted for the calculation of deflections in stiff-jointed frames. When the frame is stiff-jointed as opposed to pinned, bending moments are transferred from one element to another across the joint.

In the part of the frame shown as Fig 54 subject to twisting and bending, deflection due to twist $= \int Tt\,dx/GJ$. This deflection would then be added to that due to bending, obtained from the formula given in the earlier section. Usually the deflection due to axial straining of the elements is so small that it can be discounted.

A good example of a frame subject to twisting and bending is the portal shape frame of Fig 55 having loading and support, in the horizontal plane, indicated. If it is required to find the vertical deflection at D for the load P applied at B, equating vertical forces and moments is first carried out to find the reactions at the supports. Next step is to break the frame up into its elements and determine the bending/ moments in each — ensuring compatibility of these and that associated loads are 'transferred' across the artificially broken joints.

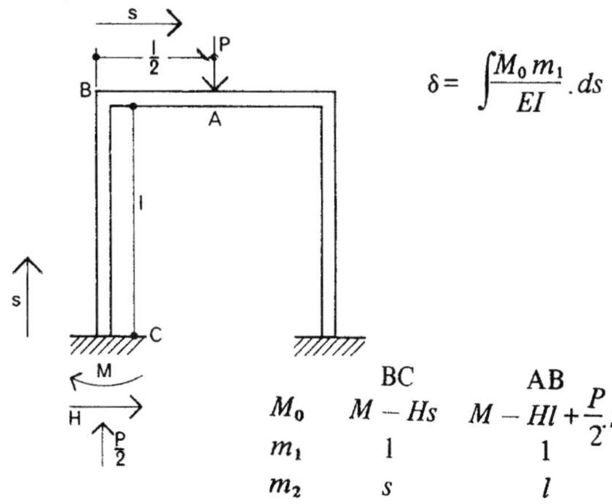

$$\delta = \int \frac{M_0 m_1}{EI} . ds$$

	BC	AB
M_0	$M - Hs$	$M - Hl + \frac{P}{2}.s$
m_1	1	1
m_2	s	l

Integrations have to be carried out for the three deflection modes, as follows:

$$\int Sb.ds/AE + \int Mm.ds/EI + \int Ff.ds/AG$$

to obtain the combined deflection. Both vertical and horizontal components can be obtained by applying vertical and horizontal imaginary unit loads, in turn, at the point where the deflection is to be determined. For the example shown in the figure, elastic and shear moduli, E and G, are 200 and 80 GN/m^2 respectively. S is axial load and F the shear load, the latter being negative in compression. The values of these, with the bending moments, are as follows:

	S	b	M	m	F	f
Vert.						
AB	0	0	Ps	1s	P	1
BC	-P	-1	1P	1.1	0	0
Hor'l.						
AB	0	1	Ps	0	P	0
BC	-P	0	1P	1s	0	1

These can readily be substituted in the expression above to determine deflections as a factor of P.

The method can also be used for 'continuum' frames incorporating large curved elements provided curvature is high enough for the engineer's simple theory of bending to be applicable. In the semi-circular arch of Fig 56, at any point along it defined by variable angle θ, bending moment is given by $Wr \sin\theta$ due to the external load and $r \sin\theta$ due to the imaginary unit load at the position and in the direction of the deflection which is required. In this case it is the same as that of the external load and the deflection due to bending is equal to the integral of the product of these between $q = 0$ and p, divided by the flexural rigidity of the section EI, product of elastic modulus and section second moment of area.

A common example of a portal frame in a horizontal plane is the front crossmember and front ends of the chassis sidemembers of a vehicle imagined 'rooted' at the scuttle. The load case of a central load on the crossmember, perhaps simulating striking a pole, can be visualised in Fig 57.

An example of a frame having both curved and straight elements is the idealised goose-neck semi-trailer shown in Fig 58. Again a table of bending moments can be written, as shown, and the vertical deflection at B calculated by using the Unit Load formula. Given the value of I for the sidemember of 0.25 x 10^4 metre^4 and E of 210 GN/m^2, the deflection works out to be 53 mm, Fig 59.

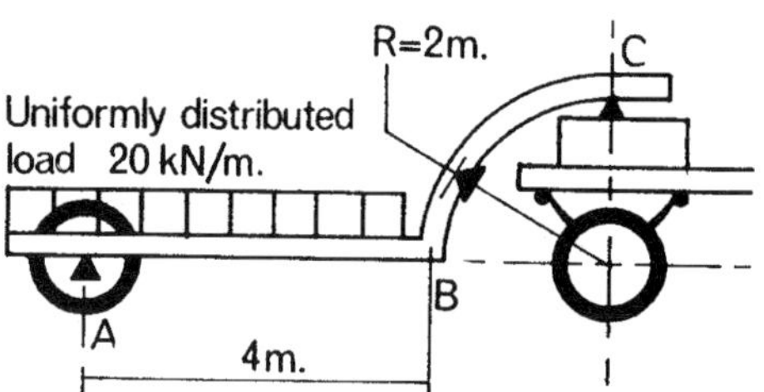

Fig 58: Goose-neck semi-trailer frame

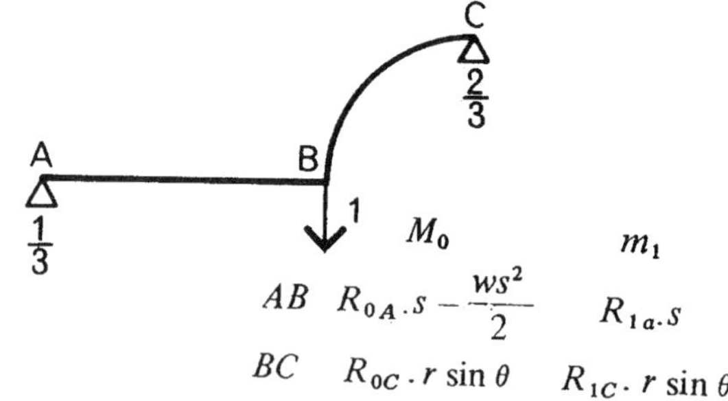

Fig 59: Bending moments for trailer frame

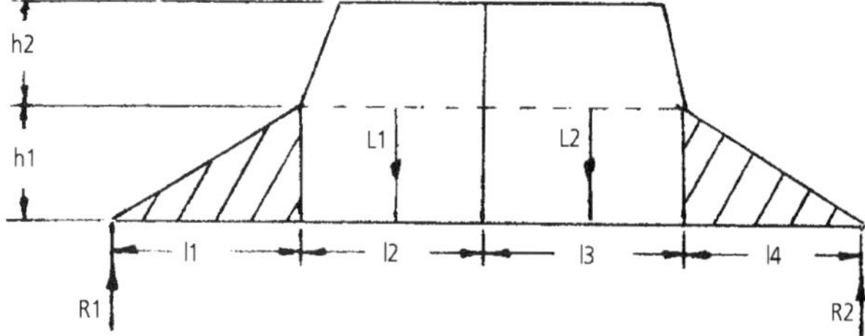

Fig 60: Simplified sideframe idealisation

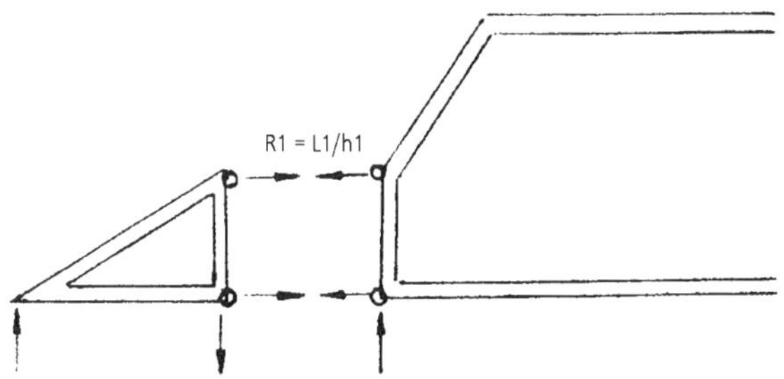

Fig 61: Reaction load referred into sideframe

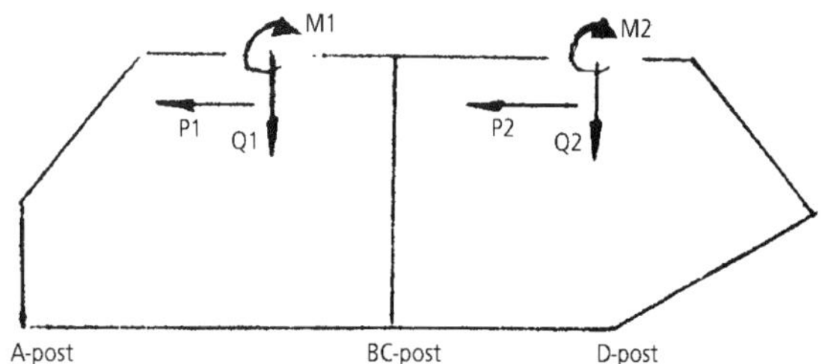

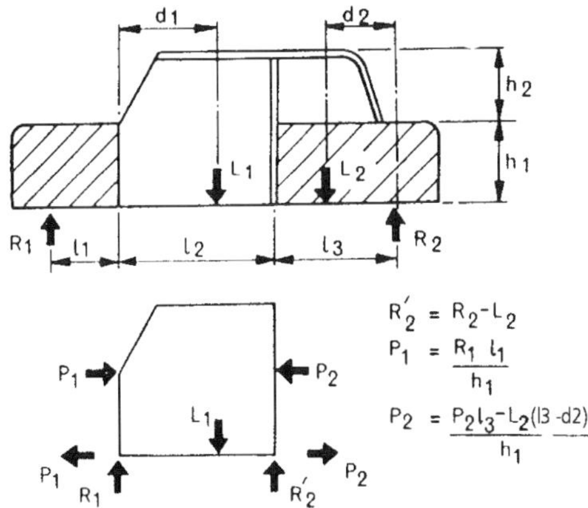

Fig 63: Redundant sideframe idealisation

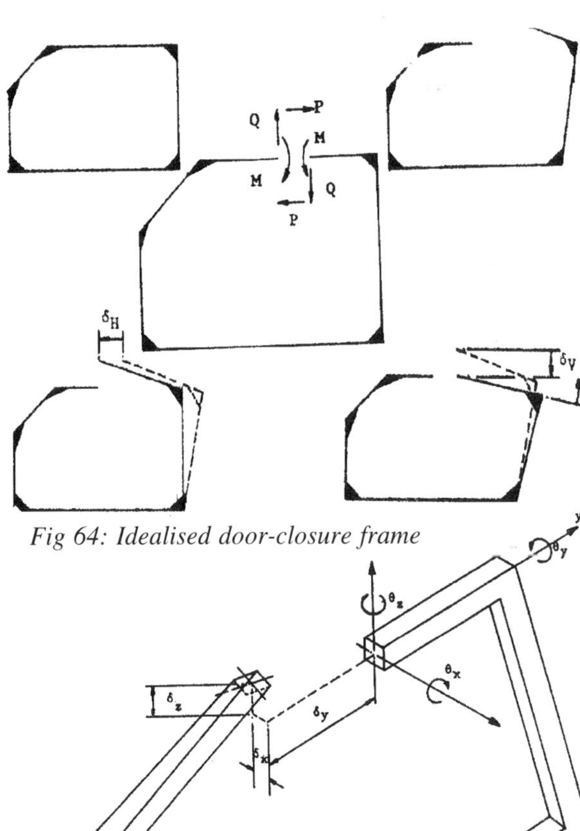

Fig 64: Idealised door-closure frame

It is generally best to integrate along the perimetric co-ordinate for curved elements and, as above; when requiring a translational deflection, apply unit force, and for a rotational displacement, apply unit couple. In the case of redundancies, remove the redundant constraint by making a cut then equate the deflection of the determinate cut structure with that due to application of the redundant constraint.

Idealised car-body side-frames

Saloon-car sideframes can be idealised into stiff-jointed frames with a relatively small number of redundancies — for analysis. Fig 60 suggests an idealisation with a considerable degree of symmetry and Fig 61 the feeding in of reaction loads. Normally a four door vehicle side frame would have six degrees of redundancy as shown in Fig 62. Various simplifications can be made to obtain an approximate solution of the structure. Pin-joints assumed at the ends of the B-C post would reduce the redundancy to 4. A strong D-post top-joint might also justify the assumption of the A-post top joint being pinned, reducing the structure to three redundancies.The overall bending moment on the frame can be replaced by compression force in the cant rail and tension in the sill. If the maximum moment is in front of the B-C post, BM at the base of the post is Ph^2. The balancing compressive force will be shared by the B-C and D-post in proportion to their relative stiffnesses. Other vehicle sideframes can be simulated by 'L' and 'U' shaped members as shown in Fig 63 and again Unit Load equations can be set up in terms of the main elements. Flexibilities can be added in series to obtain a picture of the whole structure including shear panels whose contribution will be At^2/Gt where G is rigidity modulus, A the panel area, t its thickness and t the shear flow due to unit torque. Method of assessing the degree of redundancy of stiff jointed frames such as the door closure ring shown at Fig 64 is to cut the ring and consider the forces necessary to reclose it (*centre*). Here two linear displacements and a rotation are required giving an order of redundancy of three. For a three dimensional ring structure this rises to six, Fig 65. Generally for a stiff-jointed structure of M members and N nodes, the number of rings will be $M - N + 1$. Fig 66 shows a worked example using algebraic symbols for a pinned parabolic arch to demonstrate the Unit Load method.

Torsion in box beams

As many vehicle structures can be idealised as box-beams, the behaviour of these structural elements in torsion is fundamental in body stressing and deflection evaluation. The theory is a valuable teacher in design prior to the more accurate prediction of loads and displacements by finite element techniques.

Torsion of box beams was introduced at the end of the earlier section on thin-walled structural analysis. It is useful to examine the change from a fully closed section box beam to an open section one, in Fig 67, which is really a limiting case of the closed tube as it collapses. The shear stress distribution has a basic influence on the torsional rigidity of the cross-section. Closed tubes resist torsion by the area of the cross section bounded by the mean thickness of the wall. Shear flows around an elemental section of the wall are equal and constant (Fig 68). Also by integrating along the circumference of the section, the sum of the horizontal and vertical components of tangential forces dP can be summed to zero and their tangential components equated with the externally applied torque T to give shear flow, $q = T/2[A]$ where $[A]$ is the area of the closed tube bounded by the mean thickness line of the wall.

The twist per unit length of a closed tube is found by equating the shear energy per unit volume with the work done on the twisted cross section. This gives twist/unitlength:

$$T/4GA^2 \; \Sigma ds/t$$

for shear modulus G and wall thickness t.

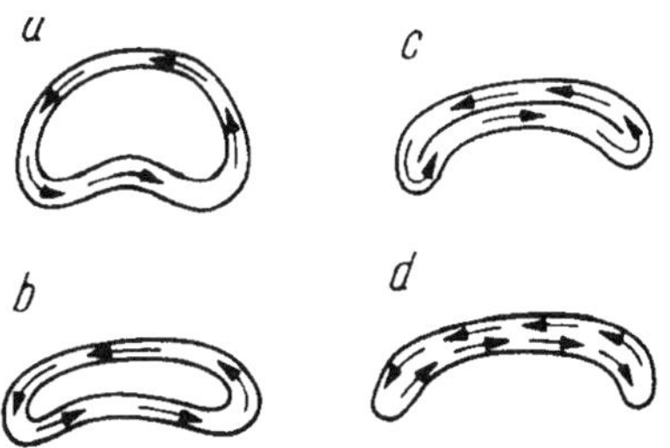

Fig 67: Closed to open section beam

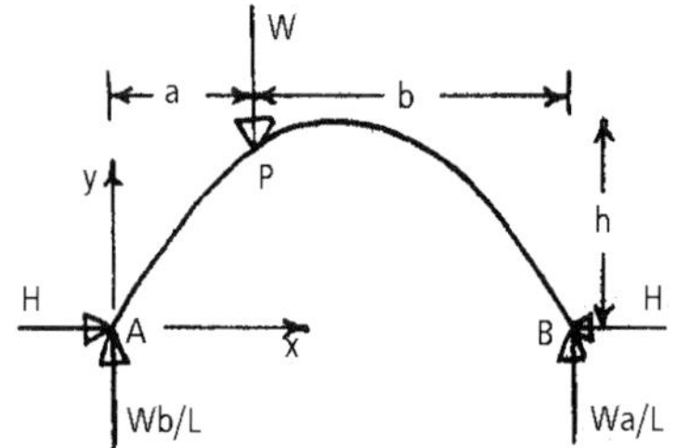

Part AP origin of x and y at A

$$\left.\begin{aligned} M &= \frac{Wb}{L}x - Hy \\ m &= -y \end{aligned}\right\} \therefore \int_A^P Mm\,ds \simeq H\int_0^a y^2dx - \frac{Wb}{L}\int_0^a xy\,dx$$

Part PB (origin of x and y at B).

$$\left.\begin{aligned} M &= \frac{Wa}{L}x - Hy \\ m &= -y \end{aligned}\right\} \therefore \int_B^P Mm\,ds \simeq H\int_0^b y^2dx - \frac{Wa}{L}\int_0^b xy\,dx$$

Adding the integrals for the two parts and equating to zero gives

$$0 = H\int_0^L y^2dx - \frac{Wb}{L}\int_0^a xy\,dx - \frac{Wa}{L}\int_0^b xy\,dx$$

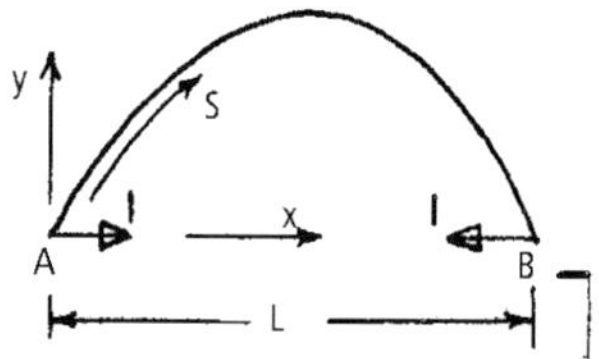

$$i.e.,\ 0 = 16h^2H\int_0^L \left(\frac{x^2}{L^2} - \frac{2x^3}{L^3} + \frac{x^4}{L^4}\right)dx$$

$$-\frac{4Wbh}{L}\int_0^a \left(\frac{x^2}{L} - \frac{x^3}{L^2}\right)dx - \frac{4Wah}{L}\int_0^b \left(\frac{x^2}{L} - \frac{x^3}{L^2}\right)dx$$

$$i.e.,\ 0 = 16h^2HL\left(\tfrac{1}{3} - \tfrac{1}{2} + \tfrac{1}{5}\right) - \frac{4Wbh}{L}\left(\frac{a^3}{3L} - \frac{a^4}{4L^2}\right)$$

$$- \frac{4Wah}{L}\left(\frac{b^3}{3L} - \frac{b^4}{4L^2}\right)$$

$$\text{Which reduces to:}\quad H = \frac{5\,Wab}{8\,L^3h}\left(4a^2 + 4b^2 - \frac{3a^3}{L} - \frac{3b^3}{L}\right)$$

Fig 66: Worked example of pinned parabolic arch

These formulae can also be used to examine the torsion in closed tubes with stringers (using the aircraft terms to describe the box beam reinforced by longitudinal bars, Fig 69). A saloon car body in torsion (Fig 70) is seen in contrast to the van type structure in Fig 71. The first illustrates the significant factor governing passenger compartment torsional stiffness is equal and opposite transverse shear forces in the roof and floor influenced substantially by the

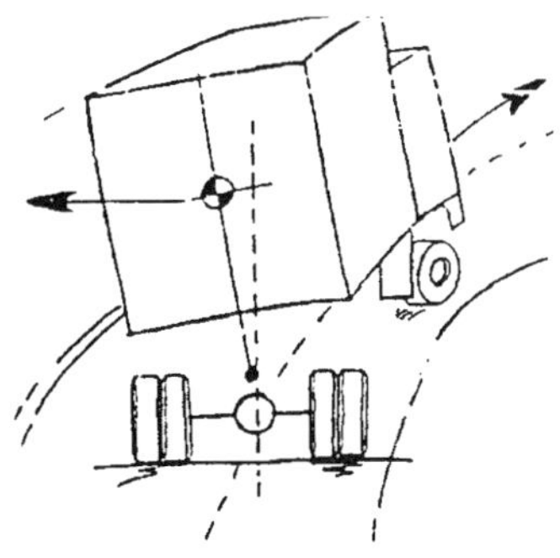

Fig 71: Van as box tube

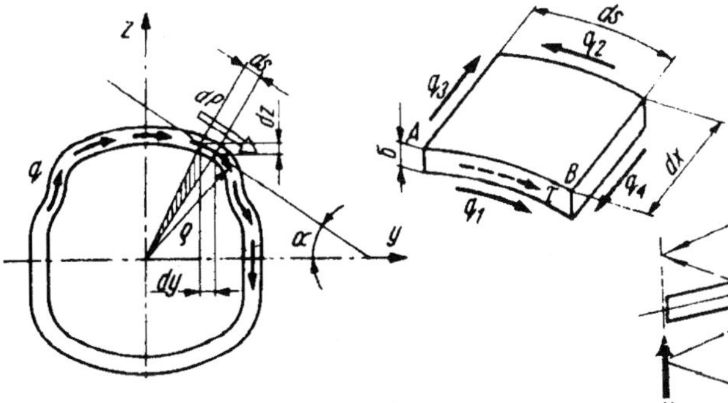

Fig 68: Shear flows in beam element

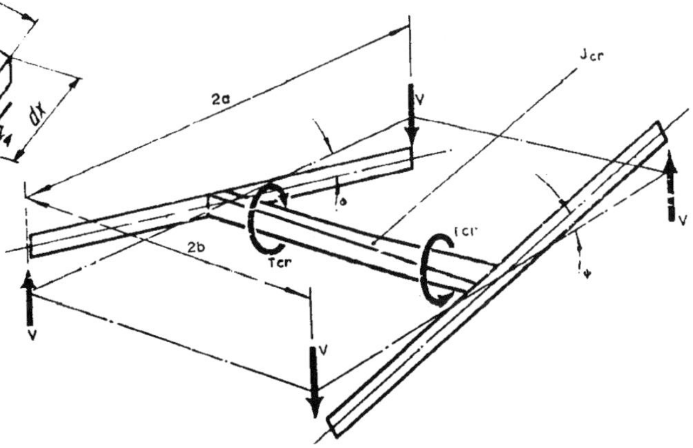

Fig 72: Underbody cross-member in torsion

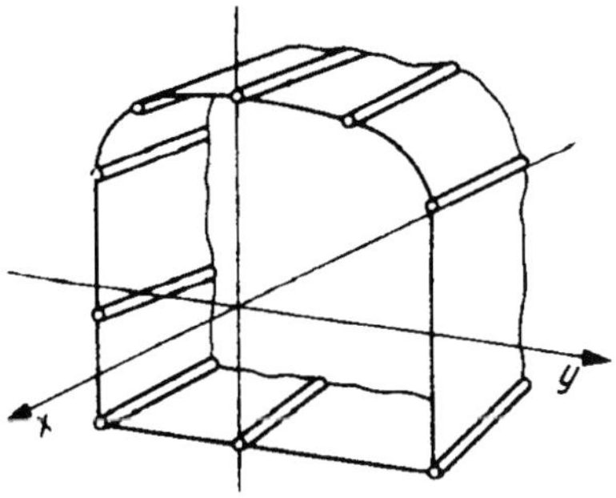

Fig 69: Reinforced box beam

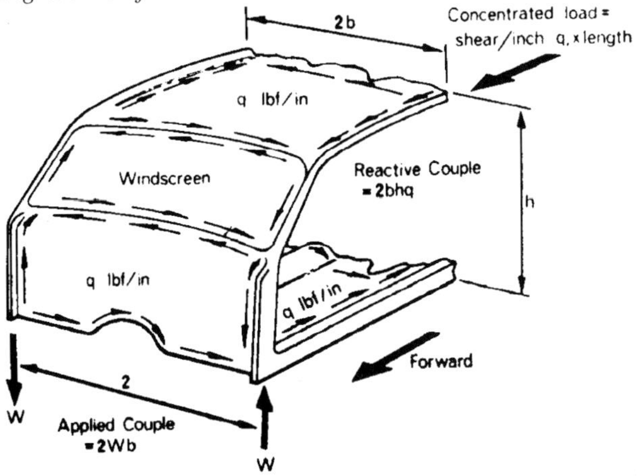

Fig 70: Saloon body as box tube

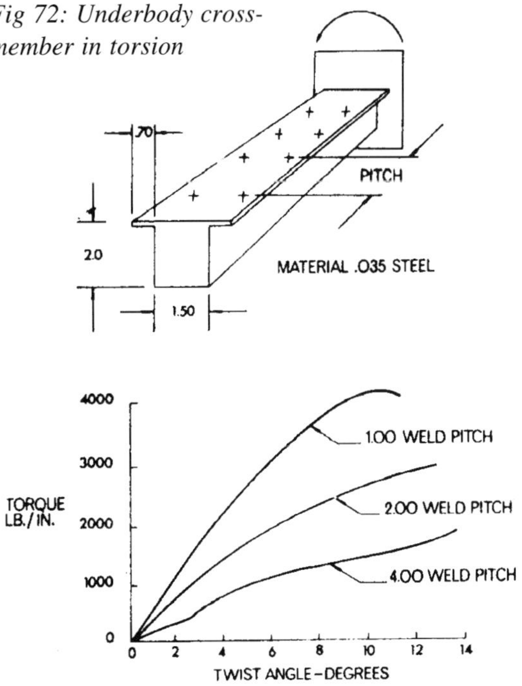

Fig 73: Spot-weld pitch effect on twisting

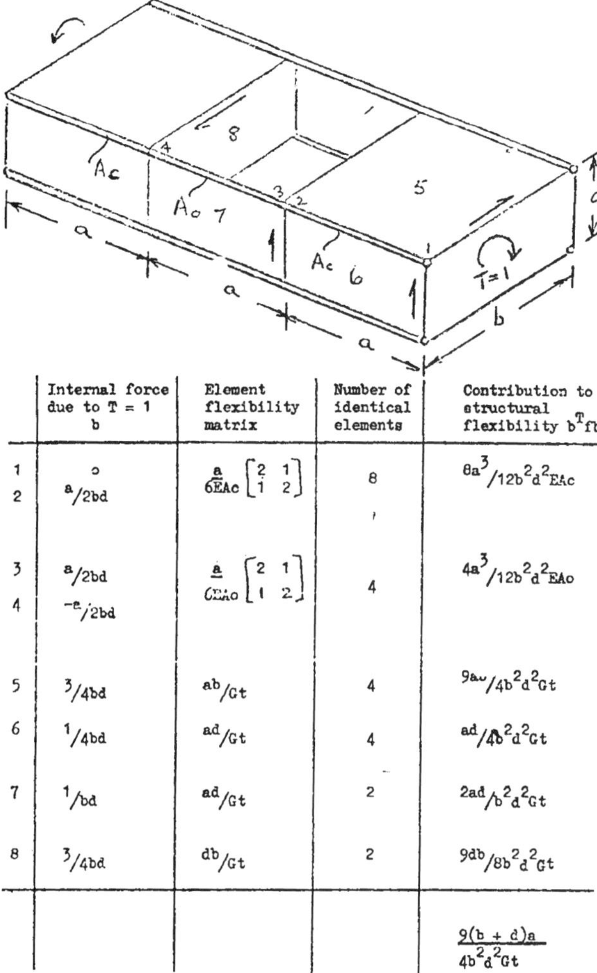

	Internal force due to T = 1 b	Element flexibility matrix	Number of identical elements	Contribution to structural flexibility $b^T fb$
1	0	$\frac{a}{6EA_c}\begin{bmatrix}2 & 1\\ 1 & 2\end{bmatrix}$	8	$8a^3/12b^2d^2EA_c$
2	$a/2bd$			
3	$a/2bd$	$\frac{a}{6EA_o}\begin{bmatrix}2 & 1\\ 1 & 2\end{bmatrix}$	4	$4a^3/12b^2d^2EA_o$
4	$-a/2bd$			
5	$3/4bd$	ab/Gt	4	$9ab/4b^2d^2Gt$
6	$1/4bd$	ad/Gt	4	$ad/4b^2d^2Gt$
7	$1/bd$	ad/Gt	2	$2ad/b^2d^2Gt$
8	$3/4bd$	db/Gt	2	$9db/8b^2d^2Gt$
				$\frac{9(b+d)a}{4b^2d^2Gt}$

Fig 74: Effect of panel removal on box tube

windscreen which closes a rectangular framework otherwise subject to lozenging deformation. There is also a contribution to torsional stiffness made by the two sills acting as a pair of torque tubes and of these acting with tranverse members to form a floor frame. Fig 72 shows torsional resistance set up in the underbody cross member due to the difference in slopes of the sills at each end, for which:

$$V = T_{cr}/2a = \phi\,(J_{cr}G/2ab)$$

where J_{cr} is the polar moment of inertia of the crossmember and 2f = pb/360a. It is important to note that spot-weld or rivet spacing has an effect on torsional rigidity as indicated by Fig 73.

Torsion is often the overall structural design criterion for vehicle bodies. Open sections of the body are torsionally much more flexible than closed ones unless the axial deflection of the ends of the open section can be restrained. In this case the torsion can be resisted by differential bending. This type of restraint might be achieved in an open car body if the front bulkhead, A-posts and dash panel form a closed torsion box. In the idealised structure of Fig 74, the effect on torsional stiffness of removing the panels from torsion boxes can be seen in the accompanying

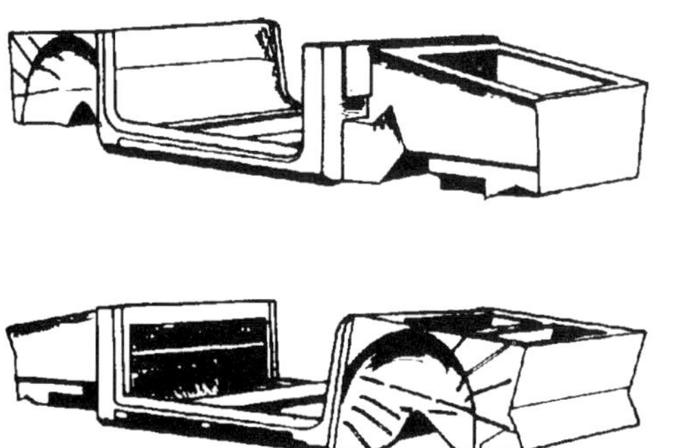

Fig 75: Car structure with enlarged sills and bulkhead

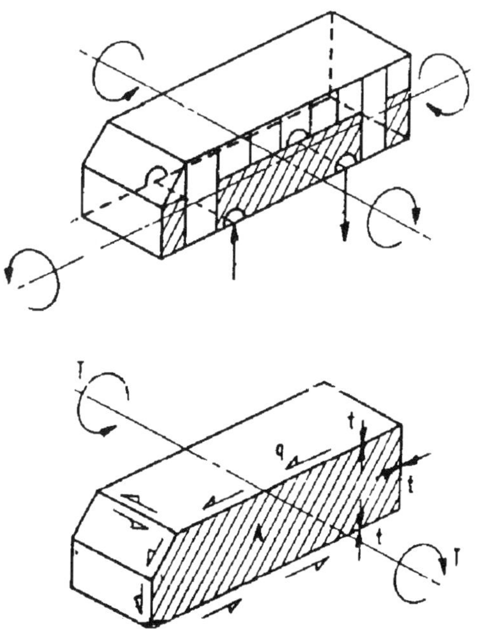

Fig 76: Coach as box tube

table, due to Dr J M Howe of Hertfordshire University. Torsional flexibility is 50% greater than that of the closed tube (or open tube with a rigid jointed frame of similar shear stiffness to the removed panel surrounding the cut-out) if the contributions of flanges and ribs are neglected.

Enlargement of the sill sections is another way to achieve the overall torsional stiffness of a vehicle body as well as partially boxing in the luggage and engine compartments in the case of the car body, Fig 75. For the example of the passenger coach structure the kind of idealisation shown in Fig 76 is useful in estimating torsional resistance. A yet more generalised idealisation for several types of semi-closed body structure is due to C J Cooke[10] of Rolls Royce Motors (Fig 77). Torsional stiffness is then given by:

$$[(bh/6a)(A+B)]/[1+(e_1 + e_2)/2a-(h/a)(k_A + k_D)]$$

due to roof and floor acting in differential shear and:

$$b^2/688F[(ab^2/J_a G) + (b^3/EI_b)]$$

due to the sills dash and rear squab acting as a rectangular frame. The terms A, B, k_A, k_D, F may be evaluated for unit angle of twist by methods given in Cooke's original paper.

The effect of joint flexibility on vehicle body torsional resistance must also be taken into account. Experimental work carried out by P W Sharman[11] at Loughborough University have shown how some joint configurations behave. The importance of adding diaphragms at intersections of box beams was demonstrated, Fig 78. Without such stiffening, the diagram shows the vertical webs of the continuous member are not effective in transmitting forces normal to their plane so that horizontal flanges must provide all the resistance. The distortions shown inset were then found to take place if no diaphragms were provided.

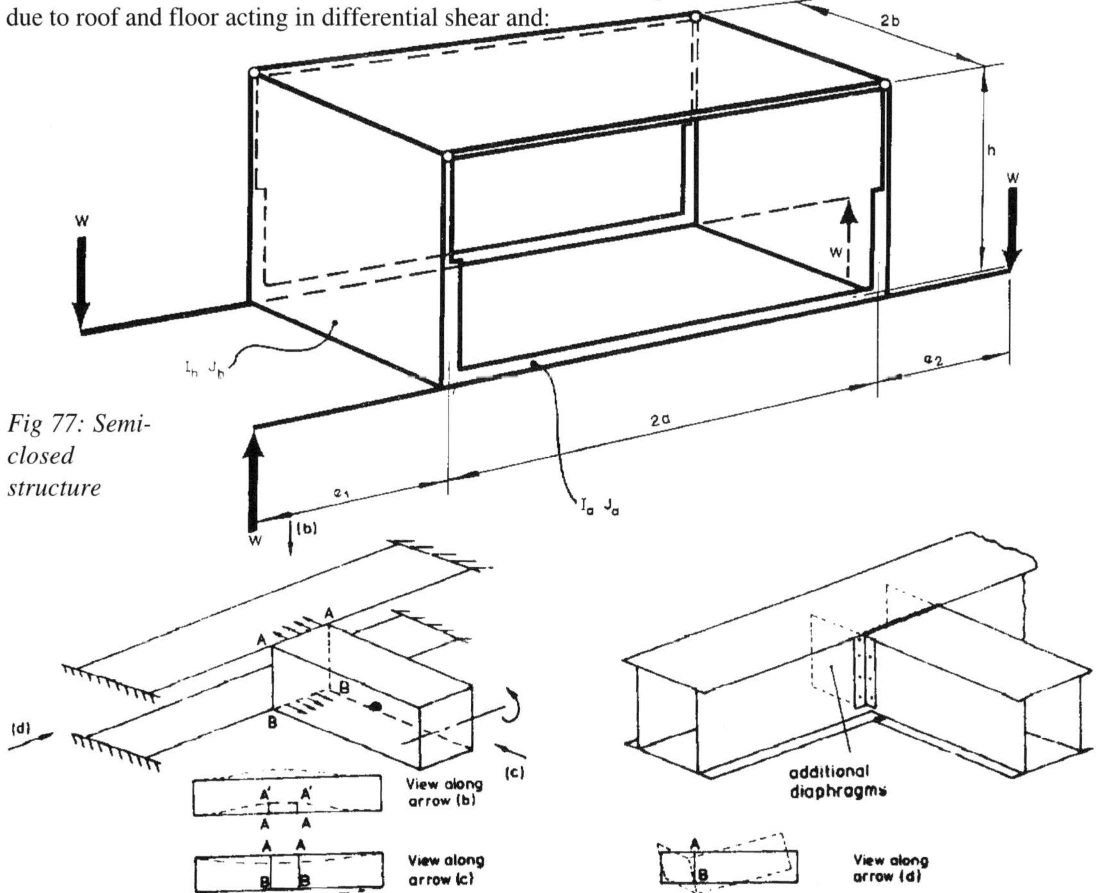

Fig 77: Semi-closed structure

Fig 78: Use of diaphragms at beam intersections

Stability of columns and plates

A key factor in the successful structural design of thin sheet panels is examination for structural instability. Too often great efforts are made to compute the predicted stresses in a vehicle body due to bending, end-load or shear without realising that in many cases the structure will suffer an instability collapse at much less than the theoretical maximum 'direct' stress due to external loading.

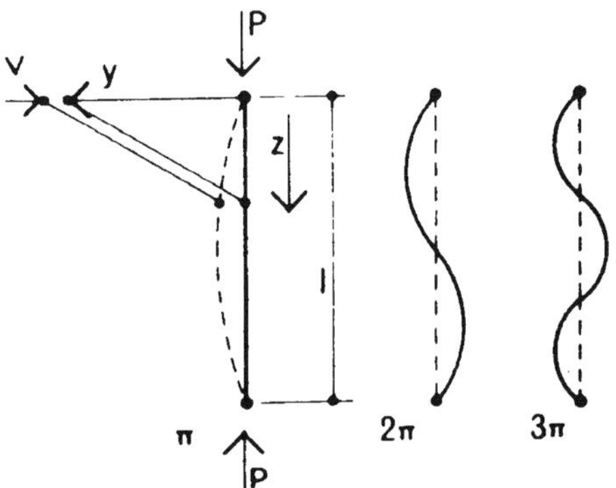

Fig 79: Strut under end-load

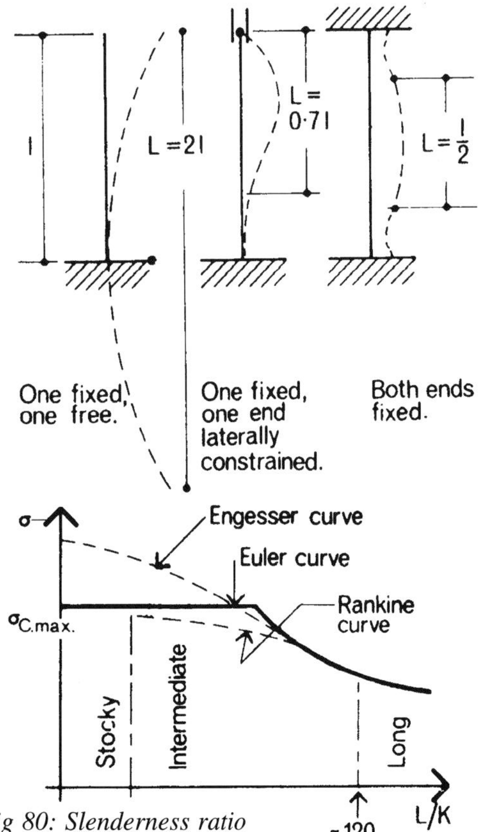

Fig 80: Slenderness ratio

The theory of panel instability is merely an extension of that applying to a narrow strut so that is an obvious starting point in the analysis. Fig 79 shows an end-loaded strut, diagrammatically, with some possible buckling modes alongside. If the strut is short enough then buckling instability will not apply; it is thus necessary to determine a critical slenderness-ratio above which the member becomes susceptible to instability failure. As the strut bows, in the single curvature mode, it is clear from the diagram that a bending moment develops in the strut due to the displacement of the deflected centre line from the line of action of the end loads. The classic analysis due to Euler involved equating this with the second derivative of lateral deflection of the strut y at any distance z along its length from one end. This is an extension of the equation of flexure considered in the section on beams earlier in this chapter. This is written as:

$$(d^2y/dz^2) = -Py/EI$$

where flexural rigidity EI is product of elastic modulus and section second moment of area of the strut. The standard solution for this differential equation, putting $a = (P/EI)^{1/2}$ is:

$$v = A\sin az + B\cos az$$

where A and B are constants which are eliminated by substituting the values of the variables at the ends of the strut. Thus at $z = 0$, $v = 0$ and so $B = 0$ while at $z = L$, again $v = 0$ so that $A\sin aL = 0$ and hence $aL = p, 2p, 3p$, successive solutions which show the theoretical possible mode shape of the strut if it were perfectly straight and the end loads perfectly in line.

Due to inevitable slight eccentricities, only the single curvature mode occurs in most practical cases, for which $a^2 = \pi/l^2 = P/EI$, a relationship leading to the formula for Euler critical load (onset of buckling):

$$P_e = \pi^2 EI/l^2$$

An 'effective length' can be substituted for l, Fig 80, when other than pinned-end conditions apply to the end-fixing of the strut. L is determined from the points of contraflexure (zero bending moment and directional change in slope) arising from the bending modes induced by the end fixing moments. Experiments show that the Euler theory is inaccurate below a slenderness ratio of *L/K = 120* (*K* is the least radius of gyration in the expression for section second moment of area $I = AK^2$ where *A* is the section area. If slenderness ratio is plotted against critical buckling stress, three broad categories of strut show themselves. The shortest, 'stocky' struts fail by direct stress without buckling while those above 120 S-R are Euler struts; between the two an 'intermediate' series exists which requires empirical formulae to allow for the fact of initial curvature or load eccentricity affecting 'Euler' critical values. Rankine's formula states:

$$1/\sigma_{crit} = 1/\sigma_{c.max} + 1/\sigma_{c}$$

from which the critical stress:

$$\sigma_{crit} = \sigma_{c.max}/[1 + (\sigma_{c.max}/\pi^2 E)(1/K)^2]$$

Classic examples of strut members in vehicle bodywork include the windscreen pillars and B-posts of saloon cars and CV cabs in the roll-over accident situation. Such a B-post section, idealised for analysis is shown as Fig 81. To determine its critical end load for buckling in the roll-over situation, its neutral axis of bending has first to be found — using a method such as the tabular one in Fig 82. Assuming the roof end of the pillar to impose 'pinned' end

Element	A	y	Ay	Ig	h	Ah²	Ig + Ah²
1.	100	½	50	100/2 = 8·3	18·5	34 200	342 083
2.	2 × 15	1½	45	30/12 = 2·5	17·5	9250	9252·5
3.	2 × 51	26½	2750	2·50³/12 = 1040	7·5	5620	6660
4.	50	51	2550	50/12 = 4·14	32	34 000	34 004·17
			Σ = 83 070				Σ = 84 124·91

Fig 82: Neutral axis determination

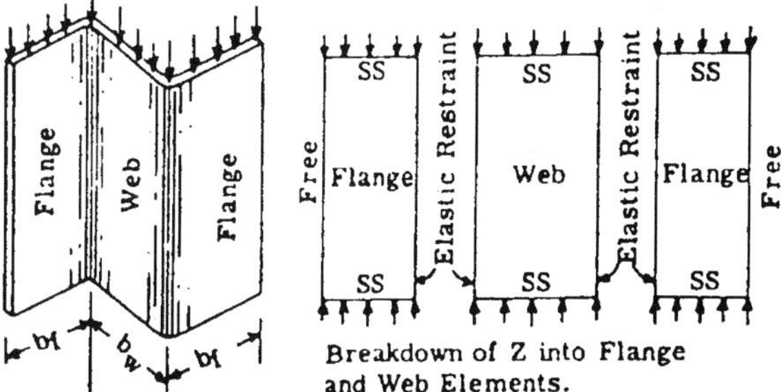

Fig 84: End-loaded zed or channel-section

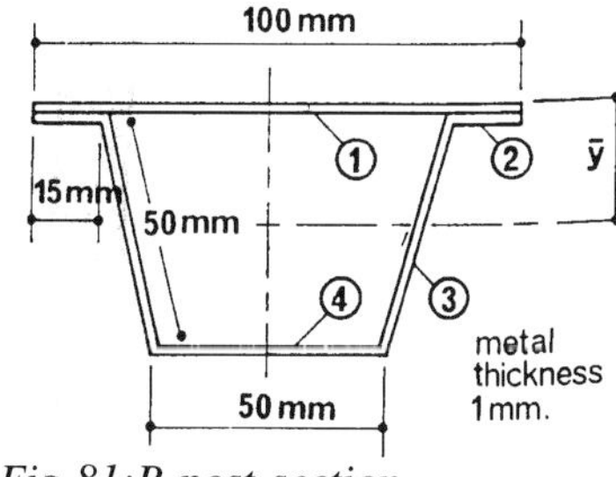

Fig 81: B-post section

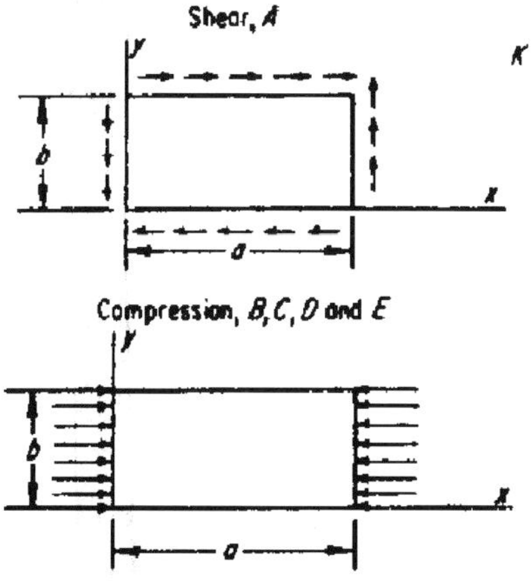

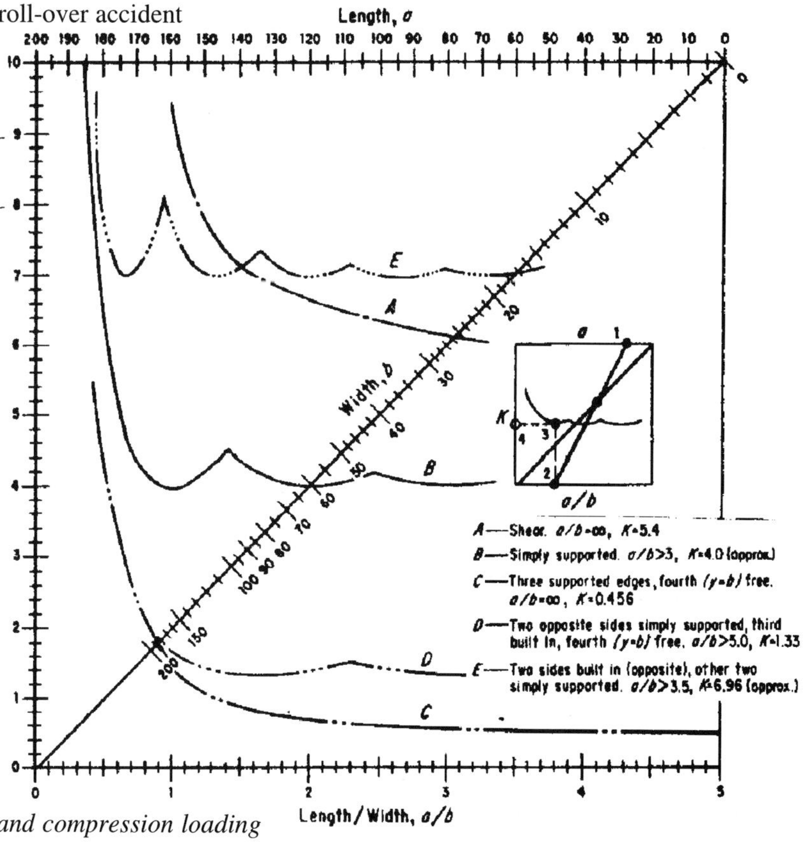

Fig 83: K-factor curves for shear and compression loading

fixing conditions (so that $L = 2l$) and that the pillar is one metre in length, then critical load is one metre in length, then critical load is $10.210.10^9.8.4.10^{-8}/22$. Taking E for steel as 210 x 10^3 N/mm^2, the stress at this load is $44.10^3/280.10^6 = 157$ MN/m^2 since $A = 250$ mm^2 — which is above the critical buckling stress. If, however, the cant rail is assumed to provide lateral support at the top end of the pillar then $L = 0.7l$ and critical buckling stress is 1280 MN/m^2 and the strut would fail at the direct yield stress of 300 MN/m^2. Other formulae, such as those due to Southwell and Perry-Robertson, will allow for estimation of buckling load in struts with initial curvature.

Plate buckling

An extension of the Euler theory can also be used to predict the buckling resistance of flat rectangular plates under end load. For a relatively narrow rectangular 'strip' it is sufficient merely to substitute strip width b in the formula for second movement of area for a rectangular cross-section of $I = bt^3/12$ for strip thickness t and then substitute this in the Euler formula. However, as the strip gets wider — to assume the proportions of a panel — then the effects of anticlastic bending (that is in the plane at right angles to the one of primary concern) have to be considered. The buckling load is then factored by $1/(1 - \nu^2)$ where ν is the Poisson's Ratio of the material. This same factoring is necessary when the panel is bending under the effect of lateral rather than end loading.

The formula for critical buckling load of a plate under end load thus becomes:

$$K\sigma\, 2Et^2/12b(1 - \nu^2)$$

with values of K, depending on edge fixing conditions taking values from the curves in Fig 83 which also gives values for buckling under shear loading, to be considered below. For a plate under end load, Fig 84, it has been found that the stress distribution across the plate is of the order of that shown in Fig 85. This has given rise to the concept of 'effective width' whereby load is assumed to be carried only by the edge strips of the plate.

Under shear loading, the critical buckling load for a flat rectangular plate is $K\pi^2 EI/b^2$. A common method of treating shear buckling is to

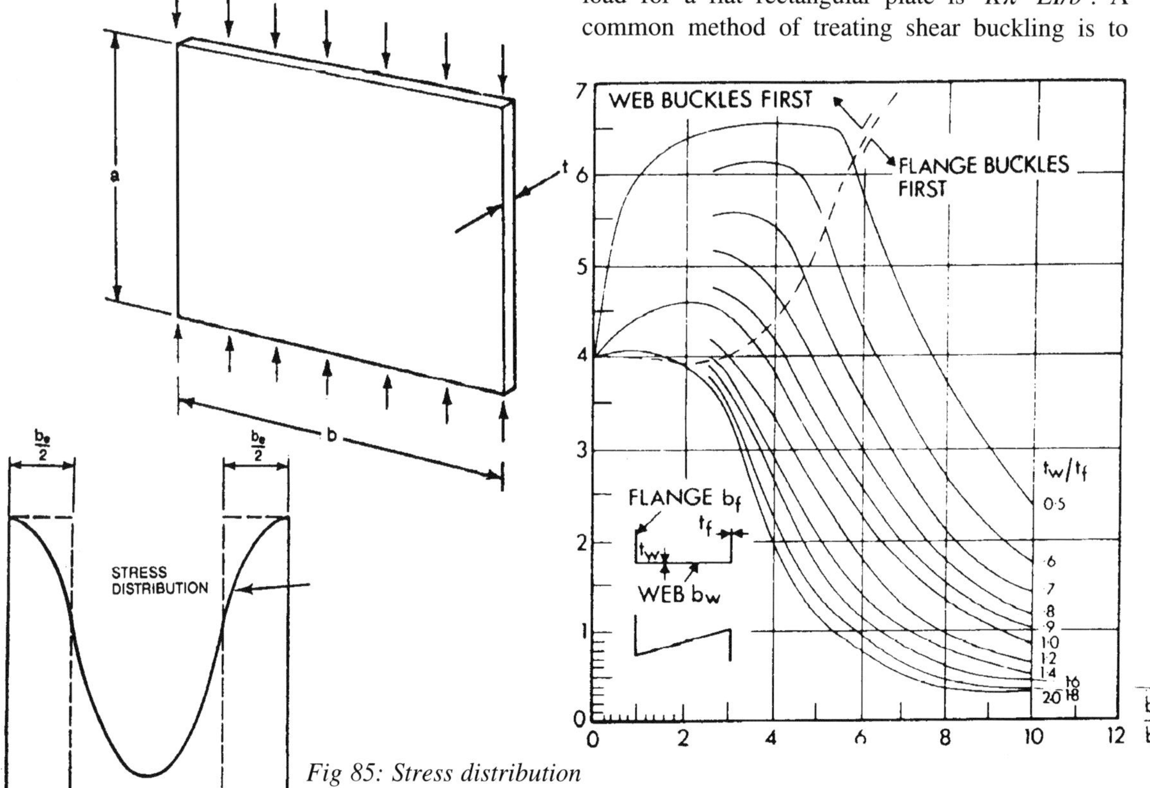

Fig 85: Stress distribution for end-loaded plate

Fig 86: K-values for zed and channel sections

represent the panel by a framework of end-load carrying bars along each edge and stabilised by two diagonals, the whole being pin-jointed at the corners. This was illustrated in an earlier section. Buckling of the compression diagonal under shear loading of the whole frame is analogous to shear buckling of an equivalent panel. The tension in the opposing diagonal is analogous to the tension fields which exist in the equivalent panel and give rise to its shear resistance. The zed or channel section under end load can also be considered to be an assembly of 'plates' subject to buckling deformation, Fig 84. The plates supply mutual elastic support for each other and curves due to Bruhn have been constructed, Fig 85, to obtain the K values for use in the buckling formulae, Fig86. As well as the overall column buckling mode the local instability of elements of such sections should also be examined. The approach is first to determine limiting stress level based on overall instability then to evaluate local instability at this level of stress.

Steel-section framed and skinned structures

High strength steels as well as mild steels can be used in structural sections but it is important to observe the so-called tension-to-yield ratio as a measure of the material's elongation. A minimum ratio of 1.08 between the yield to maximum tensile stress has been recommended for materials with limited elongation. Yield strength can of course be increased by cold forming particular elements of the section and is a funtion of the sharpness of bend between section elements.

Presence of corners in a section also permits it to carry load after buckling-collapse of the flat elements of the section. End load carrying capacity of comparatively long members is first considered in terms of overall buckling by the methods given in the previous section. The critical stress for overall column buckling is then used to evaluate the local

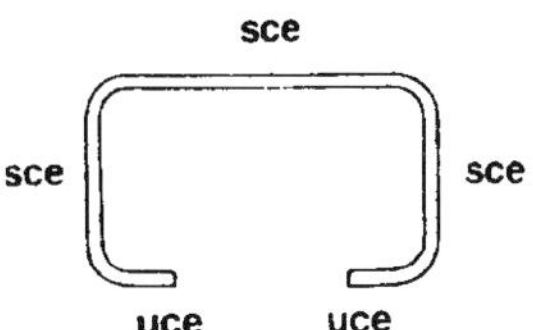

Fig 87a: Stiffened (sce) and unstiffened (uce) compression elements

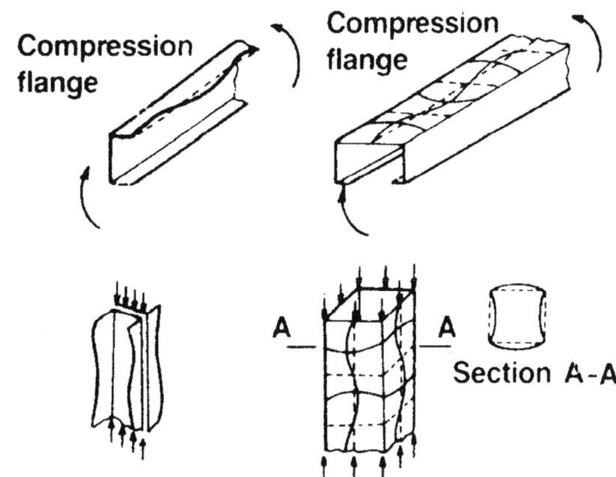

Fig 87b: Local buckling of stiffened (left) and unstiffened (right) elements

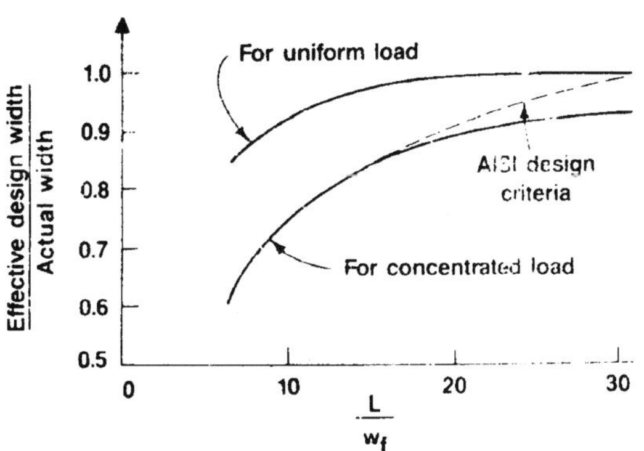

Fig 89: Shear lag effect

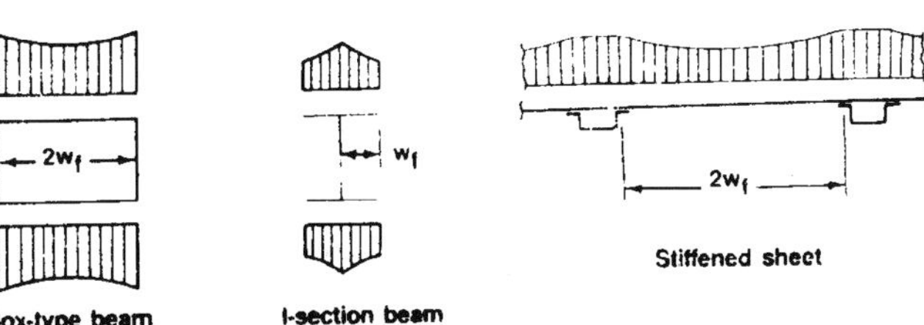

Fig 88: Effective flange width

instability of flat elements of the section. For this purpose the elements are classified as stiffened and unstiffened as shown in Fig 87a in an analytical method due to the American Institute of Steel Industries (AISI)[10].

For unstiffened elements, the overall buckling stress is reduced to account for loss of full section capacity under end load due to local buckling. The latter behaviour is assumed to be as depicted in Fig 87b as governed by the slenderness ratio of the element expressed as:

$$1.052(w/t)(\sigma/E)^{1/2}(k)^{-1/2}$$

where s is stress, *E* the elastic modulus, *k* the flat-plate buckling coefficient and *w/t* the width to thickness ratio. For unstiffened elements a maximum *w/t* value of 60 is recommended by AISI.

The bending capacity of thin wall section profiles is governed by yield strain in compression, local instability and overall lateral instability as well as the effect of local stress build-ups due to combined loading. In short beams, shear stress becomes a significant factor, as shown earlier, and the effective flange-width (Fig 88) reduces due to the effects of shear lag, Fig 89.

While it is usual to calculate web shear stress as a uniform value of *F/ht*, where *F* is shear force, the maximum shear stress at the neutral axis may be 25% larger. With web stiffeners which divide the web into rectangular panels, an increase in shear capacity is effected. The capacity is similarly increased if the web is curved to a maximum force carrying capacity of:

$$4.895Et^3/h + 0.10Et^2h/R$$

where *R* is the web radius of curvature.

Lateral instability has to be considered where section second moment area is considerably less in the transverse plane to that in the vertical plane of bending. Lateral buckling involves both bending and twisting, Fig 90, the latter due to the compression flange wanting to buckle while the tension one stays straight. As a result, torsionally stiff tubular members have little tendency to lateral buckling in bending. Rectangular section tubes have little strength reduction from this effect even down to depth to width ratios of 10. Torsional shear stress and stiffness

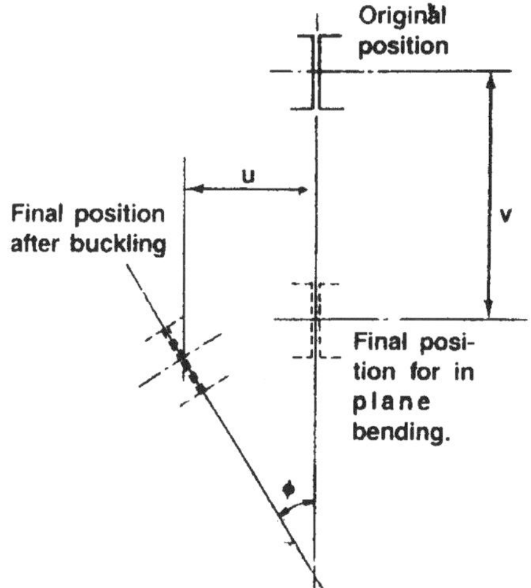

Fig 90: Lateral buckling

Cross Section	Shear Stress	Torsional Stiffness Constant J; $\phi = \frac{T}{JG}$
R, t	$v_t = \frac{T}{2\pi R^2 t}$	$J = 2\pi R^3 t$
b, h, t	$v_t = \frac{T}{\frac{\pi}{2} bht}$	$J = \frac{\pi b^2 h^2 t}{2\sqrt{2(b^2+h^2)}}$
h, t	$v_t = \frac{T}{2h^2 t}$	$J = h^3 t$
t_1, t, h, b	$v_t = \frac{T}{2bht}$ $v_{t_1} = \frac{T}{2bht_1}$	$J = \frac{2tt_1 b^2 h^2}{bt + ht_1}$ $t_1 = t;\ J = \frac{2b^2h^2t}{b + h}$

Fig 91: Torsional shear stress and stiffness

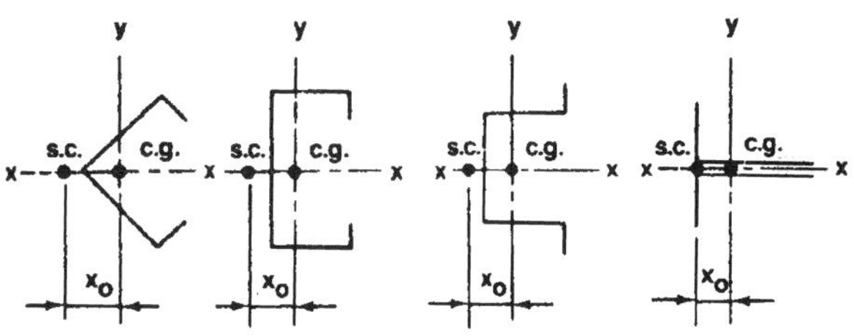

Fig 92: Sections subject to torsional buckling

formulae for closed sections are shown in Fig 91.

Sections with curved elements which are open rather than closed have different behaviour. These are not torsionally weaker than similar closed sections but have resistance made up of both shear due to lozenging tendency of the developed 'wall' (St Venant, so-called; see earlier section) and that due to axial warping. To examine the second effect, of 'differential bending', common on beams with widely spaced flanges likely to bend as well as 'lozenge' the critical factors of beam end restraint and span have to be considered.

Another consideration is torsional buckling and singly symmetric sections such as those in Fig 92 can buckle about the y-axis when the axial load is to the left of the shear centre. With loads to the right of the shear centre, failure can also occur in a torsional/flexural mode (flexurally about the x-axis). In the case of curved members, inner fibres in bending have higher stresses than determined by conventional theory and radial stresses are developed in the webs due to bending moments. Those parts of the flanges distant from the webs lose their effectiveness and tranverse bending occurs. When the ratio a/c (Fig 93) is less than 10, stress increase at the inner fibres should be considered. Stress in a curved member can be calculated from:

$$(M/Aa)[1 + y/Z(a + y)]$$

where A is section area, y the distance from the centroidal axis and Z takes a value from Fig 94, for some example sections. This is due to the neutral axis moving towards the centre of curvature as seen in Fig 93. The result is distortion of the flanges and an effective flange width can be calculated as $0.7(a_f t_f)$.

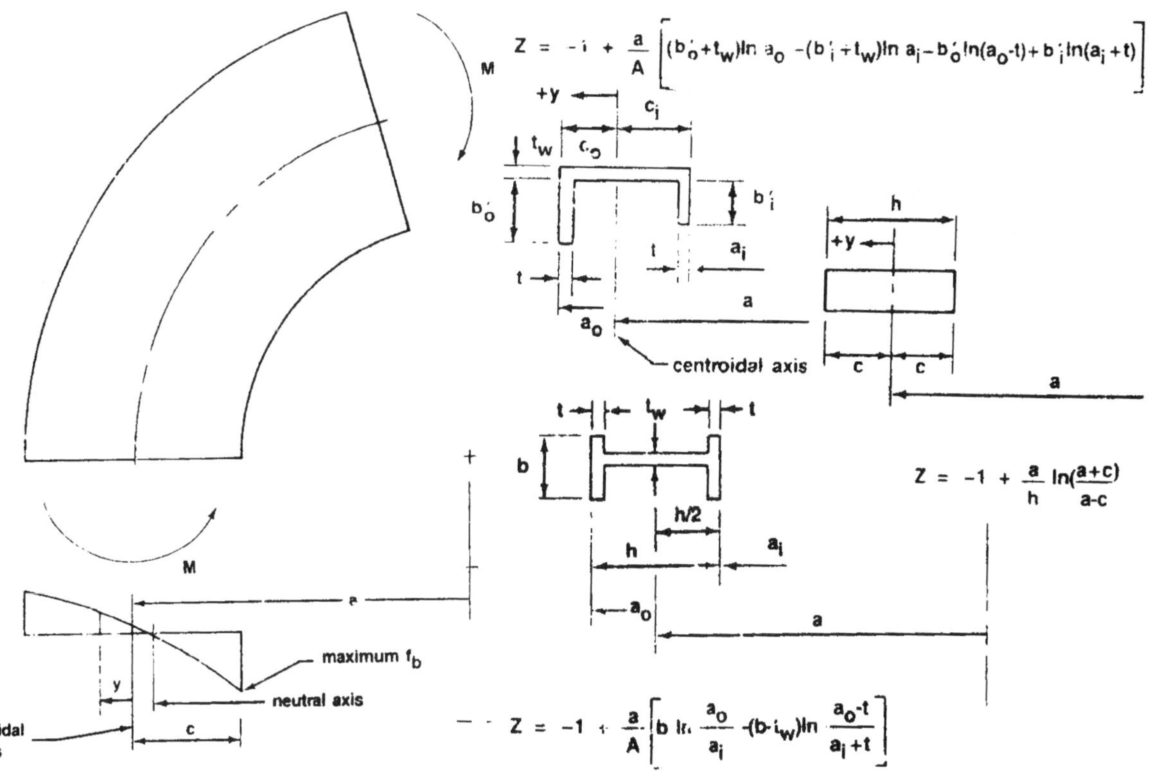

Fig 93: a/c ratio

Fig 94: Zed values for three sections

Buckling under end load

In the case of curved section elements such as those in Fig 95, both stiffened and unstiffened, but typically braced along one or both sides by a corner bend, an extension of buckling theory for flat plate elements can be applied. For flat elements, the buckling stress is given by:

$$f_{crf}=180,760(t/L)^2$$

But for most elements, which are straight over their length, the length dimension is such that it is beyond that where buckling, evaluated by the above analysis, would occur. Local buckling will be similar to that derived for a cylinder where critical buckling stress in MPa is given by:

$$f_{crc}=0.605Et/R$$

where t is thickness, L the length and R the sheet curvature radius. Elastic buckling of curved plate elements is between these two values and given by:

$$f_{crp}=[f_{crc}{}^2+(f_{crf}/2)^2+f_{crf}/2]$$

Inelastic buckling can also be obtained from the above by substituting the secant modulus for E.

There is also post buckling resistance offered by curved elements such as the ones illustrated, due to their perpendicular flanges. An effective width of plate to represent the flanges is considered, Fig 96, analogous to that assumed for a flat element. However, unlike the latter, the ineffective portion of a curved element is assumed to carry the critical buckling stress as a circular cylinder having the same thickness and radius. Instead of neglecting the centre portion, this is assumed to have an effective thickness:

$$t_e=(A_o/A)(F_y/f)t$$

where A/A_o is taken from Fig 97.

When the curved segment is tangential to the straight segment, rather than separated by a sharp bend, the evaluation is more complex. According to AISI[9] a curved segment with R/t </= 7 is normally used to stiffen a flat element and in this case the flat element is measured from the point of tangency, Fig 98. For a curved element with radius thickness ratio

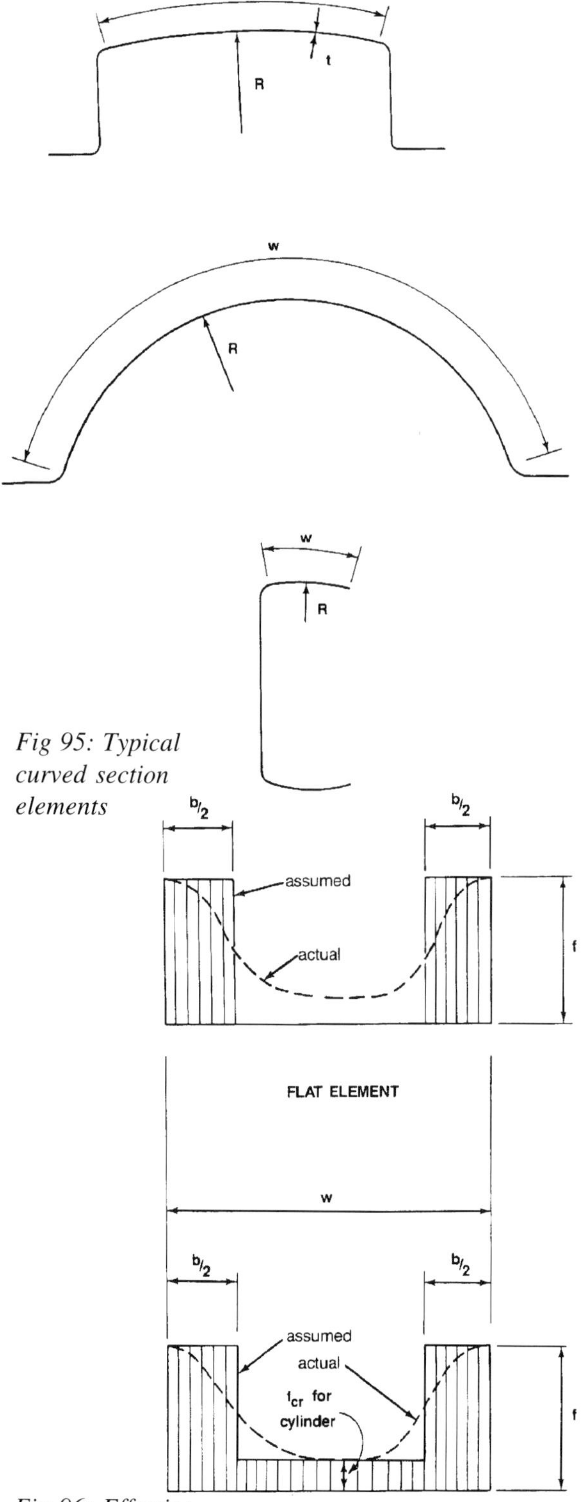

Fig 95: Typical curved section elements

Fig 96: Effective width assumption

</= 33 (assuming 345 MPa yield steel), the complete yield strength of the element can be realised before buckling. But measuring *w* from the point of tangency may not be safe, nor the assumption that the curved element provides the stiffened edge. At some dividing point the flat element is taken as half the stiffened element or even as an unstiffened element, as illustrated.

For the case of element (b), in the figure, an approach to evaluating this problem is seen by example. Here $(R/t)>7$ and the radiused portion will thus reach yield. If the flat portion has width 90 mm, thickness 1.8 mm and radius 36 mm, a new *w* is established, arbitrarily, which is larger than 90 mm, by estimating the appropriate additional amount of arc to add from $(p/4)t[(R/t)-7]$ such that w =108.4 mm. Equivalent width *b* is then calculated as 80 mm so that 40mm would be consider effective adjacent to the sharp bend and 21.6mm of the flat portion adjacent to the large radius corner would be effective. For the element at (c) where *R/t* is large enough for the equivalent area A_o of the radius is less than the actual area. The effective region near the sharp radius can thus achieve a higher stress than the large radius portion. So by changing, as before, the radius to 84 mm and using the same relationship for amount of arc, then w=146.1 mm. Fig 96 can be used to find *b* and then the stress on the section will be shown to be

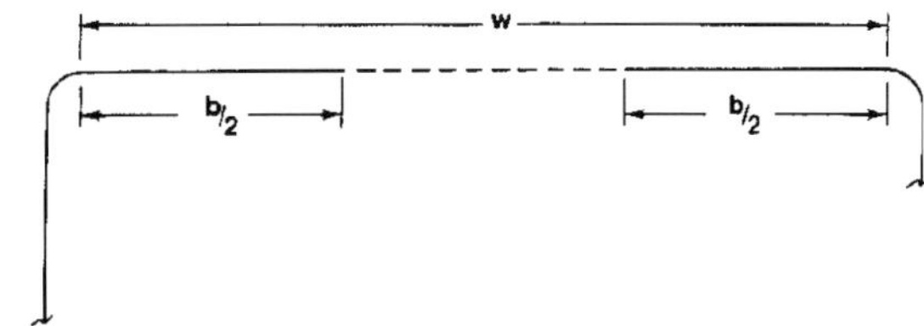

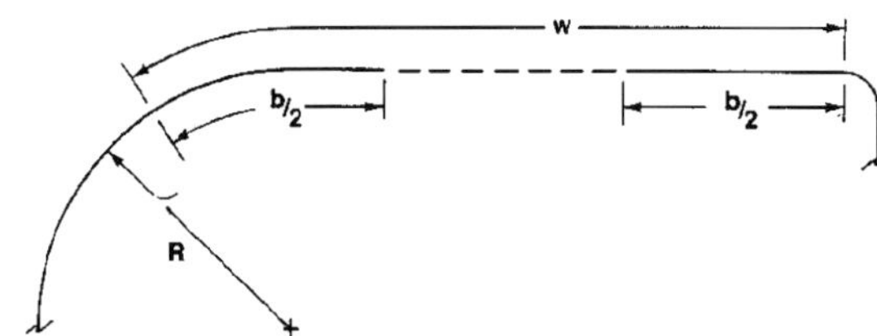

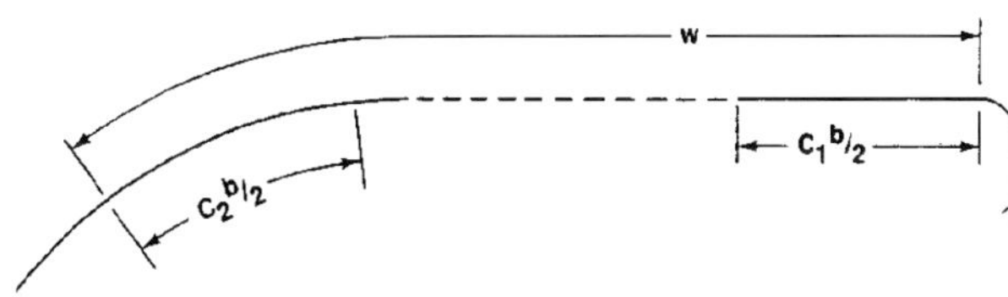

Fig 98: Combined flat and curved elements - a,b and c from top to bottom

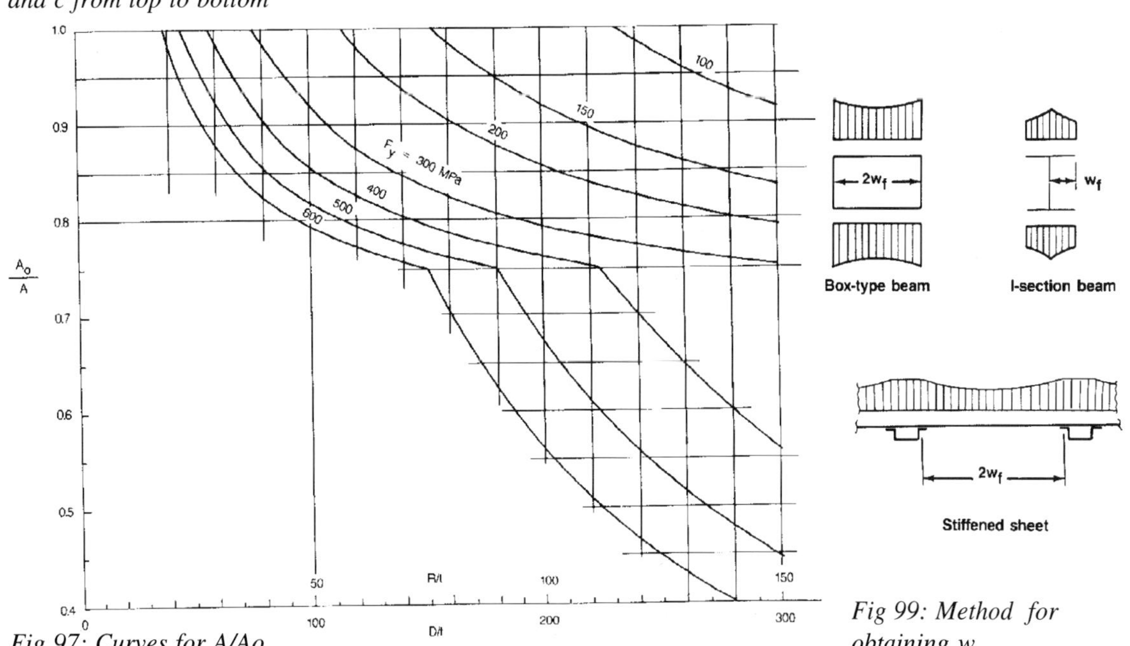

Fig 97: Curves for A/Ao

Fig 99: Method for obtaining w_f

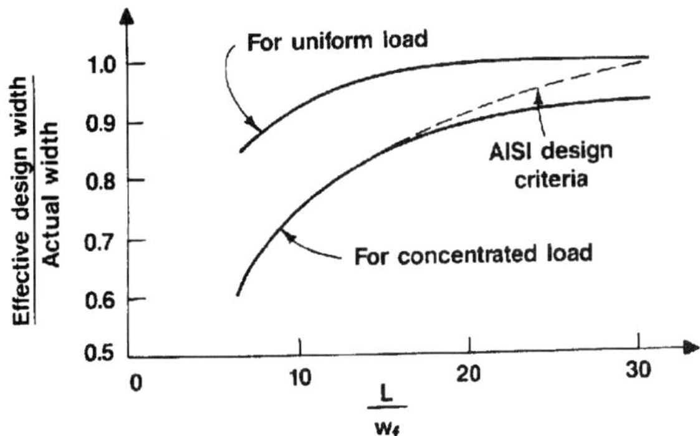

Fig 100: Curves for w_f

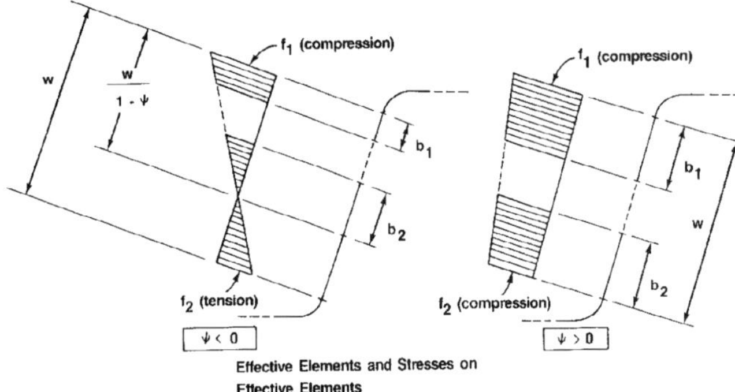

Fig 101: Stress distribution on web

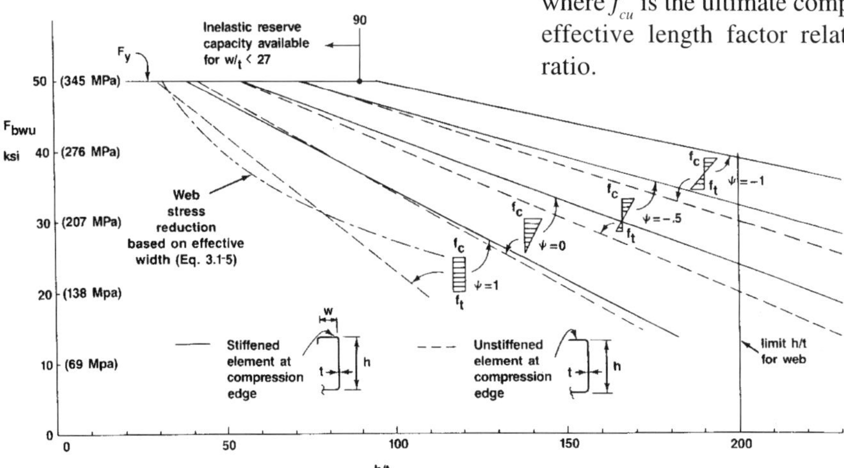

Fig 102: Element stiffened on the compression side of the web

governed by the limiting stress on the radius portion.

Sections comprising a number of flat and curved elements pose greater difficulty and requires an approach which considers all components in the section and their relationship to one another. If such a section has a sharp corner with $(R/t)<0.057E/F_y$ then post-buckling strength can be expected; if not the radius portion, or flat portion tangential to the radius, will limit capacity. If the sharp corners are not evenly spaced over the section, post-buckling strength may be useless as the section properties in that condition may produce a load less than the elastic buckling load.

Overall stability of structural members has been calculated using the total cross-section area A and a factor Q representing the ratio of effective-to-total area at yield. Now the approach is to calculate on the effective area calculated at the ultimate stress determined by lateral buckling of the member:

$$\text{For } f_e<F_y/2,\ f_{cu}=f_e;\ \text{for } f_e>f_y/2,\ f_{cu}=f_y[1-f_y/4f_e]$$

where critical Euler stress:

$$f_e=p^2E/(KL/r)^2$$

the above applying to any sections not subject to torsional-flexural buckling. Axial load capacity:

$$P_u = A_e f_{cu}$$

where f_{cu} is the ultimate compressive stress and K an effective length factor related to the slenderness ratio.

Bending of thin-walled beams

For short span thin walled beams the effect of shear lag is to cause an effective flange width which is less than the distance between the webs. The width w_f is defined for different beam sections in Fig 99. Since the load enters the flange from the webs, and develops gradually with bending moment, other factors than compressive stress and element slenderness can affect effective width which is determined from the curves in Fig 100. If the *w/t* ratio of a stiffened compression element is lesss than $(E/f_y)^{1/2}$ it may be posssible to permit inelastic stress distribution in the web so as to increase ultimate bending moment capacity beyond the yield moment.

The flat elements are the webs of the flexeral member section and these have a linear stress variation across them as in Fig 101, according to the symmetry of the section about its neutral axis of bending. Generally a web height is limited to *h/t=200* if no transverse or longitudinal stiffening is provided. If the section assymetry makes tension the larger stress, the likelihood of compression-stress control is small if *h/t<200*. When compression is the larger stress the limiting stress can be found from the expressions which follow. A stiffened element at the compression side of the web exerts a rotational constraint leading to a higher buckling load than can be achieved by an unstiffened one, Fig 102.

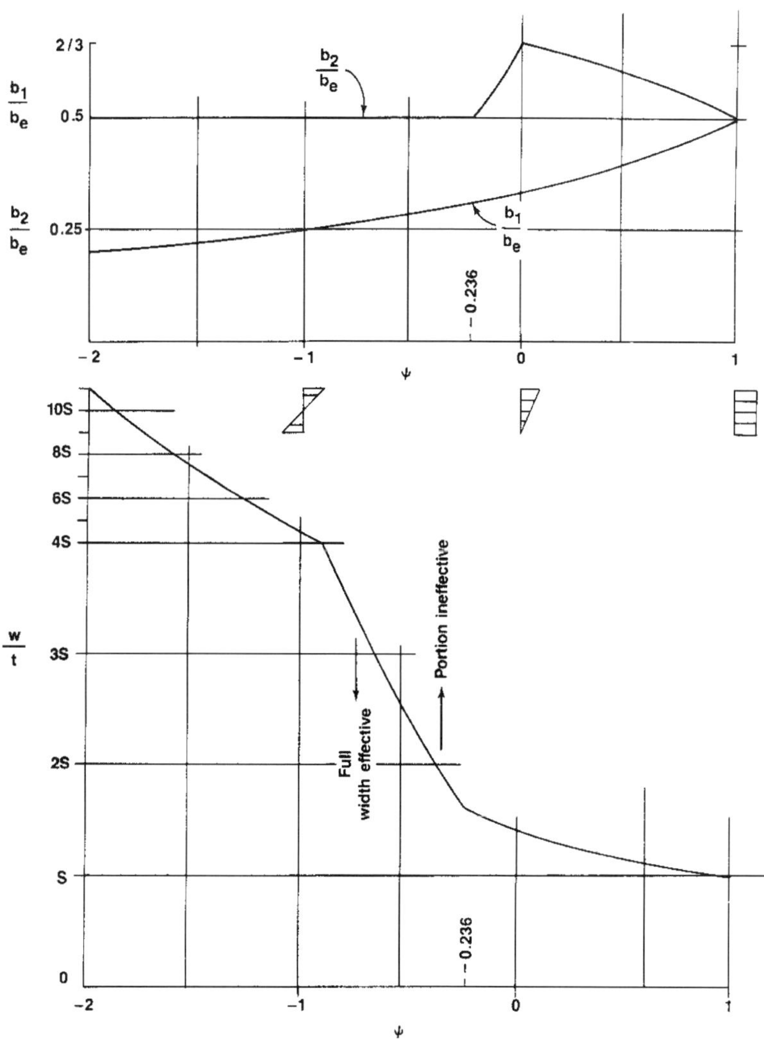

Figs 103 and 104: Effective flange widths

For the stiffened element:

$$f_{bwu}=[1.197 - 0.247(h/t)(f_y/kE)^{1/2}]f_y$$

and for the unstiffened one:

$$f_{bwu}=[1.229 - 0.374(h/t)(f_y/kE)]f_y$$

where $k=4 + 2(1-j)^3 + 2(1-j)$ for $j=f_t/f_c$ (Figs 103 and 104). k is the buckling factor as a function of variation in tensile to compressive stress ratio as plotted in the figure. When $j = 1$ (uniform stress) an h/t between 30 and 39 is where buckling begins to reduce stress limit. As compression on the element decreases and pure bending approaches, the element can achieve stress levels shown for $j = -1$. The effective width of the compression portion of the flat element is shown as b_1 and b_2 in Fig 103 with $b_1 = b_e/(3-j)$. For mainly bending stress variation ($j=/<-0.236$) width $b_2=b_e/2$ and for $j>-0.236$ $b_2=b_e-b_1$ w.

These are plotted in Fig 104 over the practical range of j, the w/t at which the flat width becomes effective being shown in the lower part of the figure.

Inelastic stress redistribution allows the section to strain beyond yield in some cases. If w/t for the flange under compression is less than $1.11/(f_y/E)^{1/2}$ and the depth to thickness ratio of the compression portion of the web is less than the same amount, the section can be strained beyond yield, provided there is no instability, shear stress is below $0.577F_y$ and web inclination is less than 30 degrees. The compressive strain is limited to three times the yield strain but there is no limit to tensile strain. The flange must also be stiffened by two webs so generally only tubular or top-hat sections can achieve this extra capacity. For preliminary design it is best to limit the compression strain to 0.005. Fig 105 shows stress distribution which can occur in bending; (a) and (b) are governed by the equations in (c).

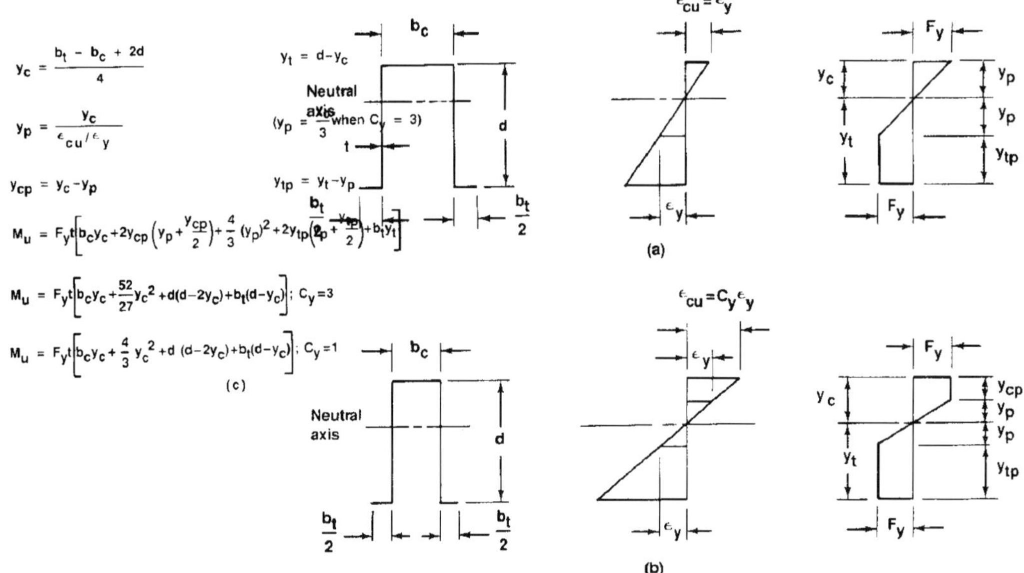

Fig 105: Stress distribution in bending

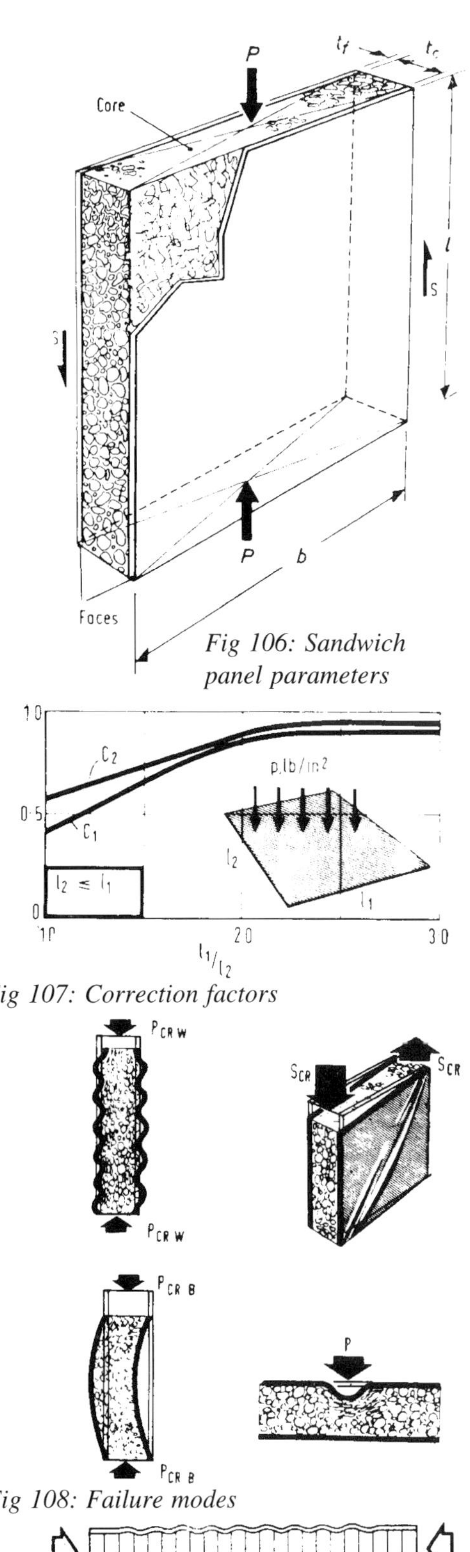

Fig 106: Sandwich panel parameters

Fig 107: Correction factors

Fig 108: Failure modes

Fig 110: Honeycomb-cored panel

Sandwich panel analysis

Analysis of forces and deflections in faces and cores as a means of optimising the main parameters is important for structural performance. Choice of skin and core modulus, density and sustainable working stress can be made so that the lightest possible panel is stable with respect to the external loading and that acoustic properties are optimised.

Principal dimensions of sandwich panels and a representation of the external forces for analysis is shown in Fig 106. Direct compressive and shear loads shown, rather than lateral bending or local indentation forces, are the ones for which the resistance of the panel to buckling, or local wrinkling, must be considered. The concept of shear buckling in unsupported skin panels has been introduced in an earlier section alongside buckling of unsupported skins under direct end load. At this stage formuli will be provided, but not fully derived, to show the critical buckling loads. The objective of a structural sandwich panel is to fully exploit the strength of the face material without premature instability of the panel. Even the most elaborately reinforced 'sheet/stringer' arrangement of panel and framework is said to be liable to failure under end load — with induced stress below 50 per cent of the maximum stress sustainable by the panel. This is because of a lack of bending stiffness in the panel and its frame. In the case of the sandwich panel, the core material, being capable of transmitting shear force between the sandwiching skins, effectively acts as the web between two 'flanges' represented by the skins. The efficiency of the sandwich panel is thus usefully defined as the ratio: maximum load per given width of panel by the weight of the given area of panel — divided by the same ratio for an equi-sized panel which has faces fully stabilised by a weightless core. This points to the key properties of the core material as specific stiffness and specific strength and the failure modes under end load being local instability of the faces or overall column buckling of the panel (or, if over stabilised, direct overstress of facings or core). Sandwich panels were first seriously analysed in the 1940s co-incident with their use in aircraft structures such as the Mosquito of World War Two fame. Researchers at the Royal Aircraft Establishment (Farnborough) used a strain-energy method to derive the critical end-load for wrinkling failure as:

$$0.63E_f^{1/3}E_c^{2/3}$$

where E is the elastic modulus for face and core. For column buckling of the panel, the parameters of panel length and edge support must also be considered — together with thicknesses of face and core plus the bending stiffness of the combination.

To calculate sandwich panel bending stiffness it is generally assumed that shear stress is uniformly distributed across the core. Central deflection for a uniformly loaded sandwich panel of moderate width, simply supported at its ends, is given by:

$$(5wa^4/384E_fI) + (wa^2/8G_cA)$$

where a is the length between supports, A is the core cross section area, I the section second moment of area and G_c the shear modulus of the core material. To allow simply supported panels to be designed without resort to complex calculation, correction factors shown in Fig 107 have been derived which, if used with simple beam theory, give good approximations in calculation. The deflection, or bending moment, is first calculated by simple beam theory, above, and then multiplied by the appropriate factor, C_1 in the case of central deflection and C_2 in the case of bending moment. The critical end load P_{cr} for unit width of panel, in overall column buckling, can then be found from the formula:

$$(1/P_{cr}) = (2a^2/\pi^2E_ft_ft_c^2) + 3/2Gt_c$$

factored by $(1 + a^2/b^2)^2$ for panel length a and width b where dimensions shown in Figs 106 and 108 apply.

For shear instability of the panel depicted in Fig 106, critical shear stress in the panel is given by:

$$(2.75K_sE_c\text{-}f\,t_c^2/b^2)/(1+ 5.5E_ft_ft_c/[b^2k_{sp}G_c])$$

where K_s and K_{sp} are factors relating to single faces and sandwich panels respectively as obtained from Fig 109. The above calculation formulae apply to isotropic core materials such as rigid foams for which it can be assumed that physical and mechanical properties are uniform in all directions. A modified approach is necessary for honeycomb cores such as those used for racing car construction, Fig 110. Formulae due to Ciba Geigy which apply to that company's Aeroweb honeycomb core material used in sandwich panels give critical buckling stress for the face skins as $(d/b)^2EK$ where d is defined in Fig

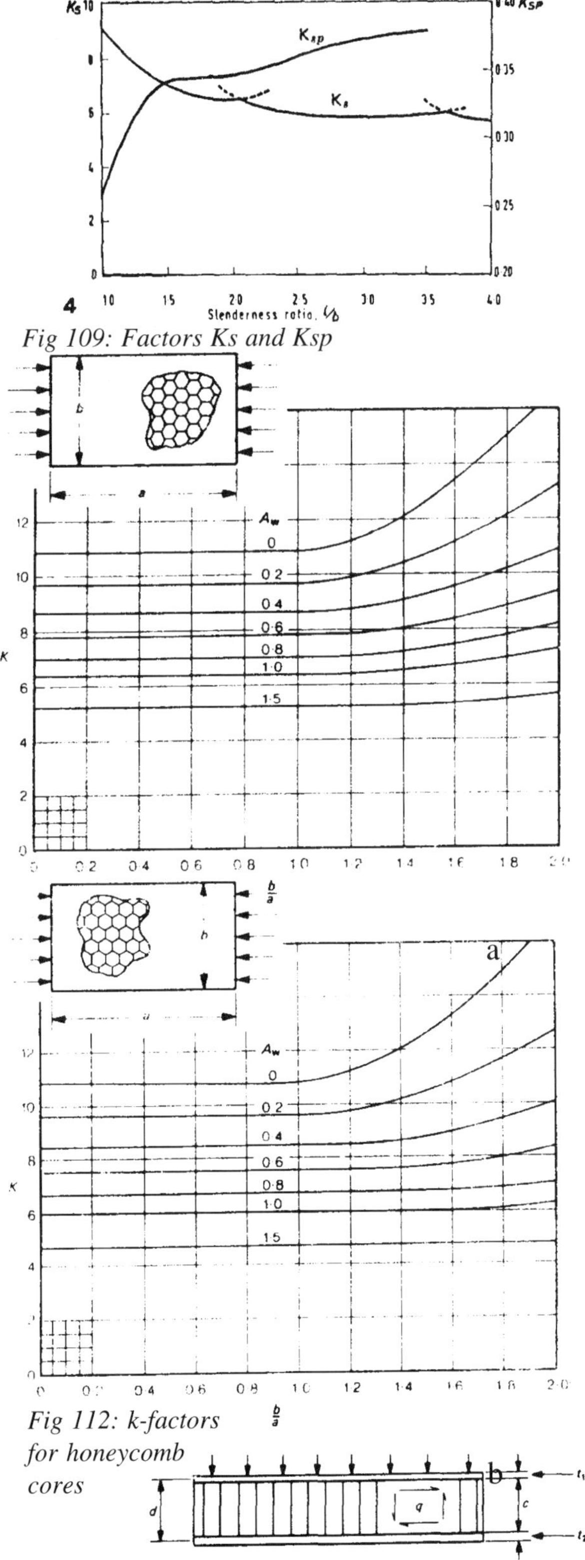

Fig 109: Factors Ks and Ksp

Fig 112: k-factors for honeycomb cores

111 and K is obtained from Fig 112a or b according to the core ribbon direction. The various failure modes for honeycomb core panels are described in the company's Construction Trends publication series (July 1989) and the critical wrinkling stress for end-loaded panels becomes $0.50(G_c E_c E_f)^{1/3}$. A further possible failure mode is inter-cell buckling shown in Fig 111. Here the critical buckling stress in the skins is given by $3E_f(t/d)^2$ where d is now the diameter of a circle inscribed within that cell.

For a rectangular honeycomb-cored sandwich-panel subject to uniform normal pressure, the maximum value of stress in the direction of the long l or short s side of the panel is given by $\sigma_{l,s} = s^2pC_2, C_3/dt_1$ where C_2 and C_3 are obtained from either Fig 113a or b depending on the core ribbon direction of the honeycomb. For the lower facing skins, the stresses are obtained by substituting t_2 for t_1 in the formula above. Shear stresses in the core under this load condition can be found from $q_{1,2} = spC_4, C_5/d$ where $C_{4,5}$ are obtained from the appropriate graph in Fig 114a and b. The central-deflection of panels under uniform side loading p can be calculated from:

$$s^4p\ C_1\ /EJ \text{ where } J = t_1t_2d_2/(t_1 + t_2)$$

and C_1 takes values from the graphs in Fig 115a and b.

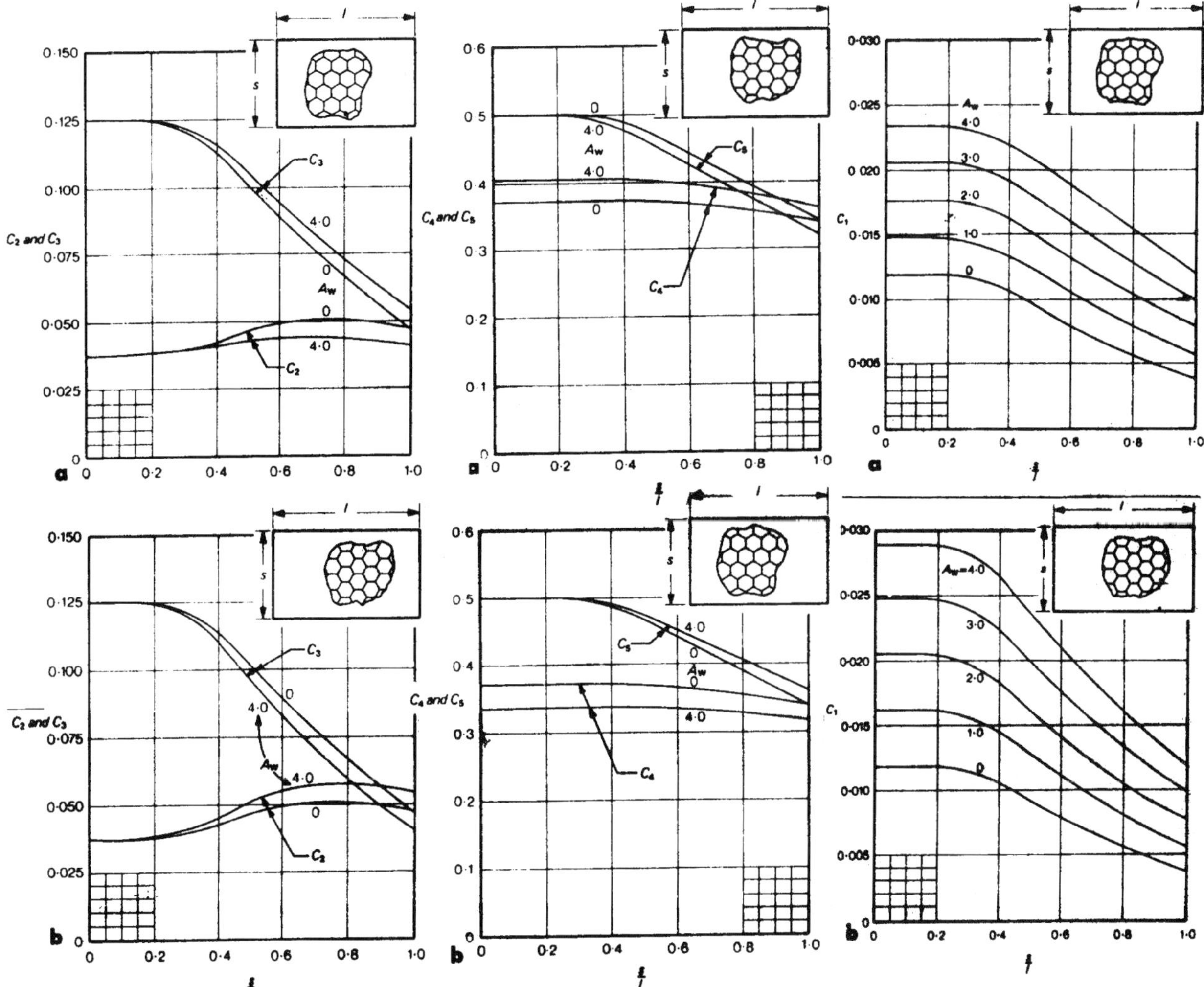

Chapter 7: Open and closed structures

Open platform structures; closed body structures; chassis-analysis; sub-assembly analysis; body shell analysis; shell structure loading, mounting and stability; stress concentration in cut-outs; finite-element analysis and FEA applied to car bodies

Flat-platform structures

Since the beginnings of the motor vehicle the prediction, at the design stage, of how the body structure behaves in terms of internal stresses and deflections, has challenged body engineers. Historically the presence of a chassis structure strong enough to react the high loads occurring at the mountings of the engine, gearbox and suspension springs ensured the adequate support of even the flimsiest of body structures. In recent times the integration of chassis/body structures in vehicles like saloon cars and road tankers has required sophisticated predictions of structural strength — aided by the arrival of computer programs to make the handling of calculations less tedious.

Even where a separate chassis frame exists, changes in the severity of vehicle operation, and the need to make bodies lighter, means that ways of assessing strength before the body is committed to metal are likely to appeal to engineers seeking to improve vehicle efficiency. Approaches to analysis vary with the category of body structure involved and in the first part of this chapter the open-top body will be considered. This applies to both cars and commercial vehicles — open, or coupé type, performance cars and platform, or dropside, bodied trucks. Such flat, or punt-like, structures are characterised by having all torsion and bending transferred through the 'floor' (Fig 1) which requires a rigidity in both these respects. In the case of a car, the floor panel is bounded by edge-members (sills) and beams (crossmembers) used to mount power-train and suspension. A useful approach to the analysis of truck body structures is due to a former Cranfield Institute of Technology visiting lecturer, Dr Ing Pawlowski[1]. It is an admirably simple and flexible technique known as the simple structural surfaces (SSS) method. It is a systematic procedure for applying simple idealisation and analysis procedures developed during the early days of monocoque aircraft structures. For the purpose of analysis, the vehicle is represented by mostly rectangular sheet panels and associated rails (or bars). These do not exactly correspond to the panels, rails and pillars of the actual final structure for they are 'representative' in the sense only of the principal internal structural load-carrying capacity of the members — shear, in the case of the panels, and end-load, in the case of the bars. This gives a very simple, yet quite accurately representative, 'idealisation' of the body structure.

Bending forces are represented indirectly in the sense that a beam element of the structure is ideal-

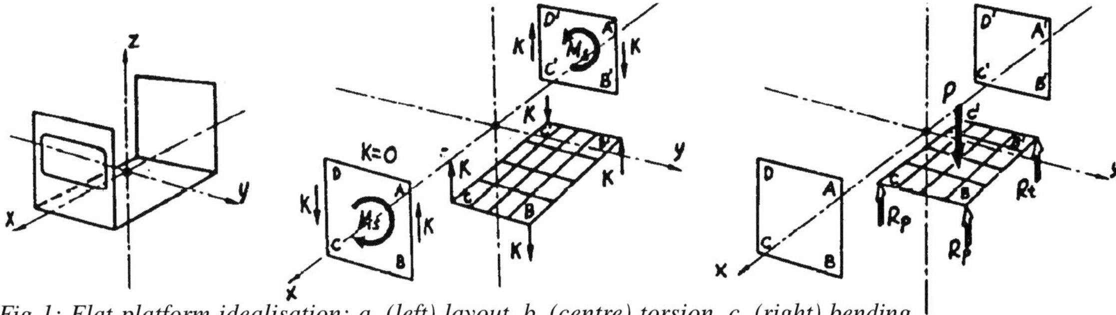

Fig 1: Flat-platform idealisation: a. (left) layout, b. (centre) torsion, c. (right) bending

ised by a panel representing the web of the beam and by bars representing the flanges (very similar to the way civil engineers analyse the behaviour of deep 'composite' I-beams). In the structure as a whole, which is divided into 'structural surfaces for analysis', all forces are assumed to act in the plane of the surface and, conversely, that all forces perpendicular to the surface (except where another surface joins it at right angles) are zero.

The simplest structural surface idealisation is a single shear panel with one bar at each edge to take the end loads, Fig 2. To help understand the validity of the idealisation, first consider a flexible hinged framework of the edge bars. This flexible frame will become rigid with respect to the load *P* by adding the shear panel. The next part of the diagram shows the actual stress distribution and the following part, the one assumed for the idealisation (in a very similar way the early monocoque airframe designers analysed their sheet/stringer constructions by assuming the stringer, for the purposes of analysis, had not only its own cross-section area but also an additional one equivalent to the thickness of the adjacent sheet for an assumed distance from the edge of the panel. As can be seen, the panels are assumed to act in 'pure' shear, that is with equal and opposite forces on their opposing edges, and that the shear force is transferred to the end-load carrying members which form the main skeleton of the structure, as seen in the final part of the diagram. The co-operation of the adjacent skin panels with the bars is accounted for in the same way as in early aircraft-structural-analysis, as mentioned above. Typically a strip of skin 20-60 times its thickness is added to the 'bar' cross-section area depending on the real section shape and the method of connection to the skin, also the curvature of the skin close to the 'bar' member, Fig 3. It is convenient to consider shear on the edge of the panel as 'flow', or force per unit length.

Truck body loading

The punishing treatment received by truck bodies in service, together with the great variety of use, and abuse, means that combinations of many load cases have to be considered in finding the worst one, for realistic use in the analysis. Considering first the downward force of the payload, in the uniformly distributed case, the following surface loadings typically apply for these five ranges of payload capacity corresponding to average platform sizes for each range:

Load capacity Tonnes	*Surface loading* kg/m
0.8-2.0	190-330
2.0-4.0	300-450
4.0-7.0	450-650
7.0-12.0	450-650
12.0 & more	600-1000

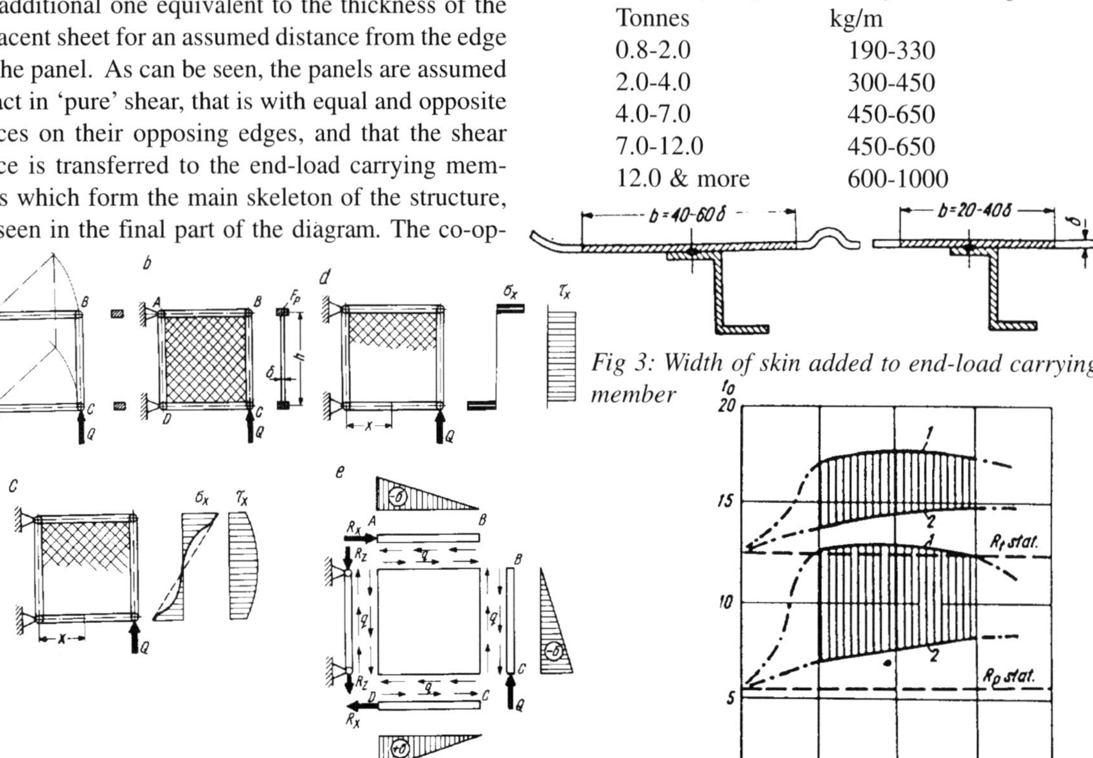

Fig 3: Width of skin added to end-load carrying member

Fig 2: Shear panel idealisation: a. pinned-frame, b. panel added, c. stress pattern (actual), d. stress assumed, e. equilibrium

These 'best-case' stress distributions can then be modified, in a general-haulage vehicle, to allow for highly concentrated loading that might occur from the legs of stillages or perhaps the mounting pads of heavy machinery. Dynamic factors should be built in to allow for travelling over rough roads, at speed, on harsh suspensions. Combinations of inertia loads due to acceleration, braking, cornering and kerbing should also be considered. One approach might be to detach a platform body of the type mounted on horizontal runners and remount it on four cornerwise load cells; by taking readings before and after static loading of the vehicle — and again, by means of suitable recording equipment, during testing over a severe trial-course. Then a good idea of the factors needed for severity of operation could be obtained. Large scale truck manufacturers will normally have large banks of such 'road-load' data.

Since it is argued that a vehicle structure which is strong enough to endure the maximum dynamic loads, which occur sparodically, will also have sufficient resistance against repeated dynamic forces at less than their average value (fatigue loading), then key factor in design is the determination of dynamic coefficients which can be used in static analysis. Fig 4 shows the results of an investigation of the vertical forces on the front and rear axles of a heavy truck with unladen weight 8 tonnes, gross weight 18 tonnes; static axle loads 5.4 tonnes front and 12.6 tonnes rear; wheelbase 4.9 metres. Curves (1) apply to driving into a cylindrical segment, radius 200mm, height 35mm, and curves (2) apply to driving over a concrete surface with joints every 10 metres. Horizontal, dashed, lines are static axle loads — front, lower, and rear, upper. Typically a co-efficient of 3 is used in design to factor up the static vertical loads, for trucks, in the absence of more detailed knowledge.

As well as the above loads which apply to two wheels riding a vertical obstacle, the case of the single wheel bump, which causes twist of the structure, must be considered. The torque applied to the structure is assumed to be 1.5 x the static wheel load x half the track of the axle. Depending on the height of the bump, the individual static wheel load may itself vary up to the value of the total axle load. If, as in Fig 5, it is assumed the body is torsionally rigid compared with the chassis of the truck (as might apply to a tipper), and similar bumps are encountered by opposite sides of front and rear axles simultaneously, the bump height is equal to the tyre deflection + spring deflection x track/spring-base, an expression which can be converted into forces by substituting tyre and spring stiffnesses. Bump heights of +/- 300 mm are reported to have been used in design.

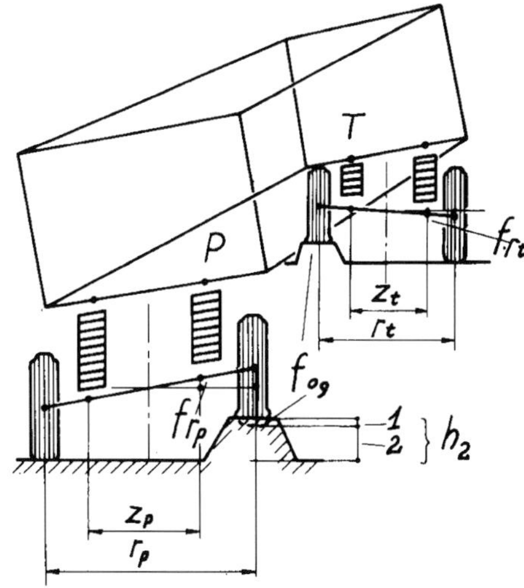

Fig 5: Twist from single wheel bump

For longitudinal forces, largest braking deceleration could be considered as seven metres/second for trucks. Higher forces can be generated by the driving of the vehicle into a kerb and higher still if the vehicle has side forces due to cornering superimposed as well. For safety in design, these combined forces would then be assumed to act simultaneously with vertical 'bending' and 'torsion' loads arising from the 'double' or 'single' wheel bump cases above. The well-known technique of constructing bending-moment and shear-force diagrams for

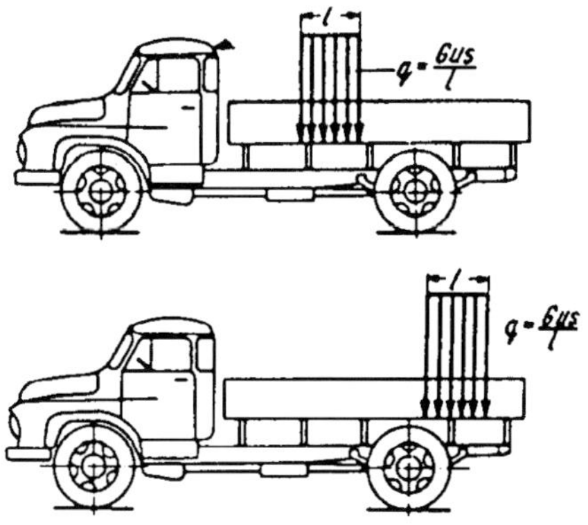

the vehicle-structure as a horizontal beam supported by its axles can then be used to determine worst-case positions of the payload for different parts of the structure, Fig 6.

Platform body stressing

Taking this particular body structure to comprise floor-plate, edge beams and longitudinal cross-members, its behaviour will depend on whether the floor-plate comprises a homogeneous sheet with bending stiffness in all directions in the horizontal plane — or corrugated/planked in which case only 'lengthwise' bending stiffness will apply. Longitudinal loads require the presence of an anti-lozenge bay as shown in the idealisation of the structure, Fig 7. This is made rigid by the addition of additional longitudinal members or cross-ties. Side loads are transferred to the chassis mounting points because the cross-member is rigid with respect to shear, Fig 8, while under vertical loading the structure reacts with internal forces as shown in Fig 9.

Without factoring for dynamic inertial effects, the reactions between body and chassis are:

$$R_l = pbl_l/2$$

$$R_x = q_x\, l_p/2$$

$$R_{xz} = q_x\, h_p$$

$$R_{yi} = q_{yi}\, b/2$$

$$R_{zi} = q_{yi}\, h_p\, b/r$$

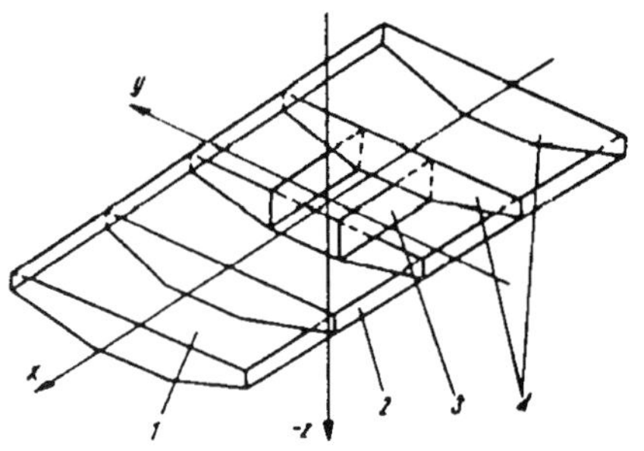

Fig 7: Platform body elements: 1. floor plate, 2. side beams, 3. longitudinals, 4. crossmember

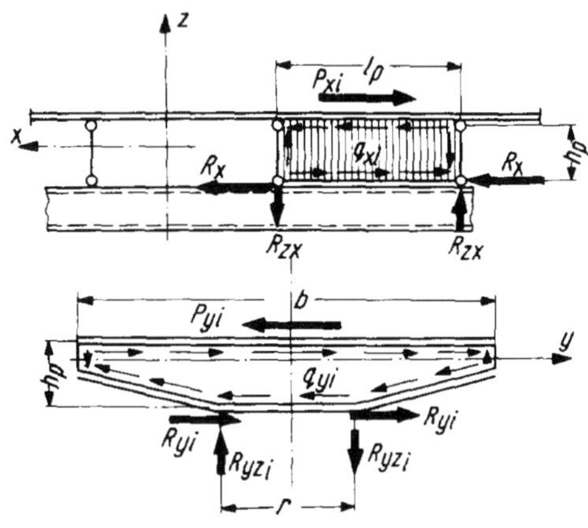

Fig 8: Floor-bearer rigid in shear

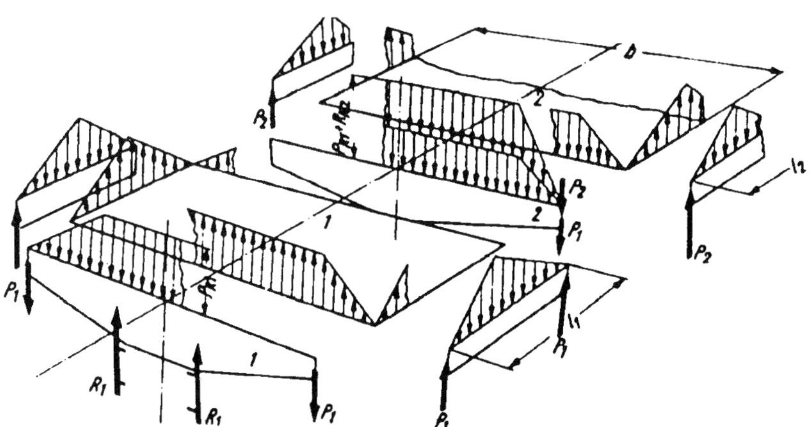

Fig 9: Internal forces, due to bending, in truck

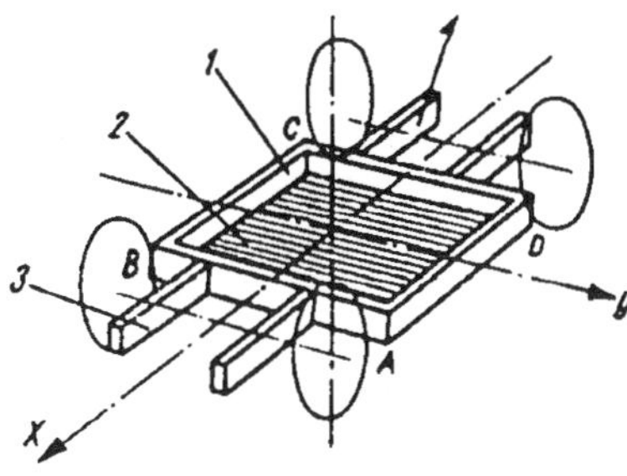

Fig 10: Car underbody elements: 1. sill, 2. floor-pan, 3. front longitudinals, 4. rear longitudinals

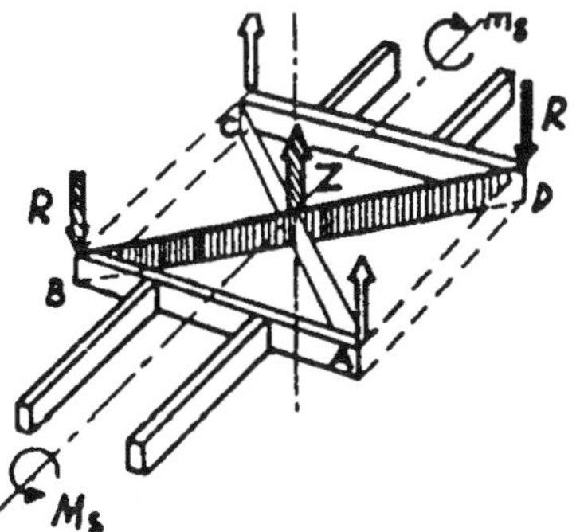

Fig 12: Cruciform-braced frame

where q = shear stress for the respective directional axes shown.

In the case of cars with self supporting underbody structures, or the ladder-frame chassis of the truck, the idealisation of the 'chassis' is shown in Fig 10. The co-operation of the remainder of the body is ensured through the joint of the sill with the sidewall — and the front and rear bulkheads with the crossmembers of the flat structure. The degree of contribution will of course depend on the number and flexibility of joints to the remainder of the body assembly.

Under symmetrical loading, overall bending, the structure comprises members in shear bending and torsion as in the case of assymmetrical loading, or torsion, when the pattern of loading shown in Fig 11 exists. If in the layout shown, and with the particular design constraints applying to the rest of the vehicle, the sidemembers, or sills, cannot be made sufficiently rigid, then an alternative layout as shown in Fig 12 is typical. The floor panel, in the case of the car, takes no part in carrying the overall bending and torsion loads, its function being to transfer vertical loads placed on it to the adjacent structural elements.

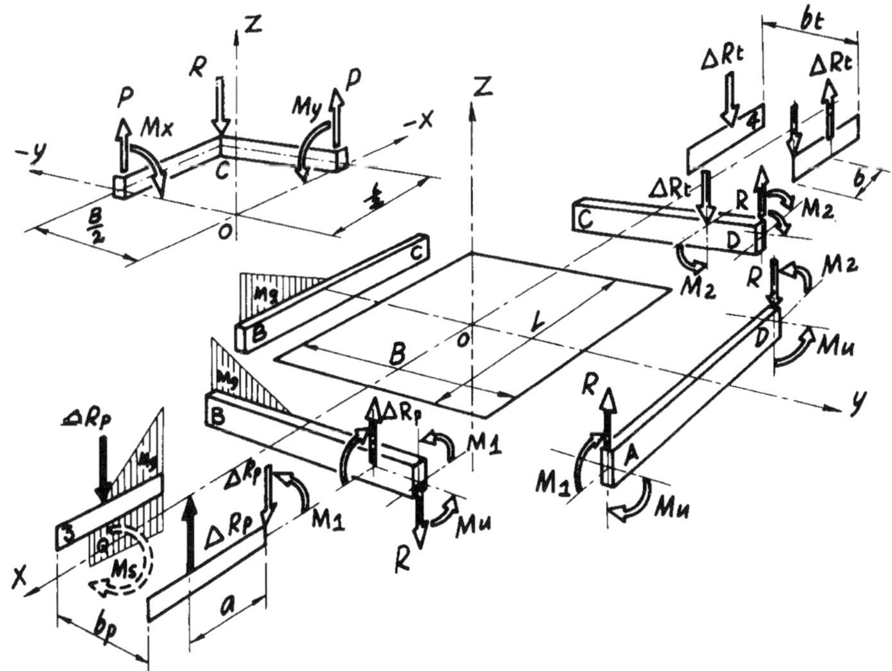

Fig 11: Internal forces, due to bending, in car structure

Closed body structures

The first part of this chapter examined the design-analysis of flat-platform structures; here we apply the same techniques to consider closed, 'box-like', structures typical of saloon-cars, vans and buses. Shown in simple, idealised, form in Fig 13, this illustration gives the general distribution of forces in the alternate torsion and bending modes. External loading is reacted by all sides, of the 'box', behaving as shear panels. If either the roof panels, or front and rear end-walls, are non-structural then the floor panel must again be rigid with respect to forces at right-angles to its plane — as in the case of the flat-platform structure considered in the first part. Fig 14 shows how typical car-shell and bus structures are idealised into 'Simple Structural Surfaces' as previously defined.

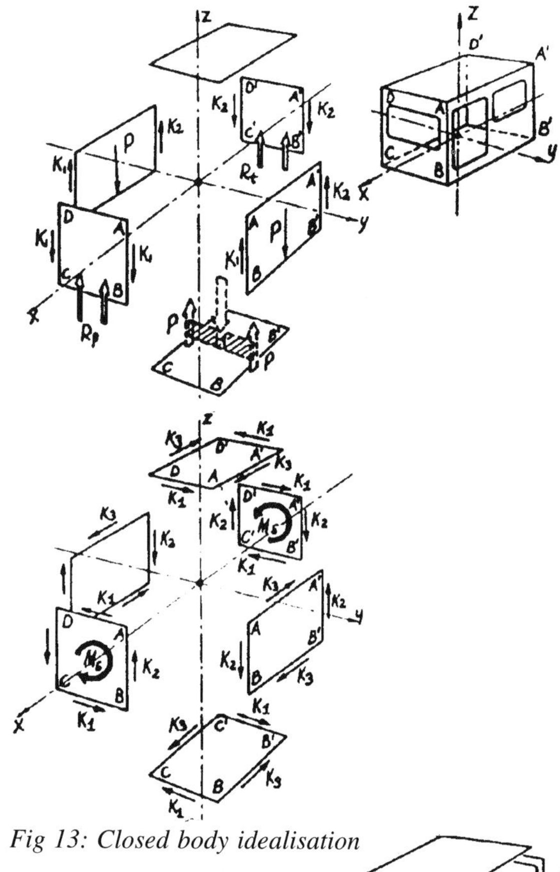

Fig 13: Closed body idealisation

In the case of an integral box van, the idealisation of Fig 15 can be considered. This assumes the frames enclosing windscreen, cab doors and rear doors are able to resist shear deformation — otherwise 'open-structure' theory has to be applied. When the force balance for all the structural surfaces is equated, Fig 16, edge forces caused by the torque Ms amount to:

$$R_p = M_s / S_p$$
$$R'_p = R_p\, a_2 / (a_1 + a_2)$$
$$R''_p = R_p\, a_1 / (a_1 + a_2)$$

The separate elements of the structure, such as the sidewall in Fig 17, can be examined by analysing them as panel and bar systems — as discussed in the previous section. The problem of van design is that the most heavily loaded structural surfaces have the largest openings cut in them. This results in high bending moments in the corners of the surrounding frameworks given by:

$$M_n = q_2\, H\, l_2 /4$$

for opening height, *H*, length.

For saloon-cars, a parallel breakdown would give the force distribution shown in Fig 18 — in this case for torsional loading. Structural surface idealisations, such as for the sideframes, require stiff-jointed beam representation. Such frames are normally redundant structures which imply the presence of more than one (statically determinate) load path

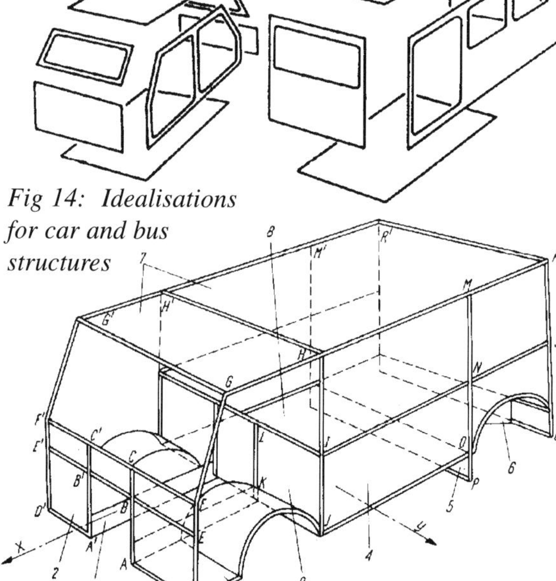

Fig 14: Idealisations for car and bus structures

Fig 15: Van structure idealisation

between external load(s) and reactions. A redundant (or statically indeterminate) structure requires knowledge of the elastic properties of the members to establish the load distribution paths. Since real structures contain an infinite number of alternative load paths they are infinitely redundant so will always require idealisation for analysis into a workable number of redundancies. The sideframe for a four door saloon, as shown in the figure, has six degrees of redundancy visualised by cutting the two ringframes and counting the fixing reactions required at the cuts. To satisfy fixing against vertical, horizontal and rotational displacement of the members would require three reactions at each cut — giving a total of six redundancies for the sideframe.

Van as box beam

A semitrailer box-van body enjoys a degree of symmetry and structural uniformity that can allow it to be considered as a box-beam in structural design — particularly so if a stiff sandwich construction is employed for its main surfaces. With such 'fully stabilised' walls, and other outer panels, shear loads can be transferred across the panels in the same way as assumed for the much smaller car and panel-van structures. In a large trailer van, loading on the horizontal floor planks is transferred by the cross-bearers to the sidepanels. The latter act as beams which pass the

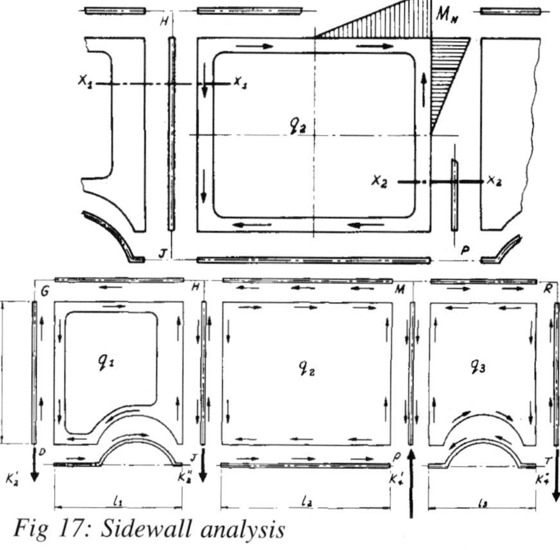

Fig 17: Sidewall analysis

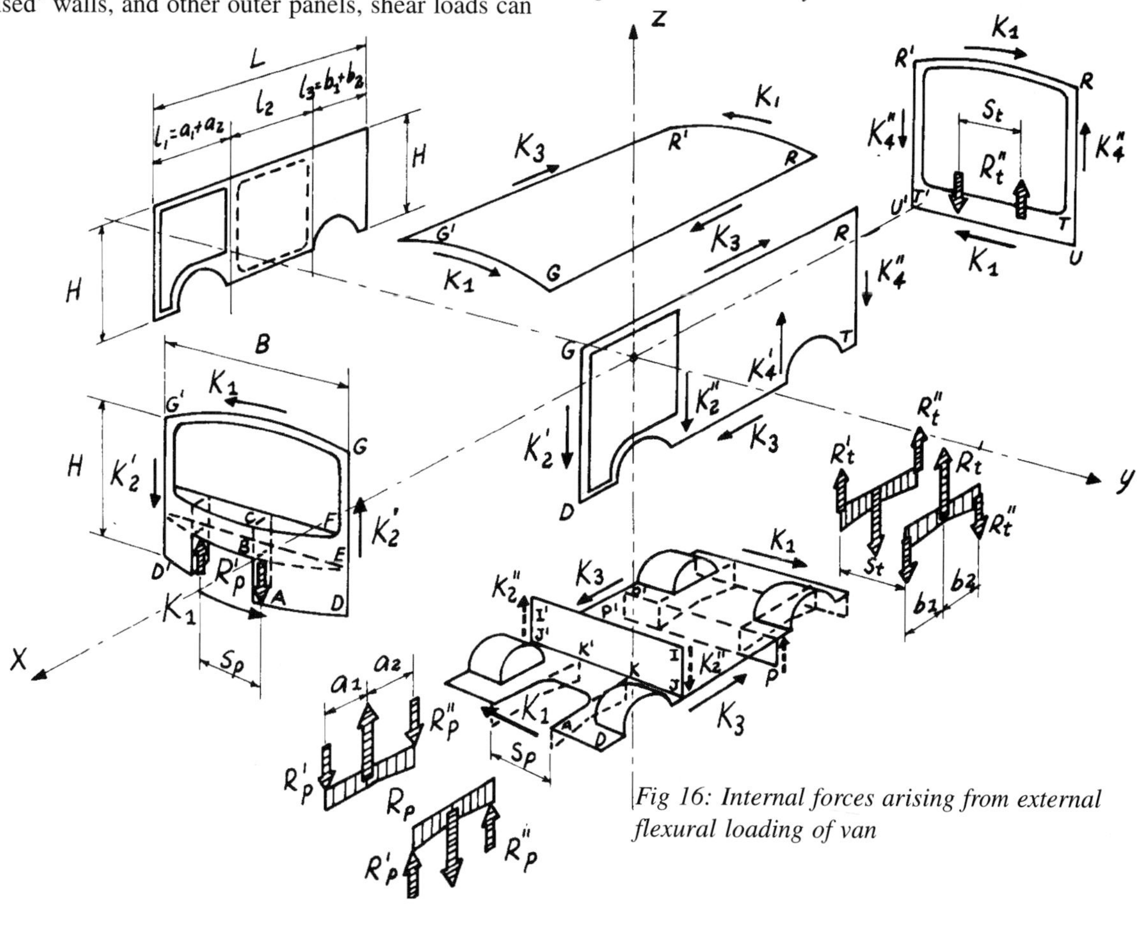

Fig 16: Internal forces arising from external flexural loading of van

load through to the coupling-gear and wheel-suspension respectively. Twisting is resisted by the entire body acting as a tube.

The bending and torsion of the box tube can be analysed directly if certain conditions are met for the application of simple beam theory. In bending, plane cross-sections must remain plane to validate the theory (there must be no axial warping) and cross-sections should also retain their planar outline (no swelling or bowing-in); there is also the proviso that the load should be applied uniformly across the cross-section. Departure from these ideals can be understood for practical box beams by considering the deformation mechanism in the box cantilever of Fig 19. Loaded as shown, there is a tendency for the 'essing' deformation of the sidewalls to take place, arising from the parabolic distribution of shear stress across the vertical panels. There is also axial warping of the horizontal panels caused by shear lag. This is due to prevention of bending forces applied at the panel edges, by shear flexibility, from being able to maintain even value across the panel width.

When fully stabilised sandwich panels are not employed, systems of pillars and rails are the rule for reinforcing side-panels. Part of the objective is to divide the sidewall into panels which are small enough to avoid the possibility of buckling. This buckling has traditionally been analysed using 'tension-field' beam theory which assumes, as the panel is sheared, a diagonal tension field, only, applies in the panel. Fig 20 nicely describes this concept, in which a single bay truss framework, top, is compared with a plane sheet panel, bottom. Under load *P* the tendency for the diagonal brace in the frame to react tension load is paralleled by the tension field assumed to exist in the panel. Bowing of the other brace is

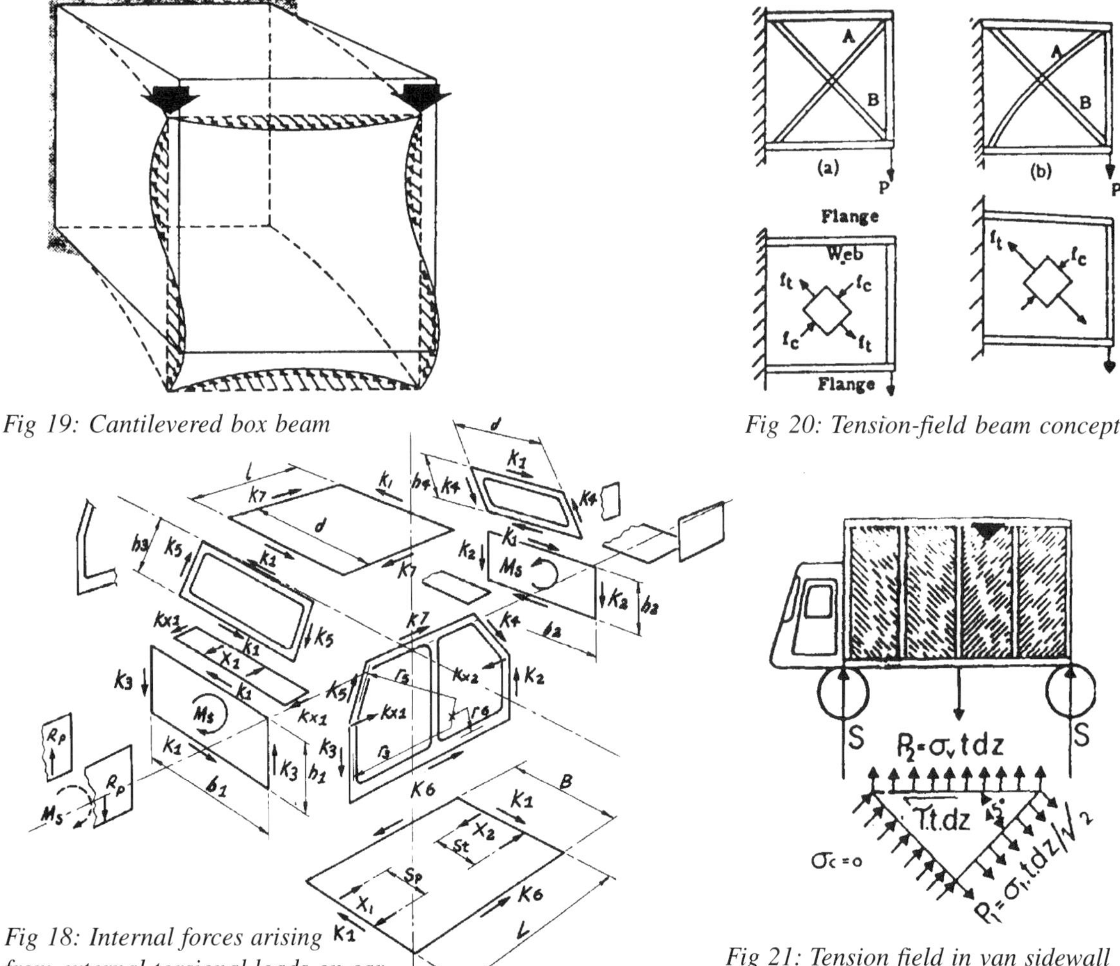

Fig 19: Cantilevered box beam

Fig 20: Tension-field beam concept

Fig 18: Internal forces arising from external torsional loads on car

Fig 21: Tension field in van sidewall

paralleled by compressive wrinkling of the panel.

In the sidewall depicted in Fig 21 assumed to be under the action of a central concentrated load 2S then tension bands, corresponding to the broken lines in the figure, can be considered. If the pillars and rails provide a condition in which pure shear of the panels takes place, then the forces acting on a triangular element of the panel can be used to obtain the relationship:

$$\sigma_v\, t\, dz - \sigma t(tdz/2^{1/2})\; 1/2^{1/2} = 0$$

for direct stress sigma and a similar one for shear stress τ — the two of which can be equated to give:

$$\sigma t = 2\tau \text{ and } \sigma_v = \tau$$

If *h* is the height between upper and lower rows of rivets securing the panel to its top and bottom rails then $\tau = S / ht$ and rivets are subjected to loading τtdz, horizontally and vertically, with resultant:

$$2^{1/2}\tau tdz$$

or rivet load per unit length of *1.41 S/h*. With rails held apart by pillars, compressive load in them is $P_l sin\ 45 = Sd/h$ for pillar spacing *d*.

Pillars and rails in body stressing The effect of horizontal intermediate rails (stringers) related to the corner rails (booms) can be analysed by early aircraft structural techniques. For the reinforced box beam shown in Fig 22 the effects of shear lag on the departure (full lines) from simple bending theory is shown in Fig 23 from which it can be seen: the longer the beam the more accurately can simple beam theory be applied.

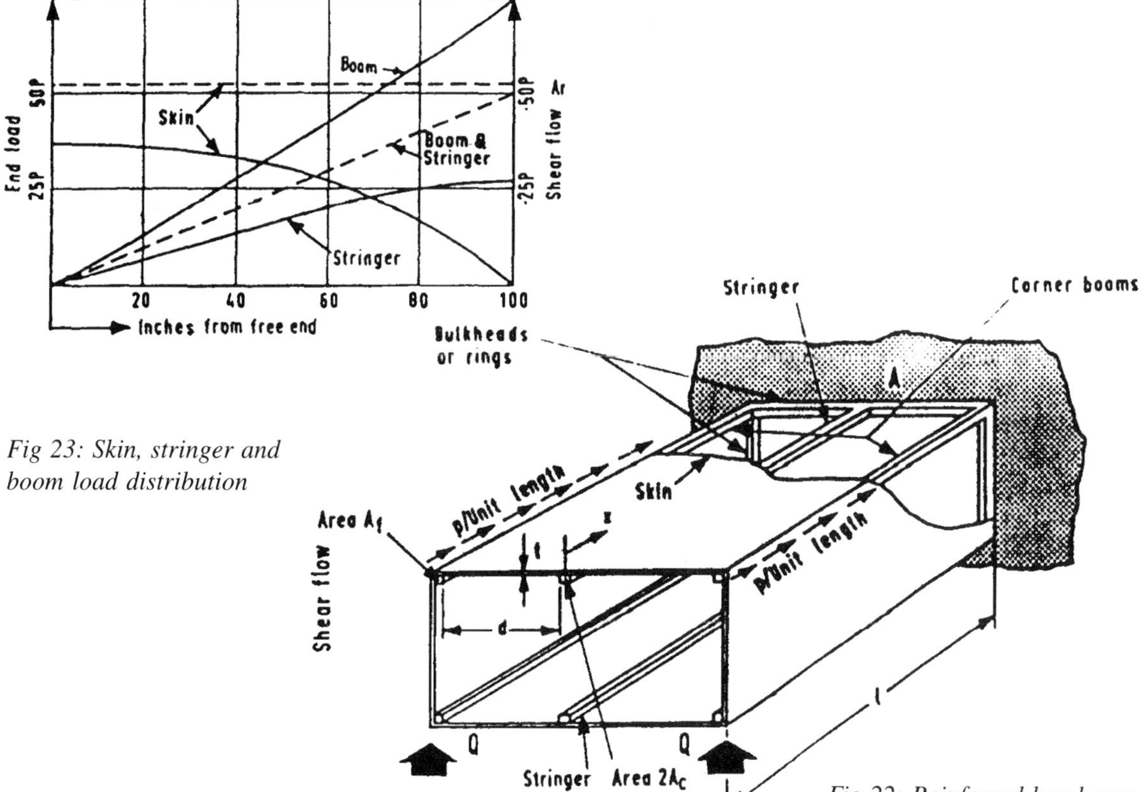

Fig 23: Skin, stringer and boom load distribution

Fig 22: Reinforced box beam

Chassis frames

The conventional commercial vehicle, and traditional coach-built car chassis is a shallow semi-stiff-jointed frame subject to bending and torsional loading and often substantial deflection. The compromise structures used on volume built vehicles in this category, to match a whole range of body types on one chassis configuration, can often be improved upon for specific specialist vehicle categories.

The goods and passenger vehicle chassis frame, in forming a mounting for a great many different body configurations, is inevitably a compromise in the torsional stiffness it provides for many body types. Because flexibility has to be provided for flat-platform trucks involving cross country travel in military operation there is little relative stiffness available to relieve a stiff box, or tanker-shell, body from wracking loads. Codes of practice stipulated in readiness for World War Two required a typical three-ton payload platform truck to be capable of having one wheel lifted from the ground without affecting the others. Bodybuilding manuals provided by truck manufacturers thus lay down elaborate systems of reinforcement for such special applications as tankers, tippers and mixers.

Specialist chassis builders, including some trailer makers, have the opportunity to tailor chassis stiffness to body type and, of course, to provide, in some instances, chassis structures integrated with the bodies. This creates the need for being able to predict stiffness at the design stage in order to avoid the high degree of structural-redundancy arising from gross differences in stiffness between chassis and body. The different ways a chassis can be loaded in torsion include an asymmetric payload on the load platform; uneven ground surface; lateral accelerations acting on high centre-of-gravity loads and severe cornering manoeuvres.

The different types of torsional load result in extremes of torsional deflection mode as shown in Fig 24. On balance it is best to have the torsionally stiffest crossmembers towards the ends of the chassis frame. Closed rather than open section members are obviously an advantage for this purpose. However, deep open sections can also exert a considerable influence on overall torsional stiffness of the chassis frame if the ends of the crossmembers are rigidly attached to the sidemembers. Remember, though, that high stresses will be built up at the interface — due to constraining the axial warping of the crossmember sections. Figs 25 and 26, due to PW Sharman of Loughborough University[2] compare stiffnesses and associated stresses for various sections.

The former Swiss company Saurer[3] built an all-welded chassis on an experimental basis to establish the desirability of greater torsional stiffness. Sufficiently soft springs had to be provided to replace the frame's traditional supplementary role of maintaining wheel/ground contact. The separate effects of

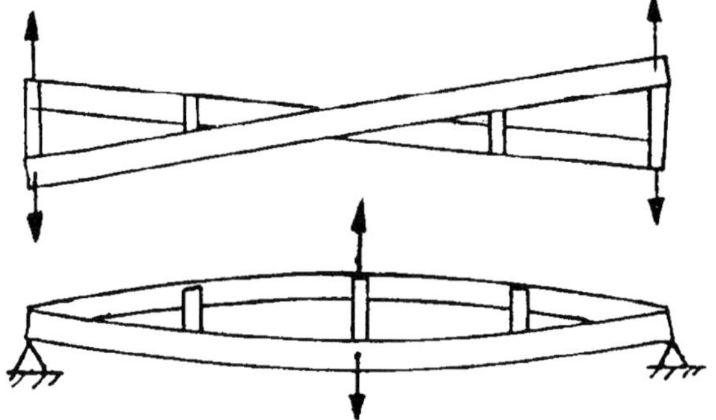

Fig 24: Twist modes for different loadings

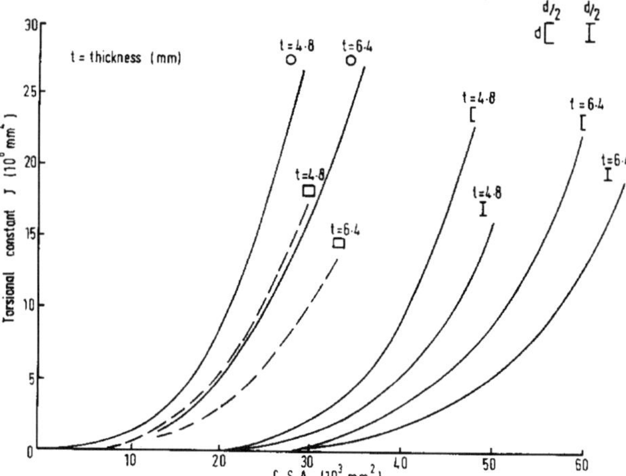

Fig 25: Stiffnesses of various sections

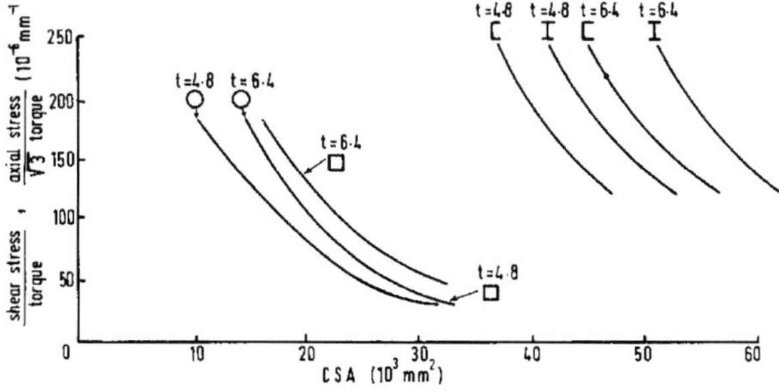

Fig 26: Related stresses in twisting

frame and suspension stiffnesses can be understood by referring to Fig 27 and using the expression for wheel deflection that will just lift the front of the vehicle given by:

$$h = F_1 f = W_v f l_g / l$$

which equals sum of flexibilities due to spring, tyre and frame stiffnesses given by:

$$(1/k_{fs})(t_{fw}/t_{fs})^2 + 1/k_{ft}\,(t_{fw}/t_{rw})^2$$
$$x[(1/k_{rs})(t_{rw}/t_{rs})^2 + (1/k_{rt})\,(t_{fw}^{\,2}/2k_{ch})(\pi/180)]$$

Frame stiffnesses measured by Saurer were 2175 lb ft (2948 Nm)/deg for a conventional frame and 10,875 lb ft (14745 Nm)/deg for the experimental welded frame with tubular crossmembers. These resulted in values of h around 6.5 ft (1.98 m) for the conventional frame and 5 ft (1.52 m) for the new frame. The theoretical results obtained are reduced by the effect of helper springs, the company point out. Fig 28 shows a comparison of frame stresses with the two frames, each twisted around 0.5 deg about the longitudinal axis.

An approximation for the torsional stiffness of a complete chassis frame can be obtained by summing individual sidemember and crossmember stiffnesses as seen in the representation of Fig 29 — using the reciprocals of stiffness as appropriate (inset). However there is an incompatibility of strain between assumed bending and twisting modes which

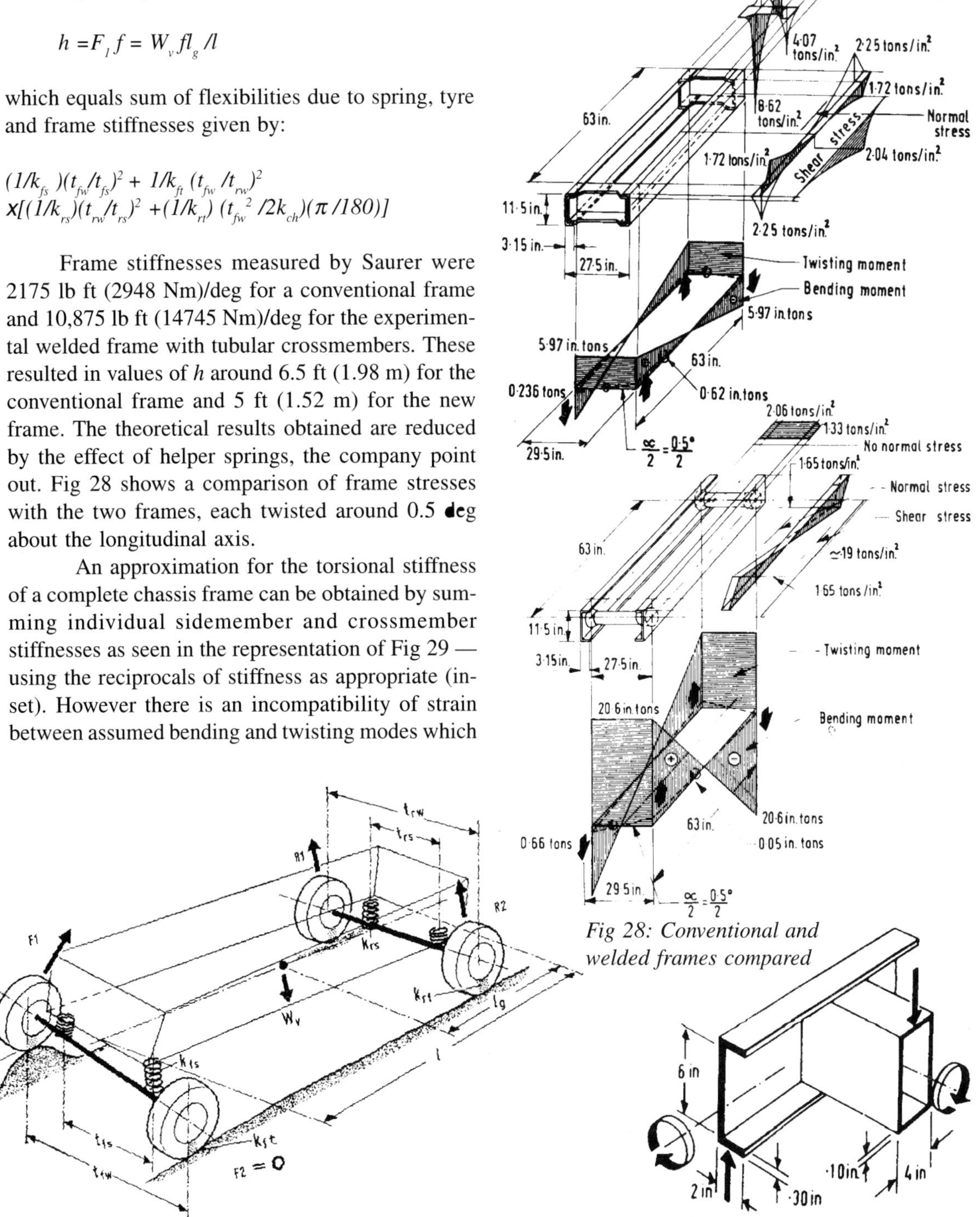

Fig 28: Conventional and welded frames compared

Fig 27: Torsional parameters

Fig 30: Example frame joint dimensions

makes the following expression for torsional flexibility α/T only an approximate one:

$$[5l-1/G(J_1+J_2)]$$
$$+10/[3l_2E\{(I_1/l_1^3)+(I_2/l_2^3)\}+(3G/2l_1)(J_1+J_2)]$$

and E, G are section second moments of area and moduli in bending and twisting respectively. For the example sections of Fig 30, a figure of 1320 lb ft/deg is obtained for torsional stiffness which compares with a figure of 1594 lb ft/deg using a more exact (but more laborious) strain-energy method[4] involving more accurate compatibility of strains shown in the deflection modes of Fig 31.

Conventional designs usually involve ladder frames of channel section members as shown in Fig 32 with variations of crossmember section and attachment as shown in Fig 33. There are other loadings on a typical crossmember due to such effects as offset fuel tank mounted to the frame; also lozenging of the frame in the horizontal plane due to effects such as asymmetric braking. These can add up to the kind of crossmember loading combinations shown in Fig 34.

Another aspect of frame torsional stiffness is the effect it has on roll moment distribution between front and rear suspensions. The ability of the suspension to transfer roll-moment on to the wheels depends then on spring rate and spring base. Rear wheels usually take a greater proportion, on goods vehicles, due to the sprung mass centre being nearer to the rear axle — resulting in both a smaller torsional compliance of the frame and a greater rear spring stiffness. There is also greater roll moment transfer at the rear of a conventional truck chassis due to the higher roll centre of the rear suspension. The combined effect can lead to a limiting oversteer condition in cornering unless the frame is stiffened, and front suspension roll stiffness increased so that a more equable roll moment distribution is obtained.

There is need too for ensuring sufficient stiffness of the chassis frame against bending in the horizontal as well as the vertical plane, to avoid differential steer effects on front and rear axles. The ride of the vehicle can also be affected by frame stiffness

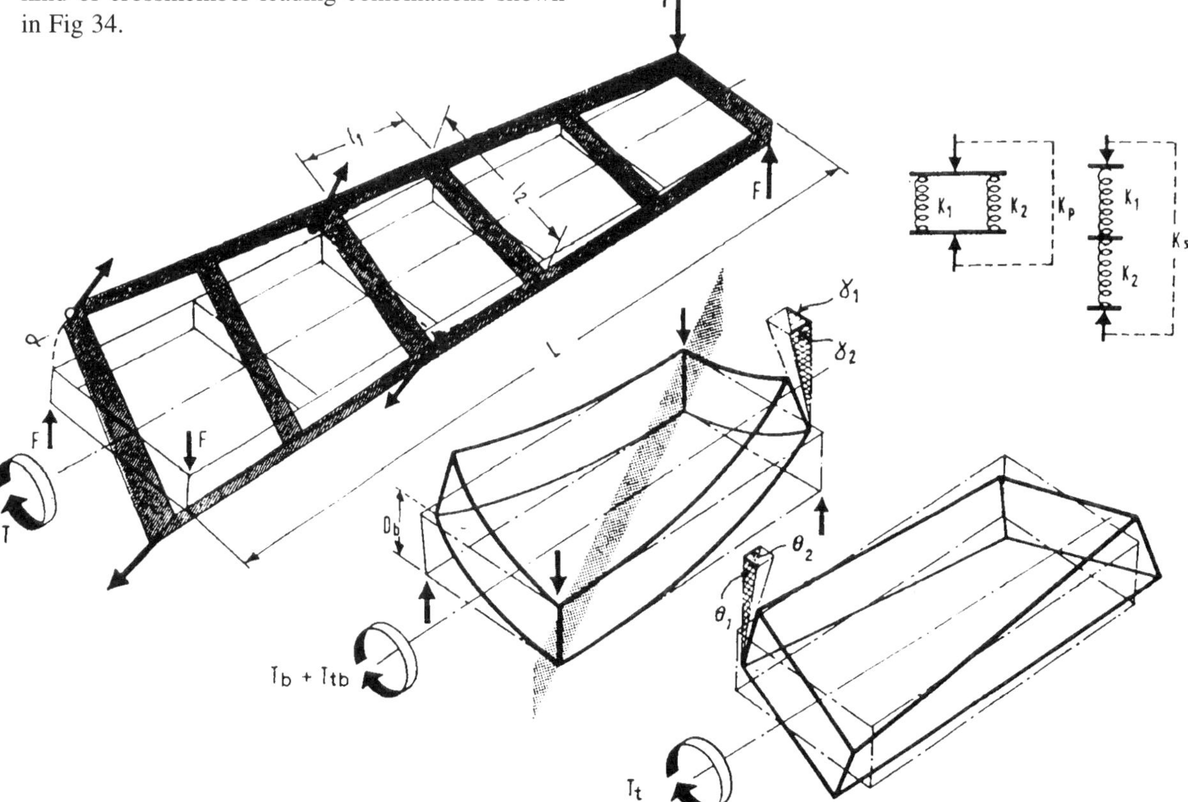

Fig 29: Simplified chassis-representation

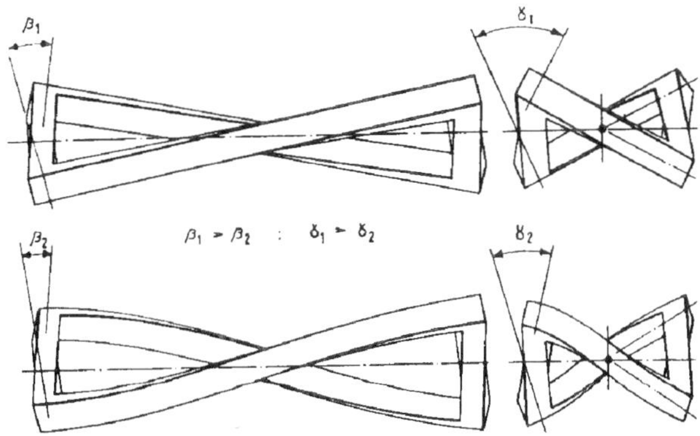

Fig 31: Correct twisting modes for strain compatibility

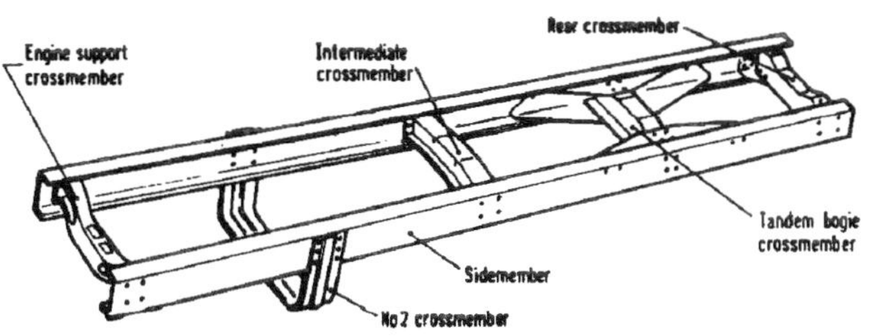

Fig 32: Conventional channel-sectioned ladder chassis

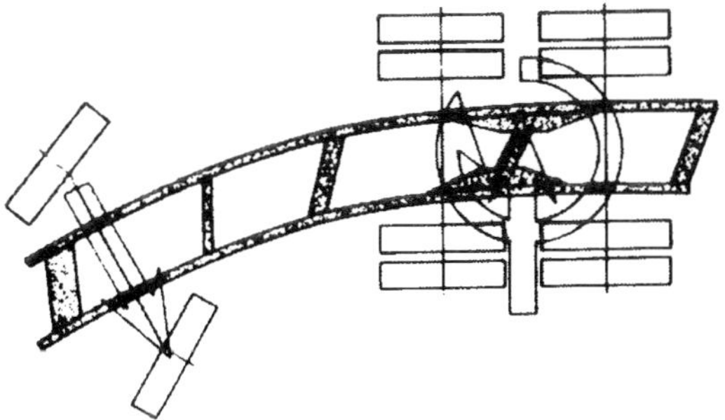

Fig 34: Lateral loading of frame

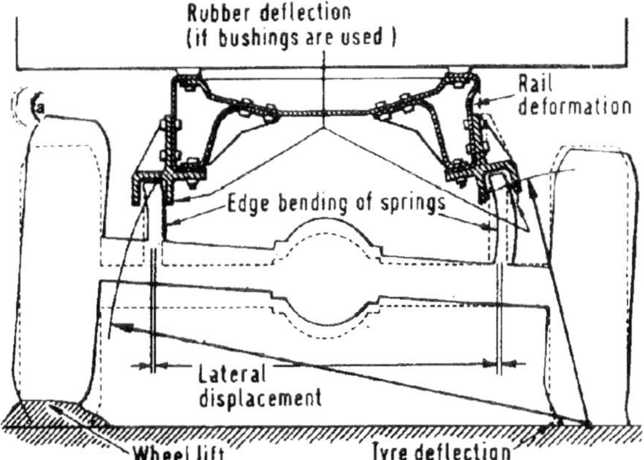

Fig 35: Diaphragming of the sidemember web

according to the bending and twisting mode frequencies excited by dynamic road inputs. Using thin walled sidemembers, very high natural frequencies can be obtained with deep webs and these can be made more economical in weight by circular or diamond shape cut-outs along the length of the web. Weight can also be saved by profiling the sidemember section so its depth of section more nearly corresponds to the bending-moment distribution.

Sidemember to crossmember attachment methods vary with vehicle type and conventional practice has seen lighter trucks with flange attachment and heavier ones attached through the webs. Spacing of crossmembers affects both lozenging resistance and the resistance to buckling of the compression flange of the sidemember under bending load. Diagonal bracing can also be employed to advantage. The frame as well as being strong enough in bending and torsion, must also have sufficient resilience (strain energy) to allow the designed deflection without over-stressing — an important criterion in crossmember to sidemember joint design. The 'alligator' type crossmember is one of the design solutions to this problem.

Frame dimensions on special purpose commercial vehicles are often set by vehicle legal dimensional limits, axle loads and spacing. SAE practice is to space sidemembers at 34 in (0.86 m) overall width and their height is often governed by suspension movement requirements. Sidemember depth follows from bending moment calculations, the allowable bending stress being conventionally taken as one third of the yield stress. The crossmember corresponding to the rear suspension anchorage sets an important design criterion. Fig 35, due to Sherman[5], shows a flange connected alligator crossmember under load due to 'single wheel bump' of the suspension. Because chassis twist is accompanied by lateral displacement of the axle, the latter exerts a high bending moment, via the spring, on the hanger bracket which can cause diaphragming of the web unless properly restrained by the crossmember.

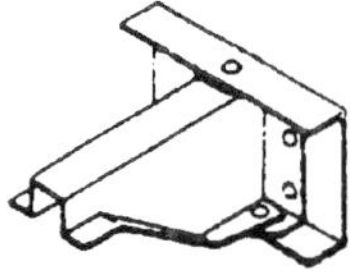

Fig 33: Typical crossmember/sidemember joints

Body sub-assembly analysis

Structural analysis of individual elements such as beams, can be combined to examine the behaviour of body sub-assemblies prior to analysis of the complete vehicle body shell. This adds to the overall 'feel' of the structure during the design process.

Examination of sections of the body as a separate entity can help the visualising of deflections influenced by remoter sections to which the first might be anchored. An example of this is the flexing of the forward extensions from the bulkhead of a saloon car. The overall distortion in Fig 36 results from the fact that the principal axis of the beams formed by the wing and valence assemblies are inclined to the vertical. The result is that a component of the applied force bending the beams about their weak axes

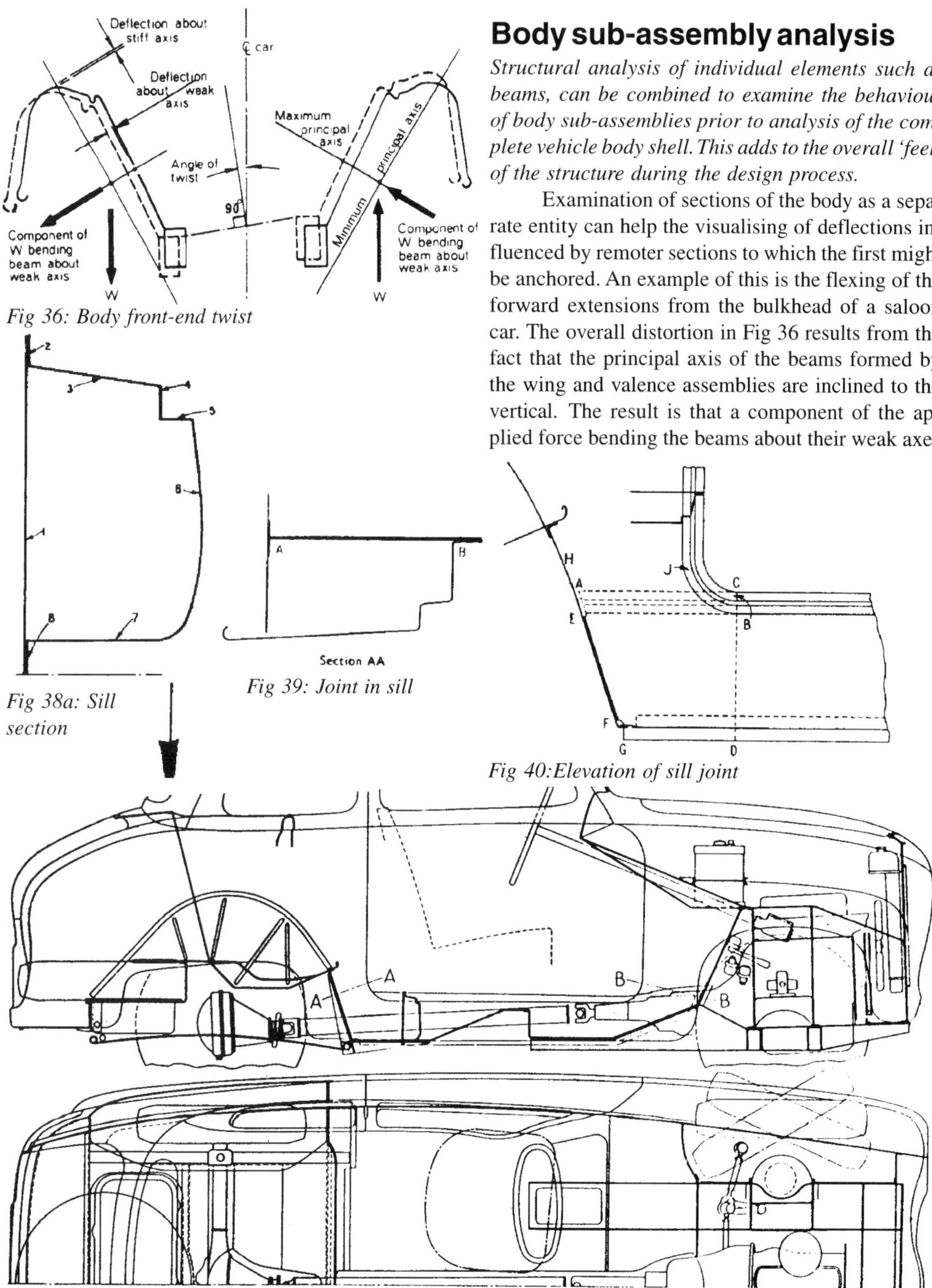

Fig 36: Body front-end twist

Fig 38a: Sill section

Fig 39: Joint in sill

Fig 40:Elevation of sill joint

Item	Size	Area	y	y²	Ay	Ay²	Local I
1	0·036 × 5·0	0·180	2·5	6·25	0·4500	1·1250	0·374
2	0·036 × 0·5	0·018	4·75	22·5	0·0855	0·4050	0·0004
3	0·036 × 2·0	0·072	4·4	19·36	0·3170	1·3950	—
4	0·036 × 0·5	0·018	4·0	16·00	0·0720	0·2880	0·0004
5	0·036 × 0·5	0·018	3·75	14·05	0·0675	0·2530	—
6	0·036 × 3·25	0·117	1·875	3·52	0·2190	0·4120	0·103
7	0·036 × 2·0	0·072	0·5	0·25	0·0360	0·0180	0·024
8	0·036 × 0·5	0·018	0·25	0·0625	0·0045	0·0001	0·0004
		ΣA = 0·513		Σ =	1·252	3·896	0·502

Fig 38b: Sill section properties

to the extent that the radiator frame is unable to stabilise the assembly. Only good continuity between wing and underbody side rail extension, via the bulkhead, ensures that the lozenging displacement of the lower A-post has a positive stiffening effect on the assembly.

The approach of analysing individual assemblies was taken by the engineer who carried out the structural evaluation of one of the UK's first integral car body structures.

The Austin A30 body shell was analysed by a former aeronautical stressman, Ken Garrett, who was one of the first to apply aircraft thin-wall structure techniques in road vehicle bodies. The complete structure is shown as Fig 37. The sill section of Fig 38a was considered as a beam with the help of the table of elemental properties shown in Fig 38b. Study of Fig 37 shows that to the rear of the sill, bending loads are taken by the outer quarter panel — also by an inner quarter panel attached to the sill inner plate and rear pillar. This inner panel extends over the rear wheel arch, which joins it and its upper edge, is stabilised by both its shape and its attachment to the outer panel; to the rear it is stabilised by joining the side of the parcel shelf and to the drip channel in the boot opening.

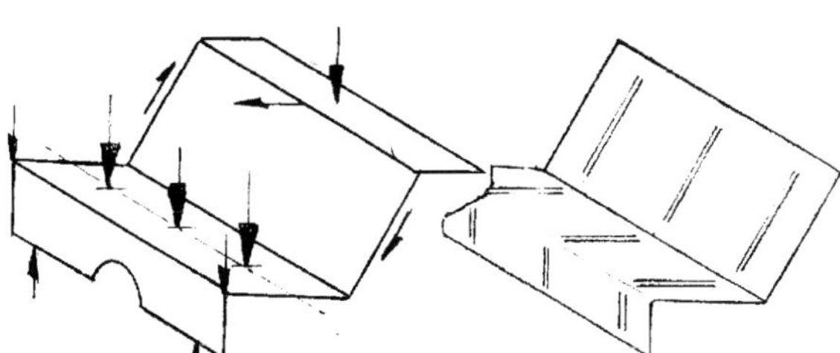

Fig 41: Rear seat pan

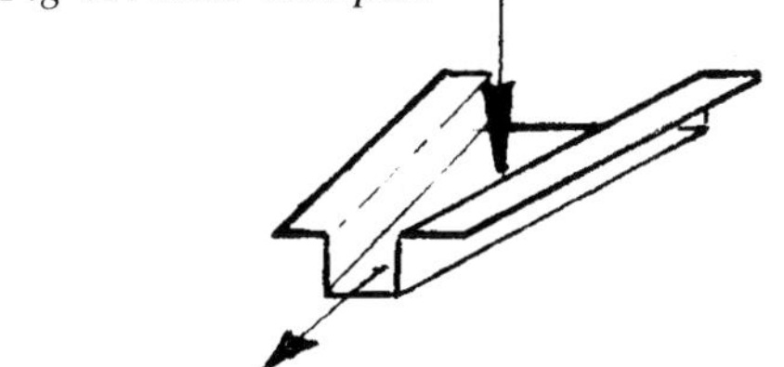

The junction of the inner quarter panel with the heelboard on Section AA of Fig 37 is similar to A in Fig 39. If there were only single thickness between A and B, it may be necessary to extend the inner panel of the sill upwards and the inner quarter panel downwards so that they overlap as shown, to double the strength of the web portion. At the rear of the two panels one is flanged inwards and the other outwards so as to connect both with the heelboard.

The importance of structural continuity, in order to transfer loads from one part of a structure to another is clear from the example shown in Fig 40. Here it is necessary to transfer both shear and bending moment between the sill and the rear end of the vehicle. The sill outer panel is joggled under the rear quarter panel and there is a vertical joint line CD. At the rear end the sill is closed by flanging outwards of the outer panel and its connection to the heel board at EF. End loads to be transferred from the top and bottom of the section to the rear quarter panel are found by dividing the section depth into the bending moment. These loads will govern the number of spot welds required at the connection. The direction of the loads will determine whether a 'kink strut' is required at the joint, its dimensions being found by resolving the forces along the line of the strut.

The rear seat and heelboard structure of a saloon car can also be considered as a separate assembly for analysis if the assumption is made that its edges are rigidly rooted in the vehicle sidewalls, Fig 41. In this example, interior vehicle width is assumed to be 1.5 metres and the four constituent panels are rectangular and set at right angles to one another. The shorter sides measure 0.2 (parcel-shelf), 1.0 (squab), 0.5 (seat) and 0.2 metres (heelboard). The latter has a cut-out of 0.15 metre radius and the resulting necked portion of the heelboard is given an inward folded lower flange. This is boxed into the the seat panel with a zed section reinforcing member.

A load system can be used to simulate a spare-wheel carrier suspended from the parcel-shelf; shear loads of 2.5 kN on the edges of the squab to represent reaction to overall twist of the body shell; horizontal forward pull at the centre of the leading edge of the parcel shelf to simulate forward inertia loads on a seat-belt anchorage at that point; passenger loads vertically downwards on the seats assumed to act on the seat centre line at three equi-spaced intervals of 0.375 metres; finally loads of 2 kN applied in the plane of the heelboard panel to represent upward forces from the spring hanger brackets which cause four-point bending of the panel arising from the vertical downward reactions at the sidewalls.

These loads are factored for inertia effects: 3g bump assumed on the spare-wheel carrier; 1.5g bump assumed on the squab side-loads; 10g impact load assumed for the forward pull of the parcel-shelf. This order of load could be used in selecting panel gauge, joint configuration, swaging patterns or additional reinforcement as necessary. A maximum working stress of 100 MN/m^2 could be assumed for the mild steel panels and a 0.5 kN working load for each spot weld. A reinforcement for the parcel shelf could be in the form of Fig 42 where a common member both diffuses the restraint force across the width of the panel and avoids bending of the panel in its plane by transferring vertical support loads to the edges. Since there will be diagonal tension fields set up in the squab due to the alternating shear reactions in body twist, either diagonal or vertical swaging is probably appropriate for this panel. Vertical ones could be used, in conjunction with parallel ones on the seat, to give both panels continuity of end load and flexure resistance within their own planes.

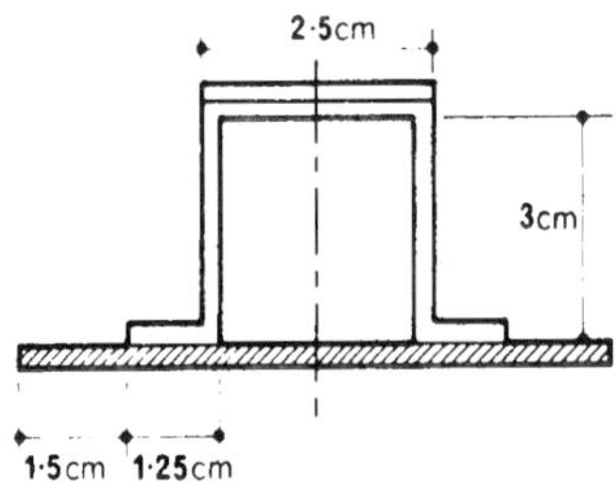

Reinforced top-hat beam section

$$y_{n.a.} = \frac{\sum A \cdot y}{\sum A} = \frac{2{\cdot}64}{2{\cdot}15} = 1{\cdot}23 \text{ cm};$$

$$I = \sum (I_g + A \cdot h^2) = 3{\cdot}877 \text{ cm}^2$$

Fig 43: Heel-board section property calculation

	Area A (cm^2)	Centroid distance y (cm)	Ay (cm^3)	I_g (cm^4)	Centroid distance to N.A. h (cm)	Ah^2 (cm^4)
Top 'flanges'	0·5	3·2	1·60	$(2{\cdot}5 \times 0{\cdot}2^3)/12 = 0{\cdot}00167$	1·97	1·94
Webs	0·6	1·6	0·96	$(0{\cdot}2 \times 3^3)/12 = 0{\cdot}45$	0·37	0·082
Bottom flanges	0·25	0·15	0·04	$(2{\cdot}5 \times 0{\cdot}1^3)/12 = 0{\cdot}00021$	1·08	0·293
Reinf plate	0·8	0·05	0·04	$(8 \times 0{\cdot}1^3)/12 = 0{\cdot}00067$	1·10	1·11
	$\sum = 2{\cdot}15$		$\sum = 2{\cdot}64$	$\sum = 0{\cdot}4525$		$\sum = 3{\cdot}425$

Fig 44: Second-moment calculation

Distance from neutral axis to extreme fibre $y = 3{\cdot}3 - 1{\cdot}23 = 2{\cdot}07$ cm.
Given a bending moment $M = 300$ Nm, bending stress

$$\frac{M \cdot y}{I} = 300 \times 2{\cdot}07 \times 10^{-2}/3{\cdot}877 \times 10^{-8} = 160 \text{ MN/m}^2$$

Distance from neutral axis to centroid of reinforcing plate

$$\bar{y} = 1{\cdot}18 \times 10^{-2} \text{ m}$$

Given $R = 500$ N for each spot weld and assuming the flanges are held by two rows of spots

$$p = \frac{R \cdot I}{S \cdot A \cdot \bar{y}} = \frac{2{\cdot}500 \times 3{\cdot}878 \times 10^{-8} \times 10^3}{2{\cdot}0 \times 10^3 \times 0{\cdot}8 \times 10^{-4} \times 1{\cdot}18 \times 10^{-2}} = 20{\cdot}5 \text{ mm}$$

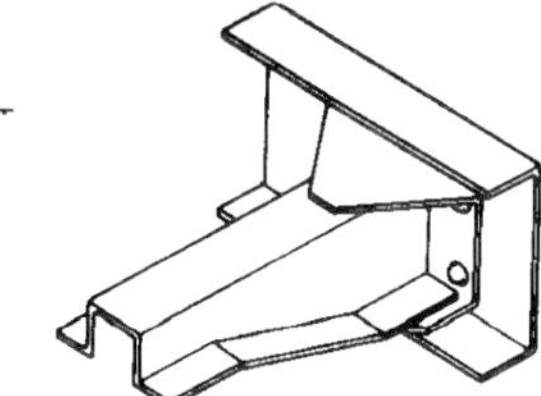

Fig 47: Crossmember/ sidemember joint for CV

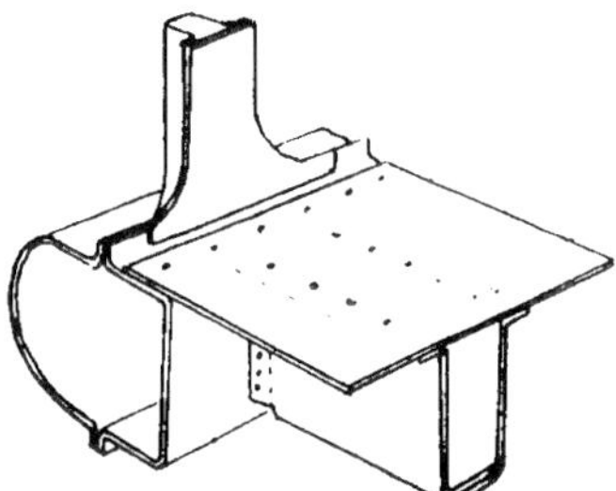

Fig 46: Crossmember/ sidemember joint for car

A critical area of the assembly will be the necked section of the heelboard with its associated box reinforcement. Here analysis of an idealised cross section, Fig 44, would be useful. Section second moment of area could be found from the tabular method in Fig 45 and the vertical deflection of this reinforcing box beam from the calculation shown in Fig 46.

Joints of sidemembers with crossmembers are critical to both car, Fig 47, and commercial vehicle, Fig 48, structures. According to Tidbury *et al*[6] the axial warping of the crossmember beam section is an important criterion in considering the intersection with the sidemember. If the axial warping of the crossmember is constrained by the intersection with the sidemember, end-loading is induced in the crossmember. When loads do not act at the centroid of the section, these cause bending moments as, of course, the loads which do not pass through the section shear centre cause torques. The axial load is conveniently expressed as a bi-moment for the sake of the analysis. This is a pair of opposite moments separated by known distance, Fig 48. Also if a bending moment is applied in a plane parallel to the longitudinal axis, it can be resolved into a moment in a parallel plane through the shear centre and a bimoment. In Fig 49, a moment is introduced to a channel member by forces on the flange as shown — resulting in a bending moment *P.d* and a bi-moment *M.h* as well as a shear force *P* and torque *P.h* over the distance d between the forces.

It is necessary to evaluate the bi-moments caused by direct loads in the crossmember which do not pass through points of zero warping. In Fig 50 this applies to forces *F3* and *F4*; the exception to this is shown of the two equal and opposite forces *F1* and *F2* whose effects cancel each other out as they are equidistant from the point of zero warping. Forces *F5* and *F6*, however, do cause both a bi-moment and a bending moment. It is necessary to add these effects to the normal sets of forces and moments which are passed across the joint when setting up finite element models of the overall structure under bending and torsion.

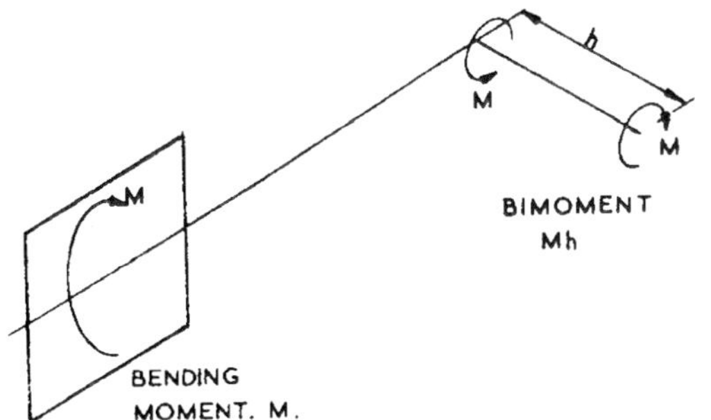

Fig 48: Bi-moment

Fig 49: Force system on channel sidemember

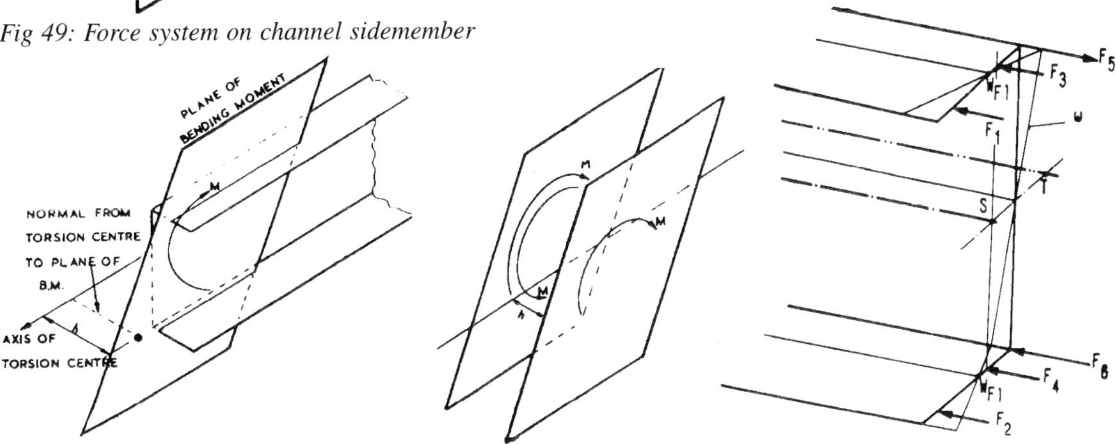

Body shell analysis

The bending and torsion of idealised box beams has been covered in earlier sections. Here the body shell is considered in a less idealised form to see, by examining the likely load paths, how elements of the structure are best represented for analysis.

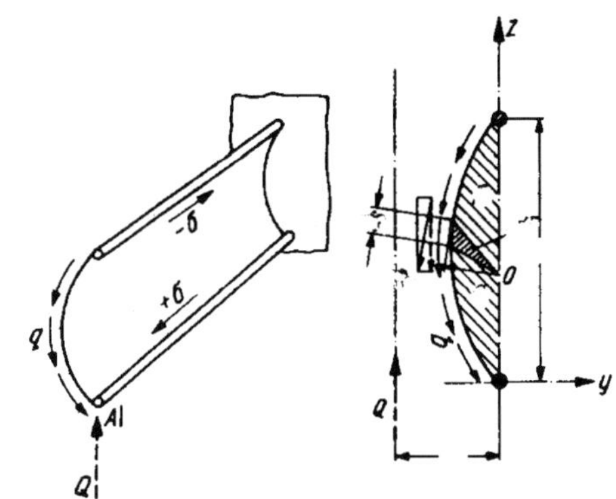

Fig 52. Idealisation of curved panel

Janusk Pawlowski argued that his Simple Structural Surfaces method of analysis, in which structures are represented by surfaces comprising shear panels with end-load carrying bars at their periphery, depended on classifying parts of the structure into the appropriate combinations of bars and panels. The simplest combination comprises two bars and a panel as shown in Fig 52. Shear flow q in the panel is found from the sum of the projections on the related reference axis. Thus on the x and y axis respectively it is

$$\int_0^h q ds.\cos a \text{ and } \int_0^h q ds.\sin a$$

Shear flow does not depend on the shape of the wall but only on height of the element h which is effectively the depth of the beam which the element represents. Location of the resultant shear, through the shear centre, is found by taking moments, which transform to give $a = 2A/h$ where A is the area bound by the mean width of the panel

Cumulative combinations of this basic beam element can be made to form open 'vee' section beams, closed tubes and open channels as depicted in Fig 53. In closed shell form the multi-stringer arrangement of Fig 54 is rigid to a force Q which causes shear, bending and twisting of the structure. The bending is statically determinate but the shear not so; thus the methods are used which were introduced in the earlier section on thin-walled-structures to evaluate the flows — a figure then shown depicting the shear flow diagram for the 'cut' section alongside a diagram illustrating calculation of the unknown shear flow q_0. Full exploitation of the strength and rigidity of shells built up in this way depend upon maintenance of the cross section shape by the use of bulkheads.

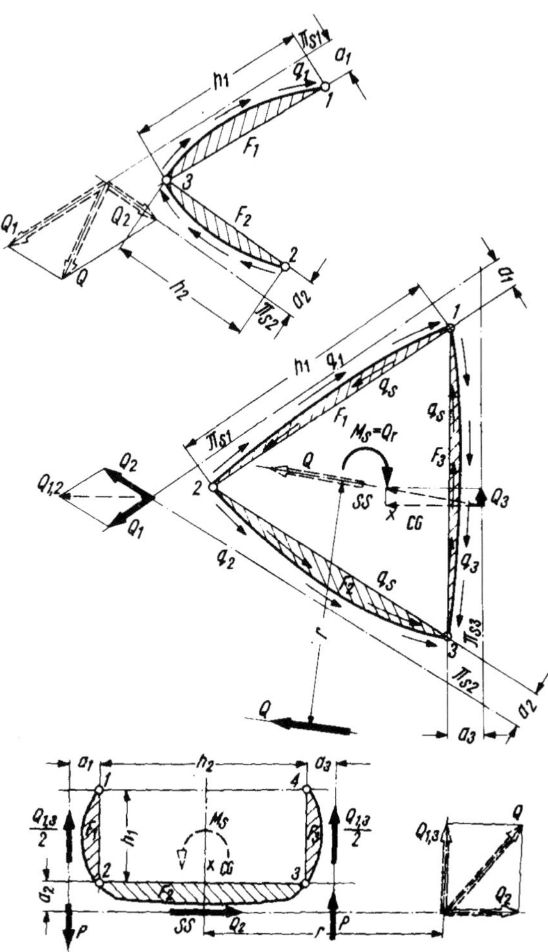

The tendency for axial distortion of the shell in twist should also be accounted for. With the box-tube shell of Fig 55, subject to torque T, there is axial defection of the corner:

$$(T/16abG)[(b/t_1) - (a/t_2)]$$

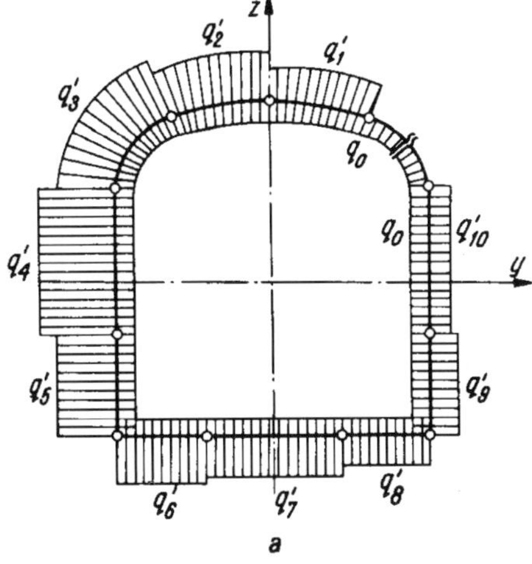

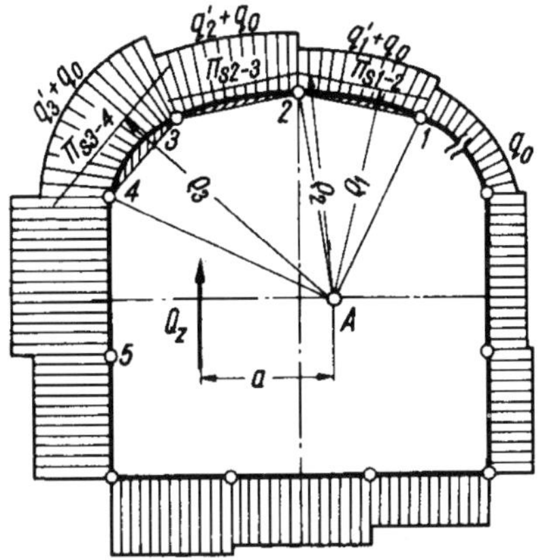

Fig 54: Multi-stringer shell beam

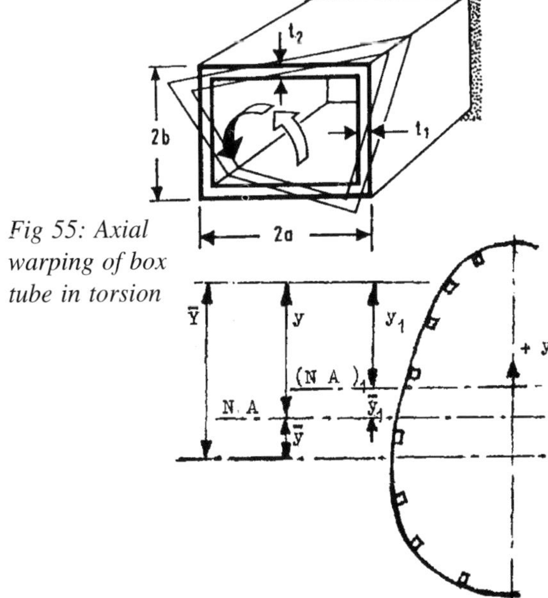

Fig 55: Axial warping of box tube in torsion

There is no axial distortion (warpless tube) when:

$b/t_1 = a/t_2$.

Bending of a multi-stringer tubular shell can be analysed relatively easily with the aid of some simplifying assumptions. For the arbitrary section of Fig 56, stringer cross-section-areas *A* are enlarged by allowing for the contribution made by adjacent panelling. For width of skin panel *b*, thickness *t* and distance of effective stringer area from neutral axis y, direct stress is notated as σ and buckling stress as σ_b. When there is no skin buckling, effective stringer area on the tension side of the shell:

$$A = A_s + (1/2)b_u t_u + (1/2)b_l t_l = A_s + b_m t$$

when $t_u = t_l = t_s$ meaning stringer; *u* is upper and *l* lower. After the onset of buckling, the effective area of the panel is assumed to be:

$$(1/3)bt + (2/3)bt(\sigma_b/\sigma)$$

and the effective stringer area is approximately:

$$A_l = A_s + (1/3)b_m t = (2/3)b_m t(\sigma_{bm}/\sigma_s)$$

where *m* is mean value between upper and lower panels. The method involves tabulating *A* and *Y* from a chosen datum and finding its distance from the neutral axis as *sAY/sA* and the section second moment of area about the neutral axis as:

$$I = \Sigma AY^2 - y\Sigma A$$

If some panels are assumed to buckle then their effectiveness will have been over-estimated and a reduction of area is necessary of $dA = A - A_l$. A new neutral axis can then be found from $y_l = -\Sigma dAy/\Sigma A_l$ and a new value of area moment $I_l = I - \Sigma dAy^2 - y_l \Sigma A l$ and the stress becomes $\sigma_l = M(y - y_l)/I_l$. Successive further approximations can also be made following the same procedure.

To simplify the determination of degree of redundancy of a shell structure, shear panels can be simulated as pin jointed frames for this purpose, in a method due to G B Bolland of Loughborough University. Fig 57 shows representations for simple beams and torque tubes. The formulae given earlier in the section on frameworks can then be used to add up the redundancies for the whole shell. Fig 58 shows a curved panel represented in this way and Fig 59 a box with framed cut-outs and longitudinal torsion members. In the latter case, the structure is regarded as three-dimensional with two-dimensional elements hung on and a general form of the equation for counting redundancies becomes:

$$R = m - (3j_3 + 2j_2 - 6)$$

where m is the number of members and j_2, j_3 the numbers of two and three dimensional joints respectively. In this case, number of redundancies $R = 8$ since $m = 74$, $j_3 = 16$ and $j_2 = 12$.

Often vehicle structures are idealised as combinations of pin- and stiff- jointed frames as in the case of the saloon car treated as a symmetrical side frame in Fig 60. Members BI and DH are pin-jointed but all others are rigid and the system has five redundancies, as a plane frame. Integral bus structures have been idealised as shells such as that shown as Fig 61 and box vans as shown, due to G H Tidbury of Cranfield University, as Fig 62.

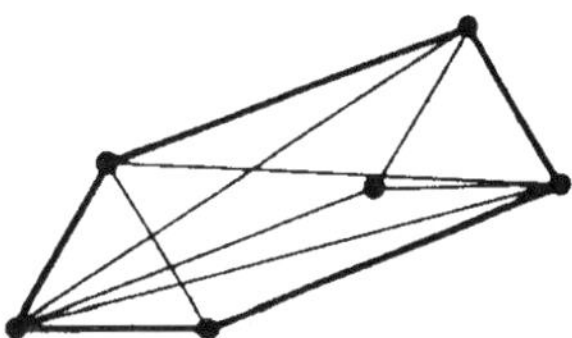

Fig 57: Frame representaton of shell

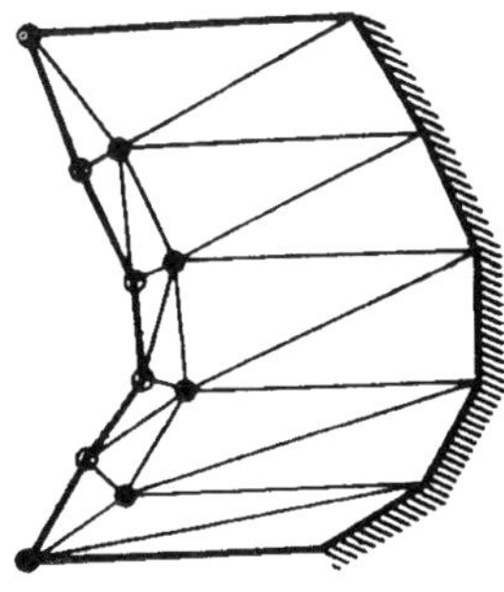

Fig 58: Frame representation of curved panel

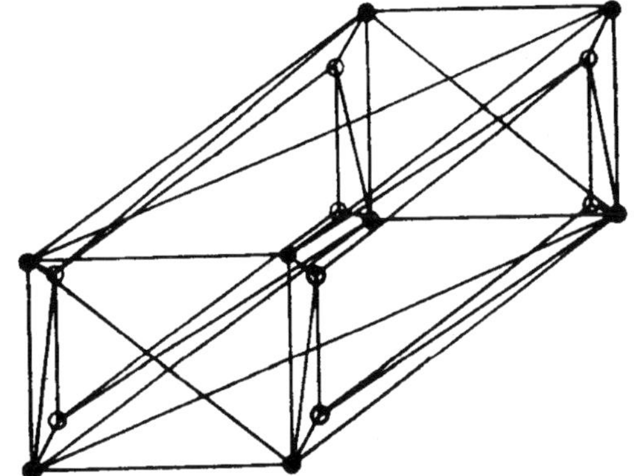

Fig 59: Frame representation of box with cut-outs

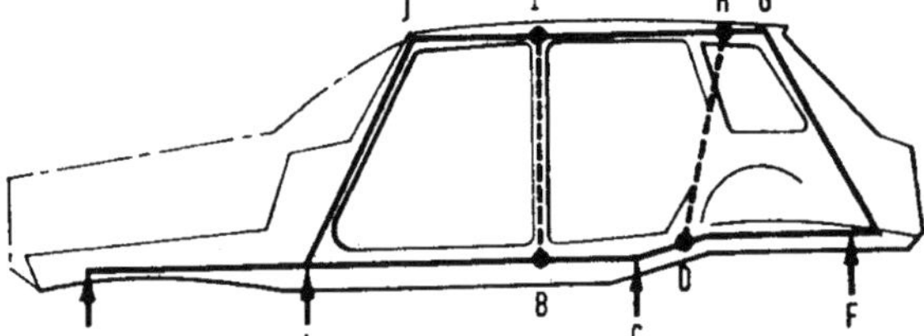

Fig 60: Pin- and stiff-jointed frame representation

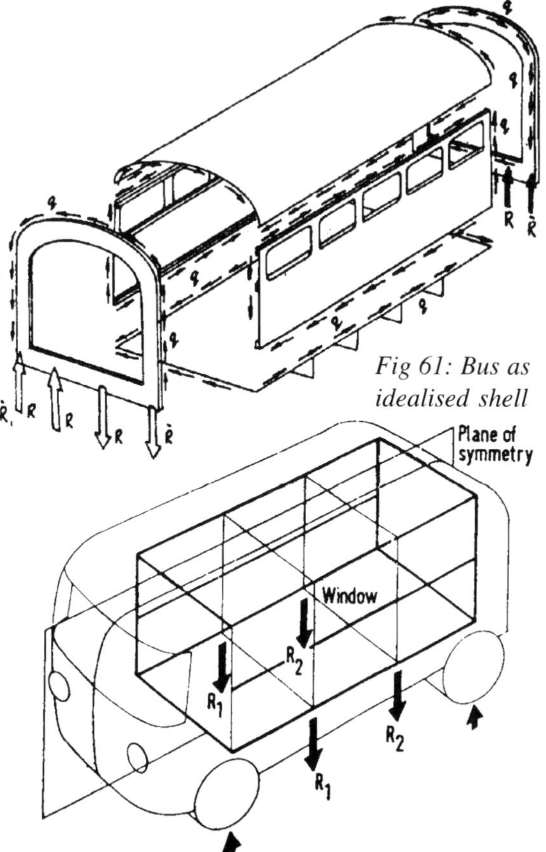

Fig 61: Bus as idealised shell

Fig 62: Van shell idealisation

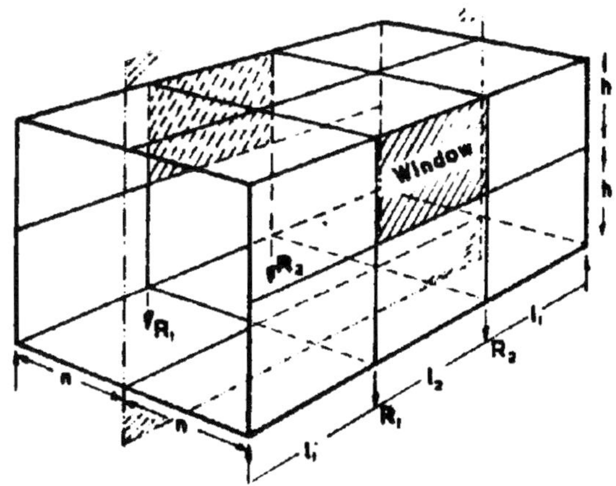

Fig 63: Van body as 3-bay box tube

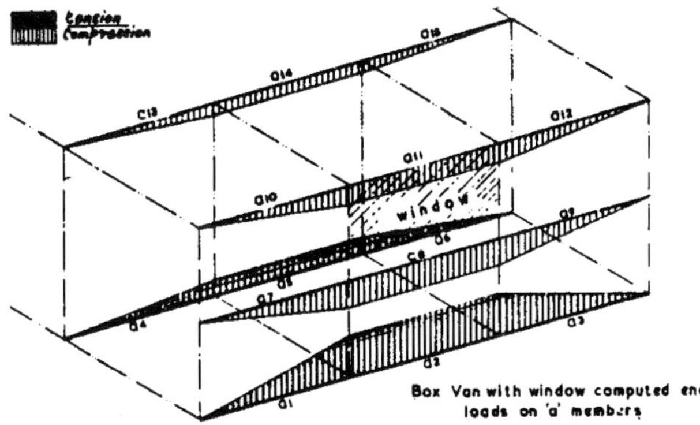

Fig 64: Stressing a 3-bay box-tube

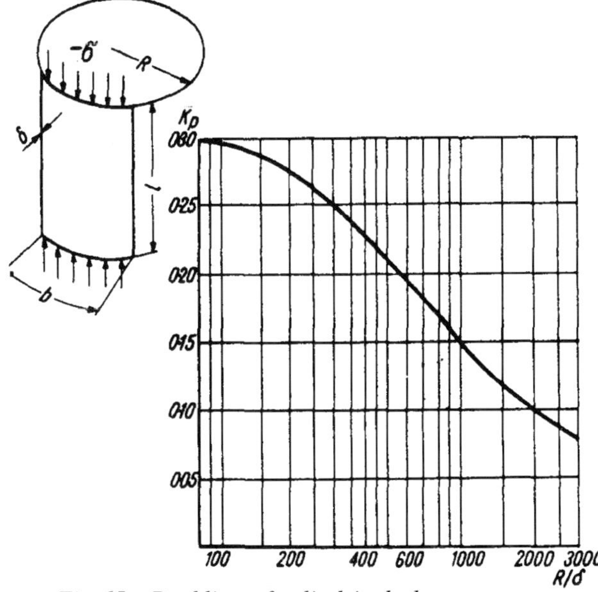

Fig 65: Buckling of cylindrical plates

Shell structure loading

In the section on finite element analysis of body structures , the construction of the mesh is shown to require a knowledge of the behaviour of the structural elements of the body shell. The effectiveness of the finite-element analysis and the economy of the meshing procedure depends on an understanding of the load paths and likely deflection modes, a number of which are discussed in this section.

Fig 63 shows the results of a matrix force analysis carried out by Tidbury[7] on the idealised van body of Fig 62. As the structure is divided into relatively few panels, a good visualisation of the structural behaviour, Fig 64, can be obtained — prior to a more comprehensive analysis using finite-element techniques. As opposed to the simpler structure analysed in an earlier section, this one required computer solution of the matrices.

Curved shell sections

For curved sections of a shell, the treatment of cylindrical plates under compression is a useful starting point. Critical stress for local buckling depends on ratio of radius-of-curvature to wall thickness. An empirical formula gives critical stress as:

$$=kE/(R/\delta)$$

where k is the experimental bucking co-efficient obtained from the graph in the lower part of Fig 65. E is the elastic modulus of the shell material and the increase in buckling strength depends upon dimension R, decrease of which substantially increases critical stress.

In considering those parts of the shell which are flat plates, the shear deformation is an important criterion. The formula discussed in an earlier section, relating shear stress and direct stress at an angle 45 degrees. In Fig 66, $\sigma_1 = \sigma_2 = \tau$ and when σ_2 exceeds a given value the plate will begin to buckle, waves appearing at right angles to the compressive stress. Critical shear stress is given by:

$$KE(B/\delta)^2$$

where K is another experimental co-efficient depending on the edge fitting as shown in the lower part of the figure. Decreasing dimension B is the most effective way of increasing critical shear stress. Another technique is by suitable choice of edge dimensions, Fig 67.

Simultaneous compression and shear stresses are given by the equation:

$$\sigma/\sigma_k = -1+(\tau/\tau_\kappa)^2$$

attributed to Pawlowski; also that between tensile and tangential stress as:

$$\tau/\tau_k = 1+(\sigma/2\sigma_k)$$

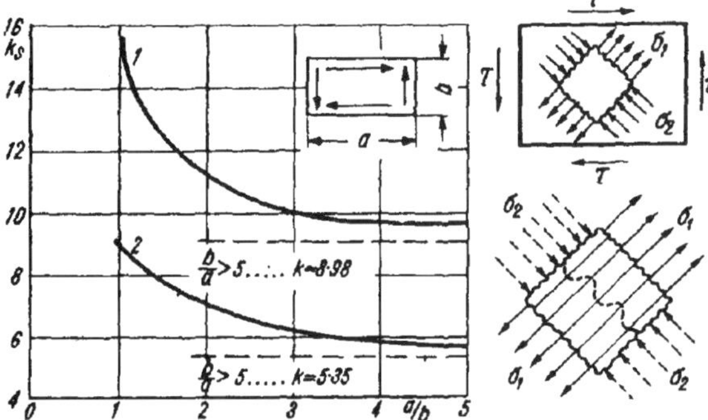

Fig 66: Shear and direct stress relationship

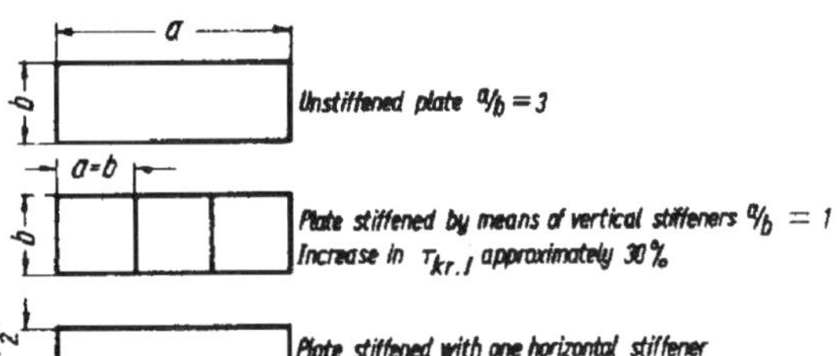

Fig 67: Stiffening of a shear panel

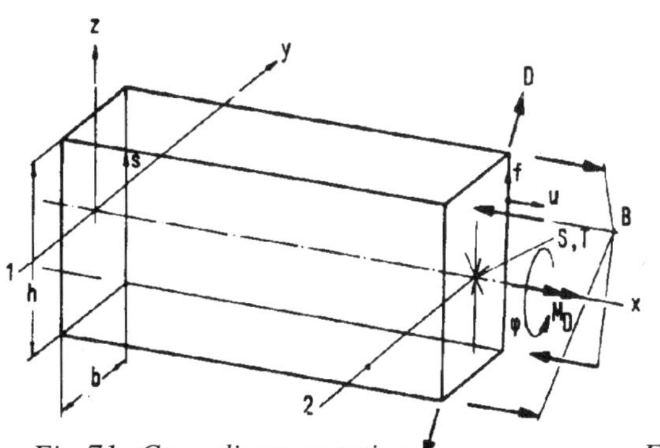

Fig 71: Co-ordinate notation

These are shown graphically as Fig 68 in which the region of stability is limited by a parabola which becomes a straight line near the ordinate and the abcissa. Points with values of t/t_k and s/s_k within this region do not buckle.

Shear lag in bending Distribution of elastic shear lag in the wide flange of a box beam is introduced in a separate section. The effect of this is to increase the overall deflection of a box-beam under load — also an increase of the longitudinal stresses at the web/flange intersection and a decrease at mid-section. In design it is convenient to replace breadth b by an equivalent breadth b_e such that the application of simple bending theory gives correct stress and deflection values, in a method described by Donald[8]. This has become standard practice as set out in BS 5400 (Part 3). Shear lag is shown to be significantly dependent on plan dimensions of the flange B/L as

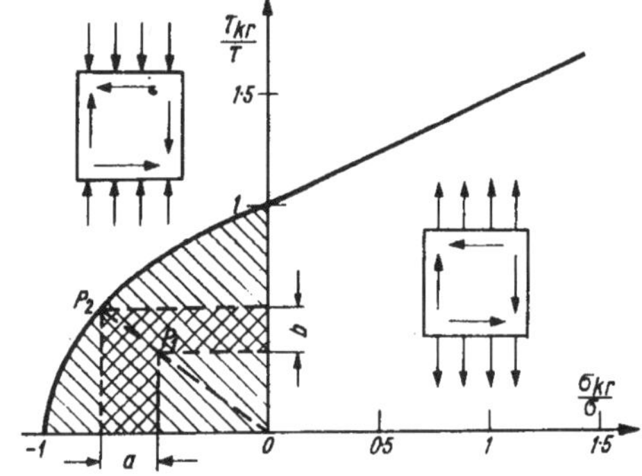

Fig 68: Critical stress diagram

(Uniform load over length of each web)

a	$\frac{B}{L}$	Stress effective breadth			Deflection effective breadth		
		Mid-span	Quarter-span	Support	Mid-span	Quarter-span	Support
0·0	0	1·00	1·00	1·00	1·00	1·00	1·00
	0·05	0·98	0·98	0·84	0·98	0·98	0·98
	0·10	0·95	0·93	0·70	0·94	0·94	0·93
	0·20	0·81	0·77	0·52	0·79	0·79	0·77
	0·40	0·50	0·46	0·32	0·48	0·47	0·47
	0·60	0·29	0·28	0·22	0·31	0·30	0·30
	0·80	0·20	0·19	0·16	0·21	0·20	0·20
	1·00	0·16	0·15	0·12	0·17	0·16	0·16
1·0	0	1·00	1·00	1·00	1·00	1·00	1·00
	0·05	0·97	0·96	0·77	0·93	0·92	0·92
	0·10	0·89	0·86	0·60	0·84	0·84	0·83
	0·20	0·67	0·62	0·38	0·62	0·62	0·60
	0·40	0·35	0·32	0·22	0·33	0·33	0·32
	0·60	0·22	0·20	0·15	0·21	0·20	0·20
	0·80	0·16	0·15	0·11	0·16	0·16	0·16
	1·00	0·12	0·11	0·09	0·12	0·12	0·12

Fig 70: Effective width ratios for simply supported box girders

seen in Fig 69, bottom half. The effective breadth of the flange in a continuous box beam can be estimated by treating each portion of the beam, between points of contraflexure, as an equivalent simply-supported span. The beam is considered to consist of an assembly of simply-supported L-section beams each having a web plate and a flange plate. Effective breadths of these are determined from the table in Fig 70 which applies to the case of a uniform load over the length of each web.

Axial warping in torsion

Another important effect in box-beams is the out-of-plane distortion of cross-sections of a box-beam known as axial warping; this is described for open-section beams in a separate section. It has also been described by Beerman[9] who points out that while inhibition of warping does not produce large torques in closed section beams, the compatibility of any warping displacement at the end connections of the beams must be taken into account. Fig 71 shows a co-ordinate notation used in analysis; *u* displacements represent warping while *f* displacements are both rotation and lozenging of the cross-section. The total deformation can be separated into three parts as shown in Fig 72.

The lozenging displacement is associated with a lateral bi-moment *D*, a set of self-equilibriating shear forces, seen in Fig 73 to be the product of shear stress τ, length of the sides and material thickness *t*; this is not to be confused with the longitudinal bi-moment *B*. Fig 74 shows the distribution of these bi-moments as well as warping and lozenging functions for two rectangular and two square sections with equal total length of sides and different thicknesses — subjected to to equal bi-moments at each end and with lozenging inhibited at these ends. Note that the equal bi-moments cause opposite internal loads at the two ends so that the bi-moment will be zero at the centre of the beam.

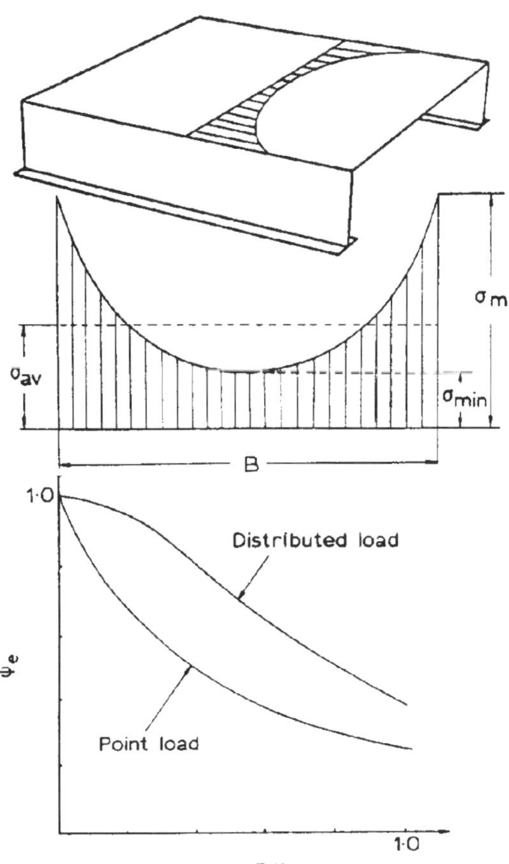

Fig 69: Shear lag effect

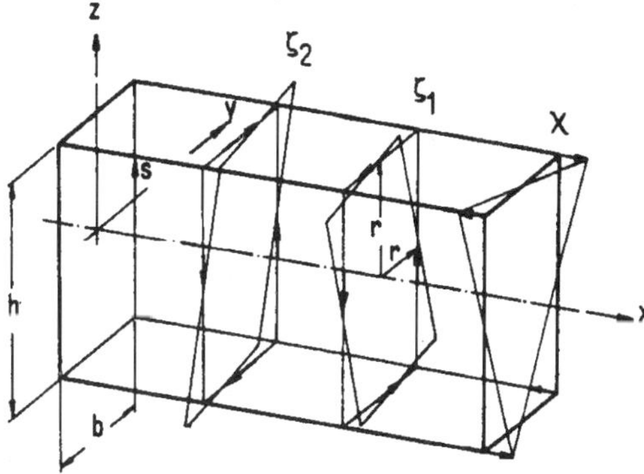

Fig 72: The three deformations

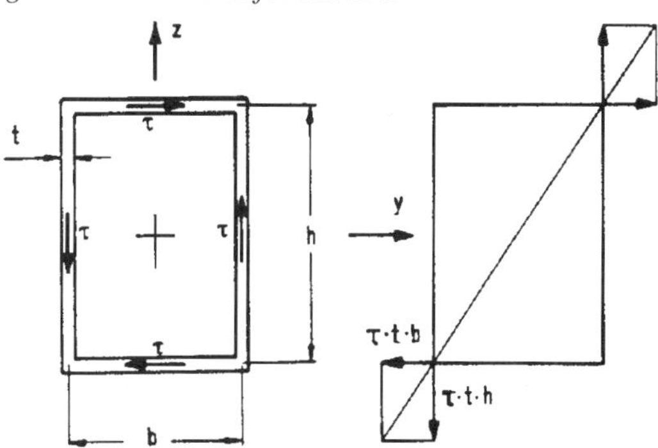

Fig 73: Force system

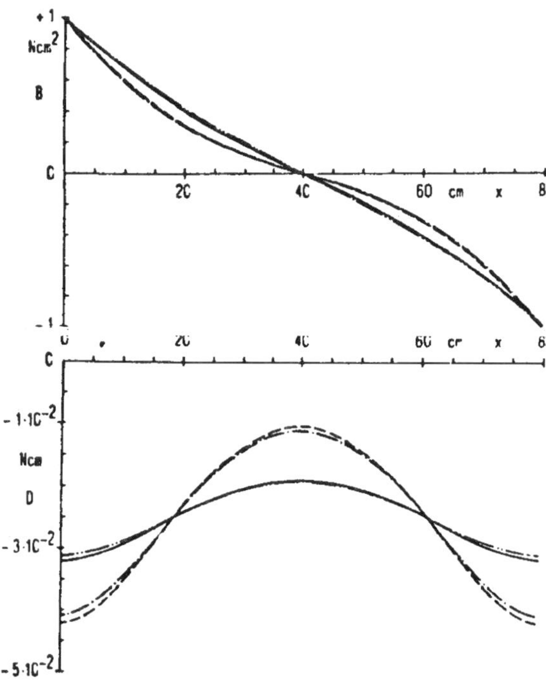

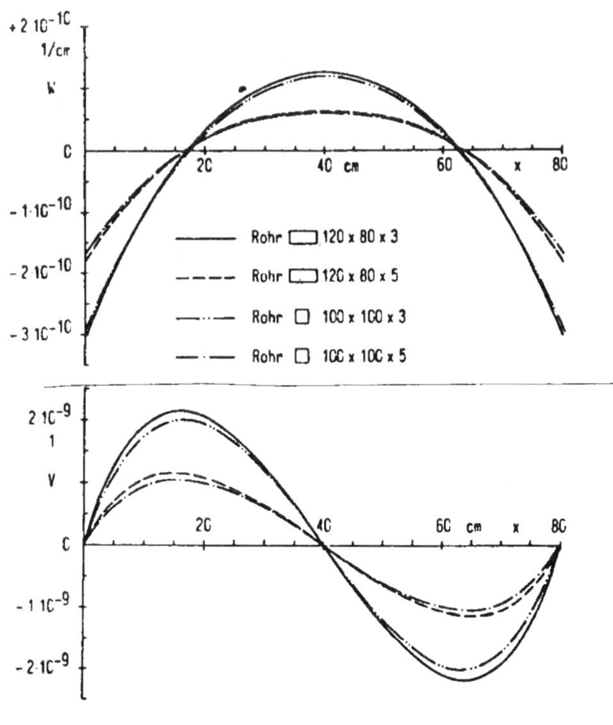

Fig 74: Bi-moment distribution

Mounting shell structures

The mounting of structures such as CV cabs and tanker shells needs particular care in design — to diffuse the concentrated loads at the mount over as wide an area as possible of the supported structure. Some of the analytical approaches are discussed here alongside an examination of some of the accepted mounting arrangements for cabs and tanks.

The loading of the shell through the mounts can also be controlled by supporting the mounts on pivoting subframes, for example, to relieve the shell of torsional loads. The subframe can also sometimes be in the form of a cradle which itself has a load-spreading effect over the shell. The cradle can then be

Fig 75: Supporting a tank shell

Fig 76: Cab-support system

trunnion mounted to a further cross-tube, Fig 75, which would be resiliently-mounted to the chassis-frame. In the case of cabs, traditional practice was to ensure relative stability for the occupants in the transverse plane, by having at least one stiff mount either side of the cab, while including further flexible mounts at the rear to allow for torsional weaving of the chassis frame.

Beyond these generalities, a much more detailed analysis is now usual in considering load-spreading, mounting spring-stiffness and a controlled degree of damping the relative motion between chassis and shell. A sophisticated damping system for a contemporary truck cab is seen in Fig 76. Four air-springs are used for the Renault AE cab — one of the tallest in production — the fronts at 4.5 bar and the rears at 3.5 bar. Longitudinal stability is obtained by Panhard rods; anti-roll and rear lateral location bars are also fitted. Three levelling valves are employed and a ride frequency of about 1 Hz was the design target.

In tankers, the approach is often one of using the bulkhead of the tank shell as the position for mounting the tank on to a saddle-bracket, Fig 77. The weight of the cargo is carried first by circumferential tension in the shell; the tension in turn creates shear stresses, resulting in an overall beaming action. The shear is a maximum at the supports, of course, where it feeds down from the shell wall via the interior bulkheads, Fig 78. By augmenting the boot at the base of the bulkhead stiffeners, it is possible to make the stiffener carry all the load and dispense with the saddle. The support reaction is then transferred directly from the longitudinal to the boot, Fig 79. If the boot is replaced by an internal longitudinal between two bulkeads, see inset, then the shear load at each end of the internal longitudinal will be transmitted to the stiffener by an appropriate welding bead.

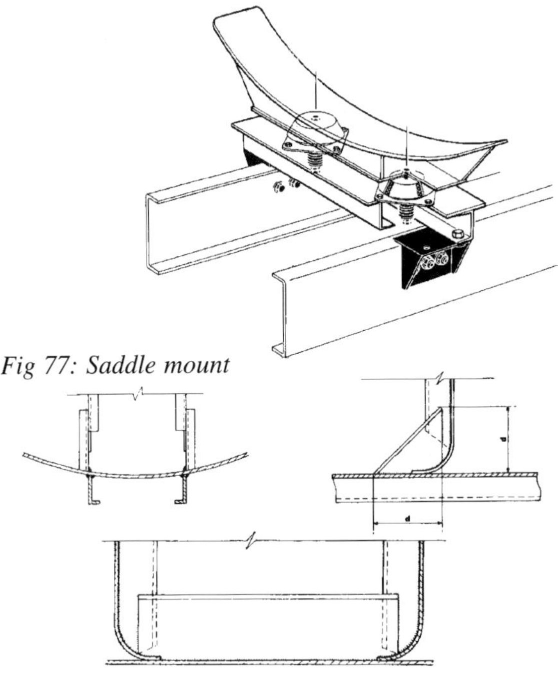

Fig 77: Saddle mount

Fig 79: Loading through boot

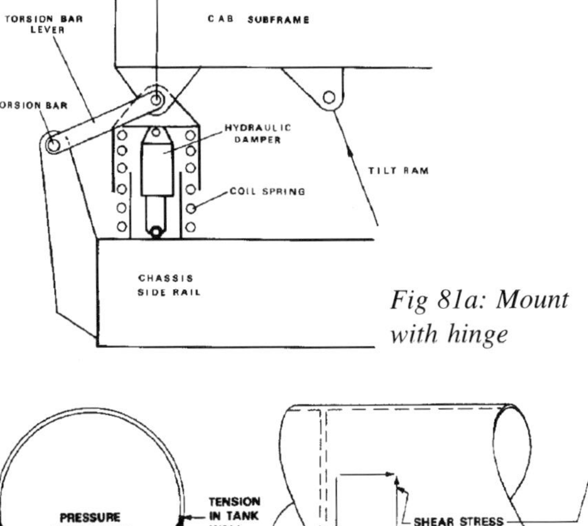

Fig 81a: Mount with hinge

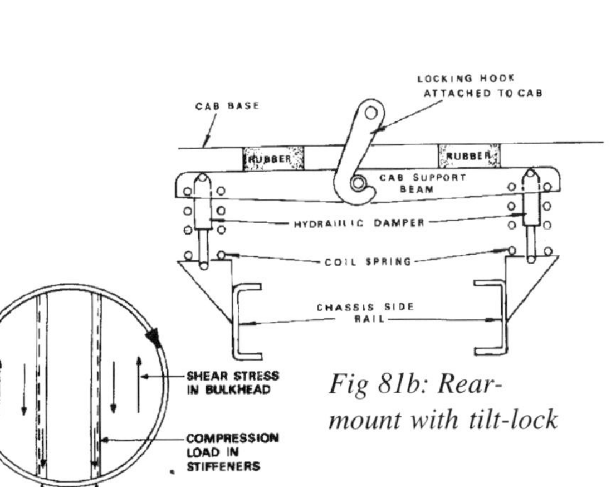

Fig 81b: Rear-mount with tilt-lock

Fig 78: Tank loading through bulkhead

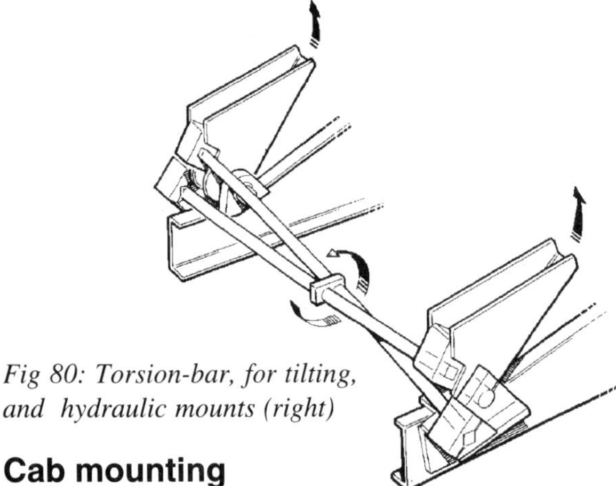

Fig 80: *Torsion-bar, for tilting, and hydraulic mounts (right)*

Cab mounting

In cab suspension the theoretical ideal, two mounts at the front and one at the rear, to allow chassis-frame weaving is often made impractical by the requirements for cab tipping. The tilting mechanism may be torsion bar or hydraulic, Fig 80, the latter being favoured for heavier cabs. One way of incorporating the front mount with the tilt hinge is shown diagrammatically in Fig 81. This arrangement minimises stresses imparted into the cab structure. A rear mount for a tilt cab is usually incorporated with a cab tilt-lock. Conventionally two rubber springs are mounted close together to simulate single-mount characteristics with respect to chassis twist.

The pattern of stress distribution into the cab arising from the load inputs at the mounts, is an ideal candidate for analysis by finite-element methods. For the Leyland C44 cab introduced on its Roadtrain truck range, the FE model was as shown in Fig 82.

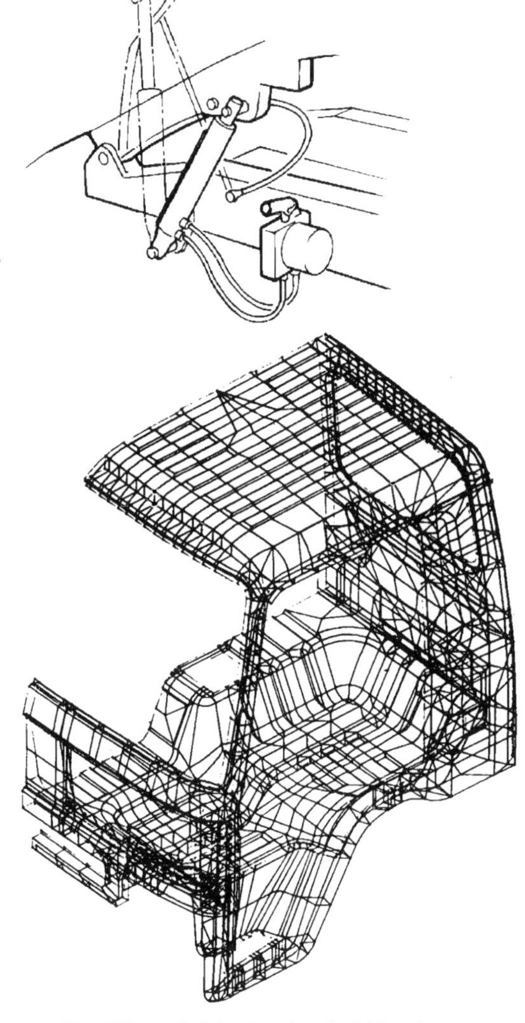

Fig 82: FE model for Leyland C44 cab

Analyzing simple shells

If some estimate is required of a shell's ability to sustain external load, without FEA, the classical spherical case provides some indication. A uniform pressure q produces a membrane stress of qa/rt where a is the spherical radius and t is the thickness of the sheet. The value of q can be increased up to a point where the shell buckles when its critical value is approximately $(E/2)(t/a)^{1/2}$ where E is the elastic modulus. If the surface is stiffened by ribs, the last equation can be factored by $(t_B / t_m)^{3/2}$ where t_m is substituted for t, being the ratio of stiffener area to stiffener spacing S_s and t_B the equivalent bending thickness evaluated from the stiffener moment of inertia I_s as $(12I_s / S_s)^{1/3}$. In the case of a concentrated load P on a spherical shell, the theoretical stiffness is given by:

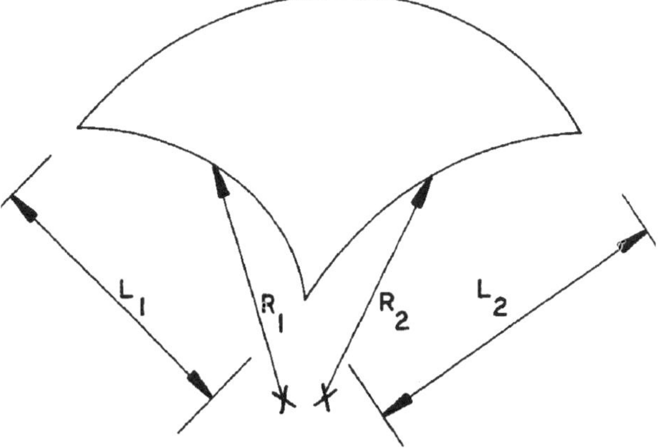

Fig 83a: Double-curved panels

where d = deflection, C is a constant. The expression is generalised for double-curved panels which are not necessarily spherical by replacing the curvature $1/a$ (Fig 83a) by:

$$[(L_1^2/R_1) + (L_2^2/R_2)]/2L_1L_2$$

The constants and a fuller description of the technique can be obtained from an SSRC publication[10].Simplified design charts are also provided by Marshall[11].

The case of box-tubes intersecting, as shown in Fig 83b, is a useful aid to design of brackets which connect mountings to shells. Here the tendency for the end-on box-tube to punch through the wall of the shell is expressed as a punching shear efficiency E_v given by maximum allowable punching shear stress divided by allowable tensile stress in the shell. E_v is shown in Fig 84 as a function of non-dimensional parameters given in the text. The first of these figure shows the yield pattern. The first localised failure is punching at the material yield stress thereafter sidewall yielding of the main shell occurs. For compression loads this is in the form of web crippling with capacities shown on the right hand side of the second figure.

Tanker shells use a variety of sandwich rubber mounts for isolation from chassis motion. A four point mounting system may have either simple plane sandwich units at each corner or vee-configurations to give additional transverse or longitudinal stiffness, as required. Conical sandwiches can be used to obtain stiffness in all directions within the horizontal plane. Another interesting approach taken by a French company for a 12 000 litre tanker was the arrangement shown in Fig 85. On each side of the rear, a tee shaped fitting was used with rubber bushes on each arm of the tee which are parallel to the longitudinal axis of the vehicle. A circular rubber mount secures the leg of the tee to the chassis frame while the two arm bushes are carried by a tank shell bracket. At the front end, the tank is connected by a single rubber-bushed pivot, also with its axis parallel to that of the vehicle, to the centre of a yoke. Each end of this yoke is connected by a rubber-bushed shackle to a bracket beneath it on the frame. Axes of the bushes in the shackles are arranged transversely in the vehicle so as to take length changes due to frame deflections.

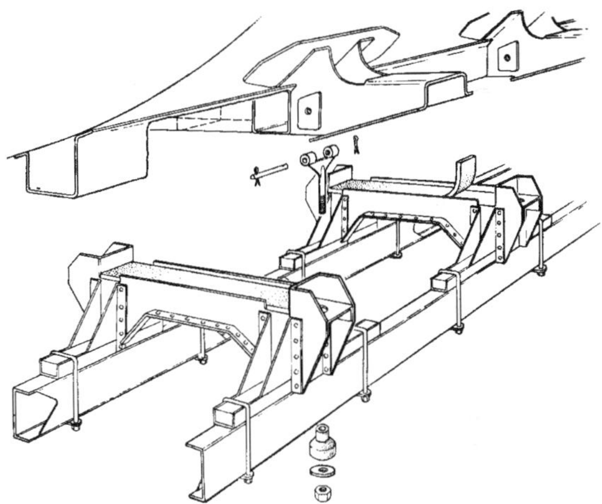

Fig 85: Special tank mounting

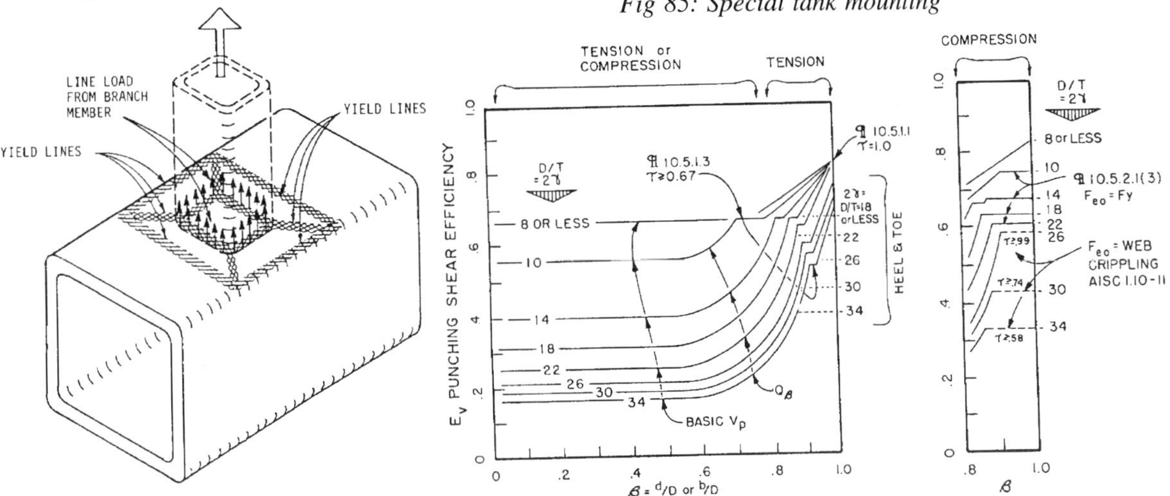

Fig 83b: Intersecting box tubes

Fig 84: Allowable 'punching' stress

Stability of shell structures

As well as considering the stability of struts and plates under end loading, shell stability is also of importance in the design of thin-walled structures under many conditions of loading and is a pre-requisite to the study cab structures, for example. A thin tube or box-beam under end loading can suffer stability failure as a buckling shell, the controlled behaviour of which can be used in the production of impact-absorbing structures for impact protection.

Another section deals with impact collapse of structures, within which plastic-hinges are introduced as an analytical tool. This theory can be extended to the consideration of thin cylinders and box-tubes under axial loading. On the assumption of a concertina-type collapse mode, the collapsing member can be idealised for analysis as shown in Fig 86. The energy dissipated in plastic bending of the four circular hinges can be summed to arrive at an overall figure for the axially loaded shell. Also to be added is the energy dissipated in stretching, under a substantially uniform tensile hoop stress, in the metal between the hinges. This will depend on whether the buckling is entirely external as shown or entirely internal, at the other extreme. For calculation purposes a mean value of energy is chosen between these extremes and P/Y becomes $= 6t_o(Dt_o)^{1/2}$

In crushing a tube under static load conditions, as each new row of buckles forms, the axial load rises to a maximum and falls to a minimum as the buckles fold flat, the load varying about a mean value as shown in Fig 87. When the tube is subject to dynamic impact, the ratio of the dynamic to the static crumpling load increases with speed. The ratio reflects changes in buckling mode due to inertia — also as a consequence of the rate-of-strain effect on the material of the tube.

Experiments with mild steel tubes preliminary to the design of railway coaches, for example, have shown a linear increase in the ratio from 1 to 2.5 with impact speed increase from 0 to 125 ft/sec. A moving tubular structure colliding with a flat stationary body will experience a resisting force $F = \sigma A$ where σ is the mean crushing stress for the structure and A its cross section area. Mass of the structure is ρAL, for density ρ and length L. Uniform retardation is therefore $\sigma/\rho L$, by dividing the two expressions above. Retardation is thus inversely proportional to length in this simplified analysis.

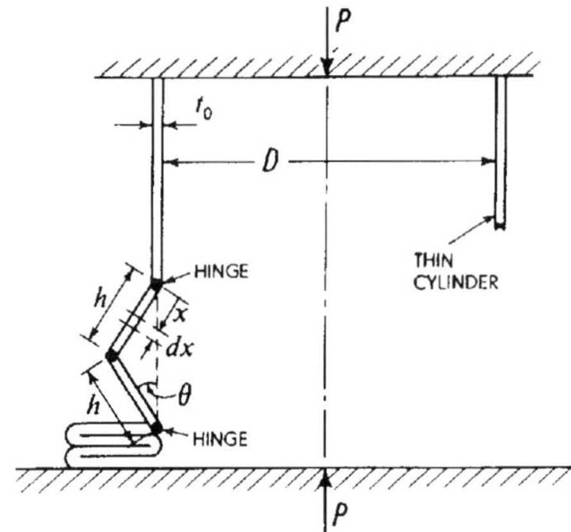

Fig 86: Collapse idealisation

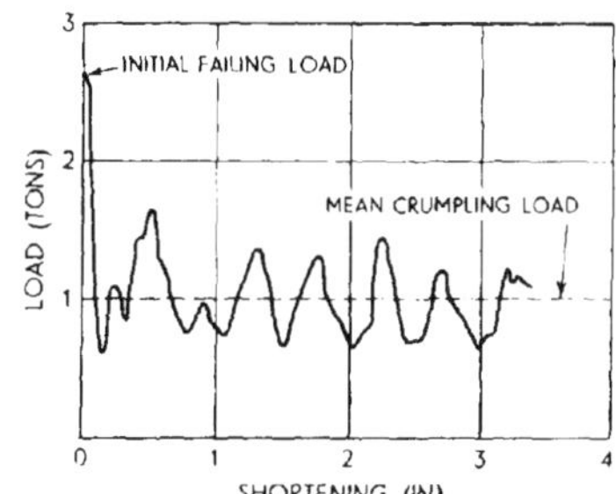

Fig 87: Static crush load trace

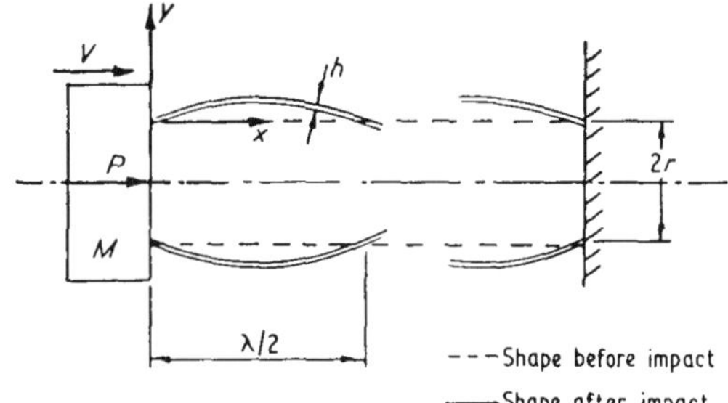

Fig 88: Pre-failure phase

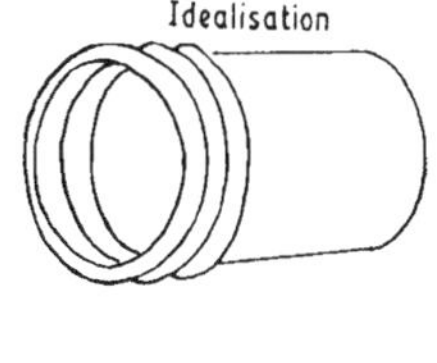

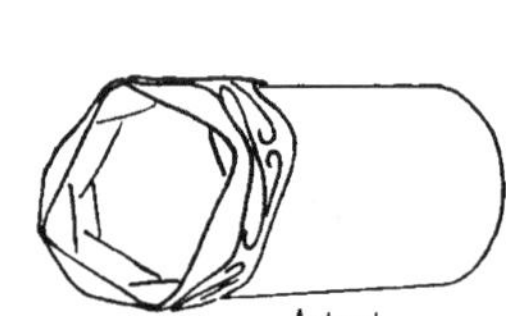

Fig 89: Cylinder collapse mode

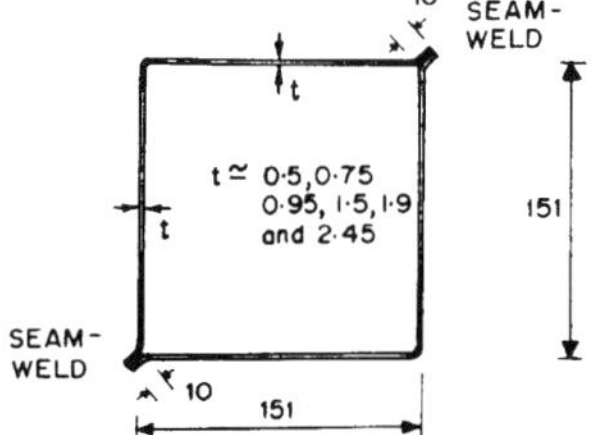

Fig 90: Box-section test specimen

Only the front part of the structure, closest to the impact area, is subject to crushing. This length x can be estimated from the assumption that all the kinetic energy is dissipated in the plastic deformation, such that $Px = Mv^2/2$ and:

$$x/L = (\pi/12)[(D/t_o)(\rho v^2 / Y)]^{1/2}$$

where Y is the yield stress of the material.

Workers at Birmingham University showed the pre-failure phase of a thin cylinder being impacted as Fig 88 and devised an expression for the failure load as $P_e = n^2\pi^2/EI/L^2$.

They showed in Fig 89 how a cylinder actually collapses compared with the idealised mode which is assumed in a simplified analysis and suggested an expression for the mean load during crumpling as $P = KYh^{1.5}(2r)^{1/2}$ where K is approximately 6 — to give reasonable accuracy for the post-failure stage of a dynamic test. It is suggested that this is valid even for cylinders with cut-outs in them provided the cut-out does not exceed 50 per cent of the cylinder circumference.

Civil engineering techniques of analysing box columns are also of potential value to automotive designers in calculating the instability of shell structures. Work at the Australian Monash University[8] has involved testing box specimens as shown in Fig 90. Others involved b/t ratios ranging from 60 to 158 and exceptionally as high as 200 and 300.

Those with low b/t developed pent-roof shaped buckles while the higher ones had so-called flip-disc

Fig 93: Roof type failure

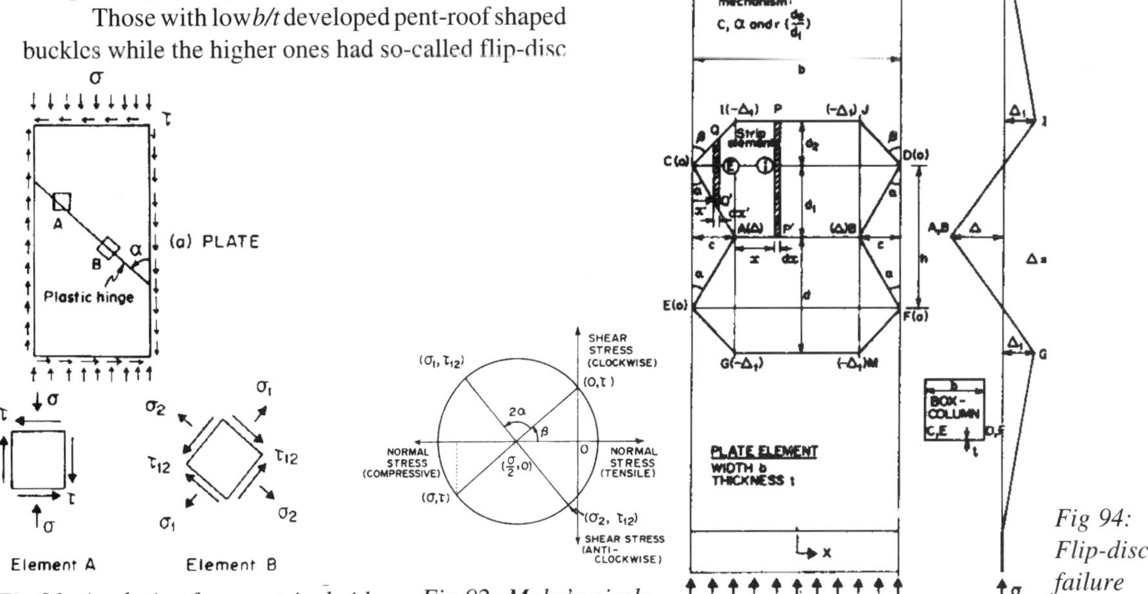

Fig 91: Analysis of symmetrical side

Fig 92: Mohr's circle

Fig 94: Flip-disc failure

shapes with general corrugations having curved tops. On increasing the applied load, specimens pass through three deformation phases: elastic, elasto-plastic and plastic. The last stage is analysed using rigid plastic theory based on observation of the test specimen failure modes.

The assumption of similar modes applying to each side of the box is made and the consideration of one side, only, in the analysis, Fig 91. The plate is considered to carry average compressive stress σ and shear stress τ, subscripts 1 and 2 applying respectively normal-to and in the direction of the plastic hinge.

Average shear stress τ_{12} is found using the Mohr's circle technique, Fig 92. Plastic moment per unit length of hinge is $M_pR[1 - (2d/t)^2]$ where R depends on the stress distribution, established in sample testing, and d is the distance from the neutral axis of the plastic hinge to the mid-plane of the plate, again found empirically. Fig 93 shows the idealised 'roof' mechanism of failure and Fig 94 the flip-disc.

In work aimed at finding the crush resistance of monocoque bus structures, Leyland and UMIST researchers[12] have examined the theory of crumpling in square tubes and suggest the Macauley formula:

$$P_v/P = n(1 + nKV/\lambda)$$

relating static to dynamic crumpling load P_v where n is a factor depending on the variation of strength with impact velocity V and 1 is a geometric scaling factor. The researchers simplified the eight tonne bus, 9 ft high x 8 ft wide x 34 ft (2.74x2.44 x10.36 m) long to a plain rectangular tube of mild steel (to use in the empirical relationship:

$$P/P_o = 3/p\ (A/A_o)^{1/2}$$

with A_o, the enclosed area of tube and the tube cross-section area, giving a so-called solidity ratio of 0.0069. $P_o = YA = 1432$ tonnes. The results shown in the table of Fig 95 compare the two equations. Because of the discrepancies, the researchers constructed test models as in Fig 96 for testing under a drop hammer — with the results shown in Fig 97.

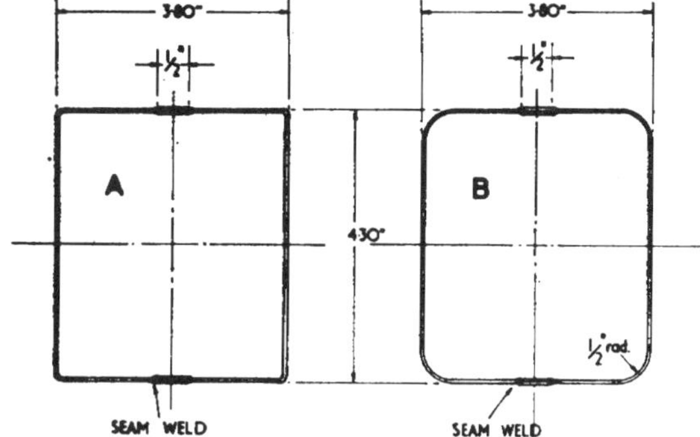

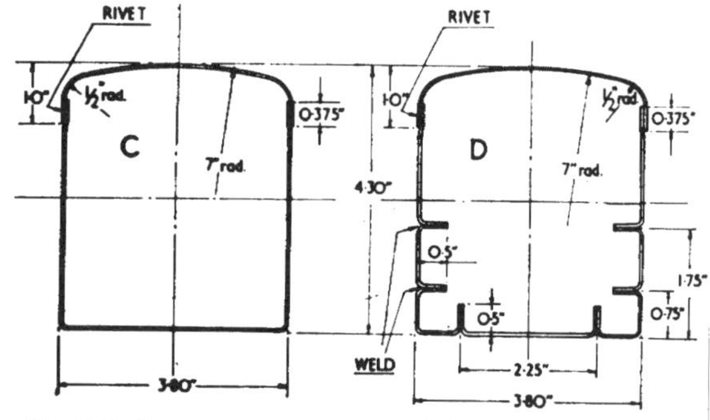

Fig 96: Bus structure test models

P/P_0	Static Mean Crumpling Load P (tons)	Dynamic Mean Crumpling Load P_v (tons)	Shortening $\bar{x}$ (inches)	Deceleration a (ft/sec²)	Deceleration k (g's)
0.0552	79.1	158.2+	8.1 (1.99%)	640	19.8
0.0794	113.8	113.8	11.25(2.75%)	457	14.2

Fig 95:Results from two formulae

COACH MODELS

MODEL NUMBER	TUBE SECTION	SIDE CUT-OUTS RIGHT HAND	SIDE CUT-OUTS LEFT HAND
1	A	PLAIN TUBE	
2	B	—II—	-II-
3	C	—II—	—II—
4	D	—II—	—II—
5	A	A	A
6	A	B	NONE
7	A	C	C
8	A	D	A+C
9	C	D	A+C
10	D	D	A+C

The models were manufactured from a mild steel EN 2c and its yield stress was taken to be 20 ton/in² throughout the calculations. The hammer weight was 77lb. Kinetic energy = ½ (mass of hammer + mass of model) V_0^2.

Model No.	Weight lbs.	Net C.S.A. 'A' in²	Enclosed C.S.A. 'A'$_0$ in²	Solidity Ratio 'A/A$_0$'	P_0 = YA ton	Static Tests: Approx Mean Crumpling Load 'P' (tons)	$\frac{P}{P_0}$	Dynamic Tests: Velocity 'V$_0$' ft/sec	Kinetic Energy lb–ft	Shortening 'x' ins	Mean Crumpling Load 'P$_v$' (tons)	$\frac{P_v}{P_0}$
1	3.75	0.829	16.32	0.0507	16.55	2.8	0.169	34.1	1455	1.5	5.0	0.302
2	3.65	0.806	16.16	0.050	16.12	3	0.186	38.5	1855	1.5	6.625	0.41
3	3.53	0.78	15.9	0.0491	15.6	2.8	0.18	34.5	1485	1.5	5.32	0.341
4	4.96	1.098	15.9	0.069	22	7.5	0.341	35.0	1560	9/16	14.85	0.675
5	3.28	0.829	16.32	0.0507	16.55	2	0.121	35.4	1560	1¾	4.78	0.239
6	3.6	0.829	16.32	0.0507	16.55	2.1	0.127	35.2	1550	1.15/16	4.29	0.214
7	3.63	0.829	16.32	0.0507	16.55	1.8	0.109	35.2	1550	2⅜	3.5	0.175
8	3.15	0.829	16.32	0.0507	16.55	1.7	0.103	36.2	1630	2⅝	3.32	0.166
9	3.07	0.78	15.9	0.0491	15.6	1.2	0.077	35.2	1540	3.5/16	2.08	0.104
10	4.04	1.098	15.9	0.069	22	3	0.136	35.0	1540	1.11/16	4.9	0.245

Fig 97: Results from model tests

Stress concentration in cut-outs

In fully-monocoque body shell structures, panel cut-outs give rise to stress concentrations which must be allowed for in design. Techniques originally developed for aerospace structures can be used to determine the reinforcements required to alleviate the effects of such stress raisers.

Structural analysis used for determining the behaviour of bars or tubes under direct or shear stress; beams, plates and shells under bending or internal pressure; all these are assumed, for simplicity of analysis, to have distribution of stress following some simple law. For direct stress in a tension member, for example, it is assumed that stress is uniformly distributed and that shear stress varies proportionally to the distance from the centre, under torsional loading. The effect of a hole in a tension member, however, is to produce the stress distribution shown in Fig 98. The high concentration of stress at the edges of the hole show that weakening effect is due to more than just loss of area.

The hole is termed a stress raiser and the ratio of the true maximum stress to that calculated by elementary theory, assuming uniform stress distribution, is defined as the stress concentration factor. Such factors for a whole variety of stress raisers such as notches, fillets, grooves and holes of many different shapes are provided by Young's classic work[13, 14,15], Fig 99.

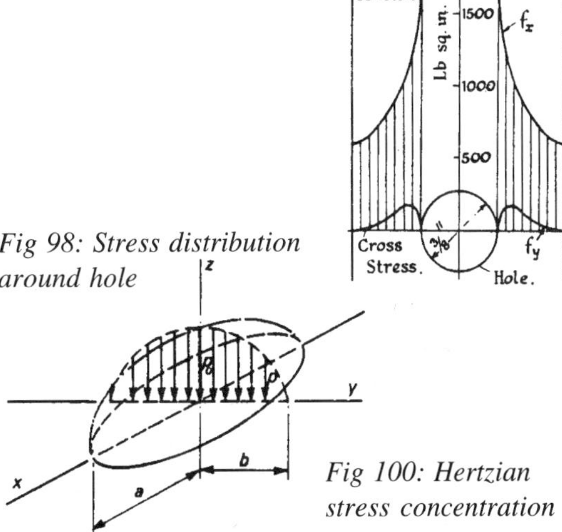

Fig 98: Stress distribution around hole

Fig 100: Hertzian stress concentration

Stress condition and manner of loading	Factor of stress concentration k for various dimensions
Elastic stress, in-plane normal stress 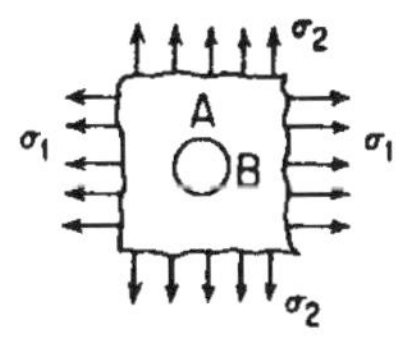	Uniaxial stress, $\sigma_2 = 0$ $\sigma_A = 3\sigma_1 \quad \sigma_B = -\sigma_1$ Biaxial stress, $\sigma_2 = \sigma_1$ $\sigma_A = \sigma_B = 2\sigma_1$ Biaxial stress, $\sigma_2 = -\sigma_1$ (pure shear) $\sigma_A = -\sigma_B = 4\sigma_1$
Elastic stress, out-of-plane bending 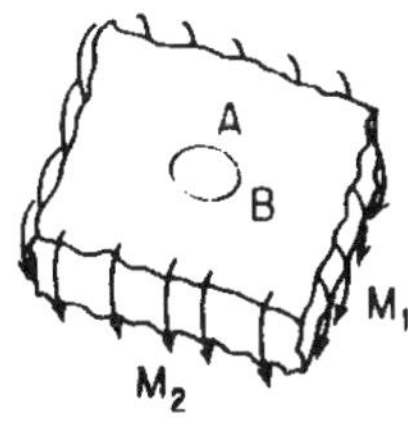	Simple bending, $M_2 = 0$ $\sigma_A = k\frac{6M_1}{t^2}$ where $k = 1.79 + \frac{0.25}{0.39 + (2r/t)} + \frac{0.81}{1 + (2r/t)^2} - \frac{0.26}{1 + (2r/t)^3}$ Cylindrical bending, $M_2 = \nu M_1$ $\sigma_A = k\frac{6M_1}{t^2}$ where $k = 1.85 + \frac{0.509}{0.70 + (2r/t)} - \frac{0.214}{1 + (2r/t)^2} + \frac{0.335}{1 + (2r/t)^3}$ for $\nu = 0.3$ Isotropic bending, $M_2 = M_1$ $\sigma_A = k\frac{6M_1}{t^2}$ where $k = 2$ (independent of r/t)

Fig 99: Stress concentration factors (Young)

These usually assume elastic behaviour and are 'safe' limits for design in that in reality plastic yielding which occurs on over-stressing mitigates the concentration effect. This is why the stress concentration at rivet holes in ductile materials under static loading can safely be ignored. Under fatigue loading, however, this may not be the case and the familiar term of 'notch sensitivity' is used to describe the effect. A separate stress concentration factor in fatigue is thus defined, being the ratio of the endurance limit of a plain member to the nominal stress at the endurance limit of the member containing a stress raiser. An expression due to Neuber defines this factor as:

$$k_f = 1 + (k-1)/[1+\pi(\rho/r)^{1/2}/(\pi-\theta)]$$

where q is the flank of the notch, or stress raiser, r the radius of curvature at the root of the notch and r a dimension related to the grain size tof the material (= 0.0189 inch for steel).

Stress concentration arises too from contact of a 'sharp' pointed body with an otherwise uniform member. Hertzian contact stresses can be estimated for the general case of two curved surfaces being pressed together, for which the pressure distribution over the contact patch is given by the ordinates of a semi-ellipsoid, Fig 100, contained by the surface of contact, such that:

$$p = p_o[1 - (x/a)^2 + (y/b)^2]^{1/2}$$

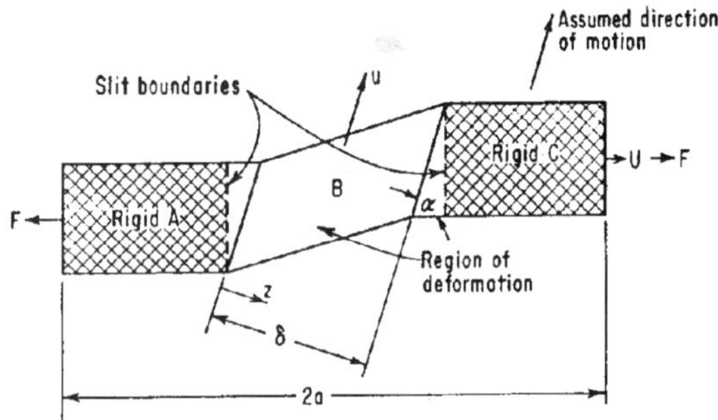

Fig 101: Slit cut-out in tension panel

Fig 102: Assumed deformation of slit panel

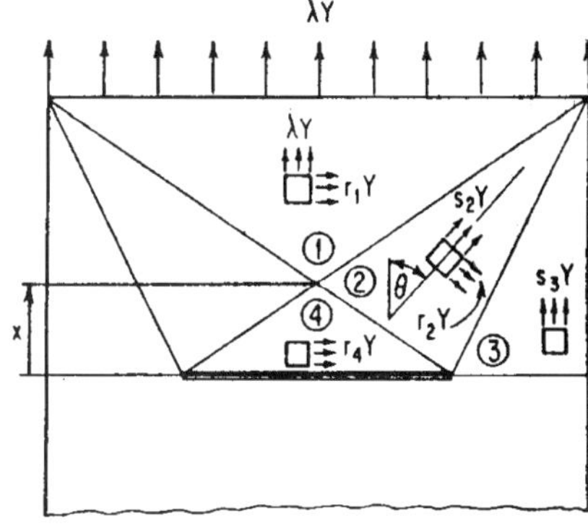

Fig 103: Analysis of segments

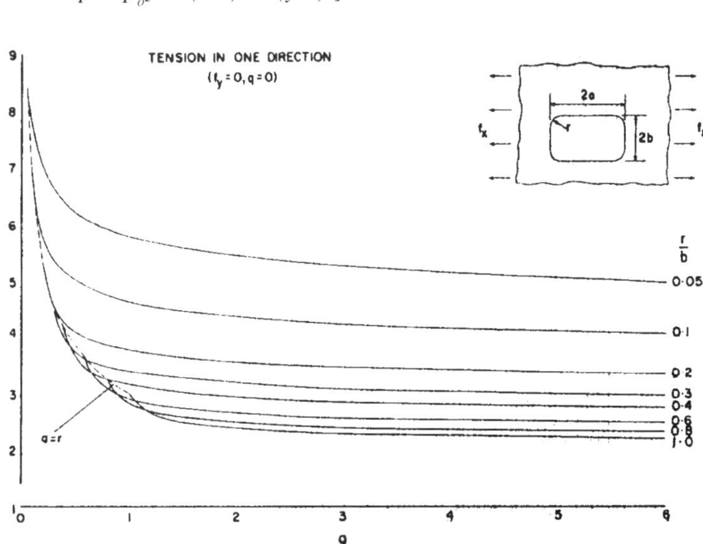

Fig 105: ESDU stress concentration factors

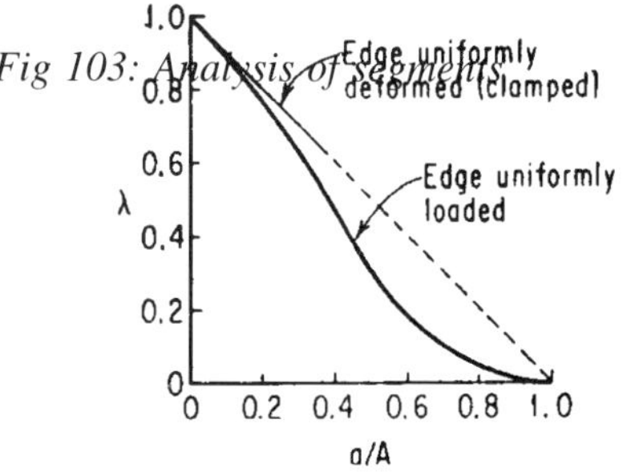

Fig 104: Cut-out factors for square panel

The total load F is represented by the volume of the semi-ellipsoid $F = 2pabp_o/3$ such that the peak pressure p_o is 1.5 times the average pressure $(F=p\ ab)$. This theory is useful for determining stress concentrations in cut-outs of stressed panels in such structures as vehicle bodies. For a large panel containing an elliptical cut-out the stress concentration factor for the points at the end of the a-axis is:

$$k_t = 1 + 2a/b$$

thus for a circular hole, $a=b$, where the local stress is three times the average stress, dropping off rapidly as distance is increased from the discontinuity. The magnitude of k_t depends on the sharpness of the discontinuity and, in the limiting case of cracks, the theory of fracture mechanics needs to be applied.

For larger cut-outs in panels of significant thickness a cut-out factor l is defined to determine the maximum edge load that can be carried by, say, a slit panel as in Fig 101. The objective is to find the weakening of the panel so that it can be properly reinforced. The assumed deformation model is seen in Fig 102 as the panel under tension kinks at the slit. Technique for analysis is to divide the panel up into segments as in Fig 103 and stresses in each determined. For a square panel of side $2A$, with a circular cut-out of radius a, for example, the approximate cut-out factor is as shown in Fig 104.

Engineering Science Data Unit (ESDU) Data Sheet 65004, available from 27 Corsham Street, London N1 6UA, UK, provides stress concentration factors for both reinforced and unreinforced rectangular holes with rounded corners in flat panels. The data is presented in the form of curves as shown in Fig 105.

Plastics and composites Stress concentration effects vary between linear-elastic and plastic materials. Fig 106 shows the case of a circular hole in an infinite flat plate for the two materials. Whereas the stress concentration factor is taken as 3 for both cases the different stress/strain curves will result in different peak stresses for the different materials.

Polymer suppliers Du Pont point out that the effect of strain rate is also important, Fig 107, for the shape of the stress-strain curve. The company's Computer-Aided Technical Service group in Geneva recommends that the allowable stress should be the smallest value of s_{ult} or SEe/S_c where S is a safety factor taken from the table in Fig 108 and S_c is the stress concentration factor discussed above.

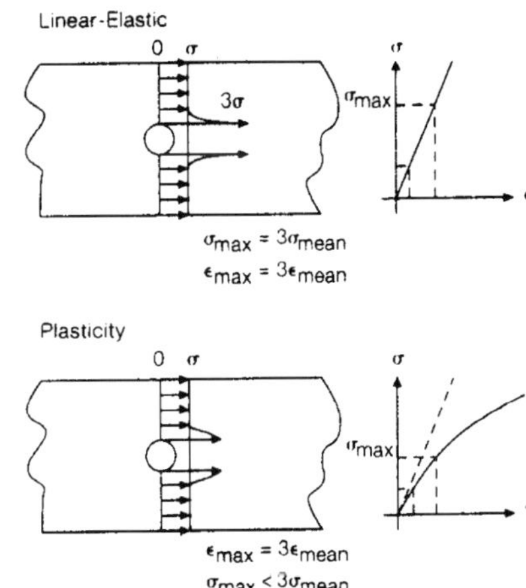

Fig 106: Effect of plasticity

Stress-Strain curves for a typical nylon 6,6 at 23°C and for 50% R.H.

Remarks: Should also be available for
• T = -40°, 80°, 120°C
• D.A.M. (In case of nylon)

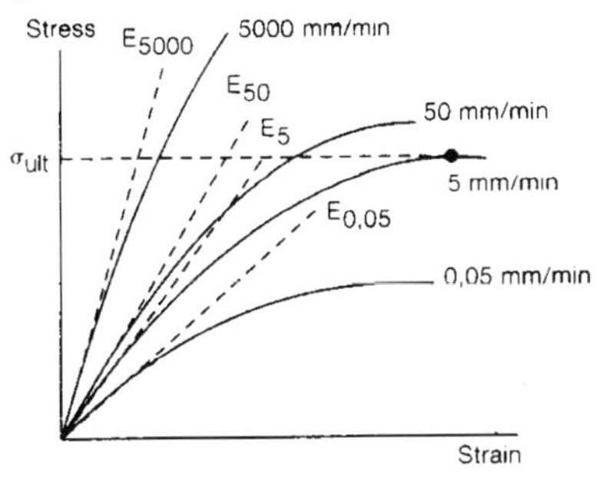

Fig 107: Effect of strain rate

s1 = Risk of Injury to People in Case of Failure
no risk = 1,0
possible injury = 0,7
probable injury = 0,5
(Example: *If an accelerator pedal fails, injury is possible; if a brake pedal fails, it is probable.*)

s2 = Processing Factor *†
For materials that are not reinforced, processing factor s2 = 1,0. For most glass-reinforced materials, where processing conditions are not known, assume fibres are not optimally oriented and calculated stresses – to be on the safe side – are: S = 0,5 for glass-reinforced "Zytel" and 0,35 for "Rynite".

s3 = Stress Calculation Accuracy
Accurate model, conservative approach = 1,0
Use of analytical formulae = 0,75
Estimates = 0,5

s4 = Material Degradation
Natural material or carbon black = 1,0

Fig 108: Safety factors

Finite element analysis

This computerised structural analysis technique has become the key link between structural design and computer-aided drafting. As well as fine-mesh analysis which gives an accurate stress and deflection prediction, course mesh analysis can give the structural feel vital in conceptual design.

One of the longest standing and largest FE software houses is PAFEC who have recommended a logical approach to the analysis of stuctures. This is seen in the example of a constant-sectioned towing hook shown in Fig 109. As the loading acts in the plane of the section the elements chosen can be plane. Choosing the optimum mesh density (size and distribution) of elements is a skill which is gradually learned with experience. Five meshes are chosen in Fig 110 to show how different levels of accuracy can be obtained.

Next step is to calculate several values at various key points — using basic bending theory as a check. In this example nearly all the meshes give good displacement match with simple theory but the stress line up is another story as shown in Fig 111. The lesson is: where stresses vary rapidly in a region, more densely concentrated smaller elements are required; over-refinement could of course, strain computer resources.

Each element is connected to its neighbour at a number of discrete points, or nodes, rather than continuously joined along the boundaries. The method involves setting up relationships for nodal forces and displacements involving a finite number of simultaneous linear equations. Simplest plane elements are rectangles and triangles and the relationships must ensure continuity of strain across the nodal boundaries. Fig 112 shows a force system for the nodes of a triangular element along with the dimensions for the nodes in the one plane. The figure shows how a matrix can be used to represent the co-efficients of the terms of the simultaneous equations.

Another matrix can be made up to represent the stiffness of all the elements *[K]* for use in the general equation of the so-called 'displacement method' of structural analysis:

$$[R] = [K].[r]$$

where *[R]* and *[r]* are matrices of external nodal

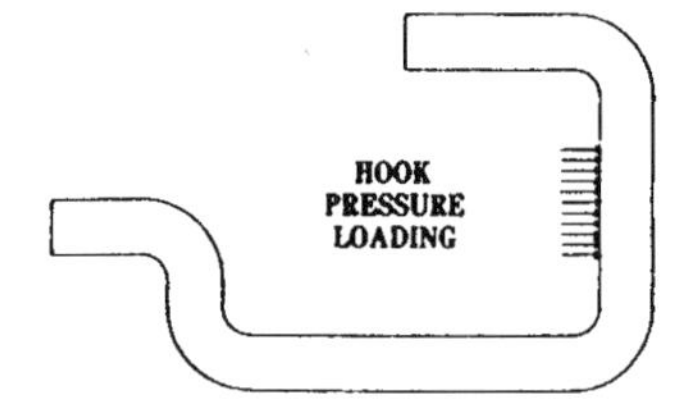

Fig 109: Towing hook as FEM structure example

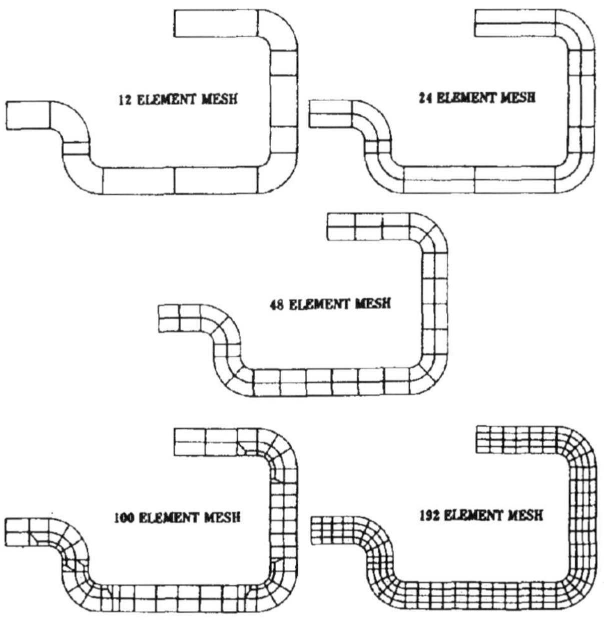

Fig 110: Various mesh densities

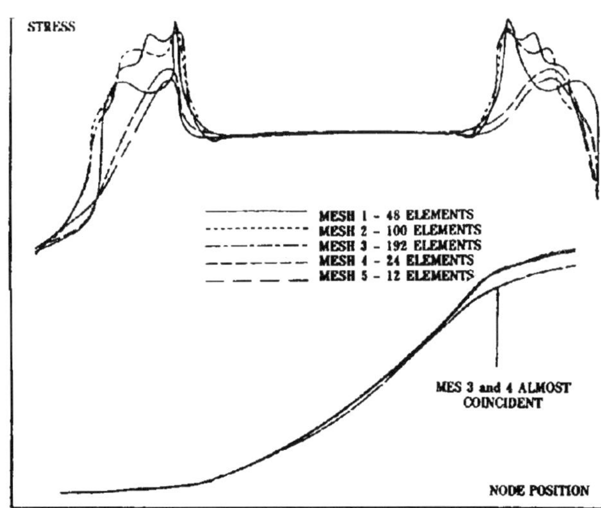

Fig 111: FEM vs elasticity theory

forces and nodal displacements; the solution of this equation for the deflection of the overall structure involves the inversion of the stiffness matrix to obtain $[K]^{-1}$. Computer manipulation is ideal for this sort of calculation.

A parallel structural method is the 'force', rather than displacement method, which uses forces rather than displacements as the unknowns in the equation. Prior to FEA, this was used with simple structural surfaces, discussed earlier, to divide the structure into idealised sections and similarly write down stress-strain relationships for matrix analysis. This approach is sometimes preferred to FEA, in the earlier stages of design, so as to obtain a better visualisation of the force patterns in the structure — in relation to the external loads.

An example might be the idealised van body shell of Fig 113 in which concentrated loads are considered to act at the centres of the lower edges of the sidewall panels and reacted at the ends of these edges. The structure comprises three bulkhead frames having pin-jointed members connected by eight sandwich panels — assumed for the analysis to be hinge-jointed at their edges. The structure is then divided into stress systems in the same manner as the redundant frameworks considered in a previous section. The determinate *[R]* system conveniently comprises the plate and four bars making up the lower half of the sidewalls while the three redundant *[X]* systems are 'self-equilibriating', Fig 114, *X1*, *X2* and *X3*.

Akin to the equation describing the stress/strain relationship for redundant frameworks, a similar one applies to the panelled structure as follows:

$$[S] =\{[b_o] - ([b_1][b_1]'[f][b_1])^{-1}\} ([b_1]'[f][b_0])[R]$$

used to evaluate matrices when unit loads are applied to the stress systems. The b_0 matrix is multiplied by *1/2l*, since for simplicity in algebraic expression of the matrix elements, the unit load due to the *[R]* system is multiplied by *2l* and the member forces are then quickly determined by simple statics.

The Unit Load theorem is used to evaluate sub-matrices for the bars and plates in the overall flexibility matrix, Fig 115. For this example, modulus *G* is taken as *E/2* and half the section area of the sandwich panel on each side of the equivalent 'a' bar considered concentrated at the bar for end-load resistance. Section area of the intermediate bulkhead

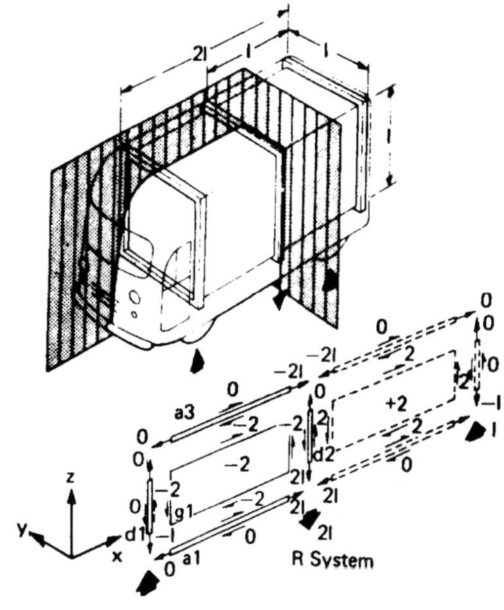

Fig 113: Box van idealised for analysis

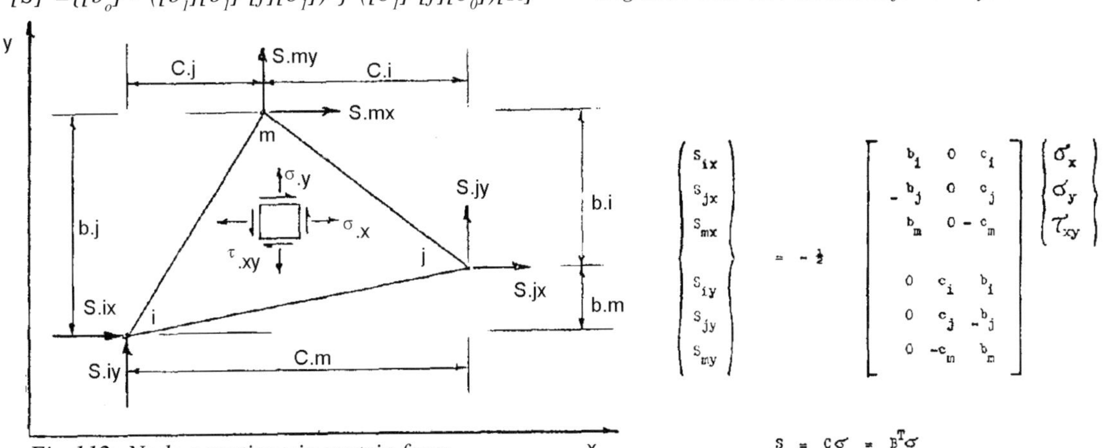

Fig 112: Node equations in matrix form

bars is taken as a quarter of that of the adjacent panel while those at the end take halves — resulting in sub-matrices for the bulkheads as in Fig 116. This leads to the matrix formulation and manipulation of Fig 117 which is about the largest one would want to undertake without a computer. The final matrix of member forces results from taking a central side load of 10 tonnes and these forces are plotted on the

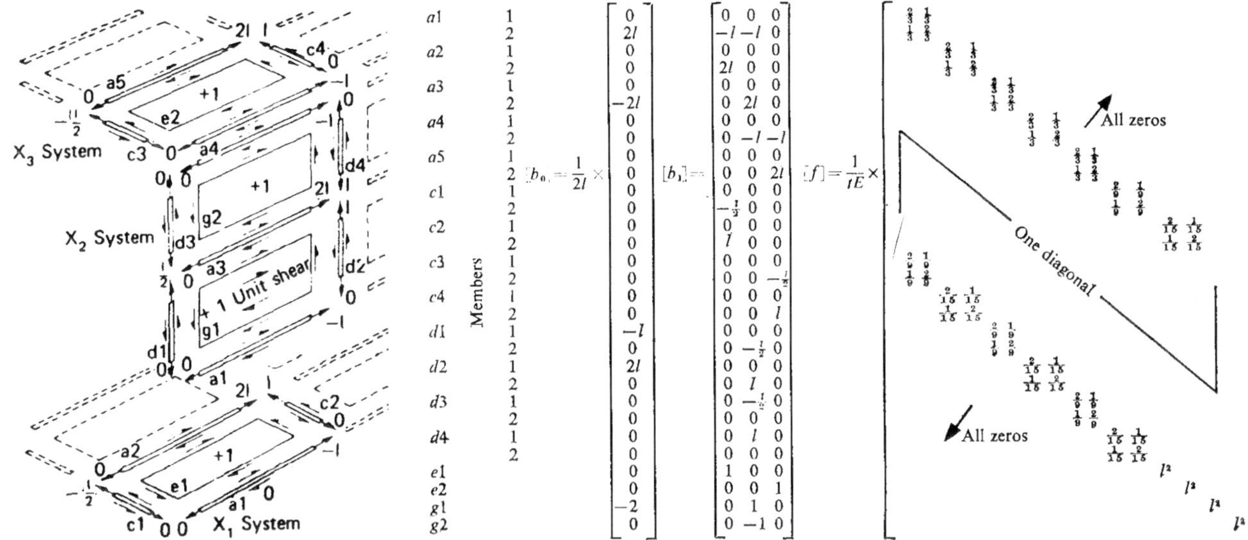

Fig 114: Self-equilibriating force system

Fig 115: Matrix relationship

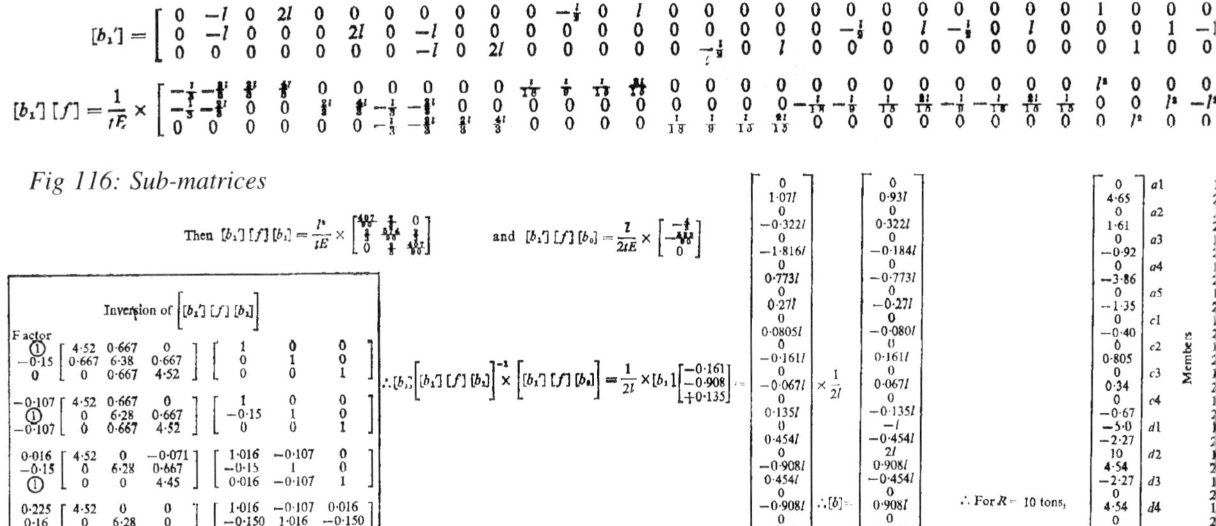

Fig 116: Sub-matrices

Fig 117: Matrix manipulation

distribution diagram for the van structure in Fig 118. This gives a good 'feel' for the structure in that, by applying the external load at the bottom of — rather than uniformly over — the cross section of the box beam formed by the van structure, it has resulted in uneven compressive and tensile forces in the top and bottom corner booms. Also a reduced loading of the intermediate top and bottom longitudinal stringers shows the action of shear lag.

By applying the matrix displacement analysis to a vehicle structure, Fig 119, in a very similar manner shows how course-mesh finite element analysis can similarly be used to obtain structural feel. The structure which represented a Land Rover-like, but monocoque, field car was evaluated some years ago by workers at Cranfield and Birmingham University, being idealised for analysis as shown. A uniformly distributed loading over the deck produced shear loading in the panel idealisations as shown — the substantial load distribution throughout the structure justifying the choice of monocoque construction.

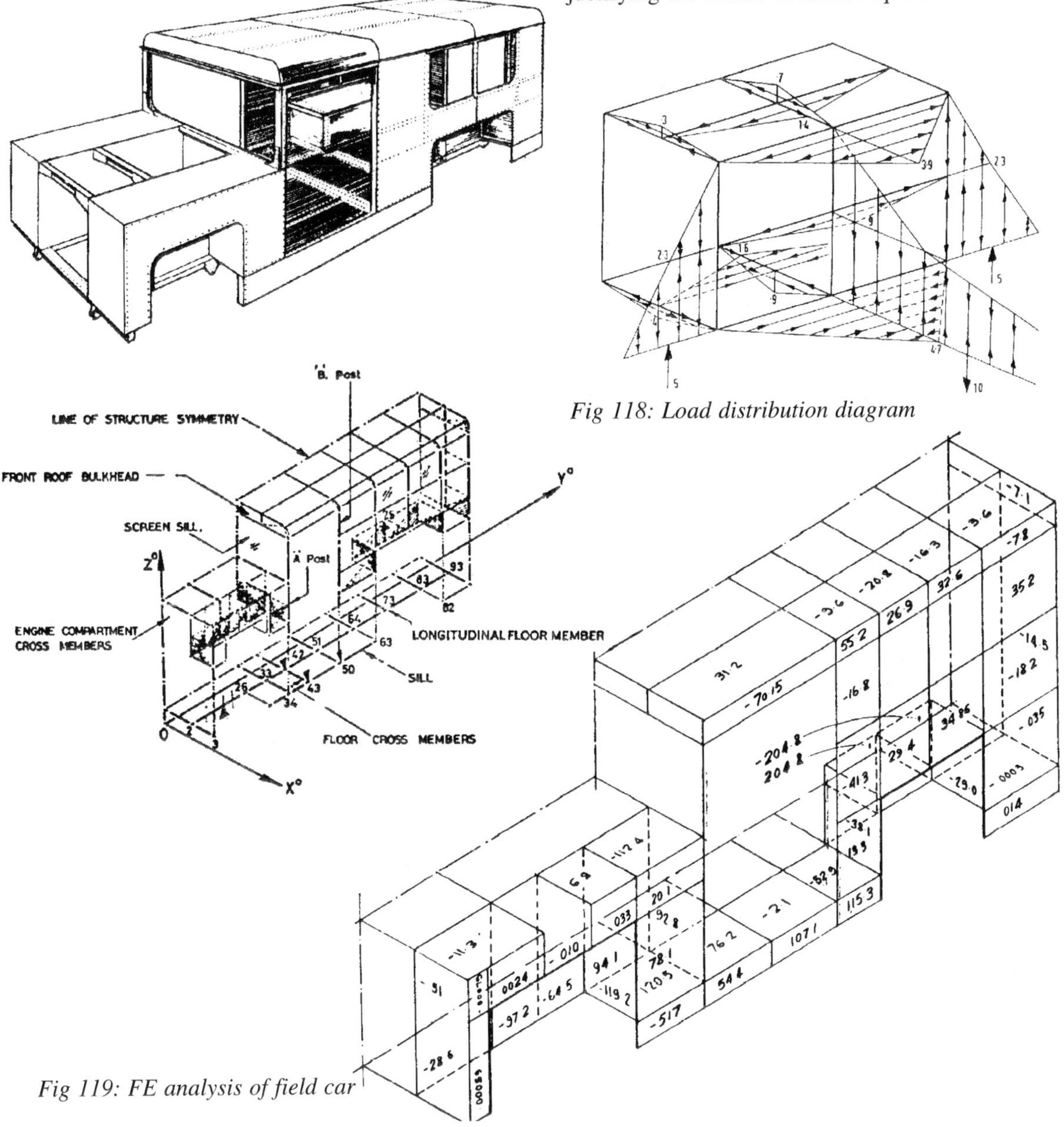

Fig 118: Load distribution diagram

Fig 119: FE analysis of field car

FEA for car bodies

The Finite Element method of structural analysis was introduced in the previous section which described its evolution from matrix methods of solution for unit-load (force) and displacement techniques of analysis. Here some of the accepted procedures are discussed for applying to body structural analysis.

As well as for loads and displacements, FE techniques, of course, cover temperature fields and many other variables and the structure, or medium, is divided up into elements connected at their nodes between which the element characteristics are described by equations. The discretisation of the structure into elements is made such that the distribution of the field variable is adequately approximated by the chosen element-breakdown.

Equations for each element are assembled in matrix form to describe the behaviour of the whole system. Computer programs are available for both the generation of the meshes and the solution of the matrix equations such that use of the method is now much simpler than it was during its formative years.

Economies can be made in the discretisation by taking advantage of any symmetry in the structure to restrict the analysis to only one half or even quarter — depending on degree. As well as planar symmetry, that due to axial, cyclic and repetitive configuration, Fig 120, should be considered. The latter can occur in a bus body, for example where the structure is composed of identical bays corresponding to the side windows and corresponding ring-frame.

Element shapes are tabulated in Fig 121 — straight-sided plane elements being preferred for the economy of analysis in thin wall structures. Element behaviour can be described in terms of 'membrane' (only in-plane loads represented); in bending only or as a combination entiled 'plate/shell'. The stage of element selection is the time for exploiting an understanding of basic structural principles; parts of the structure should be examined to see whether they would typically behave as a truss-frame. beam or in plate-bending, for example. Avoid the temptation to over-model a particular example; however, because number and size of elements are inversely related, as accuracy increases with increased number of elements.

Different sized elements should be used in a model — with high mesh densities in regions where a rapid change in the field variable is expected. Different ways of varying mesh density are shown in Fig 122, in the case of square elements. All nodes must be inter-connected and therefore the fifth option

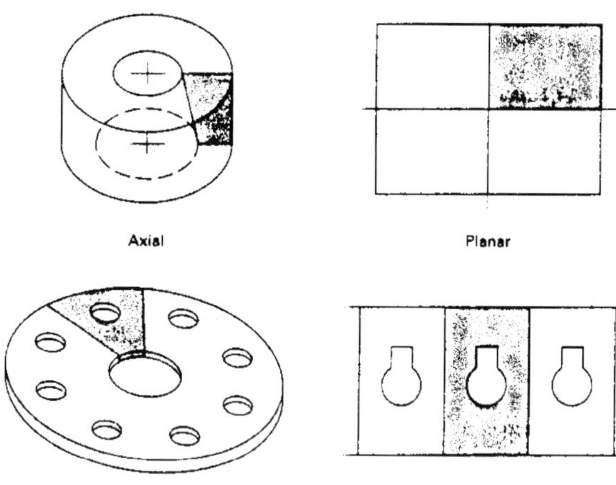

Fig 120: Types of symmetry

Shape	Type	Geometry
Point	Mass	
Line	Spring, beam, spar, gap	
Area	2D solid, axisymmetric solid, plate	
Curved area	Shelf	
Volume	3D solid	

Fig 121: Element shapes

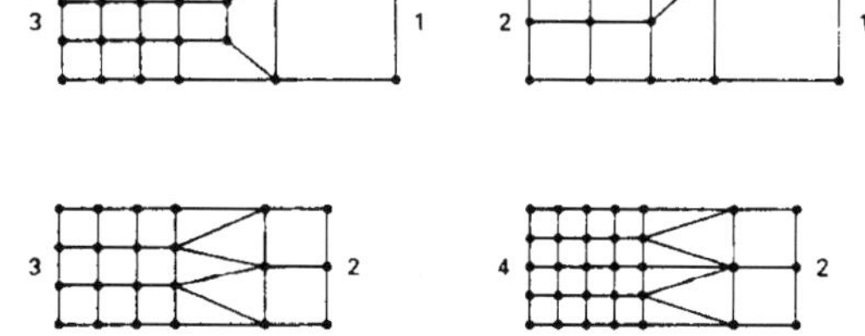

Fig 122: Varying mesh densities

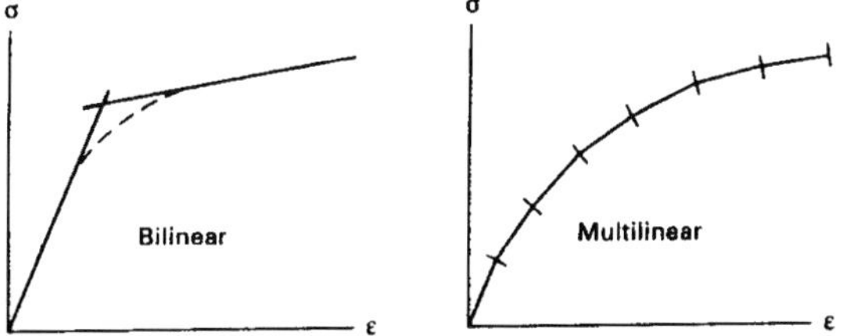

Fig 123: Stress-strain curve representation

shown would be incorrect because of the discontinuities.

As element distortion increases under load, so the likelihood of errors increases, depending on the change in magitude of the field variable in a particular region . Elements should thus be as regular as possible — with triangular ones tending to equilateral and rectangular ones tending to square. Some FE packages will perform distortion checks by measuring the skewness of the elements when distorted under load. In structural loading beyond the elastic limit of the constituent material an idealised stress-strain curve must be supplied to the FE program — usually involving a multilinear representation, Fig 123.

When the structural displacements become so large that the stiffness matrix is no longer representational then a 'large-displacement' analysis is required. Programs can include the option of defining 'follower' nodal loads whereby these are automatically re-orientated during the analysis to maintain their relative position. The program can also recalcu-

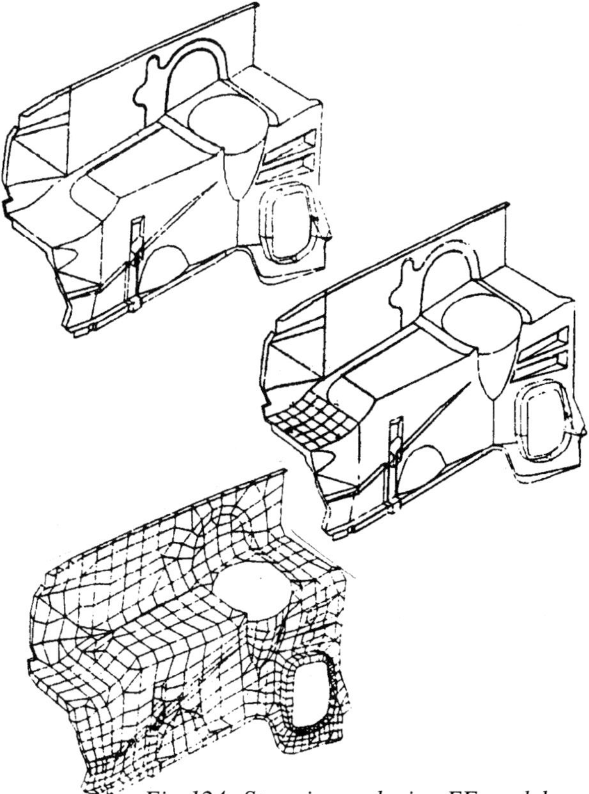

Fig 124: Steps in producing FE model

Structural Analysis.

NASTRAN	linear static analysis
ABAQUS	non-linear static and dynamic analysis
MOTRAN	body loadings from road surface
OPUS	optimisation of body
WRECKER-F	crash analysis
MENTOR	crash analysis
ANSYS	small strains on impact components
FATSUP	fatigue analysis

Pre- and Post- Processors.

TIDE	inter-face between NASTRAN and MOVIE.
MOVIE	graphics display post-processor of F.E. results.
FELS	pre processor for ABAQUS
NAVIE	pre/post processor for NASTRAN
PATRAN-G	pre/post processor for F.E.packages generally
NASTECK	NASTRAN display package for Tektronix terminals.
NASPLOT	NASTRAN plotting package.
Lundy/Prime	generation of F.E. models and display of NASTRAN stresses and deformations.

Applications.

BELTFIT	fit of seat belts
GRASP	liftgate opening geometry and efforts (gas strut assistance)
HOODLUM	hoodtop opening geometry and efforts
CALSPAN CVSP	kinematic dummy predictions (Crash Victim Simulation Program)
TMIRROR	external mirror field of view checks using Tektronix display
TENSP/TORSP	various spring programs
DOORLIFT	door swing as function of hinge inclinations
PROFTEK/PROFTAB	cross-section properties with digitised input
IMIRROR	inside mirror field of view
WIPER	wind screen wiper layout

Fig 125: FE programs

late the stiffness matrices of the elements after adjusting the nodal co-ordinates with the calculated displacements. Instability and dynamic behaviour can also be simulated with the more complex programs.

The principal steps in the FEA process are:

* idealisation of the structure (discretisation)
* evaluation of stiffness matrices for element groups
* assembly of these matrices into a super-matrix
* application of constraints and loads
* solving equations for nodal displacements
* finding member loading.

For vehicle body design, programs are available which automate these steps, the input of the design engineer being, in programming, the analysis with respect to a new model introduction. First stage is usually the obtaining static and dynamic stiffness of the shell, followed by crash performance based on the first estimate of body member configurations. From then on it is normally a question of structural refinement and optimisation based on load inputs generated in earlier model durability cycle testing. These will be conducted on relatively course mesh FE-models and allow section properties of pillars and rails to be optimised and panels thicknesses to be established.

In the next stage, projected torsional and bending stiffnesses are input as well as the dynamic frequencies in these modes. More sophisticated programs will generate new section and panel properties to meet these criteria. The inertias of mechanical running units, seating and trim can also be programmed-in and the resulting model examined under special load cases such as pot-hole road obstacles. As structural data is refined and updated, a fine-mesh FE simulation is prepared which takes in such detail as joint design and spot-weld configuration. With this model a so-called sensitivity analysis can be carried out to gauge the effect of each panel and rail on the overall behaviour of the structural shell.

Joint stiffness is a key factor in vehicle body analysis and modelling them normally involves modifying the local properties of the main beam elements of a structural shell. Because joints are line connections between panels, spot-welded together, they are

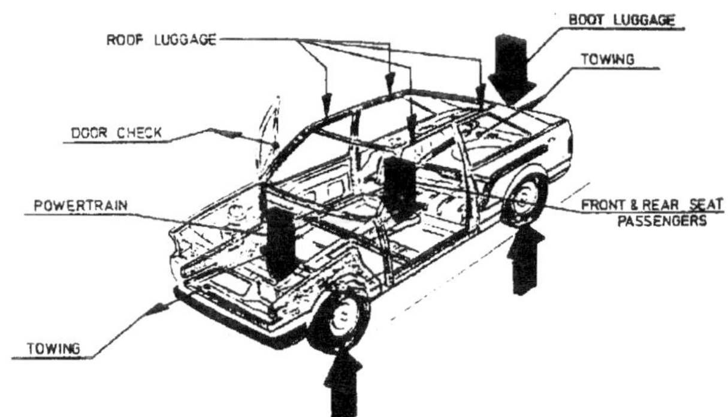

Fig 126: Load inputs

difficult to represent by local FE models. Combined FE and EMA (experimental modal analysis) techniques have thus been proposed to 'update' shell models relating to joint configurations. Vibrating mode shapes in theory and practice can thus be compared. Measurement plots on physical models excited by vibrators are made to correspond with the node points of the FE model and automatic techniques in the computer program can be used to update the key parameters for obtaining a convergency of mode shape and natural frequency.

An example car body FE analysis at Ford was described at one of the recent Boditek conferences, Fig 124, outlining the steps in production of the FE model. An extension of the PDGS computer package used in body engineering by the company — called FAST (finite element analysis system) — can use the geometry of the design concept existing on the computer system for fixing of nodal points and definition of elements. It can check the occurrence of such errors as duplicated nodes or missing elements and even when element corners are numbered in the wrong order. The program also checks for misshapen elements and generally and substantially compresses the time to create the FE model.

The authors considered that upwards of 20 000 nodes are required to predict the overall behaviour of the body-in-white. After the first FE analysis was carried out, the deflections and stresses derived were fed back to PDGS-FAST for post-processing. This allowed the mode of deformation to be viewed from any angle — with adjustable magnification of the deflections — and the facility to switch rapidly between stressed and unstressed states. This was useful in studying how best to reinforce part of a

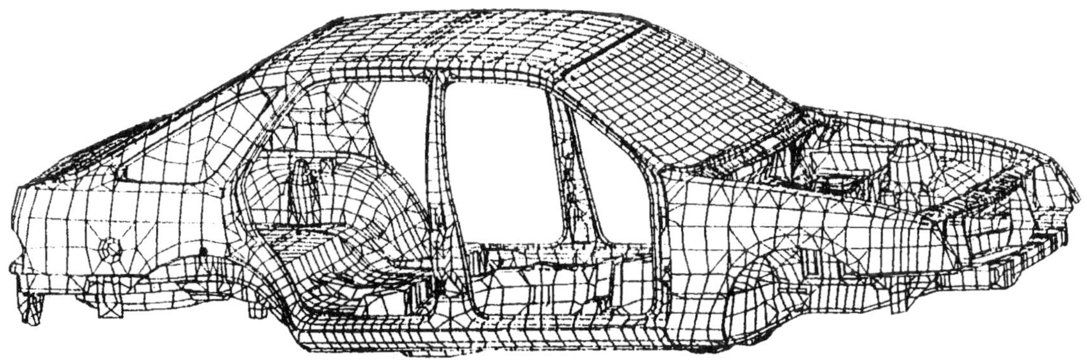

Fig 127:Global model for body-in-white

structure which deforms in a complex fashion. Average stress values for each element can also be dis played numerically or by graduated shades of colour. The authors released a list of the programs used, as Fig 125, by the company about a decade ago. Load inputs were as shown in Fig 126 and the FE model for the BIW, in Fig 127.

Chapter 8: Safety under impact

Impact protection basics; design for crashworthiness; calculating crashworthiness; laws of mechanics applied to safety; side-impact protection; controlled-collapse longitudinals; aluminium and steel crashworthiness; tailoring front-end deformation to cabin shell stiffness; structural design for seat mounts; crashworthiness of road tankers

While the need to obtain advantageous trade-off between fuel-ecomomy and exhaust emission control may be the prime mover in structural design for reduced weight, probably the main stimulus for body structural analysis has been secondary-safety design, in recent years.

Impact protection

Design for impact-protection has now become one of the key factors in the structural engineering of volume cars and the same regulations often apply to specialist cars and commercial/industrial vehicles when used on public roads. Much of the safety protection can be built in at the concept stage if predictive techniques are used.

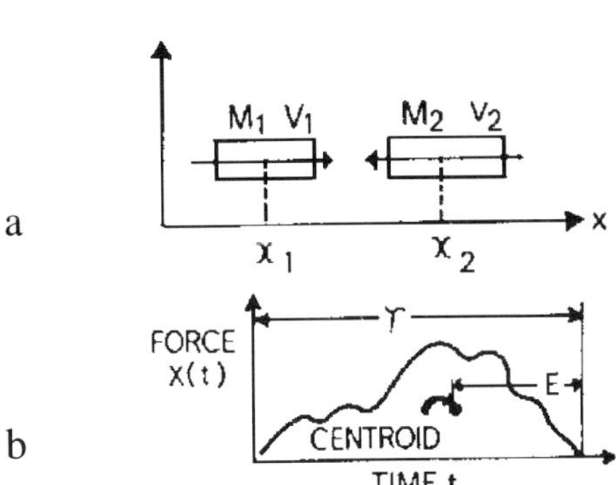

Fig 1: Collision parameters

Impact protection for vehicle occupants has been in focus since the pioneering work of Ralph Nader in the 1960s recognised, in particular, the poor protection offered by standard production cars of the day. This work has now, of course, spread to most categories of vehicle including trucks, buses and racing cars. UK investigations at that time[1] showed

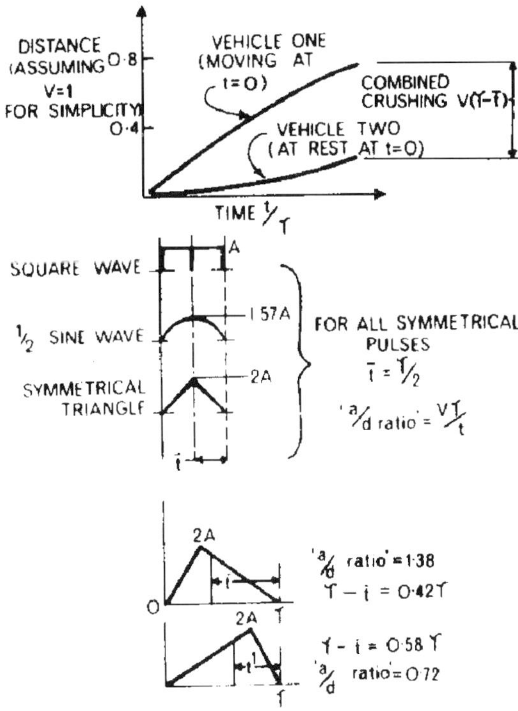

Fig 2: Idealised impact pulses

saloon cars typically offering up to 3 ft (0.91 m) crush resistance, with 0.1 second deceleration, in front-end barrier impacts at about 35-40 mph (55-65 kph). This was for average sized cars of typically 13.75 ft (4 m) long by 5.25 ft (1.6 m) wide and weighing around 2350 lb (10.5 tonnes).

When vehicles collide with one another, head-on, as represented in Fig 1a, an impact pulse might be typically as shown in Fig 1b. Each vehicle has momentum (mass x velocity) and there will be a common velocity after impact:

$$V_3 = (M_1V_1 + M_2V_2)/(M_1 + M_2)$$

and it can be shown that the combined crushing of the two vehicles is $(V_1 - V_2)(T - t)$ the shape of the impact pulse determining time t, Fig 2 showing the effect of different idealised standard pulse shapes.

Plastic collapse

Crush resistance depends upon front-end structural layout and for many vehicles, the bending resistance of the front crossmember is crucial. Bending beyond the elastic range is the case to examine. Consideration of the stress/strain curve for mild steel, for example, Fig 3a, shows that initial yield occurs at a strain of 0.001 while the plastic range continues to a strain of 0.010 to 0.020 when strain hardening commences. Because of the material's ductility, and long plastic range, increasing the bending moment beyond that at which yield is just reached results in a spread of yielding across the depth of the beam section. Distribution of bending stress across the section of a beam undergoes the transformation shown in Fig 3b, the attainment of the upper yield stress is followed by a falling off to a lower yield stress, as shown, until failure at the lower level applies across the whole of section and a 'plastic hinge' is said to develop. The corresponding plastic moment:

$$M_o = (\sigma Lbd/2)d/2 = \sigma Lbd^2/4$$

against $M_b = sBbd^2/6$ for bending in the elastic range; the ratio of the two bending moments M_o/M_b is known as the 'form factor' for the beam. It is a measure of the additional load which can be carried under part plastic strain. Typical values are 1.15 for an I-beam and 1.7 for a circular section tube.

In considering again the front crossmember of a vehicle, one legal requirement for cars in certain countries is for structural integrity to be maintained up to a certain impact velocity when striking a post-like object at mid-span. Collapse of the beam occurs in this situation when the central load reaches either $4M_o/L$ or $8M_o/L$ according to whether the crossmember 'beam', length L, is 'simply-supported' (unrestrained) at its end or embedded in fixed supports, for calculation purposes. It can also be shown that when the

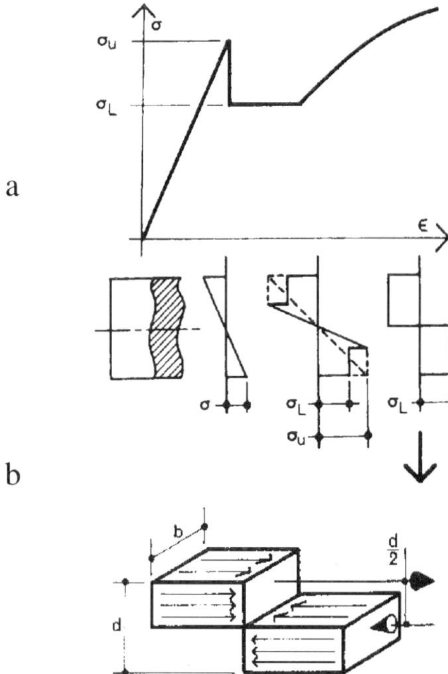

Fig 3: Plastic bending

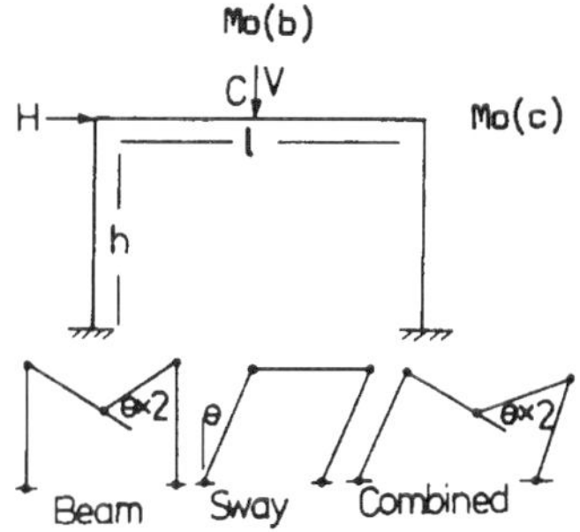

Fig 4: Front-end as portal frame

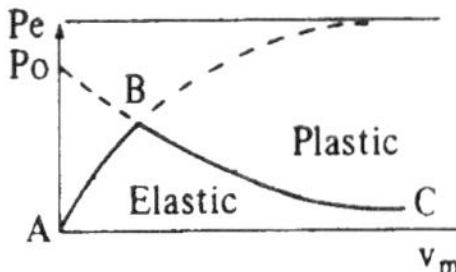

Fig 6: Buckling load vs lateral displacement

central load rises to $45.6M_o/L$, plastic hinges also form at points $0.202L$ either side of mid-span, the hinges moving inwards towards mid-span for even greater loads.

Technique in designing for impact-resistance is to use as much permanent deformation of the structure as possible without encroaching on the occupants' 'clearance zone'. Rails and pillars can be arranged so that plastic hinges occur at their intersections and subsequent deformation will be as a hinged 'mechanism'.

Collapse mode can be modified by gusseting of joints, to alter the effective linkage lengths of the mechanism and the addition of shear panels further affects behaviour. Fig 4 suggests that the front-end of a conventional vehicle, the structure around the engine bay in plan view, can be likened to a portal frame — again under the action of the central impact load. Mode of collapse is seen to depend on the geometry and relative stiffness of the structural elements, as well as the load pattern. In the beam-collapse case, work done by the central load, $VL/2$, is equal to the energy dissipated in the plastic hinges $M_o(b)2\theta + 2M_o(c)\theta$ giving $VL/4 = M_o(b) + M_o(c)$. Similarly, for sway collapse, $Hh/2 = M_o(c)$ and for combined collapse, $VL/4 + Hh/2 = M_o(b) + M_o(c)$, assuming the portal has pinned feet. If fixed feet are assumed:

$$VL/4 + Hh/2 = M_o(b) + M_o(c) + M_o(p).$$

A typical vehicle might have a front-end portal 1.5 metres wide x 1.5 high and be subject to 10 kN central impact load with a lateral component of 2.5 kN. Thus $Hh/2 = 1.875$ kN and $Vl/2 = 3.75$ kN and if pinned feet are assumed, a section modulus of $M_o / Ss(b)$ is required for $M_o(c) = M_o(b) = M_o$ where S is shape factor.

US researchers[2] have shown that the collapse rate of specially designed front-end structures can readily be predicted. For the arrangement in Fig 5, curved beam theory can be used to show controlled-collapse can be predicted for the S-shaped longitudinals. For moment arm lm of the frontal impact force the radius defining the neutral axis of the beam can be calculated from:

$$A/[(W_1 l_m r_2/r_1)+(W_2 l_m r_4/r_2)+(W_3 l_m r_5/r_3)+(W_4 l_m r_6/r_5)]$$

For $l_m = 4.7$ inches, the 50,000 lb/in[2] steel of 0.132 inch thickness can be shown to have R = 5.64 inches and a collapse load per longitudinal of 21 700 lbf. Even practically straight members subject to end load tend to bend because of unavoidable load-eccentricity creating a bending moment. In a plot of axial buckling load against lateral displacement, Fig 6, the latter grows at an increasing rate, with end load, until at B, a plastic hinge forms. The collapse load does increase under impact conditions, however, because high lateral inertia forces usually restrain the longitudinal. When thin-walled closed section mem-

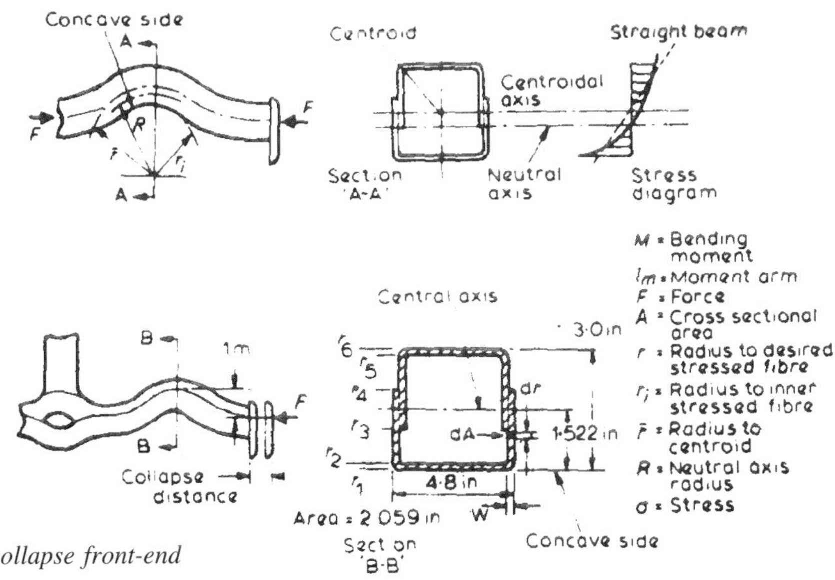

Fig 5: Controlled-collapse front-end

bers with substantial relative width are employed, crumpling under axial impact usually only occurs in the front part of the member because of longitudinal inertia effects and the formula for static crumpling has been found to describe adequately the dynamic impact condition, for which crush load is:

$$6s_y t(Dt)^{1/2}$$

and crush length:

$$MV^2/(12s_y t^{1.5}D^{0.5})$$

for a circular tube of diameter D and thickness t hit by mass M at closing speed V, where s_y = yield stress of the tube material.

When the whole vehicle is modelled as a rectangular section tube (akin to a bus structure without doors and windows) an approximation can be made at predicting the crush length of the vehicle shell as it impacts a completely rigid object as:

$$P/(V^2/2g)(1 + c)dAL;$$

$$\text{for } P/P_o = [(6A/A_o) + 0.04][1 + (v/50)]$$

for cross section area A and A_0 the area enclosed by A (A/A_0 is the solidity ratio); P_o is As_y.

Safety frames

Industrial vehicles such as farm tractors, and increasingly CV cabs, have safety frames affording a substantial degree of protection to the driver in both overturning and impact situations. Early work reported by Massey Ferguson on farm tractor safety frames was based on exploiting the considerable increase in the yield strength of mild steel when it is subject to rapidly applied loads, Fig 7. For calculation purposes the frames were modelled as links connected by plastic hinges and after applying the test load in the position shown, observe how far the point of application of the load can move before any of the links encroaches on the driver's zone of clearance. For prescribed test energy levels given by British Standards for the weights of tractor concerned, a critical 'test' load is established. This load is then applied to the same shaped frame, but under elastic rather than plastic deformation conditions, to determine member proportions and deflections.

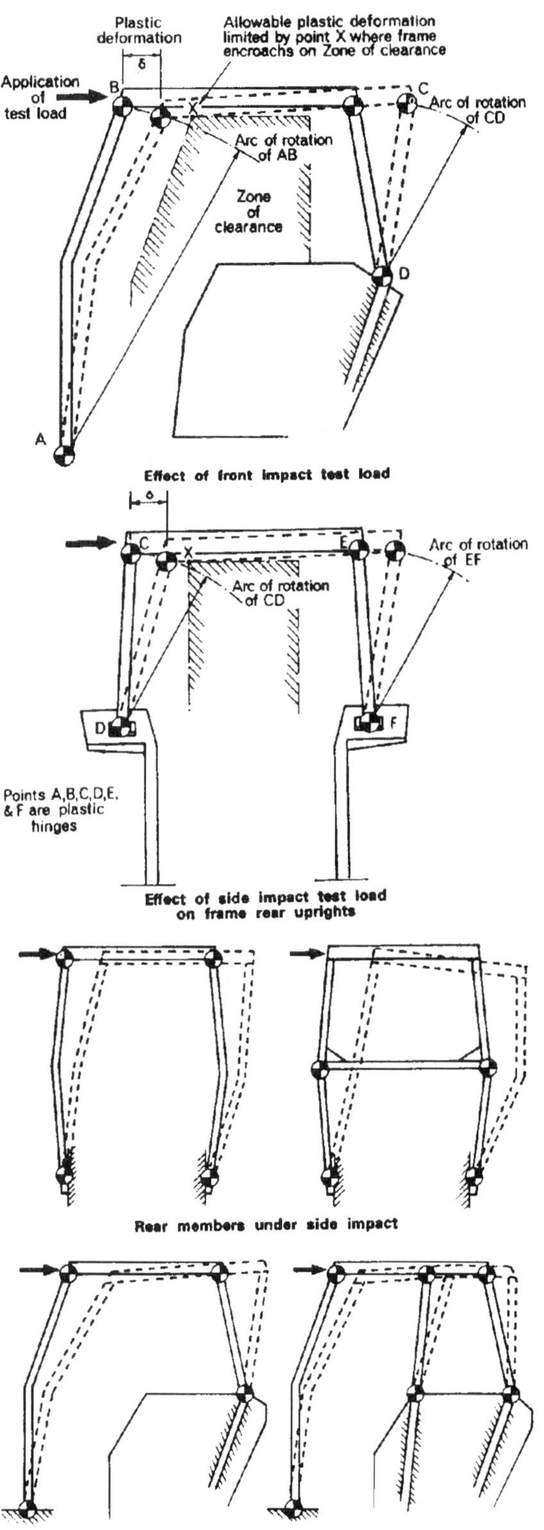

Fig 7: Tractor cab safety frame

Design for crashworthiness

The behaviour of structural elements is described in a way that can lead to a better understanding of the frontal impact resistance of vehicle structures. The previous section showed how the shape of the crash time history pulse can be altered by building in particular collapse rates into the impacting elements of the body structure; here the design of such elements is considered.

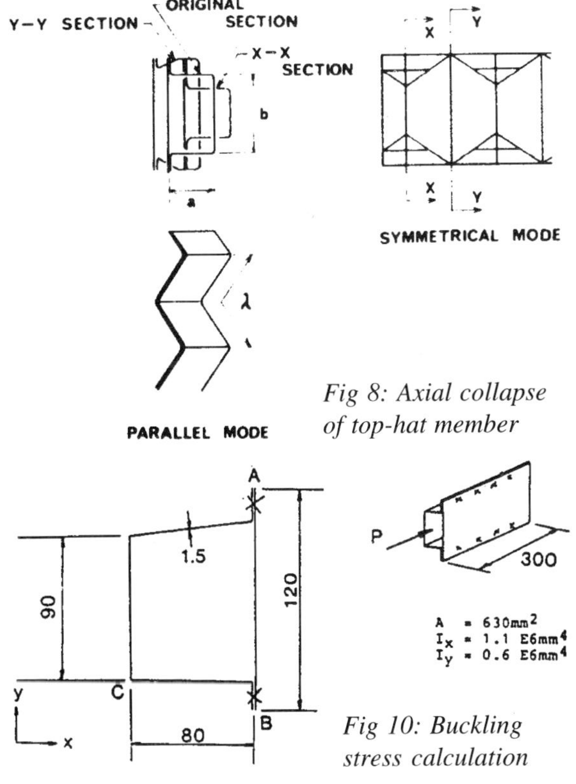

Fig 8: Axial collapse of top-hat member

Fig 10: Buckling stress calculation

It is estimated that some 100 kJ of energy is required to be absorbed by the individual front-end structures of two car-weight vehicles impacting head on at a combined speed of 30 mph. Such values need to be equated with absorption capacities of members in either bending or end-on crush mode.

Accurate predictions of crashworthiness can only be obtained by building inertia effects and strain rates into the calculations. However, prior to the linear dynamic modelling necessary to simulate these effects, some useful preliminary calculations can help understand the mechanisms of failure and provide a first guide to the required structural behaviour.

In the case of heavier commercial vehicles, very little of the crash energy is absorbed into the structure when impacting a car or lighter vehicle, therefore crashworthiness has to be built in by means of such devices as front underrun guards. These must be positioned with ground clearances no greater than 300 mm so that impact is made with the strong part of the car's front-end. The truck bumper must also be a yielding structure and the stroke of the under-run guard should be about 300-400 mm for compatibility with car front-end collapse characteristics. It should be designed to collapse at a constant horizontal force equivalent to that which a car can withstand without producing serious intrusion or too violent a deceleration — such that the car front and underrun struts collapse together.

Tests have shown that a force of 230 kN is developed when car crush typically reaches 300 mm, representing a 60 kJ energy absorption — equivalent to a light car moving at 40 km/h. In practice, progressive rates are used so that the 'bumpers' are tolerant

BEADING TYPES / BEADING PITCHES	A1	A2	A3	B1	B2	C
$\frac{a+b}{2}$ xk1						
xk2	HAT-PLATE FLAT-PLATE	O		O	O	
xk3	O		O			
xk4						
xk5						

Fig 9: Folding controlled by beading

of lower speed impacts and initial yielding is typically arranged at 100 kN or less. Struts are often made of Invertube cartridges; each of a pair would be about 450 mm long and made to compress, under a lever system by 200 mm.

In a design reportedly due to the Transport and Road Research Laboratory, the Invertubes comprised steel tubes of 100 and 80 mm diameter, the smaller of which was turned back on itself at one end and then peripherally welded to the open end of the large tube. In impact the smaller tube progressively turns itself inside out. Tubes of 16 swg each had a collapse load of about 50 kN to give a total energy absorbing capacity of the bumper of around 20 kJ.

One technique of stabilising the axial collapse of lighter vehicle front-end side rails is to make them deform into a series of folds. This is considered by some to be more favourable than allowing collapse by overall bending deformation of the siderails — as energy can be absorbed more efficiently. Fig 8 shows axial collapse of a top-hat sectioned member in which stability is achieved by initial folding near the impact end with peaks and troughs of waves perpendicular to the axis — subsequent folding progressing towards the root end. Beading is used to initiate folds and so ensure a stable collapse.

In early work carried out by Toyota, Fig 9 shows how different beading arrangements can produce different folding effects and below it the resulting folding modes. A method of calculating the absorption characteristics of front rail sections has been reported by Hawtal Whiting based on the critical buckling stress of the constituent panels. In the rail section of Fig 10, when considering the axial stability of side AB, buckling stress can be calculated from:

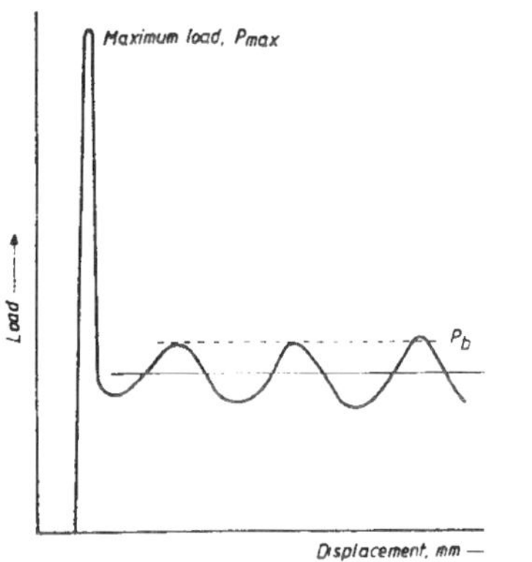

Fig 11: Box-tube crush performance

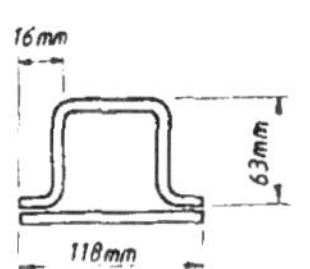

Fig 12: Actual dynamic impact pulse

Effect of weld pitch on collapse of spot welded box sections tested under 30 mile/hr impact loading (6000J energy)

Batch number	Type	Thickness, mm	Yield strength, N/mm^2	Weld pitch, mm	Reduction in sample height, mm	Energy to deform sample 10mm*, J	Mode of collapse
146	Low carbon	1.2	186	21	161, 169, 149, 170, 151, 161	375	6A
				45	184, 177, 183	330	2A, 1B
				81	226, 199, 216, 212, 201, 211	285	6A
				122	236, 208, 222	270	2A, 1B
151	HSLA	1.2	340	21	132, 137, 139	440	[illegible]
				45	139, 150, 168	395	3A
				81	168, 174, 172	350	3A
				122	198, 188	310	2A
192	Dual phase	1.2	370	21	123, 136, 121	475	3A
				45	130, 135	485	2A
				81	147, 148, 144	410	3A
				122	161, 183, 141	370	3A
193	Dual phase	1.2	405	21	100, 107, 102	580	3A
				45	107, 111, 122	530	3A
				81	120, 111, 120	510	3A
				122	137, 158, 148	405	3A
212	5182 Al	1.6	131	21	182, 186, 186, 188	325	4A
				45	201, 219, 199	290	3A
				81	258	235	1B
				122	367	165	1B

Effect of yield strength on collapse of box sections tested under 30 mile/hr impact loading (6000J energy)

Batch number	Type	Thickness, mm	Yield strength, N/mm^2	Weld pitch, mm	Reduction in sample height, mm	Energy to deform sample 10mm*, J	Mode of collapse**
146	Low carbon	1.2	186	21	170, 151, 161, 169, 161, 149	375	6A
				81	212, 201, 226, 199, 211, 216	285	6A
151	HSLA	1.2	340	21	132, 137, 139	440	3A
				81	168, 174, 172	350	3A
192	Dual phase	1.2	370	21	123, 136, 121	475	3A
				81	147, 148, 144	410	3A
193	Dual phase	1.2	405	21	100, 107, 102	580	3A
				81	120, 111, 120	515	3A
508	Dual phase	1.2	425	21	101, 106	580	2A
				81	122, 115	505	2A
212	5182 Al	1.6	131	21	182, 186, 186, 188	325	4A
				81	258	235	1B

$$\sigma_b = k.E/(1 - 0.32)^2(t/b)^2$$

where E = 208000 N/mm^2; t = 1.5 mm; b = 120 mm and k = 3.29, giving s_b = 117 N/mm^2 and P = 73 kN. A swage would be incorporated into the closing plate AB designed to make it stable until yield of the material is reached. Peak axial load capacity for the section is given by:

$$1112t_1 86b_0 14.\sigma_y 0.57.B_0.43$$

where B = 1.0 for t/b < 0.016 so that peak capacity is about 105 kN and mean load capacity is 33.5 kN — given fold initiators to trigger the crush mode. For a circular section thin tube the classic formula due to Macauley gives the collapse load as:

$$2.\sigma_y.A[10(t/r) - 0.03]$$

where r is tube radius, t the wall thickness and A the enclosed area of the cross section.

Spot welding patterns affect the crush performance of fabricated box tubes and this has been studied by TWI using a drop test rig. At crushing rates up to 25 mm/min, quasi-static loading, load/compression characteristics are typically as shown in Fig 11 with permanent deformation after peak load Pmax is reached. Subsequent 'waves' represent the generation of plastic hinges along the section, fluctuating about a mean load P_m. Area under the load/deflection curve gives the energy absorbed. For this form of loading, tests showed effect of spot weld pitch, below a value of 50 mm, to be insignificant. The shape of a typical dynamic impact curve is theoretically different from the quasi-static case, as seen in Fig 12.

Dynamic impact loading was found to give substantially different results — and these depended

Effect of spot weld pitch on collapse of weld bonded box sections tested under 30 mile/hr impact loading (6000J energy)

Batch number	Steel type	Thickness, mm	Yield strength, N/mm²	Weld pitch, mm	Reduction in sample height, mm	Energy to deform sample 10mm*, J	Energy to deform equivalent box section 10mm without adhesive*, J	% increase in energy absorption due to presence of adhesive
146	Low carbon	1.2	186	21	148, 137, 132, 140	430	375	15
				45	148, 148, 139	415	330	26
				81	154, 161, 182	360	285	26
				122	174, 199, 200, 166	325	270	20
151	HSLA	1.2	340	21	111, 109, 122	525	440	19
				45	125, 121, 133	475	395	20
				81	156, 130, 142	420	350	20
				122	156, 150, 145	400	310	29
192	Dual phase	1.2	370	21	106, 96, 99	600	475	26
				45	98, 109, 108	570	485	17
				81	144, 140, 110	455	410	11
				122	120, 126, 125, 132	480	370	30

Effect of adhesive type on the collapse of weldbonded box sections tested under 30 mile/hr impact loading (6000J energy)

Batch number	Steel type	Thickness, mm	Yield strength, N/mm²	Adhesive type	Weld pitch, mm	Reduction in sample height, mm	Energy to deform sample 10mm*, J	% increase to energy absorption due to presence of adhesive
146	Low carbon	1.2	186	No adhesive	21	161, 169, 149, 170, 151, 161	375	—
					122	236, 208, 222	270	—
				A15	21	148, 137, 132, 140	430	15
					122	179, 199, 200, 166	325	20
				A16	21	133, 119, 130	470	25
					122	205, 207, 203	290	8
				A17	21	139, 128, 136	445	19
					122	191, 178, 184, 164	335	24
151	HSLA	1.2	340	No adhesive	21	132, 137, 139	440	—
					122	198, 188	310	—
				A15	21	111, 109, 122	525	19
					122	156, 150, 145	400	29
				A16	21	110, 103, 120	540	23
					122	161, 182, 156	360	16
				A17	21	136, 120, 118	480	9
					122	179, 154, 146	410	32
192	Dual phase	1.2	370	No adhesive	21	123, 136, 121	475	—
					122	161, 183, 141	370	—
				A15	21	106, 96, 99	600	26
					122	120, 126, 125, 132	480	30
				A16	21	101, 95, 89	630	33
					122	138, 137, 134	440	19
				A17	21	89, 101, 98	625	32
					122	142, 118, 137	420	14

Fig 13: Effect of weld pitch and yield strength

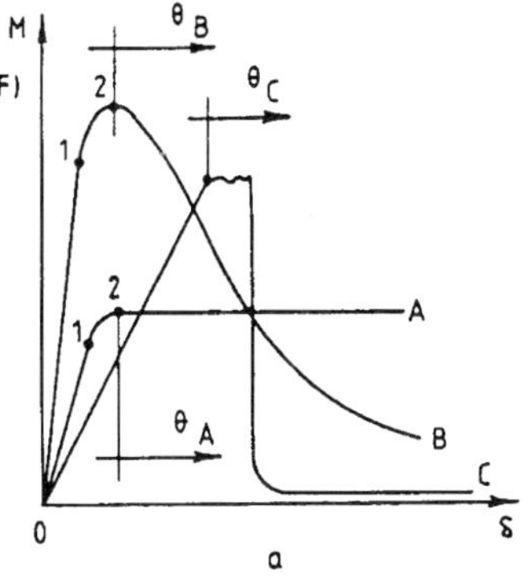

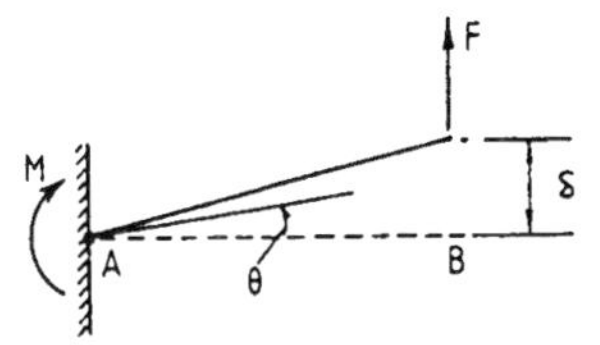

Fig 14: Constant moment curve

considerably on different jointing methods for the fabrication, such as spot-welding, weld-bonding and adhesive bonding. TWI therefore impact-tested specimens with these different jointing methods, as well as different materials and thicknesses. Tests were carried out at 30 mph impact with a drop mass of 66.8 kg giving impact energy of 6 kJ. Tables in Fig 13 show the effect of weld pitch and yield strength on the spot-welded box sections compared with weld- and adhesively- bonded types. Spot-welds were found to act as sites for fold initiation with deformation increasing for longer weld pitch. With weld-bonding, presence of adhesive increased the energy required to deform the section. With pure bonded joints a quite different failure mode was experienced.

The Cranfield Impact Centre has been prominent in the prediction of structural crash behaviour and has developed computer programs for the analysis of both structural elements and complete systems. These can deal with the plastic hinges formed in gross bending deformation of the box-beam elements. Unlike the compact section members referred to in the previous section, hinges in large thin-walled sections develop in a different way. Instead of a constant moment, curve 'A' in Fig 14, a variety of curves can be obtained depending on the location of beam and joint — also on the nature of local failure. Local buckling usually causes a moment drop-off, curve 'B', or if the material separates, the kind of behaviour in curve 'C'. A typical box section hinge is shown in Fig 15, in which the plastic deformations are concentrated along the yield lines of bending and rolling deformation. The figure represents the kinematics of the joint between about 5 degrees and the angle of 'jamming' when the two opposite sides of the compression flanges come into contact. Experimental (dotted) and theoretical (solid) load/deflection curves are compared in Fig 16, alongside their respective collapse modes, for the quasi-static (left) and quasi-dynamic, right, load cases.

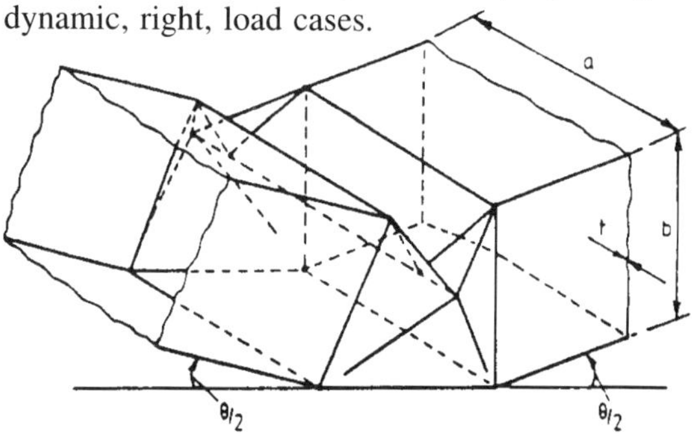

Fig 15: Box-member plastic-hinge

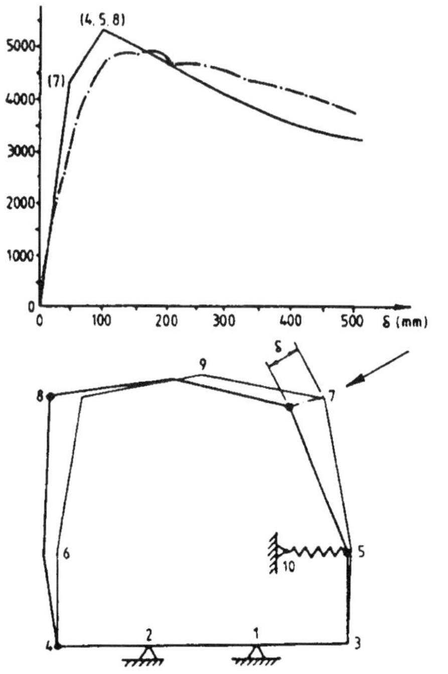

- Experimental (dotted) and theoretical (solid), quasi-static load-deflection curves of a bus ring (a) and the overall collapse mode (b)

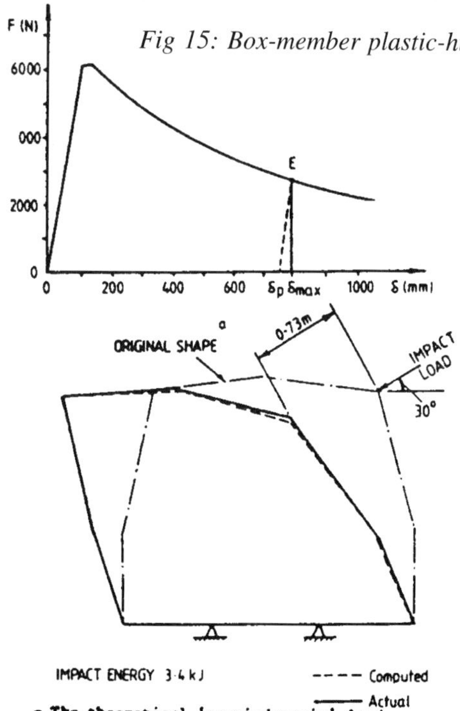

- The theoretical 'quasi-dynamic' load-deflection curve of an impacted bus ring (a) and its final deflections (b).

Fig 16: Quasi-static and quasi-dynamic performances

Calculating crashworthiness

According to researchers at HW Structures[3], determining the modes of failure and crashworthiness of a vehicle at concept stage is essential if feasible designs are to be achieved within project times scales and costs. Although application of CAD/CAE techniques have reduced the time during the detail design phase, they have not had the same influence on the concept design or vehicle layout stage. A structure has to be conceived before it can be analysed, the author argues.

For frontal impacts, energy absorption in a conventionally designed vehicle (Fig 17) is mainly achieved by deformation of the upper and lower longitudinal structural members. Depending on the overall deformation, other components such as the engine cradle may also contribute to the energy absorption process. Usually the lower side structure is responsible for the majority of the energy absorption during the early stages of the impact (say between 10 and 30 msec). Thereafter, the upper side structure will start to deform plastically. These two components will be mainly responsible for the early rise in the crash pulse so it is important, from a structural engineering viewpoint at least, that failure modes remain predictable and do not form grossly unstable mechanisms during the crash.

As the deformation progresses, and depending on the vehicle in question, other components may have a significant effect on the crash pulse. This is particularly so for vehicles with confined engine compartments or short front ends. It is essential to limit intrusion of engine or driveline components into the passenger compartment without creating high retardations detrimental to the occupant. Design crash pulses for the development of future models have been developed based on previous testing and/or occupant modelling (Fig 18).

Energy absorbing structures are typically constructed of ductile metal: they make use of the metal's ability to absorb mechanical energy by plastic deformation. For legislated impacts, the majority of vehicle structures absorb energy via crush and gross bending mechanisms, the authors explain. In a 30 mph (13.3 m/s) impact into a rigid barrier, a saloon car will expend about 105 kJ. Of this amount, about 100 kJ will be required to be absorbed by the front structure. Simple formulae (Fig 19) for mean crush force derived from earlier work can be applied to axially loaded side members. The complication arises when either the axial stability of the member cannot be maintained or the member is loaded both axially and in bending during the impact.

Quantitative approaches to impact protection design provide some insight into the stability of axial collapse modes. In vehicle structures, the complexity of loading during impact may induce instabilities derived from bending loads rather than axial buckling. Post yield stability in thin walled structures is influenced by: reduced section stiffness during fold-

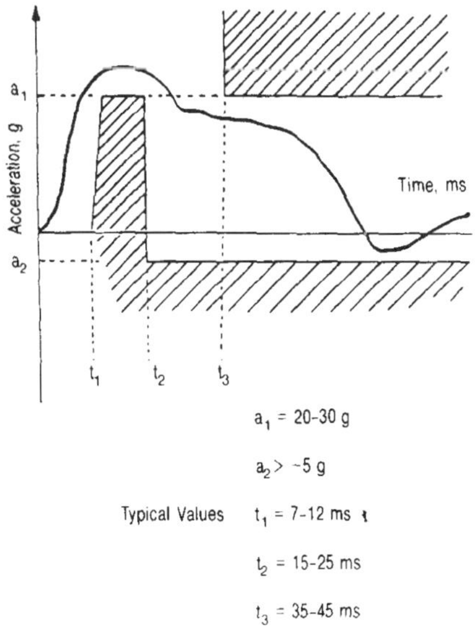

Acceleration, g
Time, ms
a_1 a_2 t_1 t_2 t_3

Typical Values
a_1 = 20-30 g
a_2 > -5 g
t_1 = 7-12 ms
t_2 = 15-25 ms
t_3 = 35-45 ms

Fig 18: Design crash-pulse

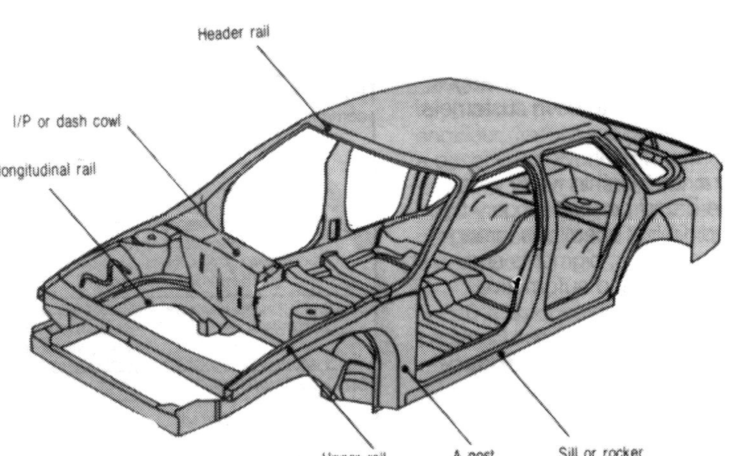

Fig 17: Key elements for impact protection

ing, column length and compactness of the column (ratio of elastic buckling to axial crush loads). If the bending loads remain small then axial stability can generally be maintained if the ratio of the lengths of the sides of a member is less than 1.5. Hinging of structural members offers a more easily controllable, albeit less efficient, method of absorbing energy. This becomes the predominant method of absorbing energy later in the impact when stability of crush mechanisms is less predictable. The overall geometry of the front-end structure (especially for front wheel drive vehicles) often means bending becomes the predominant failure mode.

It is possible though to use quasi static plastic methods to derive energy management strategies for the structure. The post yield collapse behaviour, shown in Fig 20, can radically effect the energy absorption characteristics. Clearly some guidance on designing structures with rising or falling force or moment characteristic is needed. Strain rate and inertia effects make dynamic plastic collapse estimates more difficult to understand than the corresponding quasi static situation. Strain rate can significantly influence the yield stress of the material and subsequently the collapse load. Inertia can influence the collapse mode of the structure. Early studies found that the final deformation of structures, with load-deflection characteristics that fail, is sensitive to the impact velocity. Although the post yield material behaviour is less sensitive to strain rate, the influence of progressive creasing (additional yielding of material) can alter the calculated performance of the structure.

Rigid body calculation methods are used to assess gross motion of components and vehicles during crash. The degree of energy absorbed by rotation during an impact can be significant in real accidents. Lumped mass modelling used in conjunction with quasi static methods and hand calculations provides a useful design tool. The iterative process is quick but does not provide a rigorous solution; it is however a cost effective method before design details are confirmed and more elaborate non-linear dynamic models are constructed.

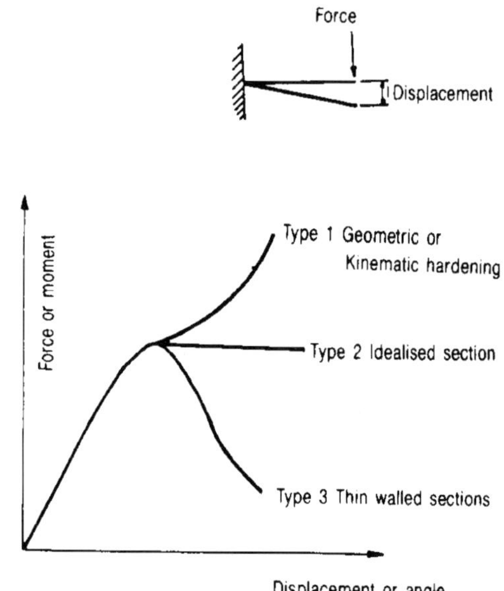

Fig 20: Energy-absorption characteristics

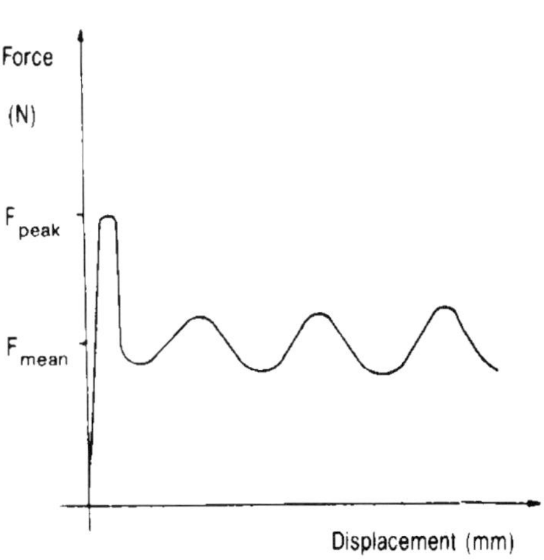

Round Tubes

$$F_{mean} = \frac{Yt^2}{4}\left(\frac{20.79(2R/t)^{1/2}+11.90}{1-0.568(t/2R)^{1/2}}\right)$$

R - Radius of Tube
t - Thickness of Tube
Y - Yield Stress

Square Tubes

$$F_{mean} = 387\, t^{1.86}\, b^{0.14}\, Y^{0.57}$$

t - Thickness of Tube
b - Width of Tube
Y - Yield Stress

Fig 19: Axial-collapse characteristics of thin-walled tubes

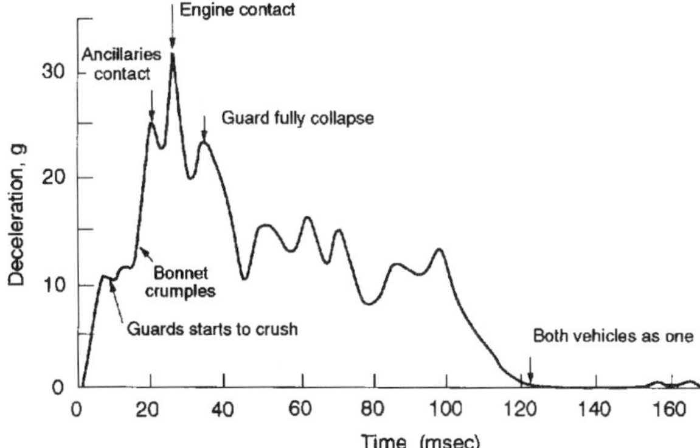

Fig 21: Representative crash pulse

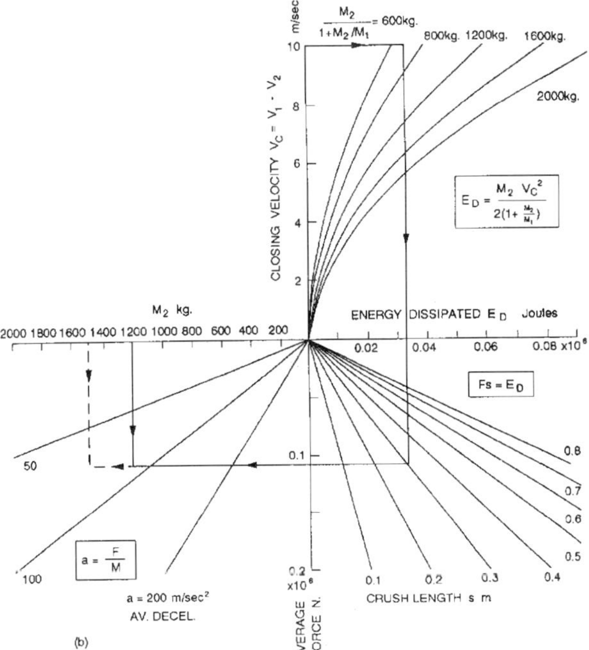

Fig 22: Relating closing velocity with energy to be dissipated

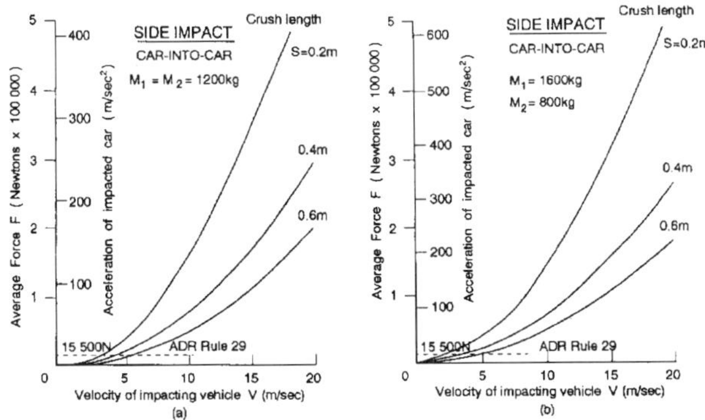

Fig 23: Average force/acceleration for various crush lengths

Laws of mechanics applied to vehicle safety

While the previous sections consider most aspects of secondary safety, it is also necesseary to examine the external force systems, and related rigid-body dynamic analysis of the accident situation. According to Professor Noel Murray[7] of Monash University in Australia (noted for its civil engineering structural research) the application of the basic laws of mechanics to the crash situation has revealed serious omissions and inconsistencies in current practice.

The author contrasts the situation of the racing driver who, more often then not, walks away unharmed, from relatively high speed crashes, with the public-roads environment where huge causality tolls are now all too familiar worldwide. Although the industry spends large sums on body structure design and manufacture to meet existing legal requirements, commensurate with the light weight structures required for low fuel consumption, perhaps not enough is done for the safety of the vehicle buying public in studying the crash behaviour of vehicles as a whole.

Murray likens the occupant cell of a typical saloon car with an unbraced Vierendeel truss structure where no triangulated bracing is supplied for structural integrity. Such unbraced frames rely on the integrity of the joints between pillars and rails, where very high bending moments can occur in certain crash situations — which lead to lozzenging deformation of the door and window openings. Joints have thus to be as least as strong as the members radiating from it. Where the joint relies entirely on spot-welding, the spacing of welds is obviously crucial and any load case which can cause unzipping of a line of spots is clearly to be avoided.

A car structure is cited in which the B-pillar is fixed to the rocker panel by only four spots and where the spot-weld pitch on the A-pillar is so wide as to induce local buckling in the flanges of the pillar between spots. The quality of production spot welds is also shown to be far below the quality obtained on laboratory test specimens in a number of cases.

The author provides an interesting analysis of a crash pulse, Fig 21, to show the effects of different parts of the vehicle (stiff and flexible) reacting to the impact. It helps make the point that in applying basic dynamic laws the average force for a particular crush length has to be considered in working out the energy to be dissipated by a particular part of the structure.

For a car impacting a stationary heavy truck with mass of order of magnitude higher than the car, and assumed infinitely rigid, the author suggests that it is reasonable to assume the work done in crushing the structure is equal to the loss in kinetic energy of the car as it is brought to rest by the truck:

$$Fs = MV^2/2$$

for car mass M and speed V, crushing force F and crush distance s. A 1200 kg car impacting at 60 kph would involve 166 667 joules of energy and if crushed by 0.4 metre, average force is 417 kN or 42 tonnes force. This would, of course, double if the crush length was only 0.2 metre. The force also is seen to rise as the square of impact speed. The average deceleration of the car can be estimated from:

$$a = Fs/Ms$$

as 347 metres/sec^2 or 35 g. Where two comparable vehicles impact these expressions are modified by putting $s1 + s2$ for s in allowing for the deformation of each structure and the author has constructed the graph in Fig 22 to solve these expressions without calculation.

In the side impact situation, when two 1200 kg cars impact so that the front of the moving one contacts the side of a stationary one, the author shows the energy dissipated to be $300V^2$ joules for a closing velocity in metres/sec. At 50 kph this amounts to 58,800 joules. If just half this is assumed to crush the side of the stationary car, then a crush distance of 0.2 metre would involve an average force of 15 tonnes and acceleration of 12.5g. A generalised graphical interpretation is given by the author in Fig 23. The adequacy of side impact protection legislation is questioned by the author in the light of these results.

In the case of frontal impact of a car against the rear of a stationary platform truck, test results obtained by Monash University laboratory are shown in Fig 24 to reveal the effects of various elements in the crushing process and leads to a plea for appropriate underrun protection legislation. A problem is to decide how much energy should be absorbed by an underrun barrier. While UK researchers are cited as recommending a figure of 20 000 joules, Fig 25 shows the effect of a 1200 kg car impacting at 50 kph; it has some 117 600 joules to dissipate. By subtracting the suggested barrier absorption figure, the car is left with 97 600 joules which would reduce impact velocity to 46 kph, probably an insufficient margin in the author's opinion.

Besides the casualty aspects of accidents, the

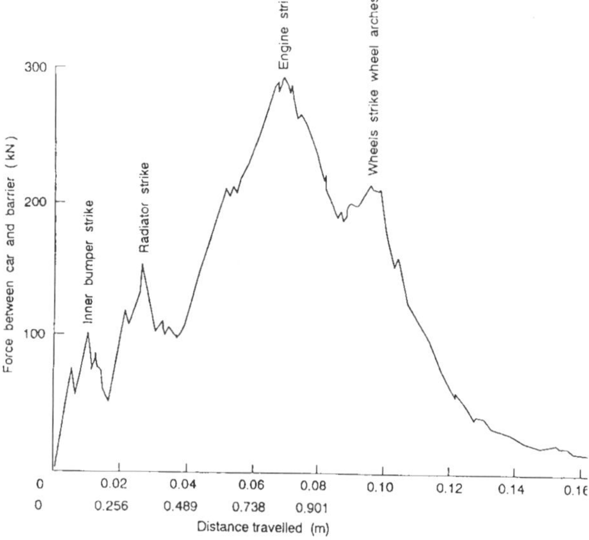

Fig 24: Underrun barrier test at Monash

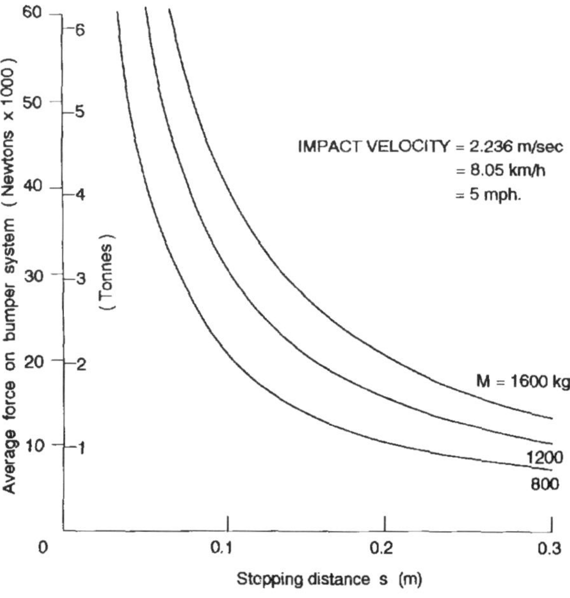

Fig 25: Bumper forces at low velocity impact

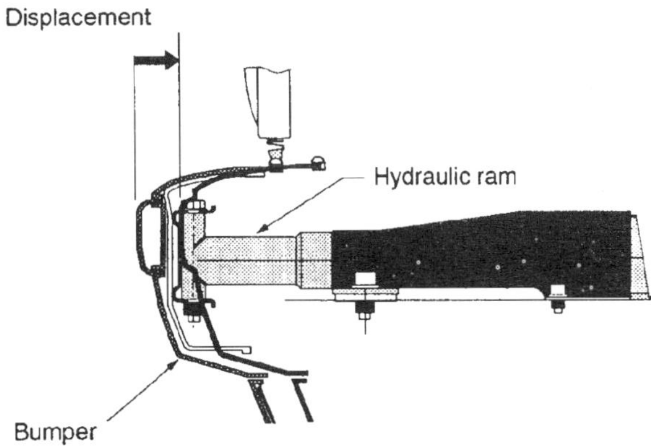

Fig 26: Hydraulic energy-absorbing device

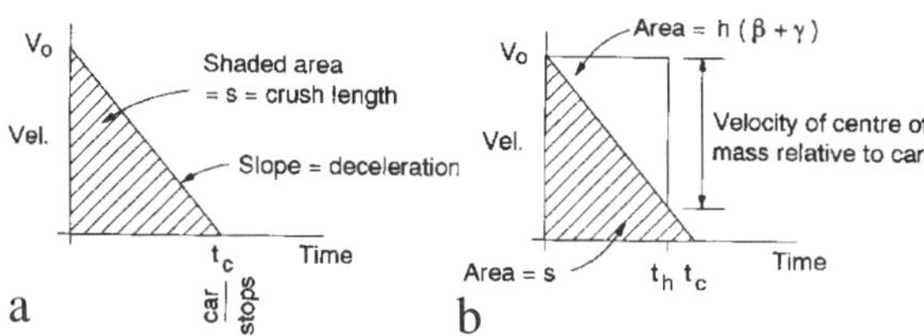

Fig 27: Crush length representation

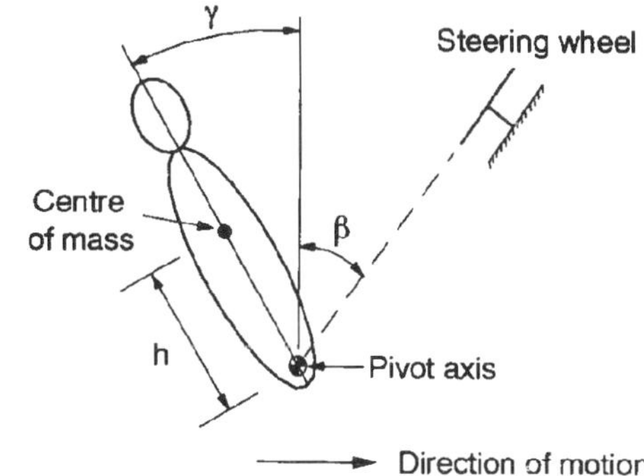

Fig 28: Occupant representation

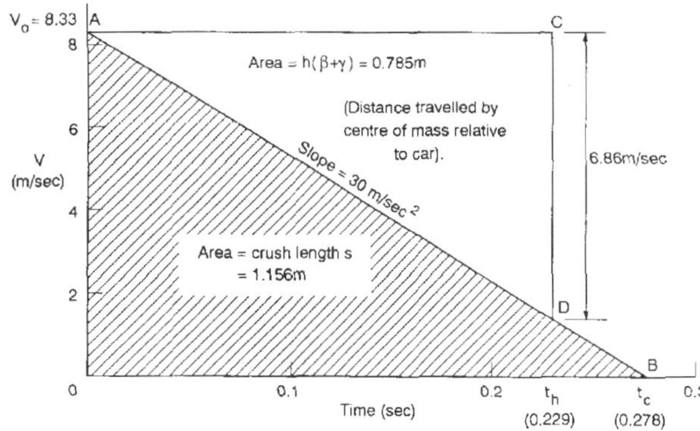

Fig 29: Construction for driver striking wheel

question of physical damage in low speed impacts is considered in the book. In particular the US requirements for bumper bars are examined. For the 5 mph impact requirement, the energy to be dissipated for a car mass M striking a stationary object is 2.5M joules. On equating this with an average force times stopping distance, the results shown in Fig 25 are obtained for three different car masses. The most striking point is how rapidly the average force increases as the stopping distance decreases. In the case of the peak force, it is argued that this is at least 50 per cent greater than the average — an important factor in design. Current bumper designs, it is argued, allow typically only 0.05 metre deflection before expensive parts of the bodywork are damaged and in fact the figure of 60 kN shown by the graph could not be sustained by typical modern bumpers. A solution might be to use an energy-absorbing bumper of the type seen in Fig 26, says the author.

A better feel for studying real-world accidents is obtained by using graphical techniques, it is argued. By expressing the time taken to bring an impacting vehicle to rest as:

$$t = 2s/V$$

This can be expressed graphically as in Fig 27a. When the driver is represented as in the diagram of Fig 28, the length of the path of its centre of mass is $h(b + g)$ and over this length it is assumed the original speed of the mass is maintained. In Fig 27b this information is added to that of Fig 27a. The unshaded are is equal to the distance travelled by the centre of mass up to the instant of head strike. In Fig 29 an example is given in which the deceleration is relatively low and parameter values are h = 0.6 metre, g = 0.524 radian, b = 0.785 radian, V = 8.33 metre/sec and d = 30 metre/sec^2 (3.06g). Having established line AB, a horizontal line is drawn from A to represent uniform speed of the centre of mass. End point t_h can be found by measuring the appropriate area and the speed of the centre of mass at impact is given by length CD. The speed which the head strikes the steering wheel is thus 13.72 metres/sec.

In his book, some estimate is made by the author of the non-uniform deceleration which occurs as the real vehicle crushes at different rates during the impact and a similar analysis then shows the head striking the wheel at around 25.8 metre/sec (93 kph!).

Side-impact protection

Differences in available survival space mark a major difference between frontal- and side-impact analysis. Computer simulation techniques are described which are tending to simulate structural deformation and occupant movement simultaneously.

In 1990 a federal standard for side-impact protection was introduced in the USA (FMVSS 214) and a proposed standard is now being discussed in the European market for 30 mph impacts. Experts consider that the restraint system cannot be considered separately from the crash pulse, as in front-impact analysis, because of the limited space available to absorb impact energy. A single analytical technique is thus favoured to simulate both structural collapse and occupant kinematics.

In a car or cab structure key parameters are door interior stiffness and topography, door to occupant initial separation, door intrusion profile after impact and door to bodyshell interaction. In determining the door intrusion profile the structural analysis requires the collapse velocity of the vehicle's side structure and the compliance of the vehicle structure adjacent to the occupant's body. It is also necessary to relate occupant contact time and distribution of contact with inner structure compliance.

Test work at the Cranfield Impact Centre has resulted in simplifying the shape of the velocity pulse at the occupant's chest area, for analysis purposes, to that shown in the straight lines in Fig 30. Before point A, rate of change of velocity with time is -35g; change in time in ms is 15 and 10, and change in velocity in m/s is 0 and 2.5, for AB and BC respectively. The resulting velocity template, obtained after testing, proved valuable in identifying the type of collapse velocity pulse likely to cause serious injuries. For a given geometry and inner wall compliance, if the side structure velocity remains below that on the template then injury level is likely to be low. Initial peak velocity influenced chest acceleration while the second peak influenced rib deformation. Tests showed that by placing an occupant next to a door, scope for improved protection through better door design and use of energy-absorbent material will be significantly enhanced.

Modelling of side-impact has been carried out by MIRA using the DYNA3D package involving distributed masses to represent the inertia of the

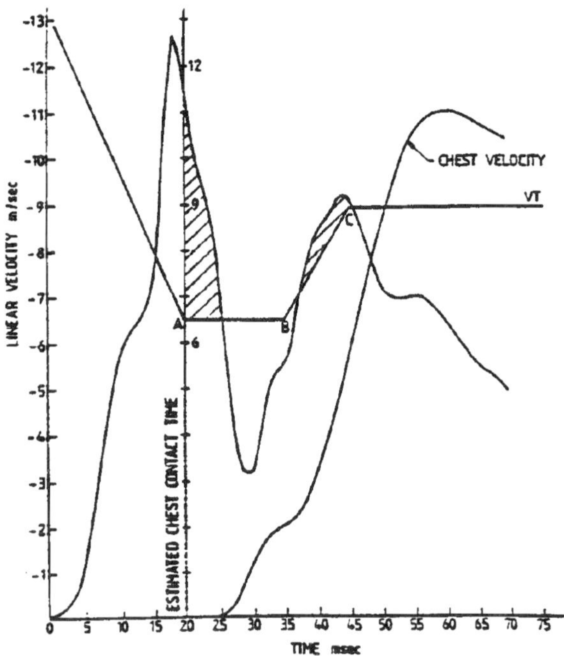

Fig 30: CIC standard impact pulse

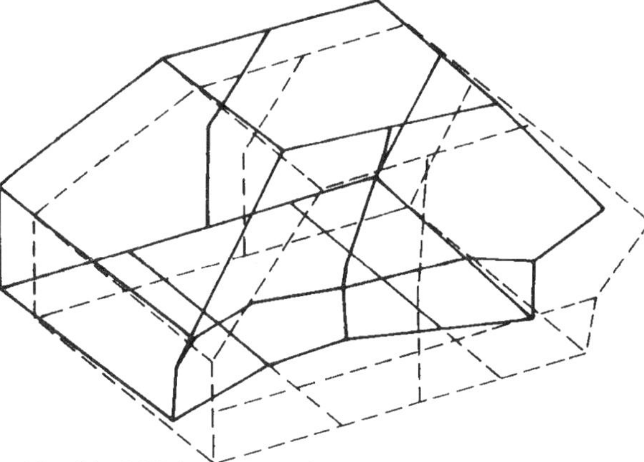

Fig 31: MIRA structural model

MODEL OUTPUT
TEST RESULT
VELOCITY
BARRIER
VEHICLE
(NON STRUCK B-POST)
TIME

Fig 32: MIRA model verification

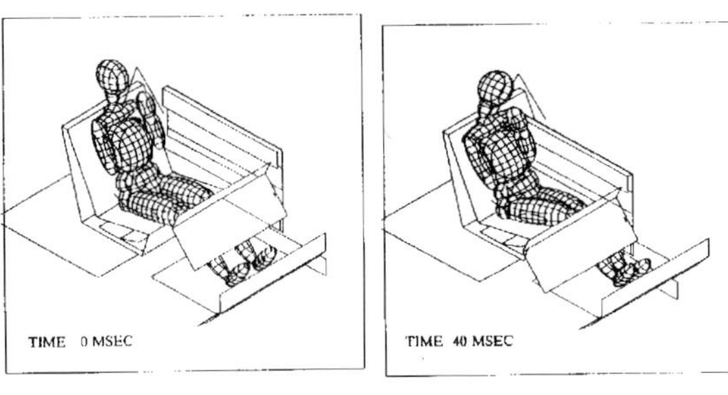

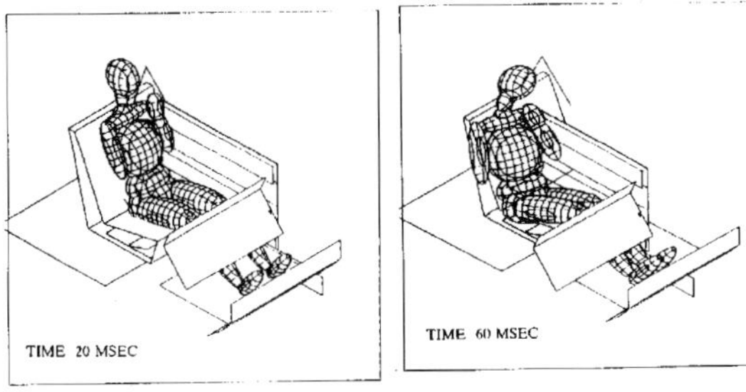

Fig 33: MADYMO simulation of occupant motion

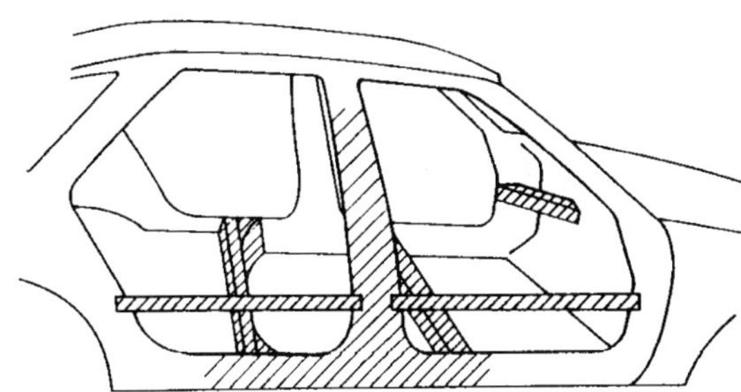

Fig 34: Elements investigated

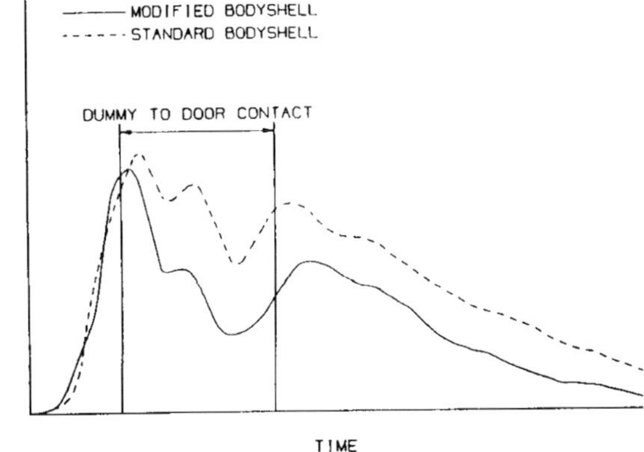

Fig 35: Before modification

structure. Fig 31 shows the simulated structure before, and 100 ms after, impact. The model included a simple representation of the Eurosid dummy. Good correlation with crash test results is seen in Fig 32. A separate detail analysis of occupant kinematics was used for a detail study of injury levels using the MADYMO package, with the time history of structural deformation and vehicle acceleration taken from the first, structural, simulation, Fig 33.

MIRA investigated a number of structural elements in detail as shown in Fig 34. The B-post has a tendency to fail at waist rail height and this was a prime candidate for reinforcement to prevent premature buckling failure. Non-linear properties of these structural elements were determined by quasi-static tests on the individual elements. Comparisons of door velocity and intrusion before and after modification are shown in Figs 35 and 36 taken from full-scale dynamic sled-testing which takes account of strain-rate effects. Note that a 30% reduction in average door velocity was achieved while the dummy was in contact with the door. Because peak injury levels have been found to take place about 15 ms after initial dummy/door contact, it is very important to reduce velocities during contact and not just at the initial instant of contact.

Earlier work by the Cranfield Impact Centre, with Ford showed how crashworthiness criteria could be included in concept design. The study started with only the interior package data for a future three or five door hatchback car for which structural behaviour of member joints and B-pillar deformation, from tests on an earlier vehicle, were used. A simulation program called SIMSIM1 was used to obtain the collapse mechanism of the structure when impacted into a standard lateral test barrier. Figs 37 and 38 show the respective three and five door vehicles and their respective structural models.

The work was able to quantify the structural contribution of each failed beam element by calculating the crash energy absorbed, Fig 39 . This showed that the bottom of the B-pillar (A) in the three and five door designs absorbed up to 25% of the crash energy, resulting in extensive deformation of the B-pillar. Next stage was to limit the side structural collapse by distributing more even loading by the impacting vehicle into as many beam elements as possible, to obtain the optimised results shown in these figures. This included improving properties of the B-pillar

(A), rear-seat pan (B) and adding a door intrusion beam (C). The figures show the percentage contribution to crash energy absorbed of each member throughout the structure, the reduced B-pillar contribution ensuring reduced deformation of the B-pillar and hence greater survival space. In fact the mid-span deflection was reduced 147 to 70 and 175 to 67 for three and five door vehicles respectively. The table in Fig 40 shows the resulting reduced injury levels to the occupants.

The SIMSIM1 program is a coarse-mesh beam-element non-linear finite-element code suited to side collisions which essentially involve beams in bending. Compression and torsion can also be simulated and the program is hybrid in the sense of being able to analyse elastic behaviour in traditional beam theory as well as non-linear material behaviour from test-derived experiments. It can simulate large deformations plus local and global collapse mechanisms. Simulation of non-linear behaviour includes both 'perfectly-plastic' (beam can maintain load-bearing capacity after failure) and 'moment drop-off' (beam's supported load falls off rapidly after failure). The finite stiffness of the joints can also be taken into account; these can otherwise significantly degrade the analysis results.

Another program employed by Cranfield Impact Centre is SIMSIM2 which can predict the rigid body motions and vehicle side deformations in a 90 degree side collision. This is achieved with a model comprising a series of unidirectional spring-mass systems representing both vehicles as they contact one another. General equations of motion, together with impacted vehicle tyre and suspension data, are used to predict the whole vehicle motions after impact.

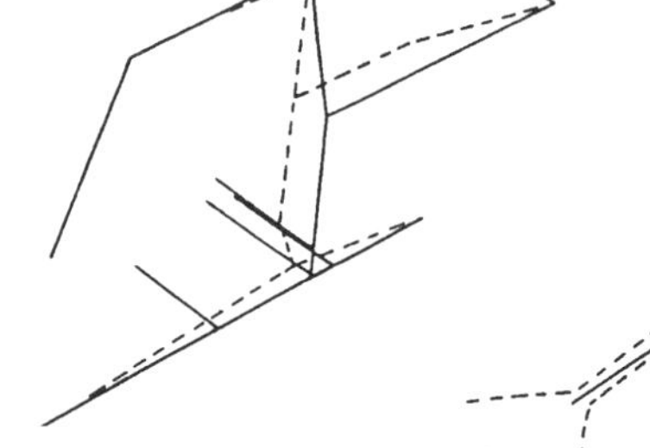

Fig 37: 3-door model

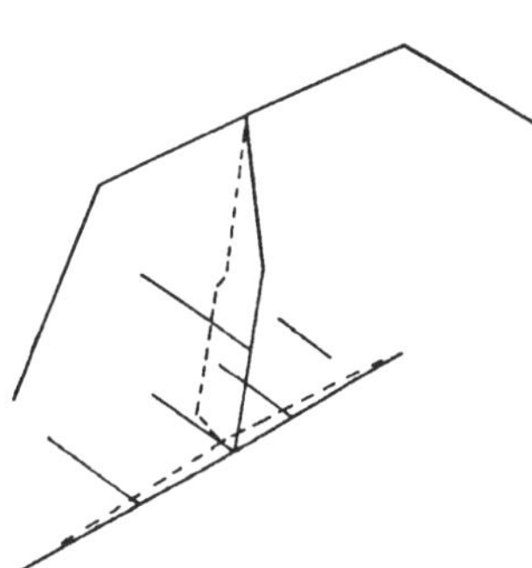

Fig 38: 5-door model

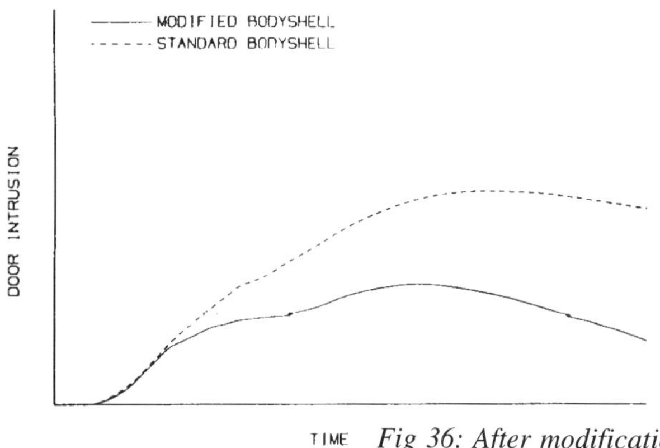

Fig 36: After modification

CVS Analysis	Initial 3 door Vehicle	Optimized 3 door Vehicle	Initial 5 door Vehicle	Optimized 5 door Vehicle
Max Velocity of Side Structure (m/s)	12.5	10.7	15.2	12.7
Velocity of Side Structure on Contact with Dummy (m/s)	12.5	10.2	15.1	12.2
Velocity of Side Structure at Chest Velocity of 0.1 m/s (m/s)	7.9	4.7	9.9	7.2
Head Injury Criteria (HIC)	138.0	67.0	224.0	94.0
Max Lateral Head Acceleration (g)	29.5	21.2	31.8	22.7
Chest Severity Index (CSI)	375.0	154.0	497.0	229.0
Max Lateral Chest Acceleration (g)	67.3	49.5	76.2	51.8
Max Lateral Chest Velocity (m/s)	10.5	7.8	12.5	8.9
Max Lateral Pelvis Acceleration (g)	110.0	80.6	117.5	83.9
Max Lateral Pelvis Velocity (m/s)	12.0	8.5	14.4	9.8

Fig 40: Reduced injury levels

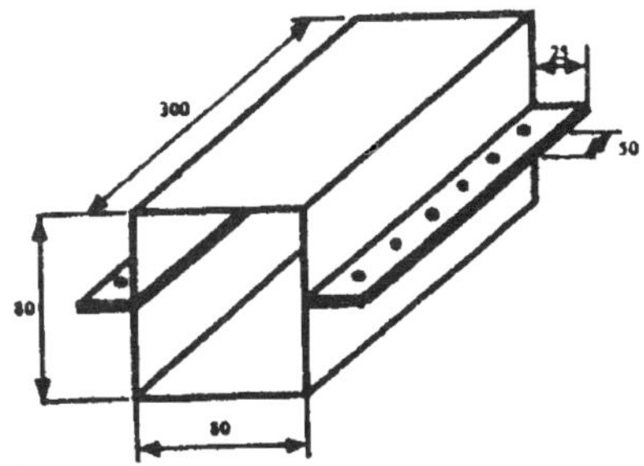

Fig 41: Specimen member

	MS th = 1.2 mm Ys = 240 MPa	HSS th=1mm Ys = 400 MPa	HSS th=1.2 mm Ys = 400 MPa
Quasi - static	4815	4755	6416
15 Km/h	7440	6190	8120
33 Km/h	7510	6600	8620

Fig 42: Test results (absorbed energy — J)

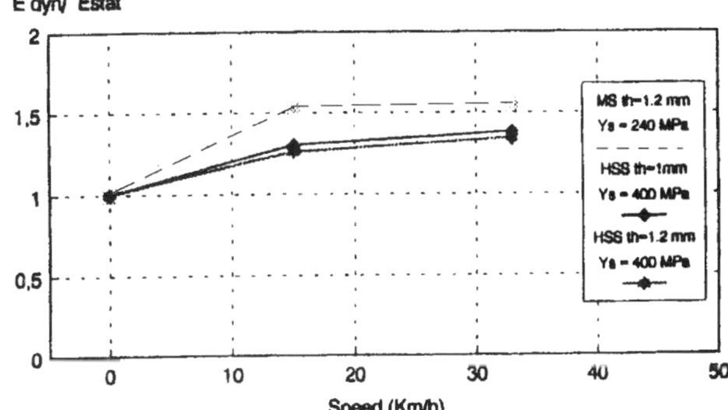

Fig 43: Absorbed energy gains

Controlled-collapse longitudinals

To evaluate the performance of high-strength steel in this context, Cockerill-Sambre researchers[4] have carried out laboratory controlled drop tests, and static crush tests, on standardised welded box sections, Fig 41, resembling those of car front longitudinals, and obtained interesting conclusions.

Specimens were subjected to 200, 300 and 1000 kg masses falling through five metres to reach impact speeds of 36 km/h. Steel thicknesses of 1.2 and 1 mm were studied in mild and HSS steels and absorbed energies were obtained as in Fig 42 which are mean values at 50 per cent deformation. From these can be deduced that a 20 per cent weight reduction can be obtained in the quasi-static case, with equivalent safety level, for a move from mild to HSS steel. When the results are plotted on an impact speed base for the dynamic case, the plotted curves have different shapes and a significant gain can be noted for HSS. This is explained by Fig 43 which plots the ratio of dynamic (dynamic absorbed) to static (crushing deformation) energies. For a mild steel the observed gain is 50 per cent between 0 and 15 km/h — and remains the same when impact speed reaches 33 km/hh. For an HSS steel the gain between 0 and 15 km/h is lower (20 per cent) but a higher gain is obtained at 33 km/h (30 per cent). Because the two curves for HH are nearly the same the phenomenon is shown to be a material effect rather than a geometric one.

At the Austrian producer, Voest-Alpine, a survey of crash test simulations was carried out. The most frequently used of all was the offset crash with 50% demolition of the body front section at 55

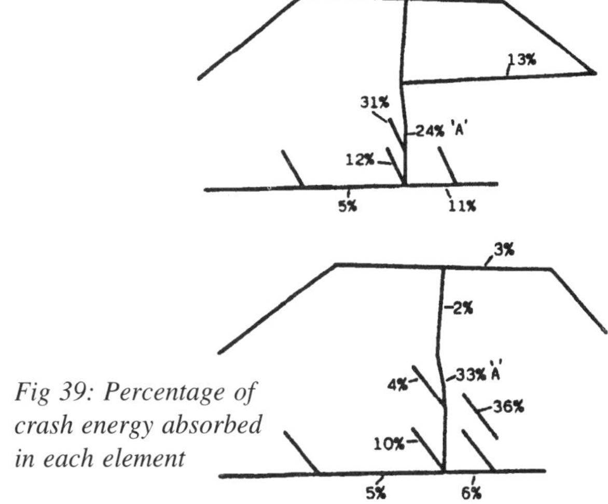

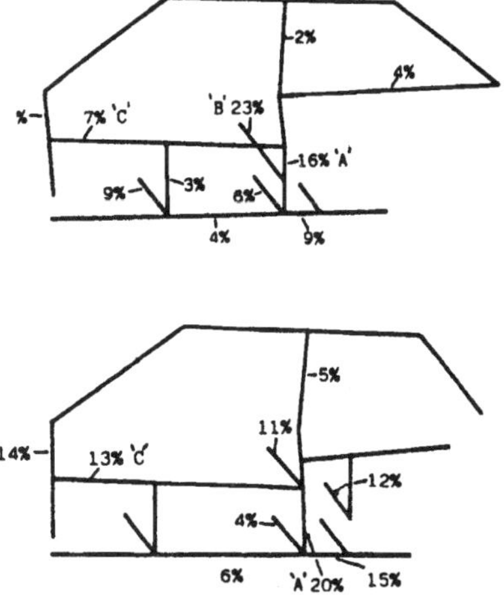

Fig 39: Percentage of crash energy absorbed in each element

km/h and again stressed the potential for weight saving in the hard worked front longitudinals. The company's FePO4 HSS material had a clear dependence of strength and forming behaviour on increasing rate of load application.

The researchers carried out simulated crash behaviour of longitudinals using octagonal sectioned box-beams of 420 mm length. In their tests, rate of load application was 45 km/h and energy input 8000 Nm. Comparative tests were carried out which included aluminium alloy specimens and the results obtained are shown as Fig 44. Spot-welding, laser welding and clinching were used for the steel materials while adhesion combined with clinching was used for the aluminium alloy ones. The AA 6010 was exposed to heat treatment similar to the BH steel but in order to achieve indicated high tensile strength values for aluminium, either higher baking temperatures than are available in the paint shop are required or distinctly longer exposure periods.

Deformation of the aluminium is found to be twice as great when compared with equally thick deep-drawing steels or 17 per cent thinner HSS steels. Advantages over aluminium are only found when steel thicknesses are reduced by 45-50 per cent compared with those of the aluminium. Energy absorption time ratios are constant for steel and aluminium but there is greater standard deviation. This, say the company, is attributable to worse deformability of aluminium, than steel, particularly after heat-treatment which leads to a 63 per cent drop of the uniform elongation before reduction in area.

Aluminium alloy Alcan Automotive Structures[5] have presented the technology for aluminium alloys in impact, particularly AA5000 and AA6000 materials. At the time of its presentation the company explained that axial collapse was well understood but bending

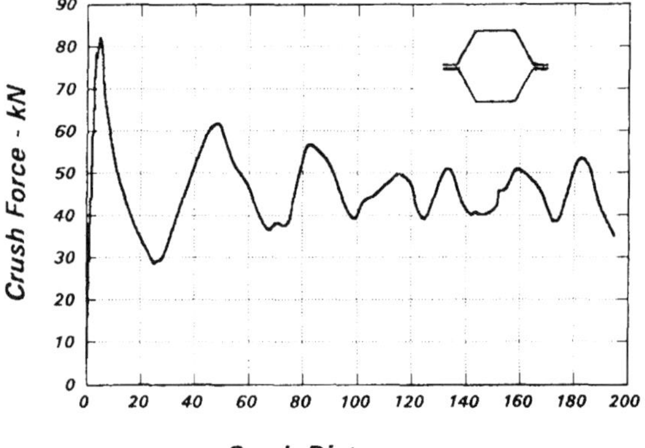

Fig 45: Force response

Average Collapse Force(kN)

(Based on 2mm 5754-0)

STANDARD EDITOR	33.85	64.8	55.4	76.0
ALUMINIUM EDITOR	28.09	53.2	41.9	66.65
EXPERIMENT (Q/S)	26	49.6	39.8	70.1

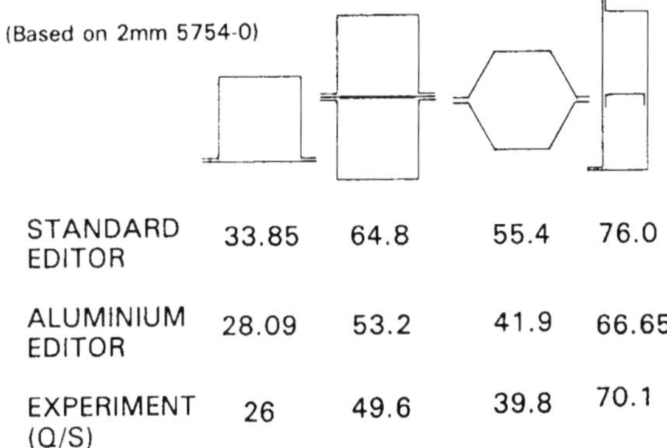

Fig 46: CRASH-CAD force predictions

material	thickness mm	Rp0,2 N/mm²	Rm N/mm²	AG %	A80 %	n4-6	r18
Fe P04	1,50	181	309	21,8	44,2	0,229	2,100
AA 6010	1,50	296	322	7,0	10,5	0,074	
THP 220	1,45	254	367	21,4	39,4	0,195	1,978
Alform 34 K	1,25	384	471	15,1	24,5	0,172	0,970
Fe P04	0,85	184	311	21,6	42,7	0,218	2,319
Alform 220 BH 170°C/20min	0,75	276	347	16,6	34,1	0,158	2,223
Alform 34 K	0,75	373	453	15,7	26,5	0,156	1,073
Alform 260 P	0,70	286	407	18,3	28,8	0,197	1,740
AA 6082	0,70	161	269	20,5	27,1	0,256	0,823

Fig 44: Data for materials handled

collapse technology was still being researched. The progressive axial buckling of thin tubular members into wave-like folds provides the force response characteristic typified by Fig 45 for the octagonal section shown. The behaviour of various sections has also been predicted using CRASH-CAD code adapted to the non-linear stress-strain relationship of aluminium. Comparison of predicted and test results is shown in Fig 46. The company also say that a quick method of assessing material type and thickness is to use the relationship:

$$P_{ave} = GST^{1.6}(DF)(JF)$$

where P_{ave} is average crush force, G section shape factor. S the strength term $(S_y + s_{uts})/2$, T the thickness and with DF = dynamic factor and JF = joint factor. The dynamic factor relates to the ratio of dynamic to static energy absorption, typically 1.1 for aluminium sections. Joint factor would be 1.15 for bonded aluminium, reflecting the superiority over spot-welding. A wide range of forms have now been tested as indicated by Fig 47. Objective in design is to avoid fracture caused by wrong alloy selection and inappropriate heat-treatment and to obtain reliable and regular fold patterns within a tolerance of loading eccentricity. The company point out that increasing the sheet thickness to attain a performance target could result in adverse affect on corner foldability. Collapse force requirements are said to vary between 50 and 100 kN for mid-range vehicles.

Inclusion of a central diaphragm, Fig 48, can sometimes improve performance by encouraging the member to fold uniformly. Careful design can also enable collapse to be triggered at an initial force level no higher than that producing the folds, Fig 49. A tapered octagonal member has been built by the company for this purpose. This was found to improve section stability but required additional swageing on the wide wall side.

Material	Average Collapse Force-kN	Mass-Specific Energy Absorption (kJ/kg)
AA5754-O sheet	85	29.9
AA5754-O sheet	80	29.5
AA5754-O sheet	95	28.8
AA5754-O sheet	60	22.5
AA5754-O sheet	82	24.8
AA6063-T4 extrusion	34	16
AA6063-T6 extrusion	58	28

100mm by 90mm

100mm by 50mm

Fig 47: Performance of 2.6 mm gauge sections

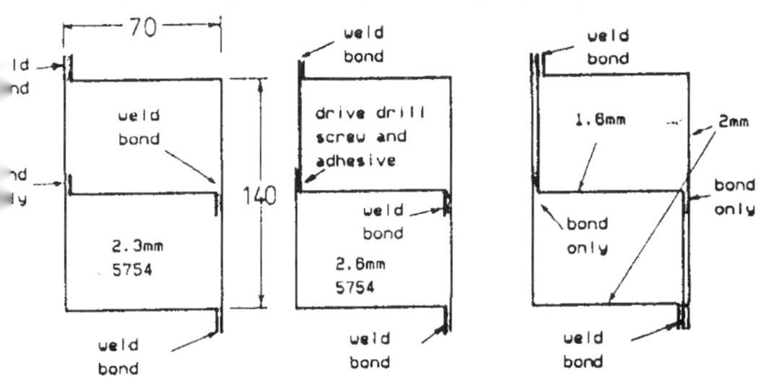

Fig 48: Internal diaphragm

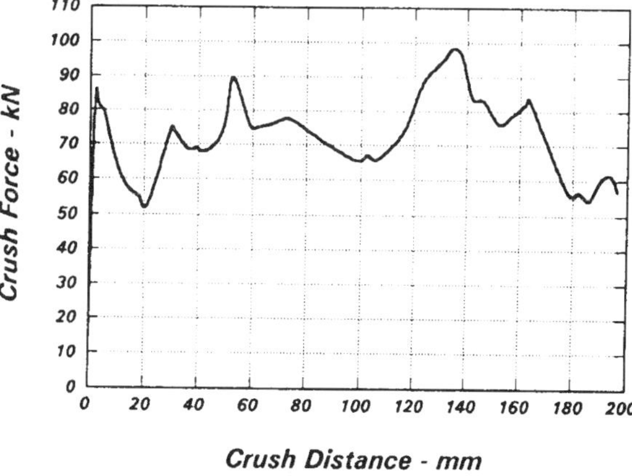

Fig 49: Response for triggered member

Folding-modes of aluminium alloy Work carried out by Alusuisse-Lonza in conjunction with MS ME[6] shows that good correlation is now possible between simulated and test results for tubular members subject to impact forces. The need to optimize the energy absorption capability of aluminium alloy tubes used as vehicle front longitudinals for controlled impact-collapse has led to analyses of heat treatment specification. By exhaustive trials the company has been able to establish a heat-treatment which provides the foldability of soft annealed aluminium alloy with the absorption capability of artificially aged alloys.

Further investigation has gone into accurate simulation of the folding behaviour of aluminium alloy tubes under impact compression so that accurate performance prediction can be carried out. Existing techniques have failed to take into account the importance of wall thickness relative to the tube section size which is seen as important in the compressive fold sequence of Fig 50. A variety of section shapes have also been tested including circular, square and diagonally braced rectangular sections, with the results shown in the table of Fig 51 and a collapse mode shown by the FE-analysis simulation in Fig 52 for the circular and diagonally-braced sections. The company has also carried out quasi-static three-point bending tests on tubular members to simulate the behaviour of side-impact protection beams. Using ABAQUS/Implicit non-linear FE analysis good correlation was obtained between experimental and calculated results. Fig 53 shows a simulated bending deformation. The importance of using true stress-strain curves is emphasised by the report. While the company has observed no strain-rate sensitivity with aluminium alloy, standard material data based on standard tensile stress-strain tests were unreliable. Fig 54 shows the effect of using true against conventional stress-strain curves. The true curves are presumably obtained by carrying out quasi-static compression tests in the laboratory.

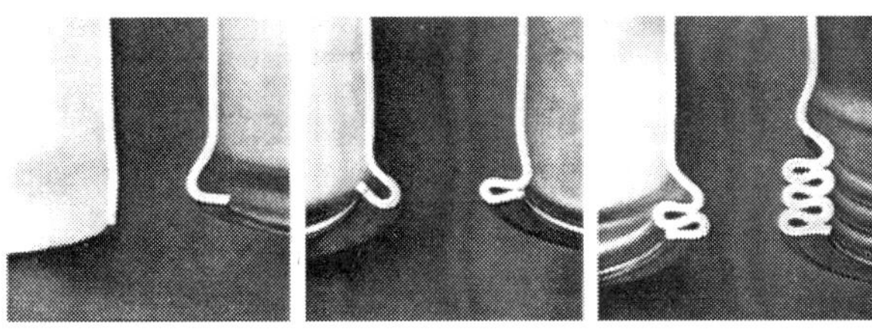

Fig 50: Folding sequence detail of circular tube

Cross Section (mm)	Alloy	Remark	Energy Absorption Capacity (kJ/kg)			
			Aged Artific.	Soft Ann.	Heat Treat.	Opt. Heat Treat.
Circular 60/1.5	AA 6060	–	34**	18	33	43
Circular 100/1	AA 6016	welded	–	11	26	29*
Square 55/2	AA 6060	–	26*	13	–	–
Rectang. 45/25/2	AA 6060	–	38*	18	–	–
Zigzag Stiffened	AA 6063	1)	6**	4	7	–

* Local fractures ** Totally failed 1) Loaded in transverse direction, see Fig. 7

Fig 51.: Energy absorption at different heat treatments

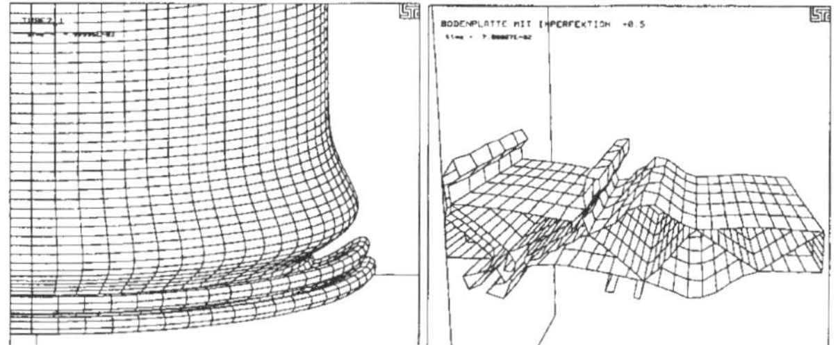

Fig 52: FEA simulation of circular and zig-zag stiffened rectangular tubes

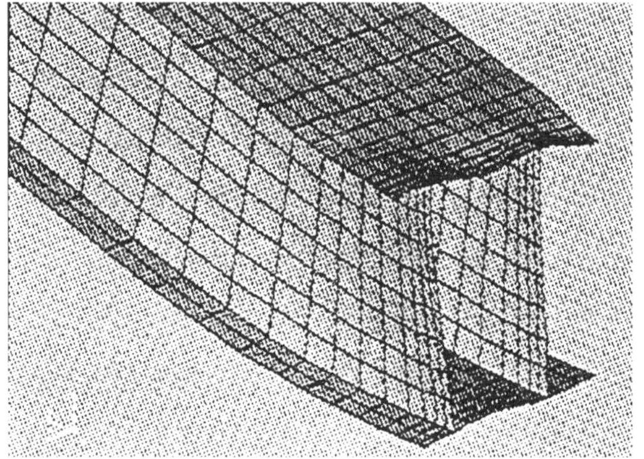

Fig 53: FEA simulation of bending deformation

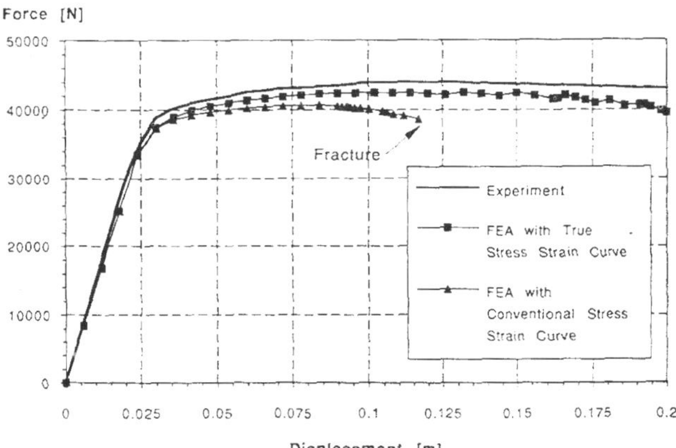

Fig 54: Experimental vs FEA results of three-point bend test on side collision beam

Tailoring deformation and cabin-shell stiffness

In the trend to ultra light weight vehicles the structural behaviour of both deformable front-end, and the safety cage of the cabin shell, both need separate attention, prior to combining their behaviours in an analysis of the complete structure.

A limitation of current US Federal test criteria (FMVSS 208) for car crash testing is that there is no requirement for compatibility in collisions between cars of different masses, particularly in head-on impacts. Fuel-efficient low-mass vehicles (LMV) can therefore look bad in crash injury statistics unless specific efforts are made to design-in protection against heavier cars. Researchers from the Working Group on Accident Mechanics at the Universities Zurich[7] have shown that well-defined properties in the crush zone, used in conjunction with appropriate occupant-restraints, can make ultra-light cars sufficiently crashworthy.

Because of the compact packaging of LMVs, available crush space is shorter than average and it is necessary to optimize the collapse-structure so that it generates the near ideal rectangular shape crash-pulse (deceleration x time). Tendency for all crash legislation to be framed around barrier impact tests in which the car is effectively crashing against a mirror image of itself means that a light car would have a barrier impact pulse inappropriate to median weight car with which it would be statistically likely to impact in the real world. Mean barrier decelerations of average cars are around 20 g while the values for heavy cars, by comparison, are about 15 g. LMV values are typically 25 g and relatively 'soft' crush

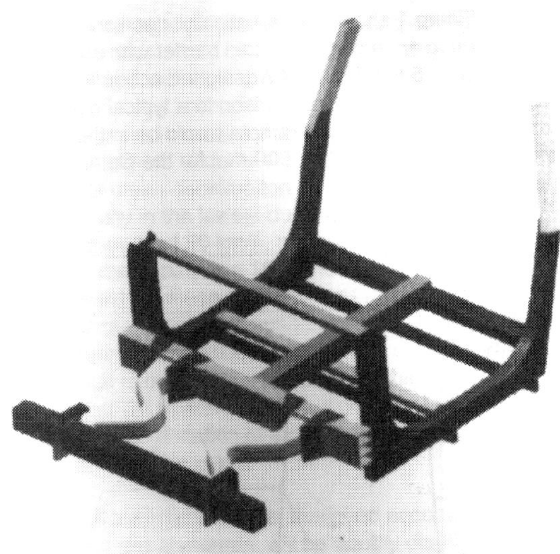

Fig 56: Experimental LMV front-end structure

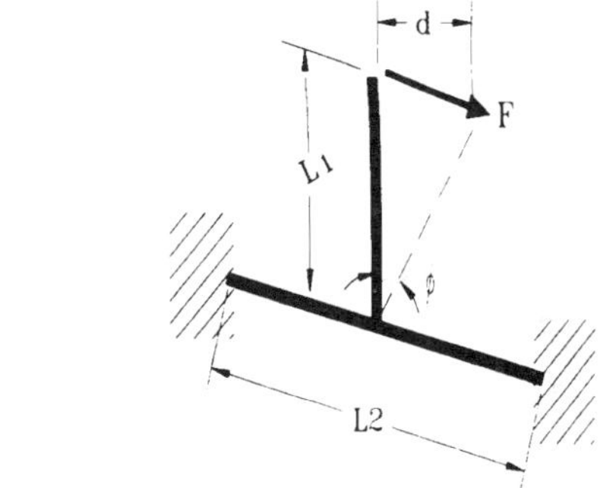

Fig 59: Joint boundary condition

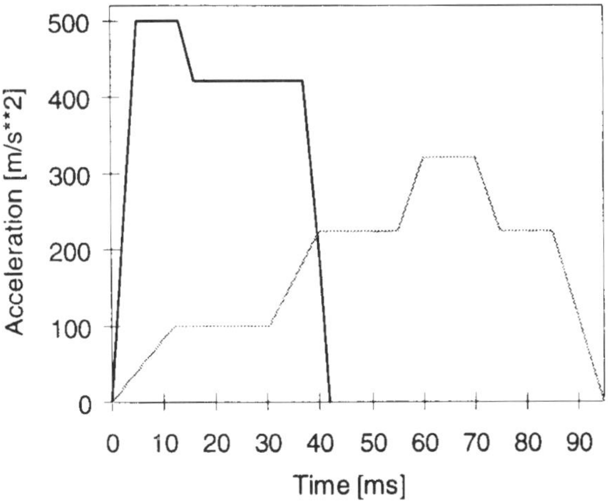

Fig 55a: Impact pulses for LMV and compact cars

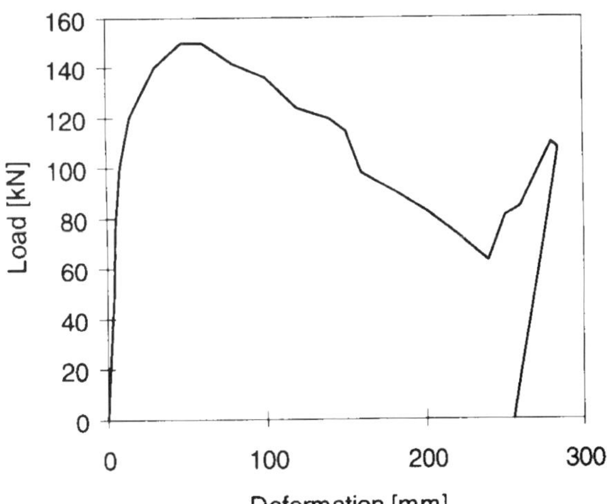

Fig 55b: Static crush test for one deformation element

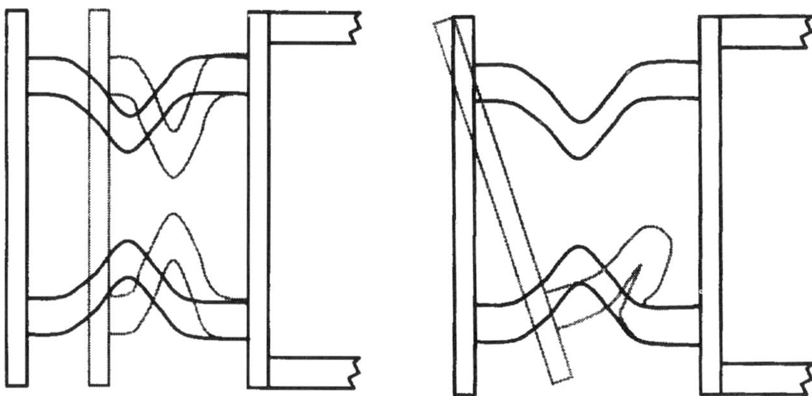

Fig 57: Deformable front-end structure under 100 and 50 % overlap collisions

structures are typical. In fact, for impacting the heavier median-weight car, which also imparts greater momentum energy on impact, slightly higher than normal crush stiffness is required of the LMV.

A critical parameter for the design of the occupant restraint is the mean impact-deceleration level, which for an LMV, the authors argue, should not exceed 40-50 g. Fig 55a shows barrier-impact deceleration pulses for an LMV, higher curve, impacting at 15.5 m/s, and a typical compact car, the lower curve. The deformation would be 300 mm for the LMV against 800 mm for the compact. With such a pulse for the LMV additional 'ride-down' space within the survival space of the car interior. The restraint system should be designed to become operative as soon as possible and its components ideally deform at a constant load level.

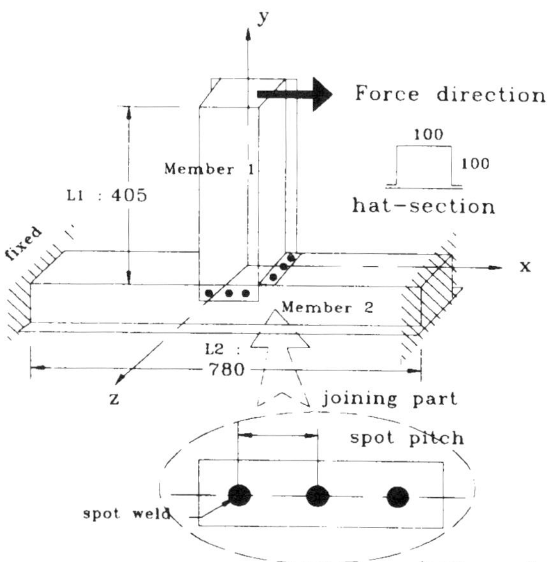

Fig 58: Geometry and dimensions of T-joint

An ideal design of a deformable front-end structure for an LMV, the authors maintain, is to distribute all its energy absorption to its structural elements even in a 50 per cent overlap situation. A stiff transverse beam just behind the front bumper, over the full width of the LMV, helps to react either point or distributed impact loads and can react most overlap situations. In the authors' experimental LMV the beam is mounted at 440 mm bumper height and is 120 mm deep; the beam transfers loads on to curved longitudinal beams as seen in Fig 55b. The extra stiff cabin structure does not deform until the deformation elements are completely crushed, the interface being another stout transverse beam.

The deformation elements are rectangular sectioned thin-wall beams forming a stiff-joined frame with the two transverse beams, designed to transfer longitudinal as well as transverse loads. Both elements absorb energy by plastic bending and folding of the tube cross-section at three locations each, in 100 per cent overlap, but in only four locations in 50% overlap collisions, Fig 56. A static crush pulse is shown in Fig 57 for one deformation element, set up to correspond to 100 per cent overlap and the dynamic deformation curve was estimated to be 10 per cent higher than the static curve.

Cabin shell joint stiffness

Researchers from Kia Motors[8] have carefully studied the contribution of joint design in the body structure for aluminium-shelled vehicles, to show the benefits to overall structural efficiency of joint reinforcement, vis-a-vis steel construction. With stiffness of

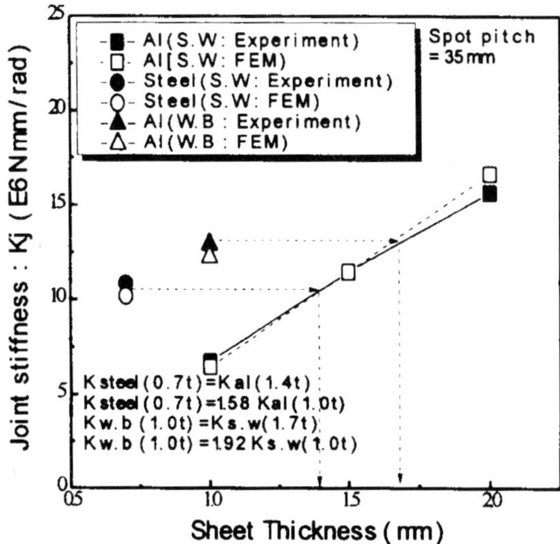

Fig 60: Effect of sheet thickness on T-joint stiffness

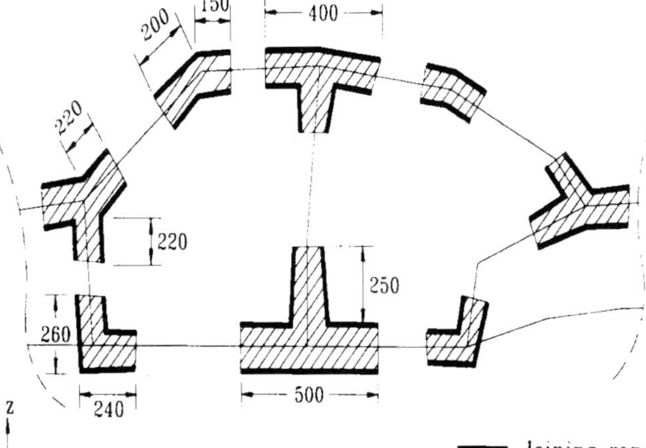

Fig 61: Representation of joints in BIW shell

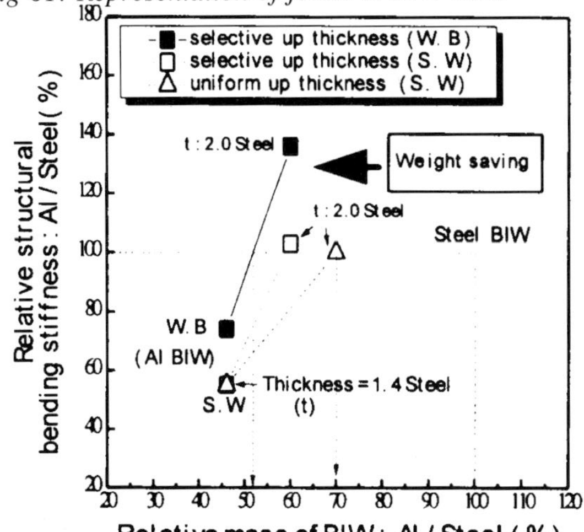

Fig 62: Effect of sheet thickness on BIW shell

sheet material proportional to the cube of thickness, and directly proportional to modulus, equivalent steel to aluminium stiffness was achieved with chosen aluminium thickness 1.4 times thicker than steel (0.7 and 1.0 mm), resulting in about half the weight. Sheet specimens were used to obtain parameters from simple joint configurations and load cases such as single-lap with shear, direct and fatigue loading. More complex T-, Y- and L-shaped vehicle structural joints were then considered and the tee-joint (sills to B-posts) was particularly focused as requiring higher stiffness and strength than the others. The joint was then analysed using both FE simulation and practical testing for different sheet thicknesses and spot-pitches. Geometry, dimensions and boundary conditions are seen in Figs 58 and 59. The authors described the overall joint stiffness as a combination of in-plane bending stiffness, out-of-plane bending stiffness and torsional stiffness, or the in-plane bending case, the stiffness of the three elements were added reciprocally:

$1/k_t = 1/k_j + 1/k_1 + 1/k_2$ where global stiffness

$k_t = FL_f/f$ and rotational stiffness of member 1

$k_1 = EI/L_1$ and rotational stiffness of member 2

$k_2 = 8EI/L_2$, the suffix j implies the joining part

Effect of spot pitch was found to have little effect on the results but sheet thickness effect was as shown in Fig 60 for the in-plane bending case, spot-welded (SW) and weld bonded (WB).

Next stage was to examine real rather than idealised vehicle body joints, the analysis showing local reinforcement of sheet as the most rewarding method of increasing joint stiffness. An effort to determine the effect of joint stiffness on the overall body structure was based on the representation seen in Fig 61. Firstly FE analysis was performed with thickness of aluminium parts increased uniformly up to 1.4 times those of steel, the result indicating weak joints at sill to B-pillar, A-pillar to roof rail and sill to A-pillar. After subsequent correction of sheet thickness on these joints the result obtained is seen in Fig 62, comparing the steel and aluminium shells. The weight/stiffness trade-off advantage for weld-bonded joints is particularly clear.

Case study: crashworthiness of tanker shells

In applying crashworthiness theory to tankers, assessing the several failure modes and crash scenarios, involved in tanker accidents lends itself to computer simulation as real-life tests would prove to costly. This is the conclusion of AR Duffin[9] of Frazer-Nash Consultancy in work on the impact resistance of road tankers based carried out for the Health and Safety Executive.

The finite element code DYNA3D was shown to be used for simulation of four scenarios, rear-impact, side-impact, roll on to side and rolled tank sliding impact with kerb. The models used in the simulations were arranged to investigate the effect of changing material type, tank thickness, baffle spacing and side protection within the accident scenarios depicted in Fig 63. Failure criterion was onset of rupture of the tank. The effect of side protection in the form of lengths of motorway guard rail attached to the tank sides was also examined.

Materials and thicknesses employed for the different cases are listed in the table of Fig 64. The representative tanker was a MAX-section design, of 2 + 3 axle articulated configuration with a trailer length of 11.4 metres. Flat tank baffles were assumed and a full liquid load with 58. Duffin, A, ISATA, 1993 ullage. Fig 65 shows the finite-element mesh, the 5 mm thick aluminium alloy being assumed to have yield stress of 190 MN/m^2, UTS of 250 MN/m^2 and plastic strain at failure of 58. Duffin, A, ISATA, 1993 . The FE model was refined to suit each impact case by increasing density of the struck area. Rupture was considered to have occurred when plastic strain at failure was exceeded in over four elements in the mesh. The kinetic energy of the impacting body (the tanker itself in the roll-over cases) at the point of failure was compared with the initial value to obtain an index of resistance. Five cases examined in detail are shown in the table of Fig 66. In Case 1 only part of the trailer length was modelled (the two rear compartments) the remainder of the trailer and load being represented by distributed masses at the non-struck end. The resulting greatest depth of the deformed profile, due to the impacting corner of the ISO container, was 110 mm. The container was taken to weigh 38 tonnes and impact at a speed of 15 metres/sec assumed. This velocity was found to have dropped by 0.012 m/s (0.15 % drop in KE) and indicated serious rupture under this localised high energy impact. For Case 2, the results of the simulation were presented in the form of von-Mises stress plots on the deformed profile as seen in Fig 67. Greatest stresses can be detected in the area of folding which occurs in the upper part of the tank. Location of initial failure was predicted to be at the fold closest to impact. Again the impacting container retained most of its kinetic energy (99.3%), the velocity dropping by only 0.05 m/s and maximum depth of deformed profile was 0.35 metres. With Case 3 the roadside object was assumed to be a 150 mm high

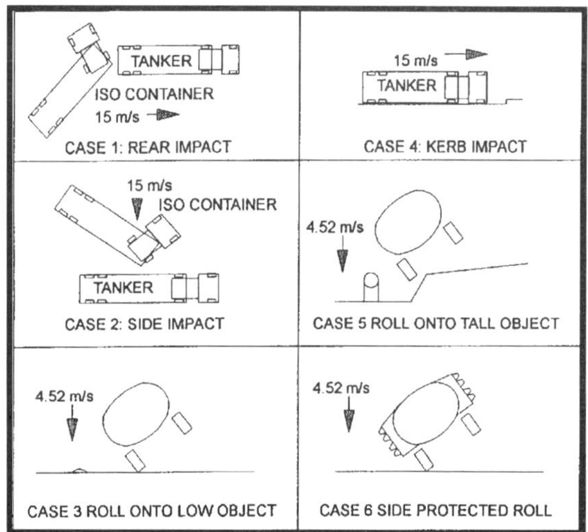

Fig 63: Accident scenarios

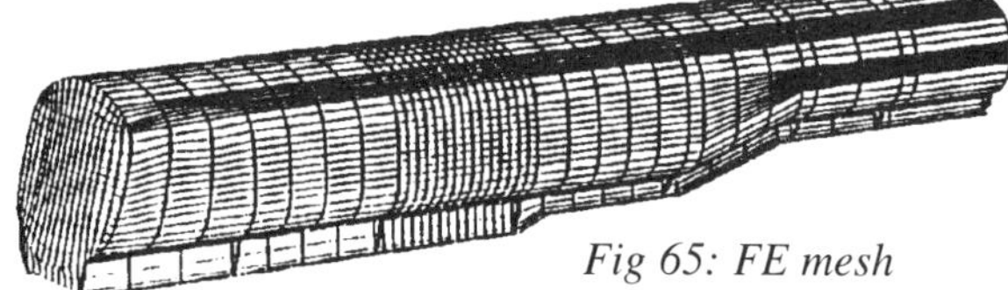

Fig 65: FE mesh

Case No.	Impact Scenario	Initial Velocity	Final KE	Failed Y/N
1	Rear impact	15 m/s	99.9%	Y
2	Side impact	15 m/s	99.3%	Y
3	Roll onto low object	4.52 m/s	N/A[1]	Y
4	Rolled, sliding into kerb	15 m/s	99.8%	Y
5	Roll onto tall object	4.52 m/s	88%[2]	Y

Fig 66: Cases examined in detail

spherical hump and the tanker rolled on to it at a velocity of 4.52 m/s. In this case failure did not occur at the primary impact location which just wrapped around the object until the rest of the tank struck the ground. At that point baffles began to fold.

For Case 4 a rolled tanker was assumed to be sliding at 45 degrees into a 150 mm high kerb to impact, without rotation, at 15 m/s. A greatly refined mesh was required at the highly localised impact area of the tank top corner striking the top corner of the kerb. Failure occurred soon after impact with little tank distortion and negligible change in tanker velocity. Case 5 differentiated from Case 3 in that tank failure was caused at the local impact site prior to the tank main body hitting the ground. The object has height of 1.4 m and diameter of 0.4 m to simulate this condition. Fig 68 shows the considerable tank side deformation which resulted in 0.28 m/s velocity change, involving 88% of KE remaining in the trailer. The Case 5 simulation was rerun to test the

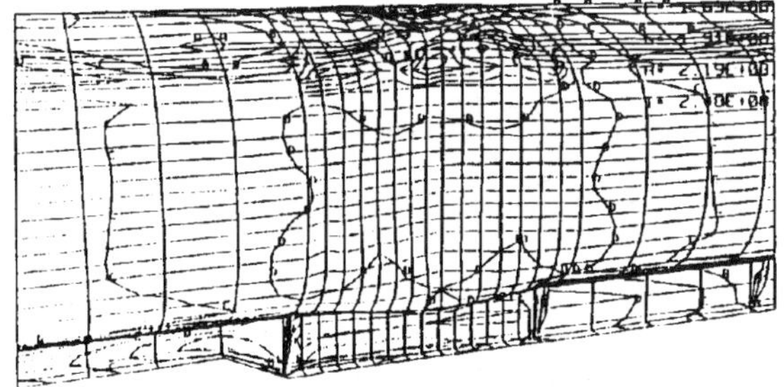

Fig 67: Case 2 impact

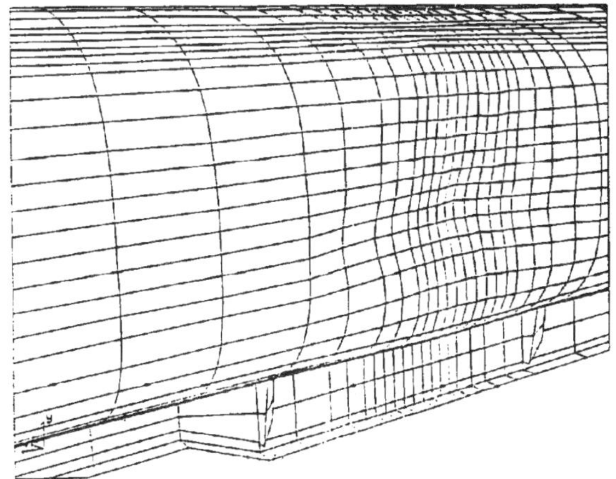

Fig 68: Case 5 impact

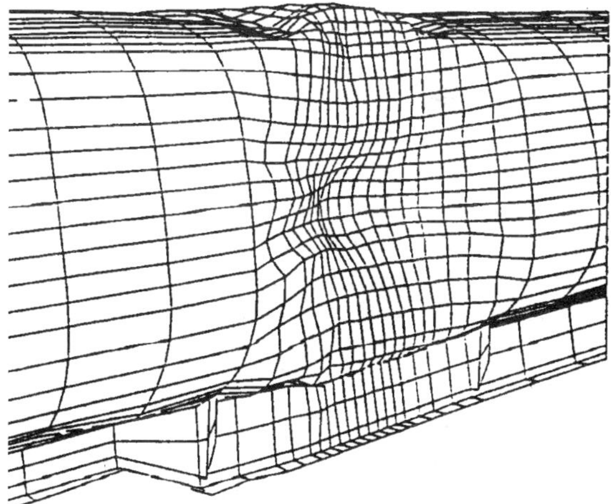

Fig 69: Case 5 repeated with parametric changes

Case Number	Impact Scenario	Tank Material[1]	Tank Thickness
Case 1	Rear impact by ISO container	Al	5 mm
Case 2.i	Side impact by ISO container	Al	5 mm
Case 2.ii	Side impact by ISO container. Tanker with side protection	Al	5 mm
Case 3	Roll onto low roadside object & ground	Al	5 mm
Case 4	Tanker rolled & sliding into kerb	Al	5 mm
Case 5.i	Roll onto tall roadside object	Al	5 mm
Case 5.ii	"	Al	7.5 mm
Case 5.iii	"	St	3 mm
Case 5.iv	"	St	5 mm
Case 5.v	"	St	10 mm
Case 5.vi	"	SSt	2.5 mm
Case 5.vii	Tanker with narrow baffles rolling onto tall object	St	3 mm
Case 6	Tanker with side protection roll onto flat ground	Al	5 mm

Fig 64: Materials and thicknesses used

effect of parametric changes, the results being tabulated as Fig 69. Because low ductility of the baseline aluminium alloy leads to low energy absorption, mild steel to BS 1501-225 490 and stainless steel to 316 S62 were examined for comparison, the increased reduction in the kinetic energy being seen in the table. With the stainless steel model wall thickness could be halved and Fig 70 shows the considerably greater deformation available to absorb impact energy. A modified mesh was used to model the effect of additional side impact protection but this showed little value against localised high energy impacts, the sideguards merely flattening and bending. The mesh was further modified and refined to test sideguard effectiveness in pure rollover.

Case No	Material[1]	Thickness (mm)	Failed (Y/N)	Velocity Change[2] (m/s)	Remaining Kinetic Energy
5.i	Al	5	Y	0.28	88 %
5.ii	Al	7.5	Y	0.39	83 %
5.iii	St	3	Y	1.24	53 %
5.iv	St	5	Y	2.52	20 %
5.v	St	10	N	4.52	0
5.vi	SSt	2.5	N	4.52	0
5.vii[3]	St	3	Y	0.77	69 %

Fig 70: Effects of parametric changes

Chapter 9: Occupant protection/restraint

Occupant and cargo restraints; passive restraint systems; responding to occupant protection legislation; bumper systems; energy-absorbent-foam analysis and systems; new-departure impact-absorbing systems and analysis; integrating absorbers into trim; design for pedestrian protection; structural design for seat mounts

Occupant and cargo restraints

The physical restraint of occupants and cargoes on road vehicles is vital in the accident situation and, in the case of cargoes, is increasingly important in normal running as traffic densities increase. It is a case where the 'village-blacksmith' approach to vehicle design of yesteryear has less and less relevance.

Hunch assumptions about inertia loads on passengers and cargoes during violent braking nearly always tend to be wrong and the efforts spent in a few simple calculations will always be rewarded. Many times have truck operators forgotten the huge inertia effects of a moving load and have made the assumption that the the heavy static weight of a payload can keep it in place on the open deck of a lorry-body. Equally remiss was the assumption, before the age of safety belts, that it was only necessary for an occupant to brace him/herself against a car bulkhead for protection in head-on collision.

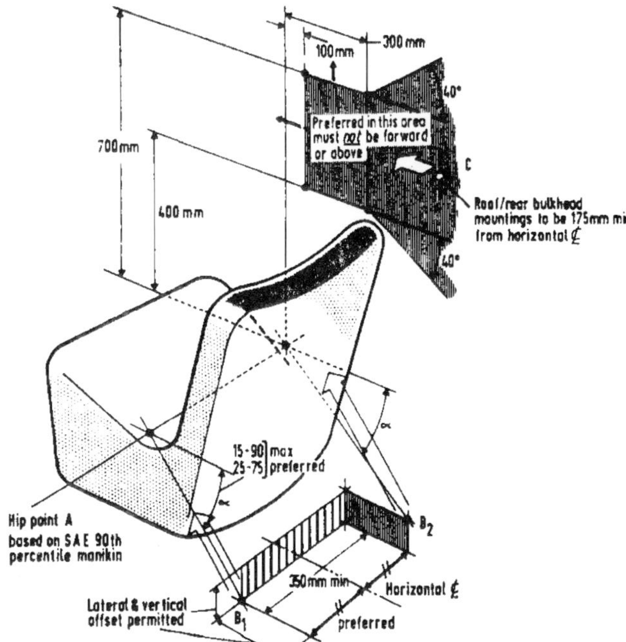

Fig 1: BS seat belt anchorage points

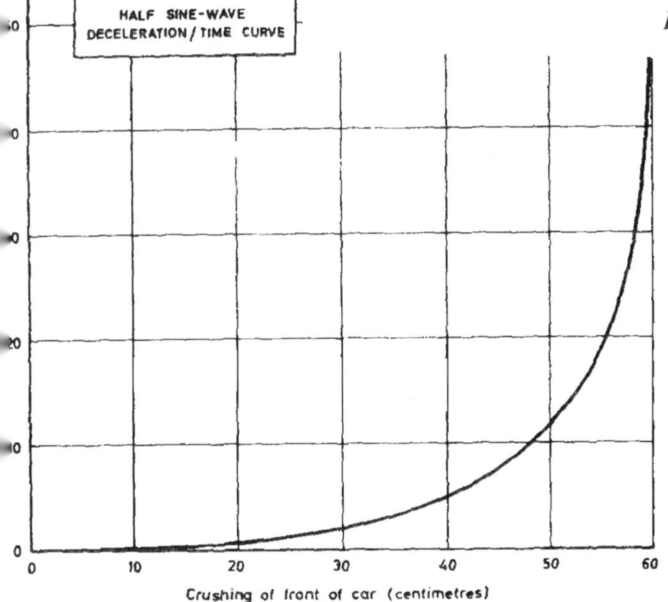

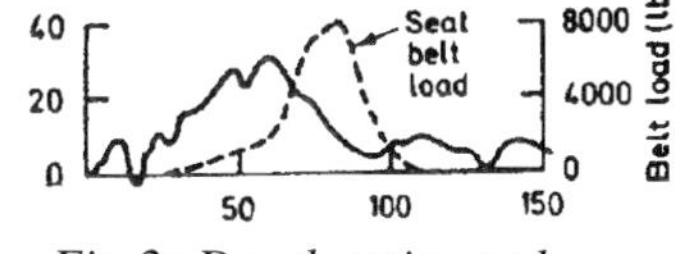

Fig 2: Deceleration pulse

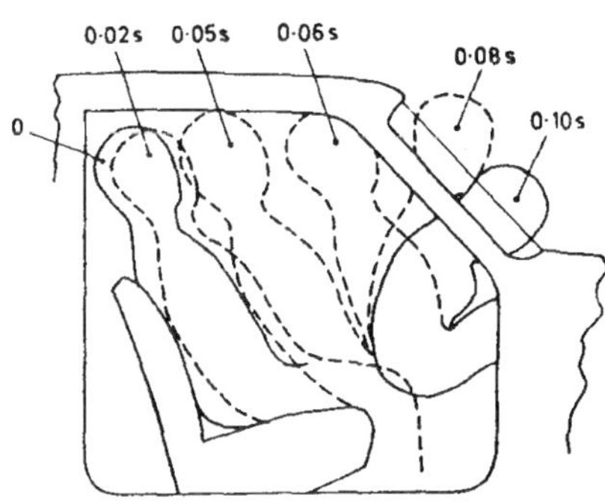

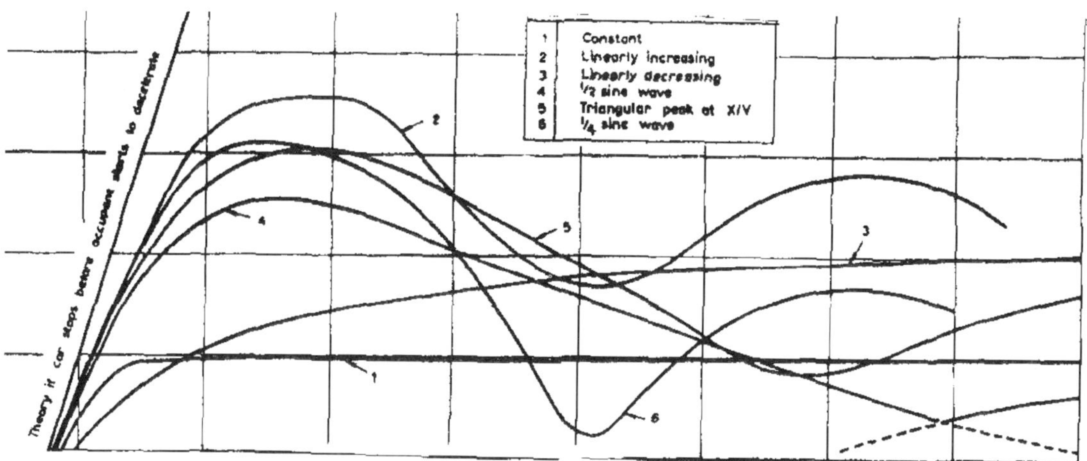

Fig 5: Belt stretch for different crash pulses

Even at modest town speeds of 30 mph, striking a rigid object such as a concrete lamp standard will cause the car to be brought to rest in a little more than one tenth of a second over a front-end crush distance of perhaps two feet. An unrestrained occupant will strike the car interior, in secondary impact, at virtually the initial speed of the car — with minimal protection even given from a padded dash.

The effect of the seat belt is not just to restrain the occupant away from projections which can imposed high localised forces but also to reduce the forces which act on the whole body. Stretching of the belt in impact allows the body to move forward about eight inches and because the body is tied to the car some of the crushing distance is added to the stretch of the belt in what is known as a 'ride-down' effect. A resulting stopping distance of some 12 inches for the body could typically be assumed in the 30 mph 'rigid' impact situation. Belts have, of course, to comply with standards applying to the country of operation. In the UK British Standards define the belt

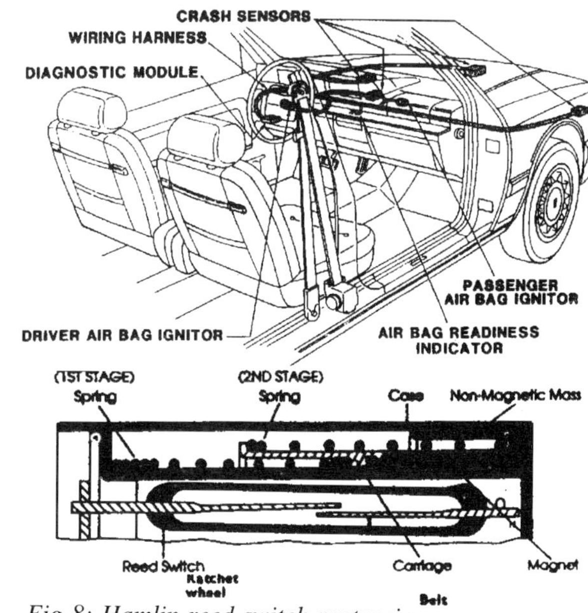

Fig 8: Hamlin reed-switch pretensioner

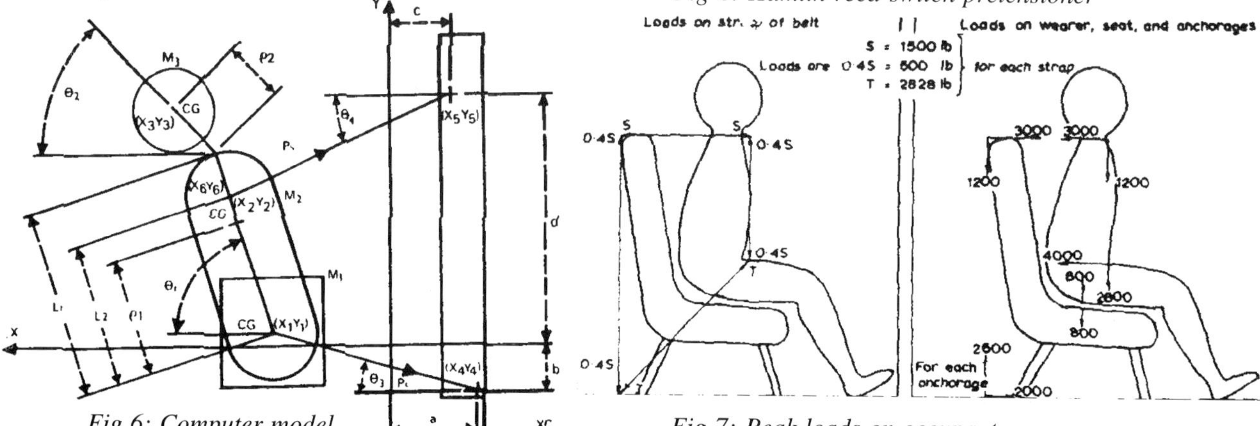

Fig 6: Computer model

Fig 7: Peak loads on occupants

and its attachments as well as prescribing positions of the anchorage points as shown in Fig 1.

The three-point belt with retractor reel has become almost the accepted norm. The retractor mechanism is such that when a certain vehicle deceleration value is reached a fast-response interlock inhibits the webbing roller. Extended protection can be obtained using belt pre-tensioner devices and even air bags sensitive to different levels of deceleration. Very quick responses are necessary for successful operation and typically a bag must be inflated by the time the occupant has moved forward eight inches — say, 30 milliseconds. The pre-tensioner would need to be fired after about 10 milliseconds. Sensors may be electro-magnetic or electronic.

Pioneering work by the Transport and Road Research Laboratory was important in setting the standards for passenger restraint. Fig 2 shows measurements taken of head-on car impacts which included traces of seat belt loading during the accident situation. The 0.1 second or so duration of impact is sufficient to use the energy absorption of the structures to the benefit of occupant survival. An earlier section suggested approximations for the shape of crash impact pulses. A deceleration time curve of a half-sine wave is found to be a reasonable approximation in calculating the movements of a restrained occupant in an impact situation.

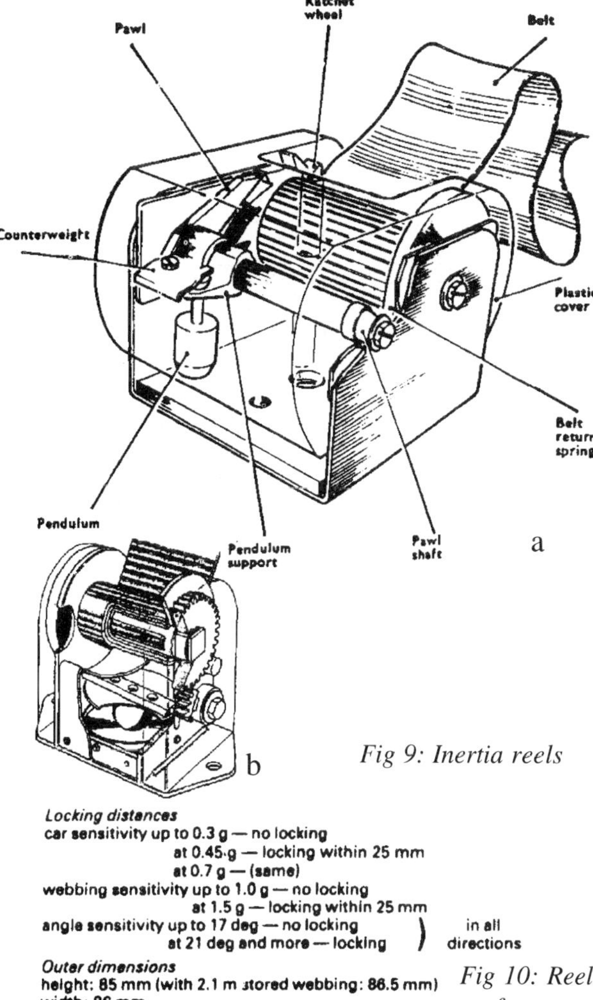

Fig 9: Inertia reels

Locking distances
car sensitivity up to 0.3 g — no locking
at 0.45 g — locking within 25 mm
at 0.7 g — (same)
webbing sensitivity up to 1.0 g — no locking
at 1.5 g — locking within 25 mm
angle sensitivity up to 17 deg — no locking
at 21 deg and more — locking } in all directions

Outer dimensions
height: 85 mm (with 2.1 m stored webbing: 86.5 mm)
width: 86 mm
depth: 62mm (with 2.1 m stored webbing: 67 mm)
Webbing capacity 2.1 m (webbing thickness 1.2 mm)

Fig 10: Reel performance data

Fig 3 shows calculation results for the motion of an unrestrained dummy. The relation of occupant movement to front end crushing of the vehicle is shown in Fig 4 based on such calculations. In the case of a seat belt with 1 in (25.4 mm) slack, at this amount of forward movement of the occupant, the car has crushed by about 13 in (330 mm) and 11 in (279.4 mm) remains — which is added to 14 in (356 mm) stretching of the seat belt. Body deceleration was found to be about 17 g — only about one quarter of that imposed by even a well-designed padded instrument panel. Even a collapsible steering column on its own was found to impose an acceleration of some 35 g after 10 in (254 mm) deformation.

There is always scope for improved design of the vehicle interior, with reference to the projected movement of the occupants on impact, to allow the belt as much stretch as possible. There is also scope for retractor mechanism designs which would result in a constant belt force during impact rather than a rising one. At the time of the research it was suggested that such work could result in belt protection for occupants with impact speed raised from a limit of 35 to 50 mph. The calculations were based on assumed occupant deceleration of 30 g and a 17 inch survival space in front of the occupant — together with a vehicle front-end crush distance of 40 inches. There is also great scope for improved comfort and convenience for the vehicle occupant and recent trends by such makers as Mercedes Benz, to incorporate belt anchorages into the seat structure, for open-roadster applications, could also have important general vehicle usage spin-offs.

The TRRL carried out work to establish whether belt systems designed to BS 3254 could be improved upon. By simplifying assumptions to make vehicle crush and belt stretch stiffnesses both linear a convenient analysis can be carried out. Such was the basis for the characteristics obtained in Fig 5 for different assumed impact pulses shown. Fig 6 shows the kind of computer model used to predict impact levels between occupant and vehicle interior when the survival space is insufficient or to demonstrate the reduced impact loads arising from the use of the seat belt. Subsequent analysis has been carried out which allow for the non-linearities in the various stiffnesses. Fig 7 shows peak loads on occupants calculated on the basis of a 50 g occupant deceleration impact. A recent example of a system introduced for pre-tensioning of seat belts or actuating an air bag uses the Hamlin 59500 series sensor based on a reed-switch, Fig 8. It incorporates a coil spring and torroidal shaped ferrite core. Under a shock force, its magnet deflects against the spring until positioned over the reed switch. Magnetic force then actuates the switch to deliver a signal when a shock force of a particular magnitude is reached. Operate time and threshold sensitivity can be preset. The development of inertia reels is seen in the examples of Fig 9. At (a) is an early Kangol system showing the separate belt and vehicle inertia mechanisms acting in the webbing shaft cog. A more recent design is the Britax Kolb system seen at (b) which has spindle locking built into both sides of the frame yet uses the minimum number of moving parts, to improve reliability. Performance data is summarised in Fig 10.

Recent design of air-bag restraint is exemplified by the Honda system of Fig 11. This is of the so-called progressive-power type and comprises four sensors to detect impact in frontal collision; a compact propellent-gas generator in the steering wheel hub with the folded air-bag. There is also a subsidiary sensor, Fig 12, in the passenger compartment which fires at 2 g rather than 12 g for the frontal sensors. Relative settings are based on the crush resistance of the vehicle to allow for time delay between front-end and interior response to impact. The inflater comprises an igniter (zirconium potasium nitrate plus a heater) and sodium nitride to generate hydrogen. The 60-litre air bag is of nylon cloth with a rubber-coated inside and it inflates in 0.2 seconds.

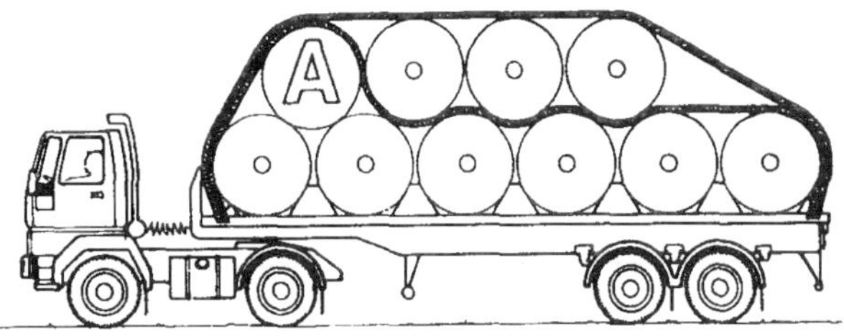

Fig 15: Roll retention

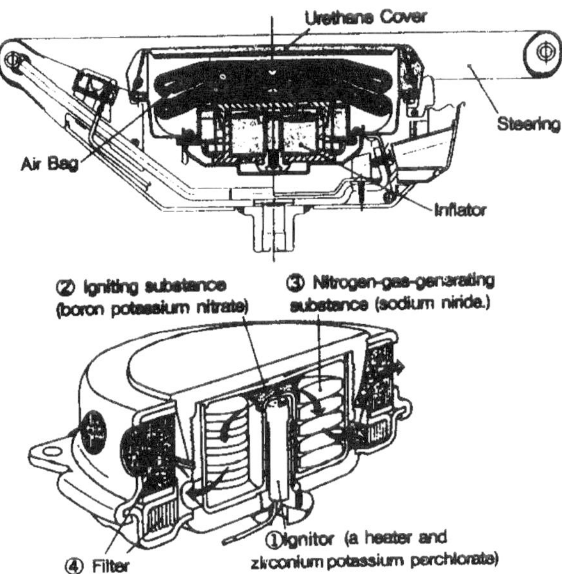

Fig 12: Sensors for air-bag firing

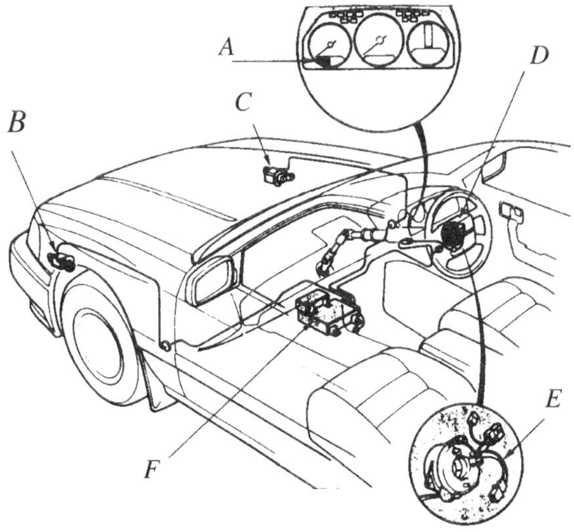

Fig 11: Honda air-bag restraint — A. warning light, B. front service (left), C. front service (right), D. air bag inflator, E. cable unit, F. diagnostics unit

Cargo restraints

The high masses associated with many goods cargoes result in very high inertia loads when sudden braking takes place and in special-purpose vehicles particularly, with greater design freedom to match the vehicle to the payload, there is considerable scope to make loads totally secure, even in the collision situation. The importance of lateral acceleration on turns and even horizontal acceleration in sudden 'take-offs' should also be accounted for. Any shifting of the load, en route, can also create serious stability problems due to the alteration of inter-axle load distribution. In the UK alone there are said to be 4500 successful prosecutions for unsafe loads per year — a figure that might rise to 10 000 when undetected cases are included.

The scale of the problem is such that legislation could well force the design of special purpose vehicles for such potentially dangerous loads as paper rolls and even steel girders. Specialist coil carriers are already widely used for heavier rolls of steel strip and horrifying reports have been heard of items like steel bar piercing the back of driver's cabs on open platform lorry-bodies, under emergency decelerations.

Construction and Use Regulations are clear in the legal requirement to ensure safety and stability of payloads. One of its clauses insists on suitability of use for purpose. While the interpretation of this is clear for such loads as liquids and bulk 'solids', it is less obvious in the case of awkwardly shaped loads — other than abnormally large ones requiring transport under Special Types orders.

Of considerable help is the Department of Transport publication 'Safety of loads on vehicles' which provides a code of practice for construction requirements of load retention equipment. The requirement is to withstand inertia forces equal to the load weight, forwards and to half the load weight sideways and backwards. Reliance on the headboard of an open-platform truck may not be sufficient and it is extremely difficult to have a standard design to meet a wide variety of payload requirements.

The code recommends varying the number of anchorage points for different types of payload. Some guidance is given in Fig 13 for reasonably symmetrically shaped cargoes according to overall dimension. Useful advice is also given on sheeting, ropes, chains, strapping and nets for general cargoes. More specific practices apply to rolls and coils (Figs 14 and 15), sacks and bricks.

A specific British Standard (BS 5073) covers the special case of loads inside containers while the Health and Safety Commision have published codes, available from HMSO, for dangerous substances. While many of these are specifically operator requirements, some fall squarely in the court of the specialist vehicle designer. An important case in point are low-loading machinery transporters. The particular case of load restraint in light vans has been covered in MIRA research reports made public in recent years. These include a code of practice covering bulkheads, load anchorages, storage bins and restraint systems.

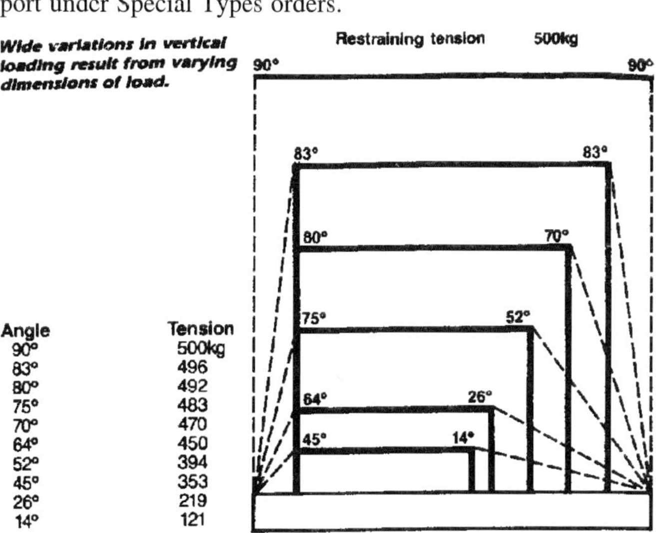

Fig 13: Anchor points for cargo restrait

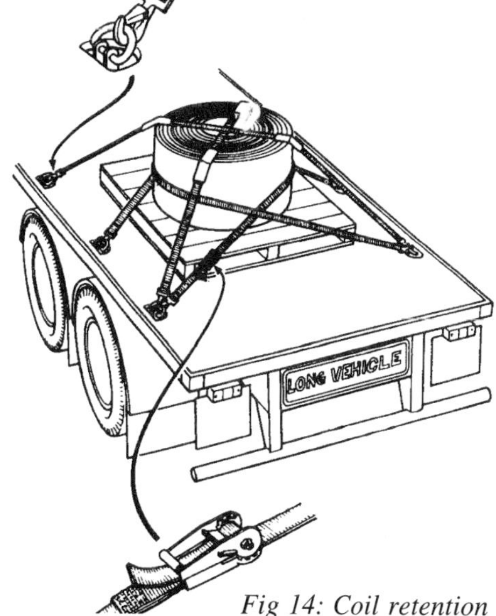

Fig 14: Coil retention

Passive restraint systems

Long experience with passive restraint systems for vehicle occupants in the USA has led automotive engineers to be wary of public reaction to a variety of restraint systems which either cause inconvenience or add considerable extra cost to the vehicle. Legislation is likely to increase for all categories of vehicle as the years roll by so it is prudent to examine the rival systems in some detail.

The requirement for passive restraint systems on 1987 model-year cars in the USA stimulated the considerable development of different solutions and brought into focus the rivalry between air-bags and passive seat-belt systems. At the time of the impending legislation, GM in the USA examined a variety of seat-belt systems which were 'passive' in the sense of not requiring a conscious effort by the user when entering or leaving his or her vehicle. These are described below after consideration of different air-bag solutions.

A representative air-bag system is that developed by Honda and known as the SRS in which a sensor unit is attached to each side of the body front-end. These comprise a roll-spring, roller and base, enclosed in a sealed casing, Fig 16. With impacts over 12 g (equivalent to a frontal- barrier impact of 16 km/h), the roller actuates the switch. Two cowl sensors, of similar design, are enclosed in a diagnostics unit on the tunnel structure between front-seat occupants. These sensors switch at 2 g typical of a deceleration of this part of the structure after the front-end has collapsed.

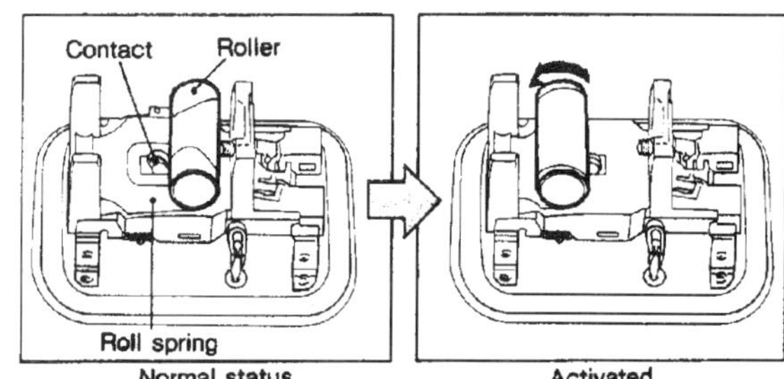

Fig 16: Honda SRS sensors

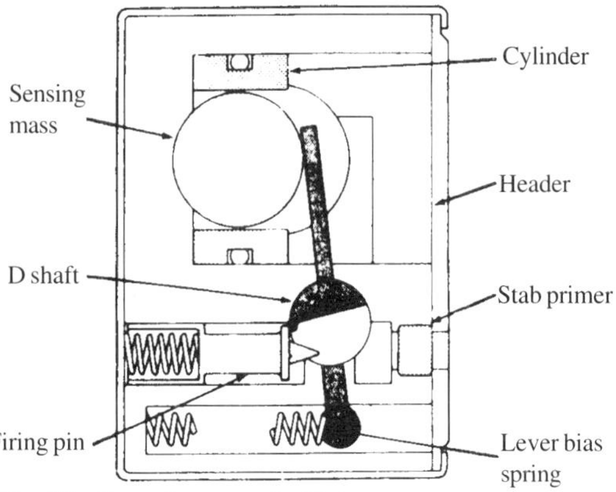

The steering-wheel hub inflator was introduced and illustrated in the previous section; it

Fig 17: Breed sensor/initiator

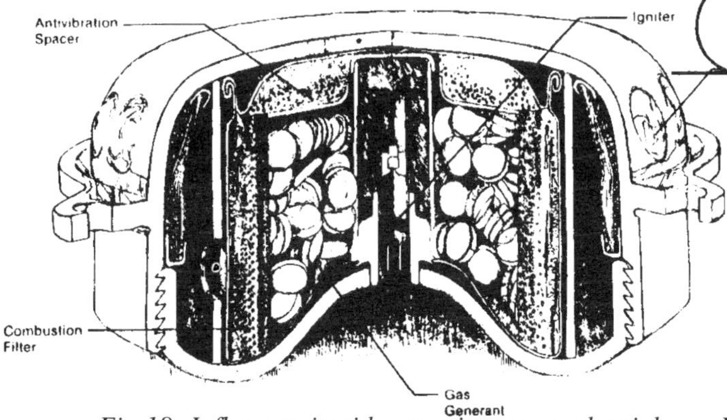

Fig 18: Inflator unit with gas exit ports on the right

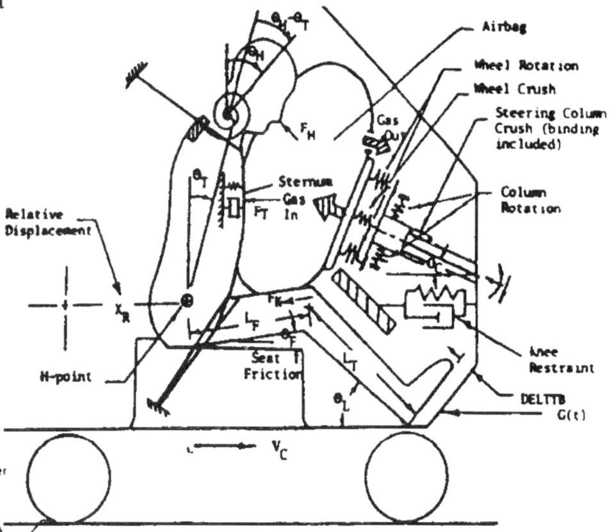

Fig 19: Computer model

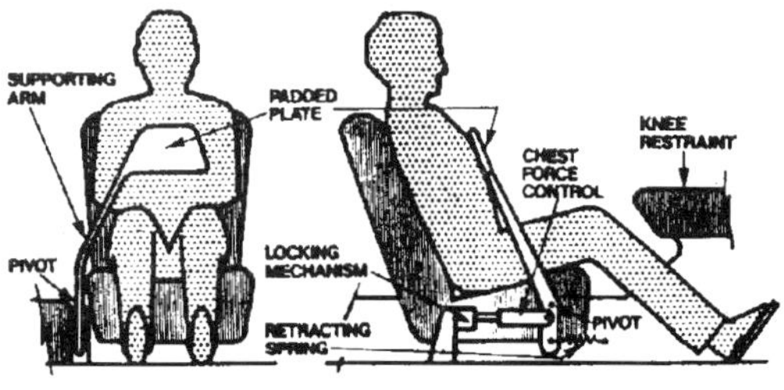

Fig 23 (below): Cushion restraint

consists of an igniter (heater and zirconium potassium perchlorate) and a nitrogen generator (sodium nitrate). When both front-end and cowl sensors record impact, the inflator activates and the chemicals 'burn' to inflate the air-bag. The bag is of nylon cloth with an inside rubber coating. The nitrogen fills the 60 litre bag, two vent holes releasing gas to prevent too high an impact on the driver's face.

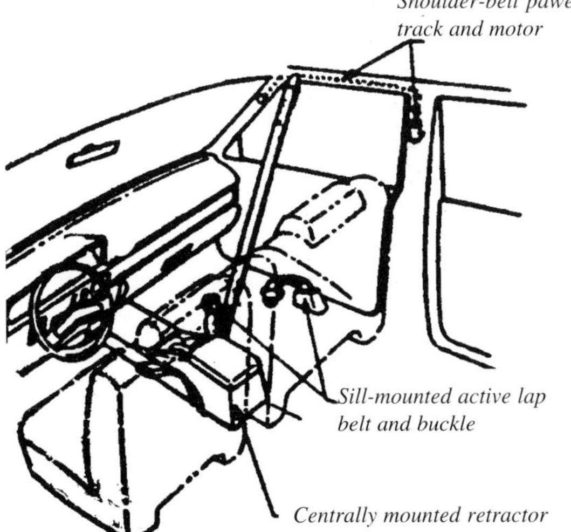

Fig 20: 2-point passive belt

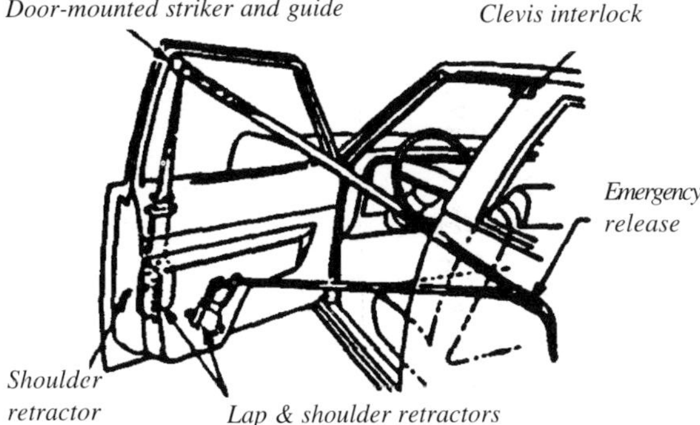

Fig 21: Door-mounted 2-point belt

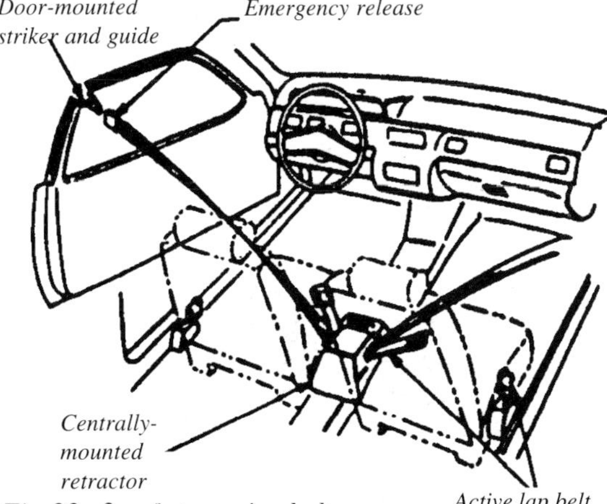

Fig 22: 3-point passive belt

Simplicity in design and operation is obviously the key to guaranteed working of air-bag systems. The equipment is only effective if the timing of the bag inflation is appropriate to the events of the very fast impact cycle which occurs in collisions. This has led to the use of computer simulation of both occupant movement and dynamic structural deformation of the vehicle in a wide variety of accident situations so that the air-bag effectiveness can be optimised.

This procedure was used in the development of the US Breed Corporation air-bag system whose crash sensor initiator is shown as Fig 17. If a deceleration of over 4 g occurs the ball-sensing mass moves forwards against the spring. The ball will push the bias lever far enough to release a cocked firing pin to initiate the gas generating material for inflating the air-bag. Two such sensor-initiators are built in to each system to provide high statistical reliability. The system differs from conventional ones in that, at the time of its introduction, most of the latter used electrical switches in the crush zone of the body, with associated electrical-circuits and diagnostics, to signal inflation of the bag.

Pyrotechnic sensors for air-bags have been developed for US systems by Morton Thiokol Inc. Through a percussion mechanism a thin-layer explosive (TLX) transfer line is initiated to transfer a pyrotechnic shock wave to the inflator at a speed of some 4500 ft (1372 m)/sec. The sensor is mounted to the air-bag module housing next to the inflator, and on the opposite side is a cavity containing the energy storage capacitors. Power from the horn circuit is used to charge the capacitors and when the sensor responds to a crash impact it closes a set of electrical contacts to let current pass to the squib which fires the inflator, Fig 18.

Considerable debate still takes place about the choice of electronic or pyrotechnic firing, also about the best position for the sensor in the steering wheel. A number of companies have developed computer models which help in the selection of such design criteria. Fig 19 shows the Volvo VDRACR model which is a relatively simple two-dimensional representation using lumped masses, the air-bag being modelled as an ellipsoid.

Non air-bag solutions The body-mounted two-point passive seat-belt system, Fig 20, uses a single retractor mounted to the centre of the vehicle with the upper, shoulder, belt anchorage located on three upper frame or B-pillar. Knee bolsters are required with the system and the upper anchorage requires a means of powering itself along a guide track to facilitate entrance into the vehicle. The door-mounted two-point system, Fig 21, again uses a single retractor mounted on or near the central tunnel — and has an emergency release latch mounted in the door area. Knee bolsters are required, too, and the upper anchorage is door-mounted to clear the entrance area as the door is opened. A considerable challenge presents itself with both two-point systems in providing protection for a wide variety of occupant sizes.

The alternative is a three-point door-mounted system having two door-mounted retractors for both the shoulder and lap belts, Fig 22. This arrangement allows a third, central, seating position with just the two outer occupants having the passive belt facility. Knee restraints are usually not required and dynamic performance is similar to an 'active', manually fastened, belt system given proper location of the anchorages.

A third major category of passive restraint is the automatic cushion restraint comprising an arm hinged from the central tunnel or console. This supports a chest pad held by light spring pressure against the occupant and used in connection with a knee bolster. The pad allows free movement in normal travel but locks up on impact. Fig 23 shows an early system developed by Accles Britax in which an energy-absorbing collapsing tube device is used to control the movement of the arm and resulting chest loading when locked-up above 6 g deceleration.

An interesting derivative of this idea is the integrated child safety seat devised by Volkswagen which the company designed into one of its concept cars. The right hand rear seat of the Futura car has a leading cushion-edge which can be lifted upwards and positioned in front of the child to act as an adjustable safety restraint, Fig 24. This transverse bolster can be moved forward and backward and is unlocked with a lever underneath the seat cushion to spring back to its original position.

Occupant protection techniques responding to changing regulations

Body interior protection against injury in impact; side impact protection; steering column penetration and structural collapse in bus roll-over are all areas in which structural designers are finding new solutions to continually updated regulations. The year 2002 will see all cars sold in the USA having to comply with FMVSS 201 which covers occupant protection provided by upper vehicle trim such as on A-pillars, B-pillars, front, rear and side header rails, and the roof. Tests for compliance will involve use of a Hybrid 111 Free Motion Headform fired at designated target areas. MIRA engineers[1] have reproduced the target areas in Fig 25 and point out that the headform can be fired from any point in the vehicle providing it follows designated trajectories. The proving ground at Nuneaton now has the facilities for performing such tests.

Trim performance is measured through analysis of the resolved deceleration and consequent Head Injury Criterion (HIC) analysis. The following formula is used to analyse resolved deceleration:

$$\{[1/(t_2 - t_1)]\int_{t1}^{t2} a\,dt\}^{2.5}(t_2 - t_1)$$

where a is resultant acceleration in multiples of *g* and t_1 and t_2 are time points, during the impact, separated by not more than 36 milliseconds, Fig 26. Further analysis allows deformation characteristics to be

Fig 24: VW child seat

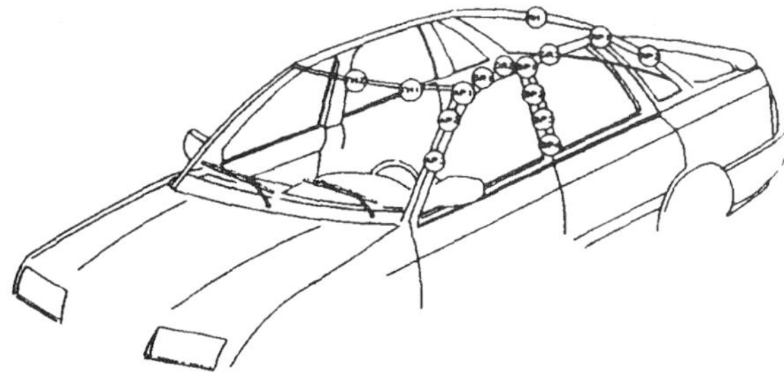

Fig 25: Impact locations — AP 1-3 A-pillar; SR 1-2 front side rail; SR3 other side rail; FH 1-2 front header; BP 1-4 B-pillar; RH rear header; RP 1-2 rearmost pillar; UR upper roof (not shown)

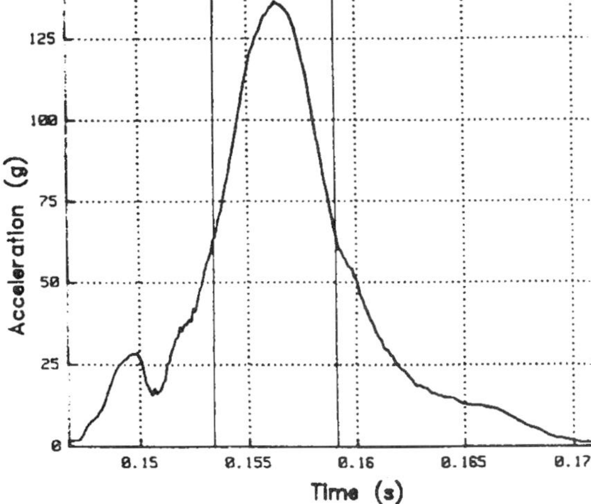

Fig 26: Typical resolved deceleration vs time traces for which HIC value = 674 (HIC[d] value 675); HIC calculation times t1 = 0.1535a, t2 = 0.1590a; CAC = 750g; CFC = 1000 Hz

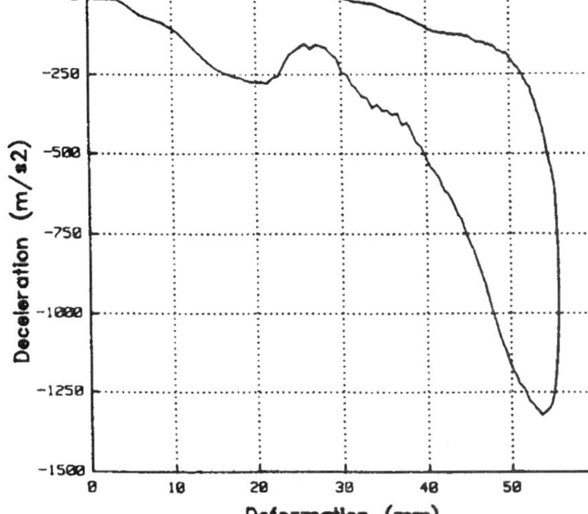

Fig 27: Deceleration vs displacement trace for which impact velocity is 6.80 m/s.

obtained, Fig 27 illustrating energy absorbed by analysing the deformation curve.

Preliminary testing at MIRA has shown A and B pillars to be particularly difficult areas for protection treatment in achieving the legislative HIC(d) of below 1000. Conventional trim materials such as PPE and ABS offer minimal energy absorption, the authors argue, particularly if a large air-gap separates them from the skin panels, as on the upper roof. Usually an appropriate padding layer is required in between. Shaped honeycombs are considered by the authors to be particularly valuable when applied to problem areas such as pillars.

Side-structure integrity in side impact Researchers at Engineering Technology Associates and co-workers at Kia Motors[2] have used FE analysis to study vehicle side structures in relation to FMVSS 214. The test requirements represent an impact between a vehicle travelling at 15 mph hit by another at 30 mph moving in a perpendicular direction to the struck vehicle. Lack of survival space at the sides of the occupant, and absence of effective controlled-deformable side structure, led the researchers to examine the possibility of using the seat structure itself as an energy-absorbent in side impact. Floor structure is important in this context, in transferring loads to the sidemembers of the vehicle and wheelbase, too, is important in that small to medium size cars tend to have the seat located behind the B-pillar so that contact occurs between the two after side-impact.

The authors analysis simulated the effect of including an optimized seat structure, Fig 28, from a side-impact protection standpoint, to a conventional design. A cruciform stiffener, of 32 mm diameter tubes with 3.2 mm wall thickness, was added to the seat-back. The seat back frame tube was raised from 22 to 32 mm diameter. Stiffeners were also added to strengthen the seat-pan and all part thicknesses of the existing seat design were doubled. High-strength steel was substituted for mild steel and the floor centre tunnel console was stiffened to provide better load path to the non-struck side of the vehicle.

Prior to performing the analysis, earlier impact case studies were used to establish the deformed shape of the B-pillar at dummy-to-door impact point in the crash pulse, with the deformed shape of the bullet test-vehicle also found at this same point, alongside the key velocity profile for B-pillar, front and rear doors, at the belt-line. After the analysis it was found that the first major component to benefit from the improved seat structure was the B-pillar, shown as Case 3 in Fig 29a, 24 msec into the crash pulse cycle. This figure also shows that the centre console has more lateral deflection, indicating more load being transferred to the side of the vehicle. The way front and rear door assemblies are affected is seen in Fig 29b. Sections are cut where the doors impact the side-impact-dummies. Both these figures, and plots of barrier face deformation on the bullet vehicle which established much greater energy absorbed, proved the effectiveness of the structural modification.

Passive restraint for minivans and LCVs For the 1999 model year, changes in FMVSS 204 mandates that upper end of the steering column must not displace horizontally rearwards more than five inches in a 30 mph fixed barrier test while at the same time obeying the requirements of Standard 208 with respect to head injury criteria. Researchers at Hawtal Whiting and GM[3] have reported an instrument panel structure redesign in magnesium alloy to meet these requirements, as well as eliminating steering column shake at engine idle by raising structural resonant frequency. In this last respect stiffness objectives were 550 N/mm vertical and lateral with modal objectives of 27.7-29.6 Hz for the fist mode and 32-34 Hz for the second. Another requirement of the design was to isolate the instrument panel cross-beam from the front of the dash, as it deformed rearward, so as to reduce the mean acceleration of the beam relative to the occupant during impact. Also some bending of the beam had to be allowed during impact to take advantage of the space available between the beam and front of the dash.

Several forces act on the beam during impact; internal forces include air-bag deployment and external ones include the mass inertia of the beam and the components mounted to it during deceleration, Fig 30. Direction of loads on the beam also add up to a pitching moment which must not reflect in undue steering column upward-rotation. The cross-beam was designed to maintain the AM60B magnesium alloy die-casting below its ultimate stress, in impact, shown by Fig 31. An FE analysis was carried out, Fig 32, to meet the correct stiffness criteria and dedicated reinforcement ribs were added, Fig 33, and material thicknesses increased, to increase column vertical

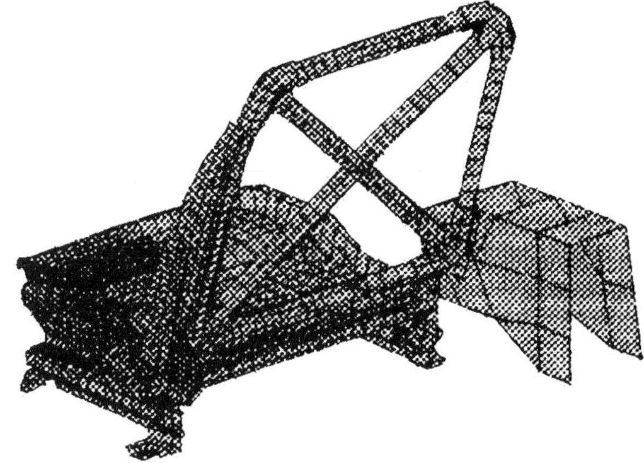

Fig 28: Improved seat structure, above, against conventional specification, opposite

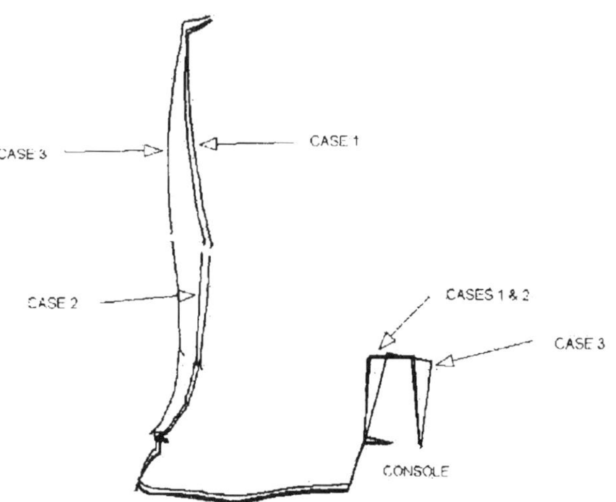

Fig 29: Deformation profiles of B-pillar (a, above) and doors (b,right)

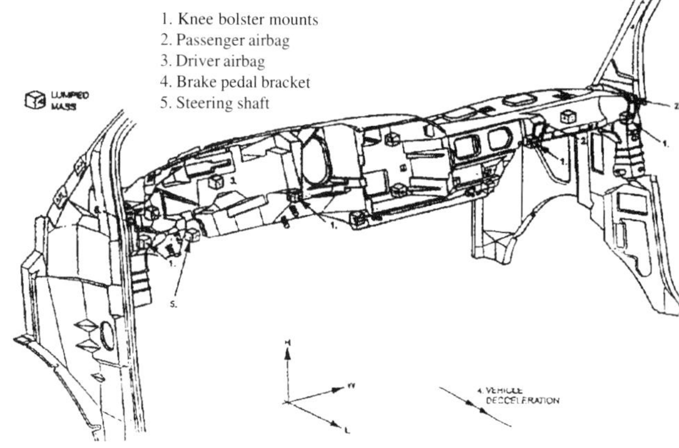

Fig 30: Cross-vehicle beam loading in frontal impact

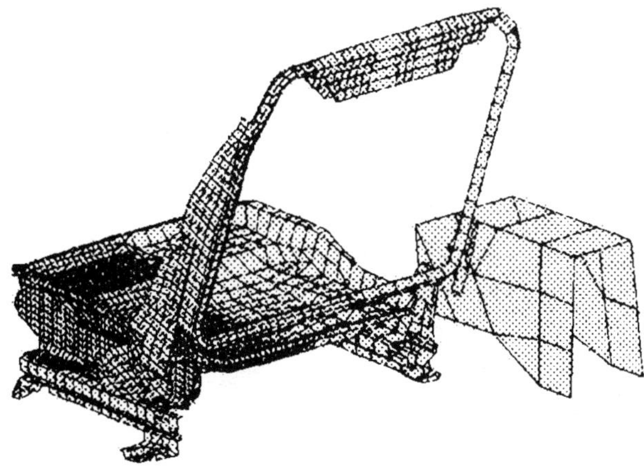

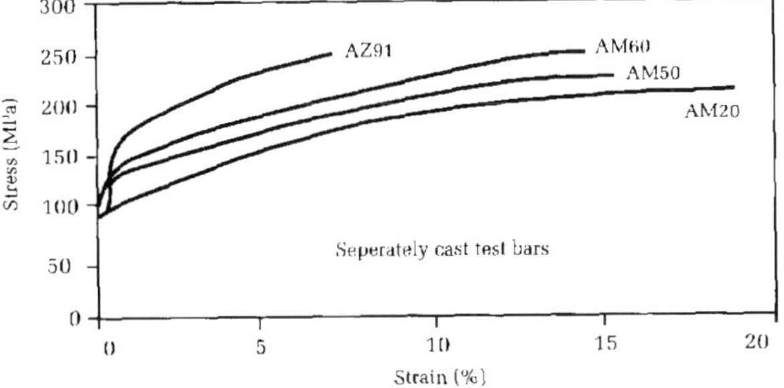

Fig 31: Material characteristics for magnesium alloys

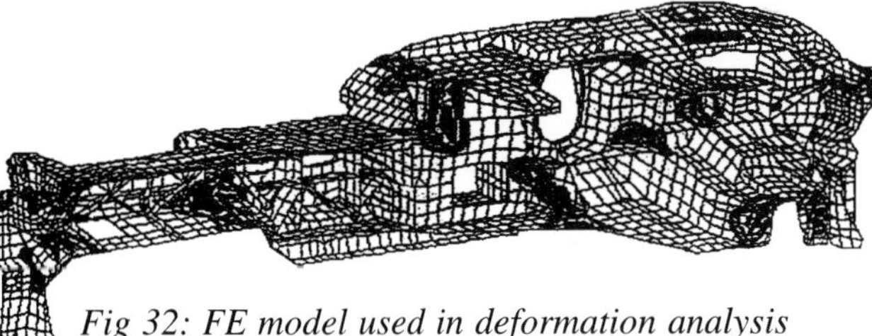

Fig 32: FE model used in deformation analysis

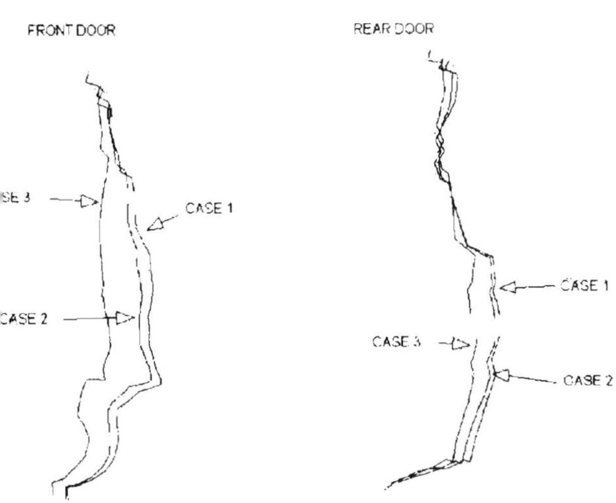

and lateral resonant frequencies. Impact energy absorption takes place in three areas: firstly, airbag compression, knee bolster deformation and surface foam crush; secondly, elastic and plastic deformation of the cross-beam and, thirdly, vehicle front-end crush.

Protection in bus roll-over Reducing occupant injury in bus roll-over has been found to depend on structural resistance, restraint systems and seat anchorages in that order of importance. Researchers at IDIADA[4] have therefore devised a simulation procedure and carried out structural-element verification tests, Fig 34, for structural resistance in roll-over which allow examination for compliance with ECE R-66 without carrying out full vehicle testing.

The ECE procedure requires the bus/coach to be impacted in the corner intersection between side and roof then ensure minimum encroachment into the survival region. IDIADA's simulation is based on an ABAQUS FE code, Fig 35, which allows experimental data to be used for defining the behaviour of the bending collapse so all relevant elements are tested and the data fed into the simulation, including moment *vs* angle effects for out-of-plane bending. The chassis presence is also included in the model of the body shell, using lumped and distributed masses. Loading of the shell involves tilting it off a 0.8 metre high platform and impacting the surrounding ground, according to the regulatory procedure. Subsequent analysis is in two parts: up to the point of and after impact with the ground, only the second being relevant to the regulations. The results of the analysis

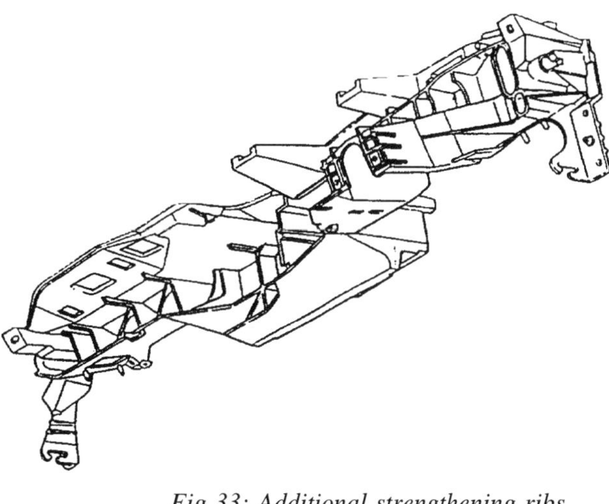

Fig 33: Additional strengthening ribs

have shown that many different types of deformation may occur but all are neglected except bending collapse since the structure absorbs all the necessary energy in this mode. The roof must not be too stiff or it will cause unduly high deflections in the pillars, Fig 36, and a balance of structural resistance between the two planes has to be met. As different collapse mechanisms can occur in the beam element corner joints there must be proper structural continuity in those between pillars and roof and no longitudinal member must interfere with this continuity. Appropriate crossmembers are required where the pillars intersect the floor and these must correspond with roof-sticks to make up ring-frames, Fig 37. All joints between pillars, roof and floor must be reinforced to a level that only elastic, and not plastic deformation occurs.

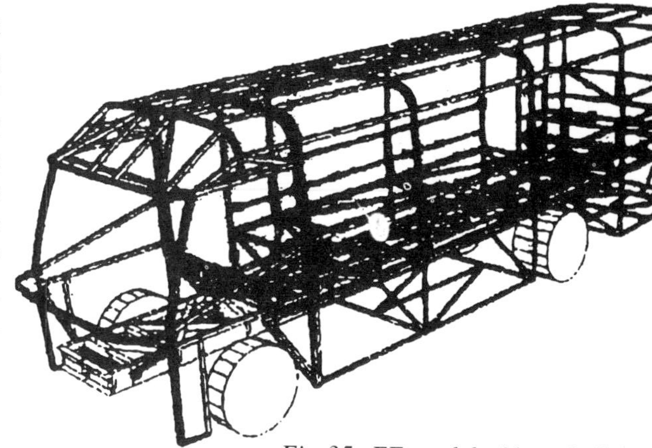

Fig 35: FE model of bus shell

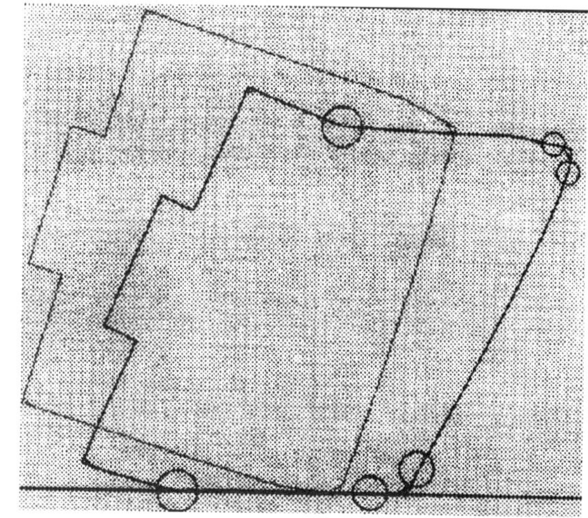

Fig 36: Undue deflection of side pillars

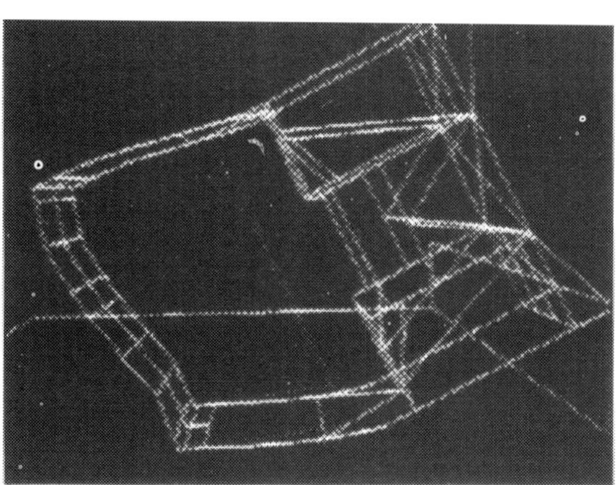

Fig 34: Physical model verification

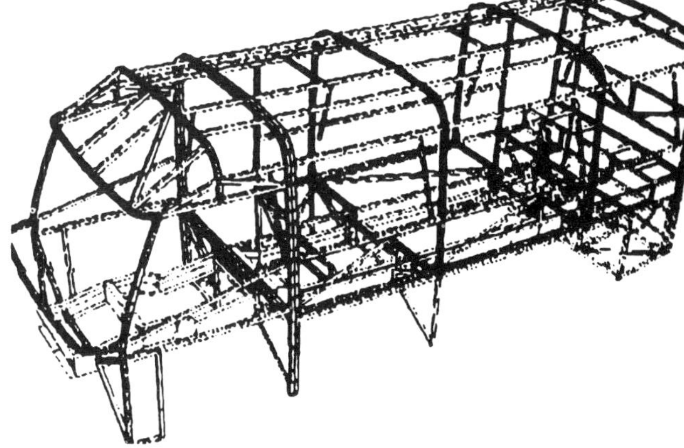

Fig 37: Ring-frames

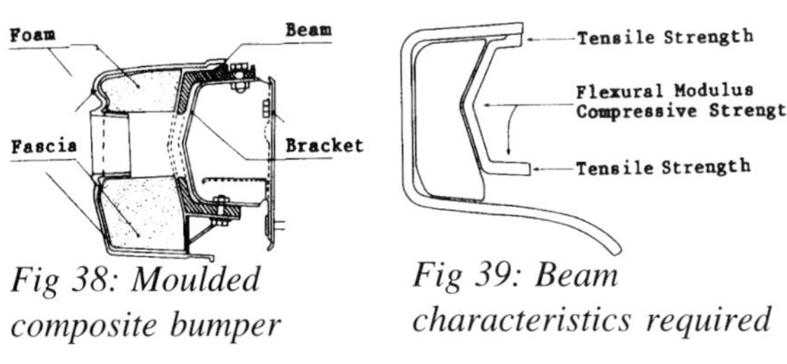

Fig 38: Moulded composite bumper

Fig 39: Beam characteristics required

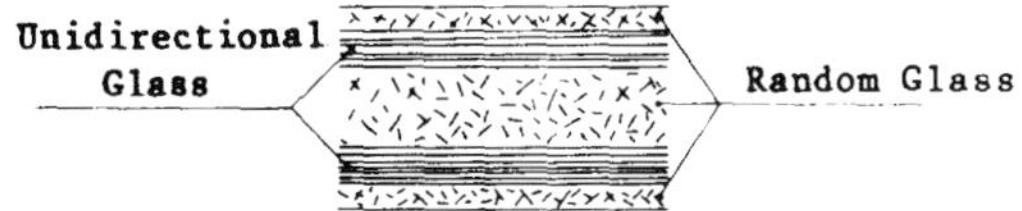

Fig 40: Reinforcement system

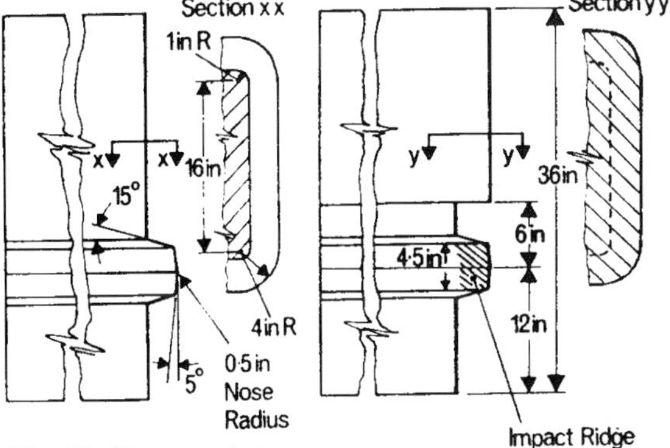

Fig 41: Test pendulum

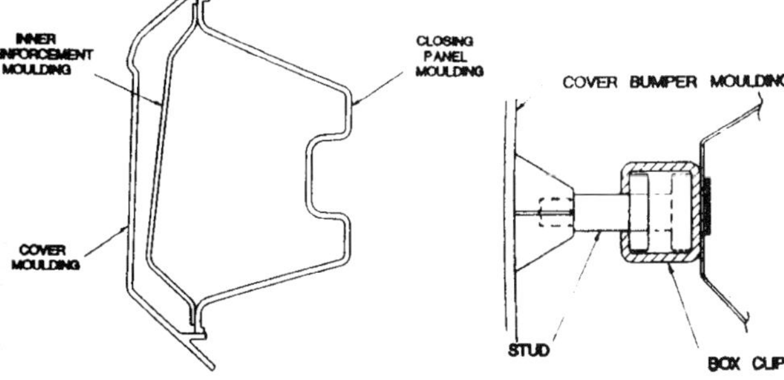

Fig 42: Cross section

Fig 43: Side fastening

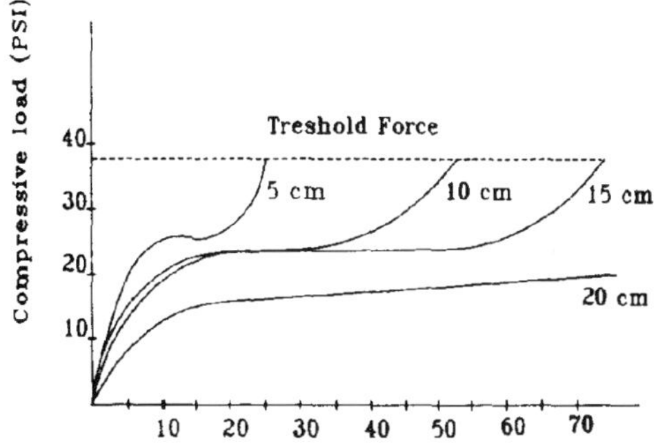

Fig 44: Foam deflection curves

Bumper systems

Large frontal impacts account for two-thirds of occupant injuries in motor vehicles and any contribution the bumper system can make to lessening this incidence is clearly worth striving for. Increasingly plastic-foam systems are being used to fulfil the major energy-absorbing role. Here a technique of optimising foam performance is suggested and an interesting way of combining foam with metal structures to deal with different parts of the impact speed spectrum.

While direct energy absorption capability of the bumpers is limited, careful design can ensure the protection in minor impact of those more important controlled-collapse structures, both at the front and rear of the vehicle, which do make a major contribution. They can also offer protection to other safety systems such as lamps, locks and latches for the various cover panels whose correct working could otherwise be jeopardised by minor impact. The prescribed dimensional envelopes for the positioning of lights on vehicles, plus the requirements of the engine cooling air inlets and rear hatch openings restrict the space available to fulfil these functions and of shielding such items as fuel filler pipes and lighting harnesses. Another factor which affects design is the incidence of minor damage to the sides of the vehicle as streets become more congested and parking more hazardous. This is leading to increased consideration of all-round bumper systems. Front and rear fascia panels are a further trend in which the bumper is integrated with 'soft' end structures for the vehicle. Design of such front/rear ends requires particular attention to the minimum legal impact absorption requirements. The 5 mph capability set initially by the US market, in FMVSS part 81, has been important in this respect.

Usually shock absorption may be accomplished by such materials as moulded foam, or honeycomb cellular matrices, attached to the back-up beam — historically made from steel. Hydraulic dampers in the place of these materials are another solution. Attempts have, of course, been made in volume-produced cars to use moulded composites in place of the steel bumper beam, Fig 38. Such a system is designed to absorb energy by the foam interposed between beam and fascia, the required characteristics for the beam being shown in Fig 39 and a typical reinforcement scheme for the composite being seen in Fig 40. An earlier US standard, FMVSS 215, gave guidance on the dimensioning of bumpers to restrict damage to lights, and prevent over-riding of bumpers, causing vehicles to become interlocked in an accident situation and leading to punctured coolant radiators. Associated with this standard was the introduction of pendulum tests. Fig 41 shows the type of pendulum head shape required for simulating an impacting vehicle. The test involved three 5 mph impacts at 20 in (508 mm) above the ground with type-b and three with type-a at 16-20 in (406-508 mm) from the ground. Corner impacts at 3 mph were also mooted at the time of the code's introduction. Pendulum mass is equal to the unladen weight of the vehicle.

In Europe, ECE 42 sets the legal requirements for bumper design which specifies 4 km/h pendulum impact tests. An example of a system designed to that standard is the Rover Montego unit which broke new ground in being integrated with a spoiler and being finished in body colours. Weight requirements limited the thickness of the combined cover moulding to 3 mm and the part had to withstand two hours at 130 C in a stoving oven; material chosen was Bayer Pocan modified polybutylene-teraphthalate. The centreline cross section is shown in Fig 42, the structure comprising a box section simply supported at its longitudinal mountings corresponding to the main longitudinal members of the body. The elements of the box beam were assembled together using a copper/polyester sacrificial tape heated to form the weld. The wrap-around portions of the bumper are fixed to the sides of the vehicle by means of the fastenings seen in Fig 43, involving a boxed clip, supporting a long shanked stud on the bumper. This allows an expansion of 25 mm on the full width of the bumper. For the longitudinal fixings at the front of the vehicle, these are pre-loaded by a double coil spring washer and a shouldered bolt. A nylon washer plate is sandwiched between the steel mounting bracket and the PBT mounting surface of the bumper closing panel, allowing the surfaces to move over one another during expansion. In developing polypropylene foams for bumper cores, Montedison-Bollate, Via Principe Eugenio 1, 1-20155 Milan, have established a calculation technique for the isotropic cellular structure of the foam. The elastic modulus E relates to the foam density ρ, with m a constant, according to $E = K\rho^m$.

Fig 44 shows load deflection curves for differ-

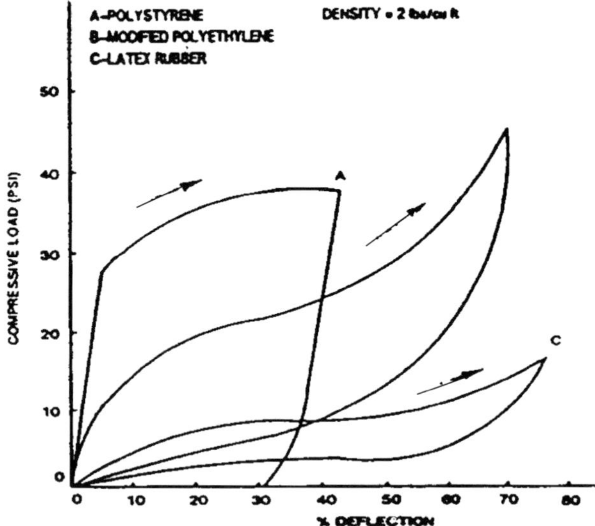

Fig 45: Foam characteristics compared

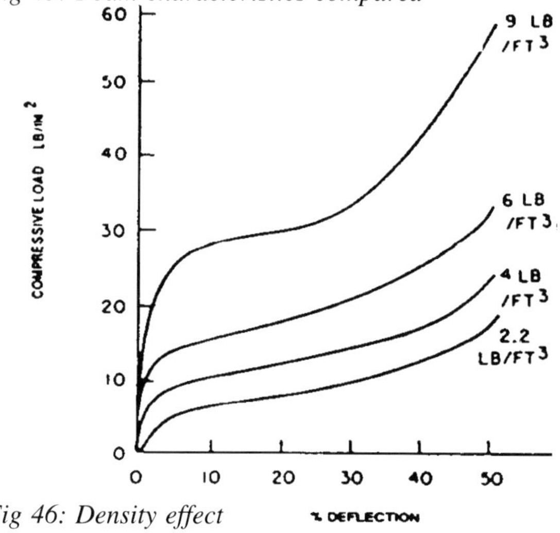

Fig 46: Density effect

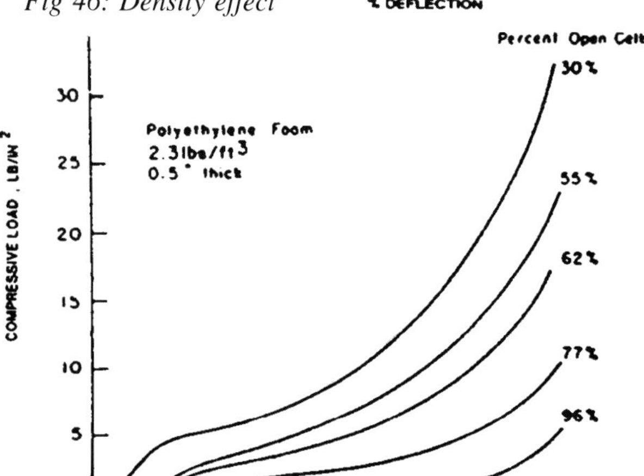

Fig 47: Grain effect

ent plastic foam thicknesses; the threshold line represents the requirements of a particular bumper system showing the compromise that has to be made between choosing thickness to obtain constant load for a given (plastic) deflection and sufficient percentage defection (and recovery).

Fig 45 plots plastic foams and latex foams at the extremes of these requirements against a good compromise modified polyolefine foam. Obtaining the compromise behaviour is a function of foam density, open-cell content and polymer phase composition as seen in Figs 46 and 47. The objective is to obtain a rectangular cellular structure of gas-filled closed cells. Fig 48 gives a mechanical model of the foam and the effect of cell size while Fig 49 shows the wide variety of characteristic curves which can be obtained. Compressive stress in a foam, stress-strain rate for which is nearly independent of the rate of deformation, is a function only of strain-energy required W_d to deform a sample, being given by the area under the loading curve, Fig 50: $W_d=\int_0^\varepsilon \sigma(\varepsilon)d\varepsilon$ for stress σ and strain ε. Velocity of the impacting masses is given by:

$$mv^2/2=mv_o^2/2-\int_o^\varepsilon \sigma(\varepsilon)d\varepsilon$$

for impacting object mass m and v_o,v velocities before and after impact. Total energy lost during impact is predicted from the hysteresis W_h (area between loading and unloading curves):

$$W_h = m(v_o^2 - v^2)/2 = \int_o^\varepsilon [\sigma(\varepsilon)-h(\varepsilon)]d\varepsilon$$

Montedison have used these techniques to develop a new-concept bumper for high speed collisions (16 km/h impacts to give no damage). Fig 51 shows how energy absorption increases rapidly with speed. The design was based on a two-stage energy management system: flexible foam to provide absorption at impacts below 2.5 mph and a bow-structure replacing the conventional steel bumper beam, for higher speed impacts.

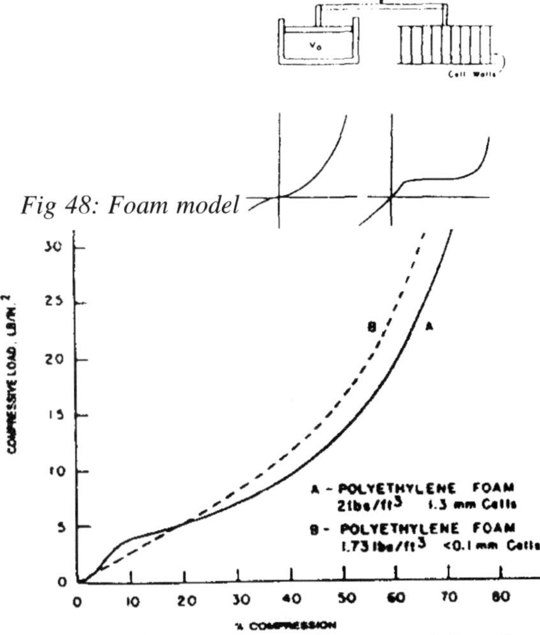

Fig 48: Foam model

Fig 50: Different foam performances, kilojoules

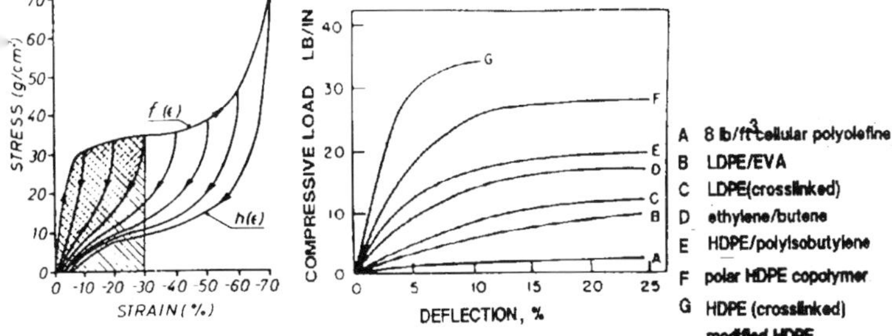

Fig 49: Speed effect (left) and loading curve

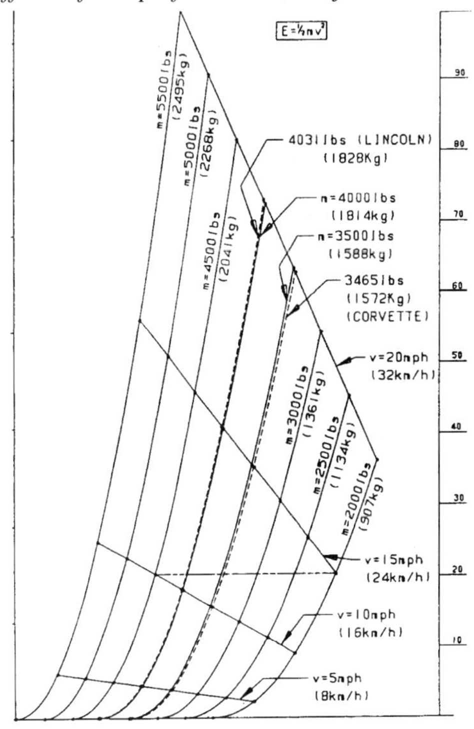

Fig 51: New departure bumper structure

Energy-absorbent foam analysis

An understanding of the different types of expanded plastic foams is necessary for their successful use in energy-management. Rigid, semi-rigid and elastomeric PUR foams are compared here alongside bead foams from expanded polystyrene

ICI's USA polyurethanes operation is addressing the NHTSA regulations for occupant safety and the company argue that PU systems are increasingly used for impact energy management. A new range of EM foams has been developed to meet FMVSS 202 for head-impact and FMVSS 208 for knee impact. The phasing in of side impact requirements (FMVSS 214/216) over the next few years is also involving development work for applications which can involve impact energies of 150 to 450 ft lbs (203-610 J).

Foams classified as rigid, semi-rigid and viscoelastic are each being developed. The first absorbs energy by a non-recoverable crushing of the foam-cells whereas the second and third create a viscous dissipation of energy followed by a slow recovery. Typical processing parameters for all three types are shown in Fig 52.

Dynamic impact testing of rigid EA foams produces a force-deflection curve such as that in Fig 53, which reveals much about energy absorption characteristics. The area within the rectangle 1+2+3 is the maximum which the foam could possibly absorb — a measure of the foam's ability to minimise high peak loads. Area 1 is unrealised energy, area 2 that absorbed by the sample and area 3 the total rebound energy. The ideal efficiency is percentage of maximum energy and input efficiency the ratio of energy absorbed to the total energy input, the latter giving a measure of the energy absorbed by the foam *vs* the energy rebounded.

Typical impact curves are seen as Fig 54 a,b,c and d. As impact energy increases, penetration into the foam, as well as the peak load, also increases. While the foam exhibits fairly uneven deceleration of the impacting body, as the impact energy increases

Component Temperatures (°F)	75-95
Component Mix Pressures (psi)	1500-2000
Demold Time (min)	2-3
Machine Throughput (g/sec)	100-350
Mold Temperature (°F)	120-140
Mold Release Agent	Wax/Water-based

Fig 52: Processing parameters

penetration and peak load also increase. Effect of surface skin is seen as a sudden increase in initial load followed by a fall-off — a useful characteristic. The figure shows that the foam crushes evenly until

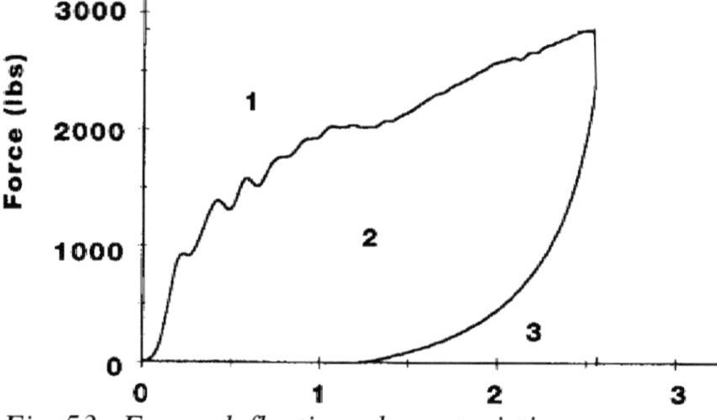

Fig 53: Force-deflection characteristics

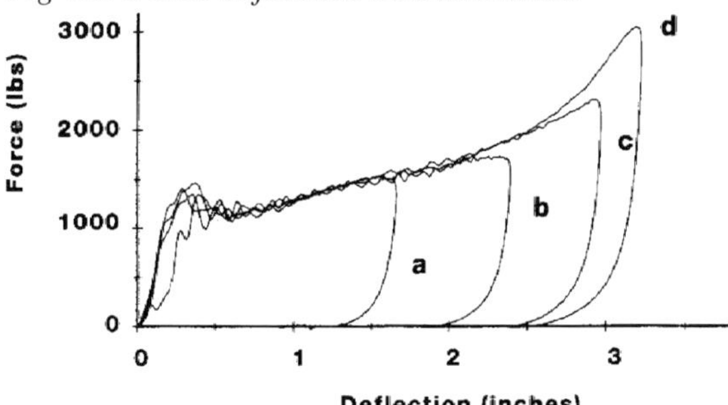

Fig 54: Typical impact curves

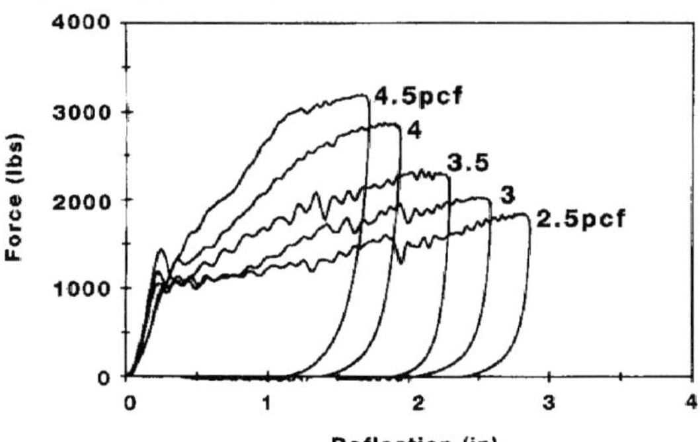

Fig 55: Effect of density increase

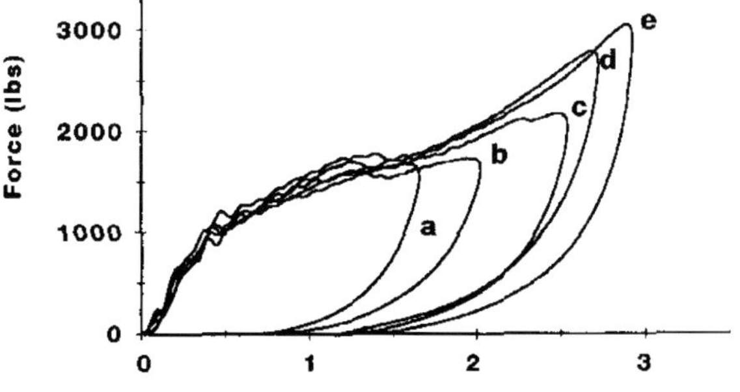

Fig 56: Semi-rigids

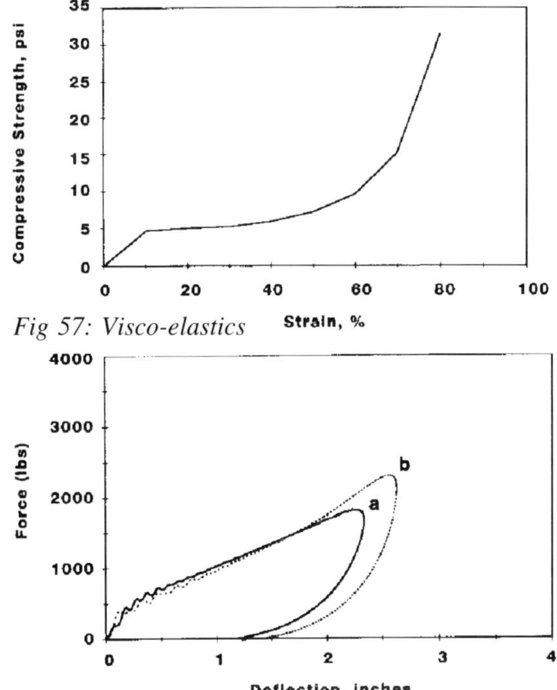

Fig 57: Visco-elastics

Fig 58: Dynamic impact curves

it begins to stack up under the impacting body. Analysis of test results has shown the efficiency of the foam ranges from around 50-70 per cent for 2.5 lb/ft^3 (0.04 g/m^3) material. Fig 55 shows the effect of increasing foam density, keeping the impact energy constant. Both density and formulation of the foam can be altered to tailor the foam for a particular application.

Semi-rigid foams are moulded at higher densities than rigids, generally. Characteristics for a 4.5 lb/ft^3 (0.072 g/m^3) system are seen in Fig 56. The increase in peak loads at higher energies is more pronounced because a flat impacting surface was used to obtain these particular results. If the striker has a rounded surface, representative of a head or a knee, the force tends to ramp up more evenly with increase in area during impact.

In the case of visco-elastic foams, a recovery is registered but it is not immediate. Static compressive strength properties are seen in Fig 57. Compressive strength is much lower than rigids and semi-rigids but it remains stable up to 60 per cent strain. The dynamic

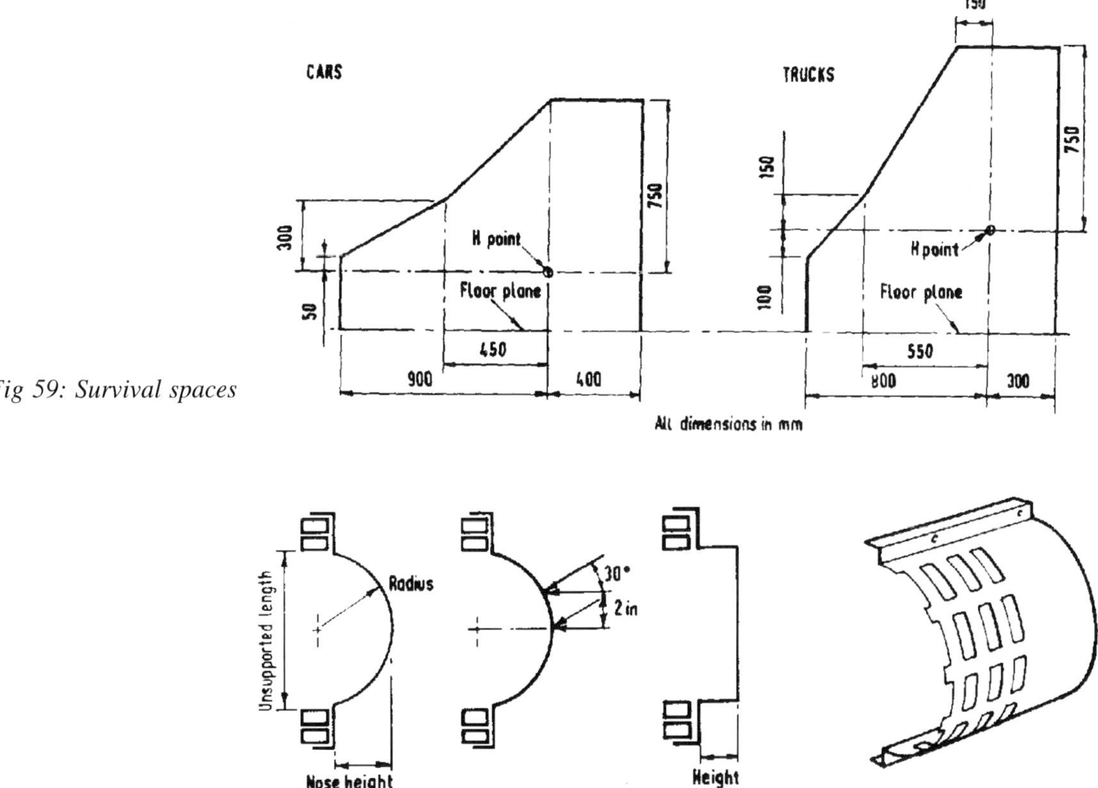

Fig 59: Survival spaces

Fig 60: Armature shapes

impact curves of Fig 58 show that the foam absorbs up to 85 per cent of the input energy with a relatively low level of rebound but does not achieve the 'plateau' situation of the rigids. At lower levels of impact energy it does show 45-50 per cent efficiency and its unique properties make it useful for such applications as anti-submarining ramps.

The design of crash pads involves the use of survival space envelopes, typical ones for cars and trucks being shown as Fig 59 for belted occupants. Armature test shapes are shown in Fig 60 and these can be designed to absorb energy in a way that relieves the foam of all the impact duty. Thinner layers of foam can then be used and the purpose of the foam becomes that of efficiently spreading the load over the surface of the armature. Human tolerance of impact loads is crucial to design studies and the maximum energy which can be absorbed is the product of that load and the total deflection of the padding.

Thermoplastics which are foamed at low densities while in the form of small particles can produce expanded bead foams which have been successfully applied to automotive interior components. One of the best known is expanded polystyrene (EPS), used in packaging, which is of course below the strength requirements for crash pads. Expanded high-heat copolymers such as styrene-maleic anhydride from ARCO Chemical are suitable candidates, however. The material is sold as solid plastic particles having a bulk density of 40 lbs/ft^3 (6.4 g/m^3). These contain the blowing agent pentane and the material transforms to a foam, of between 2 and 10 lbs/ft^3 (0.032-0.16 g/m^3) on heating. This foam is introduced into a mould and exposed to pressurized steam, the secondary heating causing further expansion and fusing together of the foam particles.

Knee bolsters are a common application in US cars where FMVSS 208 requirements call for a 30 mph vehicle crash into a fixed barrier with a maximum occupant femur or upper leg compressive load of 2250 lbs (1 tonne). At such vehicle impact speeds the knee typically impacts the bolster at about 12-15 mph with about 3000 in lbs (339 J) of energy per knee being dissipated by the bolster. Tests have shown S-MA foams to have 63 per cent efficiency under these conditions. The Eagle Premier was the first full vehicle-width knee bolster made from the material, Fig 61. It consists of a steel back-up rail, the moulded foam section, and a PUR foam and vinyl skin.

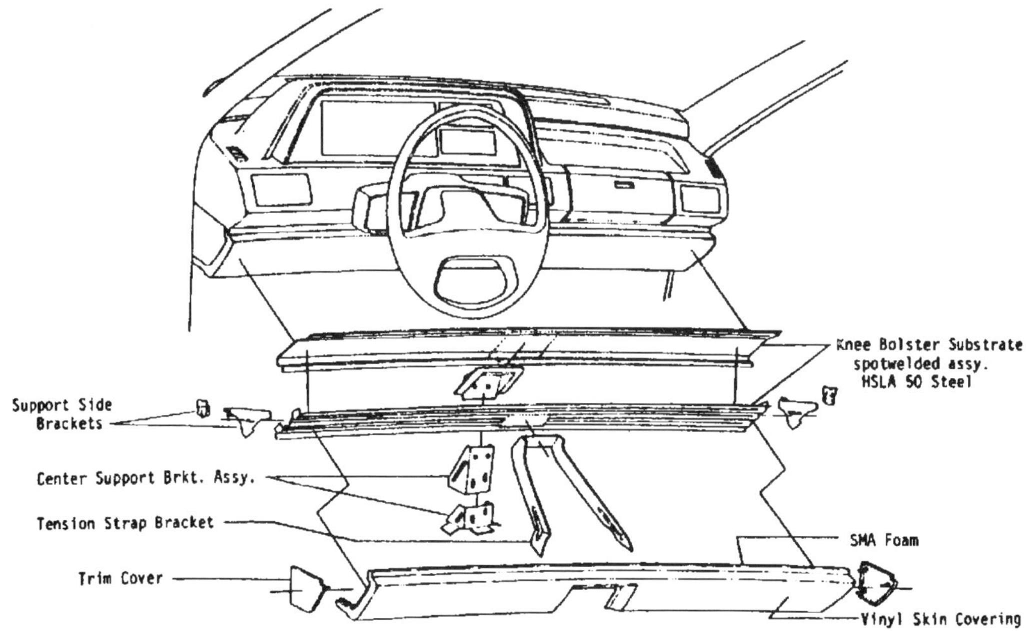

Fig 61: Knee bolster assembly

Impact-absorbing foam systems

Recent research[5] has shown how composite energy management systems are based on the inclusion of low density light-weight fillers, comprising specially selected hollow spheres, of ceramic, glass or plastic composition, with predictable compressive strengths. The sizes of hollow spheres range from 10 to 200 microns in diameter, with a normal wall thickness of between 1 to 2 microns. Hollow spheres are incorporated within a resin matrix, thermoset or thermoplastic, to enhance the properties of the final ultra-lightweight composite material.

The enhanced properties include: electrical and thermal properties; density; compressive and shear strengths; also rheology. The hollow spheres produce a final material which has less stress concentrations and as a result provides improves the impact and fatigue strength. The stress concentrations within the resin system are symmetrical, uniform and predictable around the spheres. Therefore the hollow spheres act as shock absorbers, preventing crack formation by dissipating the energy of impact loads. The crack arresting properties of the spheres also adds the impact characteristics of the final composite.

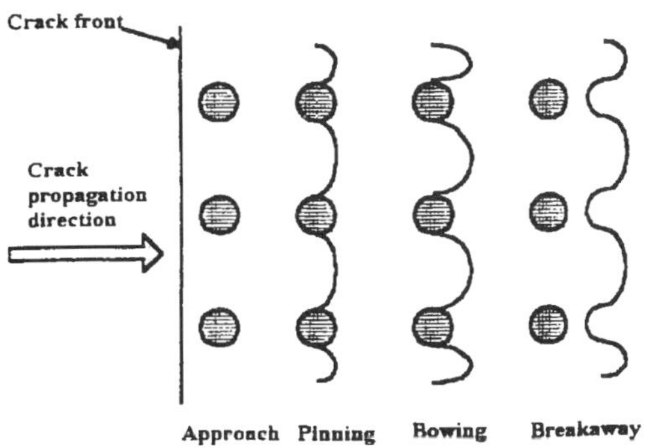

Fig 62: Crack pinning mechanism

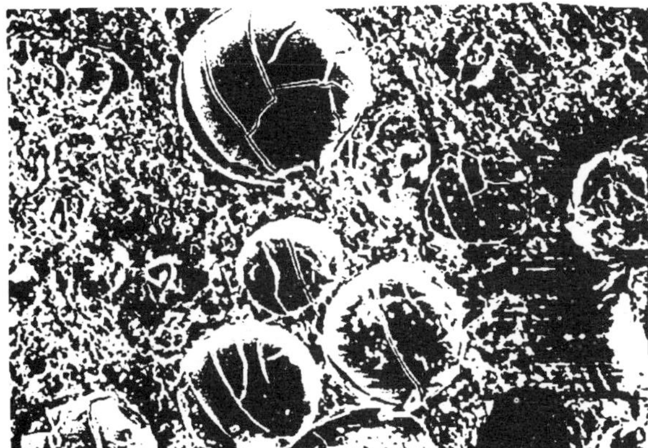

Fig 63: Sphere failure mechanism

Plastic foams are good energy absorbing materials, he argues, and are used in applications which require shock and vibration damping. They can undergo large compressive deformations and absorb significant amounts of energy during a deformation cycle. With plastic materials there are two classes of impact failure behaviour, these are: brittle fracture (the specimen breaks completely into two or more pieces or it may crack extensively; there is either no sign of yielding or only a little indication of yielding) and ductile fracture (the specimen will initially yield and will then crack). When a plastic undergoes ductile failure it exhibits yielding or permanent deformation which surrounds the region or separation under the applied load and this is followed by cracking. Yielding is identified by stress whitening of the material in the failure region. The stress whitening effect is caused by micro-voiding.

When a particulate filler is added to the relatively brittle resins the particles inhibit the crack growth and increase the fracture energy dramatically. A crack in the filled resin is pinned by a row of particles. Particles that pin the crack do not themselves fracture under the stress in the system. Also larger particles are more effective than smaller ones in increasing the fracture energies for a given interparticle spacing by pinning the crack front, allowing it to bow and increase in length before breaking away. For composites containing larger particles its fracture energies are larger.

The addition of particulate fillers has several different micro-mechanical processes that control the fracture energy. These processes include: interaction between the dispersed phase and the crack front; increased plastic deformation in the matrix due to the dispersed phase; energy absorption in the dispersed phase itself; increased surface area of the fracture faces due to an increase in surface roughness.

An investigation of the properties of composite materials has been made into their behaviour under impact. Most composite material structures may experience various kinds of impacts in real service conditions differing in term of the energy, velocity, geometry of the impact head, number of impacts and so on. In these cases it becomes questionable to base the material selection on standard impact test since there are a number of impact tests. The instrumented drop-weight impact test procedure was developed from recommendations by NASA as one of the standard tests for toughened resin composites. This test equipment was manufactured in accordance with the ASTM 3029-84 specification. The computed impact energy of the drop weight system was equivalent to 14.22 Joules. The complete system required the development of a series of separate components, including; an impact absorbing system; a progressive failure mechanism and an energy management system.

Impact absorbing materials The initial research involved developing a material which would be suitable for an impact absorbing system. This was achieved by incorporating hollow spheres of diameters 177 microns into an epoxy resin; and curing the systems with a flexible catalyst. Experimental analysis with different volume loadings were evaluated using the falling weight impact measurement procedure. An analysis of the impact measurements demonstrated that the impact absorbing material developed had undergone a ductile failure, with the development of micro-cracks present within the resin matrix. However, most the hollow spheres where still intact. A micrograph of the sample showed that almost all of the micro-cracks had been pinned by the hollow spheres, Fig 62. The impact absorbing system was therefore capable of performing in several distinct application areas. As the initial impact absorbing mechanism the system developed was capable of operating as relatively thin impact absorbing sheet or skin materials.

A progressive failure mechanism was developed by extruding a cellular tube structure from a thermoplastic polyester resin containing various loadings of hollow spheres. The range and type of hollow spheres used is dictated by the failure mechanism of the system being designed for. In instances where the failure mechanism needs to occur at varying rates the structure of the cellular tube can be modified or special inserts incorporated.

Where the progressive failure mechanism is embedded within the energy management system, the failure mechanism has been developed such that maximum energy is transmitted throughout the system. The impact absorbing system was therefore capable of performing in several distinct application areas.

The initial research involved developing a material which would be suitable for an energy management system. This was achieved by incorporating hollow spheres into a resin system; and curing the systems with a flexible catalyst. Experimental analysis with different high volume loadings were evaluated using the falling weight impact measurement procedure.

An analysis of the impact measurements demonstrated that the energy management material developed had undergone a failure mechanism in which the resin is used as a binder through which the energy is passed directly to the hollow spheres. The loading of the hollow spheres is such that the interspatial distances are minimal and under high energy transmissions there is a tendency for the spheres to collide, Fig 63.

New-departure crash-unit

Hoogovens Group[6] have examined a variety of known techniques for absorbing energy in crash units positioned between the bumper and the main structure of the vehicle. After the evaluation of the ideas phase it was decided to develop concept crash-units, based on the four collapse modes of the crash unit, inversion out-in, inversion in-out, pulverising and buckling deformation modes.

All the concept crash-units had the following characteristics: based on an impact extrusion production process, which offers part and cost reduction; the crash-unit meets all requirements set by the car manufacturer with reference to space, deformation behaviour, price and weight; it is bolted on to the crash zones, guaranteeing easy replacement ; it is connected through a hinge with a stiff bumper beam. In total, five concept crash-units were developed, namely: traditional, cylinder, two step, die and pulverisation. The concepts differ in the amount of energy they can absorb and the space they need for deformation and assembly. It was found that each concept, except the concept die, was feasible in both steel and aluminium.

The traditional concept absorbs the impact energy through buckling and is illustrated in Fig 64.

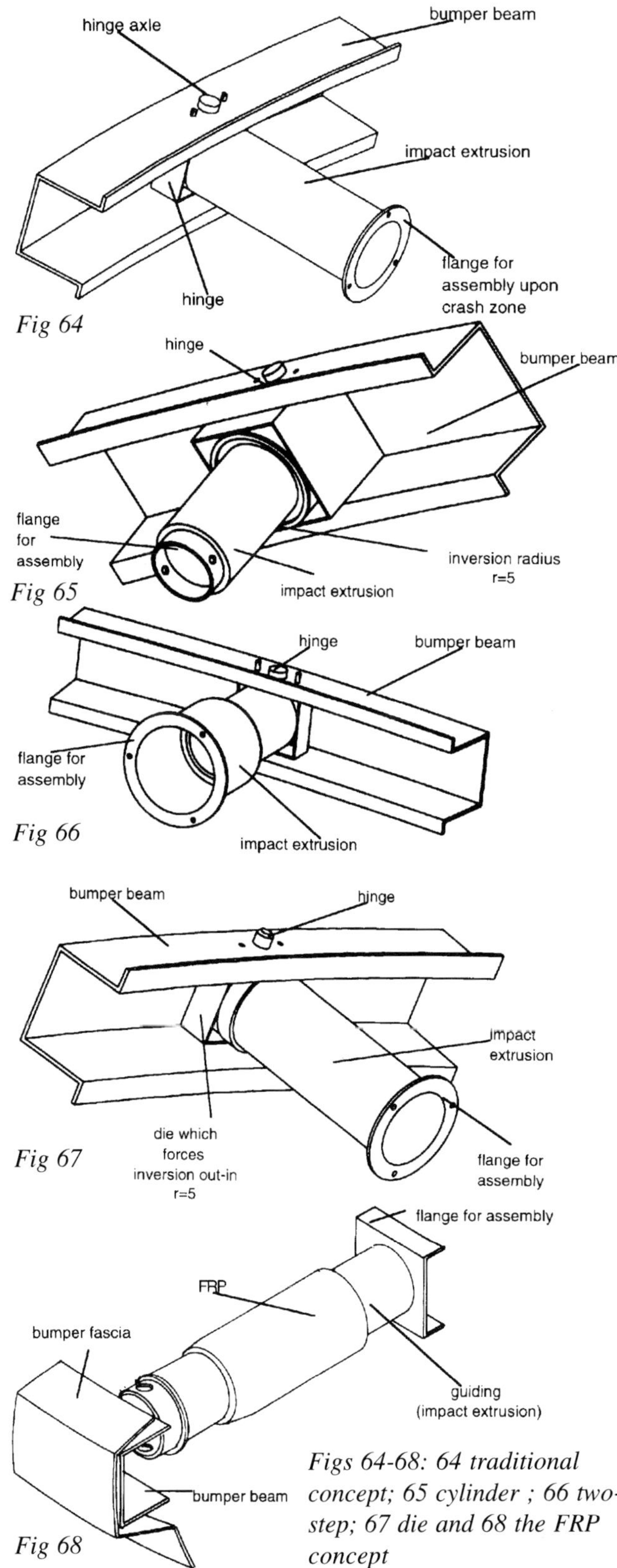

Figs 64-68: 64 traditional concept; 65 cylinder ; 66 two-step; 67 die and 68 the FRP concept

Because of load fluctuations, the mean load during deformation must be 15 per cent less than in the case of the other solutions and thus less impact energy can be absorbed. Additional measurements have to be taken for lowering the initial peak load and this can be done by pressing initiating ribs in the crash-unit, subsequent to the impact-extrusion process. Due to the simple configuration, the assembly is cost-effective and combination with any bumper system or crash zone is feasible.

The cylinder concept is based on the inversion in-out deformation mode, Fig 65. During deformation the crash-unit disappears into the bumper beam, which increases the total deformation distance and thus the amount of impact energy absorbed. A reinforcement is placed in the bumper beam to ensure that deformation into the bumper beam is possible in all impact conditions. The reinforcement adds extra weight, but also functions as a hinge. Assembly of the concept proves cost-effective, because only two parts are used. For the reinforcement, extra space in the bumper beam is needed.

The two-step concept absorbs impact energy through the inversion out-in deformation mode, Fig 66. The die radius is integrated in the crash-unit itself, thus making a die superfluous and reducing weight. The deformation distance is maximised by disappearance of the crash-unit into the crash zone, during impact. Due to the simple configuration, assembly in combination with any bumper system or crash zone is feasible, but the dimensions of the crash-unit must be in line with the internal dimensions of the crash zone. Production of the impact extrusion is complicated and therefore could influence cost-effectiveness.

The die concept absorbs impact energy through the inversion out-in deformation mode with a die, Fig 67. The crash-unit and the die are connected with a bolt, which breaks when deformation starts. The crash-unit disappears during impact into the crash zone, thus maximising energy-absorbing capacity. Due to the straightforward dimensions assembly in combination with any bumper system or crash zone is feasible. Care should be taken to ensure that the crash-unit can disappear into the crash zone. The concept can only be produced in the aluminium version, because the steel tube does not show stable deformation in combination with a die. Cost-effectiveness is not optimal, because two impact extrusions are used.

The FRP concept absorbs impact energy through the deformation mode pulverisation of a FRP cylinder, Fig 68. Because the bending stiffness of an FRP cylinder is worse (low E-modulus), a combination with a guiding impact extrusion in steel or aluminium is necessary. During deformation the impact extrusion disappears into the crash zone, while totally crushing the FRP tube. Due to the simple configuration, assembly in combination with any bumper system or crash zone is feasible. However, care should be taken to ensure that the guiding can disappear into the crash-zone. Because both an impact extrusion and an FRP tube must be produced, the solution does not seem cost-effective.

The table in Fig 69 shows the results of the company's drop-test evaluation of the different systems. The two-step concept fulfilled the requirements and wishes of the car-manufacturer co-operating in the project in the best way and therefore it was

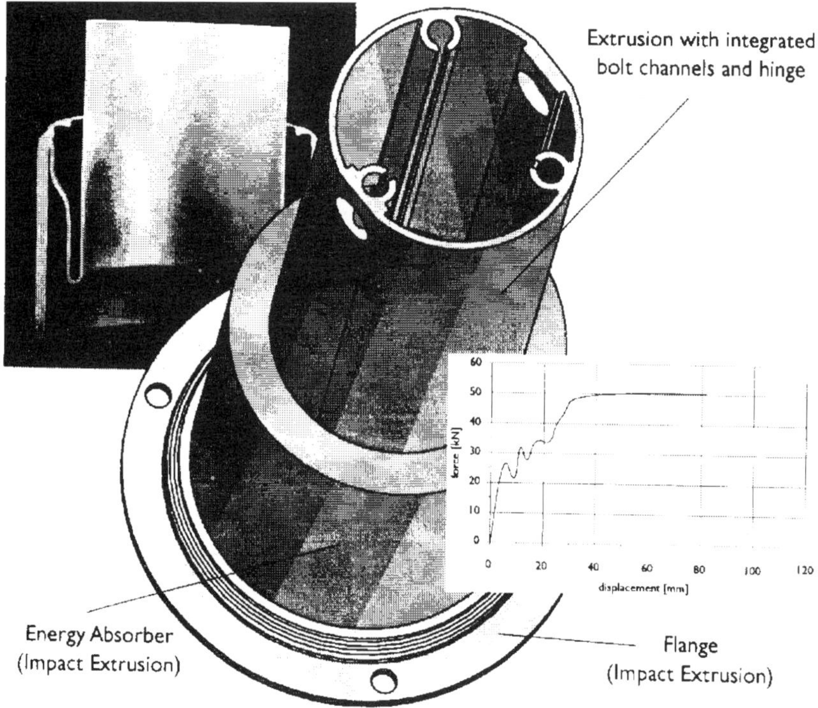

Fig 70: The aluminium step-concept crash-unit

CRITERIA	TRADITIONAL	CYLINDER	DIE	TWO STEP	FRP
ENERGY ABSORPTION					
Amount of energy	-	+	+	+	+
Peak load	-	+	+	+	-
Stable	+	±	+	+	+
Shattered parts	+	+	-	+	+
Sub Total A	0	3+	2+	4+	2+
Other					
Longitudinal beam	+	+	+	±	±
Bumperbeam	+	-	+	+	+
Amount of parts	+	±	-	+	±
Open/Close section	-	+	+	+	+
Weight	±	-	+	+	-
Repair	±	+	+	+	+
Sub-Total B	2+	+	4+	5+	2+

Fig 69: Evaluation of the concept crash-units (right)

decided to engineer the two-step concept in both steel and aluminium. Although with the two-step concept all requirements set by the car manufacturer were already met, a further optimization step was possible, regarding cost-effectiveness and function. The cost-effectiveness could be further improved through simpler produce-ability, alloy choice and assembly. Function could be further improved with regard to stabilisation of the deformation mode, crash situations (angular) and control of dimensions.

The final design of the two-step aluminium concept is illustrated in Fig 70. After discussion with Amefo, an aluminium impact extruding company in Hoogovens Groep, it was decided to redesign the crash-unit into three parts, two impact extrusions and one extrusion. The main reason is that the flange needed for assembly cannot be integrated into one part without causing very high costs. Splitting into three parts would also favour controlled dimensions, necessary for constant functioning through lifetime.

The three parts are assembled with three bolts and a hem-flange operation. Bolt channels and flanges are integrated in the parts. Welding is avoided, because it would influence the mechanical properties and thus the deformation behaviour of the crash-unit. The crash-unit is mounted upon the crash zone with three bolts, making adjustment and replacement possible. In the case of a steel crash zone, a rubber ring is placed in between to prevent galvanic corrosion. The large impact extrusion absorbs impact energy through the inversion out-in deformation mode. The wall thickness is 2.1 mm. The crash-unit is forced into the inversion out-in mode through the aluminium extrusion, produced by Hoogovens Aluminium Profiltechnik. During deformation the crash-unit disappears into the crash zone. After the crash-unit is fully collapsed the crash zone will deform around it.

The weight of the aluminium two trap design is 369 grams (excluding bolts). The crash-unit can be used unpainted, which reduces costs. In the case of the minimal deformation distance of 140 mm, 8400 kJ of impact energy can be absorbed with the aluminium two-step crash-unit. A comparable and applicable crash-unit deforming by buckling would weigh 322 grams, has a maximum deformation distance of 108 mm and would absorb 5400 kJ.

The steel two-step concept could not be economically produced as one impact extrusion (in the case of steel usually called cold forging). To improve the cost-effectiveness, the concept is divided into three parts, one impact extrusion, one high strength steel tube and a high strength steel clip, which are welded together. In the finished part, holes have to be drilled for assembly in the crash zones. To meet the corrosion tests, the crash-unit must be painted, adding extra costs. The large impact extrusion absorbs the impact energy through the inversion out-in deformation mode and is forced into this mode by the steel tube. The wall thickness of the impact extrusion is 1.40 mm. Flanges and bottom are thicker, due to the production process. The total weight of the crash-unit is 866 grams (excluding bolts) and can, when a deformation distance of 140 mm is applied, absorb 8400 kJ. In general it can be said that assembly of a painted steel crash-unit will not cause any corrosion problems. A comparable steel crash-unit deforming by buckling weighs 915 grams, has a deformation distance of 108 mm and can absorb 5400 kJ.

Computer simulation of impact-protection devices

MIRA researchers[7] have given an example of how CAE models were used to lead the development of an airbag restraint system for the Morgan sports car. Details are given on the design of the knee bolster using DYNA3D and its validation using static and dynamic test methods.

According to the authors, whilst FE analysis is recognised as a valuable technique for early assessment of a structure it is not always a panacea to the body engineer's problems. It would be unwise to embark on a vehicle test programme with a structure that had been based on theoretical predictions alone. Unless the FE model has been validated against a suitable physical test, predictions will inevitably suffer inaccuracies through assumed material behaviour, boundary conditions or a local effect not anticipated.

An FE model should therefore be validated as early as possible to improve confidence in the results it provides. Usually a hybrid approach involving part analysis and part test is more suitable in expediting a solution, especially if delays would affect programme timing. The analysis and validation process of an FE model used to develop a knee bolster within a passive restraints system is here used to exemplify the approach.

For its sports car, Morgan's objective was to meet the requirement of FMVSS 208 for unbelted occupants and thereby maintain valuable exports to the US. Improved safety would also provide them with a competitive advantage over other sports car manufacturers in the European market. The cost of developing a passive restraint system was a critical consideration for a small company. Significant savings were thus made by utilising existing production components for the airbag and steering column system. Also development costs were kept to a minimum by extensive use of CAE models to predict occupant and structural behaviour. It was therefore vital to have confidence in the CAE predictions from an early stage.

Occupant kinematics

Analysis of the restraint system under a 30 mph frontal impact was performed using MADYMO. The concept model allowed the performance of the collapsible steering column and airbags to be studied. A theoretical stiffness characteristic for the knee bolster was determined by running a number of iterations to achieve target injury levels. A baseline HYGE sled test was performed to validate the concept MADYMO model. Parametric studies were then undertaken to optimize the column stroke load, bolster stiffness, airbag stiffness and deployment angle.

The knee bolster design was based on two thin-walled tubes held in position with flat end-plates. The upper tube was expected to provide the principal energy dissipation so consequently was set at a thicker gauge than the lower tube. A thin panel was placed over the tubes to better distribute the load on the occupants lower legs as they slide forward during impact.

Higher forces are often generated on the driver side when the bolster contacts the steering column during deformation. In the Morgan, however, the column angle is very low making interaction with the bolster unlikely. Consequently a common bolster design was used for both passenger and driver. A maximum load of 10 kN per femur is specified in FMVSS 208. To improve the protection level over this criterion and also to provide a measure of contingency a load of 8 kN per femur was set as the performance target of the bolster.

CAE structural model and its validation An FE model of the bolster was used to convert the theoretical stiffness characteristic, identified from the MADYMO analysis, into a practical design. The FE model was generated using Hypermesh and analysis was performed using the dynamic, impact code OASYS DYNA3D. A surface model of the knee and lower leg was generated and represented as a rigid body. Contact surfaces were used to map its interac-

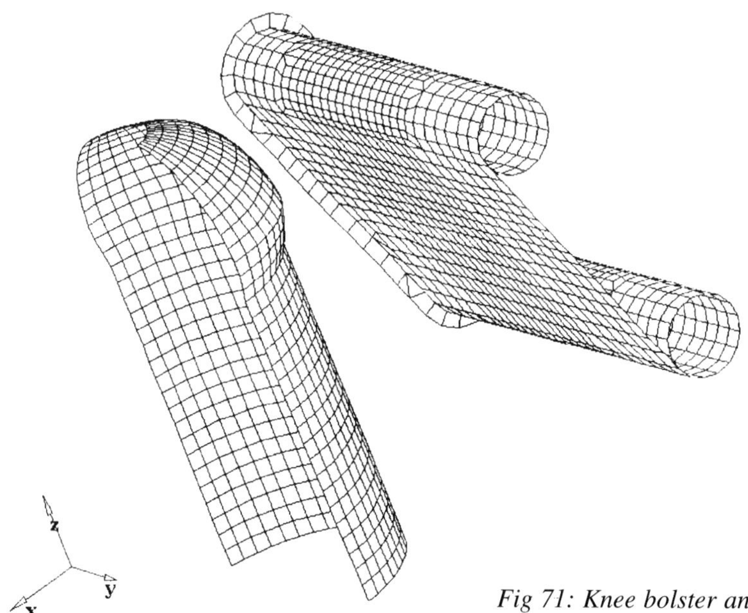

Fig 71: Knee bolster and lower leg models

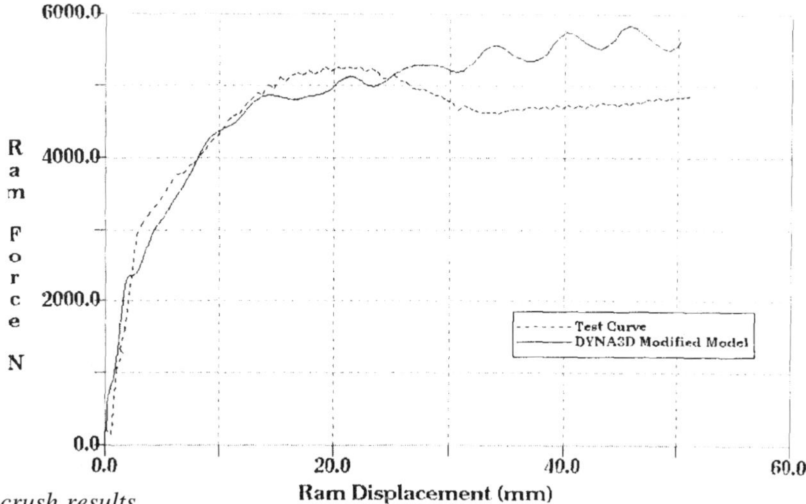

Fig 72: Quasi-static crush results

tion with the bolster. The motion of the lower leg-form was prescribed using results from the MADYMO analysis. The knee bolster and lower leg models are shown in Fig 71.

Results of the DYNA3D analysis indicated that the upper tube buckles at the knee impact locations providing relatively soft contact. Tensile forces, developed in the tube due to the large deformation, bend the end-plates inwards which further limits the peak load of the structure. A number of iterations were performed to determine the tube diameters and gauges necessary to meet the performance target. Having specified the details for the concept bolster design it was then necessary to validate the predictions. In this case, prototype bolsters were manufactured, tested and results used to update the FE model. This process is sometimes referred to as 'Hardware-in-the-loop'.

A quasi-static crush of the bolster was first performed using a Denison test machine. The purpose of this test was to eliminate the complexities inherent with dynamic impacts but still produce a realistic deformation mode. The bolster was held at an angle of 30 degrees to the horizontal in a rigid frame and a 150 mm diameter cylindrical ram was used to apply the load. The ram force and displacement were continuously monitored and at the end of the test the lateral deformation of the end-plates was also measured.

The leg form in the CAE model was substituted for a model of the cylindrical ram and the boundary

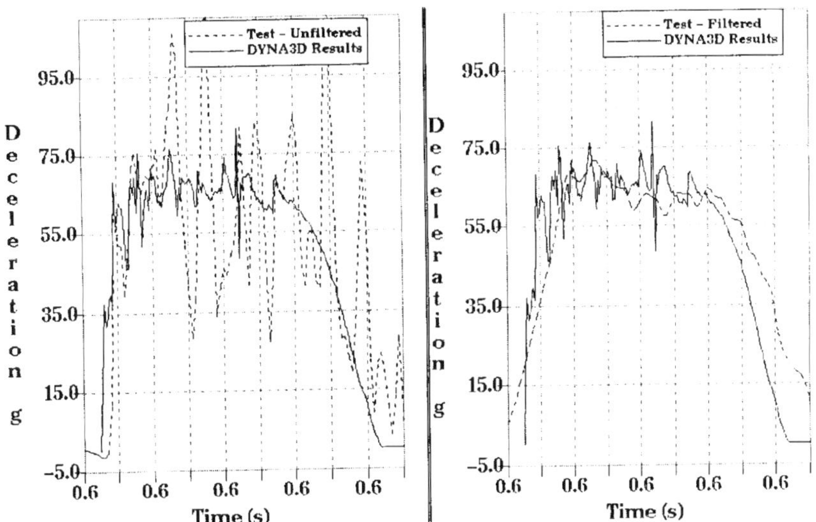

Fig 73: Drop test ram-deceleration

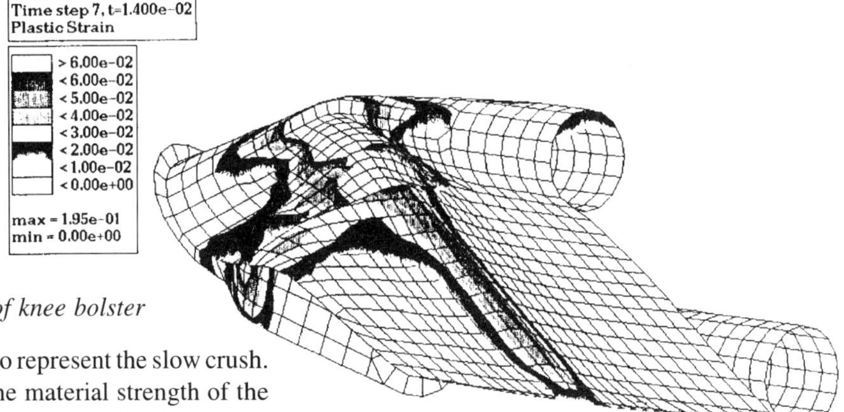

Fig 74: Deformed shape of knee bolster

conditions were modified to represent the slow crush. This simulation enabled the material strength of the tubes, panel and end-plates to be set. Nominal material gauges were also replaced with measured values. A comparison of the FE predictions with the test results is shown in Fig 72. A more representative test was then performed using a drop test rig. The bolster was again held at a angle of 30 degrees in a rigid frame on the anvil of the rig. The cylindrical ram was ballasted to a mass of 9.5 kg and dropped from a predetermined height to give an impact velocity of 9 m/s. An accelerometer was fixed to the ram to record the dynamic performance.

The FE model used to simulate the quasi-static crush was then simply converted to simulate the drop test. Constants for the Cowper-Symonds equation were added to the material description to take account of the strain-rate effects. The analysis results are compared with the test results in Fig 73. A C60 class filter was used to remove the high frequency oscillations from the filtered plot.

Results of the bolster validation exercise indicated that the concept design would need to be softened to achieve the performance target. The upper tube was consequently reduced in gauge, to be the same as the lower one, and the cover panel was also down-gauged. Further softening was achieved by detaching the cover panel from the end-plates. Results of the modified design were fed back into the MADYMO model to ensure that occupant injury criteria were still met. The deformed shape of the bolster during impact is shown in Fig 74 and dynamic force levels for the original design and modified design are shown in Fig 75.

The authors say that a sensitivity analysis of the bolster is planned, as a next stage, to investigate its performance under upper and lower bounds of knee contact velocity and also for out-of-position occupants. Further MADYMO analysis is proposed to determine the performance of the restraint system under alternative crash scenarios. Simulations of 5th percentile female and 95th percentile male occupants will also be carried out.

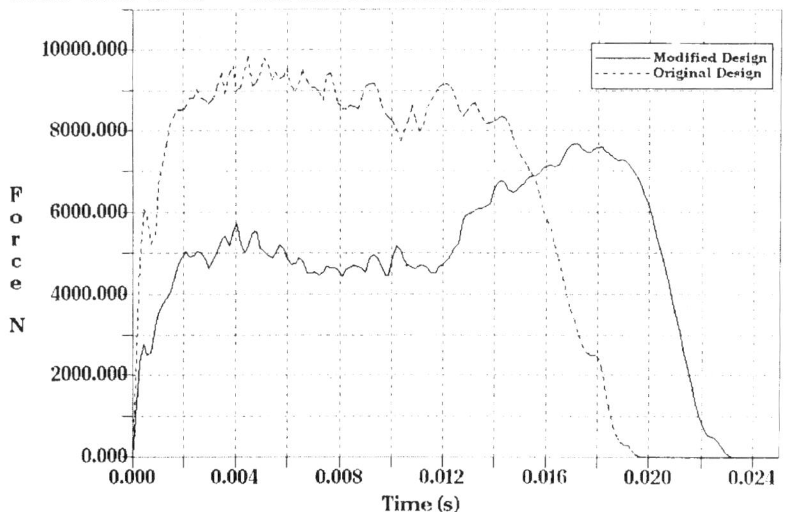

Fig 75: Dynamic force levels

Integrating energy absorbers into trim panels

A collapsible energy absorber moulded into the door panel is a new approach to side impact protection (demanded by FMVSS 214) suggested by Dow Chemical[8]. The test standard involves a moveable deformable barrier travelling at 30 mph on an angle of 27 degrees to the stationary test vehicle.

Typically the response of an average door system is seen in Fig 76. The velocity of the vehicle centre-of-gravity measured at the prop-shaft tunnel midpoint is slow to increase due to the compliance of the side-structure and the structural collapse mechanisms. As the door collapses, the outer skin crushes against the door components and inner sheet metal, and the interior trim panel accelerates against the occupant.

The occupant's velocity increases to that of the penetrating structure, occupant contact occurring 20-30 ms after impact initiation while the complete impact lasts 50-60 ms. Acceleration of the occupant has to be monitored at the points seen in Fig 77, the maximum allowable acceleration values being shown in Fig 78. In designing the conformable door inner panel, three areas involving collapse initiators and absorbers are critical during the development for side impact protection: the armrest, the belt line and the lower torso area.

Increasing compliance of the armrest can be achieved by removing material in localised areas and designing in notch lines. Plastic hinges are formed by reducing material thickness at the notch lines and during impact local instability collapse of the immediate structure occurs. For the belt-line and lower torso areas, polyurethane foam blocks are typically used, sandwiched between the door panel and the inner sheet metal.

Fig 76: Impact response of typical door

Manufacturing complexity is considerably reduced by integrating these foam absorbers in the injected moulded door inner panel as shown in Fig 79. Ribs of the 'egg-box' structure may be continuous or broken at the intersections, the latter permitting uncoupled collapse behaviour of the walls and therefore lower loads. Continuous intersections can also be waisted so that the structure can be tuned for shear loading at these points. Ideal response is a square wave force-deflection curve and this result can be approached by using some ribs of lower height as seen in Fig 80.

Choice of thermoplastic material revolves around the idealized true stress-strain response depicted in Fig 81. While the modulus controls the structural stiffness during the first stages of loading, once instability is reached plastic yielding criteria take over. The material then must reach high strain levels before plastic flow, without risk of fracture. The greater the energy absorbed the higher is the required plastic stress point at which plastic flow

PARAMETER	VEHICLE	FMVSS No. 214
TTI	4 - Door	85 g
	2 - Door	90 g
A_B	4 - Door	130 g
	2 - Door	130 g

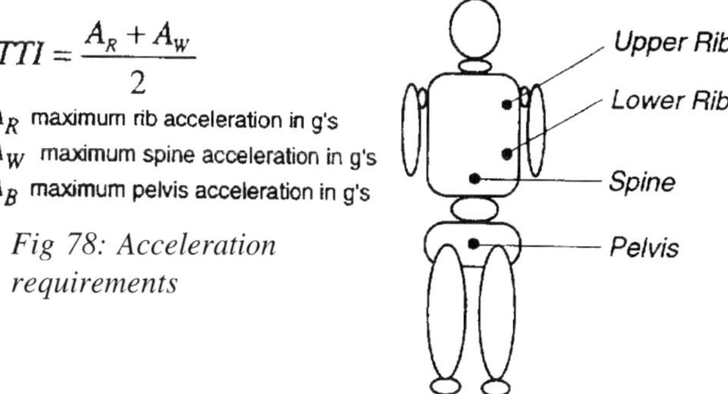

$$TTI = \frac{A_R + A_W}{2}$$

A_R maximum rib acceleration in g's
A_W maximum spine acceleration in g's
A_B maximum pelvis acceleration in g's

Fig 78: Acceleration requirements

Fig 77: FMVSS 214 acceleration measuring points

occurs. Plastic elongivity is also important. However, the yield stress controls the maximum cut-off load at which the absorber collapses.

Besides side impact, protection is also needed against impact with the upper components of the body interior, such as pillars and header rails, as prescribed in the new FMVSS 201. The test requirements are for impacting a 10 lb headform travelling in free flight over a minimum of 25 mm at a speed of 15 mph. An expression for head injury criterion defines the parameter involved in maintaining its value below 1000.

Fig 82 is typical of a rail padding system used to approximate the desired square-wave impact profile. The trim moulding is intended to deform locally to prevent the head sliding laterally and hitting other things. Foam core thickness and mechanical response are the prime parameters affecting energy absorbency, however. Mechanical fastening is the preferred means of attaching the foam to the rails but a 2-3 mm spacing is required to accommodate manufacturing tolerances and thermal expansion.

Again Dow has considered the alternative approach of integrating the absorbing system into the main design of the trim moulding, as in Fig 83, and similar arguments are advanced for optimizing the design of the 'egg-box' structures.

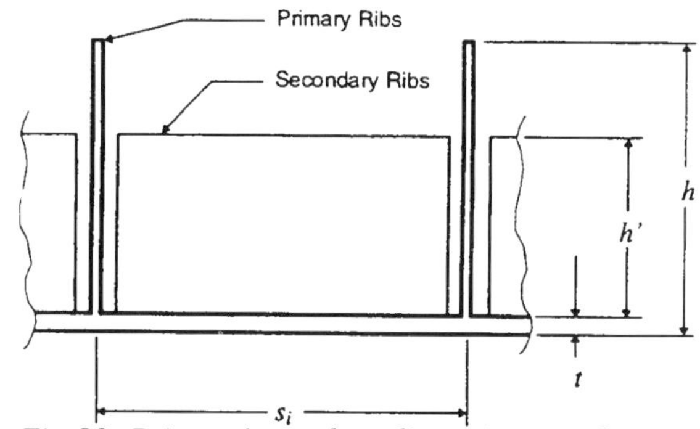

Fig 80: Primary/secondary discontinuous ribs

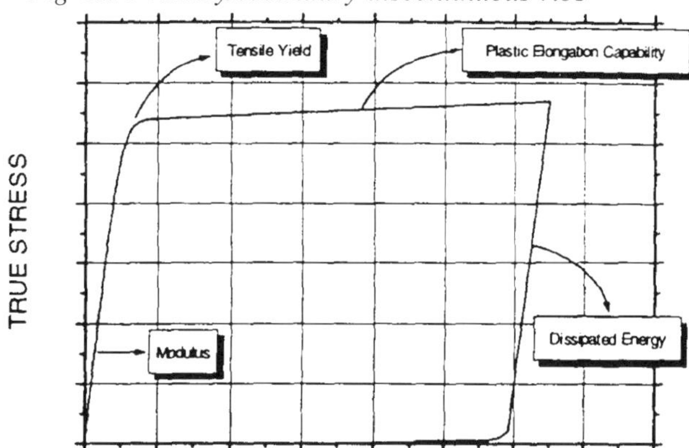

Fig 81: True stress-strain curve for thermoplastic

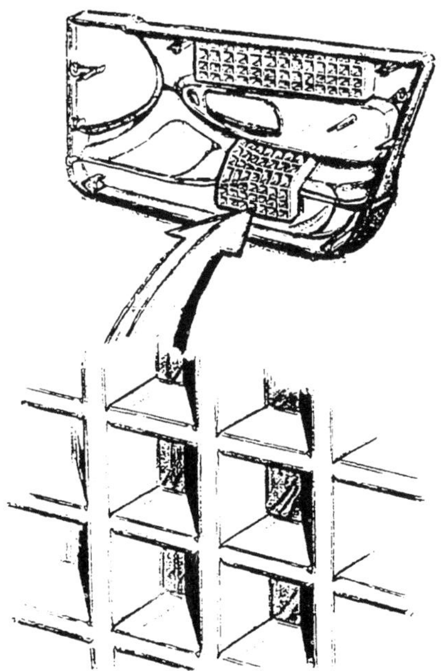

Fig 79: Integrated absorber in door panel

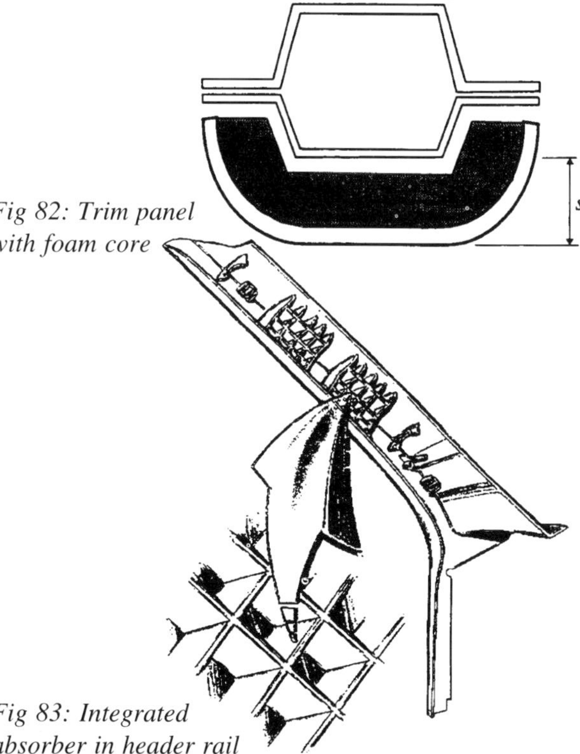

Fig 82: Trim panel with foam core

Fig 83: Integrated absorber in header rail

Structural design for seat mounts

The accident toll involving minibuses has raised awareness for the necessity of seat-belts in such vehicles. The Motor Industry Research Association has applied body engineering techniques to provide safe floor structures for mounting seats with integral belts. The recent opening of its enhanced HyGe impact testing facility, Fig 84, which offers clients the prospect of "non-destructive" crash testing by using inertial loading to test a variety of occupant-protection systems[9]. As a result of development work on the rig, the Association has produced an under-floor bracing framework which prevents any intrusion into passenger space while allowing replacement seats with integral safety belts to be fitted.

The two most significant crash scenarios for minibuses and coaches are frontal impact and roll-over; the former results in occupants impacting the rear face of the seat in front while roll-overs often result in ejection of passengers from the vehicle. Once ejected, the likelihood of severe injury is increased due to being crushed by the vehicle or striking hard objects. The adoption of seat belts is thus a primary requirement.

Research has shown that a two-point belt could adequately restrain an occupant and the seat in front could be used as an impact surface for the head whilst the upper torso rotated forwards. By optimisation of the seat stiffness characteristics and the seat spacing, 2-point belts can provide a reasonable level of protection for a 50th percentile male. Complications can arise when children and large adults are considered, as the optimum seat spacing may be different for these cases. Also, an incorrectly fitted two-point belt, for example one left loose, can result in abdominal injuries as the occupant is suddenly restrained once the slack in the belt has been taken up and the occupant's body is incorrectly positioned for the appropriate restraint around the pelvis. MIRA work has focused on the use of three-point seat belts that are integrated into the seats. An initial study was conducted to establish current practice tor the installation of seat belts in both new and used minibuses. The lack of quality standards in the evaluation of seat belt installations led MIRA to producing a report that outlined recommendations for the installation of seat belts in minibuses and coaches. The test standard recommended was the adoption of ECE 14.03 for M2 (minibuses) or M3 (coaches) vehicles.

Fig 84: Preparing a Hy-Ge sled

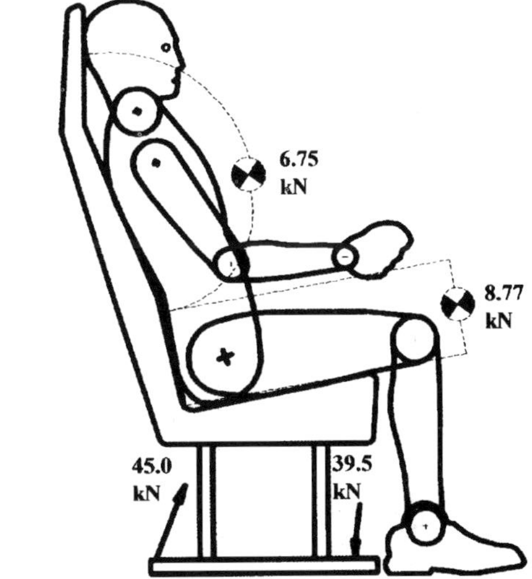

Fig 85: Loads on seat and floor mountings

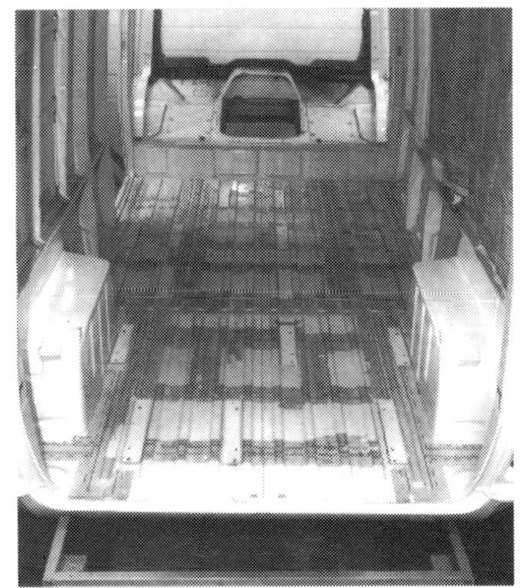

Fig 86: Position of seat mountings

Re-engineering the floor

Van floors are often constructed of longitudinally swaged thin sheet steel, spot welded to the main chassis rails and cross-members. The majority of minibus conversion companies rely on bolting the seats through the steel floor, spreading the load with large washers or plates. The loads experienced on seat mountings are illustrated in Fig 85 for a double seat, attached to the floor through two legs. The load case was taken from the European Regulation 14, for M2 vehicles. Although minibuses in the UK and Europe do not need to meet this (or any other) seat belt anchorage strength requirement, it was considered to be an appropriate design target. To re-engineer the floor, the initial step taken was to establish the strength and stiffness that was required to resist tearing the seats from the floor or cause excess deformation, when subjected to the test load.

A simple method of providing adequate strength is to construct a framework of square section tube that provides pick-up points for all the seat mountings. This framework is placed over the existing floor and is bolted through the floor using straps to clamp around the main chassis rails. As well as raising the centre of gravity of the seats the disadvantage of this method is that the weight of the framework, typically 120 to 180 kg, results in a loss of seating capacity because the maximum vehicle weight is limited by legal considerations. To produce a weight optimised solution, it was necessary to make the maximum use of the existing structure. This was achieved by using new, additional underfloor structural members that were attached to the main vehicle chassis rails and crossmembers.

Four minibus versions were to be developed for this range of vehicles. To reduce the number of variations certain 'worst case' seat layouts were chosen for development. 'Worst case' was determined by assembly which floor areas were given the minimum level of support by the main chassis rails. Additionally, areas that featured a spot welded seam where two floor sections were joined together were identified as being prone to tearing. The other areas of the floor were left to be proven in final validation tests.

The spacing of minibus seats is partly governed by the minimum required seat pitch to give sufficient leg room. However, for this project the seats were placed such that the optimum load path could be achieved between the seat mountings and the main vehicle structure (Fig 86). In particular, it was necessary to avoid bolting into the closed box created by the chassis components, since the drilling of any large holes in these key members could weaken the structure.

Basic calculations provided an initial indication of structural requirements but as they indicated that the failure mode may involve buckling and tearing of the underfloor, finite clement analysis was undertaken to account for the non-linear behaviour after the onset of plastic deformation. To minimise weight, plastic deformation of the seat mounting system was acceptable but a progressive failure was required, to prevent a sudden fracture and loss of load carrying capability. The non-linear dynamic finite element code OASYS DYNA 3D was used to

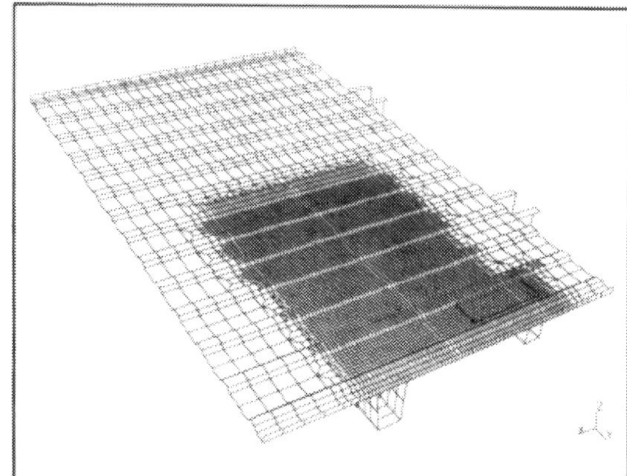

Fig 87: FE Model of floor structure

Fig 88: Small reinforcement plates

analyseinitial design proposals (Fig 87). Optimum section sizes, material specification and thickness were predicted and the solution that provided the best performance for the least weight was chosen for evaluation. The vehicle manufacturer did not want the additional structural members to be welded in place because this process would disturb the PVC anti-rust coating on the underbody. Also, control of welding quality is difficult with fully built up vehicles and a quality assured design solution was essential. Zinc coated steel 'pop' rivets were selected to attach the new sections to the main structure working in a shear mode. These fasteners provided adequate strength with no corrosion problems and allowed rapid assembly. The front mounting of each seat is loaded downwards under test and the design of the floor reinforcements was optimised to account for this load case. To prevent fatigue failure around the seat mounting holes, load spreading plates were specified for the top side of the floor, bonded in place with polyurethane adhesive to provide a good shear connection.

The form of the kit varied significantly from one area of the floor to another (Figs 88 and 89). This was because in well supported areas adjacent to a structural member, a small fixing was adequate.

However, for the first two rows of seats, steel box sections were needed to stiffen the vehicle and provide adequate strength. The finite element analysis model indicated the areas that required a higher level of reinforcement.

Fig 89: Box-section reinforcements

Chapter 10: Noise, Vibration, Harshness

Noise & vibration basics; noise analysis applied to vans; advanced noise analysis techniques; acoustic control with polyurethane foam; active noise-control; anti-noise enclosures; body structural vibration; structural dynamics applied; chassis beaming vibration; designing against fatigue; rubber as an isolator; CV-body mounting

NVH inputs: a classification method

Classifying road-induced NVH in terms of both mechanical and acoustic vibration, has been offered by classic work from Pirelli researchers[1]. In considering these random vibrations, which have a very wide frequency spectrum, this author has perfected a system of 'percentage classification' based on the production of a continuous frequency spectrum of (amplitude)^2 values.

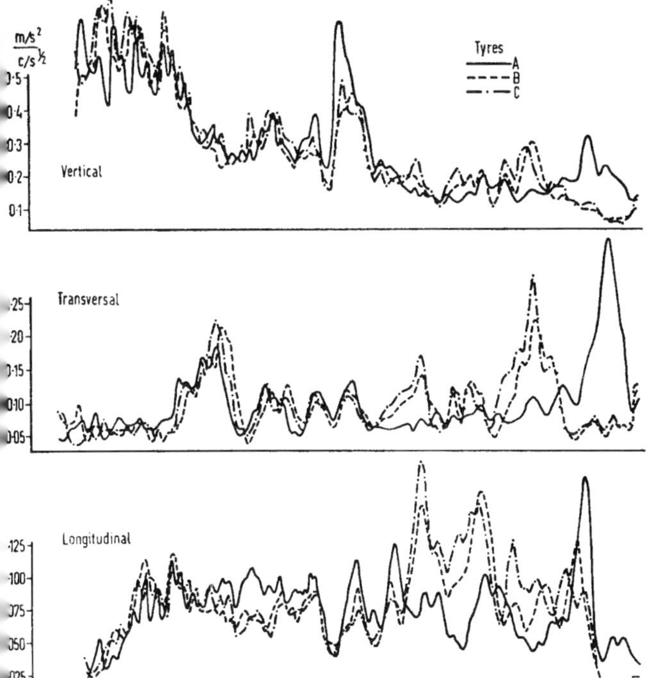

Fig 1: Vertical (top), transverse (centre) and longitudinal (bottom) car vibrations (metres/ sec^2 per cycles/sec$^{1/2}$ against cycles/sec base)

In analysing the trace recordings from strain-gauge accelerometers (linear response, 0 to 250 c/s) and a microphone mounted on the vehicle, very narrow band filters (1/15 of an octave wide) had to be used, owing to the complex spectral outlines and acuteness of the resonances. The recording was fed through the analysing equipment — in closed loop form — at eight to 16 times the recording speed. After filtering, the signal was squared by an electronic multiplier and integrated to obtain the mean square value throughout the recording interval.

The author explained that, in general, the objects of such work were either to find which parts of the vehicle produced an oscillation or to compare different suspensions and tyres in terms of comfort evaluation. Subjective impressions gave:

0.5 to 3 c/s — overall rigid movements of the human body (may lead to 'sea-sickness')

10 to 40 c/s — inner displacements of individual parts of the human body

50 to 60 c/s — skin and acoustic effects.

Mechanical considerations in the vehicle gave:

low frequency — rigid oscillation of the vehicle body on its main suspension (mainly vertical and sometimes transverse, on tyres)

middle frequency — resonant oscillations of the unsprung mass, due mainly to tyre vertical stiffness (also the engine on its mounts, and other component resonances)

high frequency — structural and component vibrations and 'distributed vibrations' of the tyres.

Procedure was as follows:

1. Single-out bands, showing characteristic differences, from the frequency spectra
2. Evaluate area between spectrum and frequency axis (in linear scale) for each band
3. Divide areas by bandwidths to give root mean square densities

4. Express densities in percentage terms in relation to the average value.

An example given concerned two cars weighing 1100 and 1400 kg and having differing mechanical structures, each in turn being fitted first with cross-bias-carcase and then radial-ply tyres of differing belt materials. Speeds of 70 and 40 km/h were adopted on rough and very rough road surfaces. Only the body front-end and acoustic spectra are reproduced in Fig 1, for one of the cars on the rough surface at 70 kph, about which the following general observations were made by the author:

Vertical: bounce, pitch, roll and coupled oscillations superimposed predominated at 3-4 c/s. Transmitted oscillations of unsprung mass, engine vibrations, etc. appeared in the middle range, and tyre distributed vibrations in the higher range.

Transverse: Spectrum very similar to that of unsprung mass.

Longitudinal: Similar to transverse generally, but increased importance of high over medium range.

Acoustic: Energy concentrated between 20 and 100 c/s. Evident correlation with longitudinal vibrations.

Noise and vibration basics

Audible noise and uncomfortable high frequency vibrations are close to one another on the frequency spectrum and can often be reduced by similar design treatments. The classification of noise sources is dealt with here alongside measures for isolation and/or insulation /damping.

Those making attempts to design against noise inside vehicles must recognise the parallel paths of air-borne and structure-borne noise when making predictions of likely noise levels. The orders of sound level attenuation need also to be appreciated before embarking on relative design solutions. Experience has shown the possibility of lowering engine noise transmission, to the vehicle chassis/body structure, by 15-25 dB, using well designed soft-mount systems. The power-train is of course the main noise source — with transmission gearbox and final drive being up to 20 dB lower than engine noise level. Lower still are levels of noise transmitted by wind and road. However the level of all these sources added together can be amplified by 20-30 dB due to the effect of structural and cavity resonances within the car-body, truck-cab or bus saloon.

In general terms acoustic power radiated by vibrating panels is proportional to the square of their area. Fig 2 shows typical sound pressure levels inside a vehicle body relative to the excitation frequency of different noise sources. The 'components' of sound need to be extracted from such results in order to determine the relative annoyance levels of different sources. As well as weighted sound level, dB(A), measurable by portable meters, 'loudness', in sones, and 'articulation' are further important criteria. Sones provide an approximate measure of auditory sensation while articulation index measures the ease with

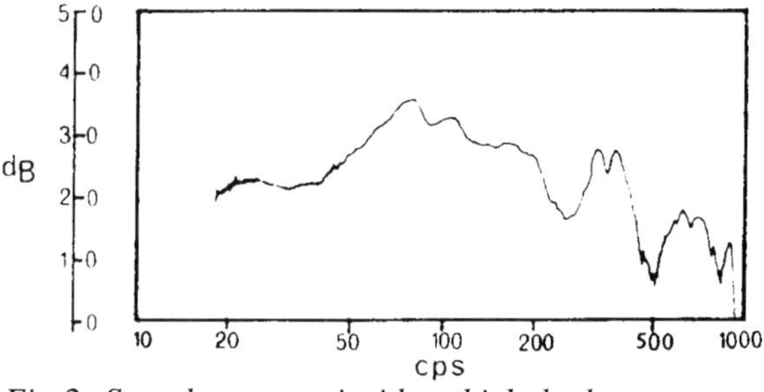

Fig 2: Sound pressure inside vehicle body

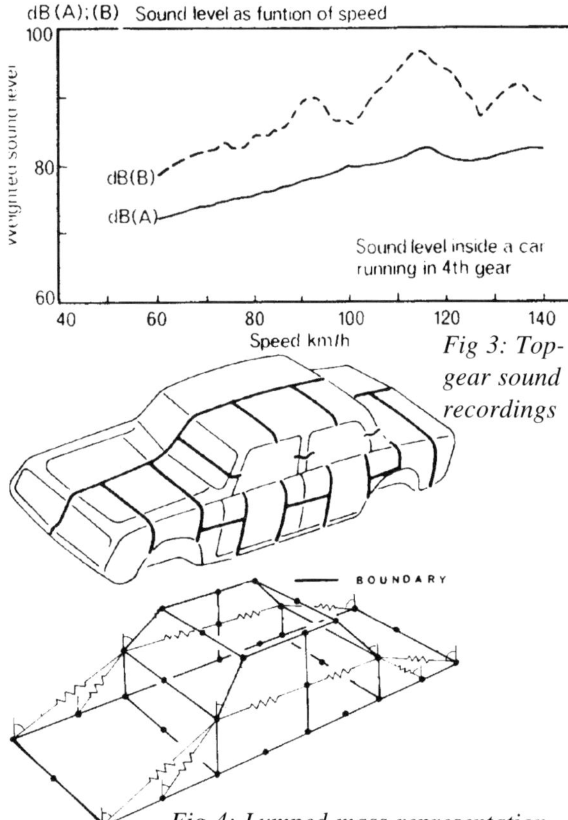

Fig 3: Top-gear sound recordings

Fig 4: Lumped-mass representation

which two people at close quarters can converse.

Also helpful in appreciation of the problem are sound level *vs* top-gear vehicle speed records such as that in Fig 3. Here dB(A) is the normal 'comfort' sound level while dB(B) gives a good picture of behaviour at low frequencies. Panel vibrations are generally responsible for high frequency annoyance while low frequency effects come from overall torsion and flexure of the structure in the so-called 'shake' mode. One approach to analysing the latter effects is to represent the structure, as in Fig 4, by 'lumped' masses of different parts of the structure concentrated at points connected by springs representing stiffness of panels, rails and pillars.

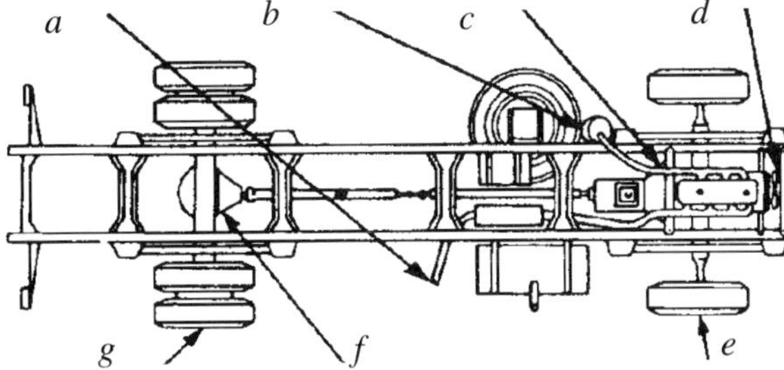

Fig 5: CV noise component intensities — a. exhaust, b. air intake, c. engine/gearbox, d. cooling fan (all $N^{4.5}$); e. front tyres (V^3W^3), f. rear axle (V^2), g. rear tyres (V^3W^3)

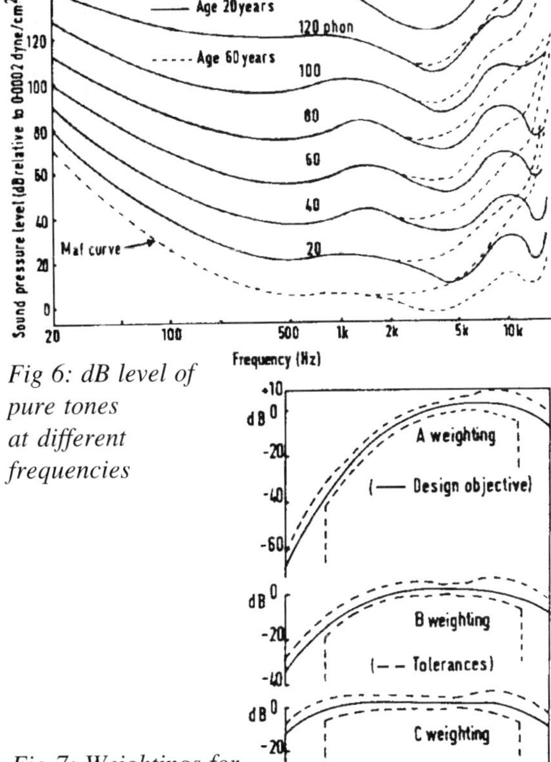

Fig 6: dB level of pure tones at different frequencies

Fig 7: Weightings for different dB levels

The decibel measure of sound pressure level is $20\ log_{10}(P/P_{ref})$ where P_{ref} is a reference pressure of 0.0002 microbars. When making a measurement it is usual to insure the background noise level is 10 dB below that of the object being measured. When considering noise levels inside a body it must be remembered that there is also a 'cavity resonance' effect akin to a stationary sound wave inside a tube. A body has a natural 'boom' frequency at which this resonance is excited at a particular road speed. A 'stationary' wave is one in which outgoing and returning waves add to produce a wave of zero intensity while a 'standing' wave is one where power is absorbed at one end and therefore the intensity is not zero. Approximations for the noise levels of various 'components' of the power train are shown in relation to vehicle weight *W*, speed *V* and engine speed *N* — useful as a starting point in designing sound attenuation and shielding systems in commercial vehicles, Fig 5. Sound pressure levels, being logarithmnic, cannot be added arithmetically and, for example, two pressure levels of 80 dB would combine to give an output of 83 dB. In general the total sound pressure level is given by:

$$10log_{10}[(N_1/10^{10})+(N_2/10^{10})+(N_3/10^{10})+\ etc]$$

where N_1, N_2 etc are the levels being added.

Loudness is not just a function of sound pressure as the ear is not equally sensitive to all frequencies. One measure of loudness is the phon which is based on comparing the sound source being measured with another pure sine wave of 1000 Hz coming from directly in front of the observer. When the two are judged equally loud the pressure level of the reference tone is the loudness in phons. Fig 6 shows the decibel level of pure tones at various frequencies required to produce equal sensations of loudness — based on the average of a number of observers.

This work was carried out at the National Physical Laboratory in the course of establishing British Standard 3383. Weightings for the different decibel levels are shown in Fig 7. As the phon scale gives no indication of the change in noise necessary to, say, double the loudness of a sound, another measure, the sone, had to be defined, Fig 8. Loudness level P in phons relates to loudness S in sones by:

$$P = 40 + 10\log_2 S$$

Airborne noise enters the vehicle interior through inadequately sealed holes and closures. Parallel noise transmission through the structure may be difficult to detect if it is about at equal level with the airborne noise. Decoupling the noise generating source from the structural path is the prime method of structural noise reduction while a combination of hole sealing and panel lining can be used to prevent ingress of exterior airborne noise as well as that generated by vibrating panels on the inside of the body. A designers-guide[2] published by the US Lord Corporation is available from Metzeler and Lord Gimetall. This provides the basic theory of decoupling using spring/damper mounting systems. Fig 9 shows the conditions for controlling the free vibration by providing a predictable spring rate and damping coefficient in the mounting. This determines the natural frequency of the system and the amplitudes of successive vibrations (x_1 and x_2) of the mass. Transmissibility curves for damped mounting-springs can be obtained as in Fig 10 to define the ratio of input to output amplitude. For an undamped spring, transmissibility is $1/[(f_d/f_n)^2 - 1]$ for values of f_d/f_n less than or equal to $2^{1/2}$ where f_d is the disturbing frequency and f_n is the spring natural frequency. The figure reflects this formula and its modification for different damping ratios is given by $(C/C_c) = (1/2)K''/K'$. In considering overall noise and vibration levels in a vehicle a distinction should be made between the periodic excitations originating from the rotating parts and the random nature of road input vibrations. Successful evaluation of road input effects is usually based on the examination of in-service input data collected on suitably instrumented vehicles. As well as for ride sensitivity, the vibrations of the vehicle can also be classified in three groups corresponding to known human sensitivities. In the 0-3 Hz frequency range, motions are evoked by the suspension springing and levels of discomfort worsened by pitching and rocking motions overlaid on the basic bounce motions. In the 3-30 Hz range, vibrations of the sprung masses come into effect while from 300 to several hundred Hz drive train and audible vibrations from the tyres and wind come into play. In the 40-70 Hz sector, unpleasant boom resonances occur just on the audible threshold.

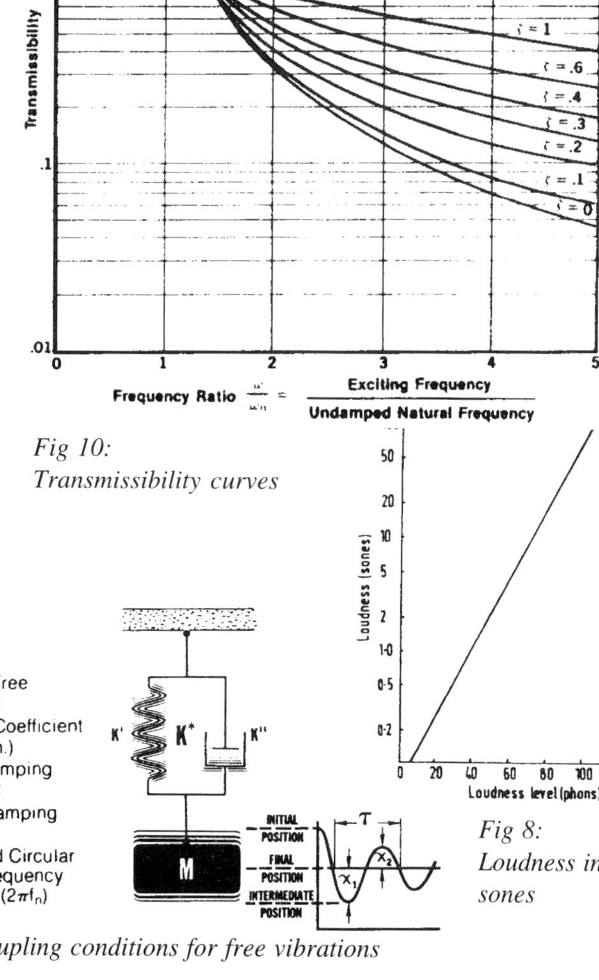

Fig 10: Transmissibility curves

Fig 8: Loudness in sones

$$f_n(\text{Hz}) = \frac{1}{2\pi}\sqrt{\frac{K'}{M}} \approx 3.13\sqrt{\frac{1}{d_{stat}}}$$

$$\tau\,(\text{sec}) = \frac{1}{f_n}$$

$$\frac{K''}{K'} \approx 2\frac{C}{C_c} = \frac{2}{\omega_n \tau} \ln \frac{x_1}{x_2}$$

where τ = Period of Free Oscillation
C = Damping Coefficient (lbs.-sec./in.)
C_c = Critical Damping Coefficient
$\frac{C}{C_c}$ = Viscous Damping Ratio
ω_n = Undamped Circular Natural Frequency in rad/sec. ($2\pi f_n$)

Fig 9: Decoupling conditions for free vibrations

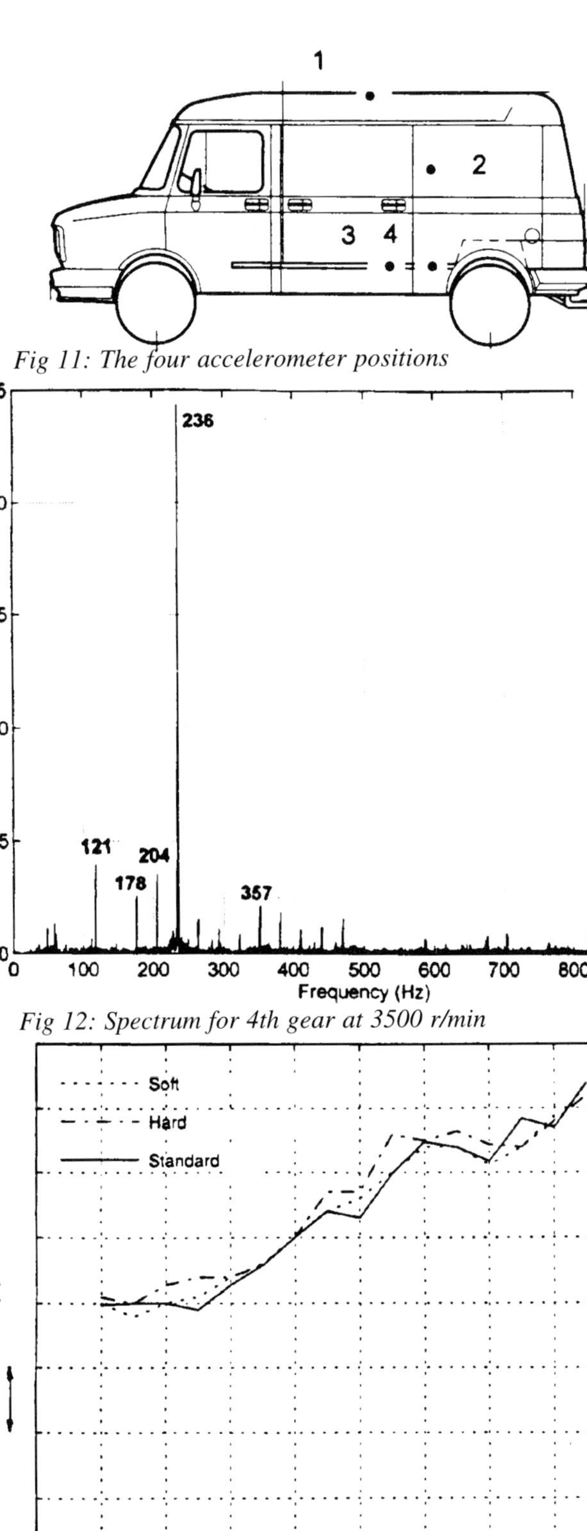

Fig 11: The four accelerometer positions

Fig 12: Spectrum for 4th gear at 3500 r/min

Fig 13: Effect of gearbox rubber mounting

Van interior noise example

Researchers at University of Central England[3] have described measures to reduce boom noise in a light commercial vehicle. The noise, which occurred in the 100 to 300 Hz range, resulted from the forces transmitted through the structure from the vibration of the engine on its mounts. A three-dimensional acoustic model was employed to compute the cavity modes of the load compartment. The noise was diagnosed as being primarily caused by panel vibrations exciting the first and second vertical acoustic modes of the load compartment. The panel and acoustic mode frequencies, as well as agreeing with each other, also coincided with measured engine orders present in the noise. The solution was to reduce coupling to the cavity by applying several longitudinal stiffeners to the inside of the roof pressing. The noise in the compartment, as measured at the passenger's position, was reduced by 4 dB(A) as a result. In addition, mount softening made an overall improvement of 2 dB(A).

Track tests were conducted to determine the dominant frequencies of the interior noise at all points in a typical light commercial vehicle's speed and gear range. Then the natural frequencies and mode shapes of the vehicle body panels were measured using a laboratory rig. Next, an acoustic model was developed to find the frequencies of the acoustic modes of a 3-dimensional space representing the vehicle interior. The predicted frequencies were compared with those measured in the road tests and near agreement taken as effective identification of the acoustic modes responsible for the noise. Further agreement with panel frequencies was taken to indicate a mechanism of acoustic mode excitation.

An eight point system comprising four microphones and four lightweight accelerometers was set up as in Fig 11 and data recorded using a portable digital computer-based data acquisition system. Data from the on-board system was processed for frequency information using Matlab software. Fourier analysis using FFT and auto-scaling in Matlab provided definition of noise resonance peaks.

An untrimmed vehicle body was mounted on soft air springs in the laboratory to approximate free-body conditions and provide the maximum isolation of the body from its support. An electromagnetic shaker was attached to the front offside corner of the body shell via a stinger and force transducer. Frequency Response Functions (FRFs) were measured for four separate panel areas of the body side-wall using a laboratory analyser and a roving lightweight accelerometer. By applying a random white noise signal and measuring the FRF at different points on each of the four panels, the main resonant frequencies were determined. The mode shape at each resonance was then found by measuring the FRFs at the resonance frequency at numerous discrete points in the panels and extracting the height of the resonant peak from each one.

A model to predict the acoustic mode shapes and frequencies of the loadspace cavity was produced based on rigid wall assumptions. The frequency of an acoustic mode with exactly *m*, *n* and *p* half wavelengths in 3-dimensions is given by:

$$f_{mnp} = (v/2)[(m/a)^2+(n/b)^2+(p/c)^2]^{1/2}$$

The lengths of the sides of the cavity, denoted by *a*, *b* and *c*, were obtained from the vehicle body dimensions. For reliable predictions, the dimensions of the cavity are critical and were tuned slightly to give the best fit to experimental results.

Fig 12 shows the large noise resonances of the vehicle tested and, by comparison with the predicted mode frequencies in the table of Fig 13, reveal the predictability of the resonances. As examples, the resonances near 121 and 240 Hz are caused by the 1st and 2nd wave modes of the air in the vertical dimension of the van's interior cavity. Engine orders 1, 2, 3, 3.5, 4 and 6 are all present, and coincide with some of the predicted cavity mode frequencies at this engine speed. The excitation of the resonances coincides with the engine orders, but not just the expected 2nd order. The interior noise problems are essentially structural in terms of setting the natural frequencies but engine-related in terms of providing the harmonic forcing input.

Fig 14 shows the effect on the interior noise of substituting a softer gearbox rubber mounting. The greatest improvement, however, is achieved by the roof panel strengthening, Fig 15. The table in Fig 16 gives the panel mode frequencies for parts of the side wall obtained in the body tests and these also fall in the range of track test noise problems. Figs 17 and 18 show the mode visualisation for just one of these at 178 Hz, which is a direct match with a track test resonance. It is interesting to note the position of the

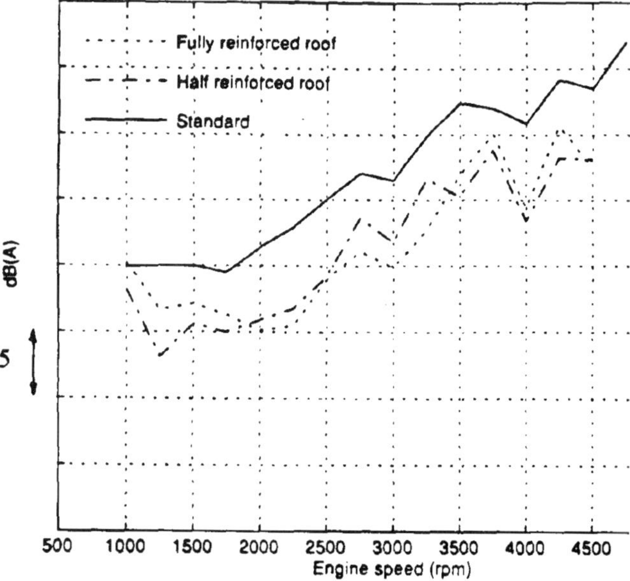

Fig 14: Effect of roof reinforcement

Combination of m and n				*Value of P*
-m	n	p=0	p= 1 .	p=2
0	0	0.0	121.0	243.0
0	1	106.3	161.0	265.0
0	2	212.5	244.0	322.0
1	0	47.2	130.0	247.0
1	1	116.0	168.0	269.0
1	2	217.0	249.0	326.0
2	0	94.4	153.0	260.0
2	1	142.0	187.0	281.0
2	2	232.0	262.0	336.0

Fig 15: Frequency eigenvalues (Hz) for low order acoustic modes in a rigid-wall rectangular box idealisation of the vehicle cavity (m = longitudinal mode number, n = lateral mode number and p = vertical mode number)

Panel Number	*Frequencies (Hz), various modes*			
1b	80	135	185	
2b	90	146	178	
3b	79	115	138	190
4b	78	144	186	

Fig 16: Panel resonance frequencies for panel modes encountered in body testing

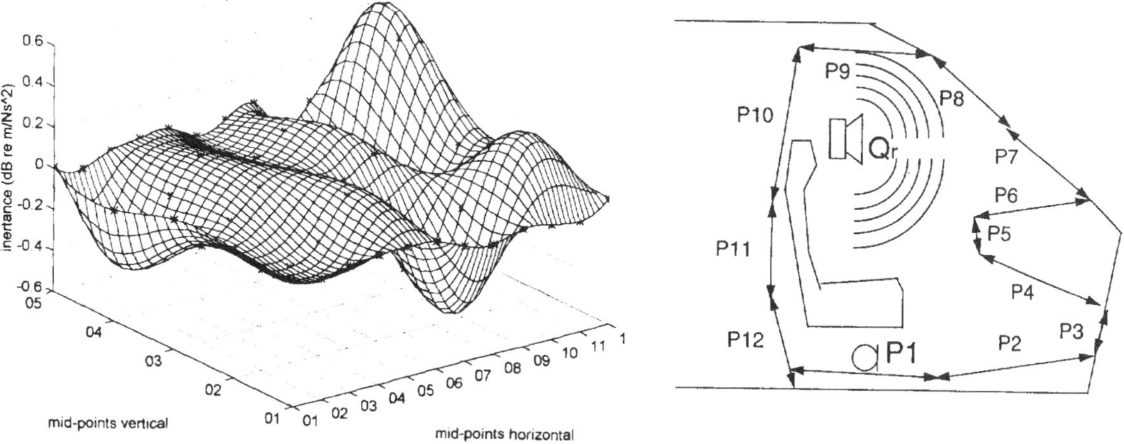

Fig 17: 3d mesh visualisation of sidewall panel at 178 Hz

Fig 19: Reciprocal measurement for vehicle interior

damping member behind this panel, the authors suggest, and to speculate as to whether its design could be improved to reduce this mode at 178 Hz.

Sources of airborne vibration According to LMS and Renault researchers[4], the method of Airborne Source Quantification (ASQ) is useful in analysing the contributions of body panels to interior noise, providing both phase and amplitude information. The possibility is shown to exist of identifying dominant contributions as well as the absorbent contribution of panels. The experimental approach makes use of reciprocal acoustic transfer function measurements alongside those for the panel surface acceleration during service. The paper describes application to a 4-cylinder diesel van and tuning of the panels to reduce noise.

Essentially the method involves break down of interior noise into contributions of panels or parts of panels, source strength of which is quantified in terms of volume velocity, while the body and cavity are simplified to a source transfer system. The latter quantifies the effect of trim absorption and cavity modes between each panel and the sound pressure at the occupant's ear. Thus the acoustic transfer describes the pressure response at a certain position in the cavity due to the volume velocity of the defined panels.

Because of the huge effort required to ensure transfer functions to build volume velocity sources into each panel, use is made of the reciprocity relation

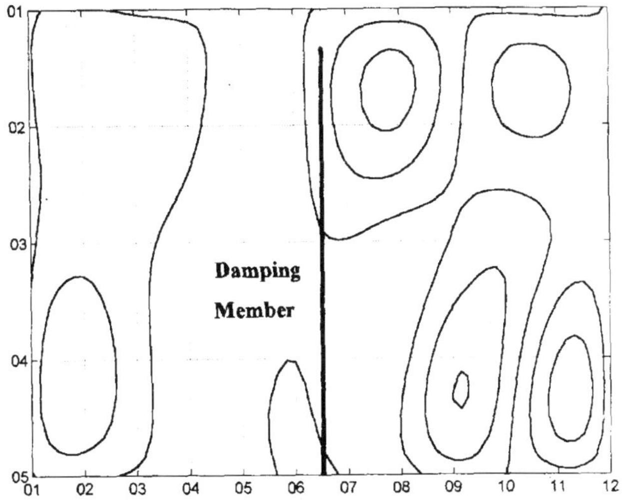

Fig 18: Sidewall contour at 178 Hz

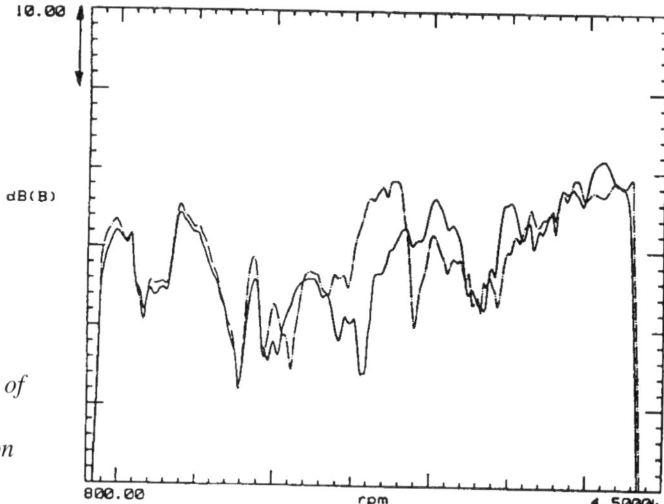

Fig 20: Second order component of sound pressure at driver's ear (solid line) and passenger position (dash-dot line).

of the transfer functions and a sound source placed at the ear position while small microphones are placed at the panel locations, Fig 19.

For the diesel van in question the interior noise spectrum has critical states at two engine speed ranges. Under full load, the second order component of engine noise becomes dominant at 1400 r/min throughout the cab and at around 2600 r/min on the driver's side, Fig 20. While the second order components of the engine forces had already been optimized, for the two critical speed ranges, noise transfer functions at the force input locations have maxima at the corresponding frequencies of 45 and 90 Hz, Fig 21. The objective was to identify panel contributions to interior sound pressure over the entire engine speed range and then devise modifications to reduce body sensitivity.

Examples of measured pressure over volume velocity transfer functions are seen in Fig 22. An additional acoustic modal analysis was made on the basis of the same acoustic isolation, as seen in Fig 23. All panel contributions were then calculated and summed, their sums being compared to two direct measured 2nd order sound pressure curves. Fig 24 is an example of identified contributions under third gear full throttle acceleration. The panels could be out of phase with the principal exciting ones and so acted as noise absorbers.

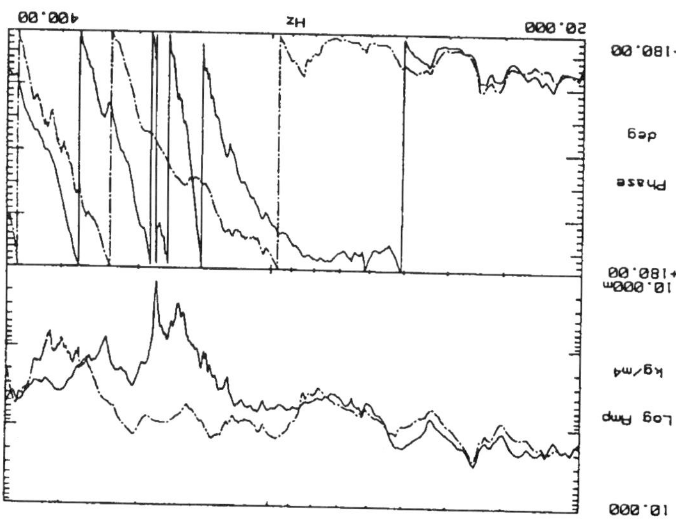

Fig 23: Acoustic FRFs from upper (dash-dot) and lower (solid line) rear wall to driver's ear

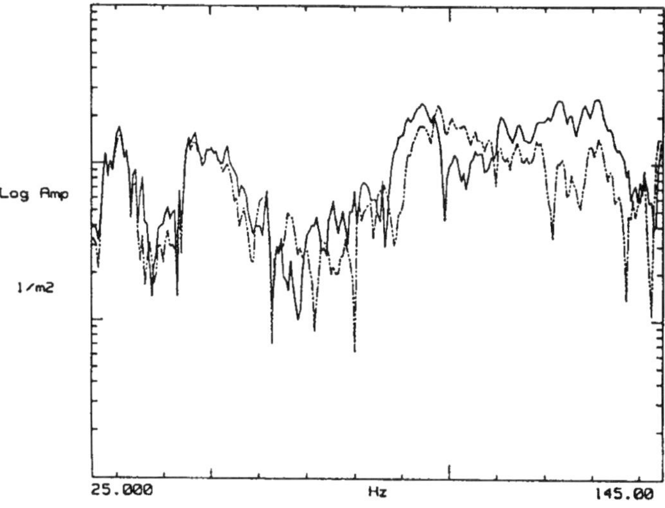

Fig 21: Vibro-acoustic FRFs from engine-mount to ear — driver, solid line; passenger, dash-dot

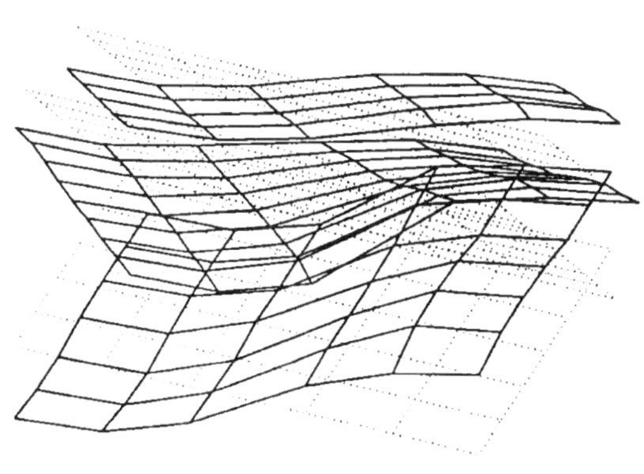

Fig 22: First acoustic mode seen from top right of cab at 91.45 Hz

At 1500 r/min engine speed, well below the first acoustic modes, the windshield was the dominant panel source, exhibiting global bending combined with deformation at the boundaries. A modification was therefore made to lower the bending mode frequency to 34 Hz which lowered the second order maximum at 1500 r/min throughout the cab but it increased the second order level around 1100 r/min, considered untypical under full load conditions. At around 2400 r/min, the first lateral acoustic mode determines cab response and left/right footwell floor pans were found to be the dominant contributors. Because these already stiff panels were difficult to modify, the first plate-bending modes of two smaller panels were tuned to 90 Hz, these being well placed near the pressure maxima of the acoustic mode and therefore acted well as acoustic vibration dampers, reducing the maximum of the second order curve as seen in Fig 25.

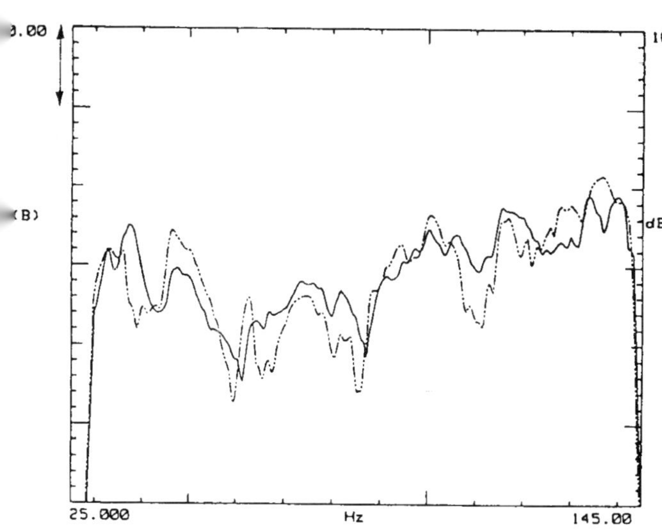

Fig 24: Effect of panel tuning on 2nd order noise: original dashed ; modified solid

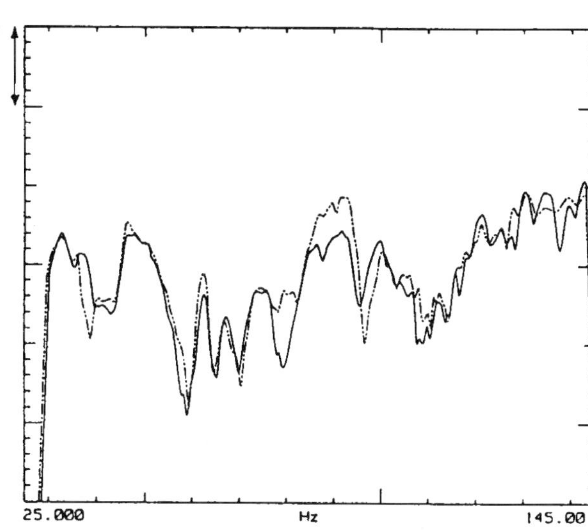

Fig 25: Effect of tuned panel absorbers

Techniques of noise analysis

Consultants have generally taken over NVH analysis from vehicle OEMs. Mascotech, for example, provide a strong capability at the design concept stage because of their accumulated expertise in predictive techniques. It was crucial, for example, to ensure that modifications to the structure for inducing controlled-collapse characteristics in impact did not, at the same time, worsen fatigue performance or alter panel vibration modes that would induce unpleasant noise. Use of predictive techniques has taken as much as one year from a five year new-model programme, the company maintain. Analytical techniques were now able to predict 90% of final product performance.

A key facility was the ability to synthesise the sound before any physical prototypes are built and present the results on a loudspeaker to the client. The effects of different powertrains in different bodies could thus be simulated. Considerable variation could even result from different tightening sequences for the powertrain mountings; the techniques could thus be used to control production as well as product engineering.

Analysis techniques were also needed so that the OEM could define the load spectrum on a suspension arm, for example, so that the tier-one supplier provided this in a form which did not otherwise adversely affect the performance of the body structure, or vehicle as a whole. The noise analysis techniques, Fig 26, also fitted in with new requirements for end-of-line 'vehicle attributes' testing by the OEM.

An important parameter in noise analysis was the noise transfer function, defined by Mascotech as the noise at the driver's ear divided by the force input to the body structure. Fig 27 shows a diagram of key noise input points to the structure, each of which can be ascribed a transfer function to allow the client a chance to listen on a loudspeaker to the alteration or isolation of particular noise inputs at the design stage. The basic cavity resonance of the body interior is very much a function of the basic vehicle shape, but some variation in noise quality could be obtained by, for example, perforating the back-of-rear seat bulkhead so that the boot cavity, on a notch-back car, could be used as a Helmholz resonator to act on the primary cavity.

Suspension and powertrain mounting bushes

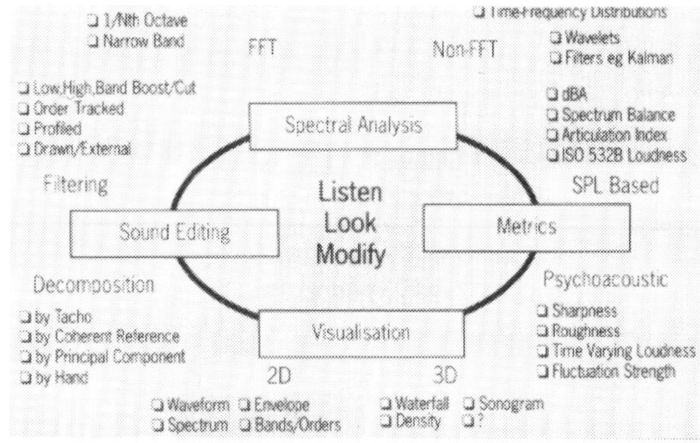

Fig 26: Mascotech approach to noise analysis

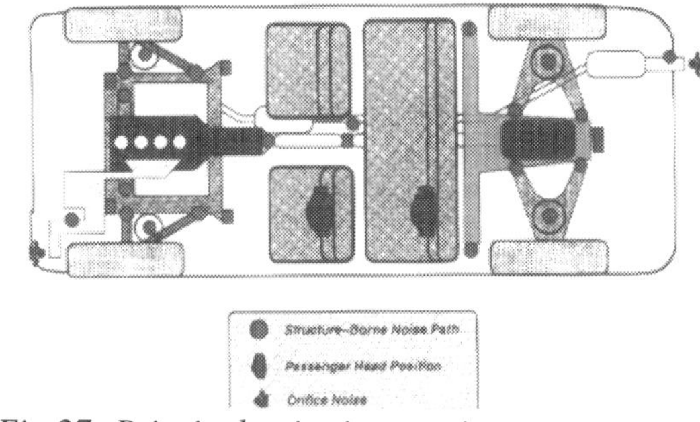

Fig 27: Principal noise input points

Fig 28a: Suspension of body shell on elastic ropes to excite load input points for modal analysis

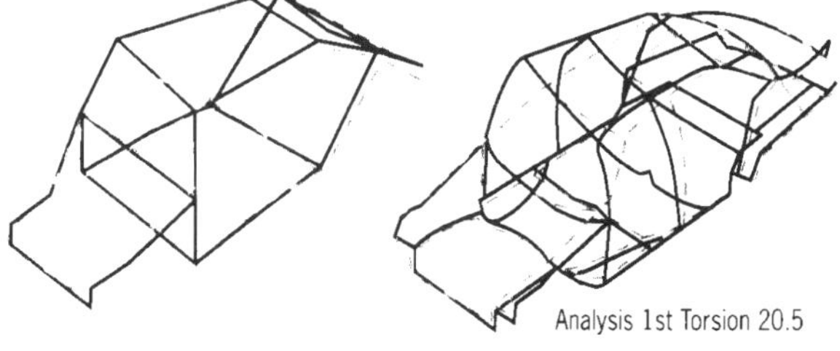

Fig 28b: Vibration modes in test and by analysis

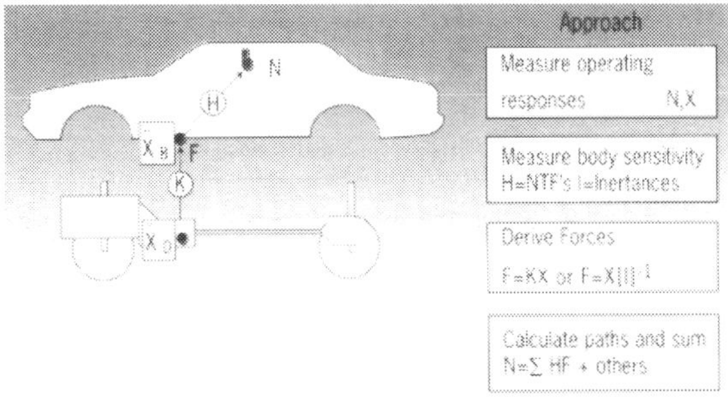

Fig 29: Building a noise-path model

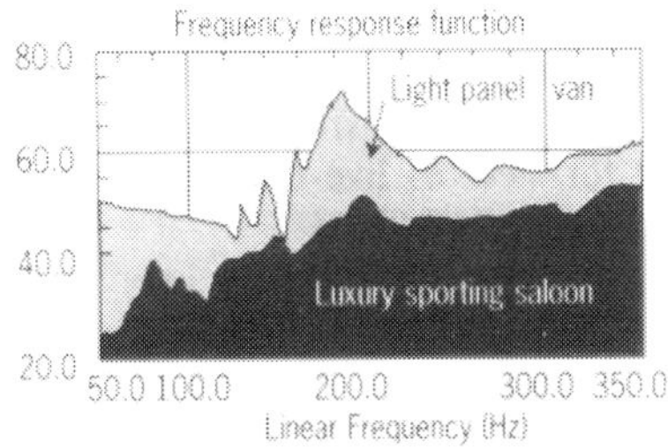

Fig 30: Noise signatures for sports saloon and van

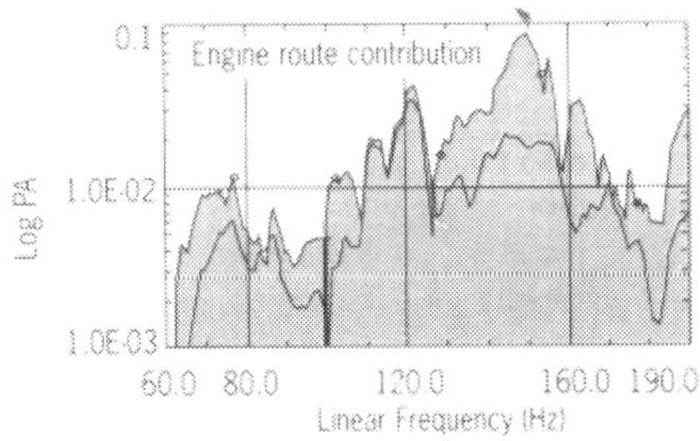

Fig 31a: Second order interior noise contribution of engine routes

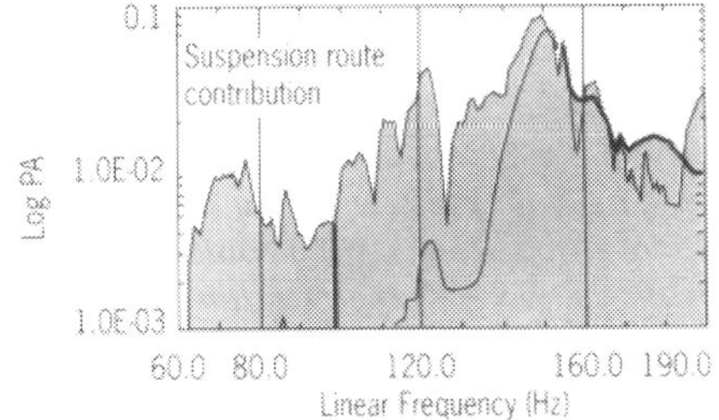

Fig 31b: Suspension contribution to total interior noise (shown shaded in both parts of the figure)

generally needed to be as large as possible, from a noise isolation standpoint, but sometimes adequate compliance could be provided on one axis for noise reduction while constraining motions in other axes that were commensurate with good handling or powertrain displacement control.

The verification of analytical simulation against experimental performance of body structures is a crucial part of the work. As well as overall modal behaviour of the body shell, carried out as seen in Fig 28a/b, localised structural responses are examined around the anchorage (input) points so that individual transfer functions can be modified. Company researchers described the general approach to sound quality analysis of the Jaguar XJ6 at a recent presentation[5] and illustrated the approach to building a noise path model from test data for the case of structureborne powertrain noise as seen in Fig 29.

The SDRC approach to route tracking of a noise is also a valuable tool and the company argue that below 500 Hz most noise is transmitted mechanically, and the first step in determining its path through the engine mountings, say, is to determine the three orthogonal forces in each elastomeric mount. This is done by fitting accelerometers to them during test runs and the signals processed to obtain displacements; these are then related to the load/deflection characteristics of the mounts.

The next step is to determine the vehicle structure's sensitivity to these mount forces. This is done by exciting the body with an electro-magnetic shaker and a microphone within the structure used to obtain a Frequency Response Function (FRF) relating force to noise computed.

Fig 30 shows a resulting noise transfer function obtained for a specialist car compared with that for a panel van. This shows a good sensitivity is in the order of 50 to 55 dB per N applied force whereas a poor example can show in excess of 70 dB per N. Forces and noise transfer functions can then be multiplied together to calculate the contribution of individual routes. Fig 31 shows contribution of engine and suspension routes to an example second order boom problem. In this case only the frequency content at the engine 2nd order (firing frequency) is used. Final stage in the troubleshooting process is to evaluate potential modifications to the vehicle by comparing recordings in the original and modified conditions.

Acoustic-control with polyurethane foam

According to ICI Polyurethanes researchers[6] the principal sources of in-car noise, over 50 per cent, are tyre/road noise, and noise from the engine compartment. Since each vehicle has its own noise 'signature', versatile methods of control are required, it is argued. The procedure is based on the use of heavy layers/carpets backed by flexible polyurethane foams - the mass/spring/dashpot principle. Key intrinsic properties of the foam need to be further tuned to the specific noise signature of each vehicle, taking into account any constraints from the foam thickness and heavy layer surface density.

The essential feature of interior vehicle noise is that it provides a broad background level on which discrete frequency bands are superimposed, Fig 32. The understanding of speech in a vehicle is measured by a single number rating called the 'articulation index.' The interior noise is weighted by an articulation function, plotted in Fig 33, which shows weighting parameters relative to 1 kHz. Annoyance, however, is caused by discrete frequency tones. The annoyance increases proportionally with differences between the band level at the discrete frequency and the broad band background. For determining the magnitude of articulation index and annoyance, the attenuating performance of the foam/carpet system must be known. In its simplest form, the system consists of a steel floorpan, foam core and heavy layer. The system parameters which have the largest influence on acoustic behaviour are listed in Table 1.

Using an appropriate sending chamber with a horizontal aperture covered by a sealed steel plate, Fig 34, the transmission loss of the panel and the panel plus carpet composite were measured in tests carried out by ICI. The steel panel was 1 mm thick and the aperture dimensions were 830 x 830 mm. A standard heavy layer, of surface density 6 kg/m^2, was used for testing, and the moulded foam thickness was always 20 mm. Each foam was also characterised in terms of its dynamic modulus and loss factor (see Table 1). These were measured using a forced vibration transmissibility test involving a random vibration input and narrow band analysis of the signal output. The arrangement of the moulded test piece is shown in Fig 35, and Fig 36 is a typical response curve. The foam modulus is calculated from the

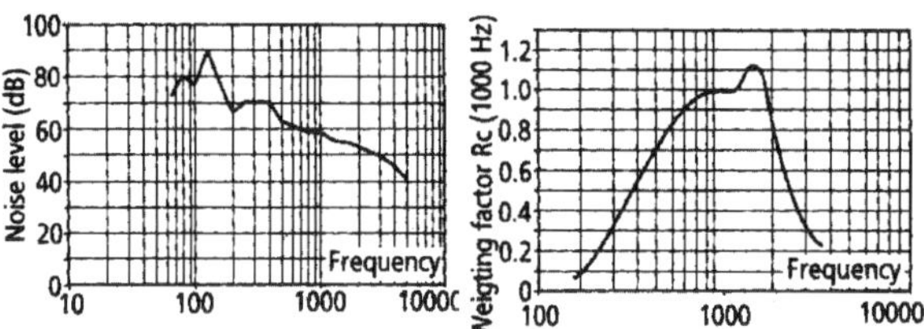

Fig 32: Typical 1/3 octave band interior noise spectrum

Fig 33: Articulation index weighting factors as a function of frequency

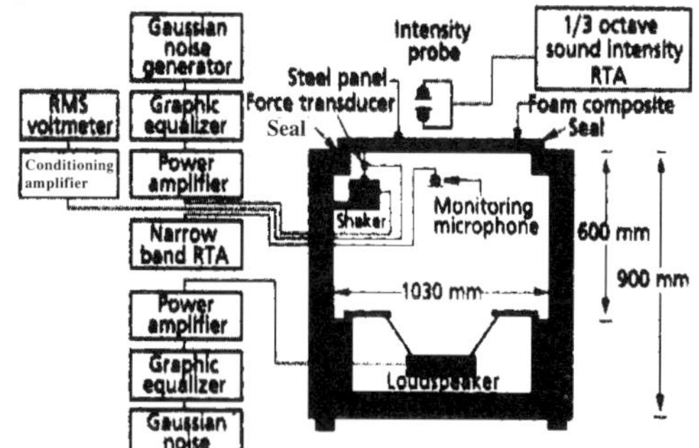

Fig 34: Laboratory test method

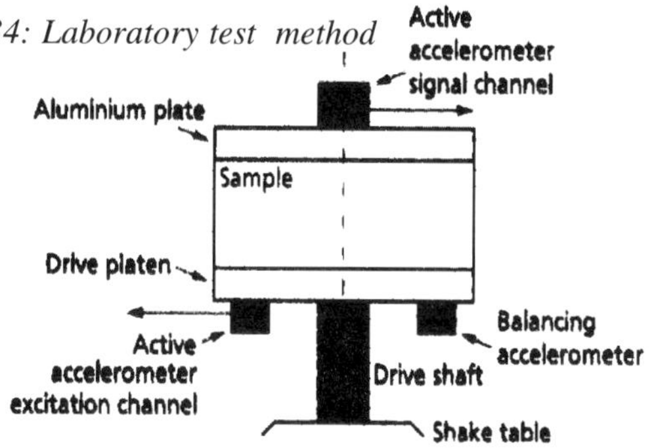

Active accelerometer signal channel
Aluminium plate
Sample
Drive platen
Active accelerometer excitation channel
Balancing accelerometer
Drive shaft
Shake table

Fig 35: Setup for measuring foam dynamic properties

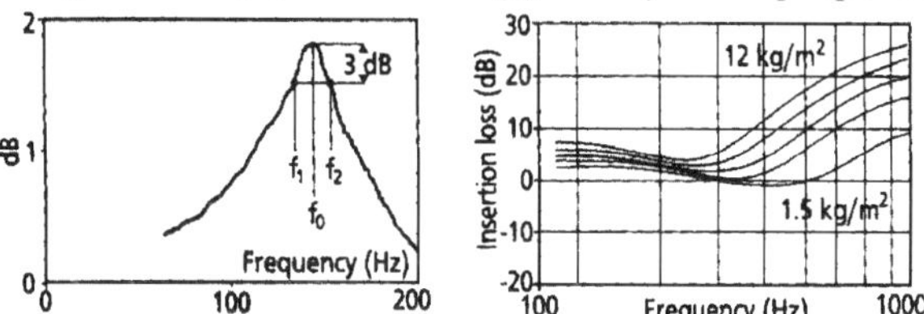

Fig 36: Response curve for foam/mass composite

Fig 37: Effect of increasing foam density

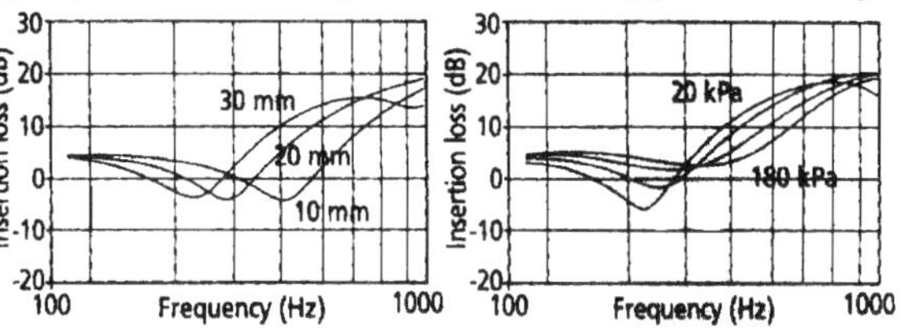

Fig 38: Effect of increasing foam thickness

Fig 39a: Effect of increasing foam dynamic modulus (h = 0.5)

resonance frequency f_O, and the loss factor η can be determined from $\eta = (f_2 - f_1)/f_R$.

The support mass for the samples was 0.46 kg and the rms acceleration was 2.2 m/s^2. Sample dimensions were 50 x 50 x 20 mm. Two series of foams were characterised. The first series, consisting of samples A, C and F (see Table 2), was designed to have increasing dynamic moduli whilst holding the foam loss factor constant and low. These are referred to as high resilience (HR) foams. The second series (P, G and J), was designed to increase modulus whilst holding the foam loss factors constant, but at a higher magnitude. These are referred to as viscoelastic foams (VEF), and are also listed in Table 2.

Figs 37, 38, 39 and 40 show the theoretical insertion loss as a function of heavy layer, foam thickness, foam dynamic modulus and foam loss factor. Apart from Fig 37, all predictions are for a septum density of 6 kg/m^2. Figs 37 and 38 show the expected result that increasing foam thickness and septum mass results in large improvements in insertion loss. In particular, increasing septum mass provides enhancement of performance in the key frequency range below 500 Hz. In practice, however, increasing these parameters is usually not a practical option, since weight and foam height increases are strictly controlled. Figs 39a and 39b predict that decreasing the foam dynamic modulus shifts the insertion loss curves towards lower frequencies. This improves the insertion loss at higher frequencies, but deteriorating it at lower frequencies. By increasing the foam loss factor, an improvement in low frequency attenuation is predicted, also shown in Fig 40. The improvement is not linear with loss factor.

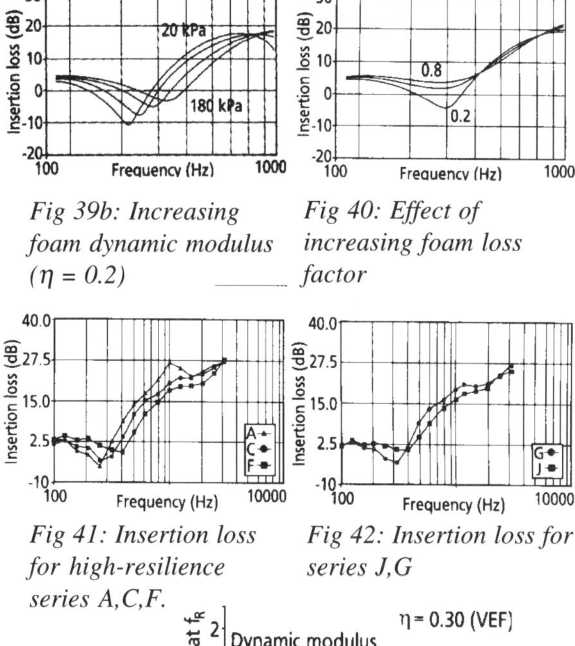

Fig 39b: Increasing foam dynamic modulus (η = 0.2)

Fig 40: Effect of increasing foam loss factor

Fig 41: Insertion loss for high-resilience series A,C,F.

Fig 42: Insertion loss for series J,G

Fig 43: Insertion loss at resonance for high resilience and viscoelastic foams

Fig 41 shows the experimentally determined insertion loss for the HR series, and good agreement with Fig 39a was obtained. Increasing the foam dynamic modulus improves the low frequency performance below 400 Hz but deteriorates the performance above 400 Hz. Fig 42 shows the insertion loss for the VEF foams, J and G. By increasing the foam loss factor from 0.16 to 0.30, the low frequency performance below 400 Hz is improved over that obtained by the HR series. This improvement is seen in Fig 43, which plots insertion loss at resonance against foam dynamic modulus. The resonance frequency point f_R is theoretically related to the composite design via the approximate equation:

$$f_R = (1/2\pi)[(K'/d)(m_1 + m_2)/m_1 m_2]^{1/2}$$

where K' = $E'_o + E'_f$; E'o = adiabatic bulk storage modulus of air (1.4 x 10^{5}Pa); E'_f is dynamic modulus of foam at f_R; d = foam thickness; m_1 =area density of

Table 1 Physical parameters that control noise attenuation:

Floor plan	Area density, m_1 (kg/m^2)
Heavy layer	Area density, m_2 (kg/m^2)
Foam core	Thickness, d (mm)
	Dynamic modulus, E' (Pa)
	Loss factors η
	Density, ρ (kg/m^3)
	Flow resistivity, R (rayl/m)

Table 3 Predicted/measured resonant frequencies:

Sample	f_R *experimental (Hz)*	f_R *theoretical (Hz)*
A	*250*	*245*
C	*250*	*267*
F	*400*	*340*
G	*315*	*284*
J	*400*	*401*

Table 2 Foam properties

Foam type	Density kg/m^3	Dynamic modulus E'x10^5 Pa	Loss factor η	Frequency of η,E' determination
High resilient foam				
A	55	0.22	0.18	40
C	53	0.52	0.16	54
F	50	1.70	0.15	100
Viscoelastic foam				
P	55	0.39	0.30	60
G	59	0.76	0.32	60
J	59	2.91	0.29	160

steel panel and m_2 = area density of heavy layer.

A comparison of the predicted values of f_R and the experimentally determined values is given in Table 3, showing the experimental results closely follow the theoretically explained behaviour. The two dominant foam intrinsic parameters, dynamic modulus and loss factor, are shown to act in different ways with respect to insertion loss.

Of the two, the most critical parameter is that of dynamic modulus, which controls the position of the resonance frequency. Since reducing the dynamic modulus improves higher frequency performance at the expense of lower frequency performance, a design compromise needs to be reached. Furthermore, the articulation index, and annoyance estimates, weight the insertion loss response differently. As a result, a particular vehicle noise input signature, requiring a defined balance between articulation and annoyance, can only be met by careful selection of an appropriate foam dynamic modulus and loss factor.

Active noise control

Although the principle of active noise control - the superposition of a sound wave of equivalent frequency, exactly 180 deg out of phase with the irritant noise - is very simple, its application to vehicle interiors is far more complex. This is because the noise sources both move and vary in frequency and amplitude during normal driving conditions. Thus any control strategy needs to both temporally and spatially match the undesired noise. Previous work has shown that the control of time-periodic and narrow band signals is far easier than that of random noise, especially if the signals appear in a broad frequency band.

Interkeller have been working on developing a modal control technique, whereby careful positioning of the loudspeakers allows them to counteract the cavity eigenmodes of the enclosure that make up the noise in the car. This technique is most applicable to enclosures with a small number of cavity eigenmodes, which are only slightly damped — limiting its usefulness to the low frequency range between approximately 60 and 250 Hz.

Of the components of noise that affect a car, wind, road and engine, the engine noise is both quasi-periodic and consists of harmonically related tones. Low-frequency ‘boom noise’, which can be traced back to a resonance phenomenon that amplifies the dominant second-order engine vibrations, is thus ideally suited to modal active control techniques. The company have been investigating the possibilities of combining active techniques for attenuating such noise, combined with passive treatment of other noise sources, and the interaction of the two.

Two types of input are required: interior noise from microphones, and engine rotational speed, which can be supplied from the engine management system. Using the engine speed for temporal matching prevents the acoustic feedback that could occur if only microphones were used. The microphones are used to collect information on the contribution of the individual modes at a particular moment. The control law must take into account the displacement of the passengers’ heads from the microphones. To achieve good results, the system uses more microphones than loudspeakers, resulting in a good spatial minimisation of the residual sound. A fast digital controller is used to generate the output for each speaker. Signal processing comprises mainly measurement and power amplifiers, A/D- and D/A-converters and filters.

Active noise control has been shown to be particularly effective in those cases where the components of the low engine orders dominate interior noise. As can be seen from Figs 44 a and b, the best results are obtained at engine speeds where the second order is predominant, namely 3500 to 4200 rev/min, and above 5000 rev/min. For back seat passengers, the second order booms are not as pronounced, and thus the system’s effectiveness is reduced.

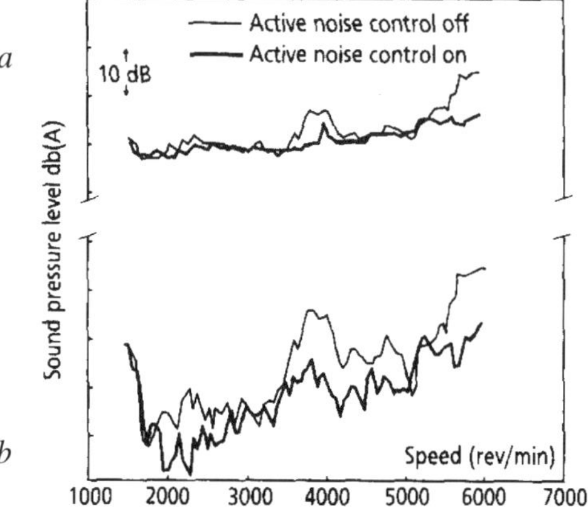

Fig 44a (top): Effect of active control on sound pressure level generally and b (bottom): effect on sound pressure level of second order component

Anti-noise enclosures

The method of isolating noise by means of 'structural enclosures' is often practised, whereby the noise source is enclosed in a box which insulates the emanation of airborne noise as much as possible. This can be the quickest and cheapest method of reducing perceived noise, which of course is airborne to the human ear, especially so if the design and prototype build is already well advanced.

The full enclosure is an extension to the use of noise shields which are sometimes preferred if the noise source is an engine which requires frequent maintenance access to it and needs airflow over its surface. Low stiffness and high damping are the requirements of such shields and it is necessary to isolate them structurally from the vibrating source so that they do not act as vibration amplifiers, typical methods of mounting being shown in Fig 45. Distance between shields and surface of the engine, say, must also be minimised in order to prevent noise build up in the intervening space. Obviously the shields do not seal in the airborne noise like an enclosure but merely interfere with the radiation of sound from the adjacent vibration surface.

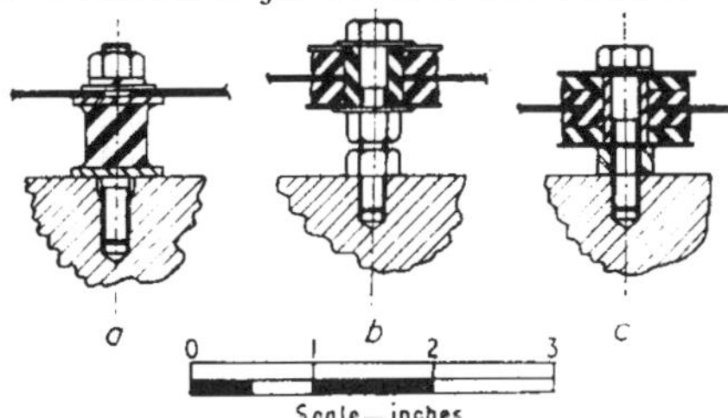

Fig 45: Types of enclosure mounting

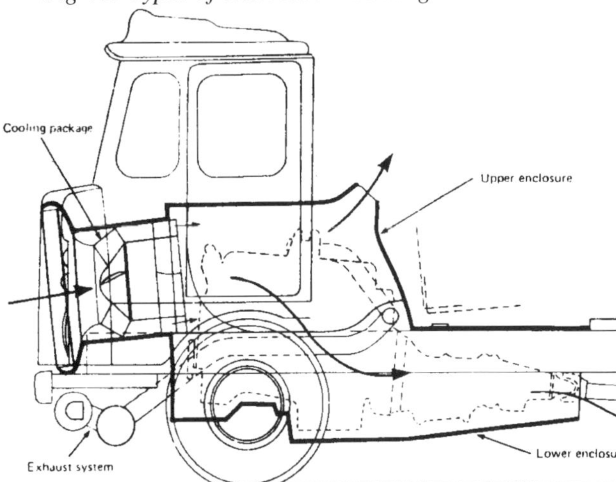

Fig 46: QHV enclosure

It is also possible to use a semi-sealed enclosure where the maintenance of airflow is vital but maintenance access less so. Fig 46 shows such an enclosure on the TRRL QHV 90 quiet heavy vehicle project where a so-called tunnel enclosure was used. This comprised an upper enclosure covering the top and back of the engine which lifted with the cab; separate panels fixed to the chassis covered the top of the gearbox and bottom of the power-unit; the complete set formed a duct through which cooling air passed.

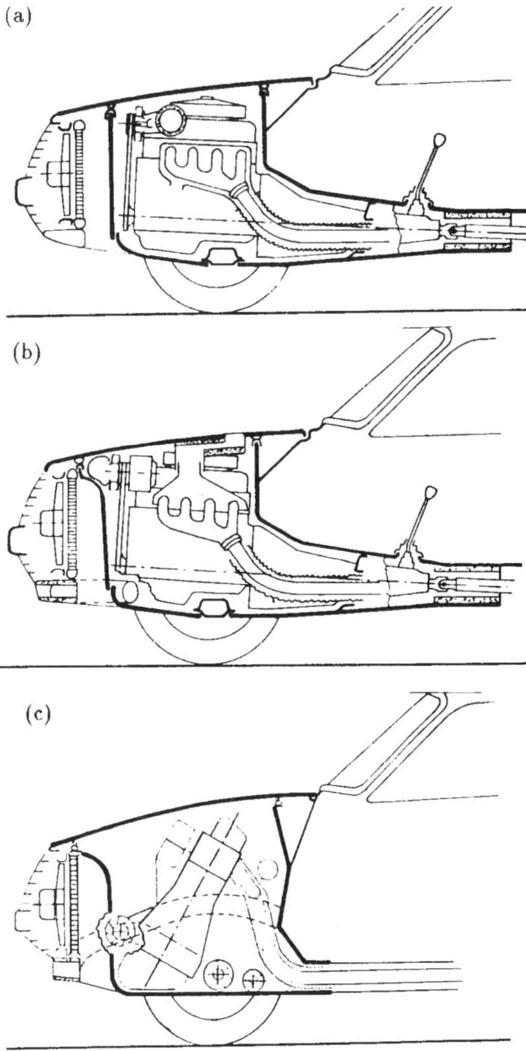

Fig 47: Car configurations to suit different engine/radiator installations

Similar enclosures have been reported by AVL for use in cars but with the cooling airflow kept out of the enclosure by arranging for separate coolant-radiator ducting. Fig 47 shows alternative solutions for different engine and radiator configurations. At (a) the radiator cooling air is forced downwards by the bonnet and leading bulkhead assisted by additional air passages through the wheelarch and bonnet as necessary. A supplementary fan on the water pump pulley provides a limited airflow within the enclosure, air entering it passing through a silencer and exiting both around the exhaust pipe and the transmission shaft. At (b) the leading bulkhead is replaced by a duct behind the radiator to achieve cost saving. Airflow in the enclosure is also simplified by replacing the encapsulation of the exhaust system by a chimney and air-silencer just beneath the bonnet air-outlets. The chimney creates a thermal draft and the exhaust system passes through the enclosure within an acoustic sleeve. This same philosophy applies to the transverse engined layout at (c) but without the chimney element, airflow inside the enclosure being assisted by the dynamic pressure head in front of the vehicle. Fig 48 shows the noise reductions obtained. In commercial vehicle noise research, reported by Ricardo, it is considered possible to reduce an engine noise source by 10dB(a) using full enclosures. Using ones which give acceptable maintenance access can also give reasonable attenuation. Fig 49 shows simple side shields having a 25 mm thick PUR coating — used together with an unlined sump undertray. The existing engine cover was used but a lining added. Shields were made from 1.6 mm thick damped steel laminate and weighed a total of 30 kg to achieve 6dB(a) sound reduction.

Antiphon Ltd are prominent in the supply of highly damped panels suitable for noise shielding. Their SA material is produced for the reduction of airborne noise and is a self-supporting absorber comprising a rigid, resinated felt-like carrier, moulded in a checked pattern with cavities and ridges. A 1 mm polyester film is heat-sealed to the panel and bonds to the ridges so that an air gap back exists between it and the base of the carrier. Noise causes the film to vibrate freely into the cavities between the ridges allowing sound waves to enter the porous carrier, the makers claim. The material weighs 0.25 lb/ft^2 (1.22 kg/m^2) and is about 1/4 in (6.35 mm) thick, Fig 50. Absorption co-efficients are in Fig 51.

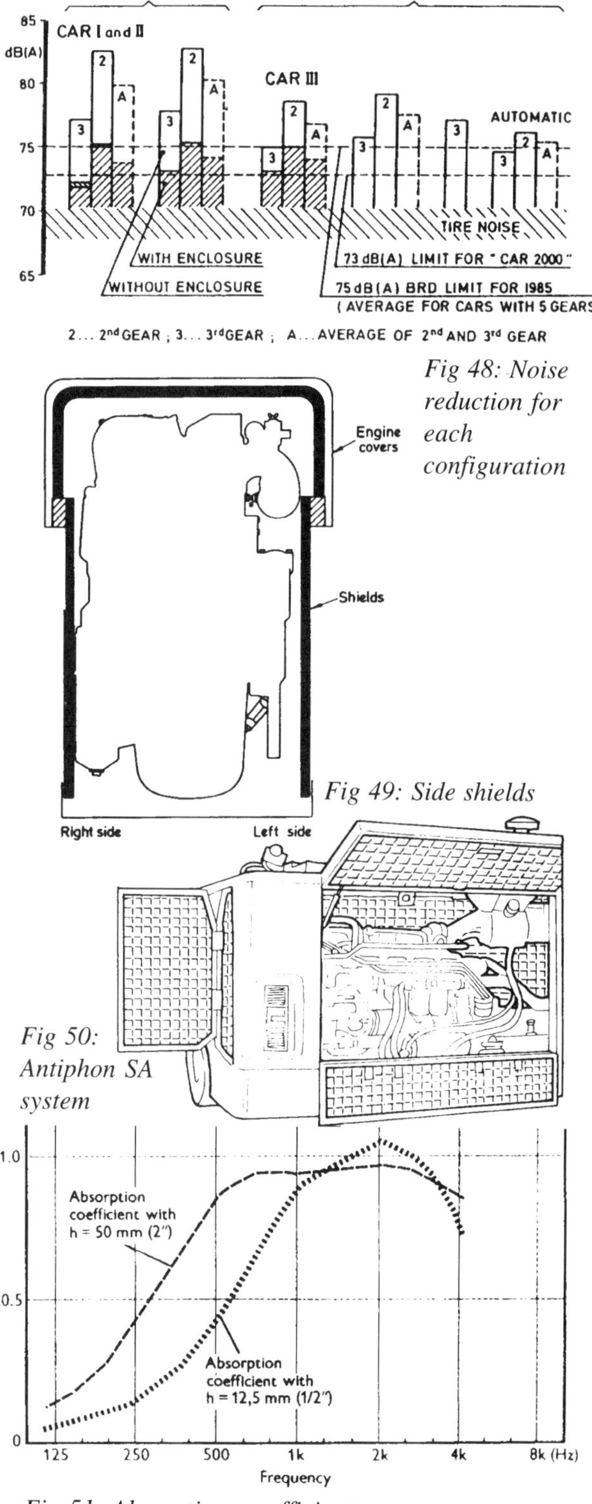

Fig 48: Noise reduction for each configuration

Fig 49: Side shields

Fig 50: Antiphon SA system

Fig 51: Absorption co-efficients

Body structural vibration

Avoidance of discomfort caused by the 'harshness' and 'shake' vibration frequencies can be achieved at the design-stage if the shape of the vehicle is virtually finalised, a complex FE analysis carried out and appropriate remedial measures taken during build such as the addition of vibration absorbers.

Without the luxury of these facilities it is possible to avoid serious vibration effects by having a thorough understanding of the possible causes as the design is built up and arranging for the structure to be outside the range of likely exciting resonances. Because road input vibrations are generally of a random nature, the variable frequency and amplitude content can excite different body panels into vibration, or cause overall bending and/or twisting motions of the structure, according to road surface and handling inputs. Generally it is the aim to design enough bending and torsional stiffness into the structure so that natural frequencies fall in a range of 20-30 Hz — to avoid the danger of excitation by suspension frequencies and the consequent loss of handling control due to movement of the anchorages. The higher road input frequencies, in the 'harshness' region of the overall spectrum, can cause difficulties which are sometimes overcome by the use of soft-mounted subframes which maintain 'chassis' and 'body' in a different vibrational environment.

Pioneer work at Birmingham University identified the vibration modes of a typical volume-built panel-van shell structure and found the associated resonant frequencies. By suspending the shell on flexible elastic cords and exciting the suspension anchorages with vibrators sweeping across a wide frequency range, localised natural frequency responses were determined. Ten measured frequencies and mode shapes are listed in Fig 52. In fact the symmetric modes were not just bending ones because of the disparity between the stiffnesses of different parts of the shell and the nature of the design which involved large cut-outs and some flexible connections. Motions involved a combination of lozenging of the load-carrying area with torsion or bending of the floor, roof and sides, Fig 53.

Subsequent studies on integral car structures have yielded vibration modes typified by those in Fig 54 for one phase condition at 140 Hz. This is obtained by affixing transducers to the surface and filtering signals to the frequency being studied. Single point harmonic forcing is also used to to obtain point responses in terms of driving force and resultant motion. The variation of point response against forcing frequency is known as 'mobility' — the inverse of mechanical impedance. The instrument used comprises a piezo-electric transducer incorporating a force and acceleration sensor on the one probe head. A sweep oscillator is then tuned to a particular resonant frequency and a roving accelerometer placed on the structure to read off RMS amplitude and phase at each grid point. The resulting quadrature amplitude response is plotted on a plan of the structure as shown in Fig 55.

Random vibration can be analysed to find its frequency and amplitude content by statistical tech-

Measured symmetric modes

Mode no.	Natural frequency (Hz)	Figure number	Description
1	24·90	15.11	Breathing mode, floor and roof in antiphase, sides in antiphase moving together while roof and floor move apart. Longitudinal bending of the floor.
2	32·56	15.12	Same as mode 1 but with greater motion of sides and front of roof.
3	38·03	15.13	Combination of longitudinal and transverse bending in both floor and roof with their motion in phase.
4	46·87	15.14	Same as mode 3 but the longitudinal and transverse motions are combined in a different sense.
5	47·55	15.15	Transverse bending of the roof and rear floor together with longitudinal bending of the front.
6	52·16	15.16	Same as mode 5 but with a reversed phase relationship between the transverse bending of the rear floor and the longitudinal bending of the front.

Measured anti-symmetric modes

Mode no.	Natural frequency (Hz)	Figure number	Description
1	20·54	15.17	Primary lozenging of the open-ended load-carrying area with the magnitude increasing towards the rear. Floor and roof moving laterally in antiphase, sides moving vertically in antiphase.
2	27·12	15.18	Torsion of the floor and roof with secondary lozenging of the load-carrying area so that the front and rear lozenge in antiphase. Lateral bending about the door opening was also present.
3	39·60	15.19	Secondary torsion of the floor but not the roof together with tertiary lozenging of the load-carrying area so the centre of the van was lozenging in antiphase with the front and rear.
4	49·36	15.20	Secondary torsion of both floor and roof with vertical bending of the sides giving a form of secondary lozenging across the load-carrying area.

Fig 52: Typical mode frequencies

niques. A relatively simple way of describing a random surface profile of a road or track is to plot the vertical and horizontal displacement from the origin of co-ordinate axes as its horizontal projection traverses at constant linear velocity, Fig 56. The key statistical 'means' are shown in Fig 57. The time in which a point lies in any narrow band about any height *x*, when divided by sample time *T*, gives the so called probability density. This can be plotted on a Gaussian distribution curve to give information on the amplitude and frequency of bumps — based on the number of times the vertical displacement variable crossed a given level. The concept of 'density' is best understood by considering a bar of constant cross-section having variable density along its length. The mass of such a bar between distances of *X1* and *X2* from end could be shown by the expression in Fig 58 and by considering distance along a frequency scale w in lieu of a length scale *X*, the spectral density *S(w)* is analogous to physical density *p(x)* and the mean square amplitude to the mass as shown in Fig 57. Panel vibrations are usually excited by engine or suspension motions transferred through their respective mounts, causing referred movement of panel edge supports. The panels are relatively flexible in bending so are easily excited into vibration if they are flat. Rectangular plates have different vibration harmonics as shown in Fig 59, the natural frequency, in the case of simply-supported plates, being:

$$\pi^2(gD/dt)^{1/2} / [(m^2/a^2) + (n^2/b^2)]$$

where *m* and *n* are the harmonics, *a* and *b* the lengths of the sides and *g* the acceleration due to gravity. Panel density and thickness are *d* and *t* respectively.

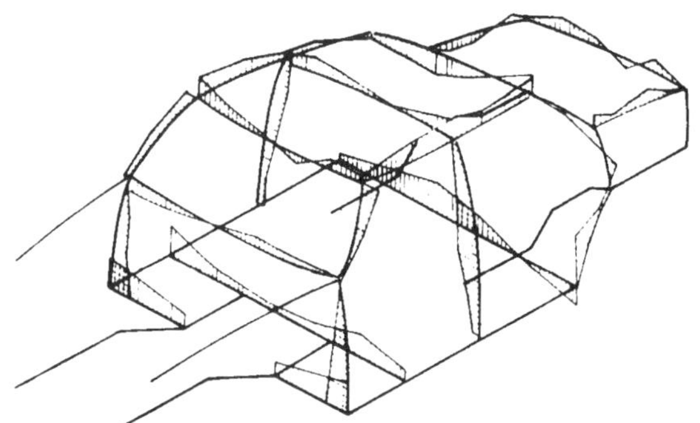

Fig 54: Car body shell vibration

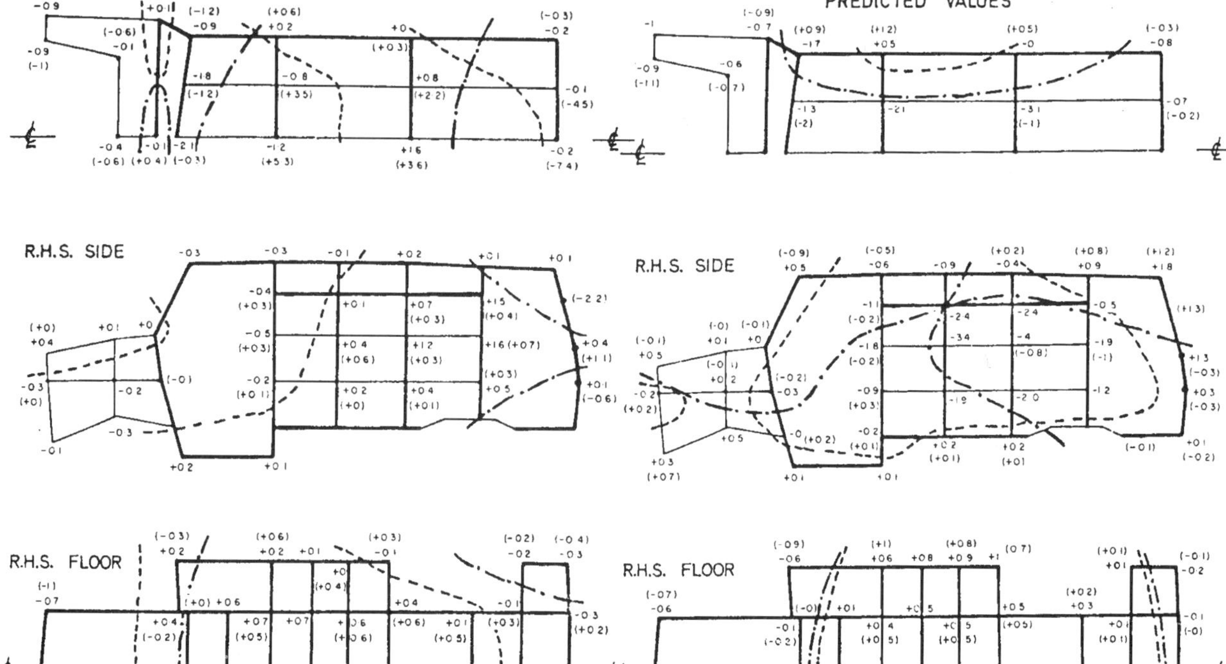

Fig 53: Combined bending and twisting motions

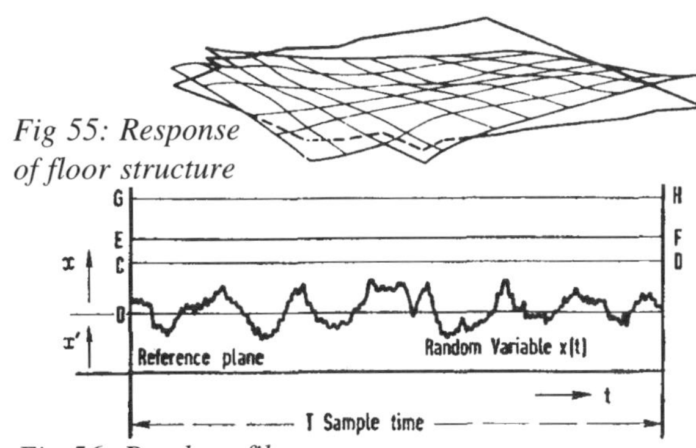

Fig 55: Response of floor structure

Fig 56: Road profile

Mean value,

$$\text{line } OO = \frac{1}{T}\int_0^T x(t)\,dt$$

Mean amplitude,

$$\text{line } CD = \frac{1}{T}\int_0^T |x(t)|\,dt \qquad \text{signified } \langle |x(t)| \rangle$$

Mean square amplitude,

$$\text{line } GH = \frac{1}{T}\int_0^T x^2(t)\,dt \qquad \text{signified } \langle x^2(t) \rangle$$

Root mean square amplitude,

$$\text{line } EF = \sqrt{\frac{1}{T}\int_0^T x^2(t)\,dt} \qquad \text{signified } \sqrt{\langle x^2(t) \rangle}$$

Fig 57: Statistical mean values

$$m_{X_1 \to X_2} = A\int_{X_1}^{X_2} \rho(X)\,.\,dX \qquad \langle x^2(t) \rangle_{\omega_1 \to \omega_2} = \int_{\omega_1}^{\omega_2} S(\omega)\,d\omega$$

Fig 58: Spectral density concept

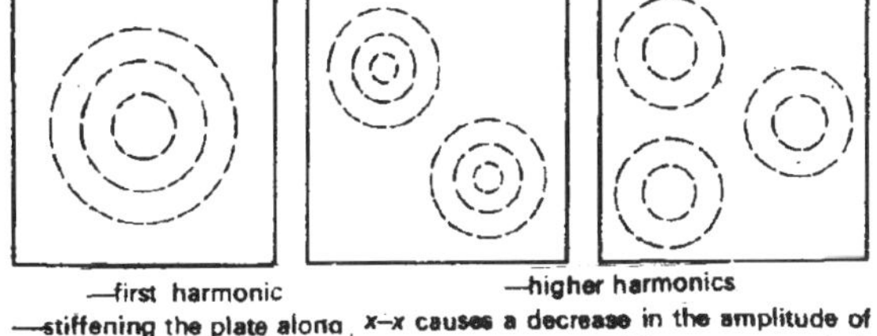

Fig 59: Flat plate harmonics

Material	Young's Modulus E (GN/m²)	Density ρ (Mg/m²)	E/ρ	$\sqrt{(E/\rho)}$	Ratio of Nat. Freqs (Plates of Equal Thickness)	Ratio of Nat. Freqs (Plates of Equal Weight)	Ratio of Thickness (Plates of Equal Weight)
Cast iron	120	7	17	4·12	1	1	1
Steel	210	7·9	25	5	1·2	1·35	0·9
Aluminium	70	2·7	26·5	5·15	1·25	3·24	2·6
Magnesium	43	1·7	26·5	5·15	1·25	5·15	4·1
Glass reinforced plastic:							
(i) random glass	5·6–10·3	1·3–1·8	5·1	2·26	0·55	2·48	4·5
(ii) bidirectional glass	16–24·2	1·3–1·8	13	3·6	0·88	3·95	4·5
Perspex	6 2	1·15	5·4	2·32	0·56	1·94	6·1

Fig 60: Panel natural frequencies

That of the first harmonic, which produces most noise, is obtained by putting *m* and *n* equal to 1. Fig 60 shows a comparison of panel natural frequencies.

No simplified method exists for estimating mode shape and natural frequencies of structures like integral saloon car bodies. A not too complex approach was developed some years ago by General Motors before finite-element methods became widespread. It has the advantage that the designer can visualise vibration effects at a very early design stage. The local masses of the structure are lumped at nodes and the properties of the weightless beams assumed between the nodes can be adjusted as experience is built up in physical model testing. As a very first estimate of calculating natural frequency, the Unit Load method can be employed assuming the body to behave as a uniform beam with the same flexibility as calculated for the front passenger load point. Then the value of *f1*, displacement due to unit load, found by applying the method to a structure such as Fig 61a can be considered within the uniform beam at 61b:

An equivalent uniform torsional stiffness *GJ* can be found from the natural frequency in torsion which can be estimated from: $(1/2L)(GJ/I)^{1/2}$ where *I* is the mean rolling moment of inertia per unit length of the body which can be assumed to roll about an axis through the mid-points of front and rear suspension mounts.

Over AB, $M1 = (b/l)x1$ and $m1 = (b/l)x1$
Over CB, $M2 = (a/l)x2$ and $m2 = (a/l)x2$ thus $f2 = (1/EI).\delta[(b/l)^2 x1^2.dx1 + (a/l)^2 x2^2 dx^2]$
from which the equivalent beam stiffness can be obtained:
$EI = a^2b^2/3lf1$
since $f1 = f2$
Assuming a uniform mass distribution *m* per unit length of the body, the first natural frequency is: $(22.4/2pL^2)(EI/m)^{1/2}$

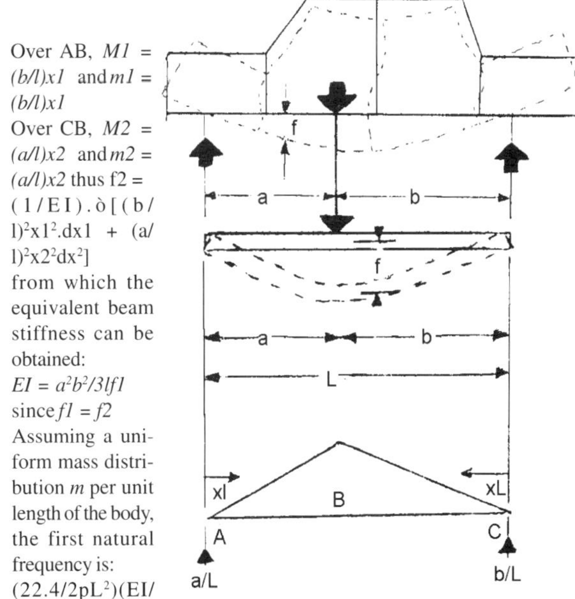

Fig 61: Car shell simplified as a beam

Structural dynamics applied

Recent computer analysis packages have the potential to automate complex structural dynamics calculations. Several recent papers are advancing the approach of integrating dynamic structural and fatigue analyses.

Optimal design changes to solve vibration induced fatigue failures can only be derived by including structural dynamics into fatigue life calculations, according to researchers at LMS International[7]. This was particularly so when desire for weight saving precludes the time-worn technique of beefing up the failed component in the area of the fatigue crack. The integration is effected by a third input to the analysis: strain histories from service records of the component.

The first step is to derive forces by representing the operating data, accelerations and strains, in the frequency domain in relation to frequency response functions (FRFs). The influence of a vibration mode can then be ascertained by synthesizing a strain signal from these forces and FRFs. Also, by applying structural modifications to calculate modified FRFs, the effect of design changes on fatigue life can be predicted.

Within the Esprit Project package 2486 DYNAMO, Fig 62, such a procedure has been demonstrated for the resonance fatigue of a car body. Design modifications can be built in to the software to test the effect of 'what if' changes in material properties and stress concentration factors, say. Resonance fatigue problems have been studied in this way by Porsche and the results confirmed in track and laboratory fatigue tests.

The approach is to drive a first test car over an endurance track until a fatigue failure occurs; then a second car is instrumented to measure strain and acceleration signals of relevance. In one case, spectral analysis of these signals identified vertical acceleration of the gearbox mounting and front differential as the most important ones which caused a fatigue crack.

These are shown in Fig 63(a), along with the critical strains close to where the crack occurred, as a function of track distance covered calculated from an integrated wheel speed signal — which indicate the four sections of the track with bad surface. Power spectra of these signals over these bad sections show a peak in the same frequency range, Fig (63b).

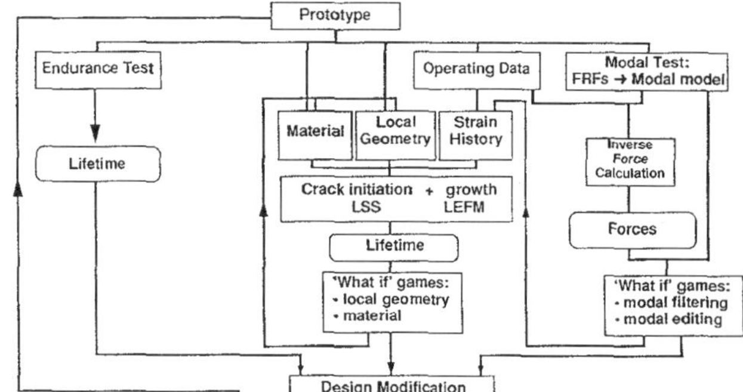

Fig 62: The LMS DYNAMO package

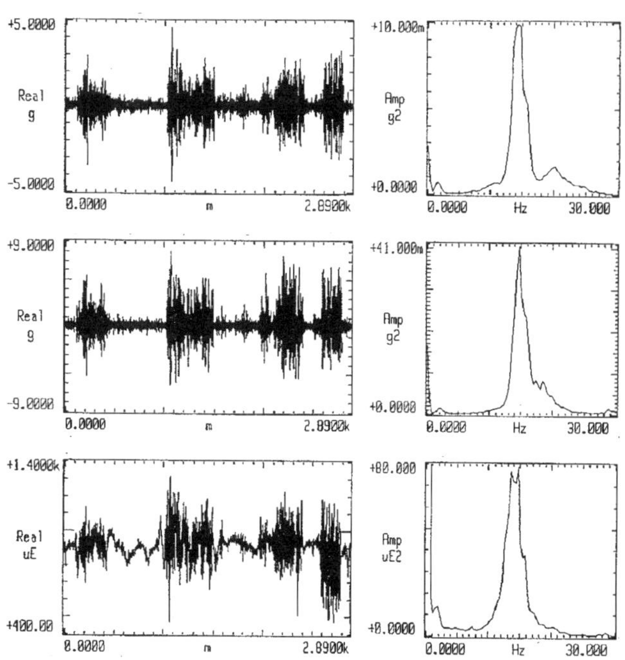

Fig 63: (a) Test recordings, left; (b) Power spectra, right

Mode No.	Reference (Hz)	Initial (Hz)	Initial Diff. (%)	Final Diff. (%)	MAC(%
7	6.77	6.77	0.01	0.01	100.
8	11.32	11.32	0.02	0.00	100.
9	17.34	17.48	0.77	0.03	99.4
10	27.21	29.95	8.76	1.22	96.0
11	33.13	33.74	1.82	0.57	70.9
12	33.53	35.39	5.53	0.98	90.4
13	41.21	43.85	6.41	0.48	97.2
14	43.21	45.71	4.13	0.31	85.4
15	47.35	49.75	5.06	1.46	53.3
16	49.17	51.01	3.73	0.84	57.0
17	53.95	54.33	0.70	0.23	51.5
18	54.24	55.61	2.52	0.37	88.8

Fig 66: Model updating procedure

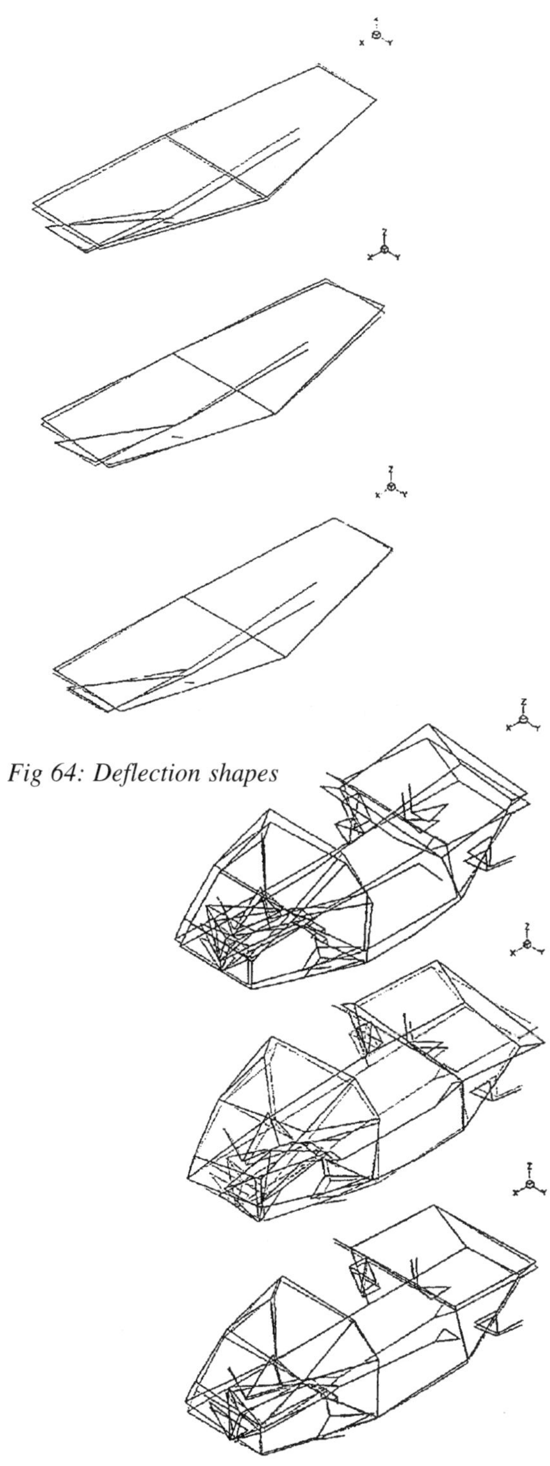

Fig 64: Deflection shapes

Fig 65: Modal analysis

Corresponding operational deflection shapes in a multi-source environment (signals from transducers throughout the structure) can be calculated by a principal-component analysis, the authors explain.

Traditional techniques for determining such shapes assume the whole structure behaves coherently with respect to one reference transducer. The coherent motion can, of course, be made up of several components when closely coupled modes are excited by driving over a test track. Principal Component analysis extracts more information from the data so as to introduce dimensionality of the operating phenomena. It results in a statement of the (independent virtual) signals called 'principal components'. By calculating cross-powers between the measurements and each of these, deflection shapes coherent to the three principal components can be displayed as a function of frequency. Fig 64 shows these at the frequency of the spectra peaks for the accelerations and critical strain. These correspond well to the three of the modes identified in modal analysis of the second test car, Fig 65:

* 1st critical component at the critical frequency corresponding to a vertical powertrain mode
* 2nd corresponding to powertrain rolling coupled to body torsion
* 3rd corresponding to a vertical front differential mode

The three modes play a key role in the structural dynamics mechanism causing the fatigue crack and the authors demonstrate good correspondence between experimentally observed and analytically predicted initiation with total life.

Sensitivity analysis

The way in which FE models can be validated and updated, using sensitivity techniques and weighted updating of physical parameters, is explained by engineers[8] of Dynamic Engineering. The technique helps the analyst to assess why a simulation model differs from a real structure. Modelling of stiffness and mass can be inaccurate, for example, because of an insufficient number of elements. The modal behaviour of the model may be taken as a basis for assessment but because of limited degrees of freedom results will only be valid in a limited frequency range.

Sensitivity-based updating is a technique for validating the FE solution automatically. It involves the calculation of a correlation criterion such as the Modal Assurance Criterion (MAC) to identify correlation between analytical modal results and those derived by Experimental Modal Analysis (EMA). Identifiable errors are those due to the FE formulation used, those due to geometrical simplification (mesh density) and those due to the input values of mass or stiffness. SYSTUNE is the name of the program which has been developed for interfacing databases from FE and EMA sources. It covers all possible situations related to the purpose of the FE model, information available on modelling assumptions and the number/type of analytical/experimental results available. Fig 66 shows the model updating procedure applied to a car body; essentially a sequence of updating runs is performed in which mass, static- and then dynamic-behaviour is corrected.

Substitution of polymer-composites

Obviously the material properties need to be accurately known to obtain an accurate FE analysis. Where properties are uncertain then approximated values can be updated in the technique, reflecting the real material as built into actual components. When several sub-structures are assembled together updating is best carried out in steps: first the components themselves and then the joints between them. Also coarse FE models can be derived from fine models and forced to adopt the same dynamic behaviour, thus saving on computer time. Such detailed effects as ribs, flanges and solid transitions can be merged into the coarse models in a dynamically correct way, to give accurate resonant frequencies and mode shapes. Fig 67 is an example of a car model for which the joints were updated by changing their rotational spring stiffnesses to obtain proper correlation, comparing actual dynamic behaviour with a reference one. No boundary conditions were considered since the structure was freely suspended. Fig 68 shows a typical torsion mode, for which the table in the earlier The Fig 69 listing applies to all modes before and after updating. Errors on resonance frequencies were reduced to an average of less than 1% and some rotational stiffnesses were reduced to zero, showing that the joints were effectively 'pinned'.

Effect of trim on structural dynamics According to researchers at Birmingham University[9], a vibrating structure can be considered analagous to a large number of simple sources distributed geometrically on the enclosure panels. Knowing the contribution of each of these, and relevant phase information, sound pressure level can be predicted at any point within the cavity. The sound pressure level also yields the vibro-acoustic relationship which leads to damping energy dissipation due to trim coverings.

When FE analysis is otherwise used to model the modal behaviour, with assumed structural damping, there is the possibility of errors, the authors maintain. With the 'simple source' technique the structure can be assessed in an absolute sense over a given bandwidth in terms of performance figures which account for effects of both resonant and non-resonant modes. Once structural problem areas are identified, the affects of damping trim pads can be evaluated. The authors define 'point mobility' as the input point velocity response normalised by the driving force. This parameter therefore describes the ability of a point in a structure to admit vibration inputs and hence is a measure of performance at an attachment point where vibration energy is introduced. Transfer mobility is used to describe the

Fig 67: Car model with modified joint conditions

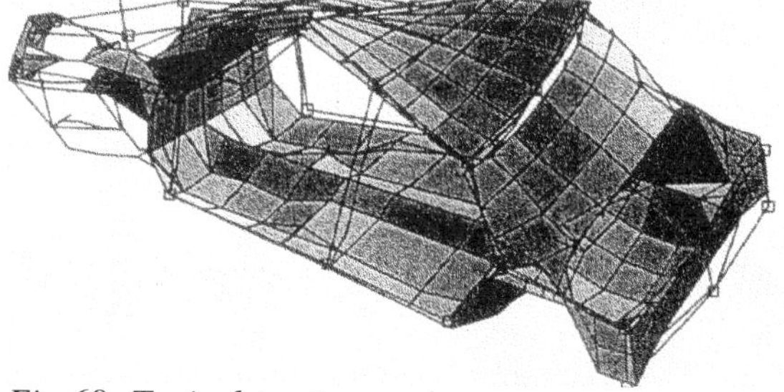

Fig 68: Typical torsion mode

velocity response of one point with respect to a vibration input at another. Modal mobility is the average transfer mobility over part of, or the whole of, the vehicle structure, as a single parameter. It allows elemental phase cancellation due to structural damping and is determined by summing the transfer mobility vectors for elemental areas dividing the structure. Thus it can be extended to provide a measure of vibro-acoustic performance in respect of structure-borne radiation.

Structural dynamic performance standards determined from a large sample of monocoque-structured vehicles over a considerable time-span are -10 and -15 for point mobility and -28 and -36 for modal mobility. The figures refer to 10Hz and 10-100Hz bandwidths, in turn — units being dB, ref 10^{-3}m/Ns. A comparison of these parameters is obtained between various stages of trim. In a test programme on five trim phases, three of them are presented in the paper, fully trimmed, partially trimmed (damping pads) and untrimmed body-in-white. The structure is hung from A-frames on soft springs. Fig 69 shows the layout of the test structure which is a luxury car monocoque carrying 156 transfer mobility measurement stations distributed over nine panels enclosing the interior cavity. Some 55 tests were carried out at the 22 input stations shown by the table of Fig 70, the right hand columns showing the results. These show that the sample structure had poor broad-band point mobility performance with a number of attachment points failing to meet the -15dB standard. The results are discussed in detail within the original paper.

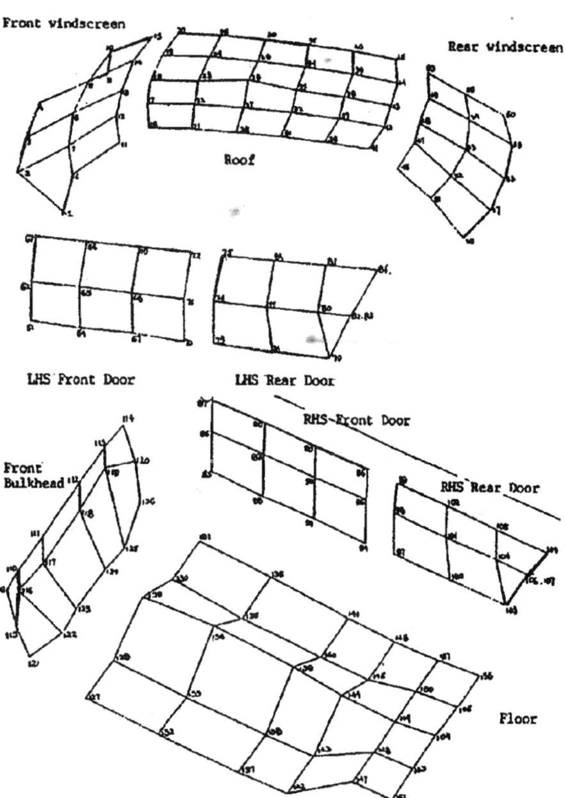

Fig 69: Layout of test luxury-car structure

Excitation Station & Station Code	Direction	Point Mobility dB, ref 10^{-3} m/Ns 10-250 Hz	Modal Mobility dB, ref 10^{-3} m/Ns 10-250 Hz
LHS Front Suspension Lower Arm Mount (R1)	Vertical	-7.78	-29.51
Front Subframe Engine Mount (R2)	Vertical	-3.71	-29.87
Rear Subframe Engine Mount (R3)	Vertical	-10.72	-29.97
RHS Engine Mount (R4)	Vertical	-19.83	-36.30
LHS Engine Mount (R5)	Vertical	4.55	-29.25
LHS Front Suspension Upper Arm Mount (R6)	Vertical	-23.47	-38.61
Bulkhead Engine Mount (R7)	Fore/Aft	-11.46	-31.66
LHS Front Suspension Tie Rod Mount (R8)	Fore/Aft	-15.83	-35.80
LHS Front Suspension Tower (R9)	Vertical	-18.98	-37.04
LHS Front Anti-roll Bar Mount (R10)	Vertical	-5.34	-29.10
LHS Rear Suspension Shock Absorber Tower (R11)	Vertical	-17.84	-39.36
LHS Rear Suspension Coil Spring Tower (R12)	Vertical	-15.55	-37.97
LHS Rear Suspension Lower Arm Mount (R13)	Vertical	-15.59	-36.46
LHS Rear Suspension Trailing Arm Mount (R14)	Fore/Aft	-22.54	-42.93
LHS Rear Anti-roll Bar Mount (R15)	Vertical	-10.15	-38.17
RHS Engine Mount (R16)	Fore/Aft	1.92	-33.00
LHS Engine Mount (R17)	Fore/Aft		
LHS Front Suspension Upper Arm Mount (R18)	Lateral	-16.40	-36.36
Front Subframe Engine Mount (R19)	Fore/Aft	-8.49	-32.50
Rear Subframe Engine Mount (R20)	Fore/Aft	-7.01	-32.67
LHS Front Suspension Lower Arm Mount (R21)	Lateral	-11.59	-34.72
LHS Rear Suspension Lower Arm Mount (R22)	Lateral	-22.45	-40.35

Fig 70: Test details and results

Chassis beaming vibration

It is possible to predict the dynamic behaviour of a chassised vehicle if the body is inherently flexible or very soft mountings are used between it and the chassis. A graphical technique can be employed to determine the modes of beaming deflection of the chassis frame and its natural frequency in bending.

While a number of analytical methods for predicting stuctural vibration behaviour depend on the assumption that mass can be concentrated at discrete points, this is usually done for complex structures, including shells, with frequent changes in section. For more regular structural members such as chassis longitudinals the distribution of the mass can

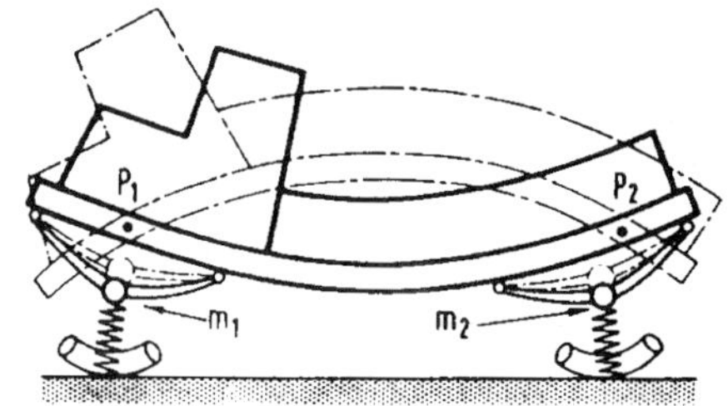

Fig 71: Beaming vibration mode

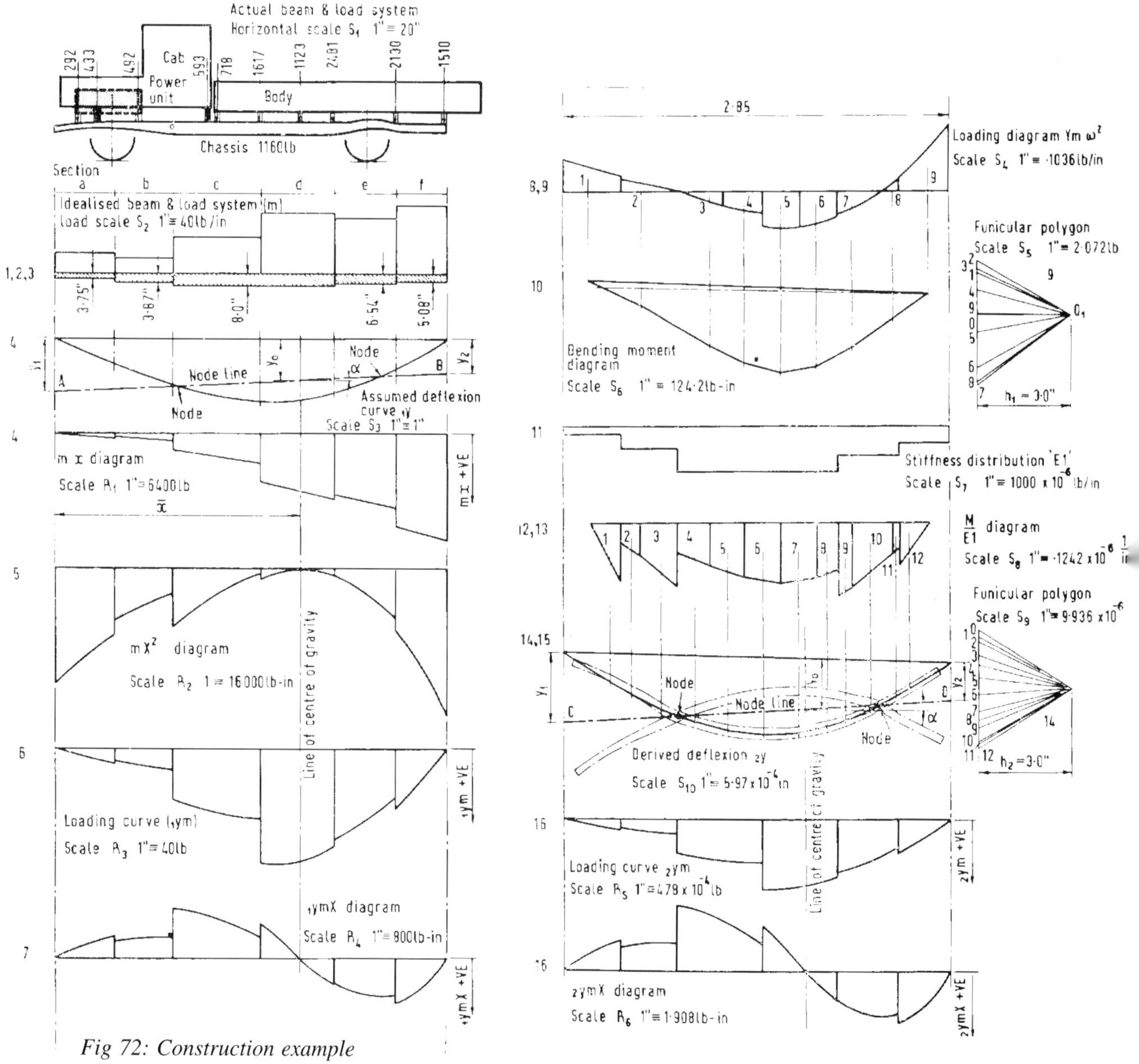

Fig 72: Construction example

be taken into account — as well of the elasticity of the member in bending. If a truck chassis, for example, is set into vibration under laboratory conditions and its vibration modes measured, the vehicle has one prominent mode as demonstrated in an exaggerated sense in Fig 71. By making the assumption that the longitudinals take up a half sine-wave shape in beaming vibration, the classical Rayleigh's method can be used to obtain a formula relating natural frequency to section dimensions and the elastic properties of the material. The strain energy of the beam at maximum deflection, and zero velocity, is then equated to the kinetic energy of the beam elements at zero deflection, and maximum velocity, as it passes through the central, non-deflected shape, within the vibration cycle. An imaginary small element of the beam is considered and its behaviour is simulated by the presence of an effective centralising spring, stiffness *k*, exerting a restoring force *ky*. The 'potential' strain energy is obtained by integrating these instantaneous forces over the full amplitude of vertical displacement. By assuming simple harmonic motion of the beam the kinetic energy of each element expressed as mv^2, can be integrated along the length of the beam and equated with the potential energy to obtain an approximate fundamental frequency of:

$$[(EI/m)\pi/4l^2\{(1/4)-(2/\pi^2)\}^{1/2}]^{1/2}$$

The avoidance of 'cab nod' is a key stimulus to the better understanding of chassis frame beaming vibration. So as to obtain more accurate vibration mode profiles a graphical method was proposed by Cranfield researchers which allowed changes in section property along the length of the beam to be taken into account. It is based on a method due to Stodola which assumes an initial deflection curve, as in the Rayleigh method above, but progressively refines it to more nearly predict the actual one. The initial curve gives an inertia loading profile based on actual variable load distribution along the chassis. The real beam is drawn out to a scale of *S1* and divided into equi-spaced ordinates. Beneath it a mass distribution curve is drawn, which will vary by stepped increments; mass per unit length *m* will have units such as lb sec^2/in^2. Below that the assumed deflection *1y* curve is plotted at a scale of *S2*. At this stage the inertia loading curve can be drawn, putting $w2 = 1$ in the load expression *1y.m.S2*. Areas between the ordinates of this curve can be replotted as a diagram of distinct inertia loads as in Fig 72, respective areas being given by an integral (*from 0 to S1*) of the above expression, to a scale *S3*. From that point the familiar technique of drawing a polygon of forces, to the same horizontal scale, is carried out. A vertical scale *h1* is chosen for convenience of drawing construction and the result leads to a bending moment diagram with ordinates *M* in scale *h1.S1.S2* in lbs per unit of measure on the drawing. The key factor in the execution of the method is carefully adhering to the scales chosen and understanding their effect at each stage. The stiffness distribution is next constructed *EI* with

1 Draw out chassis frame to tlength scale S_1, with point loads at intervals, from Fig 21.
2 Idealize the non-uniform chassis into lengths of uniform section. Draw out idealised 'distributed' load diagram m to vertical scalc S_2.
3 Write down assumed deflexion curve $_1y$ to scale S_3
4 Construct moment diagram *mx* from ordinates *m* in (2) and distance *x* from left-hand end, to scale R_1. Obtain *x* by dividing area of the resulting diagram by that of the *m* diagram (factoring for S_2 and R_1)
5 Using the c.g. as the new datum, plot the mx^2 diagram (Fig21a) to scale R_2 (area here is *I*).
6 Derive loading curve $_1ym$ as product of ordinates (2) and (3) to scale $R_3 = S_3S_2$ and find its area, *A*
7 Plot the $_1ymx$ diagram (Fig 21b) to scale R_4 and carefully find resultant area as large positive and negative parts.
8 Calculate y_1 and y_2 (Fig 21(b), and draw node line AB on curve $_1y$, measure the ordinates *Y* and multiply these by the *m* ordinates (2) to give new load $Ym\omega^2$ to scale S_4.
9 Divide resulting load diagram into small areas (loads) and plot them as the vertical line of the first funicular polygon to scale S_5, then complete the polygon.
10 Calculate y_1 and y_2 (Fig 21b), and draw bending oment diagram by normal methods, and if necessary ensure that it closes, thus overriding any small errors in calculating areas; scale is $S_6 = S_5h_1S_1$.
11 Draw the stiffness distribution *EI* to scale S_7 .
12 Divide major ordinates of BM diagram (10) bythose of diagram (11), and plot *M/EI* diagram to scale $S_8 = S_6/S_7$; divide conveniently into 11, say, areas and measure these and their centroids.
13 Plot these as the vertical line of the second funicular polygon to scale S_9 and complete it.
14 Draw derived deflexion curve $_2y$ with lines parallel to 'rays' of the polygon as in a BM diagram (lines 00_2 and $12\ 0_2$ are produced back to the ends of the beam); scale is $S_{10} = S_9h_2S_1$.
15 Ordinate $_1y_{max}/_2y$ measured at centre of beam gives the natural frequency of the loaded beam.
16 Draw $_2ym$ and $_2ymX$ diagrams to scales R_5 and R_6 so that y_1 and y_2 can be calculated and CD drawn in to give the node points.

Fig 73a: Drawing procedure

E in lb/in² and I in in⁴, to a scale $S4$, such that $S4$ = lb/in² per unit of measure on the drawing. Dividing the ordinates obtained from the last two curves will allow a further one for M/EI to be constructed. A similar technique to the one used on the inertia load curve can then be used, of integrating the areas between the ordinates, to produce a new set of ordinates which in turn will provide a more accurate beaming deflection curve for the vibrating chassis sidemembers.

The Cranfield workers essentially developed this Stodola method, covering beams simply supported at their ends, to one for free-free beams vibration mode. Fig 73a is a drawing procedure and Fig 73b an example construction.

SECTION ALONG BEAM:	LH	a		b		c		d		e		f	RH
1. LOAD DIST'N, lb/in		24·05		18·45		44·35		72·35		66·25		83·95	
2. ORD, m (S₂, 1 in = 40 lb/in)		0·60		0·46		1·11		1·81		1·66		2·10	
3. ORD, ₁y (S₃, 1 in = 1 in)	0	0·38	0·75	1·11	1·38	1·73	1·88	1·85	1·61	1·30	0·87	0·45	0
4. DIST, x (S₁, 1 in = 20 in)	0		1·87		3·67		6·35		8·70		10·62		12·25
ORD, mx (in = 6400 lb)	0		1·12,0·86		1·69,4·07		7·04,11·48		15·73,14·40		17·59,22·3		25·7
AREA (1 in² = 128,000 lb in)		0·13		0·29		1·86		4·0		3·84		4·89	
ΣA = Σmx = 15·01 in² = 1 921 000 lb in, Σm = 12,550 lb, x̄ Σmx/Σm = 153·1 in													
5. DIST, X (1 in = 20 in)	−7·65	−6·70	−5·78	−4·85	−3·98	−2·57	−1·30	0	1·05	2·0	2·97	3·79	4·60
ORD, mX² (1 unit = 16,000 lb in)	35·2	27·0	20·05,15·40	10·84	7·30,17·57	7·32	1·87,3·06	0	1·99,1·82	6·62	14·60,18·51	30·15	44·40
AREA (1 in² = 3·2 x 10⁶ lb in²)		5·10		2·01		2·28	0·13		0·07	1·40		4·96	
1 = Σmx² = 15·96 in² = 51·0 x 10⁶ lb in²													
6. ORD, ₁ym (ω²) (1 in = 0·104 lb/in)	0	0·23	0·45,0·35	0·51	0·64,1·53	1·92	2·08,3·40	3·45	2·91,2·67	2·15	1·44,1·83	0·94	0
AREA (1 in² = 800 lb in)		0·42		0·89		5·00		7·71		4·05		1·52	
ΣB = 19·60 in² = 15,670 lb in, y₀ = ΣBIM = 1·25 in													
7. ORD, ₁ymx (1 in = 3200 lb in)	0	−1·53	−2·61,−2·00	−2·48	−2·53,−6·09	−4·93	−2·71,−4·42	0	3·06,2·80	4·30	4·28,5·42	3·58	0
AREA (1 in² = 64,000 lb in²)		0·65		−1·07		−3·20	−0·08		0·42	1·93	0·38		0·95
8. D = −1·99 in² = − 126,400 lb in², α = − D/I = 0·0025													
y = 1·25 + 153·1 (0·0025) = 1·63 in, y₂ = 1·25 − 91·9 (0·0025) = 1·02 in													
ORD, Y (1 in = 1 in)													
ORD, mY (ω²)	−1·63	−1·21	−0·80	−0·40	−0·06	0·37	0·57	0·60	0·41	0·15	−0·23	−0·60	−1·02
(1 in = 0·104 lb/in)													
9. AREA = (S₉ 1 in² = 2·07 lbf)	−0·98	−0·73	−0·48,−0·37	−0·19	−0·03,−0·07	0·41	0·63,1·03	1·08	0·74,0·68	0·25	−0·38,−0·48	−1·26	−2·14
10. ORD, M		−1·36		−0·36	0·28		0·69,1·27		1·09,0·52		−0·10	−2·08	
(S₆, 1 in = 124·22 lb in)		0	0·44	0·71	1·33	1·87	2·36,2·62		2·39,2·02	1·76	0·81,0·68	0	
11. ORD, EI													
(S₇, 1 in = 1000 x 10⁶)		0·24		0·68		1·40		1·40		0·88		0·50	
12. ORD, M/EI													
(S₈, 1 in = 0·124 in⁻¹)		0	1·83,0·64	1·04	1·95,0·95	1·34	1·69,1·87	1·71	1·44,2·29,1·99,0·92		0·77,1·35	0	
13. (1 in² = 2·49 x 10⁻⁶)													
(S₉, 1 in = 9·94 x 10⁻⁶)		0·96	0·55		1·72,1·18	1·63	2·08	1·97	1·05,0·98	1·85	0·16	0·66	
14. ORD, ₂y													
(S₁₀, 1 in = 5·97 x 10⁻⁴)	0·38		0·90,1·09,1·25		1·70,1·86,2·04		2·27,2·36,2·28		2·06,1·94,1·87		1·55,1·01,0·83		0·51
15. ω² = ₁y/₂y (max) = 1335, ω = 36·8 rad/s, f = 5·85 c/s													
16. ORD, ₂ym		0,0·54,		0·50,0·78,		2·06,2·52,		4·27,3·73,		3·21,2·57,		2·12,1·07,	
(1 in = 478 x 10⁻⁴ lb)		0·29,0·66,		0·58,0·86,		2·26,2·62,		4·12,3·51,		3·10,1·67,		1·74,0	
AREA 1 in² = 0·96 lb in)		0·32		0·65		3·26		4·70		2·42		0·87	
ΣB = 12·22 in² = 11·69 lb in, ₂y₀ = 11·69/12,550 = 9·31 x 10⁻⁴/(1·56) in													

DIST. X	−7·65	−7·0	−6·15	−5·78	−5·46	−4·52	−3·98	−3·47	−2·42	−1·30	−0·20	0·69	1·05	1·26	2·05	2·97	3·28	3·88	4·60
ORD, ₂ymX	0,		−3·33,		−2·91, −3·54,			−8·20,	−6·09,		−5·55,	+2·57,	+3·37,		+5·26,		+6·30,	+4·06	
(1 in = 1·91 lb in)	−1·60,		−3·78,		−3·15, −3·41,			−7·84,	−3·41,		−0·82,	+3·68,	+3·90,		+4·96,		+5·71,	0	
AREA (1 in² = 38·16 lb in²)		−0·98			−1·51			−2·89			−1·33, −0·91		0·49,1·12			1·18,1·12			0·41
D = −3·31 = −126·0 lb in², α = −D/I = 126/51·0 x 10⁶ = 2·47 x 10⁻⁶																			
y₁ = 0·00093 + 153·1 (2·47 x 10⁻⁶) = 2·195 in, y₂ = 0·00093 −91·9 (2·47 x 10⁻⁶) = 1·18 in																			

Fig 73b: Example construction

Designing against fatigue

As well as either steady or impact loading, repetitive cyclic loading has to be considered in relation to the effective life of a structure. The aerospace industry has been central to the study of fatigue problems and is fully familiar with the concept of structural life — in, for example, the need to satisfy a 'Certificate of Airworthiness'.

Fatigue failures, in contrast to those due to steady load, can of course occur at stresses much lower than the elastic limit of the structural materials. Failure normally commences at a discontinuity or surface imperfection such as a crack which propagates under cyclic loading until it spreads across the section and leads to rupture. Even with ductile materials failure occurs without generally revealing plastic deformation. Fig 74 shows the terminology for describing stress level and the loading may be either complete cyclic reversal or fluctuation around a mean constant value. Fatigue life is defined as the number of cycles of stress the structure suffers up until failure. The plot of number of cycles is referred to as an *S-N* diagram, Fig 75, and is available for different materials based on laboratory controlled endurance testing. Often they define an endurance range of limiting stress on a 10 million life cycle basis. A log-log scale is used to show the exponential relationship $S = C.Nx$ which usually exists for C and x as constants depending on the material and type of test, respectively. The graph shows a change in slope to zero at a given stress for ferrous materials — describing an absolute limit for an indefinitely large number of cycles. No such limit exists for non-ferrous metals and typically, for aluminium alloy, a 'fatigue limit' of 5×10^8 is defined. It has also become practice to obtain strain/life (Fig 76) and dynamic stress/strain (Fig 77) for materials under sinusoidal stroking in test machines. Total strain is derived from a combination of plastic and elastic strains and in design it is usual to use a stress:strain product from these curves rather than a handbook modulus figure. Stress concentration factors must also be used in design.

When designing with load histories collected from instrumented past vehicle designs of comparable specification, signal analysis using rainflow counting techniques is employed to identify number of occurrences in each load range. In service testing of axle beam loads it has been shown that cyclic loading has also occasional peaks due to combined braking

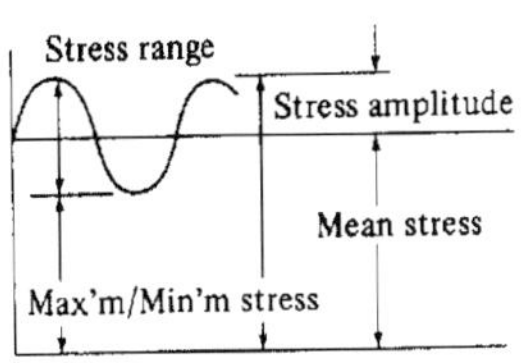

Fig 74: Terminology for cyclic stress

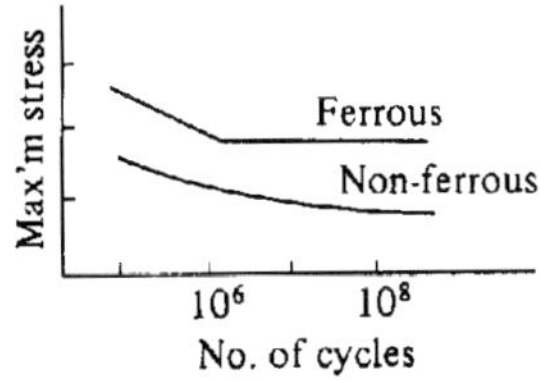

Fig 75: S-N diagram

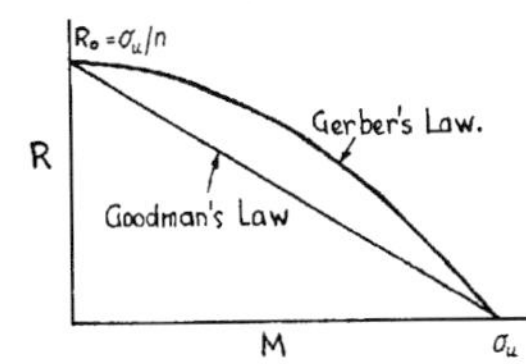

Fig 78: Fatigue limit diagrams

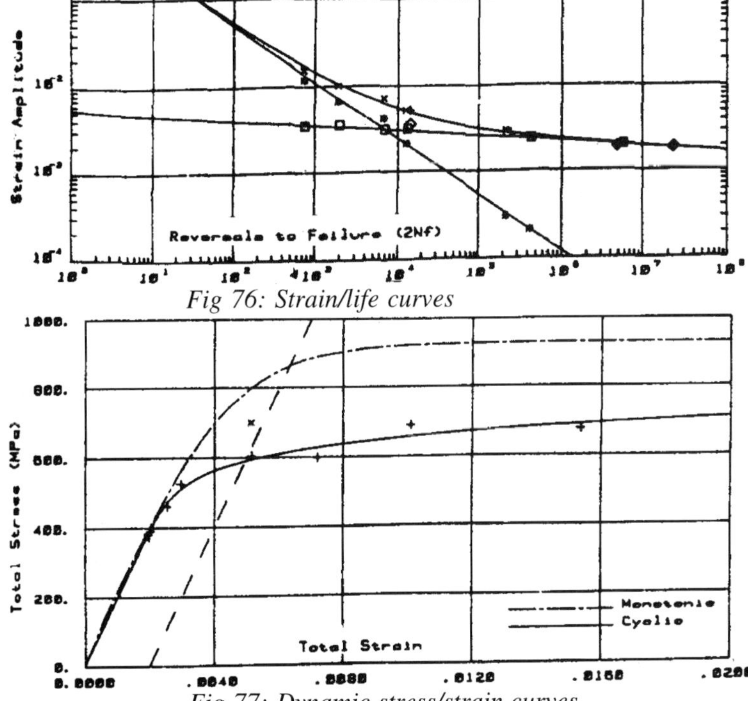

Fig 76: Strain/life curves

Fig 77: Dynamic stress/strain curves

and kerbing equivalent to four times the static wheel load. Predicted life based on specimen test data could be twice that obtained from service load data. Calculation of the damage contribution of the individual events counted in the rainflow analysis can be compared with conventional cyclic fatigue data to obtain the necessary factoring. In cases where complete load reversal does not take place and the load alternates between two stress values, a different (lower) limiting stress is valid. The largest stress amplitude which alternates about a given mean stress, which can be withstood 'infinitely', is called the fatigue limit. The greatest endurable stress amplitude can be determined from a fatigue limit diagram, Fig 78, for any minimum or mean stress. Stress range (R) is the algebraic difference between the maximum and minimum values of the stress. Mean stress (M) is defined such that limiting stresses are M +/- R/2.

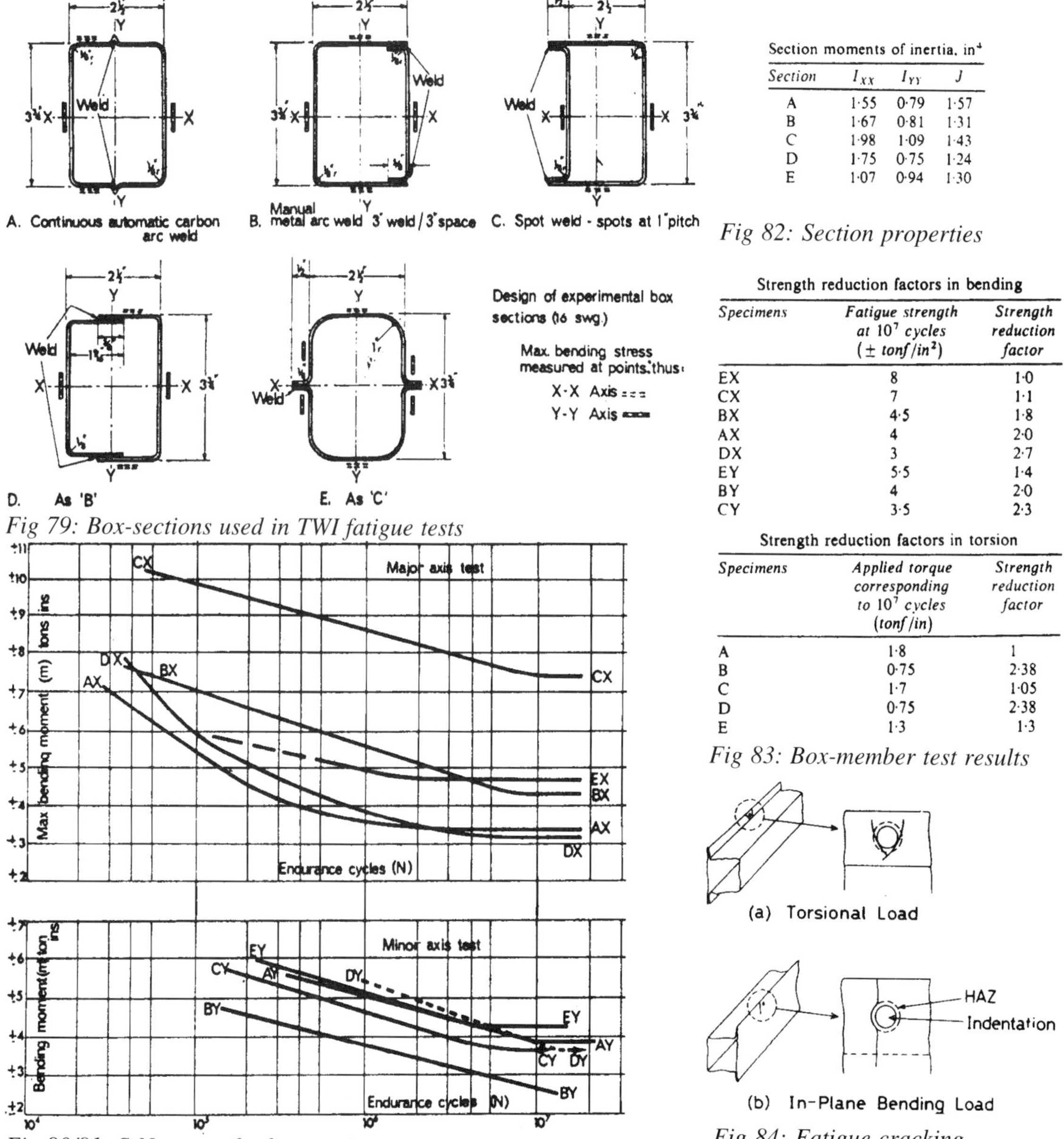

Fig 79: Box-sections used in TWI fatigue tests

Fig 80/81: S-N curves for box sections

Section moments of inertia, in⁴

Section	I_{XX}	I_{YY}	J
A	1·55	0·79	1·57
B	1·67	0·81	1·31
C	1·98	1·09	1·43
D	1·75	0·75	1·24
E	1·07	0·94	1·30

Fig 82: Section properties

Strength reduction factors in bending

Specimens	Fatigue strength at 10^7 cycles ($\pm$ tonf/in²)	Strength reduction factor
EX	8	1·0
CX	7	1·1
BX	4·5	1·8
AX	4	2·0
DX	3	2·7
EY	5·5	1·4
BY	4	2·0
CY	3·5	2·3

Strength reduction factors in torsion

Specimens	Applied torque corresponding to 10^7 cycles (tonf/in)	Strength reduction factor
A	1·8	1
B	0·75	2·38
C	1·7	1·05
D	0·75	2·38
E	1·3	1·3

Fig 83: Box-member test results

Fig 84: Fatigue cracking

Fatigue limit in reverse bending is generally about 25 per cent lower than in reversed tension and compression, due, it is said, to the stress gradient — and in reverse torsion it is about 0.55 times the tensile fatigue limit. Frequency of stress reversal also influences fatigue limit — becoming higher with increased frequency. An empirical formula due to Gerber can be used in the case of steels to estimate the maximum stress during each cycle at the fatigue limit as $R/2 + (\sigma_u 2 - nR\sigma_u)^{1/2}$ where σ_u is the ultimate tensile stress and n is a material constant = 1.5 for mild and 2.0 for high-tensile steel. This formula can be used to show the maximum cyclic stress σ for mild steel increasing from one third ultimate stress under reversed loading to 0.61 for repeated loading. A rearrangement and simplification of the formula by Goodman results in the linear relation $R = (\sigma_u/n)[1 - M/\sigma_u]$ where $M = \sigma - R/2$. Fig 78 shows the relative curves in either a Goodman or Gerber diagram frequently used in fatigue analysis. If values of R and σ_u are found by fatigue tests then the fatigue limits under other conditions can be found from these diagrams.

Where a structural element is loaded for a series of cycles *n1, n2* ... at different stress levels, with corresponding fatigue life at each level *N1, N2* ... cycles, failure can be expected at $\Sigma n/N = 1$ according to Miner's Law. Experiments have shown this factor to vary from 0.6 to 1.5 with higher values obtained for sequences of increasing loads.

Fatigue of structural box sections Spot welded box section members typical of vehicle side-rail elements have been tested by the Welding Institute and different joint configurations evaluated as in Fig 79. These were made in mild steel to EN 24 and subject to both free-free bending and torsional oscillations about both xx and yy axes at 53 to 83 Hz. S-N curves in terms of bending moment are shown in Figs 80/81 for the sections whose inertia properties are shown in Fig 82. Section E proved to be the best performer in bending and the deduced reduction factor relative to that section, for the others, is shown in Fig 83. In the torsional tests, a vibrator was used to

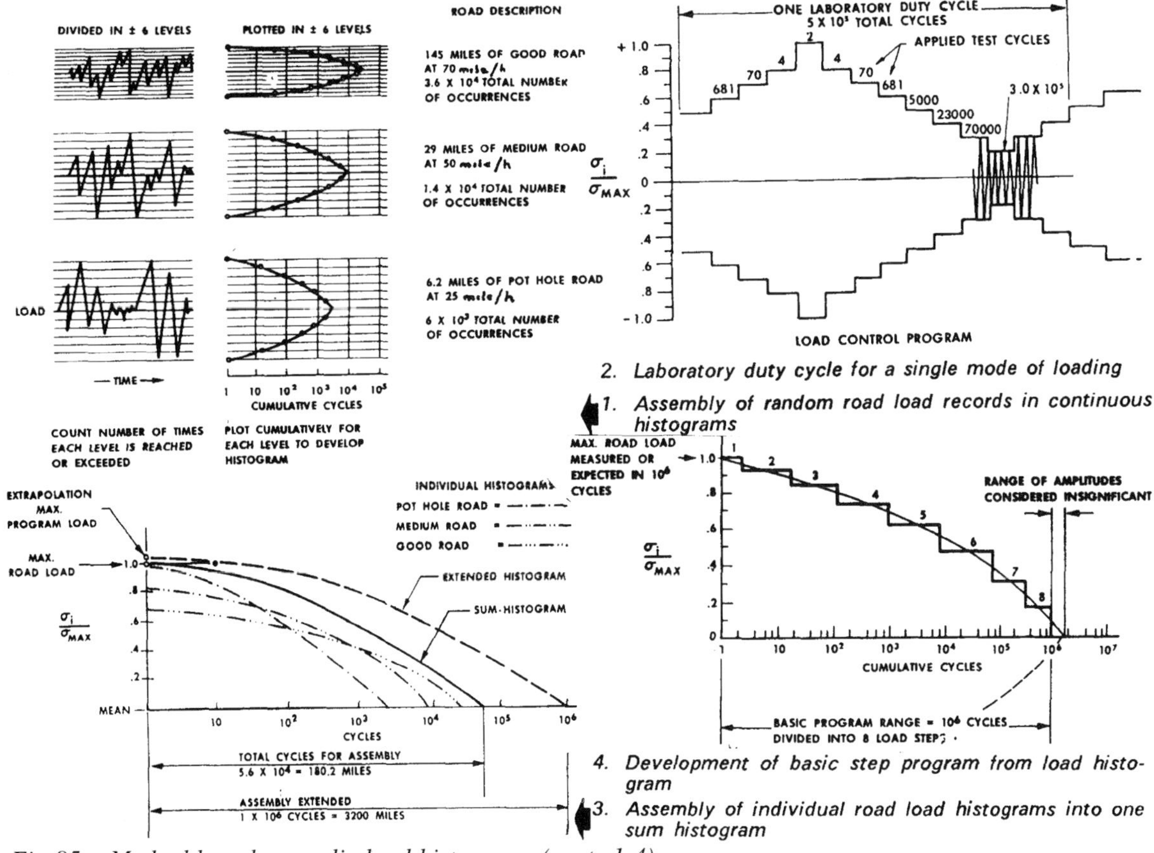

Fig 85a: Method based on cyclic load histograms (parts 1-4)

oscillate the sections at their resonant frequency given by $(60/2\pi)[GJ.12g/I_oL]^{1/2}$.

For shear modulus G = 11.7 x 10^6 lbf/in^2 (80 x 10^6 kN/m^2), length L = 36 in (0.91 m), vibrator inertia I_o = 3260 lbf/in^2 (22 MN/m^2) and torsional constant J = 1.82 , the resonant frequency is 40Hz for gravity acceleration g = 32.2 ft/sec^2 (981 cm/sec^2). In the torsional case, A proved to be the best performer with reduction factors for the others shown in Fig 83. Tests carried out by Nihon University for the Japan SAE used single and double closed top hat sections. Cracks were found to occur on spot welds close to centre span of the torsional specimens where effect of end constraint on developed stress is minimal. Cracks occurring differ from those in bending tests as shown in Fig 84 while Fig 83 shows how increased metal gauge favourably affects torsional fatigue life. Useful advice on the effects of arc welding on fatigue life is given in British Standard BS 5400 Part 10.

Service load data SAAB as aircraft maker as well as a car producer were early into the use of service load data for design life calculation. Early work on suspension parts used strain gauges to measure surface stresses from which bending moments were derived and a frequency distribution obtained for repesentative road and track routes for 10^6 cycles. The load spectrum was typically divided into half this number to obtain a mixture of high and low loads representative of operational conditions — and programmed rig-testing of the part carried out commencing with a medium load step and running with rising and descending steps until failure occurred. It was practice to scale up all stages of the test load spectrum to reduce test time — and repeat the tests at different severities. Plotting results on log-log paper produces a 'working life line' from which it is possible to read off life associated with level of bending moment experience. Another approach, Fig 85, is to obtain random load histories from service tests, by plotting load against occurrences. Several individual histograms can be combined to give a total effect and the resultant extended to 10^6 cycles by extrapolation. For test purposes the curve is approximated by a step program. Cycles with an amplitude less than 10 per cent of the maximum were omitted to achieve a 30 per cent test-time saving. Fig 87 also gives an example of the eight steps into which a curve is typically divided. Then a series of accelerated tests with different scale factors can be run to obtain the straight life line on the log-log plot which becomes a 'mixed-cycle' S-N curve for life prediction.

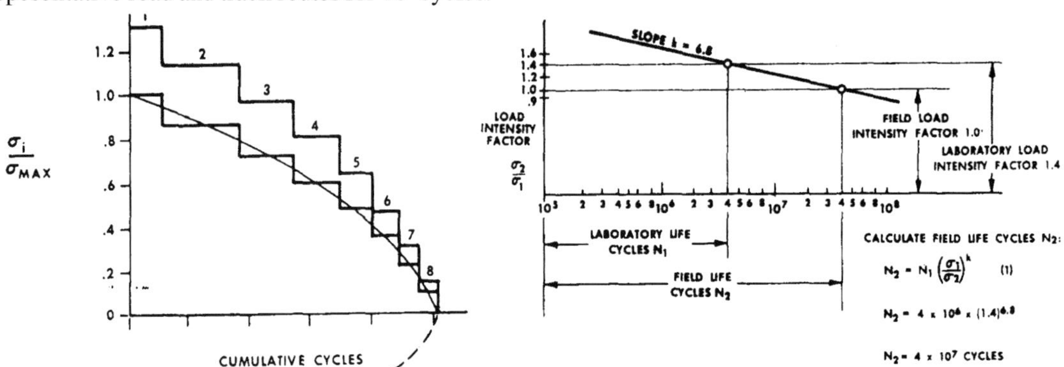

5. Development of a step program with increased intensity

6. Use of the mixed cycle S-N curve

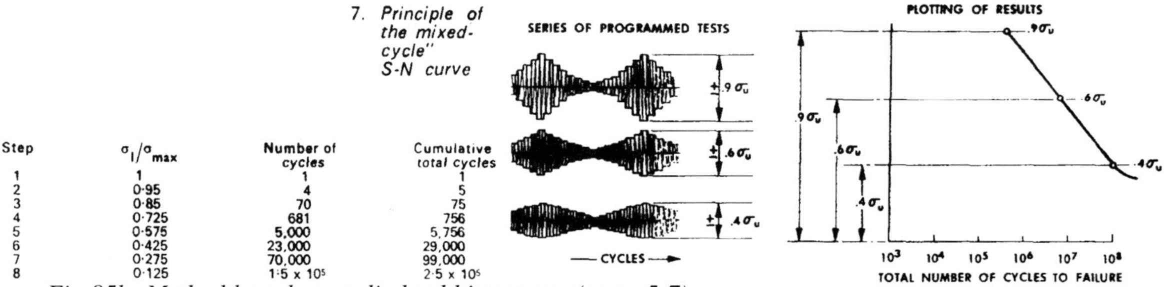

7. Principle of the mixed-cycle" S-N curve

Step	σ_i/σ_{max}	Number of cycles	Cumulative total cycles
1	1	1	1
2	0·95	4	5
3	0·85	70	75
4	0·725	681	756
5	0·575	5,000	5,756
6	0·425	23,000	29,000
7	0·275	70,000	99,000
8	0·125	1·5 x 10^5	2·5 x 10^5

Fig 85b: Method based on cyclic load histograms (parts 5-7)

Fatigue in aluminium structures

It has been pointed out[10] that as a fatigue crack progresses, the stress on the residual cross section increases so that there is a corresponding increase in the rate of crack propagation. Ultimately the stage is reached when the remaining area is insufficient to support the applied load and final rupture occurs. Due to aluminium's physical and mechanical properties, it is often used in applications where the ratio of net weight to gross operating weight is larger than that in structures built from steel. This implies that structural applications using aluminium will be governed by different ratios of minimum to maximum stress. Such applications are often found in the transportation field, which by it's very nature is subjected to fluctuating loads. Thus, it is here that all the elements of fatigue reducing design need to be applied, he explains.

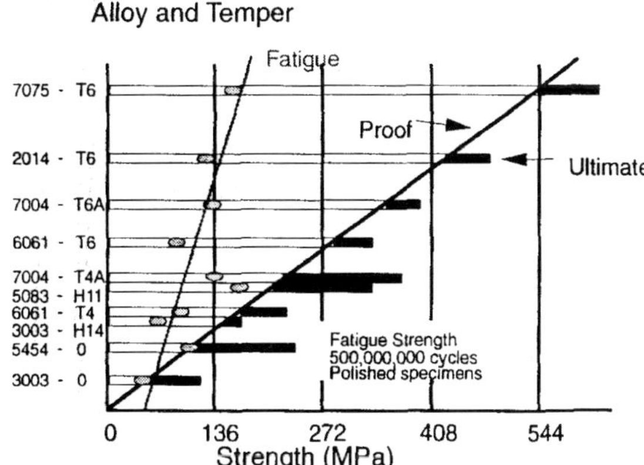

Fig 86: Fatigue, proof and ultimate tensile stresses of some wrought aluminium alloys.

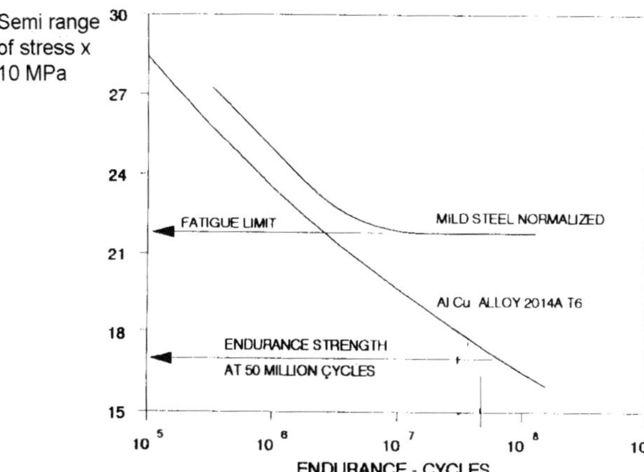

Fig 87: S-N curves for steel and aluminium alloy

Fatigue and the design It is necessary to establish as early as possible the extent to which fatigue is likely to control the design, the author maintains. In doing this the following factors important: an accurate prediction of the complete loading sequence throughout the design life; the elastic response of the structure under these loads should be accurately assessed; detail design, methods of manufacturing and the degree of quality control which have a major influence on fatigue strength need to be specified more precisely than for statically loaded structures.

Conditions for fatigue susceptibility are: high ratio of dynamic to static load; frequent application of load —slender structures or members with low natural frequencies are particularly prone to resonance and enhanced magnification of dynamic stress, even though the static design stresses are low; use of welding — some commonly used weld details have low fatigue strength — this applies not only to joints between members but also to any attachment to a loaded member, whether or not the resulting connection is considered to be structural; complexity of joint detail — complex joints often exhibit high stress concentrations due to local variations in stiffness through the load path. While these may have little effect on the static capacity of the joint, they can have a severe effect on the fatigue resistance. If fatigue is dominant, the joint shape should be selected to ensure smoothness and simplicity of design.

Potential sites for fatigue cracking The most common areas for the initiation of fatigue cracks are as follows: toes and roots of fusion welds; machined corners and drilled holes; surfaces under high contact pressure (fretting); roots of fastener threads; production and quality control discrepancies including scratches, weld arc strikes and misaligned assembly.

All of these areas are points of high stress concentration which have a far greater influence on the fatigue strength than any differences which can be attributed to the individual choice of alloy alone. This is why, other than those for very low cycle fatigue, most fatigue codes for commercial applications consider all alloys as one.

Aluminium alloy fatigue characteristics The differences in the fatigue strengths of various alloys are shown in Fig 86. From this, it is immediately apparent that the fatigue strength increases at a much lower rate than the proof and ultimate strengths as the strengths of the alloys increase. When laboratory specimens of aluminium alloys are fatigue tested the S/N curve continues on a downward slope therefore they exhibit no fatigue limit. Steel on the other hand, has a cyclic stress limit below which no failure occurs irrespective of the number of cycles applied and is therefore said to have a 'fatigue limit', as shown in Fig 87. In current design thought, where weight and costs are important this should be of little consequence. Designs are aimed to provide a specified life span while subjected to a given stress history, the author argues. The mechanism of crack growth was described in the earlier section on Designing for Fatigue. However, Cobden points out that fatigue cracks may be very difficult to detect since, unlike tensile failures there is no surface necking. Crack propagation occupies a major proportion of fatigue life especially at high cycles of loading. Final failure may be the result of the growth of one crack or of many small cracks coalescing into a final crack. The higher the number of loading cycles , the greater the production of multiple cracks.

When is a crack a crack? To most people this is when it can be observed with the naked eye although in reality it could be as small as 1000A. If no other imperfections are present the initiation of the crack begins with a slip band formation on the surface of the material, as shown in Fig 88. In pure aluminium high load cycling causes cracks to start at grain boundaries but in many commercial alloys the existence of large second phase particles (inclusions) can play a predominant role in crack generation. In engineering structures, cracks will already exist due to processes such as welding or machining.

It is possible to establish the number of load cycles, up to N, to generate an observable fatigue crack. It is usual to express the result in terms of the fraction of total life n/N. It has been shown that this ratio is 0.10 for pure aluminium and 0.40 for 2024-T6, both with a smooth surface finish. The combination of fatigue and a corrosive environment increases the speed of crack generation, while the application of compressive stresses to the surface either by shot peening or controlled stretch reduces it.

Crack growth rate The increase in crack length a for the increment DN of load cycles define the growth rate, Da/DN. This is a function of both the crack length a and the stress or strain amplitude. Observed growth rates may range between 10^{-10} m/cycle at low amplitudes to about 10^{-3} m/cycle at high amplitudes. The importance of the growth rate lies in its use to calculate the remaining life times, given a certain initial crack length a after NI load cycles. Assuming that a is a continuous function of N, the instantaneous crack growth rate da/dN can be used to give total life for a crack propagation. This ultimately leads to investigation by fracture mechanics theory. Plane strain crack growth occurs at 90 degrees to the tensile axis near the point of crack initiation. As the crack grows in length a shear lip begins to develop where the fracture surface intersects the specimen surface. This re-orientates the fracture surface to 45 degrees to the tensile axis. Crack growth rates can be expressed analytically for different crack lengths and for different loads in terms of the stress intensity factor range K. Cross sections through the crack tip at various parts of the load cycle have established that crack growth occurs by repetitive blunting and re-sharpening of the crack tip, as shown in Fig 89.

Fatigue load data In the most simple fatigue

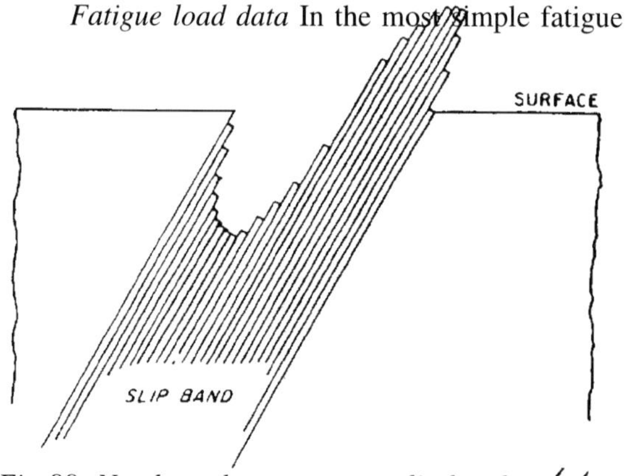

SURFACE

SLIP BAND

Fig 88: Notch-peak geometry at slip bands

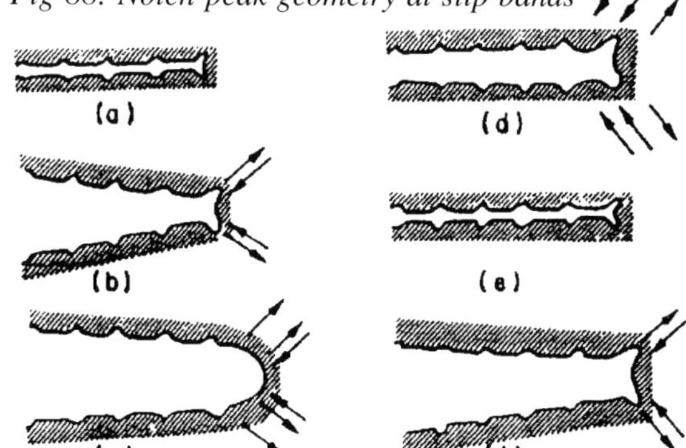

Fig 89: Fatigue crack growth by the plastic-blunting mechanism (due to Laird) — a zero load; b small tensile load; c maximum tensile load; d small compressive load; e maximum compressive load; f small tensile load; stress axis is vertical

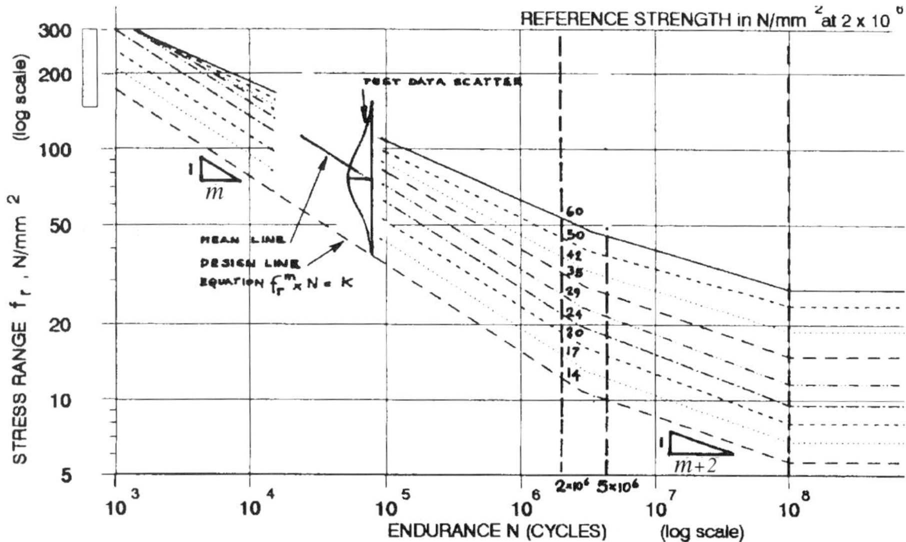

Fig 90: Design stress range x number of cycles curves for variable amplitude stress histories

loading case such as the cantilevered loaded rotating bar, the counting and loading of the cycles is a simple exercise. Variable rate tensile fatigue tests are also simple as long as the load increase slope remains positive and load decrease remains negative. In these cases when assessing fatigue three basic factors need to be known: number of stress cycles; a definition of the stress cycle and surface finish or contour shape classification. In this simple case the number of stress cycles is easily established. When defining the stress cycles there are four basic parameters: the minimum stress cycle S_{min}; the maximum stress cycle S_{max}; the mean stress:

$$S_m = (S_{min} + S_{max}) / 2 \text{ and } S_r = S_{max} - S_{min},$$

the stress range. The stress cycle is fully defined if any two of these four quantities is known.

Load spectra damage accumulation In most structural applications the load inputs are variable and it is important to take account of these. A number of methods have been devised to record load spectra and damage accumulation. Probably the simplest way to do this is to use the Reservoir Method as employed in the BS 118 standard[11].

When BS 118 was produced, 'limit state design' was used as the means of assessing stress, with fatigue being approached on the basis of stress range, as shown in Fig 90. This follows the practice which had already been used for BS 5400, 1980 (Steel fatigue).

Comparison of the aluminium and steel fatigue curves showed that a rough match was obtained if the steel data was divided by three. In the final version of 118 the aluminium data for the middle and high endurance range is generally greater than steel divided by three. More recent data from the Technical University of Munich, Fig 91, indicates widely varying fatigue strength ratios between various steel and aluminium joint details. Overall these range between 1.25 and 3.00 but the average is only about 2.3.

Detail	Aluminium	Steel range of recomm.	Steel EC3 ECCS TC6	Ratio Steel/Aluminium range	Ratio Steel/Aluminium EC 3 ECCS TC6
A1	130	116-194	160	0,89-1,49	1,23
A2	85	116-194	160	1,39-2,28	1,88
A3	95	116-194	160	1,22-2,04	1,68
A4	70	116-194	160	1,66-2,77	2,28
A5	0,9*A	87-143	0,88*A	,74-2,27	1,20-2,23
B1	55	93-194	125	1,69-3,52	2,27
B2	50	(77-90)	(90)	(1,54-1,80)	(1,80)
B3	45	76-90	80-40(?)	1,69-2,00	1,79-0,89
B4	40	52-64	71-50	1,30-1,60	1,78-1,25
B5	45	93-194	125	2,07-4,31	2,78
B6	40	(77-90)	(90)	(1,93-2,25)	(2,25)
B7	35	76-90	80	2,17-2,57	2,29
B8	30	52-64	40	1,73-2,13	1,33
B9	40	93-194	125	2,32-4,85	3,12
B10	35	69-90	90	1,97-2,57	2,57
B11	30	52-64	40	1,73-2,13	1,33
C1	60	93	---	1,55	---
C2	45	86-107	100	1,91-2,38	2,22
D1	45	77-100	100	1,71-2,22	2,22
D2	40	77-100	100	1,92-2,50	2,50
D3	35	58-80	80	1,66-2,29	2,29
E1	35	50-80	80-71	1,43-2,29	2,29-2,03
E2	23	39-64(?)	---	1,70-2,78	---
E3	35	52-90	90	1,49-2,57	2,57
E4	18	39-51	45	2,17-2,83	2,50
E5	35	50-76	50(?)	1,43-2,17	1,43
E6	23	39-51	(?)	1,70-2,22	---
E7	18	(?)	(?)	---	---
E8	23	(?)	(?)	---	---
F1	30	59-90	71	1,97-3,00	2,37
F2	25	36-39	36	1,44-1,56	1,44
F3	20	39-58	50-36	1,95-2,90	2,50-1,80

Fig 91: Comparison between steel and aluminium fatigue strengths in MPa at 2 x 10^6 cycles for 32 different structural details.

Rubber as an isolator

For alleviation of many vibration problems an understanding is needed of available rubber isolation-systems; here some of the basic physical characteristics of the material are examined and a guide given to information sources on special mixes and anchorage configurations.

Both natural and synthetic rubbers are used in anchorages and compliant bushes for vibration isolation. The key mechanical properties of vulcanised natural rubber are shown in Fig 92. This data is contained in a useful guide 'Engineering design with natural rubber' obtainable from the Natural Rubber Producers Association, 19 Buckingham Street, London WC2. The very high bulk modulus compared with the elastic (Young's) modulus means that rubber hardly changes in volume even under high loads and therefore space must be available for it to expand into on deformation. By restricting the expansion, effective increase in stiffness of a spring unit can be obtained.

Breaking strength of rubber in tension is in the range 14-28 MN/m^2 based on the unstretched cross section area. This can rise to 200 for the cross section at break — compared with compression strength at failure of 160 MN/m^2. Some typical stress/strain curves for natural rubber, depending on hardness, are shown in Fig 93. Creep and stress relaxation must also be allowed for in design. These vary linearly with the logarithmn of time as shown in Fig 94. When, conversely, rapid rates of strain are applied, rubber exhibits a substantial degree of damping, hysteresis curves being shown in Fig 95. This property varies substantially with the amount of filler (usually carbon black); left curves are rubbers with 50 parts reinforcing black and right curves for 50 parts non-reinforcing black. Dotted curves are after

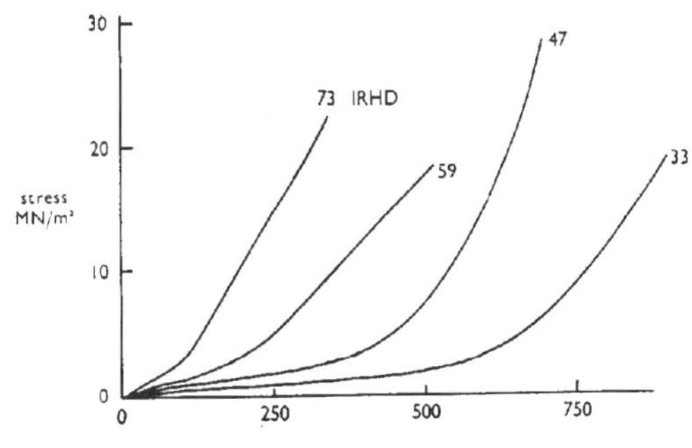

Fig 93: Stress/strain curves for natural rubber

Fig 94: Creep and stress relaxation

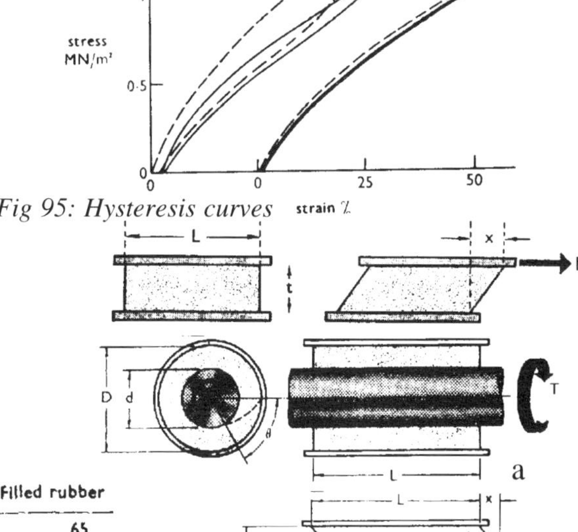

Fig 95: Hysteresis curves

Fig 96: Bonded rubber mountings

		Gum rubber	Filled rubber
Hardness	IRHD	45	65
Tensile strength (T.S.)	MN/m^2	28	21
Elongation at break (E.B.)	%	680	420
Young's modulus (E_0)	MN/m^2	1·9	5·9
Shear modulus (G)	MN/m^2	0·54	1·37
Bulk modulus (E_∞)	MN/m^2	1000	1200
Poisson's ratio		0·4997	0·4997
Resilience	%	80	60
Velocity of sound	m/s	37	37
Specific gravity		0·93	1·16
Specific heat		0·45	0·41
Thermal conductivity relative to water		0·25	0·31
Coefficient of cubical expansion/deg C		67×10^{-5}	56×10^{-5}
Electrical resistivity	ohms/cm cube	$1{\cdot}7 \times 10^{16}$	3×10^{10}
Dielectric constant		3	15
Power factor		0·002	0·1

Fig 92: Vulcanised rubber properties

one cycle and full curves after 10.

In many applications, rubber is bonded to metal; by preventing slipping at the load carrying surfaces, more predictable and repeatable performances can be obtained. In a bonded rubber shear mounting as in Fig 96a (or in circular form, 96b), the normal elasticity formulae apply. When ratio of thickness to length exceeds about 0.25, deflection due to bending should be allowed for as:

$$Ft^3 / 36AGk_r^2$$

where k_r is radius of gyration about the neutral axis $= L^2/12$ for a rectangular section. Stiffness of rubber in compression is calculated with the help of a shape factor. Provided no slipping occurs at the load surfaces, this is given by:

$$S = LB / 2t(L + B)$$

for breadth B, length L and thickness t. Variation of compression modulus with shape factor is shown in Fig 97. Synthetic rubbers include polyisoprene, styrene-butadiene, butadiene, ethylene-propylene, butyl, chloroprene, nitrile, polyurethane, chlorosulphonated polyethylene, polyacrylic, ethylene-acrylic, silicone, fluorsilicone, fluorocarbon, polysulphide, epichlorohydrin, polypropylene oxide and polynorborene. There is also a range of thermoplastic rubbers. Properties and types of most of all these are listed in British standard BS 6716.

Lord Corporation lists proprietary mounts and couplings handling loads up to 14 000 lb and powers up to 2000 bhp. Engineering theory is included in an associated Designers Guide showing different configurations which can provide the required combination of stiffness and damping, Fig 98.

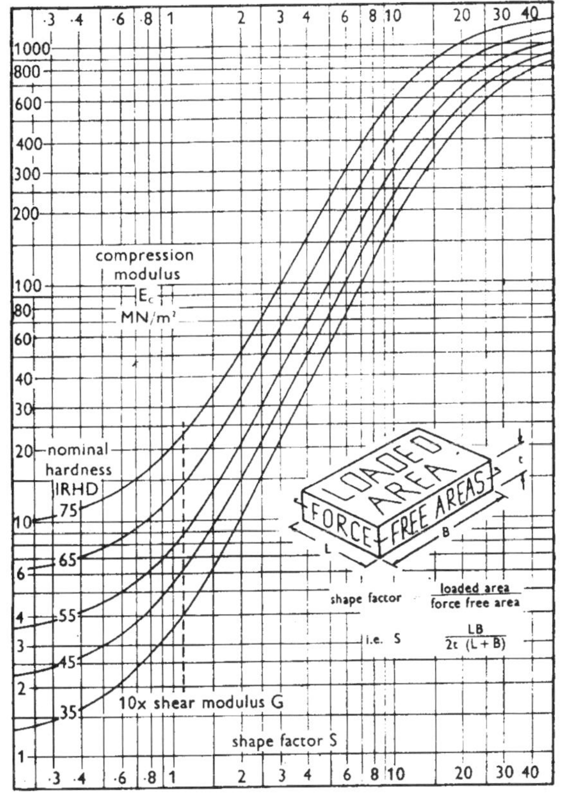

Fig 97: Compression modulus vs shape factor

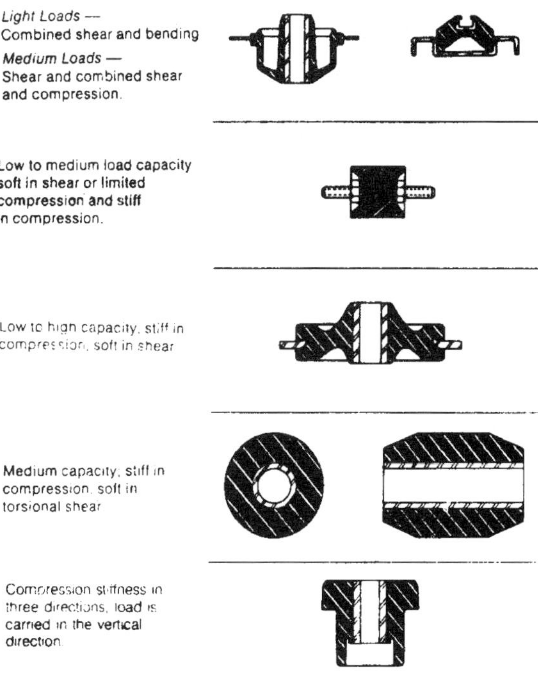

Fig 98: Isolator configurations; from top to bottom: Soft in all directions; soft in two directions, stiff in one; stiff in two directions, soft in one; stiff in two directions translationally, soft torsionally; equal stiffness in three directions

CV body mounting

The prospect of product liability legislation has brought a considerably increased application of engineering-science into the study of chassis-to-body interaction and the publications of bodybuilder codes of practice, compliance with which will make both parties less vulnerable to product-liability claims. Vehicles such as road tankers, of the non-integral type, have historically required greater attention to the chassis/body interface in order to protect the shell structures against fatigue-failure at the mounting locations. Some important lessons were learned from this experience which now benefit the mounting of box-van shells, and related configurations, which have themselves increased in torsional rigidity as construction methods have developed.

The considerable disparity between the stiffness of CV chassis and body is likely to continue as long as any one chassis design has to serve the requirements of a proliferation of body types and modes of operation. With the exception of purpose built military, and related, all-terrain vehicles, having long-travel 'soft' suspensions, truck chassis have relatively 'hard' short-travel ones. These suspensions require a torsionally flexible frame, Fig 99, to permit sufficient cross articulation of the axles for retaining wheel ground contact on rough terrains.

Chassis frame design was considered in an earlier section; besides the constraints described in it, there is also thought to be unnecessary unladen weight penalty in making chassis frames too rigid by conventional structural techniques. Truck makers

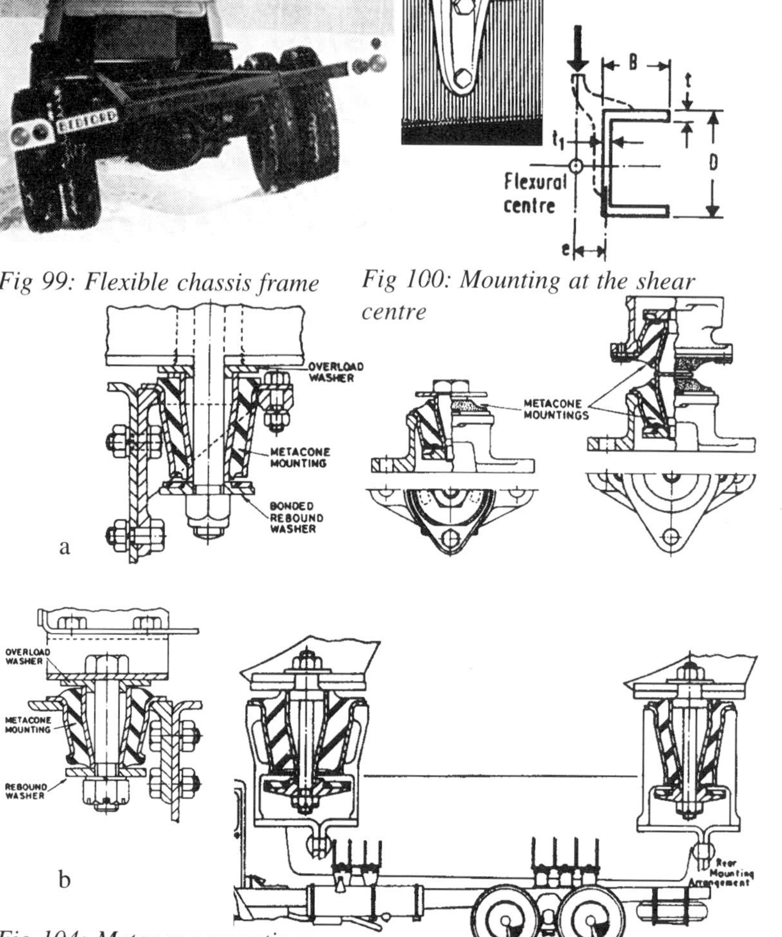

Fig 99: Flexible chassis frame

Fig 100: Mounting at the shear centre

Fig 104: Metacone mountings

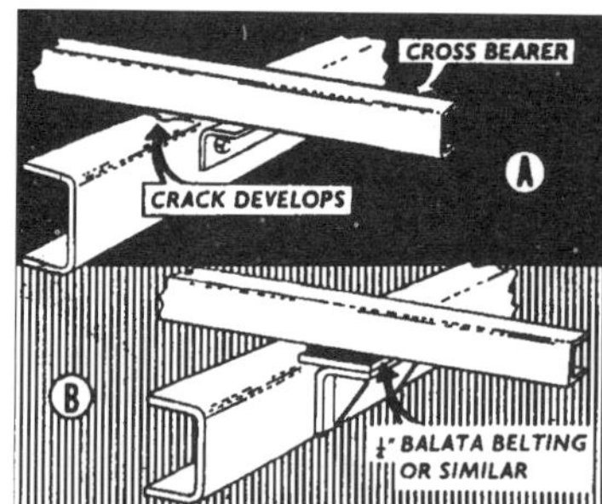

Fig 101: Protective bracket

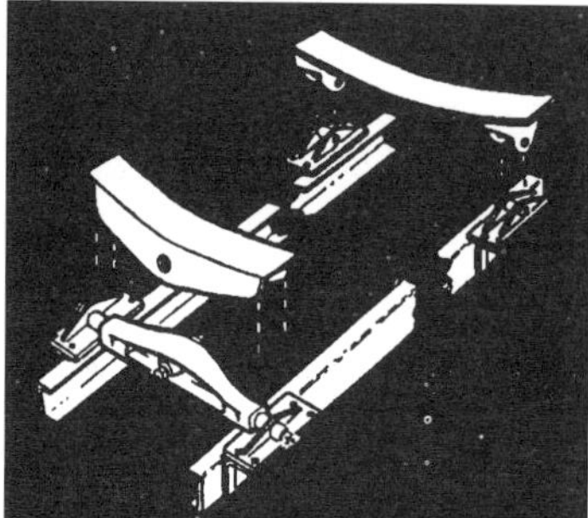

Fig 102: tanker mounts

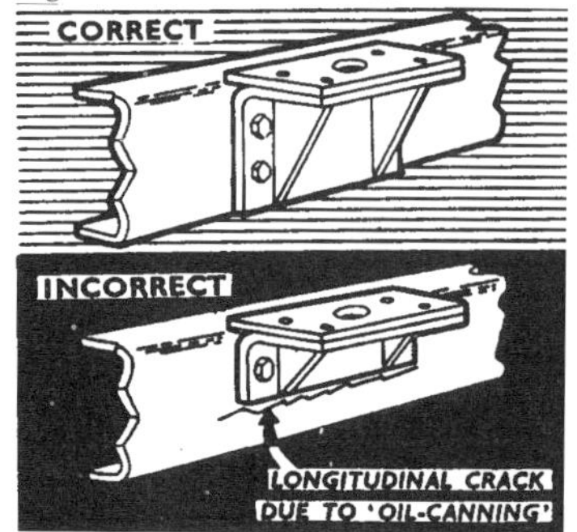

Fig 103 : Avoiding oil-canning

such as GM Bedford had considerable early experience with four-wheel-drive trucks for military application and were among the first to recommend that bodies were not direct-mounted to upper flanges of the chassis sidemembers, offering factory-supplied brackets which involved clearance between sidemembers and body cross-bearers.

This bracket design, Fig 100, was said to avoid frame cracking arising from fidgeting motions, between chassis and body, causing wear and looseness between the two, Fig 101. The distance e, defining the flexural centre, at which the line of action of vertical load gives no twist to the channel section is given by:

$$(Bt/4I)(B-t1)(D-t)(D-2t)$$

where I is section second moment of area for the channel. Lips of these mounting brackets were arranged so they did not overlap the top flange of the sidemember by more than one inch; this allowed vertical movement of the flanges during weaving caused by torsional flexure on rough ground. The object of the lip was to relieve the bracket fixing bolts from shear loading; as well as maintaining chassis/body clearance as discussed above.

The company also advised against longitudinal body runners direct-mounted to the sidemember top flange on the grounds that it is not designed for supporting the entire weight of body and payload. The increase in section depth is also said to stiffen the

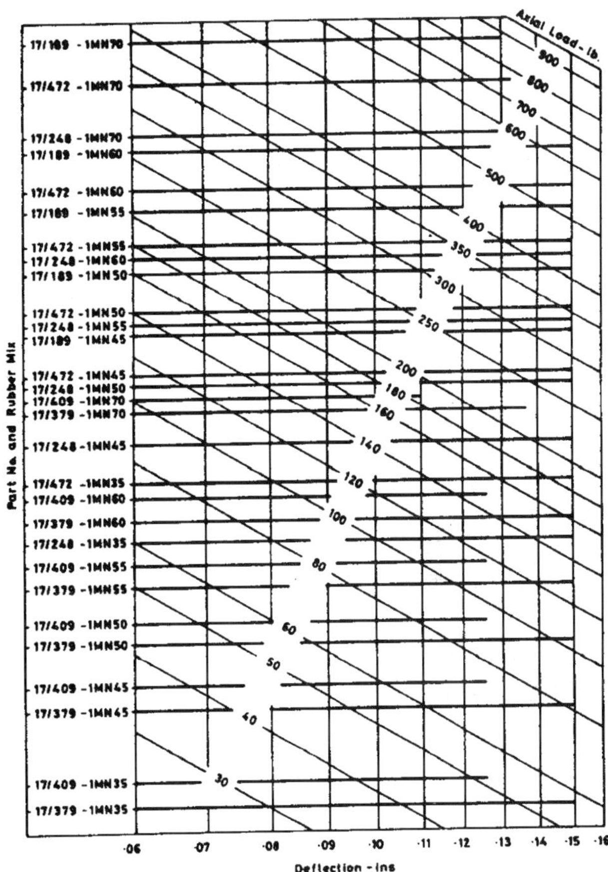

Fig 105: Metacone characteristics

frame in the area of body support such that any flexing takes place between the foremost edge of the body and the front-spring mounts — leading to possible fatigue cracking at the stress concentration point at the leading edge of the body bearer. Hence the need for brackets connecting direct to the sidemember webs, and, the company maintained, at the shear centre of the channel section, in order to avoid localised twisting moments. The company also used a U-shaped pressing between the two brackets to the rear of the back-axle, thus making any horizontal lozenging motions impossible.

For tankers, or even torsionally-stiff insulated box vans, the company recommended a layout which allows the stiff body to float on the chassis, unaffected by the latter's weaving. Three or five-point resilient mountings, Fig 102, were recommended for tanks, with a centrally located trunnion mount used to

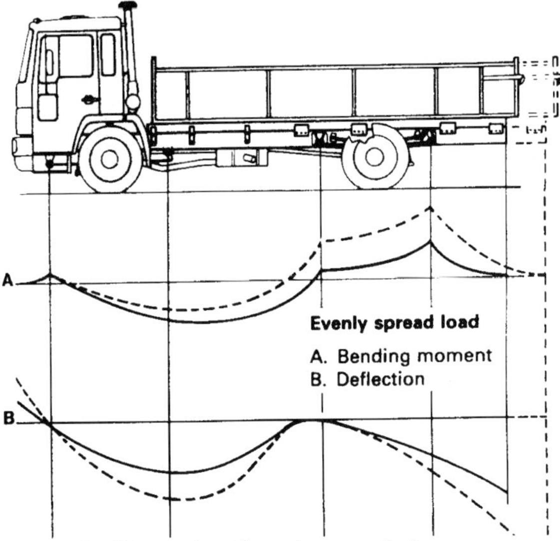

Fig 106: Chassis bending characteristics

support the front end. As fewer outrigger brackets are used, in such cases, than for conventional bodies, greater strength must be built into them and the company recommended their vertical faces extending to the full depth of the sidemember web in order to avoid 'oil-canning', Fig 103.

Proprietary resilient mounts are obtainable from such companies as Metalastik. The company recommend that mountings should be designed to give ample flexibilty over a strictly limited range of movement — beyond which stiffness of the resilient elastomer should increase rapidly and progressively. Positioning along the chassis has, of course, to be calculated so that the deflection of such units is fully utilised without overstraining. The Metacone mounting, Fig 104, is available in four types; one has integral buffers and and metal washers for bump and rebound (a). A second has integral buffer and metal washer at one end plus separate bonded rebound washer at the other (b). A third comprises mounting plus bump and rebound washers, assembled into a robust housing with two-hole fixing at the base while a fourth includes the SL derivative mounting having cut-out portions in the 'rubber' — with integral buffers and metal washers for bump and rebound. Typical characteristics are shown in Fig 105. Guidance figures for chassis frame deflections, constrained by mounted bodies, have been provided in pioneering quantitative analyses by Erz who quotes frame twist of +/- 25 deg and wheel deflections of 53 cm from ground level with maximum obstacle heights ranging from +/- 30 cm, on-road, to +/- 50 cm on road. Softer vehicle suspensions are of course continuously reducing frame deflection, with the passage of time, relative to spring deflection. One of Metalastik's approaches was to recommend stiffest mounting in the transverse plane of the payload centre-of-gravity then softening pro-rata towards the end of the body.

The main CV chassis/cab manufacturers have now issued comprehensive recommendations applying to their own chassis-frames. Volvo have provided frame bending diagrams for mounting bracket positioning; recommended bracket types for bodies of varying torsional stiffness and related bolting recommendations. Superstructure Instructions give specific advice for such variants as tippers, tankers and vans. Fig 106 is a typical bending moment and deflection plot for the FL6 'rigid' model with platform body. Reference holes are provided on the delivered chassis to facilitate measurements from the axle centrelines. Ready-punched holes (Fig 107) are also now provided for different attachment (mounting) type. Controlling dimensions are also given (Fig 108) for extra holes in the web to suit specialist bodies. Four types of regular attachments have been standardised (Fig 109) for pivoted (a), flexible one-way (b), Flexible two-way (c) and rigid

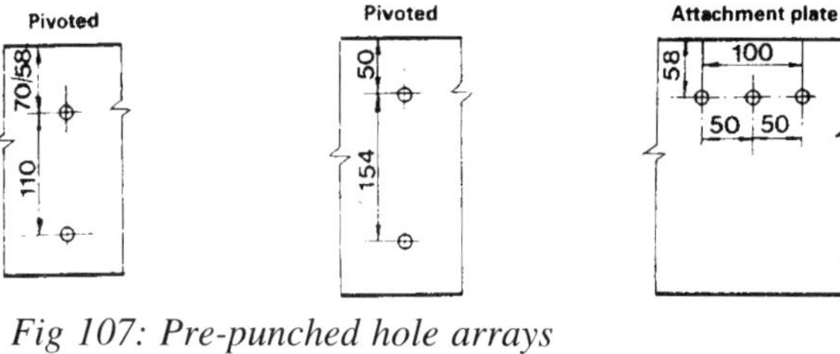

Fig 107: Pre-punched hole arrays

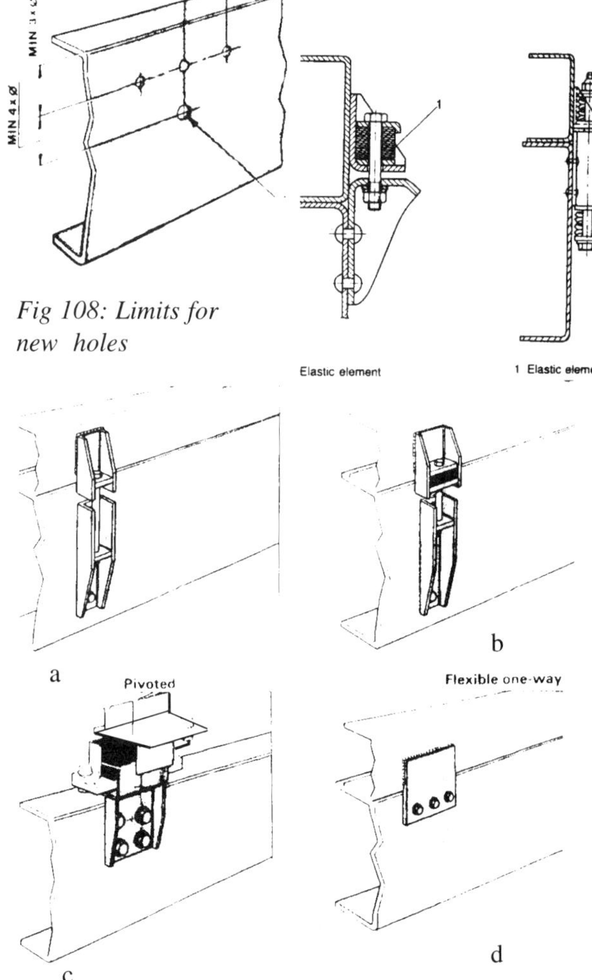

Fig 108: Limits for new holes

Fig 109: Standardised attachments

(d) according to body type (Fig 110). The Leyland DAF Superstructures manual uses the concept of matrices for giving specific recommendations according to wheelbase, axle configuration and body type. These work around the basic mounting systems shown in Fig 111 of single tie-rod attachment (a) and retainer plate (b).

Iveco Ford issue the Bodybuilders Instructions manual which recommends a number of 'auxiliary frames' and mounting instructions for distributing loads from different body types uniformly on the chassis, Fig 112. These frames may be flexible or rigid according to recommendation for a particular body type and its connection to the chassis varies accordingly. For a box-van body the minimal dimensions for the longitudinal members of the auxiliary frame are shown in Fig 113. Where the box body uses crossbearer supports brackets rising above the auxiliary frame, these need triangulated anchorages to contain lengthwise thrusts, Fig 114.

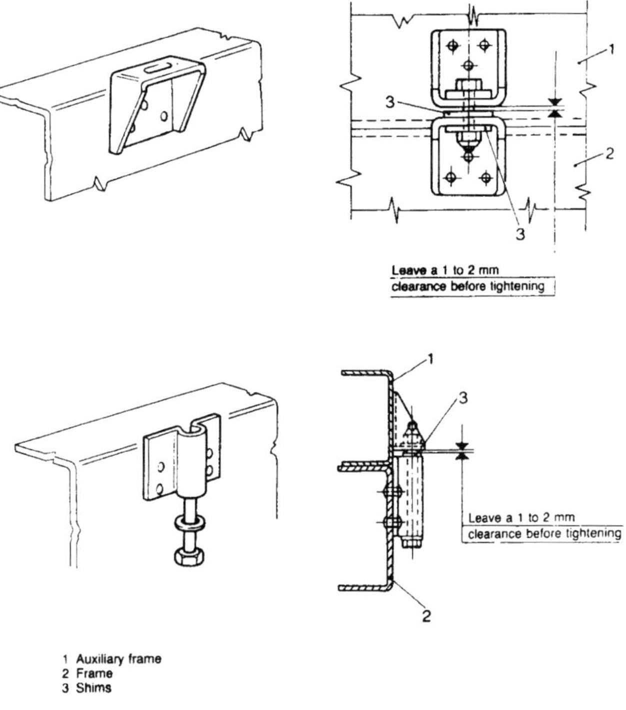

Fig 110 : Flexible attachments

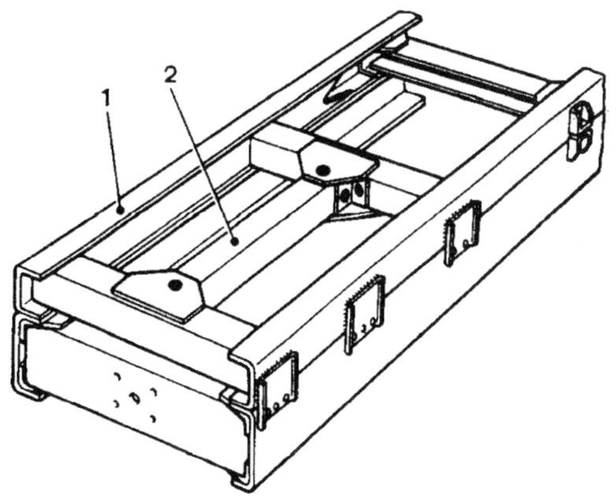

Fig 112: Auxiliary frame

Vehicle class total weight	MINIMUM SECTIONAL REINFORCING IRON	
	Moment of resistance for each section iron Wx (cm^3)	Dimensions (mm)
3 ÷ 7 ton	16	80x50x4
7,9 ÷ 10 ton	26	100x50x5
11 ÷ 33 ton	46	120x60x6

1 2 3

1 Auxiliary frame
2 Brackets
3 Securing anchorages

Fig 113: Auxiliary frame sizes

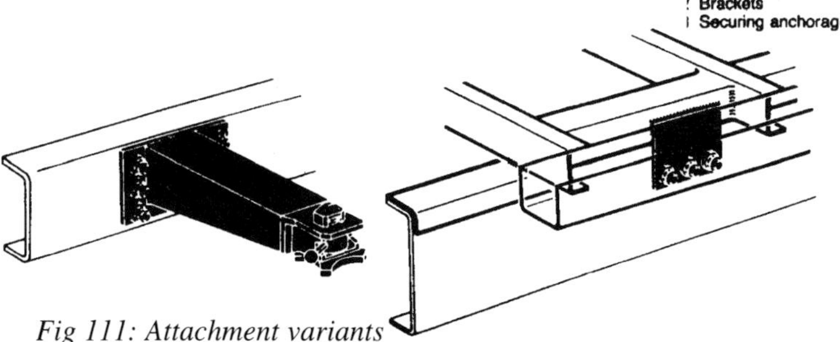

Fig 111: Attachment variants

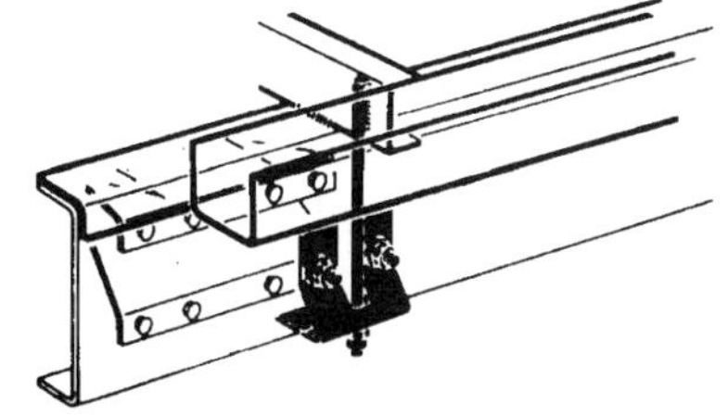

Fig 114: Triangulated anchorages

Secondary body suspension

The advances made by rubber suspensions in rail vehicles, involving steel elements bonded to elastomers, may have some lessons for heavy road vehicle body suspension. Researchers at Metalastik Vibration Control Systems have demonstrated the high load bearing capacity of rubber springs, with their ability to accommodate overload conditions without the catastrophic failures which can be associated with steel systems, might provide auxiliary suspension. Ratio of load carried to spring weight has been a historic advantage of rubber springs so too has been the unique acoustic damping property. Most important, however, has been the ability to bond rubber to metal and achieve rotational and shear displacements which do not involve sliding contact and lubrication requirements, according to Metalastik engineers.

The early problems of spring settlement due to creep of the material under load have been largely overcome by advances in compounding techniques involving natural rubbers and synthetic polyisoprenes. Care is needed in design to resolve the problems of low-creep compounds in dynamic situations. Dynamic stroking of rubber springs involves the material's internal damping effect and leads to a dynamic stiffness higher than the static one. Usually the ratio is constant over the normal frequency range of 2-20 Hz. Strain levels of 2-5 per cent also have little effect on dynamic:static ratio. However strains below 1 per cent cause much higher dynamic stiffnesses while very large strains reduce the ratio almost to unity.

There is also a temperature effect to guard against and in increase of spring standing height of some 4 per cent can be expected with a 10 C temperature rise. While the automotive market has advanced the acoustic damping applications of rubber, beyond the railway one, their may be some lessons to be learned from railway engineering in the use of sole rubber suspensions, as opposed to anchorage bushes. The ability of bogie wheel-sets to steer round curved tracks has been largely due to rubber suspension. In conical rubber springs of the metal boned type (Metacone) use has been made of air 'voids' in the rubber to give directional effects to spring stiffness — a technique that has already read over to the automotive sector.

In secondary suspensions, resilient seatings, Fig 115, are an interesting railway engineering development. These so-called Flexicoil springs have been used to produce dissimilar horizontal stiffness rates in the lateral and longitudinal directions, to lower yaw stiffness of the vehicle body by reducing end-moment constraint of the main steel coil springs.

Solid rubber bearer springs, of the type shown in Fig 116, might be effective means of achieving road-friendly heavy vehicle suspensions without resort to the complexity of air-springing. Individual spring load capacities down to five tonnes have been configured for railway application. Below 120 km/h ride quality is said to be comparable with air-springing and constant periodicity can readily be obtained.

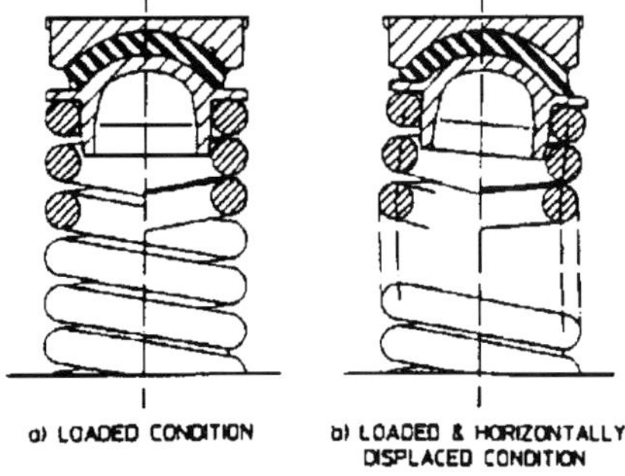

Fig 115: Flexicoil resilient spring seatings

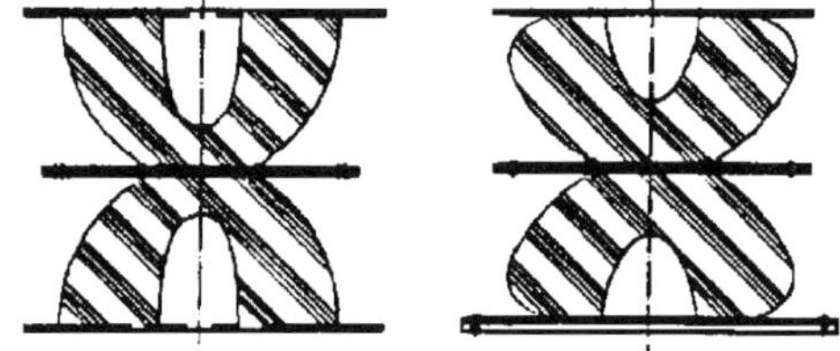

Fig 116: Solid rubber bearer springs

Chapter 11: Structural design applications

Bus structural design basics; case study in bus structural design; SPV idealisation; designing in structural-composites; analysis of composite structures; hydroformed tubular structures; bonded aluminium alloy structures; weight paring in safety-critical parts; aluminum vs steel body parts

Bus structural design

The development of design techniques is traced from the introduction of the first integral structures, through the application of regular stressing procedures for subassemblies to the overall finite element analysis of the shell. First, the approximate methods, based on considering cut sections of the shell, are examined as a means of obtaining a feel for contributions made by different elements to the overall structural performance.

Much structural analysis of bus bodies dates back to work done in East European countries who pioneered the application of aeronautical techniques to early forms of integral construction. Bus production at the Hungarian Ikarus company ran at some 12 000 units per year in the 1960s and 70s and many of the techniques were developed at the Autokut Research Institute in that country. These techniques were developed into structural design procedures at the Cranfield Structural Design Group[1]. An approximation for stresses in an integral bus structure can be obtained by assuming the bending loads are carried by the side frames. The side door apertures are critical areas as they represent large cut-outs in the structural surfaces. Shear force at the centre of each opening is assumed to be shared equally between the cant rail and sole bar according to their respective stiffnesses. These in turn cause root bending moments at the corners of the opening which can be

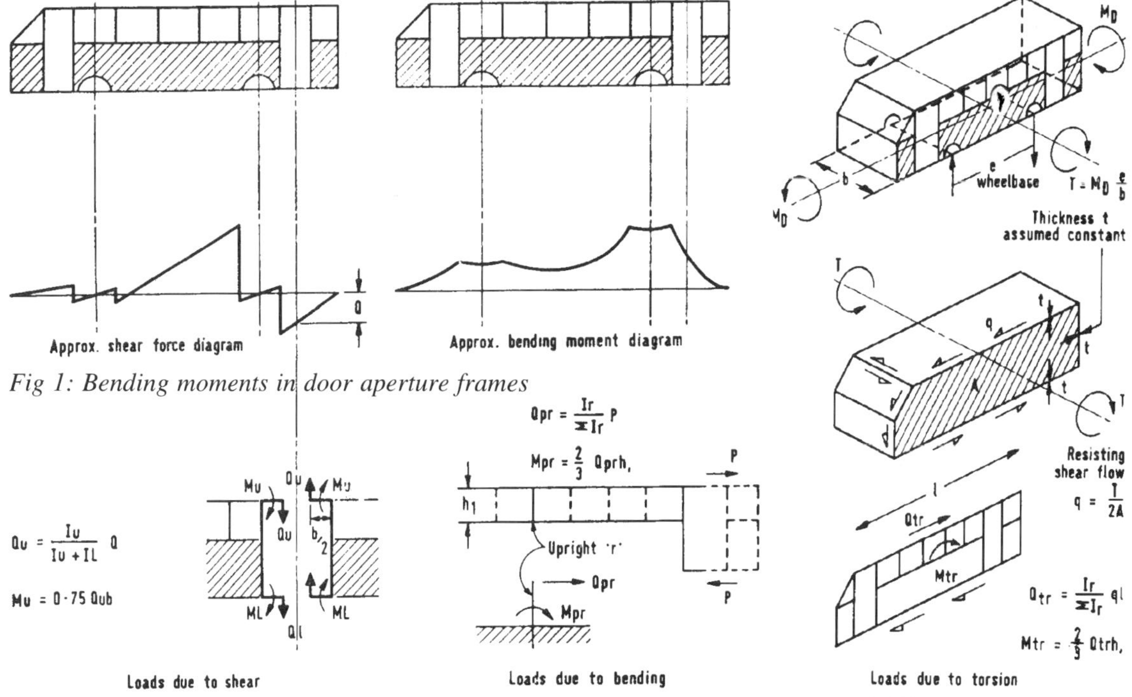

Fig 1: Bending moments in door aperture frames

Figs 2/3: Cant rail and pillar forces

Fig 4/5/6: Torsional loading

calculated as in Fig 1. Redundant-frame analysis of this ring-frame can be avoided by making the assumption of the frames being formed of a series of cantilevers, subjected to a shear force 50 per cent greater than the value obtained from the side-frame shear force diagram.

Also, the bending moment at the centre of the door will give rise to equal and opposite end loads in the cant rail and sole bar which are readily reacted by the sidewall structure up to the waist-rail. Force P on the cant rail, however, will be balanced by a sum of forces Qs developed by bending of the window pillars. These can be calculated according to the stiffnesses of the respective pillars from the formulae shown in Fig 2/3. To allow for the fact that the joints at the top of the pillars are gussetted rather than pin-jointed the bending moment in each pillar is assumed to be reduced by one third. Torsional loading of the body, Fig 4/5/6, gives rise to a torque which can be considered to act across the structure which acts as a short tube of length equal to the bus width.

The shear flow q is then given by $q = T/2A$ from the torsion formula developed in an earlier section. This shear flow is carried down from the roof by the pillars in the same way as the bending case but causing bending moments at right angles to the bending-case ones. Fig 7 shows the bending moments in the windscreen pillars for example. To compensate for the poor resistance of the side pillars to shear loading on the side frame, an anti-lozenge panel, Fig 8, can be provided.

Semi-integral construction vehicles with negligible stiffness in the pillars can be analysed as trough structures, Fig 9, with the side wall up to the waist rail carrying the substantial part of the bending loads. These will be carried by both chassis and body, in proportion to their relative stiffnesses. Whether the body is timber, aluminium-alloy or steel-framed makes a dramatic difference to the contribution made towards stiffness if it is assumed that the frame reacts only bending and the skin-panels only shear. The body contribution might be higher in those coaches which have no central door openings on the sideframes and which have ring frames made up of the chassis/

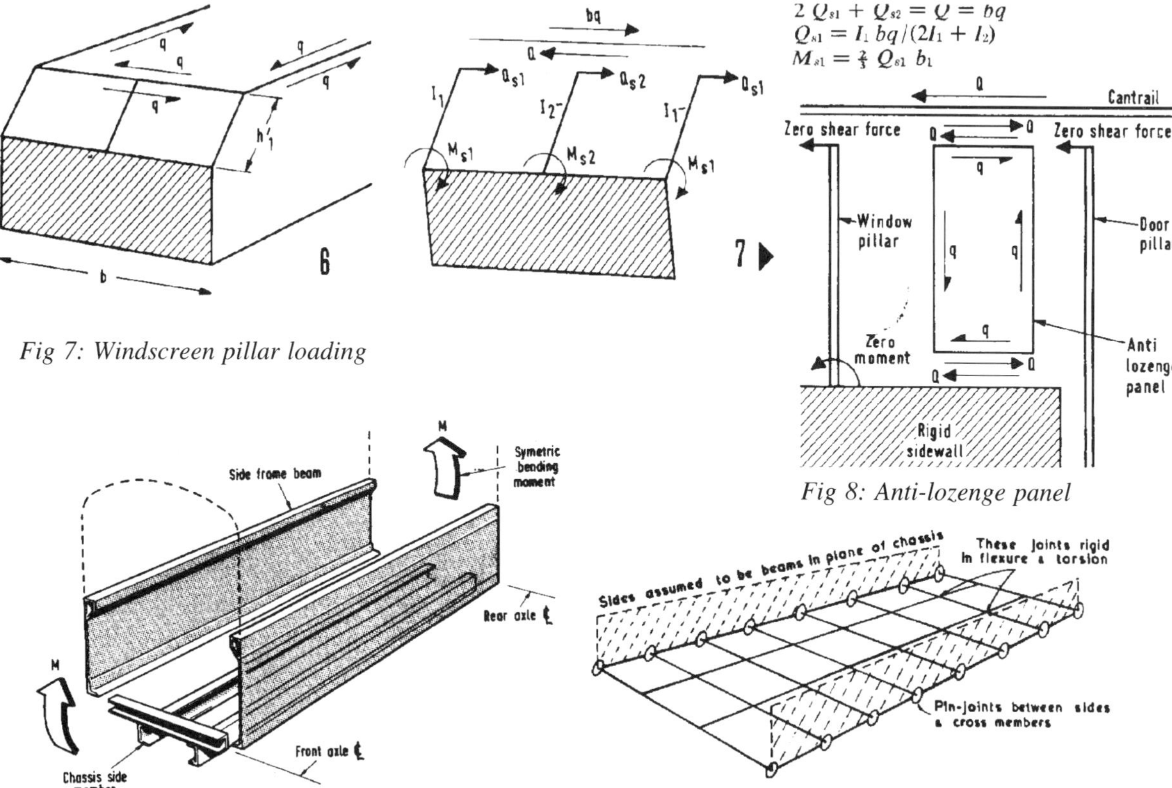

Fig 7: Windscreen pillar loading

Fig 8: Anti-lozenge panel

Fig 9: Trough structure

Fig 10: Trough idealisation

body floor cross-bearers, side pillars and roof sticks.

Fig 10 shows the 'trough' representation and Fig 11 a model test that was carried out at Cranfield to compare the stiffnesses of sideframes with and without door 'cut-outs'. The deflections of the two frames are compared in Fig 12. When the structure is made assymmetric by having a door cut-out(s) only on one side the two sides of different flexibilities are joined by a series of rings whose shear flexibility will affect the overall structural flexibility. If the sides of the rings are very stiff, shear will be carried by the roof-sticks and floor-crossbearers, as by the cant-rails and door-sills at the cut-out.The effect of one such ring, at mid-span, is calculated in Fig 13. Its flexibility, (f3) can vary widely with the assumed stiffness of the sides. Applying this analysis to a composite structure has suggested ratios of body to chassis bending stiffness of about 3.5:1.

In the pioneering Leyland National bus structure (Fig 14), galvanised steel coated with 18 microns of zinc was used in thicknesses up to 5 mm on the underframe. Epoxy powder coating from 50-75 microns was used to overcoat the zinc. Main assemblies were Avdel-riveted together, these comprising floor, roof, sideframes, front and rear assemblies of spot-welded steel pressings. Target life of 15 years was the design criterion and normal working stress was made to be one third the material proof stress. The structure was given twice the torsional and beam stiffness of traditional built structures and had also to meet rollover requirements of the American Federal School Bus Roof Load Test. Overall loading criteria included 3g vertical intensity generally, with 4g in some areas; 2g vertical plus horizontal braking loads; 2g vertical plus lateral 1g 'radial' loading and torsional loading of 50,000 ft lbs (67.8 kNm) at the axles.

Individual elements and joints such as pillars, transverse roof sticks, underbody crossmembers, wheelarch risers and underframe joints were first analysed. Next stage was the evaluation of large structural systems such as wheelarch groups, door-frames, step and engine beam assemblies. To prove these techniques a structural module bay was constructed and tested physically.

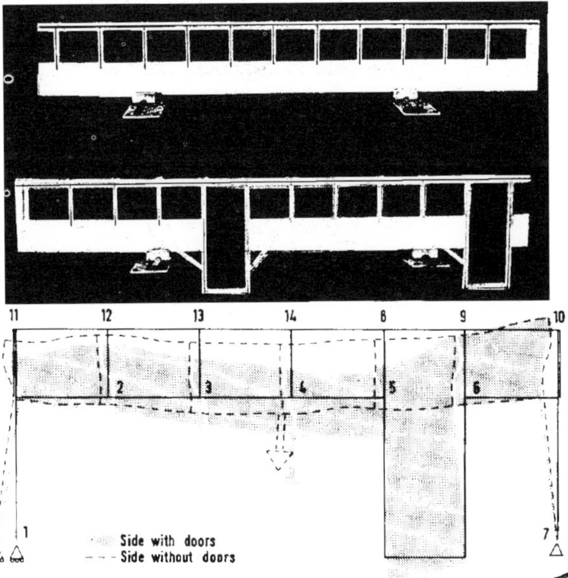

Fig 11: Model sideframe tests and computer simulation (below)

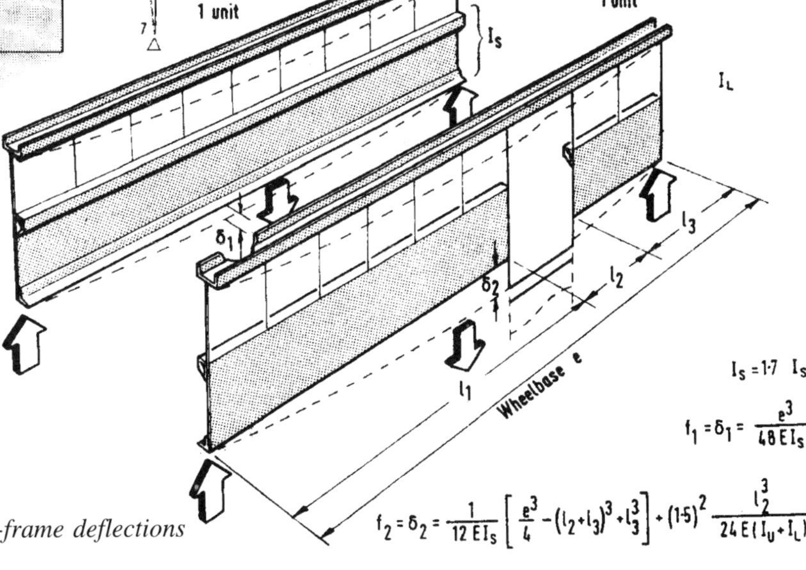

Fig 12: Side-frame deflections

CV structures case study

The Pawlowski Simple Structural Surface method for open-platform and closed structures, introduced in earlier sections, is here applied to bus structures. The closed integral-bus structure is considered as one of the most suitable for this method and the advantage of providing a structural feel for concept designs is apparent.

In such structures use is made of sidewalls and roof; Fig 15 shows the classic idealisation in which the following basic elements play the major role: cross-members, sidewall pillars and roof arches form the ring-frames which are connected by floor plates, side and roof panels. The shear deformation of the window area is crucial to the analysis. Since shearing occurs of the window cut-out frames, which depends on the bending stiffness of the window pillars, anti-lozenge panels are needed to stabilise the openings, as discussed in the earlier section.

Edge forces for the structural surfaces in the bending mode reduce the loading of the side-wall, Fig 16. Payloads P_{zi} are transferred to the lower parts of the ring frames by the floor-plates while the cross-members transfer load to the wall pillars and therefore to the side panels. There is also a portion of the roof panel, width b_w , which is included in the cant rail. Two reaction forces $0.5R_t$ correspond with the

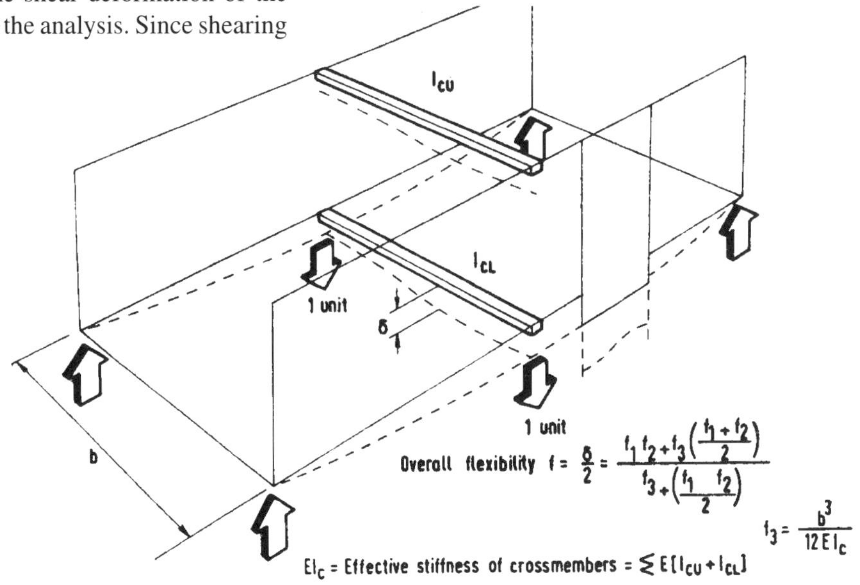

Fig 13: Ring-frame effect

Fig 14: Leyland National structure and bay model

spring hanger beams. Structural surfaces 'O' make no contribution in the bending mode.

Analysis of these forces, Fig 17, in the separate structural surfaces would normally involve a statically indeterminate problem. By treating the sidewall, in a shear panel design, as a beam, sufficient accuracy can be obtained for the purpose of a qualitative estimation. The doorless wall in the top half of the figure is the simpler one and where the pillar F_0-D_0 is loaded with force P_{z0}, this is balanced by panels E_0-D_0-D_1-E_1 and E_0-F_0-F_1-E_1 exerting shear flow:

$$q_0 = P_{z0}/2(h_0 + h_s)$$

Pillar F_1-D_1 is in equilibrium with this shear flow, together with force $P_{z1}/2$, , reaction $R_{pl}/2$ and shear flow:

$$q_1 = [(R_{pl}/2 - P_{z1}/2)/(h_0 + h_s)] - q_0$$

Shear flows $q_2 ... q_6$ are obtained similarly from the equilibrium equations and all cause bending moments in the walls which are weakened by window cut-outs. These reach a magnitude, at the window corners, of:

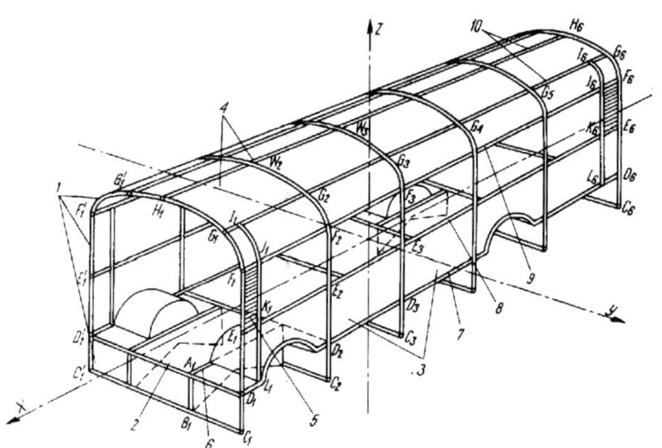

Fig 15: Main elements of monocoque bus

$$M_g = q_i l_i h_0/4$$

In the end-load carrying members, F_0- F_6 and D_0-D_6, direct forces P_n arise from the shear flows in the panels and have a magnitude at any point x in a cross-section of:

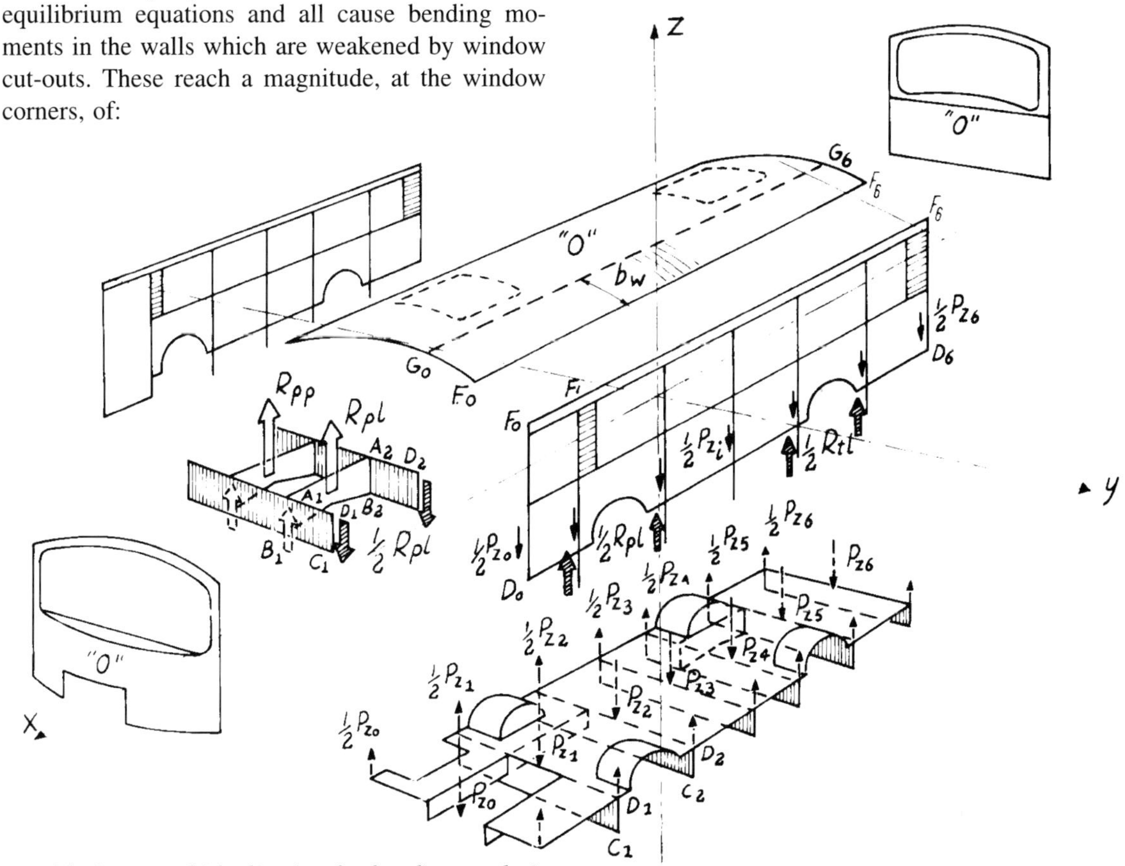

Fig 16: Structural idealisation for bending analysis

$$P_{nx} = \Sigma^{\xi}_{0}\theta_{1}\lambda_{1}$$

In the case of a sidewall with doors, reinforcements are required to make up for the loss in bending rigidity. In the lower half of the figure, the effect of part of the roof acting as cant rail is examined. Force $P_{z0}/2$ loads the pillar G_0-D_0 and is balanced by shear flow q_i in the panel G_0-F_0-F_1-G_1 where:

$$q_0 = 0.5P_{z0}/h_p$$

Since this pillar is heavily loaded by the reaction force it should be given a large cross-section so that even a high shear flow q_1 will not seriously bend the corners. This flow is calculated from the equilibrium equation of the pillar G_1-D_1:

$$q_1 = (R_{pl}/2 - P_{zl}/2 - q_0h_p)/(h_p + h_0 + h_s),$$

further shear flows $q_2 ... q_6$ being similarly obtained from the equilibrium equations of the pillars. Use of part of the roof as a cant rail has also reduced bending at the corners, by the ratio: $-\Delta M_g = h_p/h_0$. It is possible thus to reduce pillar bending moment by some 30 per cent, with appreciable effect on the strength of the body.

Torsional loading of van body

Twisting of the bus body by asymmetric vertical loading of the wheels produces edge forces shown in Fig 18. Moment M_s is transferred from the spring hanger beams A_a-B_1-B_2-A_2 etc. to the bottom of the main ring and hence, as the reactions R_1 and R_2, via side pillars D_1-C_1, D_2-C_2, D_4-C_4 and D_5-C_5, to the side

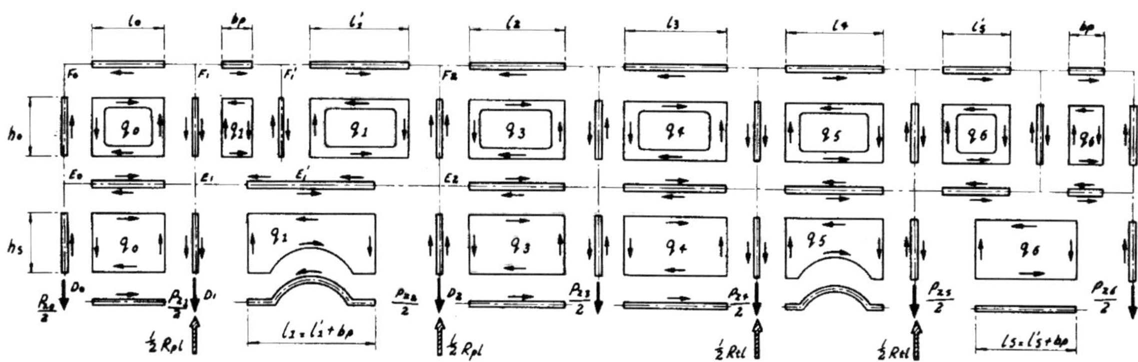

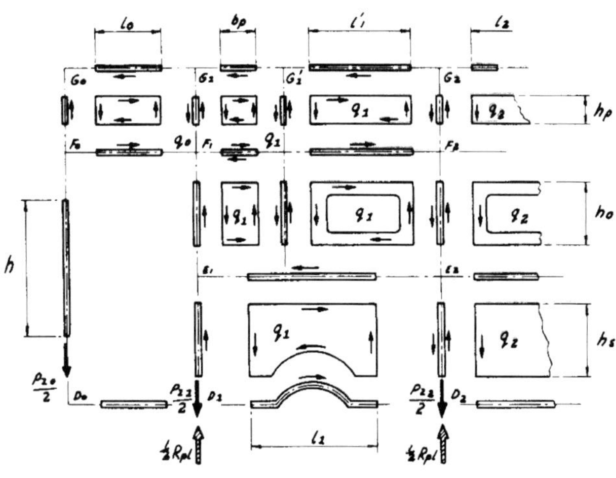

Fig 17: Analysis of forces in structural bending

wall, producing edge forces K_1, K_2 and K_3. Again the magnitude of the edge forces may be determined from a system of equilibrium equations from the separate structural surfaces:

$$K_1L + K_2h + R_1(2l_1 + l_2) - R_2(2l_1 + 2l_2 + 2l_3 + l_4) = 0$$

$$K_1B - K_3H = 0; \; K_2B - K_3L = 0$$

During twisting of the structure the most heavily loaded elements are the window pillars. These are in bending, under force K_2, and deflecting according to their individual rigidity, the force bending the *i*th column being:

$$P_i = K_2 I_i / \Sigma_1^n I_i$$

where I_i is the pillar second moment of area and K_2 the edge force. Moment at the base of the column is:

$$M_{gi} = P_i Z[_{ho/2}^{ho}$$

This formula underlines the importance of the anti-lozenge areas in a side-wall under torsion; typically two anti-lozenge panels are worth considerably more than 5-6 normally dimensioned pillars. The moment is also dependent upon the kind of fixing to the roof. If there is little bending resistance, Fig 19, the moment-arm of force P_i is the full height h_0 of the opening. If on the other hand the cant rail can transfer the fixing moment to the same extent as the panel below (c), then the moment arm length is reduced by half. Corner pillars are particularly heavily loaded since force in two planes, caused by K_2 and K_3, occurs in them; this is dangerous when the tendency is to reduce cross-section for the sake of greater visibility. Total bending moment for the pillar F_0-D_0 is:

$$M_{gn} = K_2(I_0/\Sigma_1^n I_i)(h_0/2) + K_3h_0/4$$

When the direct load on the pillar due to the force P_z, and the weakening due to the lack of panel in the door opening takes effect, large stresses are involved, given by:

$$\sigma = (MK_2/W_y) + (MK_3/W_x) + (P_{zo}/F)$$

where MK_2 is the moment caused by the force P_0 as part of the edge load, MK_3 the moment caused by the force P, P_{zo} the vertical force in the column, W_y and W_x the stiffnesses with respect to y and x axes and F the cross section area of the column. The corresponding load diagrams for bending and torsion for a low-floor integral coach are seen as Figs 20/21. Here the payload and tare load F_{zi}, together with the reactions of the suspension R_f and R_r are contained in the area bounded by crossmembers, sidewalls, longitudinals and front/rear end frames. The effects of step loadings in both modes are shown inset.

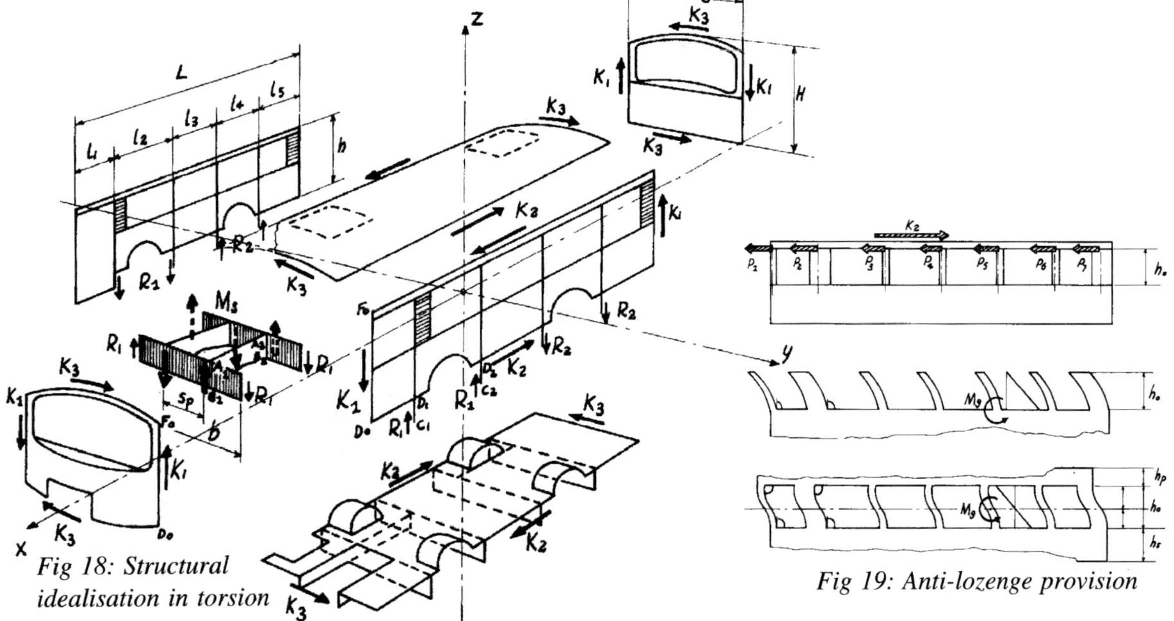

Fig 18: Structural idealisation in torsion

Fig 19: Anti-lozenge provision

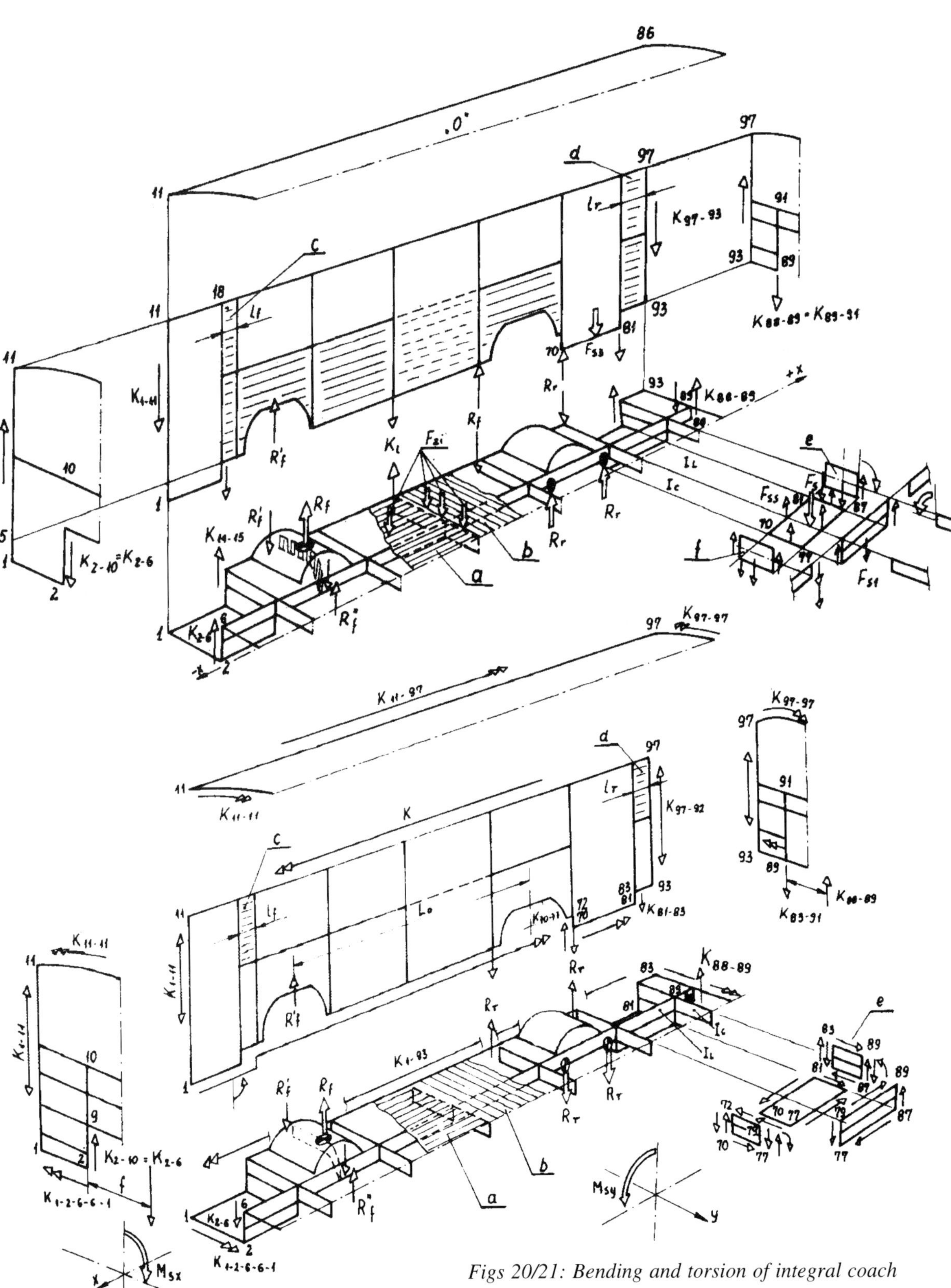

Figs 20/21: Bending and torsion of integral coach

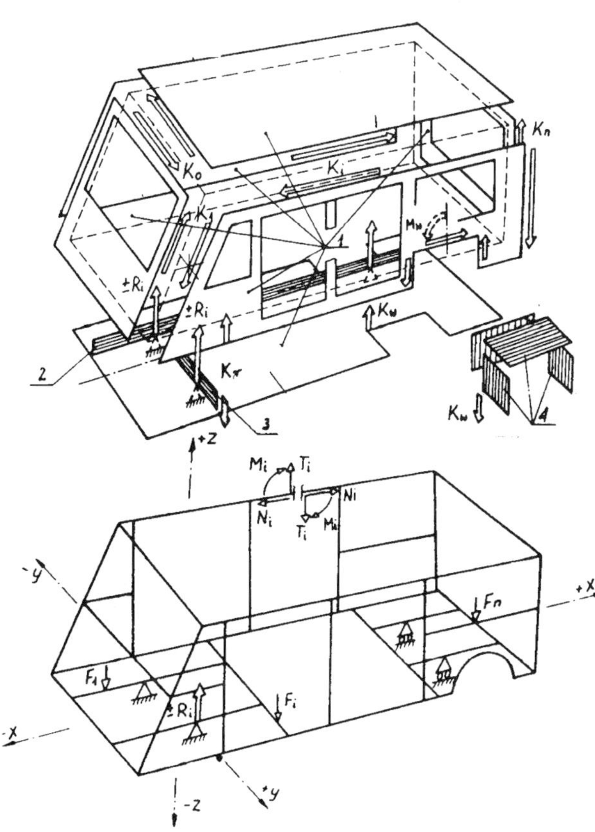

Fig 22: Beam representation for panel van

Idealising van/LCV structures

An earlier section dealt with the idealisation of regular car and CV structures by the Pawlowski Simple Structural Surface method. It introduced the former aeronautical structural technique of representing a real panel by an idealised one comprising a flat rectangular panel taking only shear load bounded by uniform section end-load carrying bars. Here more irregular shaped structures are considered typifying a van and LCV.

The vehicle structural analysis pioneer Janusz Pawlowski considered that the finite element techniques now widely used in structural design were less suitable at the initial concept stage because they depended on fairly precise modelling of the shapes and related mass centres which were unlikely to be fixed at this stage. By using the SSS method, the designer could progress from a relatively simple beam model, step by step, into one which represented the final structure more closely — and in so doing arrive at a more efficient structural solution. By contrast, he considered, the FEM modelling route was 'passive' rather than interactive.

The SSS approach was one in which local stiffnesses could be arranged to meet the requirements of external load inputs; the general elasticity depends on the characteristics of the whole body under the action of service loads and partial plasticity

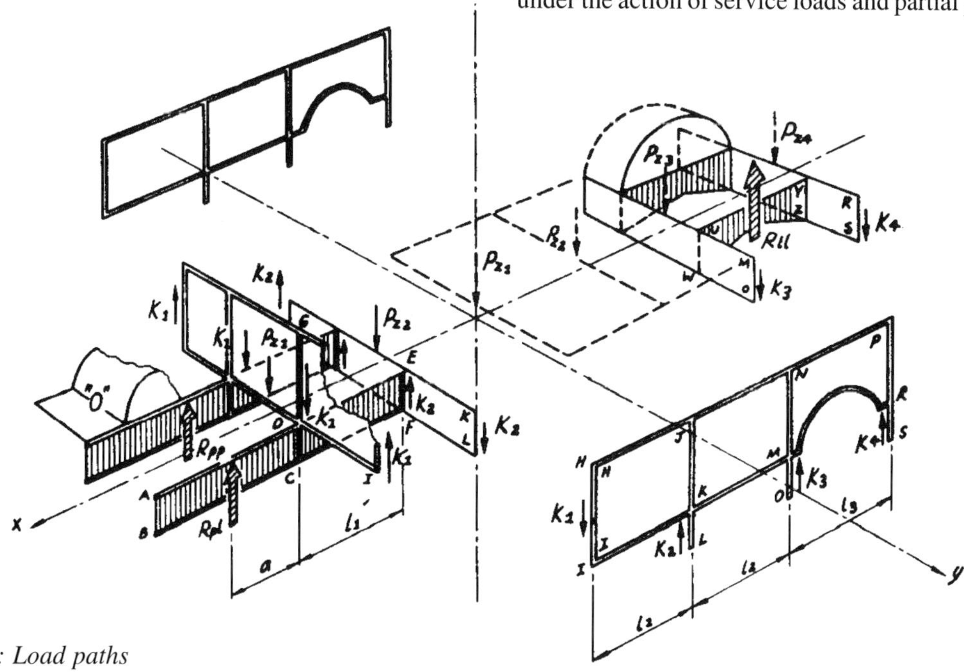

Fig 24: Load paths

of the body meets the energy absorption requirements of crash impact. This leads the designer, at an early stage, to direct loads into the thin wall structure in the planes of the thin walled elements. The preliminary structural model in Fig 22 starts as a beam representation with 15 levels of redundancy and then progresses to an SSS model comprising six main structural surfaces plus an adequate number in the underbody group depending on the support conditions. Edge loads in each group of SSSs are obtained from the equilibrium conditions applying to each one. Interaction between SSSs occurs as edge loads such as *K1*. On top of these is superimposed loading cases due to bending, tension and planar shear for the whole structure to obtain a set of linear equations which can easily be solved.

SSSs occurring in one vehicle body can also be included in a generalised description of others in a family so that the same structural efficiency can be translated to derivative models. Such a vehicle might be a pick-up truck designed for ultimate structural efficiency with an integral load-bearing cab. In the case where the cab does not participate in reacting structural loads, the structure would be reduced to that shown in Fig 23; this is reduced to front longitudinal members, wheel and cover plates, bulkhead, rear crossmembers, sidewalls and floor plate as the main load bearing elements.

Such a structure is sufficiently rigid for vertical symmetrical loads P_z. The front reactions R_{pp} and R_{pl} are introduced into the corner post H-L by the first crossmember E-F-K-L, Fig 24, and the bulkhead is in equilibrium under the edge force *K2* of the sidewall. Forces P_{z1} and P_{z2} are also introduced into these elements and from equilibrium:

$$K_2 = (R_{pl} a/l_1) + P_{z2}/2$$

$$K_1 = R_{pl} + K_2 - (P_{z1}/2) - (P_{z2}/2)$$

Note that the large load *K1* on the bulkhead, which is transferred through the corner post H-I to the sidewall; force *K2* is also introduced into the sidewall through the post K-L, which is common for the crossmember and the wall.

In order for the sidewall to be in equilibrium, edge forces *K3* and *K4* must exist in the pillars M-O and R-S:

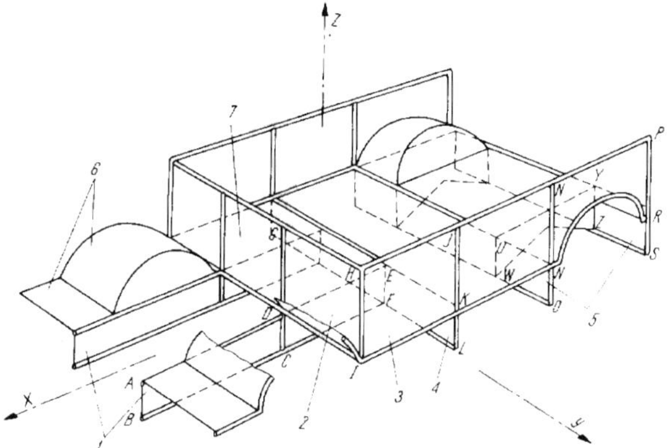

Fig 23: Non-participating cab

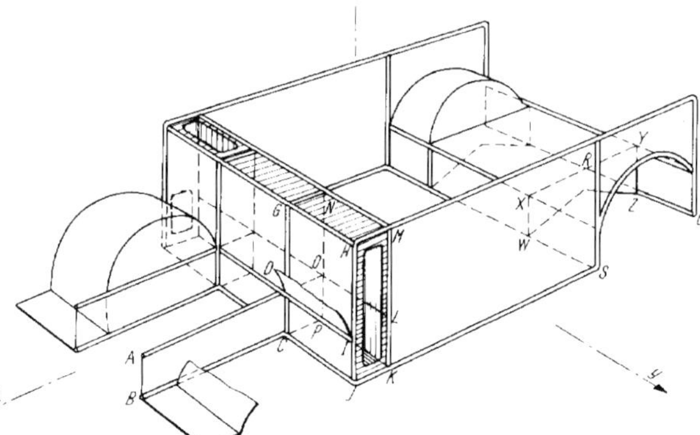

Fig 25: Pick-up with box bulkhead

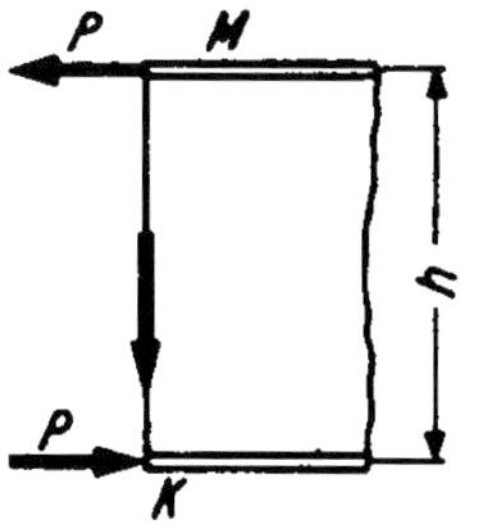

Fig 26: Load paths through bulkhead

$$K_3 = [K_1(l_1 + l_2 + l_3) - K_2(l_2 + l_3)]/l_3$$

$$K_4 = K_1 - K_2 - K_3$$

Force *K3* is usually larger than *K4* and both act on crossmembers M-O-W-U and R-S-Y-Z respectively. From these the forces are transferred to the spring beams U-W-Y-Z and the spring supports.

Normally a pick-up structure is not sufficiently rigid with respect to torsion since the torque M_s can only be absorbed through the lower members in bending — leading to large deformations. This disadvantage, along with the impossibility of lowering the floor level of the load platform, because of the height of the auxiliary longitudinal members, can be overcome by use of a box such as H-M-K-J formed from the bulkhead parts of the sidewalls and a closing wall; it can also be used to house the fuel tank and spare wheel. The analysis of such an arrangement, Figs 25 and 26, is made on the assumption of the front support beam A-B-C-D being a cantilever fixed in the box — transferring the root moment to the sidewall as a torque.

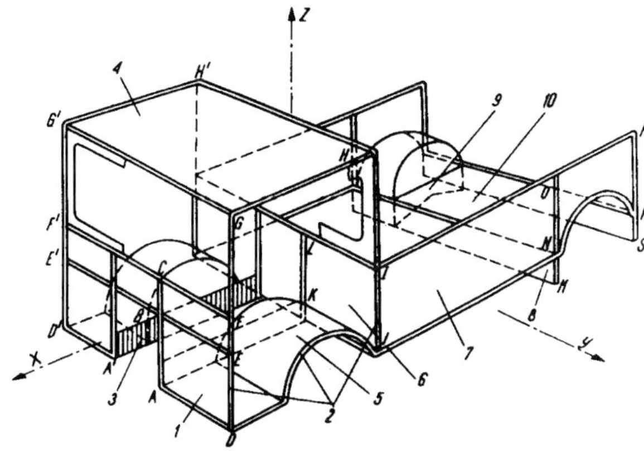

Fig 27: Integral cab structure

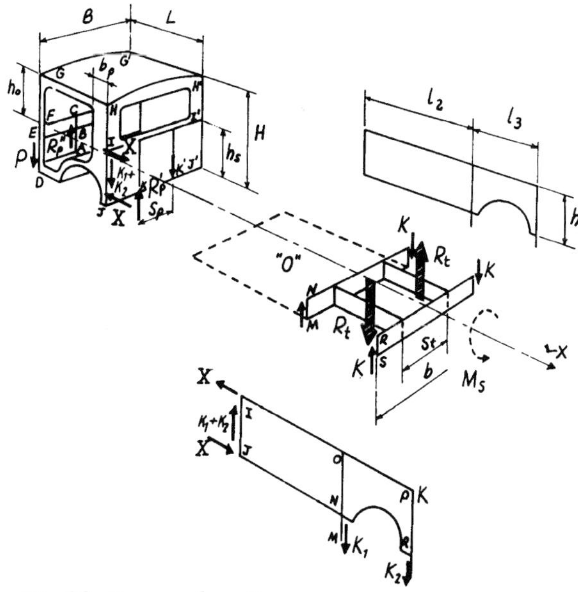

Fig 28: Rear-end section

With addition of the integral forward-control cab, Fig 27, the pick-up becomes a structure with good resistance to bending and torsion, with its rear part fixed in the box formed by the crew cab, Fig 28. The twisting moment M_s, introduced from the wheels into the suspension by the couple $R_t S_t$, is balanced with the edge forces *K1* and *K2*; the latter arise from the action of the sidewalls as a result of their being fixed in the cab with the forces *X* and *(K1+K2)*:

$$K_1 + K_2 = M_s/b$$

$$X = [K_1 l_2 + K_2(l_2 + l_3)]/h_s$$

The cab is supported by the two front spring beams, separated by a width S_p, and joined to the rear wall of the cab at point K and to the front wall at A. The force *(K1 + K2)* is transferred via the rear bulkhead to the points K and K' in the form of the reaction P on the front wall. The latter, which is fixed to the front spring beams at A, acts on the springs with a couple $4R_p$''. Assuming that the lower force *X* acts at the level of the cab mounts, these forces amount to:

$$R'_p = (K_1 + K_2)B/S_p$$

$$P = Xh_s/L$$

$$R_p'' = PB/S_p$$

The forces R'_p and R''_p are transferred from the front spring beam to the spring supports as the couple R_p, which gives a moment M_s on a lever arm S_p.

All six SSSs which form the cab have forces in them and if the sidewall of the cab has the form of a door frame, then the beam H-J should be substantial in order to absorb force *X*.

Structural composites

While racing and sports cars are currently the principal users of structural composites, it is likely that other categories will soon follow as polymer to metal price ratios change. Special techniques of analysis are required to allow for directional properties which follow the reinforcement configurations.

Structural composites can be based on a variety of resin and reinforcement systems, the properties of some of the main ones being given by Phillips[2] in Fig 29. The effectiveness of fibre reinforcement depends on the fibre aspect ratio of length/diameter which should be a minimum of 100. Most commercial fibres are less than 20 microns in diameter but the area of each available for adhesion to the resin is relatively high compared with the compressed bulk of the individual fibres. Resins, in general, are an order of magnitude lower in strength/stiffness than fibre reinforcements.

Fig 30 provides a useful relationship between fibre and resin matrix to show percentage strength against ratio of moduli. Fig 31, however, provides a more detailed examination from which basic calculation formulae can be derived. Between the two extremes of submerged depth shown, for the fibre inside the resin, there is a critical length L_c where fibre failure may be either pull-out or tensile fracture and the failing load is balanced between shear and tension. These can be equated to give: $\tau = \sigma_f r/2l_c$

At the end of the fibre tensile stress is zero and rises to a constant maximum once l_c is exceeded.

In the case of carbon-fibre reinforcement, low molecular weight epoxy resins have become the favoured matrix. Properties of commercial grades are shown in column 3 of the table in Fig 32, the first two

Fibre/material	Specific gravity density (g/cm³)	Tensile strength (MPa)	Tensile strength (× 10³ lbf/in²)	Young's modulus (GPa)	Young's modulus (× 10⁶ lbf/in²)
E-glass	2.54	2410	349	69	10
S-glass	2.49	2620	380	87	12.6
carbon type 2	1.75	2410	349	241	35
aramid (high modulus)	1.44	3450	500	124	18
boron	2.63	2760	400	379	55
alumina	3.30	3000	435	297	43
asbestos (chrysotile)	2.40	1490	216	183	26.5
wood	0.50	69	10	7	0.95
light alloy	2.69	476	69	72	10.50
steel (structural)	7.85	413	60	207	30
titanium alloy	4.52	711	103	117	17

Fig 29: Reinforcement properties compared

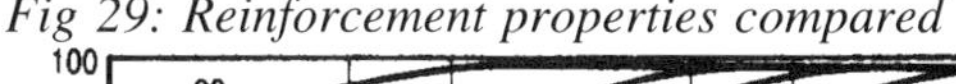

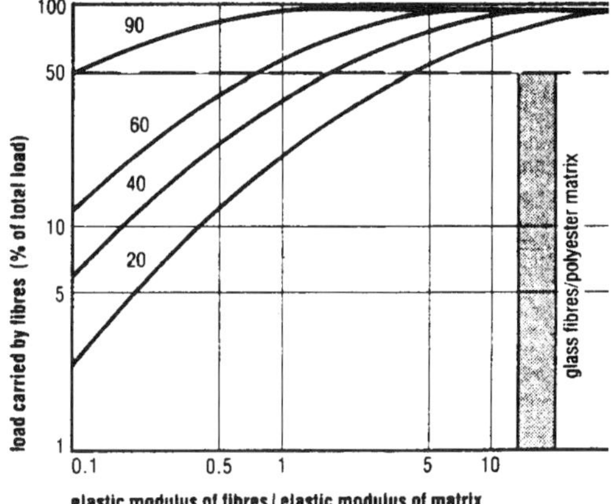

Fig 30: Percentage strength vs modulus ratio

	Tensile strength (MPa)	Tensile strength (× 10³ lbf/in²)	Young's modulus (GPa)	Young's modulus (× 10⁶ lbf/in²)	Specific gravity	Elongation of break (%)
lower-modulus fibre	2,650	384	59	8.6	1.44	4
higher-modulus fibre	2,650	384	127	18.4	1.45	2.4

Fig 33: Aramid fibre properties compared

Fig 31: Embedded fibre analysis

Fibre value \ Fibre type	1		2		3	
Young's modulus, lbf/in² × 10⁶ (GPa)	50	(345)	35	(241)	29	(200)
ultimate tensile strength, lbf/in² (GPa)	290,000	(2.00)	350,000	(2.41)	320,000	(2.21)
specific gravity	1.92	—	1.75	—	1.70	—
Unidirectional composite (V_f = 60%)						
Young's modulus, lbf/in² × 10⁶ (GPa)	26	(179)	17.5	(121)	14.5	(100)
ultimate tensile strength, lbf/in² (GPa)	145,000	(1.00)	200,000	(1.38)	175,000	(1.21)
specific gravity*	1.72	—	1.55	—	1.52	—

* Specific gravity of epoxy resin taken as 1.25

Fig 32: Carbon fibre grades compared

columns showing potential properties available when more exotic types 1 and 2 carbon-fibres are used. The table of Fig 33 shows the properties available with the available forms of aramid fibre. Large panels and shells are now commonly moulded in autoclaves — cylindrical pressure vessels which can generate several atmospheres as well as a means of producing vacuum at one side of an air-tight membrane. Fig 34 shows a typical construction.

Tensile modulus of a composite laminate is:

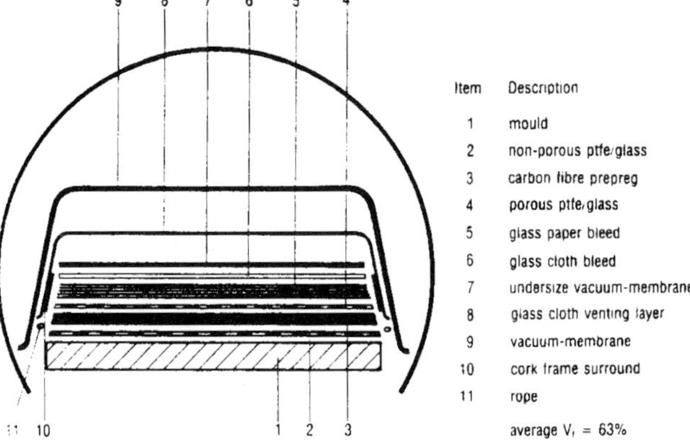

Fig 34: Autoclave construction

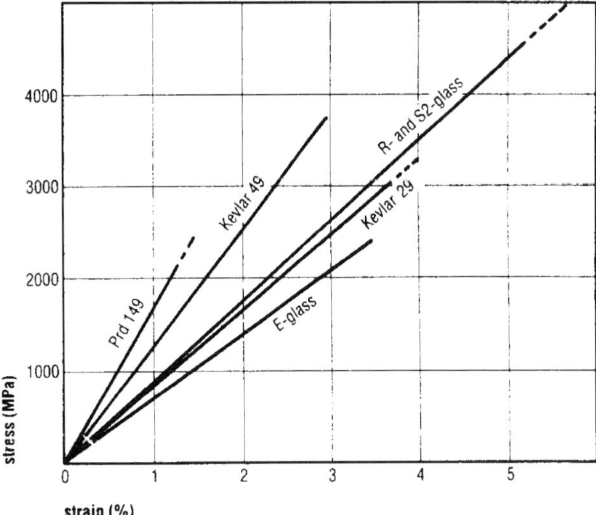

Fig 35: Stress/strain curves for aramids

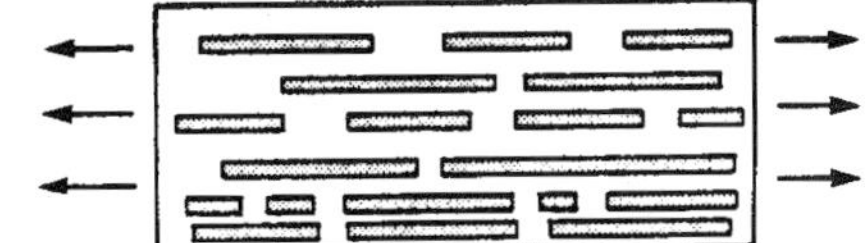

Fig 36: Fibre geometry effect

$$E_k = E_f V_f B + E_m V_m$$

where E is modulus, k composite, f fibre and m the matrix, an expression due to Hollaway[3]. V_m is the volume fraction of the matrix and the lay-up factor B = 1, 1/2, 3/8 for unidirectional, bidirectional or random (in-plane) orientation respectively. Stress/strain curves for glass and aramid fibres are shown in Fig 35, with effect of fibre geometry on stress/strain curves in Fig 36. Jointing configurations are, of course, crucial in design with the critical effect of adherend thickness being shown in Fig 37. Weakest joints are those where failure is limited by the interlaminar failure of the adherend or peel of the adhesive. Next strongest are those in which load is limited by adhesive shear-strength and strongest of all are those which fail outside the joint area at a load equivalent to the strength of the adherend.

The compressive strength and stiffness of laminates is difficult to predict and analysis is usually based on the buckling of slender columns on an elastic foundation. The two idealised buckling modes which have been suggested are given in Fig 38; these are such that adjacent fibres buckle either in or out of phase with each other. The critical compressive stresses are given by:

$$\sigma_c = G_m /(1 - V_f)$$

for shear mode in phase buckling and

$$\sigma_c = 2V_f [E_f E_m V_f /3(1 - V_f]^{1/2}$$

for extension-mode out-of-phase buckling.

Stress/strain relationships for uniaxially aligned discontinuous fibres stressed in parallel to the fibre direction can also be estimated by calculation, Fig 39. There is a portion at the end of each finite length which is stressed to less than the maximum stress of a continuous fibre. This is known as the transfer length and depends on the maximum fibre stress and the effective shear strength of the fibre-matrix interface σ_{xy} and is given by:

$$l_l = \sigma_f d_f /4 \sigma_{xy}$$

where d_f is fibre diameter and σ_f the maximum stress in fibre length l_l. Critical transfer length in which fibre stress drops from σ_{fmax} to zero is l_c. A quantity α

is defined as the area under the stress distribution curve over the length $l_c/2$; then the fibre end of the length l_c may be considered as supporting a reduced average stress $\alpha\sigma_{fmax}$ or it may be reduced to length $\alpha\, l_c/2$ and subjected to σ_{fmax}. Referring to the figure, average axial stress $\sigma_{f.av}$ for fibre lengths not less than l_c can be deduced from:

$$\sigma_{fmax}\,(1 - l_c/2l)$$

and for those less than the critical length as

$$\sigma_{fmax}\; l/2l_c$$

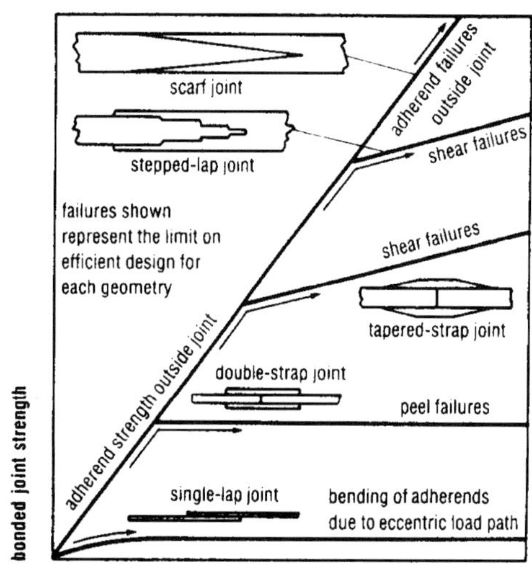

Fig 37: Joint configurations

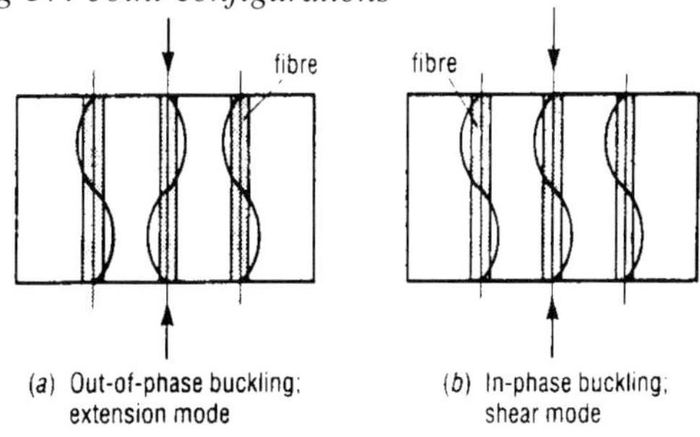

Fig 38: Idealised buckling modes

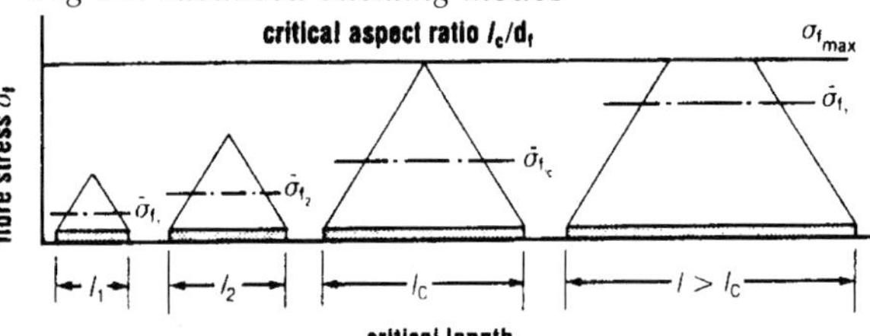

Fig 39: Fibre transfer length

Composite construction techniques

The structural evaluation of the Panther Solo car structure has been described by Parry-Williams[4]. Fig 40 shows the body shell components, notable for the two-piece roof structure. Major loadings on the shell are seen in Fig 41. Uni-directional fibres are used in strip form around all the apertures in the primary structure with aluminium inserts positioned where threaded anchorage holes are needed. A torsion test was carried out between front and rear suspension anchorages. Deflections were measured at ten places, at stations a, b, c, d and e, and torque applied at the three levels shown by the table in Fig 42. Dial gauge readings, in mm, taken at 750 mm from the vehicle centreline are shown.

Application of advanced composite materials in racing cars has been described by Dr HA Potts[5] of Advanced Composites Group Ltd. At ACT two in-house tooling systems are used for mould making — gel-faced HT5C Precom or LTM Prepreg form of the same material. The former is a room-temperature

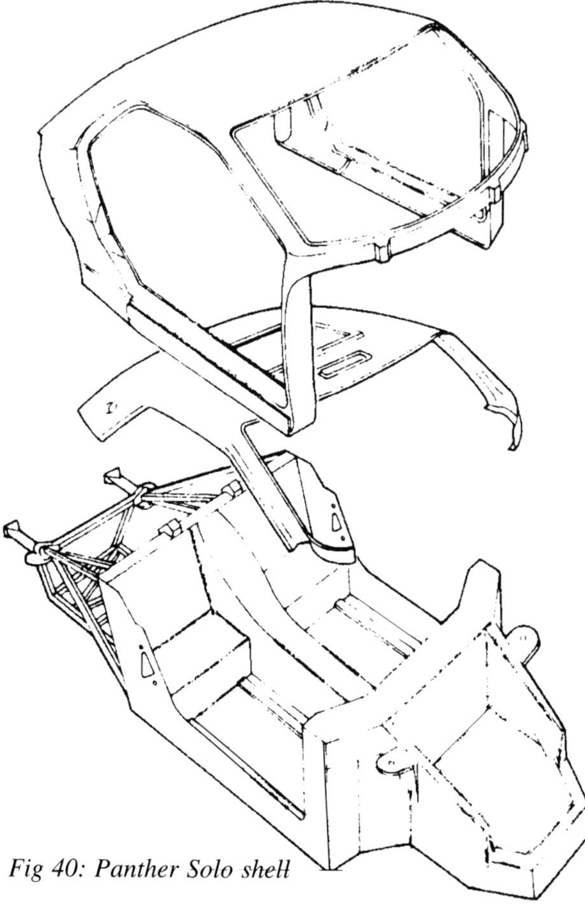
Fig 40: Panther Solo shell

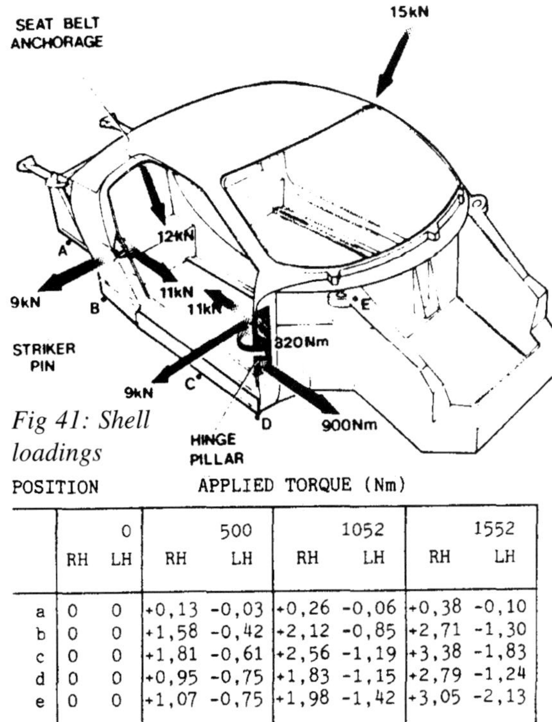

Fig 41: Shell loadings

POSITION	APPLIED TORQUE (Nm)							
	0		500		1052		1552	
	RH	LH	RH	LH	RH	LH	RH	LH
a	0	0	+0,13	-0,03	+0,26	-0,06	+0,38	-0,10
b	0	0	+1,58	-0,42	+2,12	-0,85	+2,71	-1,30
c	0	0	+1,81	-0,61	+2,56	-1,19	+3,38	-1,83
d	0	0	+0,95	-0,75	+1,83	-1,15	+2,79	-1,24
e	0	0	+1,07	-0,75	+1,98	-1,42	+3,05	-2,13

Fig 42: Deflections measured during tests

cure system which is subsequently post-cured to around 150 C under a vacuum bag at 0.5 bar while the latter requires full autoclave pressure at 20-50 C. Near zero shrinkage carbon fibre mould tools are utilised to ensure absolute accuracy of anchorage dimensions. Warping and twist of the moulds is prevented by a rigid backing structure also in carbon-fibre laminate.

Following polishing of the moulds, and the application of release agents, the first skin is laminated in stages, with intermediate debulks using either vacuum or autoclave pressure. On completion, the lay-up is cured under full autoclave pressure of 90-100 psi. When cured, the bag, breather and peel ply are removed and the laminate surface lightly abraded and degreased. A resin film ply is then fitted to the skin and preshaped honeycomb pieces added. Next the many anchorage inserts are fitted and the second stage of the lamination process takes place, usually at full vacuum in an oven. After this cure cycle, core and insert detail is carefully checked for any movement prior to the third stage in the lamination, the inner skin; here less than full autoclave pressure is used in order to avoid crushing of the core splices.

		Abbreviation	Cost £tonne^{-1}	Density kgm^{-3}	UTS* MNm^{-2}	Modulus GNm^{-2}	Impact Strength Jm^{-1}	Fatigue Strength MNm^{-2}	Friction Coefficient	Maximum Service Temperature °C	Expansion Coefficient 10^{-6} C^{-1}	Thermal Conductivity Wm^{-1} °C^{-1}	Specific Heat	Volume Resistivity ohm cm	Dielectric Strength kVmm^{-1}	Dielectric Constant	Power Factor	Refractive Index	
Alkyd	– Asbestos Filled		1500	1600	45	16	15			230	30	0.6		10^{14}	14	4	0.02		A3
Epoxy	– Cast		2500	1200	60	3.5	20			150	50	0.2	0.4	10^{14}	16	4	0.02	1.57	A7
	– Glass Filled		3000	1800	100	25	600			200	20	0.3		10^{14}	14	4	0.01		A3 A1
	– Glass Laminate			1820	270	19	370			200									A5
	– Glass Wound(60% unidirectional)			2000	1500	40		300		260	6	0.3	0.24		16	5	.025		A6
	– Carbon Fibre(")			1600	1400	120		840		70									
	– Aluminium Whisker (14% vol)				750	40													
Ester	– Cast	UPR	1000	1300	60	3	15		0.35	120	70	0.2	0.5	10^{15}	18	3.5	0.02		A7
	– Dough Moulding Compound	DMC	1000	2000	50	7-14	600			170	30	0.25		10^{15}	15	5	0.01		A3
	– Sheet " "	SMC	1000	2000	65	7-16	750			170	25	0.25	0.22	10^{15}	15	5	0.02		A3
	– Glass Matt Laminate	GRP		1600	150	12	800	100		175	12	0.25		10^{14}	24	5	.015		A5
	– Glass Wound(60% unidirectional)			1800	1000	30		280											A6
Imides	– Cast			1400	120	3	35			260	26	0.32	0.3	10^{17}	22			1.70	A3 A5
	– Glass Filled			1400	190	22	800			260	15	0.48	0.27	10^{16}	20	4.7	.005		A3 A5
Melamine	– Cellulose Filled	MF		1500	50	9	15		0.2	100	40	0.6	0.4	10^{12}	15	7.5	0.03		A1 A3
	– Glass Laminate			1900	300	18	500			140	100	0.5		5×10^{11}	8	14	0.1		A5
Methacrylate	– Glass Matt Laminate				150	12				200									A5
Phenolic	– Asbestos Filled	PF		1800	40	11	15		0.3	250	20	0.6	0.3	10^{10}	8	10	0.1		A1 A3
	– Glass Filled		200	1800	40	18	500			250	12	0.4	0.3	10^{13}	10	5	0.02		A1 A3
	– Asbestos Laminate			1720	80	16	190			250					4	15	0.3		A5
	– Glass Laminate			1650	160	10	340			270	180				24	5.5	0.03		A5
Urea	– Asbestos Filled	UF		1500	60	10	15		0.2	75	30	0.3	0.4	10^{12}	15	8	0.03		A1 A3
	– Cellulose Filled			1520	65		13			75				10^{12}	9	0.3	0.18		
Urethanes	– Reaction Moulded	RIM		1100	22	0.3				120	130								A1 A3
	– " " (reinforced)	RRIM			25	1					45								A1 A3
	– " " (foam)			500	15					90	72	0.06		3×10^{14}	10	2.4	0.007		A1 A3
Amides	– " "	RIM		1130	40	1.1	50			180									A1 A3

Fig 43: Basic mechanical properties

Analysis of composite structures

Design calculation techniques for composite structures have generally been based on pseudo-elastic methods. These have increased in value as users build up life-time confidence in engineering light-weight polymer composites, the advanced end of which is, of course, the racing car structure.

West[6] has provided basic data for thermosets in the table shown as Fig 43. He maintains that pseudo-elastic design is still the most commonly used method, in which classic elasticity theory is applied to polymer composites but that time-dependent material elastic properties are used in design formulae. The procedure is thus to specify service conditions such as life-duration and operating temperature, and select the worst combination in design, then substitute properties appropriate to the conditions.

Fig 44 shows the example of a ribbed plate subject to a uniform pressure *p* of 5 kPa, required to sustain a minimum life of 10^5 hours. Sheet thickness *t* is 2 in (50.8 mm) and 0.32 in (8 mm) width ribs are employed. Here, grid spacing is required for two materials, unreinforced PVC and polyester dough moulding compound reinforced with chopped glass. The unsupported area *a x a* is considered as a plate with built-in edges for which maximum stress is given by $0.308\ a^2\ p/t^2$. This gives values of a as 100 for UPVC and 200 for DMC.

A thin-walled rectangular box-beam made from long-fibre reinforced polymers, and having diaphragm stabilising webs along its length, would be designed against the criterion of buckling collapse. In the example shown, it is required to determine the maximum centre-point loading for a 2-cell beam, Fig 45, having uni-directional fibres on the top and bottom flanges with randomly-oriented fibre for webs and diaphragms at 50% by volume and a polyester type resin.

Composite elastic modulus in the fibre direction is :

$$E_{C1} = E_R V_R + E(1-V_R) \text{ and } E_{C2}=E_R E/[EV_R(1-V_R)]$$

for the transverse directon. Corresponding Poissons ratios are:

$$n_{C1}= n_R V_R + n(1-V_R) \text{ and } n_{C2}= n_{C1}E_{C2}/E_{C1}$$

Composite shear modulus in the fibre direction is:

$$G_{C1}=G_R G/[GV_R + G_R(1-V_R)]$$

where G_R and G are for fibre and matrix respectively and

$$G = E/[2(1 + n)]$$

Using values from Fig 1, results are seen in the table of Fig 4 and critical buckling stress is:

$$s_B = (4p^2/b^2 t)[(D_L D_T)^{1/2} + H)$$

where D_L, D_T and H are the section second moments of area corrected for Poissons ratio effect.

Du Pont have applied pseudo-elastic design methods for such structures as rectangular section seat backs, Fig 46 using the expression for breadth:

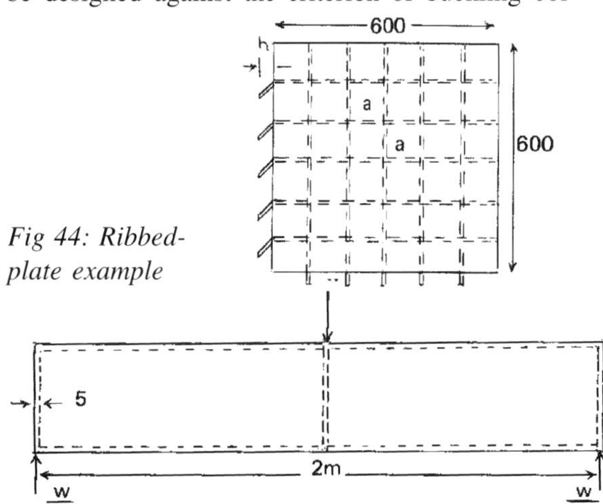

Fig 44: Ribbed-plate example

Fig 45: Two-cell box beam example

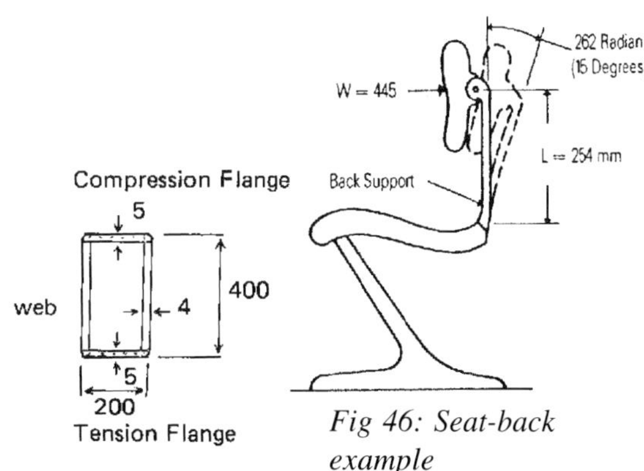

Fig 46: Seat-back example

Property	Compression flange	Tension flange	Web and diaphragm
Modulus			
– Longitudinal E_{C1} GNm^{-2}	36.5	36.6	19.75
– Transverse E_{C2} "	5.75	5.75	19.75
– Shear G "	2.14		7.53
Poissons ratio			
– Longitudinal ν_{C1}	0.275	0.275	0.312
– Transverse ν_{C2}	0.043	0.043	0.312
Strength			
– Longitudinal GNm^{-2}	$0.42(X_{C1})$	$1.26\ (X_{T1})$	0.395
– Transverse "	$0.060(X_{C2})$	$0.025(X_{T2})$	0.395

Fig 47: Derived "elastic" properties

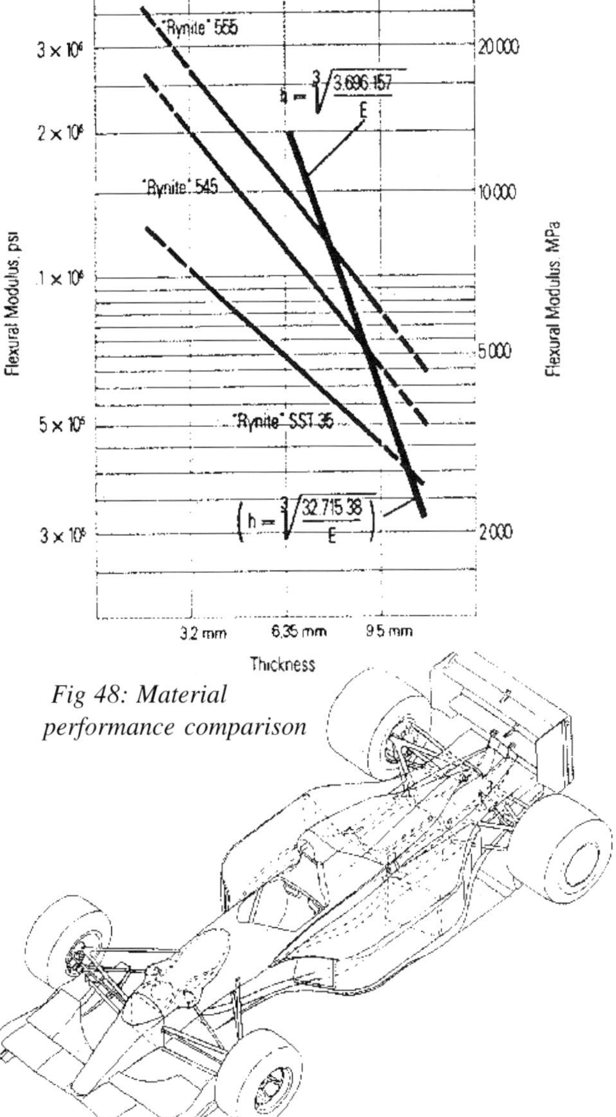

Fig 48: Material performance comparison

Fig 49: Williams F1 car

$$b = 6WL^2/Eqh^3$$

Putting various thicknesses and the relevant moduli from Fig 47 for different materials into the equation gives a series of widths; from these the amounts of material used and an idea of relative part costs is obtained. Thinner sections will have shorter moulding cycles, important in determining part cost.

For the stiffest resin, Rynite 555, the flexural modulus at a thickness of 3.2 mm is = 17.9 x 10^3 MPa, giving breadth of:

$$\frac{[16 \times 445 \times (254)^2]}{[17.9 \times 0.262 \times (3.2)^3]}$$

= 1121 mm and the c/s area — proportional to the mass of material used — is 3.2 x 1121 = 3587 mm^2.

This width is obviously too great, so Du Pont repeat the exercise with some thicker sections/smaller widths which look more acceptable: for h = 4.8 mm, E= 13.8 x 10^3 MPa then b = 431 and c/s area is 2069 mm^2, and so on for h = 6.4, 8.0, 9.5 mm, and by extrapolating the experimental data slightly, for h = 11.1 mm. All the results are :

h mm	E MPa x10^3	b mm	area mm^2
3.2	1 7.9	1121	3587
4.8	1 3.8	431	2068
64	10.3	244	1559
8.0	8.1	159	1269
9.5	59	130	1235
11 1	4.5	107	1186

To find the thickness to satisfy the 15 degree (0,262 radian) deflection criterion, using the same equation, but this time solving for h, to determine the two unknowns h and E we insert the known values for load, deflection, length and width then plot this on Fig 48, — for example, by using two values of E and the corresponding calculated values of h (E = 13800 MPa, h = 6.4 mm; E = 4140 MPa, H = 9,7 mm) and connecting the points by a straight line. This intersects the modulus/thickness line for Rynite 555 to give the solution for h as 7.8 mm.

The lines for Rynite 545 and SST 35 intersect the equation plotted at greater thicknesses, and since these resins can take more impact than Rynite 555, one has the choice of playing safe — at the cost of more resin in the back support. But the stress in the thicker section would be lower, and this could be especially important. Keeping stresses low in the back support will help to extend fatigue life, reduce creep, and ensure that the part can safely take overload several times the design load.

Racing car structures

Advanced composite materials are of course widely used in race-car structures such as the Williams FW14B seen in Fig 49. O'Rourke[6] explains that FISA load cases apply to specific portions of the structure. The structure also has to sustain inertia loads of 4.5*g* laterally and 4*g* under braking as well as 10*g* instantaneous bump load. Downforces may be as much as 5kN and a combined load envelope would be as in Fig 50. The main 'chassis' comprises five principal components, Fig 51, the outer shell being reduced to two, Fig 52.

The structural composite materials used at Willliams comprise a range of carbon and aramid fibre-reinforced epoxy resin prepregs, aluminium and Nomex honeycomb core materials, and epoxy adhesives and fillers in film and paste forms. The majority of components are made from a 120 C-curing, modified epoxy prepreg but where mechanical property retention at temperature is important a 175 C-curing resin system is employed. Carbon fibre is used in both woven and uni-directional tape forms, intermediate modulus type carbon fibre being chosen where components are of a fully structural nature and high strength type for the remainder of applications. Ablative materials are incorporated around the exhaust outlet where temperatures of 500 C are typical.

Prepreg lay-up is accomplished by hand and plies are laminated to the orientations specified in an appropriately configured mould; a vacuum bag is applied and curing takes place in an autoclave to the recommended cycle. Total factory floor space given over to composites activities amounts to 21500 ft^2 (2000 m^2) whilst curing facilities include 3 autoclaves, the largest of which is 2.5 metres in diameter and 5 in length. Structural performance is such that the levels of chassis torsional stiffness that are now attained are in excess of 250% of those measured on the last generation of metal structures, for no increase in the mass of the component.

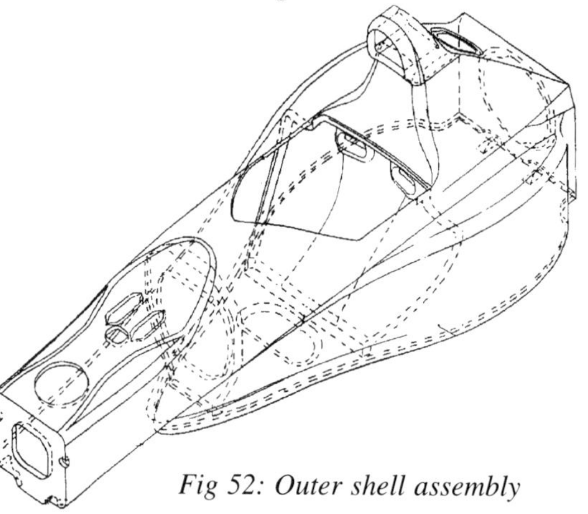

Fig 52: Outer shell assembly

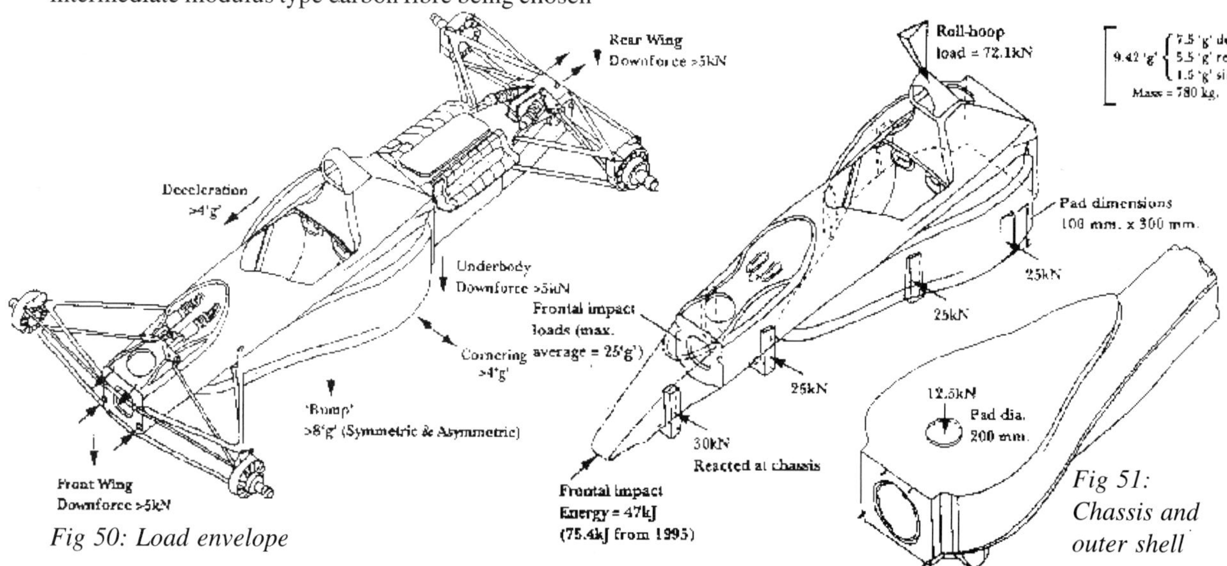

Fig 50: Load envelope

Fig 51: Chassis and outer shell

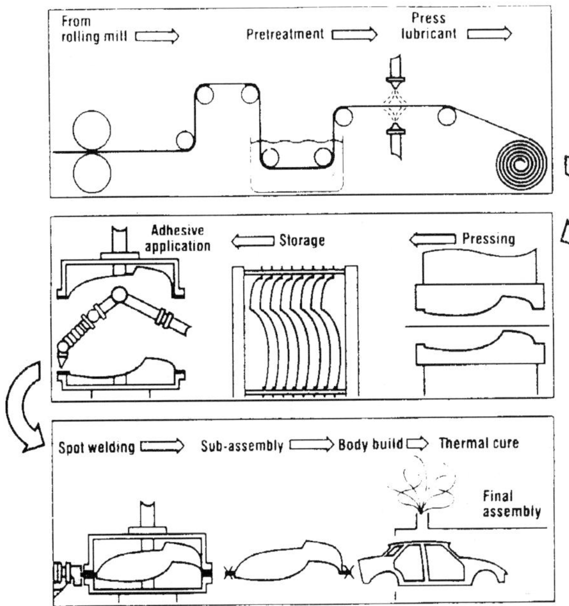

Fig 53: Production process for aluminium car bodies

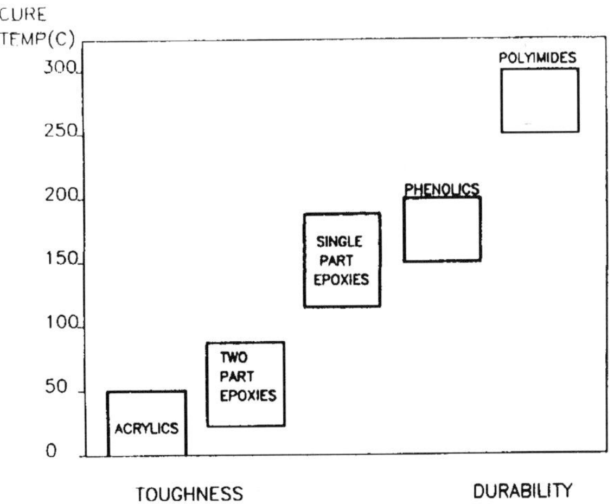

Fig 54: Ranking for structural adhesives

Joint Type	Static Strength (N/mm)	Fatigue Strength** (N/mm)
Bonded Aluminium (25% Fillet), 2mm 5754-O	68.4	24.4
Spot-Welded Steel (1.16mm CR2) 25mm pitch	28.95 (723.8)*	4.6 (115.0)*
Spot-Welded Aluminium, 2mm 5754-O 25mm pitch	30.4 (760.0)*	2.0 (50.0)*

Fig 55: Static and fatigue strengths compared

Bonded aluminium structures

Recent work at Alcan[7] has shown how adhesives could be the key to producing aluminium sheet monocoque body structures in modest volumes. The paper also refers to earlier work, reported here, in which a formal procedure is proposed for ensuring the integrity of joints.

Fig 53 shows the production process used to produce six Metro cars over a decade ago. Earlier work on Alcan's Aluminium Vehicle Technology (AVT) project had determined the necessary sheet pre-treatment for making the surface compatible with adhesives and more amenable to resistance spot welding. This was important for the weld-bonding process which the company believed to be the best way to fabricate body structures.

A suitable drawing lubricant had also to be developed which allowed panels to be pressed and adhesive applied subsequently without the need to clean down the panels after pressing. Specialised automotive sheet materials were also developed, these being typically formed in AA5754 AlMg alloy, a medium strength material offering good formability and corrosion resistance. For surface skins the AA6111 material is preferred, an AlMgSi alloy which combines good corrosion resistance with high strength after paint bake.

Fig 54 shows commercially available structural adhesives ranked according to cure temperature, an important consideration being that the adhesive must not cure during the assembly stages. Single part epoxies become the main contenders on this basis; also their peel strengths are high above ambient temperatures. A modified form of epoxy is needed, however, to provide an adequate impact strength throughout the range of vehicle operating temperatures, even at the critical sub-zero level. An advanced formulation developed with Ciba-Geigy solved this and the problem of high-temperature creep performance so as to eliminate joint failure at the hot spots on the vehicle.

For structural optimization on monocoque vehicles, Alcan developed a method for joint design of weld-bonded structures which allowed for geometric variations, manufacturing variability and complex joint loading. This involves finite-element modelling with bond line generation software involving joint-line elements, modelled joint stiffness and joint design envelopes, for the 50% of the car's structural joints in which adhesives formed a crucial element. First car to benefit from this treatment was the Jaguar XJ220.

McGregor *et al*[8] had earlier described the joint-design approach in detail, in which the table of Fig 55 compares static and fatigue strength of various aluminium joining techniques which demonstrated the superiority of the bonded joint. Fig 56 gave the static strength of various bonded joints with the special epoxy adhesive, a trend which is similar for spot-welded joints. In Fig 57, the aluminium yield line is the load per unit width for a given gauge at which the whole of the cross-section of the specimen is at yield ($90N/mm^2$). For the spot-welded joint it is seen that the aluminium adherend has not fully yielded before the spot-weld fails.

For bonded joints, however, Fig 58 shows that single lap joints fail at a load much greater than that required to yield the aluminium. Comparing the figures also shows there is much greater influence of the gauge on the strength of the spot-welded joint than on the bonded joint. The table of Fig 59 gives the influence of bondline (adhesive layer) thickness on the performance of T-peel joints. The most critical parameter has been found to be the fillet size, Fig 60. Its influence is shown in Fig 61 where fillet size is increased from 20 to 100% to give a joint strength increase of four times. By examining the stress distribution at the bondline the reason for this becomes clear, Fig 62. Most of the load transfer is seen to take place in the fillet, with almost no load in the flange of the specimen.

The manufacturing process must therefore be designed around this important criterion. The three stages in the Alcan procedure for joint design are:

Bondline Thickness (mm)	Failure Load/Width (N/mm)
0.22	141.1
0.30	140.0
0.95	145.0

2mm AA5754-O Aluminium, Fillet Size: 75%

Fig 59: Joint failure loads

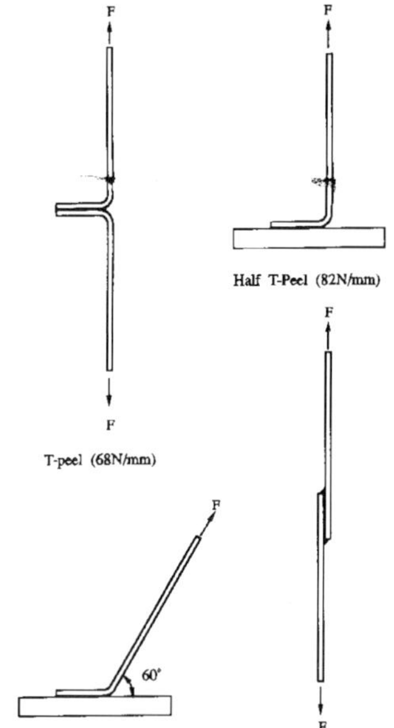

Fig 56: Different joint configurations

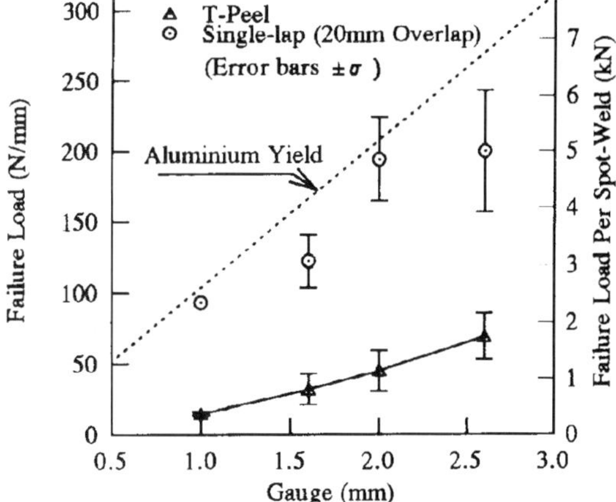

Fig 57: Effect of gauge on strength of spot-weld joints

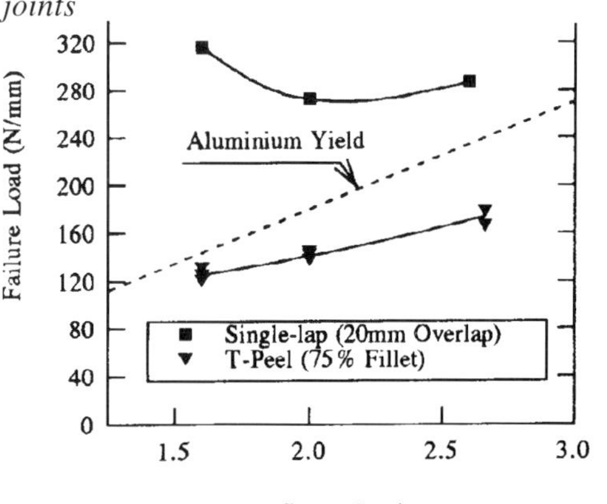

Fig 58: Gauge effect on bonded joint strength

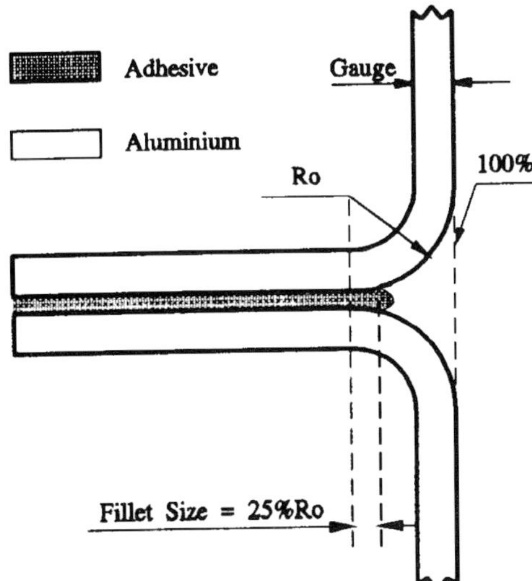

Fig 60: Defining the joint fillet size

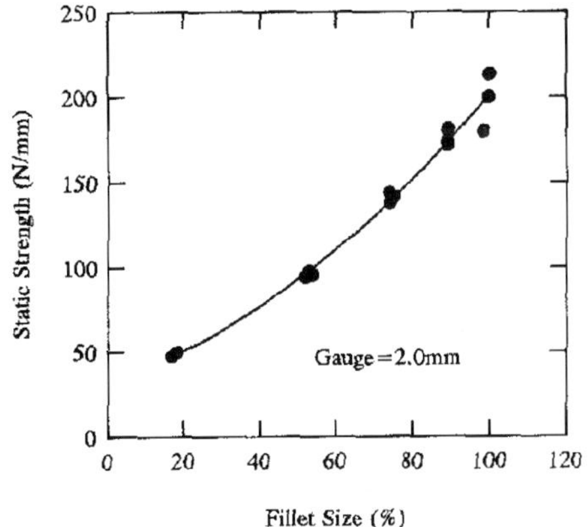

Fig 61: Effect of fillet size on static strength

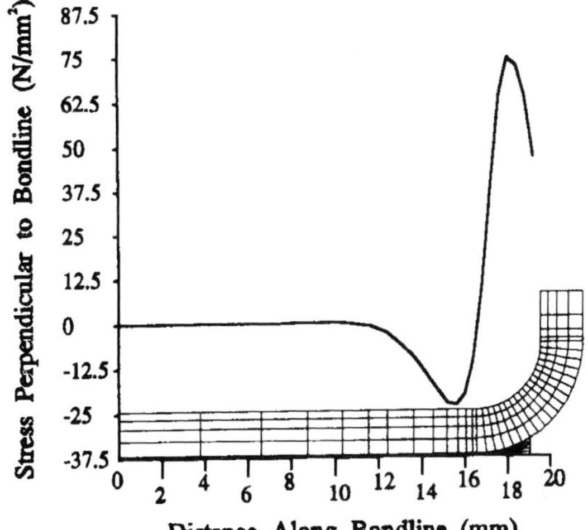

Fig 62: Stress distribution in the bondline

* modelling of the joint-line areas within the full-scale FE structural model of the vehicle
* extraction of the important loads acting on the joint from this model
* comparison of the loads on the joints with the joint failure envelopes, which provide crite ria based on detailed modelling and tests for each configuration.

By representing the joints within the structure in a simplified way, complexity of the overall model is reduced. Such a model normally involves shell elements representing its thin-walled parts and the joint design approach has a row of special shell elements along the joint lines known as Joint Line Elements. These have the special feature that all local co-ordinate systems are oriented so that the y-axis always points along the joint line, the element tension force per unit length in the y-direction *Ty* representing joint-line tension. The *Ty* value is an element output value is one which can be obtained from most shell elements. These joint line elements are also arranged to contain modified physical and geometrical properties, representational of real joints, such as tension and bending stiffnesses. Software has been developed to produce these automatically.

The Joint Line Generator accepts as input the FE mesh data including nodes and elements of a structure then creates new elements at the boundaries. These are oriented in relation to the joint line for post-processing of jointline forces. It can incorporate several of the most common joint modelling methods

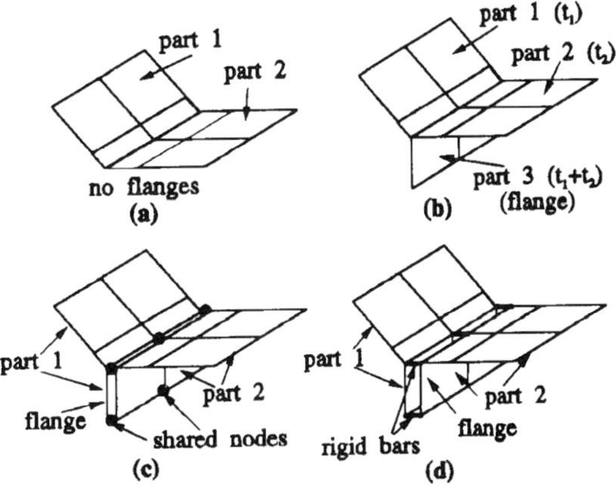

Fig 63: Joint modelling methods

including, Fig 63, no flanges, flanges modelled as one set of elements with combined thicknesses and those represented by two sets of elements but sharing the same set of nodes, also flanges modelled as two sets of elements and connected by rigid bars. Fig 64 shows an example of the use of the Generator in a four part structure.

Correct prediction of tension force and bending moment distribution depends on stiffness modelling. The table in Fig 65 compares predicted stiffnesses of weld-bonded and spot-welded T-peel joints, under tension and bending, with aluminium sheet stiffness, based on detailed FEA. This shows the bending of the joint to be much closer to the sheet performance than the tension.

In determining the critical loads on joints, FE modelling has come up with the stress distribution shown in Fig 66 based on the analysis of a 3-D model of a half-peel joint using 20-node solid elements. The figure shows the tension force produces the largest stresses in the joint with stress concentrated in the adhesive fillet while shear loading produces a concentration further into the main bondline. A critical

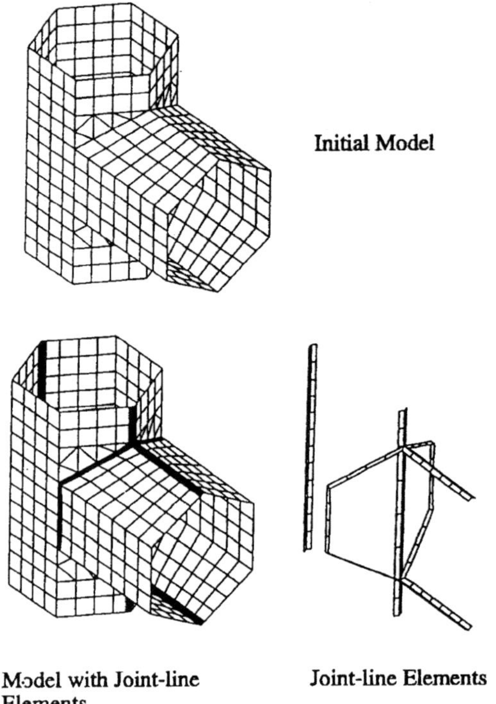

Fig 64: Use of joint line generator

TENSION STIFFNESS per UNIT LENGTH [(N/mm)/mm]			
Gauge (t) (mm)	Sheet	Weld-Bonded	Spot-Welded
1.2	8400 (100%)	954.6 (11.4%)	29.2 (0.35%)
1.6	11200 (100%)	1134 (10.1%)	57.5 (0.51%)
2.0	14000 (100%)	1337.0 (9.5%)	94.9 (0.68%)
BENDING STIFFNESS per UNIT LENGTH [(Nmm/rad)/mm]			
Gauge (t) (mm)	Sheet	Weld-Bonded	Spot-Welded
1.2	1131 (100%)	1051 (92.9%)	551 (48.7%)
1.6	2681 (100%)	2430 (90.6%)	1282 (47.8%)
2.0	5237 (100%)	4628 (88.4%)	2447 (46.7%)

Fig 65: Predicted stiffness of wel-bonded and spot-welded joints

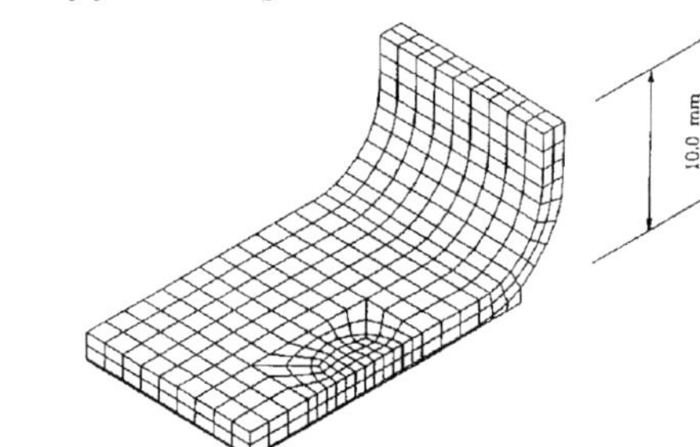

Fig 66: FE model of weld-bonded joint

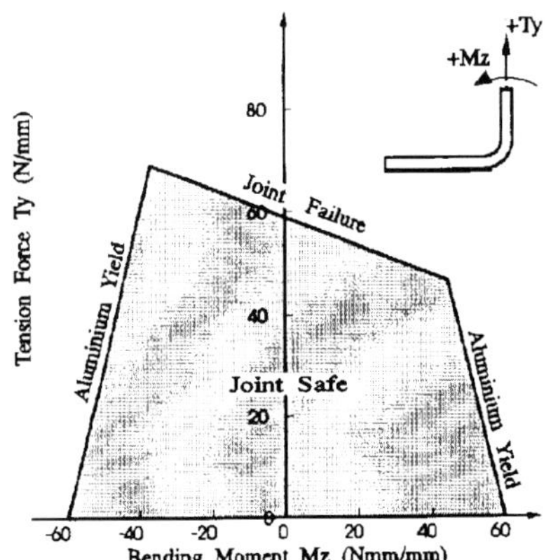

Fig 67: Typical design envelope

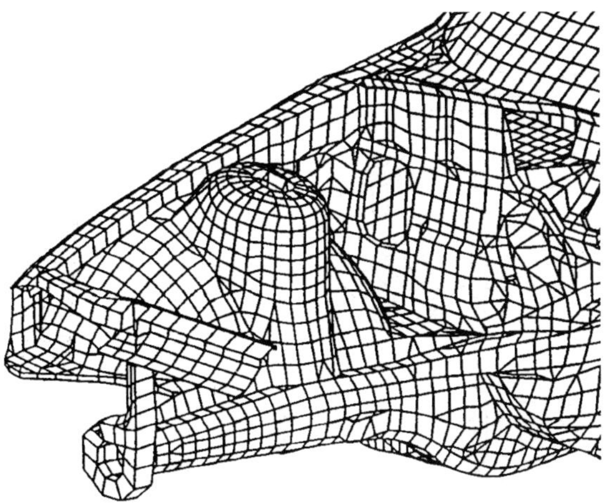

Fig 68: FE model of body front-end

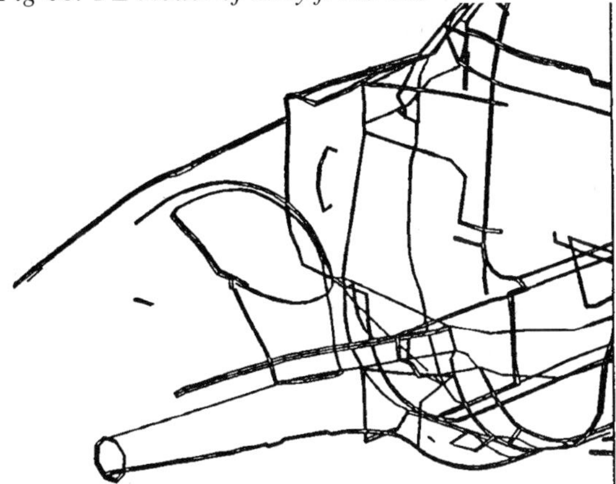

Fig 69: Joint line elements for above model

Configuration	Stiffness (kNm/deg)
Bonded Joints (fillet size = 75%)	11.04
Spot-Welded Joints (30mm pitch)	9.44
Continuous Joints	10.71

Fig 70: Torsional stiffness of aluminium vehicles

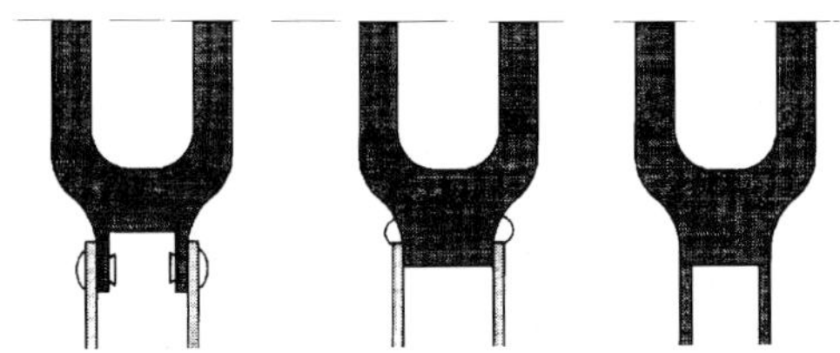

Fig 71: Attachment alternatives

combination of tension and moment loading was therefore chosen as the joint failure criterion in the design method.

Joint design envelopes were developed as a logical part of the design approach, the forces obtained from the full scale FE model being checked against them. The envelope, Fig 67, defines a safe region and every point represents a tension and moment combination. In this case the top boundary, representing adhesive failure, was obtained by a combination of FE modelling, tensile tests of peel joints and FE simulation of the test to obtain the exact combination of tension and moment forces.

The joint design procedure has been used on a number of full vehicle structures. In that shown modelled for analysis in Fig 68, the half-structure contains 16 500 elements and 14 700 nodes, involving some 71 different pressed parts. The Joint Line Generator was used to introduce 4 mm high joint line elements which led to 3280 further ones and 3382 more nodes, increasing the model size by 20%, Fig 69. Loads used for the analysis were fatigue ones measured at the suspension mounts, the structure being of aluminium gauges ranging from 1.0 to 3.0 mm weighing 132 kg. The assumption that all the joints were peel ones gave a highly conservative estimate of performance and normally this would be followed by the assumption of a more sensible joint configuration. On the basis of tension loads only, no moments, the table in Fig 70 summarises the static and fatigue strengths of various techniques of T-peel jointing which compares the modelled torsional stiffness of the structure.

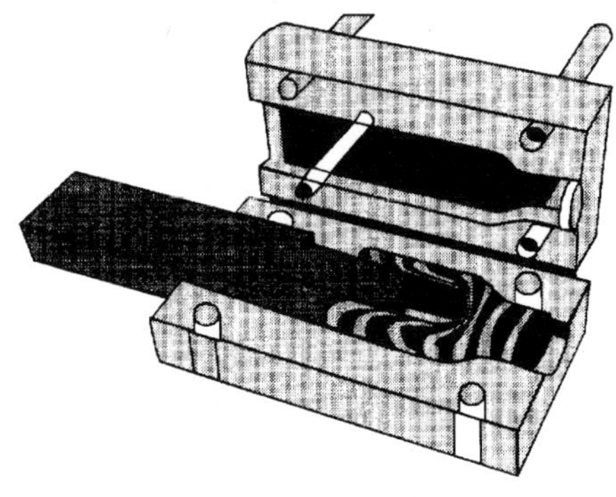

Fig 73: Model testing

Weight reduction for safety-critical components

Use of light alloys in safety critical components obviously requires special care in design and manufacture. These two case studies show an integrated design/manufacturing approach involving novel production techniques

Steering column element As part of the Aluminium Federation's TALAT initiative[11], engineers at Raufoss A/S have described light-weighting procedures, in the substitution of aluminium for steel, for the upper casing of an automobile steering column now adopted for Volvo and Saab models. The part has to transfer 160 Nm steering wheel torque, facilitate length adjustment, be collapsible on impact and allow angular articulation for torque transfer. Tube dimensions were specified on the design brief as 30 mm maximum external diameter and 42 mm equivalent for the yoke, the part having to sustain +/- 40 Nm at 5 x 10^5 cycles fatigue loading.

The first potential design solution involved a tubular member for torque transfer but with a faceted interior to allow sliding motion, the yoke allowing angular articulation. Fig 71 shows alternative solutions for attaching the yoke to the tube, riveting, welding or an integral casting/forging, all of which have disadvantages in conventional steel construction. By choosing aluminium, an extruded tube could be joined to a hot-forged, cast, extruded or cold-forged yoke. While welding could be considered, quality-control would be even more difficult then steel. Because of the high ductility of aluminium, cold-forging in one piece is a strong contender and safety consideration make it the first choice.

Four alternative methods of cold-forging are considered in Fig 72 and Fig 73 shows a technique of model testing to ensure proper material flow in the process for this safety-critical component. Preferably the initial blank is annealed before processing and then the finish part will be solution heat-treated and aged. However the cold-forging process itself improves tensile strength and hardness by the action of cold working, particularly so with 7000 series alloys containing Zn, Mg and Cu. Also grain flow follows the surface contour of the component, improving fatigue strength. However, because of their higher ductility and proneness to corrosion 7000 series alloys are slightly less preferable to the 6000 series

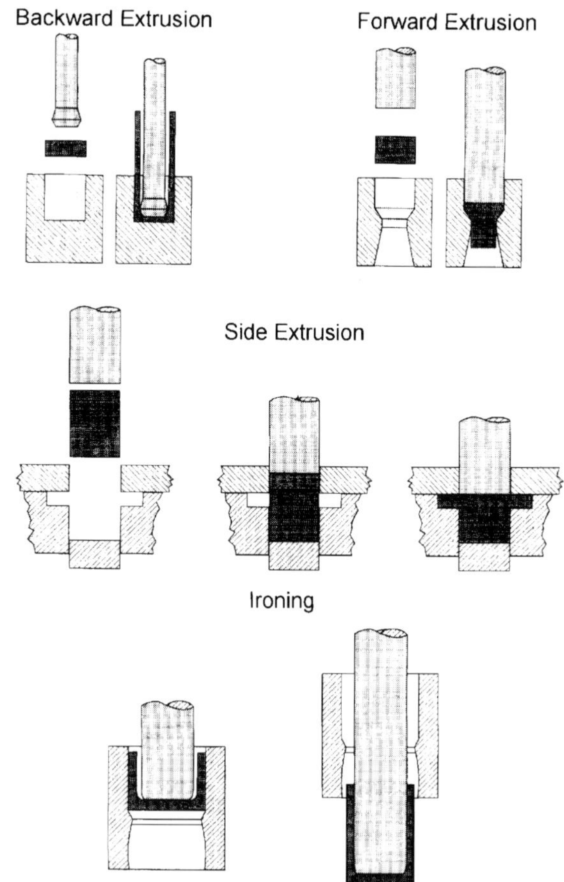

Fig 72: Cold-forging alternatives

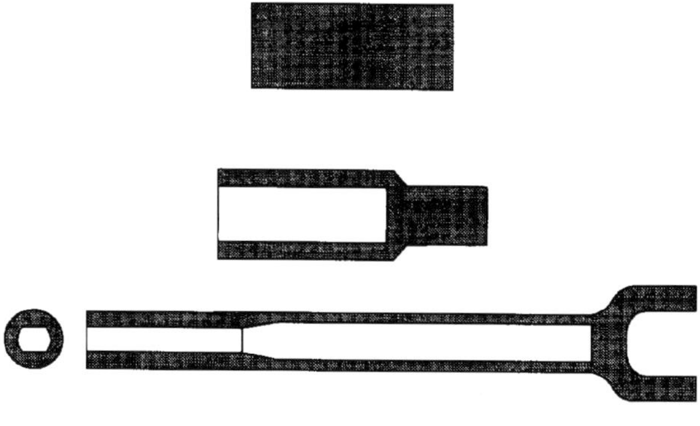

Fig 74: Stages in manufacture

from which 6351 is chosen by consideration of the following table of properties:

Alloy	*Formability*	*Strength*	*Ductility*	*Corrosion resistance*
7000	Poor	Excellent	Varies	Poor
6005	Good	Ok	Brittle	Ok
6351	Good	Ok	Ok	Good
6082	Good	Ok	Ok	Good
6063	Good	Insufficient	Ok	Ok

Thickening of the wall of the tube in the faceted section ensures torque transmission from the telescoping steel shaft causes no overstress condition. Fig 74 shows the stages in cold-forging.

Suspension sub-frame Another recent example of a safety critical component in aluminium alloy is the Alfa Romeo Spyder rear-axle's subframe described at the 1996 FISITA congress. According to Michele Spina of Centro Ricerche Fiat a highly integrated design and manufacturing process was involved in this 7 kg part which has to support the multilink suspension anchorages.

A compound structure comprising two cast lateral supports connected by extruded bars, was proposed, Fig 75, with configuration shaped by the requirements of the longitudinal and transverse loads to be reacted from the suspension, Fig 76. Reducing the structure to five components involved complex multi-function lateral support castings. Thixocasting of the A357 alloy for these elements was the chosen route, which only required an ageing process for the final heat-treatment, without risk of distortion otherwise involved in precipitation hardening. By following this approach straight constant section transverse bars could be used, TIG-welded to the castings.

The extruded bars have to transfer bending and torsional moments from the suspension loads. The central bar reacts loads from the springs ; the front bar reacts braking loads and the rear one loads from turning manoeuvres. An integrated CAD-CAE package used in the design allowed a multibody CAE analysis to proceed during the CAD development, using course and fine FE meshes, respectively. Elasto-kinetic analysis was also carried out to verify the accuracy of the suspension geometry under simulated loading.

Control of the thixocasting process was crucial to the manufacture of the structure, and the 18 kg, 5 inch diameter, billets for each lateral support were the largest ever used by the process at job-1 date. Maintenance of uniform temperature up to the commencement of melting was critical and so too the uniformity of solid/liquid ratio in the semi-solid state prior to pressing. The manufacturing facility is innovative in its ability to provide guaranteed heating capacity for 180-200 kg material per hour at 530 C to a tolerance of +/- 5C. A Buhler 1800 C press is used and a device which prevents any vortex formation in the metal flow is involved.

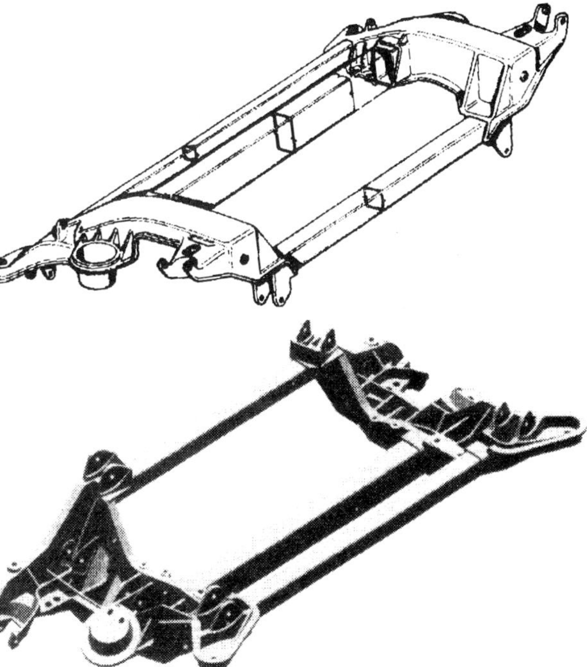

Fig 75: Subframe configuration

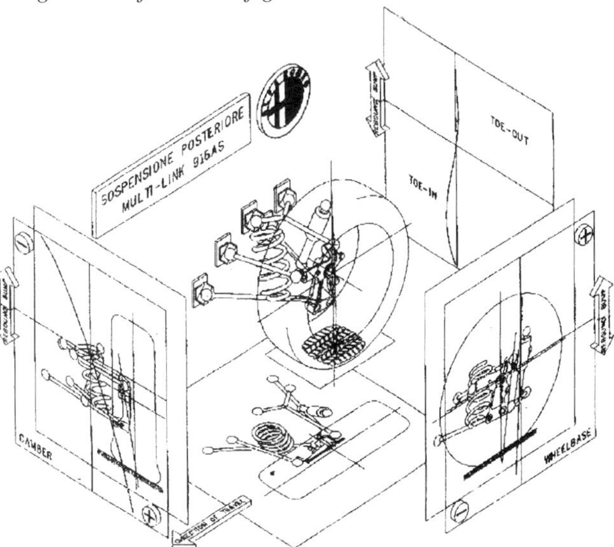

Fig 76: Suspension layout

Aluminium *vs* steel for body structures

The increasingly important criterion for vehicle body structures is safety in impact, to be considered in the next chapter. In the long raging battle between steel and aluminium body panel material suppliers, some recently reported investigation work on aluminium crash performance at Raufoss has shown the material to have better potential for body crashworthiness than was commonly considered. However, British Steel have recently shown the goal posts to be continually moving, in their report on the international Ultra-light steel car body project (ULSAB) with which they are involved.

According to researchers at Raufoss Automotive Structures[11], there are a number of different structural requirements for vehicle bodyshell materials. In normal road use, bending and structural stiffness are the key parameters. However, in low-speed impact (4 - 9 kph) elastic deformation energy is most important; in a mild crash, elastic deformation energy is the key whereas in a severe crash, structure integrity is all-important, with no breakage or fragmentation occurring.

In comparing structural performance with steel the author argues that cost per unit volume rather than per unit weight should be considered; even so, he admits that aluminium is unlikely to meet the application cost of steel in a body-structural situation. However, he reminds us that examination of individual parts of the structure can still show some surprises. Most beam elements, for example, are subject to bending and torsional moments, as well as shear and axial loads. The response-displacement characteristic of the beam element, to these moments, is a measure of the contribution of the element to global stiffness and strength, while contribution from shear and axial loads are minor in comparison.

With a straight substitution for steel in a sill member, Fig 77a, and assuming equivalence of yield strength, there is a one third weight saving, offset by a two-thirds reduction in stiffness; however, the beam profile will absorb three times more elastic energy before permanent set. When optimized for bending, Fig 77b, there is now more than 1/3 the stiffness but three times the displacement and more than three times the elastic energy absorbed. There is also more elastic-plus-plastic energy absorbed. When optimized for torsion, Fig 77c, again there is more than 1/3 the stiffness and three times the displacement and more plastic-plus-elastic energy absorbed. However, when optimized for stiffness, generally, Fig 77d, there is 40-50% weight saving, stiffness is

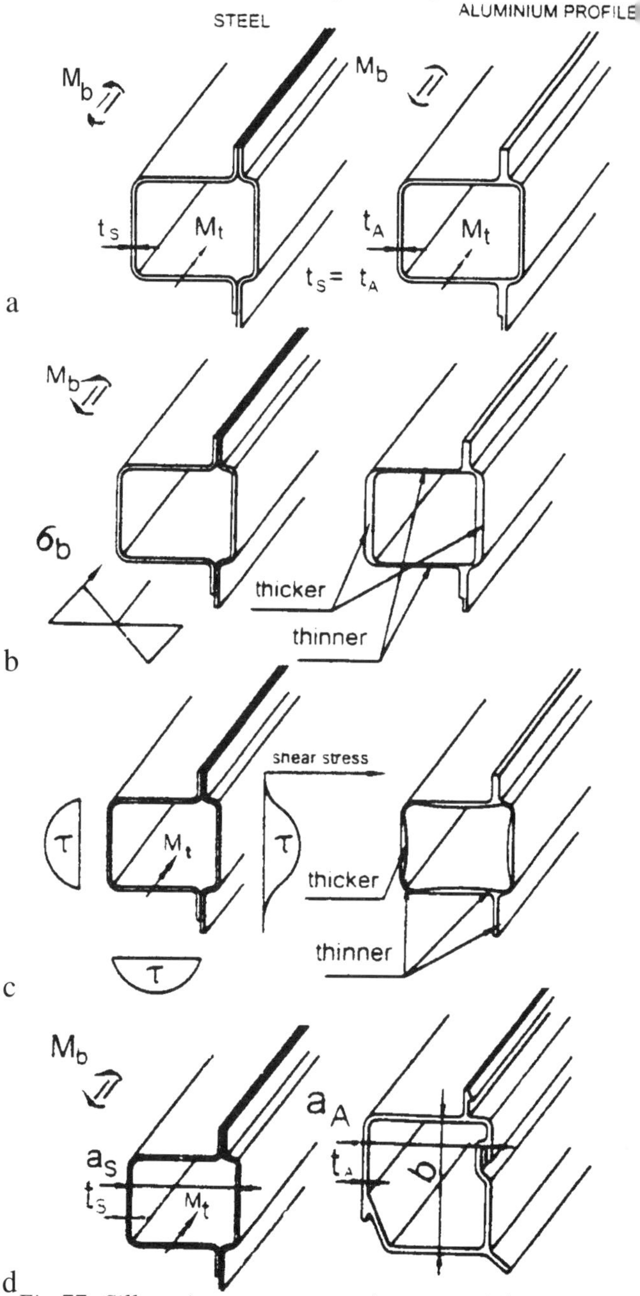

Fig 77: Sill section parameters when a. straight substitution of aluminium for steel; b. optimized for bending; c. optimized for torsion; d. optimized for stiffness

the same as steel and approximately 2.5 times the displacement applies. There is also much more elastic-plus-plastic energy absorbed, Fig 78.

Significant gains can only be made when profile dimensions are increased until the section moment of inertia is three times that of steel. But as the moment of inertia increases in relation to the cube of the depth of section, an increase in beam thickness of 30-40%, together with somewhat thicker walls and increased height, will achieve the optimum section properties. As section moment of inertia increases, the section modulus increases by the square of the amount and there is much more moment-carrying before permanent set.

Typically a car crash takes place within 0.01-1 second high-speed strain rate occurs, Fig 79. Material specimens are, of course, traditionally tested at static strain rates and material properties presented thus in charts. But it should be born in mind that some steel yield strengths change with strain rate. Tests carried out by Raufoss on Type 7000 aluminium alloys has shown elongation, yield strength and ultimate strength not changing significantly between static and crash-situation strain rates, Fig 80.

True fracture strain is considered by the author to be the parameter giving a better indication of the cracking proneness of an alloy at high deformations. By over-ageing, and moving from a T6 to a T7 temper condition, this parameter can be improved, although yield strength will drop somewhat. Overall, though, more energy will be absorbed in the crash. Examples of structural parts produced by Raufoss in 7000 aluminium alloy are shown in Fig 81.

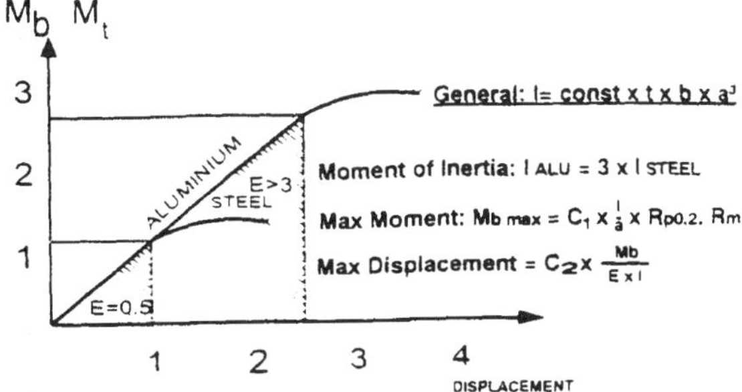

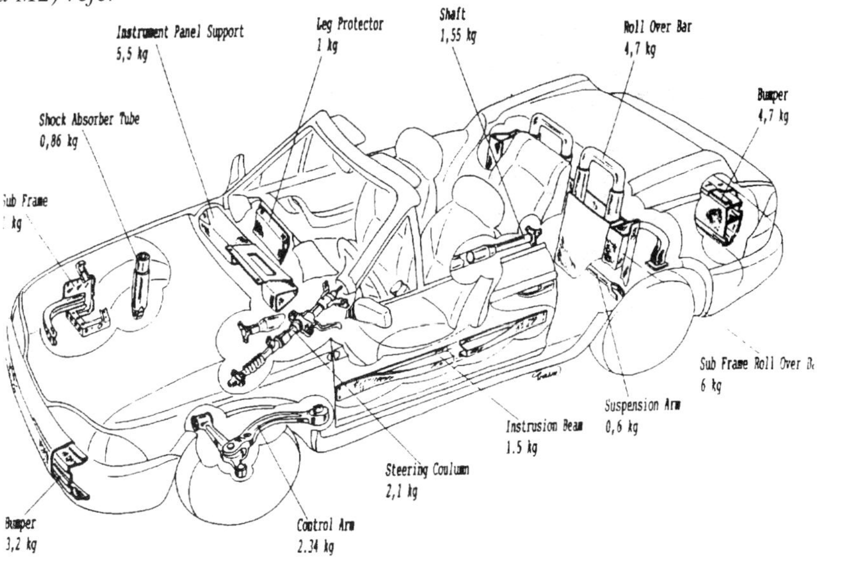

Fig 78: Applied moments vs displacement for sill member optimized for stiffness

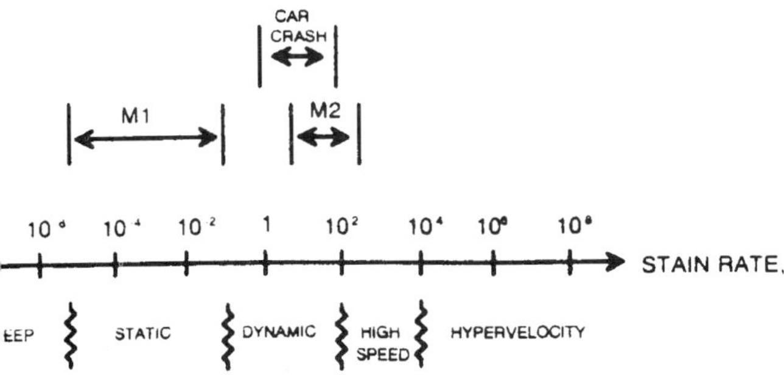

Fig 79: Strain-rate definitions (M1 and M2) refer to Raufoss testing machines

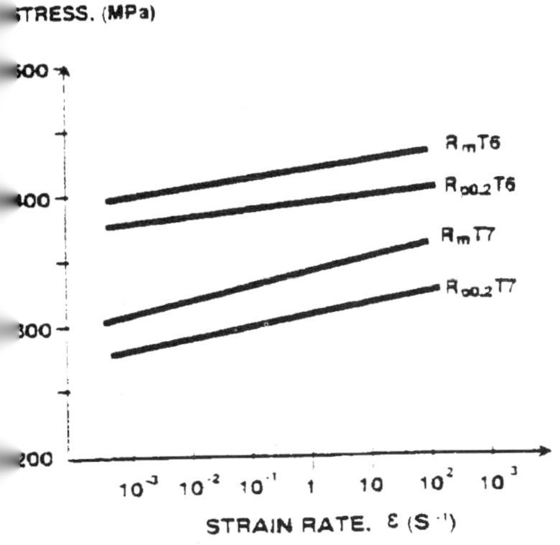

Fig 80: Yield and ultimate strength vs temper and strain rate

Fig 81 Structural parts made in aluminium by Raufoss

Steel *vs* aluminium in impact

The monocoque structure, according to British Steel[12], is the one best suited to volume construction and the company quotes Rover's Engineering Director in claiming that while steel bodies could be adequately modelled for structural and fatigue analysis, this was not yet the case with aluminium alloy. The suggestion also seemed to be given by the author that energy absorbing properties of aluminium alloy had been misrepresented against those for steel as like-for-like structures had not been compared.

In British Steel's tests on longitudinal crush box-members made from weld-bonded top-hat sections of equal weight, these showed that steel was clearly the superior material in 30 mile/h impact tests, aluminium alloy sections exhibiting overall instability failures rather than local progressive wrinkling.

Dramatic changes in sheet steel alloying had also taken place in the last decade, with carbon-manganese steels being substituted for HSLA structural members and the use of bake-hardening steels for skin panels. There was also the problem, with aluminium alloys, of prices having almost doubled in the last decade and until service load envelopes on road vehicles were as well understood as those on aircraft, the unclear fatigue limit of aluminium would work against the metal (however, it was often not mentioned that aircraft had much more efficient structures than volume-car body shells and that more structurally efficient car design could probably result in better exploitation of light alloys). It was also now argued that the amount of greenhouse gases released in production of aluminium alloys outweighed those for steel to such an extent that reduced emissions form lighter aluminium alloy vehicles would not in fact be realised.

Aluminium structural extrusions According to Alcan International[12] only Audi, Honda and Jaguar have as yet offered medium-rate production al-alloy structures to the market but its author points out that there are substantial opportunities for bolt-on al-alloy components for steel-shelled vehicles. A surprisingly wide range of alloys could be extruded, as seen in Table 1. Choice of alloy and temper can be tailored to application so, for example, an impact absorbent part could require a high-deformability material like AA6063 at an average temper of T7 while a part requiring considerable reshaping could involve working the extrusion in the solution treated T4 condition, the final strength being achieved by heat treatment to the TS condition.

In bumper armature design, the use of a spring-mass analogue is suggested for determining the com-

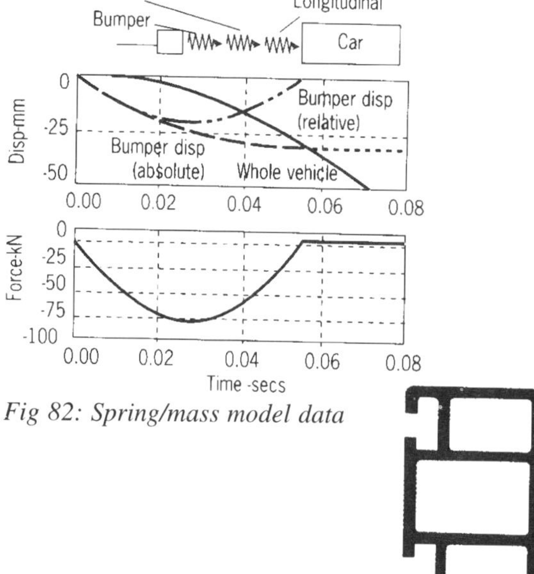

Fig 82: Spring/mass model data

Fig 83: Bumper section example

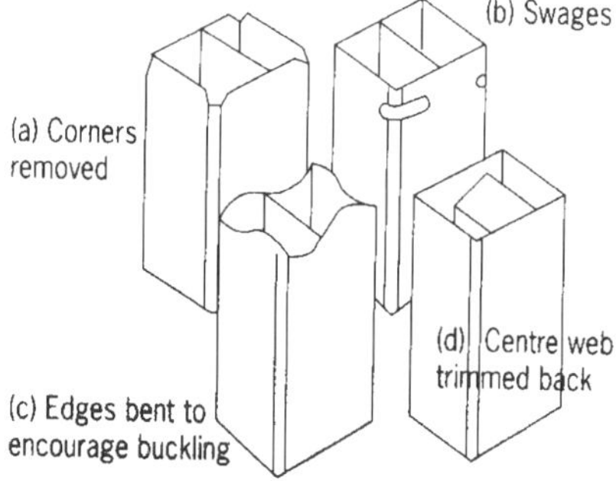

Fig 84: Crush can initiators

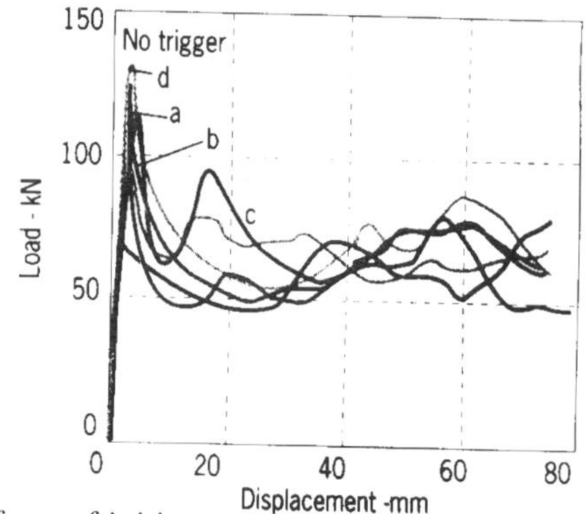

Fig 85: Effects of initiators

Table 1. Typical mechanical properties of extruded alloys

Alloy	Temper	Yield strength (N/mm^2)	UTS (N/mm^2)	Elongation (%)
6063	T4	90	160	21
	T5	210	245	12
	T7	175	205	13
6082	T4	170	260	19
	T5	310	340	11
7020	T5	335	380	13

bined effect of bumper, support-can and body longitudinal displacements, Fig 82, with respect to time. Relative displacement of the bumper can also be obtained by subtracting absolute displacements. Aluminium alloy extrusion allows the possibility of using closed section armatures in place of the usual open section rolled steel type, Fig 83. This can lead to optimized metal placement and attachment features which improve the design.

For the crush-can used for bumper mounting, fold initiators can be introduced, Fig 84, to obtain stable and predictable deformation. Instead of the usual swages or 'bird-beaks' the alternatives illustrated can be employed to give corresponding displacement characteristics shown in Fig 85.

For side intrusion an alternative to heavy and difficult to form high-tensile steel systems can be obtained with an aluminium alloy extrusion with potential weight savings of 4-6 kg per door. Beams having trapezoidal sections shown in Fig 86 are said to be particularly advantageous, having the compression face thicker than the tension face.

MODEL CONFIGURATION	TORSIONAL STIFFNESS Nmm/Degree (lbft/degree)	BENDING STIFFNESS N/mm (lbf/in)
Initial Design	2.1×10^6 (1.6×10^3)	6.1×10^3 (35.0×10^3)
Initial Design plus Rigidised Cpts	5.0×10^6 (3.7×10^3)	
Minor Modifications	2.8×10^6 (2.1×10^3)	
Final Design	4.2×10^6 (3.1×10^3)	

Fig 86: Stiffness predictons

Initial Design	2.1×10^6 (1.6×10^3)	2.0×10^6 (1.5×10^3)
Minor Modification	2.8×10^6 (2.1×10^3)	
Final Concept 1st Prototype)	4.2×10^6 (3.1×10^3)	

Fig 87: Torsional stiffness correlation

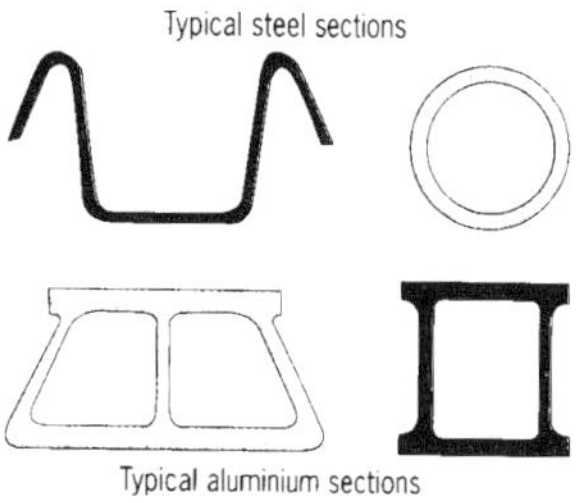

Typical steel sections

Typical aluminium sections

Fig 86: Intrusion beam sections

Analysing open car structures

For the open sports-car a greater structural challenge is set for the designer. For the latest Scimitar car, SDRC Engineering Services[13] were sought to carry out a full computer-aided concept-design in a determined effort to overcome the low-frequency dynamic behaviour problems which have dogged many open-car structures. The steel underframe was designed around a central armature to react torsional loads — with a peripheral beam structure to carry the running units plus the vehicle occupants and luggage. The structure was notionally divided into shell and beam regions, the beam section properties being factored to allow for joint inefficiencies.

The resultant analytical model obtained is shown alongside the system responses later in the section. The central tunnel was subjected to a finite element analysis using shell elements while the remainder used beam elements. So as to retain 'designer-feel' during the mathematical analysis, a preprocessor with an advanced graphics capability was used to allow the model to be presented to the designers at regular review sessions. When the model was felt to truly represent the structure it was subject to analysis using SUPERB FE code. Test loads were applied to derive overall torsion and bending stiffnesses which were compared with those of known vehicle designs. Areas of high distortion were then located on the basis of elemental strain energy densities.

Such areas were artificially 'stiffened' and the program re-run to evaluate the effect and that of strain re-distribution through the remainder of the structure. Results of this preliminary analysis are shown in the table in Fig 86, from which an unacceptably low torsional stiffness resulted. The weak points found are annotated in the computer responses which showed that the peripheral frame structure was failing to strain the central tunnel sufficiently.

This led to the design modifications shown in

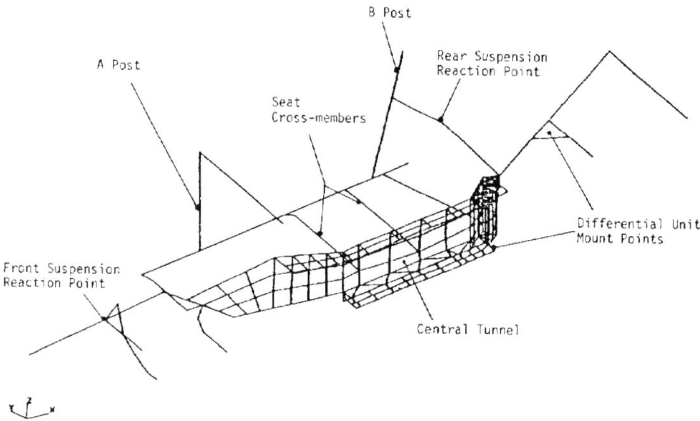

Fig 88: Analytical model

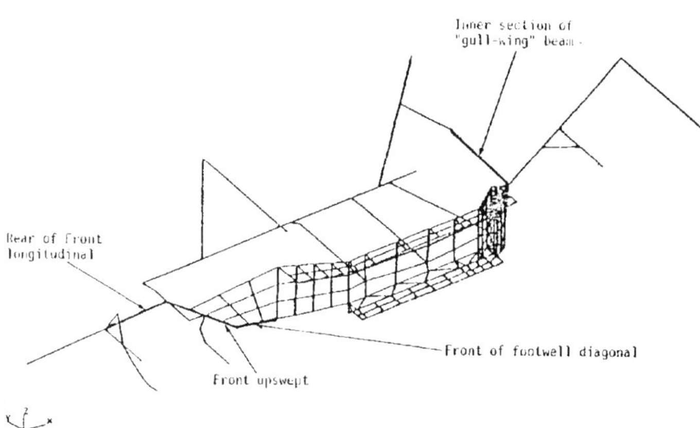

Fig 89: Weak beams made rigid for study

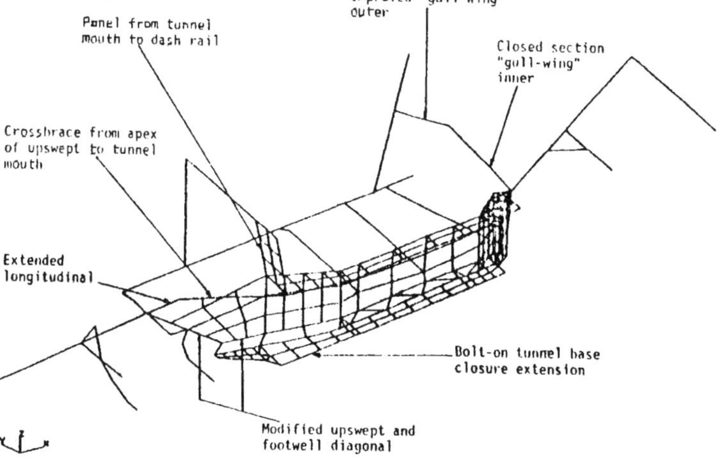

Fig 90: Design modifications in first revision

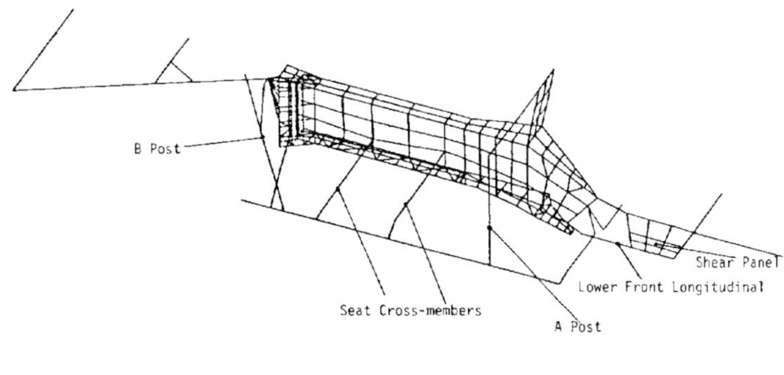

Fig 91: Sub-structure redesign

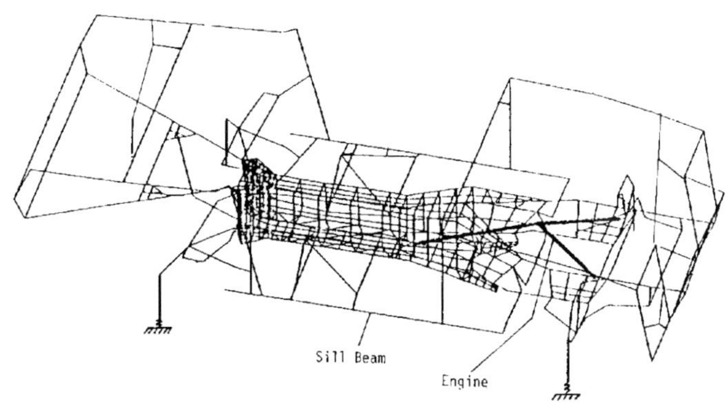

Fig 92: Full dynamic model

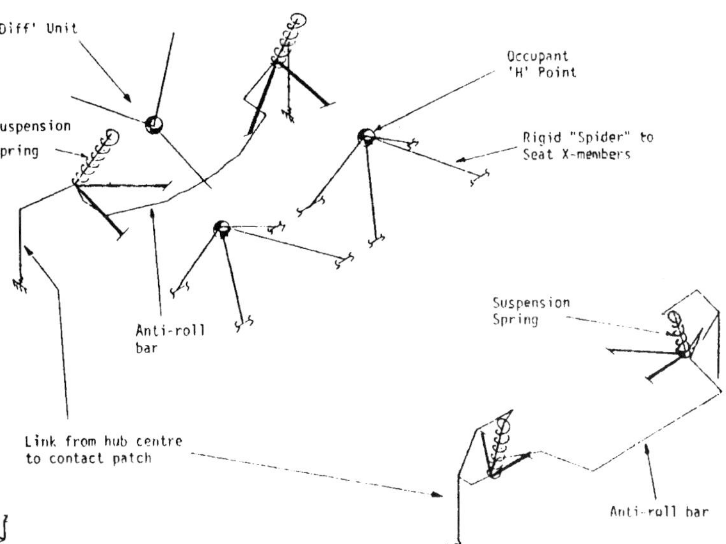

Fig 93: Dynamic model: suspension, occupants and differential

later in the analysis which proved adequate with the exception of the front longitudinals which could not be sufficiently stiffened — hence a redesign of the front-end. To achieve greater vertical bend stiffness between front-suspension reaction point and the tunnel, a lower longitudinal was added, together with a shear-panel with vertical edge stiffeners front and rear. Their addition to the model is also seen in the computer simulation and the table in Fig 87 shows that the predicted improvement to torsional rigidity closely aligned with the measured value on the first prototype.

The computer analysis commenced with a description of the original design, Fig 88, followed by an annotation of the weak points, Fig 89, a description of the necessary rectifications, Fig90, and the model redesign, Fig 91. For dynamic analysis it was necessary to add to the dydnamic model, Fig 92, the mass of vehicle components, Fig 93. This was used to study vibrational behaviour of the structure and its associated connecting masses so as to minimise the coupling paths between known dynamic forces and the vehicle occupants. An examination of the vibrational mode shapes led the designers to conclude that the critical fundamental torsional mode occurred at 18Hz — high enough to avoid interaction with the wheel-hop modes and low enough to avoid excitation by the engine at idle.

References

Chapter 1

1 Hartley, J, The changing scene in car manufacture, *Automotive Engineer*, April 1977

2 Burman, R, Achieving world class in body fabricatoin, *Automotive Engineer*, February/March 1994

3 Psarouthakis, J, Better makes us best, Productivity Press

4 Robinson, P, Body/tool engineering matrix organisation, IMechE conference paper C462/7/052

5 Burman, R, Manufacturing management, principles and systems, McGraw-Hill

6 Jubb, J, Design for spot welding, *Automotive Design Engineering*, 1966

7 Dittlo *et al*, Paper SIA9506B01, EAEC 1995

8 Hibben et al, Paper SIA9506B10, EAEC 1995

9. Houldcroft, P, Which Process?, Abington,1990

10 Gibson/Smith, Basic welding, Macmillan, 1993

11 Hicks, J, A guide to designing welds, Abington, 1989

12 *Job knowledge for welders* series, TWI, 1996/7; contact e-mail: wlucas@twi.co.uk

13 Pitchford, N, Adhesive Bonding for Aluminium Structured Vehicles, IMechE seminar, Materials — Fabricating a Novel Approach,1993

14 McGregor et al, The development of a joint design approach for aluminium automotive structures, SAE Paper 922112, 1992

15 Pearson, I, Adding welded/mechanical fastening to adhesive-bonded joints, *Automotive Engineer*, Aug/Sept 1995

16 Debell, Hartstock and McLean, Paper 5 of the Ceramics in the Automobile session, Autotech 1987

Further reading

Hartley, J, A worldwide review of manufacturing trends, Economist Intelligence Unit, 1992

Roberts, D, Designing vehicles with an economic framework. *Automotive Design Engineering*, vol 5, 1966

Timings, R, Manufacturing Technology, Longmans, 1993

Sawhill and Sawdon, A new mechanical joining technique for steel compared with spot-welding, SAE paper 830128

Snap-fits and press-fits, *DuPont review Engineering Design*, E-96323

Handbook for disc springs, Adolf Schnorr Kg, Maichingen bei Stuttgart, Germany

Structural Adhesives in engineering, IMechE conference report 6, 1986

Ed Sadek, M, Industrial applications of adhesive bonding

James, P, Isostatic pressing technology, Applied Science Publishers, 1983

Hicks, J, Welded joint design, BSP Professional Books, 1987

Redford and Chell, Design for assembly, McGraw Hill, 1994

Wines, B, Adhesive and mechanical clinch bond strengths,SAE paper 890517, 1989

Chapter 2

1 Kawka et al, Advnaces and trends in sheet metal forming simulation in Japanese automotive industry, SAE paper 970432, 1997

2 Siegert, K, Advances and trends in sheet metal forming processes, SAE paper 970436, 1997

3 Tebbutt, E, Design for presswork, *Automotive Design Engineering*, August 1966

4 Grainger and Morse, Design for steel pressings, *Automotive Design Engineering*, April 1971

5 Cole, A, Design for economical press tooling, *Automotive Design Engineering*, December 1968

6 Pearce, R, Sheet Metal Forming, Adam Hilger Publishing

7 Waller, J, Press tools and presswork, FMJ International Publications

Further reading

Cole A, Paper at conference of Institute of Sheet Metal Engineering, Design for tooling, *Sheet Metal Industries*, 1968, p 812.

Johannisson, T, Flex-forming in the automotive industry, *ABB Review*, 5/90

Bull et al, The application of a knowledge-based CAD system to press tool design, SAE paper 900797

Wright and Neate, Cold extrusion, *Automotive Design Engineering*, Feb 1965

McRobert and Watson, Design optimization of forgings, Drop Forging Congress, Cologne, June 1953

Pearce, R, Superplasicity and automotive engineering, *Automotive Design Engineering*, May 1971

Chapter 3

1 Hill, T, Manufacturing Strategy, Macmillan, 1985.

2 Shingo, S, Study of Toyota Production System from Industrial Engineering Viewpoint, Japan ManagementAssociation, 1986.

3 Shingo, S, Zero Quality Control: Source Inspection and the Poka-Yoke System, Productivity Press (USA), 1986.

4 Shingo, S, A Revolution in Manufacturing the SMED System, Productivhy Press (USA), 1985.

5 Monden, Y, Toyota Production System, Insthution of Industrial Engineering Managers (USA).

6 'Sweden's solution', *The Engineer*, page 23, 21 .1 .88.

7 'An affordable strike', *Spectator*, page 5, 13.2.88.

8 Wickens, P, The road to Nissan, Macmillan

9 Hale *et al*, Quest for quality, Tennant Company, Minneapolis, Minnesota, USA

10 Schonberger, R, World class manufacturing casebook, Collier Macmillan.

11 Shingo, S, Key strategies for plant improvement, IFS (Publications)

12 Singh and Kiangi, Risk and reliability appraisal on microcomputers, Chartwell-Brett.

13 Manton, S, Engineering for quality, Proc IMechE, Vol 202, No 19, 1988
14 Pirsig, P, Zen and the art of motorcycle maintenance, Corgi, 1987
15 Institute of Materials conference Moving forward with steel automobiles, 1993
16 Design Optimization of Forgings, GKN Technology Ltd, Birmingham New Road, Wolverhampton, UK.
17 Bloom, P, Using sintered materials, *Automotive Design Engineering*, July 1968
18 Langdon, R, Powder metalurgy progress, *Automotive Engineer*, August/September 1992
Further reading
Ed Baboian, R, Automotive corrosion by de-icing salts, US National Association of Corrosion Engineers, 1981
Fenton, J, Car body structure design for corrosion prevention, paper 3 at *ADE*/Hatfield Polytechnic conference on Nature and prevention of vehicle corrosion, 1992
Corrosion of motor vehicles, IMechE conference report, 1974
McArthur, H, Motor vehicle corrosion prediction and prevention on vehicles, Expert Books, 1990
Child, P, Vehicle fine finishing, BSP Professional Books, 1987
Loop, F, Cathodic automotive electro-deposition, SAE paper 780189
Johnson, B, Development of an automotive paint film concept, SAE paper 890352

Chapter 4
1 Mann, R, Automotive plastics and composites, Elsevier Advanced Technology
2 West, GH, Manufacturing in Plastics, PRI
3 UK DTI pamphlet on reinforced thermoplastics available from the Design Council, 28 Haymarket, London SW1Y 4SU
4 Voy and Morsch, SAE Paper 860278, 1986
5 Basu, A, SAE Paper 860283, 1986
6 Doring, E, SAE Paper 860275, 1986
7 Messenger et al, Paper C498/18/086 , Autotech, 1995
8 Bradbury, L, Thermoplastic brake pedal housing, a case study, Proc IMechE/IOM conference Plastics on the Road, Nov 1996
9 Iiduka et al, Development of damping SMC and its application as material for a rockercover, from SAE SP-1166
10 Goldbach, H, IBCAM Boditek conference, 1991
11 Wardill, G, The stabilised core composite, Autotech 1989
12 Designers Guide to Rotational Moulding, T&D Rotomoulding, Victoria Street, Pontycymmer, Bridgend, CF32 8LR, UK
Further reading
Hartley, J, The materials revolution in the motor industry, Economist Intelligence Unit, 1993
Ed Young and Shane, Materials and processes, Marcel Dekker, 1985
Gutman, H, New concept bumper in plastic, SITEV conference, 1990
Jutte and McCluskey, Development of SC for automotive body panel use, SITEV conference, 1983
Thermoplastic matrix composites, Profile series, Materials Information Service, DTI
Wootton, A, Development of a novel RP suspension system, Soc of the Plas Ind/Composites Inst conference report, 1989
West, G, Engineering Design in Plastics, Plastics and Rubber Institute
Chant P, Reservoir moulding technique: a new process for glass-reinforced resins. Reinforced Plastic Conference, 1972, paper 15, British Plastics Federation Reinforced Plastic Group
Owen M, How fatigue affects glass and carbon-fibre composites. *OEM design*, 1972, vol. 1, no 4
Wood, R, Automotive engineering plastics, Pentech Press, 1991
Design in composite materials, IMechE conference report 2, 1989
Maxwell, J, Plastics in the automobile, Woodhead Publishing, 1994
Plastics and rubbers, engineering design and applications, Mechanical engineering publications, 1989
Macosko, C, Reaction injection moulding, Hanser Publications, 1989
Limits of plastics uses in vehicle construction, VDI-Verlag, 1981
Plastics on the road, PRI/IMechE conference report, 1988
Sabol et al, A design procedure for thermoplastic bumpers, SAE paper 870109
Data for design in Propathene polyurethane, Tech Service Note PP110, ICI
Neal, P, Comparison of steel and reinforced plastics in body engineering, Inst. Body Engrs., 1967

Chapter 5
1 Ashley, C, Weight saving in steel body structures, *Automotive Engineer*, December 1995
2 For a new automobile technology, a new steel industry, SITEV conference paper, 1985
3 British Standard 8118. 1991: *Structural Use of Aluminium, Parts 1 and 2.*
4 Vaschetto et al, A significant weight saving application of magnesium to car body design, Paper SIA9506B02, EAEC congress, 1995
5 Technical Note to IMechE Proceedings (Part D, Vol 208), 1995
6 Guyen *et al*, EAEC conference paper, 1993
Further reading
Engineering steels, Profile series, Materials Information Service, DTI

Nardini and Seeds, Structural design considerations for bonded aluminium structured vehicles, SAE paper 890716/7
Ruden et al, Design and development of a magnesium/aluminium door frame, SAE paper 930413
User manual for 3CR12 steel, Cromweld Steels
Cowie, G, The AISI automotive steel design manual, SAE paper 870462
Using copper alloys, Profile series, Materials Information Service, DTI
Commercial vehicle body-building in aluminium, British Aluminium Company, 1965.
Mazurek, Y, Chrysler XLV truck cab has been designed for low-volume production overseas, SAE paper 650023.
Bouge, L, Semi-trailer van-design consideration, SAE paper 660077.
Reidelbach, W, Zur I-'rage der Gestaltfestigkeit von Karosserie-Oberbauten. *Automobiltechnische Zeitschrift*, 1964, N.3, SS.63-66.
Schreiner et al, Heat treatment processes for the reduction of wear in vehicle components, Mechanical Engineering Publications, 1991
Schiek, H, The design of aluminium alloy castings, *Engineers' Digest*, March 1966

Chapter 6

1 Campbell, C, The sports car, its design and performance, London, Chapman and Hall, 1959
2 Platt, M, Structure of automobile, Proc IMechE, AD no. 13, 1959/60
3 Short, M, Possibilities of aircraft structures in ground vehicles, SAE Transactions, 1945.
4 Peery, D, Aircraft structures, New-York, McGraw-Hill, 1950, 566 pp.
5 Analytical method for chassisless vehicle design. *Automobile Engineer*. 1953, vol. 43, pp. 103—111; Structural design Part 2: The front end structure. *Automobile Engineer*, 1953, vol. 43, pp. 152—157.
6 Megson, T, Structural and stress analysis, Edward Arnold
7 Megson, THG, Increasing semi-trailer torsional stiffness could enhance operational safety, *Automotive Design Engineering*, May 1970
8 Costin and Phipps, Racing and sports car chassis design, Batsford, 1962.
9 British Standard Code of Practice CP118:1969, The structural use of aluminium (also BS8118, 1991)
10 Cooke, C, Torsional stiffness analysis of car bodies. Proc IMechE, 1965—66, Vol 180, part 2A, no. 1
11 Sharman, P, Optimum stiffness-weight design of peripheral and ladder frames, Proc IMechE, Vol 182, Part 2A, No 3, 167-68
12. Automotive steel design manual, American Iron and Steel Institute, 2000 Town Centre, Suite 1900, Southfield, MI 48075-1138, USA.

Further reading

Giles, C, Body construction and design, Iliffe, Automotive technology series vol 6, 1971
Framework analysis and stiff-jointed frames
Goodwin, E, The unit load method of structural analysis, Draughtsmen's and Allied Technicians Association booklet, session 1966/7
Costin, Competition car chassis design, *Automotive Engineer*, Dec 1983
Bolland, G, The static indeterminacy of a shell structure, *Automotive Design Engineering*, Sep 1966
Buckling of plates, Engineering Science Data Unit ESDU data sheets 71005/71014/71015
Williams, Structures Report No 2, Royal Aircraft Establishment, Farnborough, UK
Timoshenko and Gere, Theory of elastic stability, McGraw Hill
Nichols, Stability of honeycomb sandwich panels in shear, College of Aeronautics thesis, Cranfield, UK, 1955
Timoshenko, Theory of plates and shells, McGraw-Hill
Reports 1505 and 1583, US Forest Products Laboratory
Barwell, R&M 2467, Aeronautical Research Council, UK
Williams, D, Sandwich construction — a practical approach for the use of designers. RAE Report no. 2, Structures, Report and Memoranda 2466, 1948, Aeronautical Research Council, UK
Williams, Leggett and Hopkins, Report and Memoranda 1987, Aeronautical Research Council, UK
Williams D, An introduction to the theory of aircraft structures, Arnold, 1960
Faupel, Engineering Design, Wiley, 1964
Pippard and Baker, The analysis of engineering structures, Cambridge University Press.
Timoshenko and Gere, Theory of Elastic Stability, McGraw Hill, 1961
Hargreaves, T, Introduction to plate buckling, *Engineering Materials and Design*, 25 Aug 1972
Lutz, L, Structural design using the automotive steel design manual, SAE paper 870463
Muller, G, TECHNA — a concept car to challenge automotive engineers, SAE paper 690267, 1969
Load bearing sandwich panels, Rubber and Plastics Age, 40, 654, 1959
Nanba et al, Application of aluminium honeycomb sandwiches and extrusions in a convertible, SAE paper 870147/8, 1987

Chapter 7

1 Pawlowski, J, Vehicle body engineering, Business Books, 1969
2 Sharman, P, Torsion of chassis frames, *Automotive Engineer*, Oct/Nov 1976
3 Seitler, W, Utility vehicle frame developments, *Technische Rundschau*, No 16/62

4 Cooke, C, Some problems in car body design, Vehicle structural analysis seminar, IMechE, 15 May 1968
5 Sherman D, Stresses and deformations in truck side-rail attachments, SAE paper 690175
6 Tidbury et al, Torsional stiffness of commercial vehicle chassis frames, FISITA conference 1968
7 Tidbury, G, Stress Analysis of Vehicle Structures, ASAE Note 1, College of Aeronautics, Cranfield, 1965
8 Ed. Donald, I, Thin-walled structures, Elsevier, 1990.
9 Beerman, H, The analysis of commercial vehicle structures, Mechanical Engineering Publications, 1989
10 Guide to stability design criteria for metal structures' published by Wiley (1988) for the Structural Stability Research Council
11 Marshall, P, Design of welded tubular connections, Elsevier (1992)
12 Lowe, W, et al, Impact behaviour of small scale model motor coaches, *Journal of Automotive Engineering*, January 1972.
13. Young, W, Roark's formulas for stress and strain, McGraw Hill
14. Timoshenko and MacCulloch, Elements of strength of materials, Van Norstrand
15. Peterson, R, Stress concentration design factors, Wiley 1

Further reading

Allwood and Norville, The analysis by computer of a motor car underbody structure, Proc IMechE, Vol 180, Pt 2A, No 7, 1965-6
Sharman, P, Optimum stiffness-weight design of peripheral and ladder frames, Proc IMechE, vol 182, part 2a, no. 3, 1967-8
Dean-Averns, R, Automotive Chassis Design, Iliffe.
Erz K. Uber die durch Unebenheiten der Fahrbahn hervorgerufene Verdrehung von Strassenfahrzeugen. Tell I, *Automobiltechnische Zeitschrift (ATZ)*, 1957, N. 4, SS. 89—96. Tell 2, ATZ, 1957, N. 6. SS. 163—170. Tell 3, *ATZ*, 1957, N. 11, SS. 345—347. Tell 4, *ATZ*, 1957, N. 12, SS. 371—375.
Fenton, J, Truck chassis design, *Automotive Engineer*, Jun/Jly 1976
Carver, G, Application of variable depth side-rail to heavy truck frames. SAE Transactions, 1969, vol. 78, Paper 690174.
McNitt, L, Truck frame side-rail buckling stresses. SAE Transactions, 1969, vol. 78, Paper 690176.
Megson, T, Increasing semi-trailer torsional stiffness could enhance operational safety, *Automotive Design Engineering*, May 1970
Sidelko, W, An objective approach to highway truck frame design, SAE Transactions, vol. 75, Paper SP 276, 1966
Sherman, D, Stresses and deflections in truck side rail attachments, SAE paper 690175, 1969
Ali et al, The application of finite element techniques to the analysis of an automobile structure, Proc IMechE, 1970-71
Bruhn, E, Analysis and design of flight vehicle structures, USA, Tri-State Offset Co., 1965.
Peach, R, Design to prevent buckling in rectangular plates, *Machine Design*, 1963, vol 35, no. 30
Fagan, M, Finite element analysis, Longman, 1992
Myers, R, Body stress analysis, SAE paper 660005
Kirioka, K, An analysis of body structures, SAE paper 979A, 1965
Swallow, W, Unification of body and chassis-frame, Proc IAE, 1938-39
Argyris and Kelsey, Energy theorems and structural analysis, Butterworth
Lucas, A, The Argyris method of structural analysis using matrix algebra, *Automotive Design Engineering*, 1963, vol 2, no. 6
Wang, C, Matrix methods of structural analysis, International Textbook Co
Amos, R, Structural design analysis of commercial vehicle cabs, IMechE conference paper C134/84
Fenton, J, Shell beams and the Argyris method, *Automotive Design Engineering*, March 1963
Norville and Mills, An applied finite element displacement analysis of a simple integral construction vehicle body shell, Intl J Mech Sci, Pergamon, 1972, Vol 14, pp 809-821
Curle and Watson, Structural analysis, body design and development, IBCAM Boditek conference, 1985
Emmerson, W, C.A.D. in the motor industry, *Computer Aided Design*, Vol 8, No 3, July 1976
Rees, C, Design of tanks and cylindrical fluid containing vessels, *The Engineering Designer*, December 1966

Chapter 8

I Barley, G W, Mills B, A study of impact behaviour through the use of geometrical similar models, IMechE conference on Body Engineering, 1970, Paper N3
2 Baracos and Rhodes, Ford CC9 frame, SAE paper 690004
3 Belfield and Lyle, Preliminary calculation methods for crashworthiness, IBCAM Boditek conference, 1985
4 Cockerill-Sambre, 1993 ISATA conference
5 Alcan presentation, SAE Congress, 1992
6 Jaccard and Brucker, The crashworthiness of aluminium components evaluated by experiments and FE analysis, Paper SIA9506B14, EAEC conference, 1995
7 Frei et al, Crashworthiness and compatibility of low mass vehicles in collisions, SAE SP-1231
8 Lee et al, A study on the improvement of the structural joint stiffness for aluminium BIW.
9 Duffin, A, ISATA, 1993

Further reading

Grime and Jones, Car collisions— the movements of cars and their occupants in accidents. Proc IMechE, (AD) vol

184, part 2A, no. 5, 1969/70
Miura and Kawamura, Analysis of deformation mechanism in head-on collision, SAE paper 680484
Emmerson, W, The application of computer simulation in vehicle safety, TRRL conference Towards safer road vehicles, 1972
Finch, P, Vehicle frontal barrier impacts, TRRL conference Towards safer road vehicles, 1972Bacon, G, Crash testing of cars, IBCAM Boditek conference, 1987
United Nations World Papers 29/372, 373 463, 477.
Murray, N, When it comes to the crunch, ISBN 981-02-2096-0, World Scientific Publishing
Searle and Brabin, The Invertube, Motor Industry Research Association Bul. 1970, no. 2
Johnson, W, Impact strength of materials, Edward Arnold, 1972
Grime, G, Handbook of road safety research, Butterworth
Ed Murthy and Brebbia, Structural design and crashworthiness of automobiles, Springer-Verlag, 1987
Vehicle and road design for safety, IMechE conference report, 1968
Yamaguchi et al, Efficient energy absorption of automobile side rails, Tech Session 3, Experimental safety vehicles, TRRL hosted Oxford conference report, US Dept of Transport, NHTSA, 1985
Ed Dorgham, M, Vehicle safety, Inderscience Enterprises, 1986
Kipp, W, Impact without rebound: an advanced tool for laboratory crash simulation, SAE paper 700406
Cornell Aeronautical Laboratory report YB-2684-V-7 on engine underrun deflectors for crash impact.
Hardy and Suthurst, Simulations to assess the influence of car lateral impact characteristics on occupant kinematics, Proc IMechE, Vol 199, D2, 1985
Kamal, SAE paper 700414 on dynamic impact coefficients
Progress towards safer passenger cars in the UK, IMechE conferencc report, Jun 1980
Parsons, D, Predicting crash performance, IBCAM Boditek conference, 1987
Haman, J, Proc FISITA 1972, paper on `target/bullet' impacts
Rivett and Riches, Drop weight impact testing of spot-welded, weld-bonded and adhesively bonded box sections, Welding Institute research Report 304, 1986
Wolff, Application of composite materials for medium speed collisions, Renault Research Report, 1994
Butcher and Rock, Side intrusion impact analysis for design, IBCAM Boditek conference, 1985
Kecman and Miles, Application of the finite element method to the door intrusion and roof crush analysis of a passenger car, Vehicle Structural Mechanics Conference, Troy, Michigan, 1979
Meadows et al, Aluminium crash members in axial and bending colloase, SAE paper 922113, 1992
Stephens and Bacon, The development of a method for dynamic simulation of side impacts, Experimental safety Vehicle conference, 1991

Chapter 9

1 Whiting and Brown, American legislation covering occupant protection in interior impacts, IMechE/I0M Plastics on the road conference, 1996
2 Farahani *et al*, Effect of front seat stiffness in Federal Dynamic Side Impact, page 71 et seq, SAE publication SP-1231, 1997
3 Nehan & Maloney, Magnesium AM60B instrument panel structure for crashworthiness FMVSS 204 and 208 compliance, SAE Paper 960419
4 Roca *et al*, Development of rollover-resistant bus structures, page 151 et seq, SAE publication SP-1233, 1997
5 PJ Rowberry, Impact Management and Passenger Safety, Paper C498/18/045, Autotech 1995
6 Kragtwijk and van Veldhuizen, Paper C498/18/016, Autotech 1995
7 McKie and Dixon, Frontal impact crashworthiness development using hybrid analysis and test techniques, MIRA Research Report, 1994
8 SAE papers 960151/2
9 Dickison and Buckley, Body engineering considerations to improve occupant safety in minibuses and coaches, MIRA paper 96-S5-0-13

Further reading

Rodger, M, Occupant protection in side impact, TRRL report, 1972
Wolfe, J, Use of models in automobile impact research, TRRL 1972
Coombs, G, Safety frames for tractors, *Farm Machine Design Engineering*, Dec 1971
Dept of Transport Code of Practice, Safety of loads on vehicles, HMSO, 1984
Pan et al, The use of a computer simulation program to predict occupant motion and injury levels in a coach front impact, Cranfield Impact Centre report
Gutman, H, New concept bumper in plastic materials for elevated speed collisions, SITEV forum, 1990
White and Tidbury, The simulation of the roll and tilt stability of buses, SITEV paper 865115
Kecman, D, Safety of buses and coaches, Autotech paper C498/32/243, 1995

Chapter 10

1 Chiesa and Oberto, Influence of tyres on car vibrations studied with a new classification method, ASAE Cranfield third technical symposium, Noise and Vibration, 1964
2 Industrial Products Catalogue, Mezeler / Lord Gimetall
3 Cornish and Atkinson, Reducing interior noise in light commercial vehicles, Paper C498/26/004, Autotech conference, 1995
4 Van der Linden and Varet, Paper C498/26/212, Autotech 1995

5 Williams, R, et al, Autotech 1995
6 Cunningham and Duggan, UTECH conference, 1994
7 Liefooghie et al, Autotech 1993 Paper C462/31/043
8. McCulloch and Dascotte, Autotech 1993 Paper C462/17/141
9 Cheng and Olatunbosun, Autotech 1993 Paper C462/17/022
10 Cobden, R, Fatigue in aluminium structures, IMechE seminar on Using Aluminium to Advantage, 1995
11 British Standard Code of Practice 8118. 1969: *The Noise and vibration basics*

Further reading

Cornish et al, Paper C498/26/004, Autotech, 1995 ?
McDonald and Quinn, Adaptive noise control, Lotus Engineering Research Report
Priede and Jha, Low frequency noise in cars, *The Journal of Automotive Engineering*, Jul 1970
Guignard, J, Human response to intense low-frequency noise and vibration, Proc I Mech E vol. 182, part 1, no. 3,1967/8
Jha S, Noise inside and out, *Automotive Design Engineering*, 1972, vol 11
Gibbs and Richards, Stress, vibration and noise analysis in vehicles, Applied Science Publishers, 1975
Priede et al, Experimental techniques leading to the better understanding of the origins of automotive engine noise, IMechE conference paper C151/84, 1984
Brughmans et al, The application of FEM-EMA correlation and validation techniques on a body-in-white, paper 3 in IMechE Vehicle NVH and refinement conference report C487, 1994
Burton and Southall, Noise, vibration and harshness in the automotive industry, paper 3, session 3R, Design Engineering Show conference, 1982.
Miki et al, Elastic vibration analysis of passenger car bodies, SAE paper 690272
Halvorsen and Brown, Impulse technique for structural frequency response testing, *Sound and Vibration*, Nov 1977
Anderson and Mills, Dynamic analysis of a car chassies frame using the finite element method, Int J. Mech. Sci., Pergamon, 1972
Sas et al, Prediction of near field intensity patterns based on modal deformation patterns, IMechE conference paper C147/84, 1984
Newman and Coats, Investigation of fatigue strength of welded thin gauge construction box-tube, British Welding Research Association Report FE 23, no. 5, 1956
Newman R. P. Fatigue of spot-welded box beams. British Welding Research Association Reports FE 23, 1958, no. 6, 7.
Murakami, Y, The rainflow method in fatigue, Butterworth Heinemann, 1992
Designing for finite life, *Automotive Engineer*, Feb/Mar 1983
Metalastik, Body mounting, *Automobile Engineer*, March 1958
Prebble, C, Recommended methods of body mounting, Vauxhall Motors, 1963
Boden, E, Some aspects of vehicle ride, IMechE conference on Vehicle ride, 1962
Tidbury, G, Measurement of loads between the suspension and the body structure of a small car, Conference on stresses in service, Inst. Civil Engrs, 1966
Stott, T, Fatigue testing of vehicle components, ProcIMechE, No 1, 1958-9
Jones and Williams, The fatigue properties of spot welded, adhesive bonded and weld-bonded joints in high-strength steels, SAE paper 860583, 1986
McRobert and Watson, Design optimization of forgings, Drop Forgings Congress, Cologne, 1983

Chapter 11

1. Tidbury G, The structural design of bus bodies, Paper 8 at IMechE conference on Body Engineering, 1970,
2. Phillips, L, Improving racing car bodies, *Composites* 1(1)50
3. Hollaway, L, Glass reinforced plastics in construction, Surrey University Press
4. Parry-Williams, D, Panther Solo, the first production composite vehicle structure, paper 15 in IMechE conference C387
5. Potts, H, Application of advanced composite materials in racing cars, Paper 29, British Plastics Federation conference, October 1990.
6. O'Rourke, Paper 4, I Mech E seminar on Materials, fabricating a novel approach, 1993
7. Pitchford, N, Adhesive bonding for aluminium structured vehicles, IMechE seminar on Materials, fabricating a novel approach
8. SAE Paper 922112
9 Training in aluminium application technologies (TALAT) initiative, CD-ROM, Aluminium Federation, Birmingham
10 Pederson, R, Aluminium Federation conference on Aluminium in road transport, 1996
11 Dr Hewitt, Paper C49X/3 1/240, Autotech 1995
12 Meadows, D, Paper SIA9506B03, EAEC conference, 1995
13 Fothergill et al, Computer-aided concept design of a sports car chassis system, IMechE conference on Vehicle Structures, no 7, 1984

Further reading

The design, construction and operation of public service vehicles, IMechE conference report C130, 1977
Pawlowski, J, Van idealisation for analysis, *Auto-Technika Motoryzacyjna*, no. 1, 1988
Hundy and Broadstock, The use of aluminium for commercial vehicle structures, a feasibility study, Proc IMechE,

Vol 192, no. 13, 1978

Greig, J, Concept structure development through design and analysis: an automotive case study using Hydroform tubes, IMechE conference paper C462/23/105

Pressel and Strifler, Analytical comparison of integral and chassis design on buses, IMechE conference paper C144/77

Stachel, P, Structural analysis of composite structures, SITEV conference, 1985

Spina and Vasallo, Alfa Romeo GTV and Spider rear axle's aluminium subframe, FISITA paper M1803, 1996

Index

A

B

C

D

E

F

S

T

U

V

W